Applied Regression Analysis and Other Multivariable Methods

Applied Regression Analysis and Other Multivariable Methods

David G.
Kleinbaum

Lawrence L.
Kupper

Keith E.
Muller

The University of North Carolina at Chapel Hill

Duxbury Press
An Imprint of Wadsworth Publishing Company
Belmont, California

Duxbury Press
An Imprint of Wadsworth Publishing Company
A division of Wadsworth, Inc.

Library of Congress Cataloging-in-Publication Data
Kleinbaum, David G.
    Applied regression analysis and other multivariable methods.
    Bibliography: p.
    Includes index.
    1. Multivariate analysis.   2. Regression analysis.
I. Kupper, Lawrence L.   II. Muller, Keith E.   III. Title.
QA278.K58   1987   519.5'35   87-5397
ISBN 0-87150-123-6

Printed in the United States of America

   92—10 9 8 7

Sponsoring Editor: Michael Payne
Production Coordinator: Jean Coulombre
Production: Technical Texts, Inc./Jean T. Peck
Interior and Cover Design: Jean Coulombre
Cover Photo: © 1987 by Larry Lorusso/The Picture Cube
Interior Illustration: Carl Brown
Typesetting: Bi-Comp, Inc.
Cover Printing: Phoenix Color Corp.
Text Printing and Binding: Halliday Lithograph

To Professor Bernard G. Greenberg,
mentor, colleague, and friend

Preface

This is the first revision of our second-level statistics text, originally published in 1978. As before, this text is intended primarily for advanced undergraduates, graduate students, and working professionals in the health, social, biological, and behavioral sciences who engage in applied research in their fields. We also hope that the book will provide the professional statistician with some new insights into the application of advanced statistical techniques to real-life problems. The changes that we have made to the first edition, discussed in more detail below, include a third author (Keith Muller), some reorganization of the topics, expanded coverage of some content areas, some new chapters covering topics not previously discussed, and new exercises for the reader. We have attempted, nevertheless, to retain the same basic structure and flavor of the first edition.

We three authors have been teaching and consulting with students and professionals in public health, in other health and medical sciences, and in environmental and behavioral sciences for many years. Most of the examples described in this book, particularly those used to introduce and illustrate the concepts in the body of the text, are based on this experience and reflect an orientation toward research involving a wide range of social and behavioral aspects of health problems.

The development of topics emphasizes the intuitive logic and assumptions that underlie the techniques covered, the purposes for which these techniques are designed, the advantages and disadvantages of the techniques, valid interpretations based on the techniques, and the statistical calculations required. The mathematical formulae presented require no more than simple algebraic manipulations. Proofs are of secondary importance and are generally omitted. Neither calculus nor matrix algebra is used anywhere in the main text, although we have included an appendix on matrices for the interested reader.

The text is not intended to be a general reference work dealing with all statistical techniques available for analyzing data on several variables. Instead, what we feel to be the most important advanced topics are covered in considerable detail. After becoming proficient with the material in this text, the reader should be able to benefit from more specialized discussions of applied topics not covered in this text.

Among the notable features of this second edition are the following:

1. Regression analysis and analysis of variance are discussed in considerably more detail than in other applied texts.

2. Several advanced topics involving the use of linear models are discussed, and the relationship between analysis of variance and regression analysis is highlighted.

3. The connection between multiple regression analysis and multiple and partial correlation analysis is discussed in detail.

4. Several advanced topics are presented in a unique, nonmathematical manner, including maximum likelihood methods (a new topic in this edition), analysis of categorical data using linear models, two-group discriminant analysis, and variable reduction and factor analysis (a topic considerably elaborated from the first edition).

5. In some new chapters, an up-to-date discussion of issues and procedures involved in fine-tuning a regression analysis is presented. Included are chapters on confounding and interaction in regression, regression diagnostics, and selecting the best model.

6. There are numerous examples and exercises, illustrating applications to real studies in a wide variety of disciplines.

7. Representative computer results from packaged programs are used to illustrate concepts in the body of the text, as well as to provide a basis for the problems for the reader. (We have avoided providing thorough descriptions of available packaged programs, however, because these programs are frequently changing.)

8. The complete set of data for most exercises is provided, along with related computer results. This allows the instructor to assign computer work based on available packaged programs, if desired. However, if such programs are not available or if the instructional objectives involve a minimum of computer work, the computer results given can still provide the student with practical experience in interpreting computer output from statistical programs.

For formal classroom instruction, the chapters fall naturally into three clusters: Chapters 4 through 16, on regression analysis; Chapters 17 through 20, on analysis of variance; and Chapters 21 through 24, on miscellaneous advanced topics. For a first course in regression analysis, some of Chapters 11 through 16 may be considered too specialized. For example, Chapter 12 on regression diagnostics and Chapter 16 on selecting the best model might be used in a continuation course on regression modeling, which might also include some of the advanced topics covered in Chapters 21 through 24.

The following changes have been made from the first edition:

Organization. Chapters are now divided into five logical groupings: background (Chapters 1 through 3), fundamentals of regression and correlation (Chapters 4 through 7), multiple regression and correlation (Chapters 8 through 16), analysis of variance (Chapters 17 through 20), and other multivariable methods (Chapters 21 through 24). These groupings will help an instructor to determine course objectives and a course plan.

The first edition contained a chapter on comparing two straight-line models. This chapter has been deleted in the revision, although much of the material has been incorporated into our new Chapter 14, "Dummy Variables in Regression."

The chapter on polynomial regression (previously chapter 9, now Chapter 13) has been moved from a location preceding our discussion of multiple regression to one following the chapter on regression diagnostics. The reason is to allow us to incorporate issues of collinearity into procedures for reliable fitting of polynomial models.

New Chapters. "Testing Hypotheses in Multiple Regression" (Chapter 9): This chapter expands on material originally described in chapters 10 and 11 of the first edition. "Confounding and Interaction in Regression" (Chapter 11): Confounding was not discussed in the first edition; interaction was previously discussed in the old chapter 12. "Regression Diagnostics" (Chapter 12): A small portion of this material (on residuals and transformations) was contained in the old chapter 16; however, this chapter expands previous material and adds several up-to-date diagnostic techniques, including assessment and treatment of collinearity. "Maximum Likelihood Methods" (Chapter 21): This chapter is completely new and includes applications to Poisson regression and logistic regression.

Other Important Changes. Chapter 13 on polynomial regression is expanded to consider collinearity, orthogonal polynomials, and computer solutions. Chapter 14 on dummy variables has expanded coverage of methods for comparing two straight lines and two correlations. Chapter 16 is greatly enlarged to cover a variety of current approaches to selecting variables. Chapter 20 has been revised to reflect the current literature on analysis-of-variance methods for unbalanced data arrays. Chapter 24 has been expanded to describe variable reduction techniques and to consider factor analysis in a more general context than in the first edition.

Exercises. We have added new exercises, including new data sets, to practically every chapter. For the most part, we have retained exercises used in the previous edition (numbered as before, when possible). However, because of the reorganization of chapters, some exercises have been relocated.

We wish to acknowledge several people who contributed to the preparation of this text. We are indebted to John Cassel and Bernard Greenberg, two mentors who provided us with inspiration and the professional and administrative guidance that enabled us to gain the broad experience necessary to write this book. We thank Anna Kleinbaum, Sandy Martin, and Sally Muller for their encouragement and support during the writing of this revision. We also thank our many students and colleagues at the University of North Carolina and the following reviewers for their helpful comments and suggestions: Howard A. Bird, St. Cloud University; Robert Cochran, University of Wyoming; James E. Holstein, University of Missouri; Frederick O. Lorenz, Iowa State University; and Patricia Wahl, University of Washington. Finally, we thank those persons responsible for publishing this book: Michael Payne and Jean Coulombre of PWS-KENT Publishing Company for their advice and support, and Jean Peck of Technical Texts, Inc. for her help during production.

Contents

5 STRAIGHT-LINE REGRESSION ANALYSIS   41

6 THE CORRELATION COEFFICIENT AND STRAIGHT-LINE REGRESSION ANALYSIS   80

7 THE ANALYSIS-OF-VARIANCE TABLE   96

8 MULTIPLE REGRESSION ANALYSIS: GENERAL CONSIDERATIONS   102

9 TESTING HYPOTHESES IN MULTIPLE REGRESSION   124

10 CORRELATIONS: MULTIPLE, PARTIAL, AND MULTIPLE-PARTIAL   144

11 CONFOUNDING AND INTERACTION IN REGRESSION   163

15 ANALYSIS OF COVARIANCE AND OTHER METHODS FOR ADJUSTING CONTINUOUS DATA   297

16 SELECTING THE BEST REGRESSION EQUATION   314

17 ONE-WAY ANALYSIS OF VARIANCE   341

18 RANDOMIZED BLOCKS: SPECIAL CASE OF TWO-WAY ANOVA   387

19 TWO-WAY ANOVA WITH EQUAL CELL NUMBERS   416

20 TWO-WAY ANOVA WITH UNEQUAL CELL NUMBERS   457

A   **Appendix A - Tables**   643

B   **Appendix B - Matrices and Their Relationship to Regression Analysis**   664

C   **Appendix C - Solutions to Exercises**   676

Index   709

Concepts and Examples of Research

1-1 Concepts

The purpose of most research is to assess relationships among a set of variables. *Multivariate*[1] *techniques* concern the statistical analysis of such relationships, particularly when at least three variables are involved. Regression analysis, our primary focus, is one type of multivariable technique. There are also other techniques, some of which will be described in this text. Choosing an appropriate technique depends on the purpose of the research as well as the types of variables under investigation (a subject discussed in Chapter 2).

Research may be broadly classified as one of three types: *experimental, quasi-experimental,* or *observational.* Multivariable techniques are applicable to all such types, yet the confidence one has in the results of a study can vary with the research type. In most types, one variable is usually taken to be a *response* or *dependent variable,* that is, a variable to be predicted from other variables. The other variables are called *predictor* or *independent variables.*

If observational units (subjects) are randomly assigned to levels of important predictors, then the study is usually classified as an *experiment.* Experiments are the most controlled type of study and maximize the investigator's ability to isolate the observed effect of the predictors from the distorting effects of other (independent) variables that might also be related to the response.

If subjects are assigned to treatment conditions without randomization, then the study is called *quasi-experimental* (Campbell and Stanley, 1963). Such studies are often more feasible and less expensive than experimental studies but provide less control over the study situation.

Finally, if all observations are obtained without either randomization or artificial ma-

---

[1] The term *multivariable* is preferred to *multivariate.* The distinction will be explained in Chapter 2.

nipulation (i.e., allocation) of the predictor variables, then the study is said to be *observational*. Experiments offer the greatest potential for drawing definitive conclusions, and observational studies the least; however, experiments are the most difficult studies to implement, and observational studies the easiest. This trade-off between interpretive potential and complexity of design must be considered when choosing among types of studies (Kleinbaum, Kupper, and Morgenstern, 1982, chap. 3).

To assess a relationship between two variables, one must measure both of them in some manner. Measurement inherently and unavoidably involves error. The need for statistical design and analysis emanates from the presence of such error. Traditionally, statistical inference has been divided into two kinds: estimation and hypothesis testing. *Estimation* refers to describing (i.e., quantifying) characteristics and strengths of relationships. *Testing* refers to specifying hypotheses about relationships, making statements of probability about the reasonableness of such hypotheses, and then providing practical conclusions based on such statements.

This text focuses on regression and correlation methods involving one response variable and one or more predictor variables. In these methods a mathematical model is specified that describes how the variables of interest are related to one another. The model must somehow be developed from study data, after which inference-making procedures (e.g., testing hypotheses and constructing confidence intervals) are carried out for important parameters of interest. Although other multivariable methods will be discussed, regression techniques are emphasized because (1) they have wide applicability, (2) they can be the simplest to implement, and (3) many other more complex statistical procedures can be better appreciated once regression methods are understood.

1-2 Examples

The examples that follow concern *real* problems from a variety of disciplines and involve variables to which the methods described in this book can be applied. We shall return to these examples later when illustrating various methods of multivariable analysis.

Example 1.1 *Study of the associations among the physician–patient relationship, perception of pregnancy, and the outcome of pregnancy, illustrating the use of regression analysis, discriminant analysis, and factor analysis*

Thompson (1972) and Hulka and others (1971) looked at both the process and the outcomes of medical care in a cohort of 107 pregnant married women in North Carolina. The data were obtained through questionnaires completed by physicians, patient interviews, and a review of medical records. Several variables were recorded for each patient.

One research goal of primary interest was to determine what association, if any, existed between SATISfaction[2] with medical care and a number of variables meant to describe patient perception of pregnancy and the physician–patient relationship. Three perception-of-pregnancy variables measured the patient's WORRY during pregnancy, her desire (WANT) for the baby, and her concern about childBIRTH. Two other variables measured

---

[2] Capital letters denote the abbreviated variable name.

the physician–patient relationship in terms of informational communication (INFCOM) concerning prescriptions and affective communication (AFFCOM) concerning perceptions. Several other variables considered were AGE, social class (SOCLS), EDUCation, and PARity.

The perception-of-pregnancy variables were developed by use of a method called *factor analysis*, which summarized the information obtained from several questions on this subject asked of all patients. *Regression analysis* was used to describe the relationship between scores measuring patient satisfaction with medical care and the variables above. From this analysis, variables found not related to SATIS could be eliminated, while those found to be associated with SATIS could be ranked in order of importance. Also, the effects of confounding variables such as AGE and SOCLS could be considered, so that (1) any associations found could not be attributed solely to such variables, (2) measures of the strength of the relationship between SATIS and other variables could be obtained, and (3) a functional equation predicting level of patient satisfaction in terms of the other variables found to be important in describing satisfaction could be developed.

Another question of interest in this study was whether patient perception of pregnancy and/or the physician–patient relationship was associated with COMPlications of pregnancy. A variable describing complications was defined so that the value 1 could be assigned if the patient experienced one or more complications of pregnancy and 0 if she experienced no complications. The method of *discriminant analysis* was used to evaluate the relationship between complications of pregnancy and other variables. This method, like regression analysis, allows the researcher to determine and rank important variables that can distinguish between patients who have complications and patients who do not.

Example 1.2 *Study of the relationship between water hardness and sudden death, illustrating the use of regression analysis*

Hamilton's (1971) study of the effects of environmental factors on mortality used 88 North Carolina counties as the observational units, in contrast to individual patients who were the observational units for a preceding study. Hamilton's primary goal was to determine the relationship, if any, between the sudden-death rate of residents of a county and the measure of water hardness for that county. However, it was also of interest to compare the mortality–water hardness relationship among four regions of the state and to determine whether this relationship was affected by other variables, such as the habits of the county coroner in recording deaths, the distance from the county seat to the main hospital, the per capita income, and the population per physician.

Regression analysis was used in this study. The four regions could have been compared by doing a separate analysis for each region and then comparing the results. However, the comparative analysis can be performed in a single step by defining a number of artificial, or dummy, variables to represent the regions.

Example 1.3 *Comparative study of the effects of two instructional designs for teaching statistics, illustrating the use of analysis of covariance*

Kleinbaum and Kleinbaum (1976) compared through a classroom experiment two approaches for teaching probability to graduate students taking an introductory course in biostatistics. A class of 52 students was randomly split into two groups stratified on the basis of the students' fields of study. Both groups were taught in a lecture format by the same instructor. However, for one group, the control, the standard lecture method was used, in which an instructor (using chalk and a blackboard) lectured after a handout had been distributed. For the experimental group, 16 transparencies, systematically designed to fit a carefully defined set of objectives, were used along with a set of practice problems that were done and reviewed in class. Both groups were given the same pre- and posttests, as well as questionnaires to measure attitudes. They both received a copy of the objectives, the handout, and a homework assignment. The primary question of interest for this study was whether the instructional design for the experimental group was more effective than that for the control group as measured by cognitive learning and attitudes.

In comparing the experimental and control groups, it was important to take into account pretest scores so that any differences found in posttest scores could not be attributed solely to differing levels of ability or knowledge between the two groups at the beginning of the study. One appropriate method of analysis for this situation was *analysis of covariance*, which showed that posttest scores, when adjusted for pretest scores, were significantly higher for the experimental group.

Example 1.4   *Study of race and social influence in cooperative problem-solving dyads, illustrating the use of analysis of variance and analysis of covariance*

James (1973) conducted an experiment on 140 seventh- and eighth-grade males to investigate the effects of two factors, race of the experimenter (E) and race of the comparison norm (N), which is defined below, on social influence behaviors in three types of dyads: white–white, black–black, and white–black. Subjects played a game of strategy called Kill the Bull, in which 14 separate decisions must be made for proceeding toward a defined goal on a game board. In the game, each pair of players (dyad) must reach a consensus on a direction at each decision step, after which they signal the E, who then rolls a die to determine how far they can advance along their chosen path of six squares. Photographs of the current champion players (N) (either two black youths [black norm] or two white youths [white norm]) were placed above the game board.

Four measures of social influence activity were used as the outcome variables of interest. One of these, for example, was called performance output, which was a measure of the number of times a given subject attempted to influence his dyad to move in a particular direction.

The major research question focused on the outcomes for biracial dyads. Previous research of this type had used only white investigators and implicit white comparison norms, and the results indicated that the white partner tended to dominate the decision making. In James's study, it was of interest to see if such an "interaction disability," previously attributed to blacks, would be maintained, removed, or reversed if either the comparison norm and/or the experimenter was black. One approach to analysis of this problem was to perform a *two-way analysis of variance* on social-influence-activity difference scores between black and white partners to assess whether such differences were affected by either the race of E or the race of N. No such significant effects were found, however, implying that neither E nor N influenced biracial dyad interaction. Nevertheless, through use of *analysis of covariance*, it was shown that, controlling for factors such as age, height, grade, and verbal and mathematical test scores, there was no statistical evidence of white dominance in any of the experimental conditions.

Furthermore, when combined output scores for both subjects in same-race dyads (white–white or black–black) were analyzed using a *three-way analysis of variance* (the three factors being race of dyad, race of E, and race of N), it was seen that subjects in all-black dyads were more verbally active (i.e., exhibited more of a tendency to influence decisions) under black E's than under white E's; the same result was found for white dyads under white E's. This property would generally be referred to in the statistical jargon as a "race of dyad" by "race of E" interaction. This property was also found to hold up after *analysis of covariance* was used to control for the effects of age, height, and verbal and mathematical test scores.

Example 1.5 *Study of the relationship of culture change to health, illustrating the use of factor analysis and analysis of variance*

Patrick and others (1974) studied the effects of cultural change on health in the U.S. Trust Territory island of Ponape. Medical and sociological data were obtained on a sample of about 2,000 people by means of physical exams and a sociological questionnaire. This Micronesian island has experienced rapid westernization and modernization since American occupation in 1945. The question of primary interest was whether rapid social and cultural change caused a rise in blood pressure and in the incidence of coronary heart disease. A specific hypothesis guiding the research was that persons with high levels of cultural ambiguity and incongruity and low levels of supportive affiliations with others will have high levels of blood pressure and high risk for coronary heart disease.

A preliminary step in the evaluation of this hypothesis involved the measurement of three variables: attitude toward modern life, preparation for modern life, and involvement in modern life. Each of these variables was created by isolating specific questions from a sociological questionnaire. Then a *factor analysis*, based on the method of *principal components*, determined how best to combine the scores on specific questions into a single overall score that defined the variable under consideration. Two cultural incongruity variables were then defined. One involved the discrepancy between attitude toward modern life and involvement in modern life; the other was defined as the discrepancy between preparation for modern life and involvement in modern life.

These variables were then analyzed to determine their relationship, if any, to blood pressure and coronary heart disease. Individuals with large positive or negative scores on either of the two incongruity variables were hypothesized to have high blood pressure and high risk for coronary heart disease.

One approach toward analysis was to categorize both discrepancy scores into high and low groups. Then a *two-way analysis of variance* could be done using blood pressure as the outcome variable. We will see later that this problem can also be described as a regression problem.

Example 1.6 *Study of structure of adult intellectual abilities compared by age, illustrating the use of confirmatory factor analysis*

Cunningham (1980) studied the structure of intellectual ability in 198 young individuals (15 to 32 years), 156 middle-aged individuals (54 to 68 years), and 156 older individuals (69 to 91 years). Subjects were tested using 32 tests of intellectual ability. In young subjects, it was expected from earlier work that intellectual ability could be described as a function of three skill complexes: (1) verbal comprehension, (2) sensitivity to problems, and (3) semantic redefinition. *Factor analysis* was applied separately to each of the three groups' data. Each analysis yielded a set of dimensions, called *factors*, which most succinctly described the variability in performance on the intellectual ability tests. Finally, the three sets of factors were compared using statistical techniques. The basic factor pattern was the same for all three groups, although some differences were found. In all cases the most important dimensions of performance (factors) were the three expected: verbal comprehension, sensitivity to problems, and semantic redefinition. Hence, the dimensions needed to describe intellectual functioning did not change appreciably from young (15 to 32 years) to old (69 to 91 years).

1-3 Concluding Remarks

The six examples described above indicate the variety of research questions to which multivariable statistical methods are applicable. In Chapter 2, we will provide a broad overview of such techniques; in the remaining chapters, we will discuss each technique in detail.

References

Campbell, D. T., and Stanley, J. C. (1963). *Experimental and Quasi-experimental Designs for Research*. Chicago: Rand McNally & Co.

Cunningham, W. R. (1980). "Age Comparative Factor Analysis of Ability Variables in Adulthood and Old Age." *Intelligence*, 4: 133–149.

Hamilton, M. (1971). "Sudden Death and Water Hardness in North Carolina Counties in 1956–1964." Master's thesis, Department of Epidemiology, University of North Carolina, Chapel Hill, N.C.

Hulka, B. S., Kupper, L. L., Cassel, J. C., Thompson, S. J. (1971). "A Method for Measuring Physicians' Awareness of Patients' Concerns." *HSMHA Health Rept.*, 86: 741–751.

James, S. A. (1973). "The Effects of the Race of Experimenter and Race of Comparison Norm on Social Influence in Same Race and Biracial Problem-Solving Dyads." Ph.D. dissertation, Department of Clinical Psychology, Washington University, St. Louis, Mo.

Kleinbaum, D. G., and Kleinbaum, A. (1976). "A Team Approach for Systematic Design and Evaluation of Visually Oriented Modules." In *Modular Instruction in Statistics—Report of ASA Study*, ed. J. R. O'Fallon and J. Service, pp. 115–121. Washington, D.C.: American Statistical Association.

Kleinbaum, D. G., Kupper, L. L., and Morgenstern, H. (1982). *Epidemiologic Research*. Belmont, Calif.: Lifetime Learning Publications.

Patrick, R., Cassel, J. C., Tyroler, H. A., Stanley, L., and Wild, J. (1974). "The Ponape Study of Health Effects of Cultural Change." Paper presented at the annual meeting of the Society for Epidemiologic Research, Berkeley, Calif.

Thompson, S. J. (1972). "The Doctor–Patient Relationship and Outcomes of Pregnancy." Ph.D. dissertation, Department of Epidemiology, University of North Carolina, Chapel Hill, N.C.

Classification of Variables and the Choice of Analysis

2-1 Classification of Variables

Variables can be classified in a number of ways. Such classifications are useful for determining which method of data analysis to use. In this section we describe three methods of classification: by gappiness, by descriptive orientation, and by level of measurement.

2-1-1 Gappiness

In the classification scheme we shall call *gappiness*, we determine whether there are gaps between successively observed values of a variable (Figure 2-1). If there are gaps between observations, the variable is said to be *discrete*; if there are no gaps, the variable is said to be *continuous*. To be more precise, a variable is discrete if, between any two potentially observable values, there is a value that is not possibly observable. A variable is continuous if, between any two potentially observable values, there is always another potentially observable value.

Examples of continuous variables are age, blood pressure, cholesterol level, height, and weight. Examples of discrete variables are sex (e.g., 0 if male and 1 if female), number of deaths, group identification (e.g., 1 if group A and 2 if group B), and state of disease (e.g., 1 if a coronary heart disease case and 0 if not a coronary heart disease case).

When dealing with actual data, the sampling frequency distributions for continuous variables are represented differently than for discrete variables. Data on a continuous variable are usually *grouped* into class intervals, and a frequency distribution is determined by counting observations in each interval. Such a distribution is usually represented by a histogram, as shown in Figure 2-2a. Data on a discrete variable, on the other hand, are not usually grouped but are represented instead by a line chart, as shown in Figure 2-2b.

It is important to note that discrete variables can sometimes be treated as continuous variables. This is particularly useful when the possible values of such a variable, even though discrete, are not very far apart and cover a wide range of numbers. In such a case, the

FIGURE 2-1   Discrete versus continuous variables

Gaps   No gaps

(a) Values of a discrete variable   (b) Values of a continuous variable

FIGURE 2-2   Sample frequency distributions of a continuous and a discrete variable

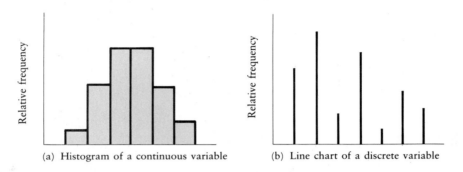

(a) Histogram of a continuous variable   (b) Line chart of a discrete variable

FIGURE 2-3   Discrete variable that may be treated as continuous

possible values, although technically gappy, show such small gaps between values that a visual representation would approximate an interval (Figure 2-3).

Furthermore, a line chart, like that in Figure 2-2b, representing the frequency distribution of data on such a variable would probably show few frequencies greater than 1 and thus be uninformative. As an example, the variable "social class" is usually measured as discrete; one popular measure of social class[1] takes on integer values between 11 and 77. When data on this variable are grouped into classes (e.g., 11–15, 16–20, etc.), the resulting frequency histogram gives a clearer picture of the characteristics of the variable than a line chart does. Thus, in this case, treating social class as a continuous variable is more useful than treating it as discrete.

Just as it may often be useful to treat a discrete variable as continuous, some variables that are fundamentally continuous may be grouped into categories and treated as discrete variables in a given analysis. For example, the variable "age" can be made discrete by grouping its values into two categories, "young" and "old." Similarly, "blood pressure" becomes a discrete variable if it is categorized into "low," "medium," and "high" groups, or into deciles.

---

[1] Hollingshead's "Two-Factor Index of Social Position," a description of which can be found in Green (1970).

FIGURE 2-4   Descriptive orientation for Thompson's (1972) study of satisfaction with medical care

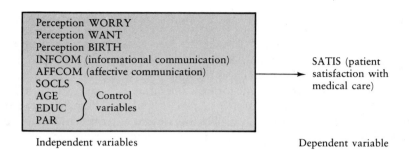

Independent variables   Dependent variable

2-1-2 Descriptive Orientation

A second scheme for classifying variables is based on whether a variable is to *describe* or *be described* by other variables. Such a classification depends on the study objectives rather than on the inherent mathematical structure of the variable itself. If the variable under investigation is to be described in terms of other variables, we call it a *response*, or *dependent*, *variable*. If we might be using the variable in conjunction with other variables to describe a given response variable, we call it a *predictor*, or *independent*, *variable*. Other variables may affect relationships but be of no intrinsic interest in a particular study. Such variables may be referred to as *control* or *nuisance variables* or, in some contexts, as *covariates* or *confounders*.

For example, in Thompson's (1972) study of the relationship of patient perception of pregnancy to patient satisfaction with medical care, the perception variables are independent and the satisfaction variable is dependent (Figure 2-4). Similarly, in studying the relationship of water hardness to sudden-death rate in North Carolina counties, the water hardness index measured in each county is an independent variable, and the sudden-death rate for that county, the dependent variable (Figure 2-5).

Usually, the distinction between independent and dependent variables is clear, as it is in the examples we have given. Nevertheless, a variable considered as dependent for evaluating one study objective may be considered as independent for evaluating a different objective. For example, in Thompson's study, in addition to determining the relationship of percep-

FIGURE 2-5   Descriptive orientation for Hamilton's (1971) study of water hardness and sudden death

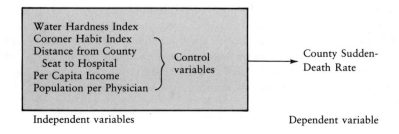

Independent variables   Dependent variable

tions as independent variables to patient satisfaction, it was also of interest to determine the relationship of social class, age, and education to perceptions treated as dependent variables.

2-1-3 Levels of Measurement

A third classification scheme deals with the preciseness of measurement of the variable. There are three such levels: nominal, ordinal, and interval.

The weakest level of measurement is the *nominal*. At this level the values assumed by a variable simply indicate different categories. The variable "sex," for example, is nominal, since by assigning the numbers 1 and 0 to denote male and female, respectively, we can distinguish the two sex categories. A variable that describes treatment group is also nominal, provided that the treatments involved cannot be ranked according to some criterion (e.g., dosage level).

A somewhat higher level of measurement allows not only *grouping* into separate categories but also *ordering* of categories. This level is called *ordinal*. The treatment group may be considered ordinal if, for example, different treatments differ by dosage. In this case we could tell not only which treatment group an individual falls into but also who was administered a heavier dose of the treatment. Also, social class is an ordinal variable, since an ordering can be made among its different categories. For example, all members of the upper middle class are higher in some sense than all members of the lower middle class.

A limitation, perhaps debatable, in the preciseness of a measurement such as social class is the amount of information supplied by the magnitude of the differences between different categories. Thus, although upper middle class is higher than lower middle class, it is debatable *how much* higher.

A variable that can give not only an ordering but also a meaningful measure of the distance between categories is called an *interval* variable. To be interval, a variable must have some sort of standard or well-accepted physical unit of measurement. Height, weight, blood pressure, and number of deaths all satisfy this requirement, whereas subjective measures such as perception of pregnancy, personality type, prestige, and social stress do not.

An interval variable that has a scale with a true zero is occasionally designated as a *ratio*, or *ratio-scale*, *variable*. An example of a ratio-scale variable is the height of a person. Temperature is commonly measured in degrees Celsius, an interval scale. Measurement of temperature in kelvin is referred to as absolute zero, and so is a ratio variable. An example of a ratio variable common in health studies is the concentration of a substance (e.g., cholesterol) in the blood.

Ratio-scale variables often involve measurement errors that follow a nonnormal distribution and are proportional to the size of the measurement. We will see in Chapter 5 that such proportional errors violate an important assumption of linear regression, namely, equality of error variance for all observations. Hence, the presence of a ratio variable is a signal to be on guard for a possible violation of this assumption. In Chapter 12 (on regression diagnostics), we will describe methods for detecting and dealing with this problem.

Similarly to variables in other classification schemes, the same variable may be considered at one level of measurement in one analysis and at a different level in another analysis. Thus, age may be considered as interval in a regression analysis or, by being grouped into categories, nominal in an analysis of variance.

It should also be pointed out that the various levels of mathematical preciseness are cumulative. An ordinal scale possesses all the properties of a nominal scale plus ordinality.

An interval scale is also nominal and ordinal. The cumulativeness of these levels allows the researcher to drop back one or more levels of measurement in analyzing the data. So, an interval variable may be treated as nominal or ordinal for a particular analysis, and an ordinal variable may be analyzed as nominal.

2-2 Overlapping of Classification Schemes

It is important to realize that the three classification schemes described overlap in the sense that any variable can be labeled according to each scheme. "Social class," for example, may be considered as ordinal, discrete, and independent in a given study; "blood pressure" may be considered interval, continuous, and dependent in the same or another study.

The overlap between the level-of-measurement classification and the gappiness classification is shown in Figure 2-6. The diagram does not include classification into dependent or independent variables, because that is entirely a function of the study objectives and not of the variable itself. In reading the diagram one should consider any variable as being representable by some point within the triangle. If the point falls below the dashed line within the triangle, it is classified as discrete; if it falls above this line, it is continuous. Also, a point that falls in the area marked "interval" is classified as an interval variable, and similarly for the other two levels of measurement.

It can be observed from Figure 2-6 that any nominal variable must be discrete, although a discrete variable may be nominal, ordinal, or interval. Also, a continuous variable must be either ordinal or interval, although there may be ordinal or interval variables that are not continuous. For example, "sex" is nominal and discrete; "age" may be considered interval and continuous or, if grouped into categories, nominal and discrete. "Social class," depending on how it is measured and on the viewpoint of the researcher, may be considered ordinal and continuous, ordinal and discrete, or nominal and discrete.

2-3 Choice of Analysis

Any researcher faced with the need to analyze data requires a rationale for choosing a particular method of analysis. Several considerations should enter into such a choice. These

FIGURE 2-6   Overlap of variable classification

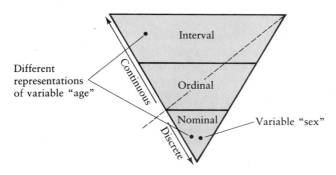

include (1) the purpose of the investigation, (2) the mathematical characteristics of the variables involved, (3) the statistical assumptions made about these variables, and (4) how the data are collected (e.g., the sampling procedure). The first two considerations are generally sufficient to determine an appropriate analysis. However, the researcher must consider the latter two items before finalizing initial recommendations.

TABLE 2-1   Rough guide to multivariable methods

Name of Method	Classification of Variables		General Purpose
	Dependent	Independent	
Multiple regression analysis	Continuous	Classically all continuous, but in practice any type(s) can be used	To describe the extent, direction, and strength of the relationship between several independent variables and a continuous dependent variable
Analysis of variance	Continuous	All nominal	To describe the relationship between a continuous dependent variable and one or more nominal independent variables
Analysis of covariance	Continuous	Mixture of nominal variables and continuous variables (the latter used as control variables)*	To describe the relationship between a continuous dependent variable and one or more nominal independent variables, controlling for the effect of one or more continuous independent variables
Discriminant analysis	Nominal	Classically all continuous, but in practice a mixture of various types can be used as long as some are continuous	To determine how one or more independent variables can be used to discriminate among different categories of a nominal dependent variable
Factor analysis	(The variables used in a factor analysis are classically continuous, but in practice may be of any type. These variables are not clearly identifiable as either dependent or independent, although the resulting factors may be used as dependent or independent variables in a later analysis.)		To define one or more new composite variables called factors from other, specially constructed or reduced variables
Categorical data analysis using linear models	Nominal	Mostly nominal, but sometimes ordinal	To describe the relationship between a nominal dependent variable and several nominal or ordinal independent variables, although applications to situations involving only dependent variables are possible

* Generally speaking, a control variable is a variable that has to be considered before any relationships of interest can be quantified; this is because a control variable may be related to the variables of primary interest and must be taken into account in studying the relationships among the primary variables. For example, in describing the relationship between blood pressure and physical activity, we would probably consider "age" and "sex" as control variables because they are related to blood pressure and physical activity and so, unless taken into account, could confound any conclusions regarding the primary relationship of interest.

In this section we focus on the use of variable classification, as it relates to the first two considerations above, in choosing an appropriate method of analysis. Table 2-1 provides a rough guide to help the researcher in this choice when several variables are involved. This guide distinguishes among various multivariable methods. It considers the types of variable sets usually associated with each method and gives a general description of the purposes of

TABLE 2-2 Application of Table 2-1 to examples in Chapter 1

Study	Name of Multivariable Method	Dependent Variable	Independent Variables	Purpose
Example 1.1	Multiple regression analysis	Patient satisfaction with medical care (SATIS), a continuous variable	WANT, WORRY, BIRTH, INFCOM, AFFCOM, AGE, EDUC, SOCLS, PAR	To describe the relationship between SATIS and the variables WANT, WORRY, etc.
Example 1.1	Discriminant analysis	Complications of pregnancy (0 = no, 1 = yes), a nominal variable	WANT, WORRY, BIRTH, INFCOM, AFFCOM, AGE, EDUC, SOCLS, PAR	To determine whether and to what extent the independent variables can be used to discriminate between mothers having and not having pregnancy complications
Example 1.2	Multiple regression analysis	Sudden-death rate for a county in North Carolina during a 5-year period, a continuous variable	Water hardness index for county, distance between hospital and county seat, per capita income, population per physician	To describe the relationship between sudden-death rate and the independent variables
Example 1.3	Analysis of covariance	Posttest score, a continuous variable	Pretest score, group designation (e.g., 1 = experimental, 0 = control)	To compare posttest scores for experimental and control groups, after adjusting for the possible effect of pretest scores
Example 1.4	Analysis of covariance	Social-influence-activity score, a continuous variable	Race of subject (e.g., 1 = white, 2 = black), age, height, etc.	To determine whether one racial group dominates the other in biracial dyads, after controlling for age, height, etc.
Example 1.4	Two-way analysis of variance	Social-influence-activity difference score between black and white partners in biracial dyads, a continuous variable	Race of experimenter (e.g., 1 = white, 2 = black); race of comparison norm (e.g., 1 = white, 2 = black)	To determine whether there is any effect of experimenter's race and of the comparison norm's race on the difference score

TABLE 2-2   (Continued)

Study	Name of Multivariable Method	Dependent Variable	Independent Variables	Purpose
Example 1.5	Two-way analysis of variance	Systolic blood pressure (SBP), a continuous variable	Discrepancy between attitude toward and involvement in modern life, categorized as "high" and "low"; discrepancy between preparation for and involvement in modern life, categorized as "high" and "low"	To describe the relationship between nominal discrepancy scores and SBP
Example 1.5	Factor analysis	None	Several scores from questions on attitudes of Ponapeans toward modern life	To reduce the information in these scores to a single composite variable summarizing an individual's attitudes toward modern life
Example 1.6	Factor analysis	None	Several cognitive test scores for three different age groups	To compare interrelationships of variables for three age groups

each method. However, in addition to using the table, one must carefully check the statistical assumptions being made. These assumptions will be described fully later in the text. Table 2-2 shows how these guidelines can be applied to the examples given in Chapter 1.

It is important to note that several methods for dealing with multivariable problems are *not* included in Table 2-1 or in this text. Among those not included are nonparametric methods of analysis of variance, multivariate multiple regression, and multivariate analysis of variance (which are extensions of the corresponding methods given here that allow for *several* dependent variables), and methods of cluster analysis. In this book, we will not cover all multivariable techniques, but simply those used most often by health and social researchers.

(Statisticians generally use the term *multivariate analysis* to describe a method in which several dependent variables can be considered simultaneously. On the other hand, researchers in the biomedical and health sciences who are not statisticians use this term to describe any statistical technique involving several variables, even if only one dependent variable is considered at a time. In this text we prefer to avoid this confusion by using the term *multivariable analysis* to denote the latter, more general description.)

It should also be pointed out that multiple regression analysis is a general technique that can be utilized with all kinds of variables. In fact, analysis of variance and analysis of covariance may be considered as special cases of regression analysis. Furthermore, even the method of discriminant analysis can be implemented using regression analysis, although the theoretical bases of the two methods are different.

Finally, factor analysis has been included in Table 2-1, since it is often used in conjunction with one or more of the other methods. Often, as when dealing with a sociological or psychological questionnaire, the researcher is confronted with a large set of basic variables that he wishes to reduce to a much smaller set of meaningful variables. In this case factor analysis can be used to combine the basic variables into one or more composite variables that summarize the essential information contained in all the variables.

References

Green, L. W. (1970). "Manual for Scoring Socioeconomic Status for Research on Health Behaviors." *Public Health Rept.*, *85*: 815–827.

Hamilton, M. (1971). "Sudden Death and Water Hardness in North Carolina Counties in 1956–1964." Master's thesis, Department of Epidemiology, University of North Carolina, Chapel Hill, N.C.

Thompson, S. J. (1972). "The Doctor–Patient Relationship and Outcomes of Pregnancy." Ph.D. dissertation, Department of Epidemiology, University of North Carolina, Chapel Hill, N.C.

3

Basic Statistics:
A Review

3-1 Preview

The purpose of this chapter is to review the fundamental statistical concepts and methods that are needed to understand the more sophisticated multivariable techniques discussed in this text. Also, through this review, we shall introduce the statistical notation (using conventional symbols whenever possible) employed throughout the text.

The broad area associated with the word *statistics* concerns the methods and procedures for collecting, classifying, summarizing, and analyzing data. We shall focus on the latter two activities here. The primary goal of most statistical analysis is to make *statistical inferences*, that is, to draw valid conclusions about a *population* of items or measurements based upon information contained in a *sample* from that population.

A *population* is any set of items or measurements of interest, and a *sample* is any subset of items selected from that population. Any characteristic of that population is called a *parameter*, and any characteristic of the sample is a *statistic*. A statistic may be considered to be an estimate of some population parameter, the accuracy of which may be good or bad.

Once sample data have been collected, it is useful, prior to analysis, to examine the data using tables, graphs, and *descriptive statistics,* such as the sample mean and the sample variance. Such descriptive efforts are important for representing the essential features of the data in easily interpretable terms.

Following such examination, statistical inferences are made through two related activities: *estimation* and *hypothesis testing*. The techniques involved here are based on certain assumptions about the probability pattern (or *distribution*) of the (*random*) variables being studied.

The key terms above—descriptive statistics, random variables, probability distribution, estimation, and hypothesis testing—will each be reviewed in the sections that follow.

16

3-2 Descriptive Statistics

A descriptive statistic may be defined as any single numerical measure computed from a set of data that is designed to describe a particular aspect or characteristic of the data set. The most common types of descriptive statistics are measures of *central tendency* and *variability* (or *dispersion*).

The central tendency in a sample of data refers to what is often called the "average value" of the variable being observed. Of the several measures of central tendency, the most commonly used is the sample mean, which we denote by $\bar{X}$ whenever our underlying variable is called X. The formula for the sample mean is given by

$$\bar{X} = \frac{\sum_{i=1}^{n} X_i}{n}$$

where n denotes the sample size; $X_1, X_2, \ldots, X_n$ denote the n independent measurements on X; and Σ denotes summation. The sample mean $\bar{X}$, in contrast to other measures of central tendency such as the median or mode, uses in its computation all the observations in the sample. This property means that $\bar{X}$ is necessarily affected by the presence of extreme X-values, in which case it may be preferable to use the median instead of the mean. A remarkable property of the sample mean, which makes it particularly useful in making statistical inferences, follows from the *Central Limit Theorem*, which states that *whenever n is moderately large, $\bar{X}$ has approximately a normal distribution, regardless of the distribution of the underlying variable X.*

Measures of central tendency (such as $\bar{X}$) do not, however, completely summarize all the features of the data. It is obvious, for example, that two sets of data with the same mean can differ widely in appearance (e.g., an $\bar{X}$ of 4 results both from the values 4, 4, and 4 and from the values 0, 4, and 8). Thus, we customarily consider, in addition to $\bar{X}$, measures of variability, which tell us the extent to which the values of the measurements in the sample differ from one another.

The two measures of variability most often considered are the *sample variance* and the *sample standard deviation.* These are given by the following formulas when considering observations $X_1, X_2, \ldots, X_n$ on a single variable X:

$$\text{sample variance} = S^2 = \frac{1}{n-1} \sum_{i=1}^{n} (X_i - \bar{X})^2$$

$$\text{sample standard deviation} = S = \sqrt{\frac{1}{n-1} \sum_{i=1}^{n} (X_i - \bar{X})^2}$$

The formula for S^2 describes variability in terms of an average of squared deviations from the sample mean, although $n-1$ is used as the divisor instead of n. This use of $n-1$ is due essentially to considerations that make S^2 a good estimator of the variability in the entire population.

A drawback to the use of S^2 is that it is in squared units of the underlying variable X. To have a measure of dispersion that is expressed in the same units as X, we simply take the square root of S^2 and call it the sample standard deviation S. A convenient computational formula for S is given by

$$S = \sqrt{\frac{\sum\limits_{i=1}^{n} X_i^2 - \left(\sum\limits_{i=1}^{n} X_i\right)^2 / n}{n-1}}$$

Using S in combination with $\bar{X}$ thus gives a fairly succinct picture of both the amount of spread and the center of the data, respectively.

When more than one variable is being considered in the same analysis, as will be the case throughout this text, we shall use different letters and/or different subscripts to differentiate among the variables, and we will modify the notations for mean and variance accordingly. For example, if we are using X to stand for age and Y to stand for systolic blood pressure, we would denote the sample mean and sample standard deviation for each variable as $(\bar{X}, S_X)$ and $(\bar{Y}, S_Y)$, respectively.

3-3 Random Variables and Distributions

The term *random variable* is used to denote a variable whose observed values may be considered as outcomes of a stochastic or random experiment (e.g., the drawing of a random sample). The values of such a variable in a particular sample, then, cannot be anticipated with certainty before the sample is gathered. Thus, if we select a random sample of persons from some community and determine the systolic blood pressure (W), cholesterol level (X), race (Y), and sex (Z) of each person, then W, X, Y, and Z are four random variables whose particular realizations (or observed values) for a given person in the sample cannot be known for sure beforehand. In this text we shall denote random variables by capital italic letters.

The probability pattern that gives the relative frequencies associated with all the possible values of a random variable in a population is generally called the *probability distribution* of the random variable. We represent such a distribution by a table, graph, or mathematical expression that provides the probabilities corresponding to the different values or ranges of values taken on by a random variable.

Discrete random variables (such as the number of deaths in a sample of patients or the number of arrivals at a clinic), whose possible values are countable, have (gappy) distributions that are graphed as a series of lines, the heights of these lines representing the probabilities associated with the various possible discrete outcomes (Figure 3-1a). *Continuous* ran-

FIGURE 3-1   Discrete and continuous distributions ($P(X = a)$ is read: "probability that X takes the value a")

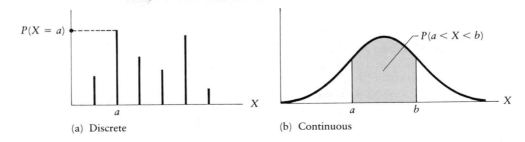

(a) Discrete   (b) Continuous

dom variables (such as blood pressure and weight), whose possible values are uncountable, have (nongappy) distributions that are graphed as smooth curves, and an *area* under such a curve represents the probability associated with a *range of values* of the continuous variable (Figure 3-1b). We note in passing that the probability of a continuous random variable taking on one particular value is 0, because there is no area above a single point.

In the next two sections we will discuss two particular distributions of enormous practical importance: the binomial (which is discrete) and the normal (which is continuous).

3-3-1   The Binomial Distribution

A *binomial* random variable describes the number of occurrences of a particular event in a series of n trials under the following four conditions:

1. The n trials are identical.

2. The outcome of any one trial is independent of (i.e., is not affected by) the outcome of any other trial.

3. There are two possible outcomes on each trial: "success" (i.e., the event of interest occurs) or "failure" (i.e., the event of interest does not occur), with probabilities p and $q = 1 - p$, respectively.

4. The probability of success, p, remains the same for all trials.

For example, the distribution of the number of lung cancer deaths in a random sample of $n = 400$ persons would be considered to be binomial if the four conditions were satisfied, as would the number of persons in a sample of $n = 70$ who favor a certain form of legislation.

The two *parameters* of the binomial distribution that one must specify to determine the precise shape of the probability distribution and to compute binomial probabilities are n and p. The usual notation used for this distribution is, therefore, $B(n, p)$; if X has a binomial distribution, it is customary to write

$$X \sim B(n, p)$$

where $\sim$ stands for "is distributed as." The probability formula for this discrete random variable X is given by the expression

$$P(X = j) = {}_nC_j\, p^j (1 - p)^{n-j}, \qquad j = 0, 1, \ldots, n$$

where ${}_nC_j = n!/j!(n - j)!$ denotes the number of combinations of n things taken j at a time. We shall return to a consideration of the binomial distribution and its extension to multinomial data in our discussion of analysis of categorical data in Chapter 22.

3-3-2   The Normal Distribution

The *normal distribution*, denoted as $N(\mu, \sigma)$, where μ and σ are the two parameters, is described by the well-known bell-shaped curve (Figure 3-2). The parameters μ (the mean) and σ (the standard deviation) characterize the center and the spread of the distribution, respectively. We generally attach a subscript to the parameters μ and σ to distinguish among variables; that is, we often write

$$X \sim N(\mu_X, \sigma_X)$$

to denote a normally distributed X.

FIGURE 3-2   The normal distribution

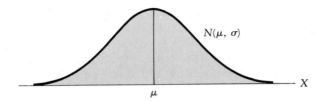

An important property of any normal curve is its *symmetry*, which distinguishes it from some other continuous distributions that we shall discuss. This symmetry property is quite helpful when using tables to determine probabilities or percentiles of the normal distribution.

Probability statements concerning a normally distributed random variable X that are of the form $P(a \leq X \leq b)$ require for computation the use of a single table (Table A-1 in Appendix A). This table gives the probabilities (or areas) associated with the *standard normal distribution*, which is a normal distribution with $\mu = 0$ and $\sigma = 1$. It is customary to denote a standard normal random variable by the letter Z, so that we write

$$Z \sim N(0, 1)$$

To compute the probability $P(a \leq X \leq b)$ for an X that is $N(\mu, \sigma)$, we must transform (i.e., *standardize*) X to Z by applying the conversion formula

$$Z = \frac{X - \mu}{\sigma} \tag{3.1}$$

to each of the elements in the probability statement about X as follows:

$$P(a \leq X \leq b) = P\left(\frac{a - \mu}{\sigma} \leq Z \leq \frac{b - \mu}{\sigma}\right)$$

We then look up the equivalent probability statement concerning Z in the $N(0, 1)$ tables.

This rule also applies to the sample mean $\bar{X}$ whenever the underlying variable X is normally distributed or whenever the sample size is moderately large (by the Central Limit Theorem). However, since the standard deviation of $\bar{X}$ is $\sigma/\sqrt{n}$, the conversion formula has the form

$$Z = \frac{\bar{X} - \mu}{\sigma/\sqrt{n}}$$

An inverse procedure for computing a probability for a range of values of X, as illustrated above, is to find a percentile of the distribution of X. A *percentile* is a value of the variable X *below which* the area under the probability distribution has a certain specified value. We denote the $(100p)$th percentile of X by X_p and picture it as in Figure 3-3, where p is the amount of area under the curve to the left of X_p. In determining X_p for a given p we must again use the conversion formula (3.1). However, since the procedure requires that we first determine Z_p and then convert back to X_p, we generally rewrite the conversion formula as

$$X_p = \mu + \sigma Z_p \tag{3.2}$$

FIGURE 3-3   The $(100p)$th percentiles of X and Z

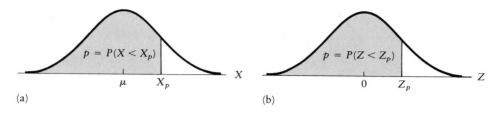

(a) $\qquad\qquad\qquad\qquad\qquad$ (b)

For example, if $\mu = 140$, $\sigma = 40$, and we wish to find $X_{.95}$, the $N(0, 1)$ tables first give us $Z_{.95} = 1.645$, with which we convert back to $X_{.95}$ as follows:

$$X_{.95} = 140 + (40)Z_{.95} = 140 + 40(1.645) = 205.8$$

Formulas (3.1) and/or (3.2) can also be used to approximate probabilities and percentiles for the binomial distribution $B(n, p)$ whenever n is moderately large (e.g., $n > 20$). The usual conditions required for this approximation to be accurate are that both $np > 5$ and $n(1 - p) > 5$. Under such conditions the mean and standard deviation of the approximating normal distribution are

$$\mu = np \quad \text{and} \quad \sigma = \sqrt{np(1 - p)}$$

3-4 Sampling Distributions of t, χ^2, and F

The Student's t, chi-square (χ^2), and Fisher's F distributions are particularly important in statistical inference making.

The (*Student's*) *t distribution* (Figure 3-4a), which is symmetric about 0, as is the normal distribution, was originally developed to describe the behavior of the random variable

$$T = \frac{\bar{X} - \mu}{S/\sqrt{n}} \tag{3.3}$$

which represents an alternative to

$$Z = \frac{\bar{X} - \mu}{\sigma/\sqrt{n}}$$

whenever the population variance σ^2 is unknown and is estimated by S^2. The denominator of (3.3), $S/\sqrt{n}$, is the *estimated standard error of* $\bar{X}$. When the underlying distribution of X is normal, and when $\bar{X}$ and S^2 are calculated from a random sample from that distribution, then (3.3) has the *t distribution with* $n - 1$ *degrees of freedom*, where $n - 1$ is the parameter that must be specified in order to look up tabulated percentiles of this distribution. We denote all this by writing

$$T = \frac{\bar{X} - \mu}{S/\sqrt{n}} \frown t_{n-1}$$

FIGURE 3-4   The t, χ^2, and F distributions

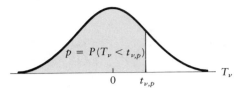

(a) Student's t distribution

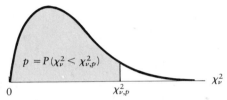

(b) χ^2 distribution

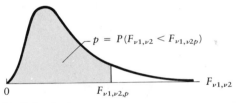

(c) F distribution

It has generally been shown by statisticians that the t distribution is often appropriate for describing the behavior of a random variable of the general form

$$T = \frac{\hat{\theta} - \mu_{\hat{\theta}}}{S_{\hat{\theta}}} \tag{3.4}$$

where $\hat{\theta}$ is any random variable that is approximately normally distributed with mean $\mu_{\hat{\theta}}$ and standard deviation $\sigma_{\hat{\theta}}$, where $S_{\hat{\theta}}$ is the estimated standard error of $\hat{\theta}$ and where $\hat{\theta}$ and $S_{\hat{\theta}}$ are statistically independent. For example, when random samples are taken from two normally distributed populations with the same standard deviation (e.g., from $N(\mu_1, \sigma)$ and $N(\mu_2, \sigma)$) and we consider $\hat{\theta} = \bar{X}_1 - \bar{X}_2$ in (3.4), we can write

$$T = \frac{(\bar{X}_1 - \bar{X}_2) - (\mu_1 - \mu_2)}{S_p \sqrt{\dfrac{1}{n_1} + \dfrac{1}{n_2}}} \sim t_{n_1 + n_2 - 2}$$

where

$$S_p^2 = \frac{(n_1 - 1)S_1^2 + (n_2 - 1)S_2^2}{n_1 + n_2 - 2} \tag{3.5}$$

estimates the common variance σ^2 in the two populations. The quantity S_p^2 is called a *pooled sample variance*, since it is calculated by pooling the data from both samples in order to estimate the common variance σ^2.

The *chi-square* (or χ^2) *distribution* (Figure 3-4b) is a nonsymmetric distribution and describes, for example, the behavior of the nonnegative random variable

$$\frac{(n-1)S^2}{\sigma^2} \tag{3.6}$$

where S^2 is the sample variance based on a random sample of size n from a normal distribution. The variable given by (3.6) has the chi-square distribution with $n-1$ degrees of freedom:

$$\frac{(n-1)S^2}{\sigma^2} \sim \chi^2_{n-1}$$

Because of the nonsymmetry of the chi-square distribution, both upper and lower percentage points of the distribution need to be tabulated, and such tabulations are solely a function of the degrees of freedom associated with the particular χ^2 distribution of interest. The chi-square distribution has widespread application in the analyses of categorical data (see Chapter 22).

The *F distribution* (Figure 3-4c), which is also skewed to the right, like the chi-square distribution, is appropriate for modeling the probability distribution of the ratio of independent estimators of two population variances. For example, given random samples of sizes n_1 and n_2 from $N(\mu_1, \sigma_1)$ and $N(\mu_2, \sigma_2)$, respectively, so that estimates S_1^2 and S_2^2 of σ_1^2 and σ_2^2 can be calculated, it can be shown that

$$\frac{S_1^2/\sigma_1^2}{S_2^2/\sigma_2^2} \tag{3.7}$$

has the F distribution with the two parameters $n_1 - 1$ and $n_2 - 1$, which are called the *numerator* and *denominator* degrees of freedom, respectively. We write this as

$$\frac{S_1^2\sigma_2^2}{S_2^2\sigma_1^2} \sim F_{n_1-1, n_2-1}$$

The F distribution can also be related to the t distribution when the numerator degrees of freedom equals 1; that is, the square of a variable distributed as Student's t with ν degrees of freedom has the F distribution with 1 and ν degrees of freedom. In other words,

$$T^2 \sim F_{1,\nu} \quad \text{if and only if} \quad T \sim t_\nu$$

Percentiles of the t, χ^2, and F distributions may be obtained from Tables A-2, A-3, and A-4 in Appendix A. The shapes of the curves that describe these probability distributions, together with the notation that we will use to denote their percentile points, are given in Figure 3-4.

3-5 Statistical Inference: Estimation

Two general categories of statistical inference, estimation and hypothesis testing, can be distinguished by their differing purposes: Estimation is concerned with estimating the specific value of an unknown population parameter; hypothesis testing is concerned with making a decision about a hypothesized value of an unknown population parameter.

In estimation, which we focus on in this section, we wish to estimate an unknown parameter θ using a random variable $\hat{\theta}$ ("theta hat," called a *point estimator* of θ). This point estimator is in the form of a formula or rule; for example,

$$\bar{X} = \frac{1}{n} \sum_{i=1}^{n} X_i \quad \text{or} \quad S^2 = \frac{1}{n-1} \sum_{i=1}^{n} (X_i - \bar{X})^2$$

tells us how to calculate a specific point estimate given a particular set of data.

To estimate a parameter of interest (e.g., a population mean μ, a binomial proportion p, a difference between two population means $\mu_1 - \mu_2$, or a ratio of two population standard deviations σ_1/σ_2), the usual procedure is to select a random sample from the population or populations of interest, calculate the point estimate of the parameter, and then associate with this estimate a measure of its variability, which usually takes the form of a confidence interval for the parameter of interest.

As its name implies, a *confidence interval* (often abbreviated CI) consists of two boundary points between which we have a certain specified *level of confidence* that the population parameter lies. For example, a 95% confidence interval for a parameter θ would consist of lower and upper limits determined so that, in repeated sets of samples of the same size, 95% of all such intervals would be expected to contain the parameter θ. Care must be taken when interpreting such a confidence interval not to consider θ as a random variable that either falls or does not fall in the calculated interval; rather, θ is a fixed (unknown) constant and the random quantities are the lower and upper limits of the confidence interval, which vary from sample to sample.

We shall illustrate the procedure for computing a confidence interval with two examples, one concerning estimation of a single population mean μ, and one of the difference between two population means $\mu_1 - \mu_2$. In each case we shall find that the appropriate confidence interval required has the general form

$$\begin{pmatrix} \text{point estimate of} \\ \text{the parameter} \end{pmatrix} \pm \begin{pmatrix} \text{percentile of} \\ \text{the } t \text{ distribution} \end{pmatrix} \cdot \begin{pmatrix} \text{estimated standard} \\ \text{error of the estimate} \end{pmatrix} \tag{3.8}$$

This general form also applies to confidence intervals for other parameters considered in the remainder of the text (e.g., those considered in multiple regression analysis).

Example   Suppose that we have determined the Quantitative Graduate Record Examination (QGRE) scores for a random sample of 9 student applicants to a certain graduate department in a university, and we have found that $\bar{X} = 540$ and $S = 50$. If we wish to estimate with 95% confidence the population mean QGRE score (μ) for all such applicants to the department, and we are willing to assume that the population of such scores from which our random sample was selected is approximately normally distributed, then the confidence interval for μ is given by the general formula

$$\bar{X} \pm t_{n-1, 1-\alpha/2} \left(\frac{S}{\sqrt{n}} \right) \tag{3.9}$$

which gives the $100(1 - \alpha)\%$ (small-sample) confidence interval for μ when σ is unknown. In our problem $\alpha = 1 - .95 = .05$ and $n = 9$, so by substituting the given information into

(3.9), we obtain

$$520 \pm t_{8,0.975} \left(\frac{50}{\sqrt{9}}\right)$$

Since $t_{8,0.975} = 2.3060$, this formula becomes

$$520 \pm 2.3060 \left(\frac{50}{\sqrt{9}}\right)$$

or

$$520 \pm 38.43$$

Our 95% confidence interval for μ is thus given by

$$(481.57, 558.43)$$

One final point is worth mentioning. If we wanted to use this confidence interval to help us decide whether 600 is a likely value for μ (i.e., we are interested in making a decision concerning a specific value for μ), we would conclude that 600 is not a likely value, since it is not contained in the 95% confidence interval for μ just developed. This helps to illustrate the connection between estimation and hypothesis testing.

Example   Suppose that we wish to compare the change in health status of two groups of mental patients undergoing different forms of treatment for the same disorder. Suppose that we have a measure of change in health status based on a questionnaire given to each patient at two different times, and we are willing to assume that this measure of change in health status is approximately normally distributed with the same variance for the populations of patients from which we selected our independent random samples. The data obtained are summarized as follows:

Group 1: $n_1 = 15$, $\bar{X}_1 = 15.1$, $S_1 = 2.5$
Group 2: $n_2 = 15$, $\bar{X}_2 = 12.3$, $S_2 = 3.0$

where the underlying variable X denotes the change in health status between time 1 and time 2.

A 99% confidence interval for the true mean difference ($\mu_1 - \mu_2$) in health status change between these two groups would be given by the following formula (which assumes equal population variances; i.e., $\sigma_1^2 = \sigma_2^2$):

$$(\bar{X}_1 - \bar{X}_2) \pm t_{n_1+n_2-2,1-\alpha/2} S_p \sqrt{\frac{1}{n_1} + \frac{1}{n_2}} \qquad (3.10)$$

where S_p^2 is the pooled sample variance given by (3.5). Here we have

$$S_p^2 = \frac{(15 - 1)(2.5)^2 + (15 - 1)(3.0)^2}{15 + 15 - 2} = 7.625$$

so

$$S_p = \sqrt{7.625} = 2.76$$

Since $\alpha = .01$, our percentile in (3.10) is given by $t_{28,0.995} = 2.7633$. So, the 99% confidence interval for $\mu_1 - \mu_2$ is given by

$$(15.1 - 12.3) \pm 2.7633(2.76) \sqrt{\frac{1}{15} + \frac{1}{15}}$$

which reduces to

$$2.80 \pm 2.78$$

yielding the following 99% confidence interval for $\mu_1 - \mu_2$:

$$(0.02, 5.58)$$

Since the value 0 is not contained in this interval, we would conclude (with 99% confidence) that there is a significant difference in health status change between the two groups.

3-6 Statistical Inference: Hypothesis Testing

Although closely related to estimating confidence intervals, hypothesis testing has a slightly different orientation. When developing a confidence interval, we use our sample data to estimate what we think is a *likely* set of values for the parameter of interest; when performing a statistical test of a null hypothesis concerning a certain parameter, on the other hand, we use our sample data to *test* whether our estimated value for the parameter is *different enough* from the hypothesized value to conclude that the null hypothesis is *unlikely* to be true.

The general procedure used in testing a statistical null hypothesis is basically the same regardless of the parameter being considered. This procedure (which we will illustrate by example) consists of the following seven steps:

1. Check the assumptions concerning the properties of the underlying variable(s) being measured that are needed to justify the use of the testing procedure being considered.

2. State the null hypothesis H_0 and the alternative hypothesis H_A.

3. Specify the significance level α.

4. Specify the test statistic to be used and its distribution under H_0.

5. Form the decision rule for rejecting and not rejecting H_0 (i.e., specify the rejection and acceptance regions for the test).

6. Compute the value of the test statistic from the observed data.

7. Draw your conclusions concerning rejection or nonrejection of H_0.

Example Suppose that we again consider the random sample of 9 student applicants with mean QGRE score $\bar{X} = 520$ and standard deviation $S = 50$. Suppose, further, that the department chairperson, because of the declining reputation of the department, suspects that this year's applicants are not quite as good quantitatively as in the previous five years, for which the average QGRE score of applicants was 600. Under the assumption that the population of QGRE scores from which our random sample has been selected is normally distributed, it is possible to test the null hypothesis that the population mean score associated with this year's applicants is 600 versus the alternative hypothesis that it is less than 600. The *null hypothesis*, in mathematical terms, is then $H_0: \mu = 600$, which asserts that the population mean μ for this year's applicants is not different from what it has generally been in the past. The *alternative hypothesis* is stated as $H_A: \mu < 600$, since there is interest in detecting whether the QGRE scores, on the average, have gotten worse. We have thus far considered the first two steps of our testing procedure:

1. Assumptions: The variable QGRE score has a normal distribution, from which a random sample has been selected.

2. Hypotheses: $H_0: \mu = 600$; $H_A: \mu < 600$.

The next step is to decide what error, or probability, we are willing to tolerate for incorrectly rejecting H_0 (i.e., making a Type I error, which is discussed later in this chapter). We call this probability of making a Type I error the *significance level* α.

(There are actually two types of errors that can be made when performing a statistical test: A Type II error occurs if we fail to reject H_0 when H_0 is actually false. We denote the probability of a Type II error by β and call $(1 - \beta)$ the *power* of the test. For a fixed sample size, α and β for a given test are inversely related; that is, lowering one has the effect of increasing the other. And, in general, the power of any statistical test can be raised by increasing the sample size.)

We usually specify α to have a value such as .1, .05, .025, or .01. Suppose, for now, that we choose $\alpha = .025$, so that step 3 is:

3. Use $\alpha = .025$.

Step 4 requires specification of the test statistic that will be used to test H_0. In this case, with $H_0: \mu = 600$, we can state that:

4. $T = \dfrac{\bar{X} - 600}{S/\sqrt{9}} \frown t_8$ under H_0.

Step 5 requires us to specify the decision rule that we will use to accept or reject H_0. In determining this rule, we divide the possible values of T into two sets: One of these sets is called the *rejection region* (or *critical region*), which consists of those values of T for which we reject H_0; the other set, called the *acceptance region*, consists of those T-values for which we do not reject H_0. The idea here is that if our computed value of T falls in the rejection region, we conclude that our observed results have deviated enough from H_0 to cast considerable doubt on the validity of the null hypothesis.

In our example we determine the critical region by choosing a point from t tables, called the *critical point*, which defines the boundary between the acceptance and rejection regions. The value we choose is

$-t_{8,0.975} = -2.306$

FIGURE 3-5   The critical point for our example

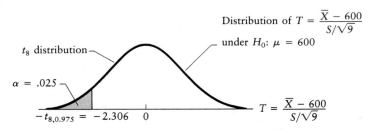

$$\text{Distribution of } T = \frac{\bar{X} - 600}{S/\sqrt{9}}$$

$$\text{under } H_0: \mu = 600$$

t_8 distribution

$\alpha = .025$

$-t_{8,0.975} = -2.306 \quad 0$

$$T = \frac{\bar{X} - 600}{S/\sqrt{9}}$$

in which case the probability that the test statistic takes values less than -2.306 under H_0 is exactly $\alpha = .025$, the significance level (Figure 3-5). We thus have the following decision rule:

5. Reject H_0 if $T = \dfrac{\bar{X} - 600}{S/\sqrt{9}} < -2.306$; do not reject H_0 otherwise.

We now simply apply the decision rule to our data by computing the observed value of T. In our example, since $\bar{X} = 520$ and $S = 50$, our computed T is

6. $T = \dfrac{\bar{X} - 600}{S/\sqrt{9}} = \dfrac{520 - 600}{50/3} = -4.8.$

The last step is to make the decision about H_0 based on the rule given in step 5:

7. Since $T = -4.8$, which lies below -2.306, we reject H_0 at significance level .025 and conclude that there is evidence that students currently applying to the department have QGRE scores *significantly lower* than 600.

In addition to the procedure above, it is often of interest to compute a *P-value*, which quantifies *exactly how unusual the observed results would be if H_0 were true*. An equivalent way of describing the P-value is as follows: *The P-value gives the probability of obtaining a value of the test statistic at least as unfavorable to H_0 as the observed value* (Figure 3-6). To get an idea of the approximate size of the P-value in this example, the approach is to

FIGURE 3-6   The P-value

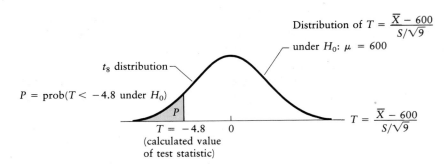

$$\text{Distribution of } T = \frac{\bar{X} - 600}{S/\sqrt{9}}$$

$$\text{under } H_0: \mu = 600$$

t_8 distribution

$P = \text{prob}(T < -4.8 \text{ under } H_0)$

P

$T = -4.8 \quad 0$
(calculated value
of test statistic)

$$T = \frac{\bar{X} - 600}{S/\sqrt{9}}$$

determine from the table of the distribution of T under H_0 the two percentiles that bracket the observed value of T. In this case the two percentiles are

$$-t_{8,0.995} = -3.355 \quad \text{and} \quad -t_{8,0.9995} = -5.041$$

Since the observed value of T lies between these two values, we can conclude that the area we seek lies between the two areas corresponding to these two percentiles,

$$.0005 < P < .005$$

In interpreting this inequality, we observe that the P-value is *quite small*. This small P indicates that we have observed a highly unusual result if, indeed, H_0 is true; in fact, this P-value is so small as to lead us to reject H_0. Furthermore, the size of this P-value means that even for an α as small as .005 we would reject H_0.

For the general computation of a P-value, it is important to point out that the appropriate P-value for a two-tailed test will be twice that for the corresponding one-tailed test. Furthermore, if an investigator wishes to draw conclusions about a test on the basis of the P-value (e.g., in lieu of specifying α a priori), the following guidelines are recommended:

1.   If P is small (less than .01), reject H_0.

2.   If P is large (greater than .1), do not reject H_0.

3.   If $.01 < P < .1$, the significance is borderline; that is, we reject H_0 for $\alpha = .1$ but not for $\alpha = .01$.

Note that if we actually do specify α a priori, we reject H_0 when $P < \alpha$.

We now look at one more worked example about hypothesis testing, this time involving a comparison of two means, μ_1 and μ_2. Consider the following health-status-change data, which were discussed earlier:

$$\text{Group 1:} \quad n_1 = 15, \quad \bar{X}_1 = 15.1, \quad S_1 = 2.5$$
$$\text{Group 2:} \quad n_2 = 15, \quad \bar{X}_2 = 12.3, \quad S_2 = 3.0 \qquad (S_p = 2.76)$$

Suppose that we wish to test at significance level .01 whether the true average change in health status differs between the two groups. The steps required to perform this test would be carried out as follows:

1.   Assumptions: We have independent random samples from two normally distributed populations. The population variances are assumed to be equal.

2.   Hypotheses: $H_0: \mu_1 = \mu_2$; $H_A: \mu_1 \neq \mu_2$.

3.   Use $\alpha = .01$.

4. $T = \dfrac{(\bar{X}_1 - \bar{X}_2) - 0}{S_p \sqrt{\dfrac{1}{n_1} + \dfrac{1}{n_2}}} \frown t_{28}$ under H_0.

5.   Reject H_0 if $|T| \geq t_{28,0.995} = 2.763$; do not reject H_0 otherwise (Figure 3-7).

6. $T = \dfrac{(\bar{X}_1 - \bar{X}_2) - 0}{S_p \sqrt{\dfrac{1}{n_1} + \dfrac{1}{n_2}}} = \dfrac{15.1 - 12.3}{2.76 \sqrt{\dfrac{1}{15} + \dfrac{1}{15}}} = 2.78.$

FIGURE 3-7   Critical region for the health-status-change example

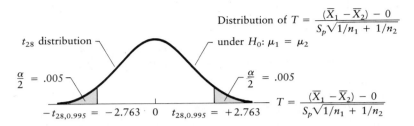

FIGURE 3-8   P-value for the health-status-change example

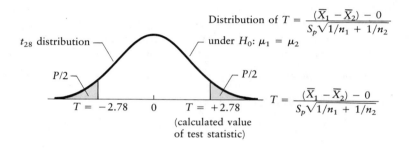

7. Since $T = 2.78$ exceeds $t_{28,0.995} = 2.763$, we would reject H_0 at $\alpha = .01$ and conclude that the true average change in health status is different in the two groups.

The P-value for this test is given by the shaded area in Figure 3-8. For the t distribution with 28 degrees of freedom, we find that $t_{28,0.995} = 2.763$ and $t_{28,0.9995} = 3.674$. Thus, $P/2$ is given by the inequality

$$1 - .9995 < \frac{P}{2} < 1 - .995$$

so

$$.001 < P < .01$$

3-7 Error Rates, Power, and Sample Size

Table 3-1 summarizes the decisions that result from hypothesis testing. If the true state of nature is that the null hypothesis is true and the decision is made that the null hypothesis is true, then a correct decision has been made. Similarly, if the true state of nature is that the alternative hypothesis is true and the decision is made that the alternative is true, then a correct decision has been made. If the true state of nature is that the null hypothesis is true but the decision is made to choose the alternative, then a false positive error has been made, commonly referred to as a *Type I error*. If the true state of nature supports the alternative

TABLE 3-1   Outcomes of hypothesis testing

Hypothesis Accepted	True State of Nature	
	H_0	H_A
H_0	Correct decision	False negative decision (Type II error)
H_A	False positive decision (Type I error)	Correct decision

hypothesis but one concludes that the null hypothesis is true, then a false negative error has been made, commonly referred to as a *Type II error*.

Table 3-2 summarizes the probabilities associated with the outcomes of hypothesis testing described above. If the true state of nature corresponds to the null hypothesis but the alternative hypothesis is chosen, then a Type I error has been made, with probability denoted by the symbol α. Hence, the chance of making a correct choice of H_0 when H_0 is true must be $1 - \alpha$. In turn, if the actual state of nature is that the alternative hypothesis is true but the null hypothesis is chosen, then a Type II error has occurred, with probability denoted by β. In turn, $1 - \beta$ is the probability of choosing the alternative hypothesis when it is true, and this probability is often called the *power of the test.*

Ideally, when we design a research study, we would like to use statistical tests for which both α and β are small (i.e., for which there is a small chance of making either a Type I or Type II error). For a given α, it is sometimes possible to determine the sample size required for the study so that β is no larger than some desired value for a particular alternative hypothesis of interest. Such a design consideration generally involves the use of a *sample size formula* pertinent to the research question(s). This formula usually requires the researcher to guess values for some of the unknown parameters to be estimated in the study (see Cohen, 1977, and Muller and Peterson, 1984).

For example, the classical sample size formula used for a one-sided test of $H_0: \mu_1 = \mu_2$ versus $H_A: \mu_2 > \mu_1$ when a random sample of size n is selected from each of two normal populations with common variance σ^2 is as follows:

$$n \geq \frac{2(Z_{1-\alpha} + Z_{1-\beta})^2\sigma^2}{\Delta^2}$$

TABLE 3-2   Probabilities of outcomes of hypothesis testing

Hypothesis Accepted	True State of Nature	
	H_0	H_A
H_0	$1 - \alpha$	β
H_A	α	$1 - \beta$

For chosen values of α, β, and σ^2, this formula provides the minimum sample size n required to detect a specified difference $\Delta = \mu_2 - \mu_1$ between μ_1 and μ_2 (i.e., to reject H_0: $\mu_2 - \mu_1 = 0$ in favor of H_A: $\mu_2 - \mu_1 = \Delta > 0$ with power $1 - \beta$). Thus, in addition to picking α and β, the researcher is forced to guess the size of the population variance σ^2 and to specify the difference Δ to be detected. An educated guess about the value of the unknown parameter σ^2 can sometimes be made by using information obtained from other related research studies. To specify Δ intelligently requires the researcher to decide on the smallest population mean difference $(\mu_2 - \mu_1)$ that is practically (as opposed to statistically) meaningful for the study.

For a fixed sample size, α and β are inversely related in the following sense. If one tries to guard against making a Type I error by choosing a small rejection region, then the nonrejection region (and hence β) will be large. Conversely, protecting against a Type II error necessitates using a large rejection region, leading to a large value for α. Increasing the sample size generally decreases β; of course, α remains unaffected.

It is common practice to conduct several statistical tests using the same data set. If such a data-set-specific series of tests is performed and each test is based on a size α rejection region, then the probability of making at least one Type I error will be much larger than α. This multiple-testing problem is a pervasive and bothersome one. One simple—but not optimal—method for addressing this problem is to employ the so-called *Bonferroni correction*. For example, if k tests are to be conducted and the overall Type I error rate (i.e., the probability of making at least one Type I error in k tests) is to be no more than α, then a rule of thumb is to conduct *each individual* test at a Type I error rate of α/k.

This simple adjustment will ensure that the overall Type I error rate will (at least approximately) be no larger than α. However, in many situations, this correction may lead to such a small rejection region for each individual test that the power of each such test may be too low to detect important deviations from the null hypotheses being tested. To resolve this antagonism between Type I and Type II error rates requires a conscientious design of the study and carefully considered error rates for planned analyses.

Problems

1. **a.** Give two examples of discrete random variables.
   b. Give two examples of continuous random variables.

2. Name the four levels of measurement and give an example of a variable at each level.

3. Assume Z is a normal random variable with mean 0 and variance 1.
   a. $P(Z \geq -1) = ?$
   b. $P(Z \leq ?) = .20$

4. **a.** $P(\chi_7^2 \geq ?) = .01$
   b. $P(\chi_{12}^2 \leq 14) = ?$

5. **a.** $P(T_{13} \leq ?) = .10$
   b. $P(|T_{28}| \geq 2.05) = ?$

6. **a.** $P(F_{6,24} \geq ?) = .05$
   b. $P(F_{5,40} \geq 2.9) = ?$

7. What are the (a) mean, (b) median, and (c) mode of the standard normal distribution?

8. A chi-square random variable can be thought as the square of what kind of random variable?

9. Find the (a) mean, (b) median, and (c) variance for the set of scores:

$$\chi = \{0, 2, 5, 6, 3, 3, 3, 1, 4, 3\}$$

(d) Find the set of Z scores for the data.

10. Which of the following statements about descriptive statistics is correct?
   a. *All* of the data is used to compute the median.
   b. The mean should be preferred to the median as a measure of central tendency if the data are noticeably skewed.
   c. The variance has the same units of measurement as the original observations.
   d. The variance can never be 0.
   e. The variance is like an average of squared deviations from the mean.

11. Suppose that the weight W of male patients registered at a diet clinic has the normal distribution with mean 190 and variance 100.
   a. For a random sample of patients of size $n = 25$, $P(\bar{W} < 180)$, in which $\bar{W}$ denotes the mean weight, is equivalent to saying $P(Z > ?)$. (*Note:* Z is a standard normal random variable.)
   b. Find the interval (a, b) such that $P(a < \bar{W} < b) = .80$ for the same random sample in (a).

12. The limits of a 95% confidence interval for the mean μ of a normal population with unknown variance are found by adding to and subtracting from the sample mean a certain multiple of the estimated standard error of the sample mean. If the sample size on which this confidence interval is based is 28, the *multiple* referred to in the previous sentence is the number ____.

13. A random sample of 32 persons attending a certain diet clinic was found to have lost (over a three-week period) an average of 30 pounds, with a sample standard deviation of 11. For these data, a 99% confidence interval for the true mean weight loss for all patients attending the clinic would have the limits (?, ?).

14. From two normal populations assumed to have the same variance, independent random samples of sizes 15 and 19 were drawn. The first sample (with $n_1 = 15$) yielded mean and standard deviation 111.6 and 9.5, respectively, while the second sample ($n_2 = 19$) gave mean and standard deviation 100.9 and 11.5, respectively. The estimated standard error of the difference in sample means is ____.

15. For the data of Problem 14, suppose that a test of H_0: $\mu_1 = \mu_2$ versus H_A: $\mu_1 > \mu_2$ yielded a computed value of the appropriate test statistic equal to 2.55.
   a. What conclusions should be drawn for $\alpha = .05$?
   b. What conclusions should be drawn for $\alpha = .01$?

16. Test the hypothesis that average body weight is the same for two independent diagnosis groups from one hospital:

   Diagnosis group 1 data: $\{132, 145, 124, 122, 165, 144, 151\}$

   Diagnosis group 2 data: $\{141, 139, 172, 131, 150, 125\}$

   You may assume that the data are normally distributed, with equal variance in the two groups. What conclusion should be drawn, with $\alpha = .05$?

17. Independent random samples are drawn from two normal populations, which are assumed to have the same variance. One sample (of size 5) yields mean 86.4 and standard deviation 8.0, and the other sample (of size 7) has mean 78.6 and standard deviation 10. The limits of a 99% confidence interval for the difference in population means are found by adding to and subtracting from the difference in sample means a certain multiple of the estimated standard error of this difference. This *multiple* is the number ____ .

18. If a 99% confidence interval for $\mu_1 - \mu_2$ is given by $4.8 < \mu_1 - \mu_2 < 9.2$, which of the following conclusions can be drawn *based on this interval*?
    a. Accept H_0: $\mu_1 = \mu_2$ at $\alpha = .05$ if the alternative is H_A: $\mu_1 \neq \mu_2$.
    b. Reject H_0: $\mu_1 = \mu_2$ at $\alpha = .01$ if the alternative is H_A: $\mu_1 \neq \mu_2$.
    c. Reject H_0: $\mu_1 = \mu_2$ at $\alpha = .01$ if the alternative is H_A: $\mu_1 < \mu_2$.
    d. Accept H_0: $\mu_1 = \mu_2$ at $\alpha = .01$ if the alternative is H_A: $\mu_1 \neq \mu_2$.
    e. Accept H_0: $\mu_1 = \mu_2 + 3$ at $\alpha = .01$ if the alternative is H_A: $\mu_1 \neq \mu_2 + 3$.

19. Assume we gather data, compute a T, and reject the null hypothesis of no difference. If, in fact, the null hypothesis is true, we have made (a) ____ . If the null hypothesis is false, we have made (b) ____ .
    Assume instead that our data lead us to accept the null hypothesis. If, in fact, the null hypothesis is true, we have made (c) ____ . If the null hypothesis is false, we have made (d) ____ .

20. Suppose that the critical region for a certain test of hypothesis is of the form $|T| \geq 2.5$ and the computed value of T from the data is -2.75. Which, if any, of the following statements are correct?
    a. H_0 should be rejected.
    b. The significance level α is the probability that, under H_0, either T is greater than 2.75 or less than -2.75.
    c. The acceptance region is given by $-3.5 < T < 3.5$.
    d. The acceptance region consists of those values of T above 3.5 or below -3.5.
    e. The P-value of this test is given by the area to the right of $T = 3.5$ for the distribution of T under H_0.

21. Suppose that $\bar{X}_1 = 125.2$ and $\bar{X}_2 = 125.4$ are the mean blood pressures for two samples of workers from different plants in the same industry. Suppose, further, that a test of H_0: $\mu_1 = \mu_2$ using these samples is rejected for $\alpha = .001$. Which of the following conclusions is most reasonable?
    a. There is a meaningful difference (clinically speaking) in population means but not a statistically significant difference.
    b. The difference in population means is both statistically and meaningfully significant.
    c. There is a statistically significant difference but not a meaningfully significant difference in population means.
    d. There is neither a statistically significant nor a meaningfully significant difference in population means.
    e. The sample sizes used must have been quite small.

22. The choice of an alternative hypothesis (H_A) should depend primarily on (choose all that apply):
    a. the data obtained from the study.

b. what the investigator is interested in determining.
c. the critical region.
d. the significance level.
e. the power of the test.

23. For each of the areas in the figure below, labeled a, b, c, and d, select an answer from the following: α, $1 - \alpha$, β, $1 - \beta$.

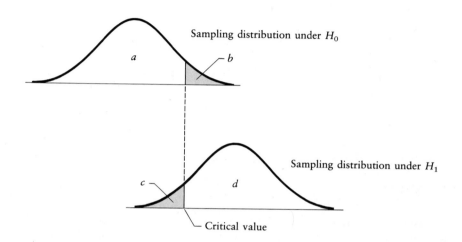

24. Suppose that H_0: $\mu_1 = \mu_2$ is the null hypothesis and that $.10 < P < .25$. What is the most appropriate conclusion?

25. Suppose that H_0: $\mu_1 = \mu_2$ is the null hypothesis and that $.005 < P < .01$. Which of the following conclusions is most appropriate?
a. Accept H_0 because P is small.
b. Reject H_0 because P is small.
c. Accept H_0 because P is large.
d. Reject H_0 because P is large.
e. Accept H_0 at $\alpha = .01$.

References

Cohen, J. (1977). *Statistical Power Analysis for the Behavioral Sciences,* 2nd ed. New York: Academic Press.

Muller, K. E., and Peterson, B. L. (1984). "Practical Methods for Computing Power in Testing the Multivariate General Linear Hypothesis." *Computational Statistics and Data Analysis, 2:* 143–158.

4

Introduction to Regression Analysis

4-1 Preview

Regression analysis is a statistical tool for evaluating the relationship of one or more independent variables $X_1, X_2, \ldots, X_k$ to a single, continuous dependent variable Y. It is most often used when the independent variables cannot be controlled, as when collected in a sample survey or other observational study. Nevertheless, it is equally applicable to more controlled experimental situations.

In practice there are several possibly overlapping situations for which a regression analysis is appropriate. Among these are the following:

Application 1   You wish to *characterize the relationship* between the dependent and independent variables by determining the extent, direction, and strength of the association. For example ($k = 2$), in Thompson's (1972) study described in Chapter 1, one of the primary questions was to describe the extent, direction, and strength of the association between "patient satisfaction with medical care" (Y) and the variables "affective communication between patient and physician" (X_1) and "informational communication between patient and physician" (X_2).

Application 2   You *seek a quantitative formula* or equation to describe (e.g., predict) the dependent variable Y as a function of the independent variables $X_1, X_2, \ldots, X_k$. For example ($k = 1$), a quantitative formula may be desired for a study of the effect of dosage of a blood-pressure-reducing treatment (X_1) on blood pressure change (Y).

Application 3   You want to describe quantitatively or qualitatively the relationship between $X_1, X_2, \ldots, X_k$ and Y but *control for the effects of still other variables* $C_1, C_2, \ldots, C_p$, which you believe have an important relationship with the dependent variable. For example ($k = 2, p = 2$), a study of the epidemiology of chronic diseases might

36

describe the relationship of blood pressure (Y) to smoking habits (X_1) and social class (X_2), controlling for age (C_1) and weight (C_2).

Application 4   You want to *determine which of several independent variables are important and which are not* for describing or predicting a dependent variable. You may want to control for other variables. You may also want to *rank* independent variables in their order of importance. In Thompson's (1972) study, for example ($k = 4$, $p = 2$), it was of interest to determine for the dependent variable "satisfaction with medical care" (Y) which of the following independent variables were important descriptors: WORRY (X_1), WANT (X_2), INFCOM (X_3), and AFFCOM (X_4). It was also considered necessary to control for AGE (C_1) and EDUC (C_2).

Application 5   You want to *determine the best mathematical model* for describing the relationship between a dependent and one or more independent variables. Any of the previous examples can be used to illustrate this.

Application 6   You wish to *compare several derived regression relationships*. An example would be a study to determine if smoking (X_1) is related to blood pressure (Y) in the same way for males as for females, controlling for age (C_1).

Application 7   You wish to *assess the interactive effects of two or more independent variables* with regard to a dependent variable. For example, you may wish to determine whether the relationship of alcohol consumption (X_1) to blood pressure level (Y) is different depending on smoking consumption (X_2). In particular, the relationship between alcohol and blood pressure might be quite strong for heavy smokers but very weak for nonsmokers. If so, we would say that there is *interaction* between alcohol and smoking. Then, any conclusions about the relationship between alcohol and blood pressure must take into account whether, and possibly how much, a person smokes. More generally, if X_1 and X_2 interact in their joint effect on Y, then the relationship of either X variable to Y is dependent on the value of the other X variable.

Application 8   You may want to *obtain a valid and precise estimate of one or more regression coefficients* from a larger set of regression coefficients in a given model. For example, you may wish to obtain an accurate estimate of the coefficient of a variable measuring alcohol consumption (X_1) in a regression model that relates hypertension status (Y), a dichotomous response variable, to X_1 and several other control variables (e.g., age and smoking status). Such an estimate may be used to quantify the effect of alcohol consumption on hypertension status after adjustment for the effects of certain control variables also in the model.

4-2 Association versus Causality

It is important to be cautious about the results obtained from a regression analysis or, more generally, from any form of analysis seeking to quantify an association (e.g., via a correlation coefficient) among two or more variables. Although the statistical computations used to produce an estimated measure of association may be correct, the estimate itself may be biased. Such bias may result from the method used to select subjects for the study, errors in the information used in the statistical analyses, or even other variables that can account for the observed association which have not been measured or appropriately considered in the analysis. (See Kleinbaum, Kupper, and Morgenstern, 1982, for a discussion of validity in epidemiologic research.)

For example, if diastolic blood pressure and physical activity level were measured on a sample of individuals at a particular time, a regression analysis might suggest that, on the average, blood pressure decreases with increased physical activity; further, such an analysis may provide evidence (e.g., based on a confidence interval) that this association is of moderate strength and is statistically significant. However, if the study involved only healthy adults, if physical activity level was measured inappropriately, or if other factors like age, race, and sex were not correctly taken into account, then the above conclusions might be considered invalid or at least questionable.

Continuing with the above example, if the investigators were satisfied that the findings were basically valid (i.e., the observed association was not spurious), could they then conclude that a low level of physical activity is a cause of high blood pressure? The answer is an unequivocal *no*!

The finding of a "statistically significant" association in a particular study (no matter how well done) does not establish a causal relationship. To evaluate claims of causality, the investigator must consider criteria that are external to the specific characteristics and results of any single study.

It is beyond the scope of this text to discuss causal inference making. Nevertheless, we will briefly review some key ideas on this subject. Most strict definitions of causality (e.g., Blalock, 1971; Susser, 1973) require that a *change* in one variable (X) always *produce* a change in another variable (Y).* This suggests that to demonstrate a cause–effect relationship between X and Y, *experimental proof* is required that a change in Y results from a change in X. Though it is needed, such experimental evidence is often impractical, infeasible, or even unethical to obtain, especially when considering risk factors (e.g., cigarette smoking or exposure to chemicals) which are potentially harmful to human subjects. Consequently, alternative criteria based on information *not* involving direct experimental evidence are typically employed when attempting to make causal inferences regarding variable relationships in human populations.

One school of thought regarding causal inference has produced a collection of procedures commonly referred to as *path analysis* (Blalock, 1971). To date, such procedures have been applied primarily to sociological and political science studies. Essentially, these methods attempt to assess causality indirectly by eliminating competing causal explanations via data analysis and finally arriving at an acceptable causal model that is not obviously contradicted by the data at hand. Thus, these methods, rather than attempting to establish a particular causal theory directly, arrive at a final causal model through a process of elimination. In this procedure, literature relevant to the research question must be considered in order to postulate causal models; in addition, various estimated correlation ("path") coefficients must be compared by means of data analysis.

A second and more widely used approach for making causal conjectures, particularly in

---

* An imperfect approximation to this ideal for real-world phenomena might be that, on the average, a change in Y is produced by a change in X.

the health and medical sciences, employs a judgmental (and more qualitative than quantitative) evaluation of the combined results from several studies using a set of operational criteria generally agreed upon as necessary (but not sufficient) for supporting a given causal theory. Efforts to define such a set of criteria were made in the late 1950s and early 1960s by investigators reviewing research on the health hazards of smoking. A list of general criteria for assessing the extent to which available evidence supports a causal relationship was formalized by Bradford Hill (1971), and this list has subsequently been adopted by many epidemiologic researchers. This list contains seven criteria, which we summarize below:

1. *Strength of association.* The stronger an observed association appears over a series of different studies, the less likely it is that this association is spurious because of bias.

2. *Dose–response effect.* The value of the dependent variable (e.g., the rate of disease development) changes in a meaningful pattern (e.g., increases) with the dose (or level) of the suspected causal agent under study.

3. *Lack of temporal ambiguity.* The hypothesized cause precedes the occurrence of the effect. Note that the ability to establish this time pattern will depend on the study design used.

4. *Consistency of the findings.* Most, or all, studies concerned with a given causal hypothesis produce similar results. Of course, studies dealing with a given question may all have serious bias problems that can diminish the importance of observed associations.

5. *Biological and theoretical plausibility of the hypothesis.* The hypothesized causal relationship is consistent with current biological and theoretical knowledge. Note, however, that the current state of knowledge may be insufficient to explain certain findings.

6. *Coherence of the evidence.* The findings do not seriously conflict with accepted facts about the outcome variable being studied (e.g., knowledge about the natural history of some disease).

7. *Specificity of the association.* The study factor (i.e., the suspected cause) is associated with only one effect (e.g., a specific disease). Note, however, that many study factors have multiple effects and that most diseases have multiple causes.

Clearly, applying the above criteria to a given causal hypothesis is hardly a straightforward matter. Even if these criteria are all satisfied, a causal relationship cannot be claimed with complete certainty. Nevertheless, without solid experimental evidence, the use of such criteria may be a logical and practical way to address the issue of causality, especially with regard to studies on human populations.

4-3 Statistical versus Deterministic Models

Although *causality* cannot be established by statistical analyses, associations among variables can be well quantified in a *statistical* sense. With proper statistical design and analysis, an investigator can model the extent to which changes in independent variables are related to changes in dependent variables. However, *statistical models* developed by using regression or other multivariable methods need to be distinguished from *deterministic models*.

The law of falling bodies in physics, for example, is a deterministic model that assumes an ideal setting: The dependent variable varies in a completely prescribed way according to a perfect (error-free) mathematical function of the independent variables.

Statistical models, on the other hand, allow for the possibility of error in describing a relationship. For example, in relating blood pressure to age, persons of the same age are unlikely to have exactly the same observed blood pressure. Nevertheless, with proper statistical methods, we might be able to conclude that, on the average, blood pressure increases with age. Further, appropriate statistical modeling can permit us to predict the expected blood pressure for a given age and to associate a measure of variability with that prediction. Through the use of probability and statistical theory, such statements take into account the uncertainty of the real world by means of measurement error and individual variability. Of course, because such statements are necessarily nondeterministic, they require careful interpretation. Unfortunately, such interpretation is often quite difficult to make.

4-4 Concluding Remarks

In this short chapter, we have informally introduced the general regression problem and indicated a variety of situations to which regression modeling can be applied. We have also cautioned the reader about the types of conclusions that can be drawn from such modeling efforts.

We now turn to the actual quantitative details in fitting a regression model to a set of data and then in estimating and testing hypotheses about important parameters in the model. In the next chapter, we will discuss the simplest form of regression model, a straight line; in subsequent chapters, we will consider more complex forms.

References

Blalock, H. M., Jr., ed. (1971). *Causal Models in the Social Sciences*. Chicago: Aldine Publishing Company.

Hill, A. B. (1971). *Principles of Medical Statistics*, 9th ed. New York: Oxford University Press.

Kleinbaum, D. G., Kupper, L. L., and Morgenstern, H. (1982). *Epidemiologic Research*. Belmont, Calif.: Lifetime Learning Publications.

Susser, M. (1973). *Causal Thinking in the Health Sciences*. New York: Oxford University Press.

Thompson, S. J. (1972). "The Doctor–Patient Relationship and Outcomes of Pregnancy." Ph.D. dissertation, Department of Epidemiology, University of North Carolina, Chapel Hill, N.C.

Straight-Line
Regression Analysis

5-1 Preview

The simplest (but by no means trivial) form of the general regression problem deals with one dependent variable Y and one independent variable X. We have previously described the general problem in terms of k independent variables $X_1, X_2, \ldots, X_k$. Let us now restrict our attention to the special case $k = 1$, but denote X_1 as X in order to keep our notation as simple as possible. To clarify the basic concepts and assumptions of regression analysis, we find it useful to begin with a single independent variable. Furthermore, researchers often find it sensible to begin by looking at one independent variable even when other independent variables are considered eventually.

5-2 Regression with a Single Independent Variable

We shall begin this section by describing the statistical problems of finding the *curve* (straight line, parabola, etc.) that *best fits* the data, closely approximating the true (but unknown) relationship between X and Y.

5-2-1 The Problem

Given a sample of n individuals (or other study units, such as geographical locations, time points, or pieces of physical material), we observe for each a value of X and a value of Y. We thus have n pairs of observations that can be denoted by $(X_1, Y_1), (X_2, Y_2), \ldots,$ (X_n, Y_n), where the subscripts now refer to different individuals rather than different variables. Because these pairs may be considered as points in two-dimensional space, we can plot them on a graph. Such a graph is called a *scatter diagram.* For example, measurements of age and systolic blood pressure for 30 individuals might yield the scatter diagram given in Figure 5-1.

FIGURE 5-1   Scatter diagram of age and systolic blood pressure

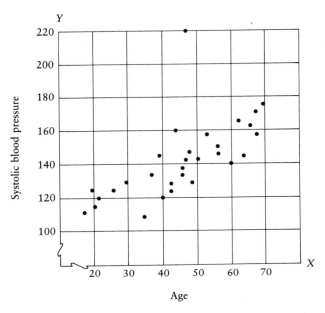

5-2-2 Basic Questions to Be Answered

There are two basic questions to be dealt with in any regression analysis:

1. What is the most appropriate mathematical model to use? In other words, should we use a straight line, a parabola, or log function, or what?

2. Given a specific model, what do we mean by and how do we determine the best-fitting model for the data? In other words, if our model is a straight line, how do we find the best-fitting line?

5-2-3 General Strategy

There are several general strategies for studying the relationship between two variables by means of regression analysis. The most common of these is called the *forward method*. This strategy begins with a simply structured model, usually a straight line, and then adds more complexity to the model in successive steps, if necessary. Another strategy is called the *backward method*. This strategy begins with a complicated model, such as a high-degree polynomial, and then successively simplifies the model, if possible, by eliminating unnecessary terms. A third approach uses *a model suggested from experience or theory*, which is revised either toward or away from complexity, as dictated by the data.

The strategy chosen depends essentially on the type of problem and on the data; there are no hard-and-fast rules. The quality of the results often depends more on the skill with which a strategy is applied than on the particular strategy chosen. In many instances, it is tempting to try many strategies and then to use those results that seem to provide the most "reasonable" interpretation of the relationship between the response and predictor variables. This exploratory approach demands particular care in order to insure the reliability of any conclusions. In Chapter 16 we will discuss the issue of choosing a strategy in detail.

For reasons to be discussed in Chapter 16, we tend to prefer the backward strategy. The forward method, however, corresponds more naturally to the usual development of theory, from simple to complex. In some simple situations, forward and backward strategies lead to the same final model. In general, however, this is not the case!

Since it is the simplest method to understand and can therefore be used as a basis for understanding other methods, we first describe the forward strategy. A step-by-step description is as follows:

1. Begin by assuming that a straight line is the appropriate model. Later the validity of this assumption can be investigated.

2. Find the best-fitting straight line, which is that line, among all possible straight lines, that best agrees (as defined below) with the data.

3. Determine whether the straight line found in step 2 significantly helps to describe the dependent variable Y. Here it is necessary to check that certain basic statistical assumptions (e.g., normality) are met. These assumptions will be discussed in detail below.

4. Examine whether the assumption of a straight-line model is correct. One approach for doing this is called *testing for lack of fit*, although other approaches can be used instead.

5. If the assumption of a straight line is found to be invalid in step 4, fit a new model (e.g., a parabola) to the data, determine how well it describes Y (i.e., repeat step 3), and then decide whether the new model is appropriate (i.e., repeat step 4).

6. Continue to try new models until an appropriate one is found.

A flow diagram for this strategy is given in Figure 5-2.

Since the usual (forward) approach to regression analysis with a single independent variable begins with the assumption of a straight-line model, we will consider this model first. Before describing the *statistical* methodology for this special case, it is appropriate to

FIGURE 5-2   Flow diagram of the forward method

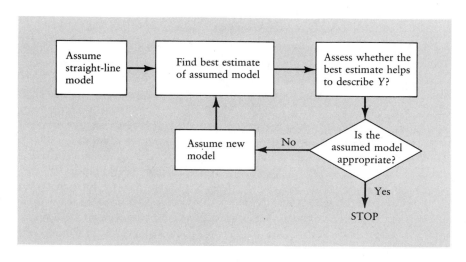

review some basic straight-line mathematics. You may wish to omit the next section if you are already familiar with its contents.

5-3 Mathematical Properties of a Straight Line

Mathematically, a straight line can be described by an equation of the form

$$y = \beta_0 + \beta_1 x \tag{5.1}$$

We have used lowercase letters y and x in this equation instead of capital letters to emphasize that we are treating these variables in a purely mathematical, rather than statistical, context. The symbols β_0 and β_1 have constant values for a given line and are therefore not considered variables; β_0 is called the *y-intercept* of the line and β_1 is called the *slope*. Thus, $y = 5 - 2x$ describes a straight line with intercept 5 and slope -2, whereas $y = -4 + 1x$ describes a different line with intercept -4 and slope 1. These two lines are shown in Figure 5-3.

The intercept β_0 is the value of y when $x = 0$. For the line $y = 5 - 2x$, $y = 5$ when $x = 0$. For the line $y = -4 + 1x$, $y = -4$ when $x = 0$. The slope β_1 is the amount of change in y for each 1-unit change in x. For any given straight line, this rate of change is always constant. Thus, for the line $y = 5 - 2x$, when x changes 1 unit from 3 to 4, y changes -2 units (the value of the slope) from $5 - 2(3) = -1$ to $5 - 2(4) = -3$. When x changes from 1 to 2, also 1 unit, y changes from $5 - 2(1) = 3$ to $5 - 2(2) = 1$, also -2 units.

The properties of any straight line can be viewed graphically as in Figure 5-3. To graph a given line, all that is needed is to plot any two points on the line and then connect them with a ruler. One of the two points often used is the intercept. This point is given by $(x = 0, y = 5)$ for the line $y = 5 - 2x$ and by $(x = 0, y = -4)$ for $y = -4 + 1x$. The other point for each line may be determined by arbitrarily selecting an x and finding the corresponding y. An x of 3 was used in our two examples. Thus, for $y = 5 - 2x$, an x of 3 yields a y of $5 - 2(3) = -1$. For $y = -4 + 1x$, an x of 3 yields a y of $-4 + 1(3) = -1$. The line $y = 5 - 2x$ can then be drawn by connecting the points $(x = 0, y = 5)$ and $(x = 3, y = -1)$. The line $y = -4 + 1x$ can be drawn from the points $(x = 0, y = -4)$ and $(x = 3, y = -1)$.

FIGURE 5-3   Straight-line plots

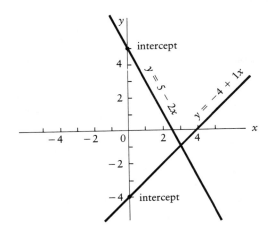

It can also be seen from Figure 5-3 that for the equation $y = 5 - 2x$, y decreases as x increases. Such a line is said to have *negative* slope. Indeed, this definition agrees with the sign of the slope -2 in the equation. Similarly, the line $y = -4 + 1x$ is said to have *positive* slope, since y increases as x increases.

5-4 Statistical Assumptions for a Straight-Line Model

Suppose that we have tentatively assumed a straight-line model as the first step in the forward method for determining the best model describing the relationship between X and Y. We now wish to determine the best-fitting line. Certainly, we will have no trouble in deciding what is meant by "best fitting" if the data allow us to draw a single straight line through each and every point in the scatter diagram. Unfortunately, this will never happen with real-life data, as we mentioned in Chapter 4. For example, persons of the same age are unlikely to have the same blood pressure, height, or weight.

Thus, we must realize that the straight line we seek is an approximation to the true state of affairs and cannot be expected to predict precisely each individual's Y from that individual's X. In fact, this need to approximate would exist even if we measured X and Y on the whole population of interest instead of just a sample from that population. In addition, the fact that the line is to be determined from the sample data and not from the population requires us to consider the problem of how to estimate unknown population parameters.

What are these parameters? The ones of primary concern at this point are the intercept β_0 and slope β_1 of the straight line of the general mathematical form of (5.1) that best fits the X-Y data for the entire population. In order to make inferences from the sample about this population line, we need to make certain statistical assumptions. These assumptions are listed below.

5-4-1 Statement of Assumptions

Assumption 1: Existence   *For any fixed value of the variable X, Y is a random variable with a certain probability distribution having finite mean and variance.* The (population) mean of this distribution will be denoted as $\mu_{Y|X}$ and the (population) variance as $\sigma^2_{Y|X}$. The notation "$Y|X$" indicates that the mean and variance of the random variable Y depend on the value of X.

This assumption will apply to any regression model, whether a straight line or not. Figure 5-4 illustrates this assumption. The different distributions are drawn vertically to correspond to different values of X. The dots denoting the mean values $\mu_{Y|X}$ at different X's have been connected to form the *regression equation*, which is the population model to be estimated from the data.

Assumption 2: Independence   *The Y-values are statistically independent of one another.* As with Assumption 1, this assumption applies to almost any regression model. Assumption 2 is sometimes violated when different observations are made on the same individual at different times. For example, if weight was measured on an individual at different times, it is to be expected that the weight at one time would be related to the weight at a later time. As another example, if blood pressure is measured on a given individual over time, one would expect the blood pressure value at one time to be in the same range as the blood pressure value at the previous or following time.

FIGURE 5-4   General regression equation

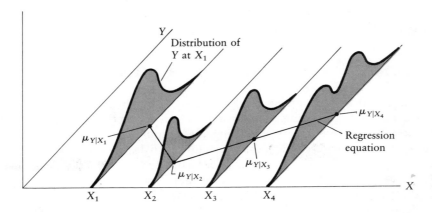

When observations are not independent, special methods can be used to find the best-fitting curve. The method chosen depends on the type of dependence and the dimensions of the problem. In some cases, multivariate linear models are appropriate. In other cases, special-purpose techniques based on weighted least squares are to be preferred. See Morrison (1976) or Timm (1975) for a general introduction to multivariate linear models.

Assumption 3: _Linearity_   *The mean value of Y, $\mu_{Y|X}$, is a straight-line function of X;* that is, if the dots denoting the different mean values $\mu_{Y|X}$ are connected, a straight line is obtained. This assumption is illustrated in Figure 5-5.
  Using mathematical symbols, we can describe Assumption 3 by the equation

$$\mu_{Y|X} = \beta_0 + \beta_1 X \tag{5.2}$$

FIGURE 5-5   Straight-line assumption

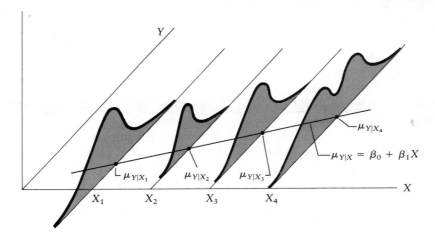

where β_0 and β_1 are the intercept and slope of this (population) straight line, respectively. Equivalently, we can express (5.2) in the form

$$Y = \beta_0 + \beta_1 X + E \tag{5.3}$$

where E denotes a random variable that has mean 0 at fixed X (i.e, $\mu_{E|X} = 0$ for any X). To be more specific, since X is fixed and not random, (5.3) represents the dependent variable Y as the sum of a constant term $(\beta_0 + \beta_1 X)$ and a random variable (E). Thus, the probability distributions of Y and E differ only in the value of this constant term; that is, since E has mean 0, Y must have mean $\beta_0 + \beta_1 X$.

(The statement "X is fixed and not random" is often associated with the statement "X is measured without error." For our purposes, the practical implication of either statement for making statistical inferences from sample to population is that the *only* random component on the right-hand side of (5.3) is E.)

Equations (5.2) and (5.3) describe a *statistical* model. These equations should be distinguished from the *mathematical* model for a straight line described by (5.1), which does not consider Y as a random variable.

The variable E describes how far away an individual's response is from the population regression line (Figure 5-6). In other words, what we observe at a given X (namely, Y) is in *error* from that expected on the average (namely, $\mu_{Y|X}$) by an amount E, which is random and varies from individual to individual. For this reason, E is commonly referred to as the *error component* in the model (5.3); mathematically, E is given by the formula

$$E = Y - (\beta_0 + \beta_1 X)$$

or by

$$E = Y - \mu_{Y|X}$$

This concept of an error component is particularly important for defining a good-fitting line since, as we will see in the next section, a line that fits the data well ought to have small deviations (or errors) between what is observed and what is predicted by the fitted model.

FIGURE 5-6   Error component E

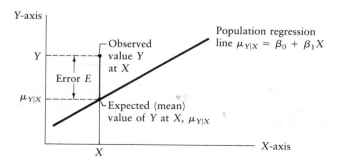

Assumption 4: _Homoscedasticity_   _The variance of Y is the same for any X._ (Homo-means "same," and -_scedastic_ comes from a Greek word that means "to scatter." An example of the violation of this assumption (called _heteroscedasticity_) is shown in Figure 5-5. In this figure the distribution of Y at X_1 has considerably more spread than the distribution of Y at X_2. This means that $\sigma^2_{Y|X_1}$, the variance of Y at X_1, is greater than $\sigma^2_{Y|X_2}$, the variance of Y at X_2.

In mathematical terms, the homoscedastic assumption can be written

$$\sigma^2_{Y|X} \equiv \sigma^2 \quad \text{for all } X$$

This formula is a short-hand way of saying that since $\sigma^2_{Y|X_i} = \sigma^2_{Y|X_j}$ for _any_ two values of X, we might as well simplify our notation by giving this common variance a single name, say σ^2, which does not involve X at all.

A number of techniques of varying statistical sophistication can be used to determine whether the homoscedastic assumption is satisfied. Some of these procedures will be discussed in Chapter 12.

Assumption 5: _Normal Distribution_   _For any fixed value of X, Y has a normal distribution._ This assumption makes it possible to evaluate the statistical significance (e.g., by means of confidence intervals and tests of hypotheses) of the relationship between X and Y as reflected by the fitted line.

Figure 5-5 provides an example in which this assumption is violated. In addition to the variances not being all equal in this figure, the distributions of Y at X_3 and at X_4 are not normal. The distribution at X_3 is skewed, whereas the normal distribution is symmetric. The distribution at X_4 is bimodal (two humps), whereas the normal distribution is unimodal (one hump). Methods for determining whether the normality assumption is tenable are described in Chapter 12.

It is important to emphasize that if the normality assumption is not _badly_ violated, the conclusions reached by a regression analysis in which normality is assumed will generally be reliable and accurate. This stability property with respect to deviations from normality is a type of _robustness_. Consequently, we recommend that considerable leeway be given before deciding that the normality assumption is so badly violated as to require alternative inference-making procedures.

If the normality assumption is deemed unsatisfactory, the observations may be transformed by using a log, square root, or other function to see if the new set of observations is approximately normal. Care must be taken when using such transformations to see that other assumptions, such as variance homogeneity, are not violated for the transformed variable. Fortunately, it often turns out in practice that such transformations usually help to satisfy both the normality and variance homogeneity assumptions.

5-4-2 Summary and Comments

One must be careful to note that the assumptions of homoscedasticity and normality apply to the distribution of Y when X is fixed (i.e., Y given X), and not to a distribution of Y's associated with different X-values. Many people find it more convenient to describe these two assumptions in terms of the error, E. It is sufficient to say that the random variable E has a normal distribution with mean 0 and variance σ^2 for all observations. Of course, the linearity, existence, and independence assumptions must also be specified.

It is helpful to maintain distinctions among such concepts as *random variables, parameters*, and *point estimates*. The variable Y is a random variable, and an observation of it yields a particular value; the variable X is a fixed (nonrandom) known variable. The constants β_0 and β_1 are parameters with unknown but specific values for a particular population. The variable E is a random variable that is unobservable. Using some estimation procedure (e.g., least squares), one constructs point estimates $\hat{\beta}_0$ and $\hat{\beta}_1$ of β_0 and β_1, respectively. Once $\hat{\beta}_0$ and $\hat{\beta}_1$ are obtained, a point estimate of E is calculated as

$$\hat{E} = Y - \hat{Y} = Y - (\hat{\beta}_0 + \hat{\beta}_1 X)$$

The estimated error $\hat{E}$ is typically called a *residual*.

Some statisticians refer to a normally distributed random variable as having a Gaussian distribution. Three reasons for this terminology are: (1) It avoids confusing *normal* with its other meaning of "customary" or "usual"; (2) it emphasizes the fact that the term *Gaussian* refers to a *particular* bell-shaped function; and (3) it appropriately honors the mathematician Carl Gauss (1777–1855).

With the above comments in mind, a useful mnemonic device for remembering the assumptions for straight-line regression analysis is the two-word German phrase "*HEIL GAUSS*," (meaning "Hail Gauss") where H stands for homoscedasticity, E for existence, I for independence, L for linearity, and *Gauss* for Gaussian distribution. Each of these terms should elicit its corresponding definition, since each word by itself is not sufficient to describe the associated assumption clearly.

5-5 Determining the Best-Fitting Straight Line

By far the simplest and quickest method for determining a straight line is to choose that line which can best be drawn by eye. Although such a method paints a reasonably good picture, this procedure is much too subjective and imprecise and is worthless for statistical inference. Two analytical approaches for finding the best-fitting straight line will now be described.

5-5-1 The Least-Squares Method

The *least-squares method* determines the best-fitting straight line as that line which *minimizes* the sum of squares of the lengths of the vertical-line segments (Figure 5-7) drawn from the observed data points on the scatter diagram to the fitted line. The idea here is that the smaller the deviations of observed values from this line (and consequently the smaller the sum of squares of these deviations), the closer or "snugger" the best-fitting line will be to the data.

Using mathematical notation, the least-squares method is described as follows. Let $\hat{Y}_i$ denote the estimated response at X_i based on the fitted regression line; in other words, $\hat{Y}_i = \hat{\beta}_0 + \hat{\beta}_1 X_i$, where $\hat{\beta}_0$ and $\hat{\beta}_1$ are the intercept and slope of the fitted line. The vertical distance between the observed point (X_i, Y_i) and the corresponding point $(X_i, \hat{Y}_i)$ on the fitted line is given by the absolute value $|Y_i - \hat{Y}_i|$, or $|Y_i - (\hat{\beta}_0 + \hat{\beta}_1 X_i)|$. The sum of the squares of all such distances is then given by

$$\sum_{i=1}^{n} (Y_i - \hat{Y}_i)^2 = \sum_{i=1}^{n} (Y_i - \hat{\beta}_0 - \hat{\beta}_1 X_i)^2$$

FIGURE 5-7   Deviations of observed points from the fitted regression
line

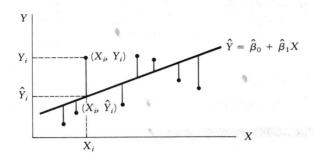

The least-squares solution is defined to be that choice of $\hat{\beta}_0$ and $\hat{\beta}_1$ for which the sum of squares above is a minimum. It is standard jargon to refer to $\hat{\beta}_0$ and $\hat{\beta}_1$ as the least-squares estimates of the parameters β_0 and β_1, respectively, in the statistical model given by (5.3).

The *minimum sum of squares* corresponding to the least-squares estimates $\hat{\beta}_0$ and $\hat{\beta}_1$ is usually called the *sum of squares about the regression line*, the *residual sum of squares*, or the *sum of squares due to error* (SSE). The measure SSE is of great importance in assessing the quality of the straight-line fit, and its interpretation will be discussed in Section 5-6.

Mathematically, the essential property of the measure SSE can be stated in the following way. If β_0^* and β_1^* denote any other possible estimators of β_0 and β_1, we must have

$$\text{SSE} = \sum_{i=1}^{n} (Y_i - \hat{\beta}_0 - \hat{\beta}_1 X_i)^2 \leq \sum_{i=1}^{n} (Y_i - \beta_0^* - \beta_1^* X_i)^2$$

5-5-2 The Minimum-Variance Method

The *minimum-variance method* is more classically statistical in nature than the method of least squares, which can be looked upon as a purely mathematical algorithm. In this second approach, determining the best fit becomes a statistical estimation problem. The goal is to find point estimators of β_0 and β_1 with good statistical properties. In this regard, under the previous assumptions, the best line is determined by those estimators $\hat{\beta}_0$ and $\hat{\beta}_1$ that are *unbiased* for their unknown population counterparts β_0 and β_1, respectively, and that, in addition, have minimum variance among all unbiased (linear) estimators of β_0 and β_1.

5-5-3 Solution to the Best-Fit Problem

Fortunately, both the least-squares and minimum-variance methods yield exactly the same solution, which we will state without proof. (Another general method of parameter estimation is called *maximum likelihood*. Under the assumption of a Gaussian distribution, the maximum-likelihood estimates of β_0 and β_1 are exactly the same as the least-squares and minimum-variance estimates. A general discussion of maximum-likelihood methods will be given in Chapter 21.)

Let $\bar{Y}$ denote the sample mean of the observations on Y, and $\bar{X}$ denote the sample mean of the values of X. Then the best-fitting straight line is determined by the formulas

$$\hat{\beta}_1 = \frac{\sum_{i=1}^{n} (X_i - \bar{X})(Y_i - \bar{Y})}{\sum_{i=1}^{n} (X_i - \bar{X})^2} \tag{5.4}$$

$$\hat{\beta}_0 = \bar{Y} - \hat{\beta}_1 \bar{X} \tag{5.5}$$

In calculating $\hat{\beta}_0$ and $\hat{\beta}_1$, one should first compute $\bar{Y}$ and $\bar{X}$, then $\hat{\beta}_1$, and finally $\hat{\beta}_0$. Although the formula given in (5.4) is mathematically correct, a computational formula for $\hat{\beta}_1$ that is often more accurate and convenient when using a hand calculator is

$$\hat{\beta}_1 = \frac{\sum_{i=1}^{n} X_i Y_i - \left(\sum_{i=1}^{n} X_i\right)\left(\sum_{i=1}^{n} Y_i\right)/n}{\sum_{i=1}^{n} X_i^2 - \left(\sum_{i=1}^{n} X_i\right)^2/n} \tag{5.6}$$

The least-squares line may be generally represented either by

$$\hat{Y} = \hat{\beta}_0 + \hat{\beta}_1 X \tag{5.7}$$

or, equivalently, by

$$\hat{Y} = \bar{Y} + \hat{\beta}_1 (X - \bar{X}) \tag{5.8}$$

Either (5.7) or (5.8) may be used to determine predicted Y's corresponding to X's actually observed or to other X-values in the region of experimentation.

(Equation (5.8) can be seen, through some simple algebra, to be equivalent to (5.7). The right-hand side of (5.8), $\bar{Y} + \hat{\beta}_1(X - \bar{X})$, can be written as $\bar{Y} + \hat{\beta}_1 X - \hat{\beta}_1 \bar{X}$, which in turn equals $\bar{Y} - \hat{\beta}_1 \bar{X} + \hat{\beta}_1 X$, which is equivalent to (5.7) since $\hat{\beta}_0 = \bar{Y} - \hat{\beta}_1 \bar{X}$.)

Example Table 5-1 lists observations on systolic blood pressure and age for a sample of 30 individuals. The scatter diagram for this sample was presented in Figure 5-1. For this data set, the best-fitting line may be calculated as follows:

$$\bar{Y} = 142.53$$

$$\bar{X} = 45.13$$

$$\sum X_i Y_i - \frac{(\sum X_i)(\sum Y_i)}{n} = 199,576 - \frac{(1,354)(4,276)}{30} = 6,585.87$$

$$\sum X_i^2 - \frac{(\sum X_i)^2}{n} = 67,894 - \frac{(1,354)^2}{30} = 6,783.47$$

$$\hat{\beta}_1 = \frac{6,585.87}{6,783.47} = 0.97$$

$$\hat{\beta}_0 = \bar{Y} - \hat{\beta}_1 \bar{X} = 142.53 - (0.97)(45.13) = 98.71$$

The equation for this line is thus given by

$$\hat{Y} = 98.71 + 0.97 X \tag{5.9}$$

or, equivalently,

$$\hat{Y} = 142.53 + (0.97)(X - 45.13)$$

TABLE 5-1   Observations on systolic blood pressure (SBP)
and age for a sample of 30 individuals

Individual (i)	SBP (Y)	Age (X)	Individual (i)	SBP (Y)	Age (X)
1	144	39	16	130	48
2	220	47	17	135	45
3	138	45	18	114	17
4	145	47	19	116	20
5	162	65	20	124	19
6	142	46	21	136	36
7	170	67	22	142	50
8	124	42	23	120	39
9	158	67	24	120	21
10	154	56	25	160	44
11	162	64	26	158	53
12	150	56	27	144	63
13	140	59	28	130	29
14	110	34	29	125	25
15	128	42	30	175	69

FIGURE 5-8   Best-fitting straight line to age–systolic blood pressure data
of Table 5-1

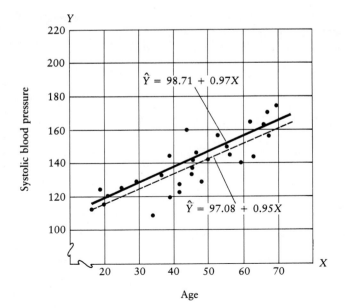

$$\hat{Y} = 98.71 + 0.97X$$

$$\hat{Y} = 97.08 + 0.95X$$

The line (5.9) is graphed as the solid line in Figure 5-8 and reflects the clear trend that systolic blood pressure increases as age increases. Notice that one point, (47, 220), seems quite out of place with the other data, and such an observation is often called an *outlier*. Because an outlier can affect the least-squares estimates, it is important to determine whether an outlier should be removed from the data. Usually this decision can be made only after a thorough evaluation of the experimental conditions, the data collection process, and the data themselves. (See Chapter 12 for further discussion concerning the treatment of outliers.) If the decision is difficult, one can always determine the effect of removing the outlier by refitting the model to the remaining data. If we do this, the resulting least-squares line is

$$\hat{Y} = 97.08 + 0.95\,X$$

and is shown on the graph as the dashed line. As might be expected, this line is slightly below the one obtained using all the data.

5-6 Measure of the Quality of the Straight-Line Fit and Estimate of σ^2

Once the least-squares line is determined, we would like to evaluate whether the fitted line actually aids in predicting Y, and, if so, to what extent. A measure that helps to answer these questions is provided by

$$\text{SSE} = \sum_{i=1}^{n} (Y_i - \hat{Y}_i)^2$$

where $\hat{Y}_i = \hat{\beta}_0 + \hat{\beta}_1 X_i$. Clearly, if SSE = 0, the straight line fits perfectly; that is, $Y_i = \hat{Y}_i$ for each i, and every observed point lies on the fitted line. Furthermore, as the fit gets worse, SSE gets larger, since the deviations of points from the regression line become larger.

Two possible factors contribute to the inflation of SSE. The first is that there may be a lot of variation in the data; that is, σ^2 may be large. The second is that the assumption of a straight-line model may not be appropriate. It is important, therefore, to determine the separate effects of each of these components since they address decidedly different issues with regard to the fit of the model. For the time being, we will assume that the second factor is not at issue. Thus, assuming that the straight-line model is appropriate, we can obtain an estimate of σ^2 using SSE. Such an estimate is needed for making statistical inferences concerning the true (i.e., population) straight-line relationship between X and Y. This estimate of σ^2 is given by the formula

$$S_{Y|X}^2 = \frac{1}{n-2} \sum_{i=1}^{n} (Y_i - \hat{Y}_i)^2 = \frac{1}{n-2}\,\text{SSE} \tag{5.10}$$

A convenient computational formula for $S_{Y|X}^2$ is

$$S_{Y|X}^2 = \frac{n-1}{n-2} (S_Y^2 - \hat{\beta}_1^2 S_X^2) \tag{5.11}$$

where

$$S_{\bar{Y}}^2 = \frac{\sum\limits_{i=1}^{n} Y_i^2 - \left(\sum\limits_{i=1}^{n} Y_i\right)^2 / n}{n - 1} \qquad \text{(i.e., the sample variance of the observed } Y\text{'s)}$$

and

$$S_{\bar{X}}^2 = \frac{\sum\limits_{i=1}^{n} X_i^2 - \left(\sum\limits_{i=1}^{n} X_i\right)^2 / n}{n - 1} \qquad \text{(i.e., the sample variance of the } X\text{'s)}$$

Example   For the data of Table 5-1, $S_{Y|X}^2$ is calculated as follows:

$S_{\bar{Y}}^2 = 509.91$

$S_{\bar{X}}^2 = 233.91$

$\hat{\beta}_1 = 0.97$

$S_{Y|X}^2 = \dfrac{29}{28} [509.91 - (0.97)^2(233.91)] = 299.77$

Some explanation is appropriate as to why $S_{Y|X}^2$ estimates σ^2, especially since, on first glance, (5.10) looks different from the formula usually used for the sample variance, $\sum_{i=1}^{n} (Y_i - \bar{Y})^2/(n - 1)$. This formula is appropriate when the Y's are independent with the same mean μ and variance σ^2. Since, in this case, μ is unknown (its estimate being, of course, $\bar{Y}$), we must divide by $n - 1$ instead of n in order for the sample variance to be an unbiased estimator of σ^2. To put it another way, we subtract 1 from n because to estimate σ^2 we had to estimate *one* other parameter first, μ.

If a straight-line model is appropriate, the population mean response $\mu_{Y|X}$ changes with X. For example, using the least-squares line (5.9) as an approximation to the population line for the age–systolic blood pressure data, the estimated mean of the Y's at $X = 40$ is approximately 138, whereas the estimated mean of the Y's at $X = 70$ is close to 167. Therefore, instead of subtracting $\bar{Y}$ from each Y_i when estimating σ^2, we should actually subtract $\hat{Y}_i$ from Y_i because $\hat{Y}_i = \hat{\beta}_0 + \hat{\beta}_1 X_i$ is the estimate of $\mu_{Y|X_i}$. Furthermore, we subtract 2 from n in the denominator of our estimate since the determination of $\hat{Y}_i$ requires the estimation of two parameters, β_0 and β_1.

When we discuss testing for lack of fit of the assumed model, we will show how to get an estimate of σ^2 that does not assume the correctness of the straight-line model.

5-7 Inferences Concerning the Slope and Intercept

To assess whether the fitted line helps to predict Y from X, and to take into account the uncertainties of using a sample, it is standard practice to compute confidence intervals and/or test statistical hypotheses about the unknown parameters in the assumed straight-line model. Such confidence intervals and tests require, as previously described in Section 5-4, the assumption that the random variable Y has a normal distribution at each fixed value of X. Under this assumption it can be deduced that the estimators $\hat{\beta}_0$ and $\hat{\beta}_1$ are each normally distributed, with respective means β_0 and β_1 when (5.2) holds, and with easily derivable variances. These estimators, together with estimates of their variances, can then be used to form confidence intervals and test statistics based on the t *distribution*.

(An important property that allows the normality assumption on Y to carry over to $\hat{\beta}_0$ and $\hat{\beta}_1$ is that these estimators are *linear functions* of the Y's. Such a function is defined by a formula of the structure

$$L = \sum_{i=1}^{n} c_i Y_i$$

or, equivalently, $L = c_1 Y_1 + c_2 Y_2 + \cdots + c_n Y_n$, where the c_i's are constants not involving the Y's. A simple example of a linear function is $\bar{Y}$, which can be written as

$$\sum_{i=1}^{n} \frac{1}{n} Y_i$$

Here the c_i's equal $1/n$ for each i. The normality of $\hat{\beta}_0$ and $\hat{\beta}_1$ derives from a statistical theorem which states that linear functions of independent normally distributed observations are themselves normally distributed.)

More specifically, to test the hypothesis $H_0: \beta_1 = \beta_1^{(0)}$, where $\beta_1^{(0)}$ is some hypothesized value for β_1, the test statistic used is

$$T = \frac{\hat{\beta}_1 - \beta_1^{(0)}}{\dfrac{S_{Y|X}}{S_x \sqrt{n-1}}} \tag{5.12}$$

This test statistic has the t distribution with $n - 2$ degrees of freedom when H_0 is true. In this formula $S_{Y|X}^2$ denotes the sample estimate of σ^2 defined by (5.10), and S_X^2 is the sample variance of the X's used in (5.11) for calculating $S_{Y|X}^2$. The denominator is an estimate of the unknown standard error of the estimator $\hat{\beta}_1$, given by

$$\sigma_{\hat{\beta}_1} = \frac{\sigma}{S_x \sqrt{n-1}}$$

Thus, the test statistic (5.12) is the ratio of a normally distributed random variable divided by an estimate of its standard deviation. Such a statistic will have a t distribution for the kinds of situations encountered in this text.

Similarly, to test the hypothesis $H_0: \beta_0 = \beta_0^{(0)}$, the test statistic used is

$$T = \frac{\hat{\beta}_0 - \beta_0^{(0)}}{S_{Y|X} \sqrt{\dfrac{1}{n} + \dfrac{\bar{X}^2}{(n-1)S_X^2}}} \tag{5.13}$$

which also has the t distribution with $n - 2$ degrees of freedom when $H_0: \beta_0 = \beta_0^{(0)}$ is true. The denominator here estimates the standard deviation of $\hat{\beta}_0$, given by

$$\sigma_{\hat{\beta}_0} = \sigma \sqrt{\frac{1}{n} + \frac{\bar{X}^2}{(n-1)S_X^2}}$$

The reason that both test statistics above have $n - 2$ degrees of freedom is that both involve $S_{Y|X}^2$, which itself has $n - 2$ degrees of freedom and is the only random component in the denominator of both statistics.

TABLE 5-2   Confidence intervals, tests of hypotheses, and prediction intervals for straight-line regression analysis

Parameter	100(1 − α)% Confidence Interval	H_0	Test Statistic (T)	Distribution under H_0
β_1	$\hat{\beta}_1 \pm t_{n-2,1-\alpha/2} \dfrac{S_{Y\mid X}}{S_X\sqrt{n-1}}$	$\beta_1 = \beta_1^{(0)}$	$T = \dfrac{(\hat{\beta}_1 - \beta_1^{(0)})S_X\sqrt{n-1}}{S_{Y\mid X}}$	t_{n-2}
β_0	$\hat{\beta}_0 \pm t_{n-2,1-\alpha/2}S_{Y\mid X}\sqrt{\dfrac{1}{n} + \dfrac{\bar{X}^2}{(n-1)S_X^2}}$	$\beta_0 = \beta_0^{(0)}$	$T = \dfrac{\hat{\beta}_0 - \beta_0^{(0)}}{S_{Y\mid X}\sqrt{\dfrac{1}{n} + \dfrac{\bar{X}^2}{(n-1)S_X^2}}}$	t_{n-2}
$\mu_{Y\mid X_0}$	$\bar{Y} + \hat{\beta}_1(X_0 - \bar{X})$ $\pm t_{n-2,1-\alpha/2}S_{Y\mid X}\sqrt{\dfrac{1}{n} + \dfrac{(X_0-\bar{X})^2}{(n-1)S_X^2}}$	$\mu_{Y\mid X_0} = \mu_{Y\mid X_0}^{(0)}$	$T = \dfrac{\bar{Y} + \hat{\beta}_1(X_0 - \bar{X}) - \mu_{Y\mid X_0}^{(0)}}{S_{Y\mid X}\sqrt{\dfrac{1}{n} + \dfrac{(X_0-\bar{X})^2}{(n-1)S_X^2}}}$	t_{n-2}
prediction band Y_{X_0} [a]	$\bar{Y} + \hat{\beta}_1(X_0 - \bar{X})$ $\pm t_{n-2,1-\alpha/2}S_{Y\mid X}\sqrt{1 + \dfrac{1}{n} + \dfrac{(X_0-\bar{X})^2}{(n-1)S_X^2}}$ [b]			

(handwritten left margin annotations: "confidence band" alongside the $\mu_{Y\mid X_0}$ row; "prediction band" alongside the Y_{X_0} row.)

Note: $\mu_{Y\mid X} = \beta_0 + \beta_1 X$ is the assumed true regression model.

$$\hat{\beta}_1 = \frac{\sum_1^n (X_i - \bar{X})(Y_i - \bar{Y})}{\sum_1^n (X_i - \bar{X})^2} = \frac{\sum_1^n X_i Y_i - \left(\sum_1^n X_i\right)\left(\sum_1^n Y_i\right)/n}{\sum_1^n X_i^2 - \left(\sum_1^n X_i\right)^2/n}$$

$$\hat{\beta}_0 = \bar{Y} - \hat{\beta}_1 \bar{X}$$

$$\hat{Y} = \hat{\beta}_0 + \hat{\beta}_1 X = \bar{Y} + \hat{\beta}_1(X - \bar{X})$$

$t_{n-2,1-\alpha/2}$ is the 100(1 − α/2)% point of the t distribution with $n - 2$ degrees of freedom.

$$S_{Y\mid X}^2 = \frac{1}{n-2}\sum_1^n (Y_i - \hat{Y}_i)^2 = \frac{n-1}{n-2}(S_Y^2 - \hat{\beta}_1^2 S_X^2)$$

$$S_Y^2 = \frac{1}{n-1}\sum_1^n (Y_i - \bar{Y})^2 = \frac{\sum_1^n Y_i^2 - \left(\sum_1^n Y_i\right)^2/n}{n-1}$$

$$S_X^2 = \frac{1}{n-1}\sum_1^n (X_i - \bar{X})^2 = \frac{\sum_1^n X_i^2 - \left(\sum_1^n X_i\right)^2/n}{n-1}$$

[a] Single observation, not a parameter.

[b] Prediction interval for a new individual's Y.

TABLE 5-3 Sample calculations of confidence intervals, tests of hypotheses, and prediction intervals for the age–systolic blood pressure data of Table 5-1

Parameter	$100(1 - \alpha)\%$ Confidence Interval	H_0	Test Statistic (T)
β_1	For $\alpha = .05$: $0.97 \pm 2.0484\left(\dfrac{17.31}{15.29\sqrt{30-1}}\right)$ or $(0.54, 1.40)$	$\beta_1 = 0$	$T = \dfrac{(0.97 - 0)(15.29)\sqrt{30-1}}{17.31} = 4.62$ Reject H_0 at $\alpha = .05$ (two-tailed test), since $t_{28,0.975} = 2.0484$ ($P < .001$).
β_0	For $\alpha = .05$: $98.71 \pm (2.0484)(17.31)\sqrt{\dfrac{1}{30} + \dfrac{(45.13)^2}{(30 - 1)(15.29)^2}}$ or $(78.23, 119.20)$	$\beta_0 = 75$	$T = \dfrac{98.71 - 75}{17.31\sqrt{\dfrac{1}{30} + \dfrac{(45.13)^2}{(30 - 1)(15.29)^2}}} = 2.37$ Reject H_0 at $\alpha = .05$ (two-tailed test), since $t_{28,0.975} = 2.0484$ ($.02 < P < .05$).
$\mu_{Y\mid 65}$[a]	For $\alpha = .10$: $142.53 + (0.97)(65 - 45.13)$ $\pm (1.7011)(17.31)\sqrt{\dfrac{1}{30} + \dfrac{(65 - 45.13)^2}{(30 - 1)(15.29)^2}}$ or $(153.09, 171.07)$	$\mu_{Y\mid 65} = 147$	$T = \dfrac{142.53 + (0.97)(65 - 45.13) - 147}{17.31\sqrt{\dfrac{1}{30} + \dfrac{(65 - 45.13)^2}{(30 - 1)(15.29)^2}}} = 2.86$ Reject H_0 at $\alpha = .10$ (two-tailed test), since $t_{28,0.95} = 1.7011$ ($.0001 < P < .001$).
Y_{65}[b]	For $\alpha = .10$: $142.53 + (0.97)(65 - 45.13)$ $\pm (1.7011)(17.31)\sqrt{1 + \dfrac{1}{30} + \dfrac{(65 - 45.13)^2}{(30 - 1)(15.29)^2}}$ or $(131.29, 192.87)$		

[a] Thus, $X_0 = 65$.

[b] Prediction interval at $X_0 = 65$.

Note: $n = 30$, $\hat{\beta}_0 = 98.71$, $\hat{\beta}_1 = 0.97$, $\bar{Y} = 142.53$, $\bar{X} = 45.13$, $S_{Y\mid X} = 17.31$, $S_X = 15.29$.

In testing either of the preceding hypotheses at a significance level α, H_0 should be rejected whenever any of the following occur:

$$
\begin{cases}
T \ge t_{n-2,1-\alpha} \text{ for an upper one-tailed test} & (\text{i.e., } H_a: \beta_1 > \beta_1^{(0)} \text{ or } H_a: \beta_0 > \beta_0^{(0)}) \\
T \le -t_{n-2,1-\alpha} \text{ for a lower one-tailed test} & (\text{i.e., } H_a: \beta_1 < \beta_1^{(0)} \text{ or } H_a: \beta_0 < \beta_0^{(0)}) \\
|T| \ge t_{n-2,1-\alpha/2} \text{ for a two-tailed test} & (\text{i.e., } H_a: \beta_1 \ne \beta_1^{(0)} \text{ or } H_a: \beta_0 \ne \beta_0^{(0)})
\end{cases}
$$

where $t_{n-2,1-\alpha}$ denotes the $100(1-\alpha)\%$ point of the t distribution with $n-2$ degrees of freedom. As an alternative to using a specified significance level, P-values may be computed based on the calculated value of the test statistic T.

Table 5-2 summarizes the formulas needed for performing statistical tests and computing confidence intervals for β_0 and β_1. Also given in this table are formulas for inference-making procedures concerned with prediction using the fitted line; these formulas are described in Sections 5-9 and 5-10. Table 5-3 gives examples illustrating the use of each formula in Table 5-2 using the age–systolic blood pressure data previously considered.

5-8 Interpretations of Tests for Slope and Intercept

Errors are often made when interpreting the results of tests concerning the slope and intercept. In this section we discuss the conclusions that can be made based on acceptance or rejection of the most common null hypotheses involving the slope and intercept.[1] In the discussion we shall assume that the usual assumptions concerning normality, independence, and variance homogeneity are not violated. If these assumptions do not hold, any conclusions based on testing procedures developed under such assumptions are suspect.

5-8-1 Test for Zero Slope

The most important test of the hypothesis dealing with the parameters of the straight-line model concerns whether the slope of the regression line is significantly different from zero or, equivalently, whether X helps to predict Y using a straight-line model. The appropriate null hypothesis for this test is $H_0: \beta_1 = 0$. Care must be taken in interpreting the result of the test of this hypothesis.

If we ignore for now the ever-present possibility of making a Type I error (i.e., rejecting a true H_0) or a Type II error (i.e., accepting a false H_0), we can make the following interpretations:

1. If $H_0: \beta_1 = 0$ is accepted (i.e., not rejected), this means one of the following:
    a. For a true underlying straight-line model, X provides little or no help in predicting Y; that is, $\bar{Y}$ is essentially as good as $\bar{Y} + \hat{\beta}_1(X - \bar{X})$ for predicting Y (see Figure 5-9a).
    b. The true underlying relationship between X and Y is *not* linear; that is, the true model may involve quadratic, cubic, or other more complex functions of X (see Figure 5-9b).

---

[1] The reader is reminded that, statistically speaking, "acceptance of H_0" really means that "there is insufficient evidence to reject H_0."

FIGURE 5-9   Interpreting the test for zero slope

Examples when $H_0: \beta_1 = 0$ is not rejected

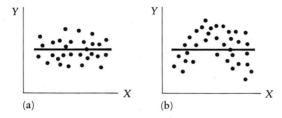

Examples when $H_0: \beta_1 = 0$ is rejected

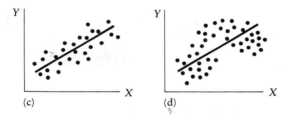

Combining (a) and (b), we can say that *accepting $H_0: \beta_1 = 0$ implies that a straight-line model in X is not the best model to use and does not provide much help for predicting Y.*

2. If $H_0: \beta_1 = 0$ is rejected, this means that:
   a. X provides significant information for predicting Y; that is, the model $\bar{Y} + \hat{\beta}_1(X - \bar{X})$ is far better than the naive model $\bar{Y}$ for predicting Y (see Figure 5-9c).
   b. A better model might have, for example, a curvilinear term (e.g., see Figure 5-9d), although there is a definite linear component.

Combining (a) and (b), we can say that *rejecting $H_0: \beta_1 = 0$ implies that a straight-line model in X is better than a model that does not include X at all, although it may very well represent only a linear approximation to a truly nonlinear relationship.*

An important point implied by the above interpretations is that *whether or not the hypothesis $H_0: \beta_1 = 0$ is rejected, a straight-line model may not be appropriate; instead, some other curve may describe the relationship between X and Y better.*

5-8-2 Test for Zero Intercept

Another hypothesis sometimes tested concerns whether the population straight line goes through the origin, that is, whether its Y intercept β_0 is zero. The null hypothesis here is $H_0: \beta_0 = 0$. If this null hypothesis is not rejected, it may be appropriate to remove the constant from the model, provided that there is previous experience or theory to suggest that the line may go through the origin and that there are observations taken around the origin to improve the estimate of β_0. To force the fitted line through the origin merely because $H_0: \beta_0 = 0$ is not rejected can sometimes give a spurious appearance to the regression line. In any case this hypothesis is rarely of interest in most studies, because data are not usually gathered near the origin. For example, when dealing with age (X) and blood pressure (Y), we are not interested in knowing what happens at $X = 0$ and we rarely choose values of X near 0.

5-9 Inferences Concerning the Regression Line $\mu_{Y|X} = \beta_0 + \beta_1 X$

In addition to making inferences about the slope and intercept, we may also want to perform tests and/or compute confidence intervals concerning the regression line itself. More specifically, we may want to find for a given $X = X_0$ a confidence interval for $\mu_{Y|X_0}$, the mean value of Y at X_0.[2] We may also be interested in testing the hypothesis $H_0: \mu_{Y|X_0} = \mu_{Y|X_0}^{(0)}$, where $\mu_{Y|X_0}^{(0)}$ is some hypothesized value of interest.

The test statistic to use for the hypothesis $H_0: \mu_{Y|X_0} = \mu_{Y|X_0}^{(0)}$ is given by the formula

$$T = \frac{\hat{Y}_{X_0} - \mu_{Y|X_0}^{(0)}}{S_{\hat{Y}_{X_0}}} \tag{5.14}$$

where $\hat{Y}_{X_0} = \hat{\beta}_0 + \hat{\beta}_1 X_0 = \bar{Y} + \hat{\beta}_1(X_0 - \bar{X})$ is the predicted value of Y at X_0 and

$$S_{\hat{Y}_{X_0}} = S_{Y|X} \sqrt{\frac{1}{n} + \frac{(X_0 - \bar{X})^2}{(n-1)S_X^2}} \tag{5.15}$$

This test statistic, like those for slope and intercept, has the t distribution with $n - 2$ degrees of freedom when H_0 is true. The denominator $S_{\hat{Y}_{X_0}}$ is an estimate of the standard deviation of $\hat{Y}_{X_0}$, which is given by

$$\sigma_{\hat{Y}_{X_0}} = \sigma \sqrt{\frac{1}{n} + \frac{(X_0 - \bar{X})^2}{(n-1)S_X^2}}$$

The corresponding confidence interval for $\mu_{Y|X_0}$ is given by the formula

$$\hat{Y}_{X_0} \pm t_{n-2, 1-\alpha/2} S_{\hat{Y}_{X_0}} \tag{5.16}$$

In addition to inferences about specific points on the regression line, it is often useful to construct a confidence interval for the regression line over the entire range of X-values. The most convenient way to do this is to plot the upper and lower confidence limits obtained for several specified values of X and then to sketch the two curves that connect these points. Such curves are called *confidence bands* for the regression line. The confidence bands for the data of Table 5-1 are indicated in Figure 5-10.

Sketching confidence bands by hand can be a painful job. However, it can be made reasonably easy if a number of helpful hints are followed:

1. Simplify the confidence interval formula (5.16) by substituting all known numbers except X_0.

---

[2] The point $(X_0, \mu_{Y|X_0})$, of course, lies on the population regression line.

FIGURE 5-10   90% confidence and prediction bands for age–systolic blood pressure data of Table 5-1

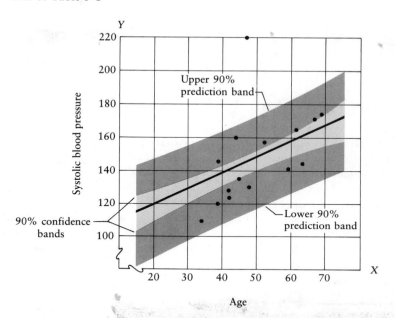

Example   For 90% confidence bands for our age–systolic blood pressure data, this simplified expression is given by

$$142.53 + (0.97)(X_0 - 45.13) \pm 29.45 \sqrt{0.033 + \frac{(X_0 - 45.13)^2}{6{,}783.48}}$$

2. Calculate the confidence limits for $X_0 = \bar{X}$. These are the easiest limits to calculate because two potentially messy terms in the preceding formula disappear. It is important to note that the minimum-width confidence interval is always obtained when $X_0 = \bar{X}$, since the second term in the square root expression of the formula above is then zero. Furthermore, as can be seen in Figure 5-10, the farther X_0 is from $\bar{X}$, the wider the confidence interval is. Thus, *we do a better job of estimating $\mu_{Y|X_0}$ near the center of the region of experimentation than in the periphery.*

Example   At $X_0 = \bar{X} = 45.13$, the formula simplifies to $142.53 \pm 29.45\sqrt{0.033}$, which yields a lower limit of 137.16 and an upper limit of 147.91.

3. Calculate the confidence limits for values of X_0 of the form $\bar{X} \pm k$, $\bar{X} \pm 2k$, $\bar{X} \pm 3k, \ldots$, for a suitable k chosen so that the region of experimentation is uniformly covered. The advantage of this approach is that calculating the confidence intervals becomes more systematic and simpler; for example, the term under the square root sign in the confidence interval formula is the same for $\bar{X} + k$ as it is for $\bar{X} - k$ (namely, $\sqrt{0.033 + k^2/6{,}783.48}$), and similarly for the other values.

Example   For the value $X_0 = \bar{X} + 10 = 55.13$, the formula becomes $142.53 + (0.97)(10) \pm 29.45\sqrt{0.033 + 100/6{,}783.48}$, which yields limits of 145.78 and 158.70. For

$X_0 = \bar{X} - 10 = 35.13$, the formula becomes $142.53 - (0.97)(10) \pm$ [same right-hand side as for $(\bar{X} + 10)$]. The limits obtained here are 126.37 and 139.28. The computations for other values are performed similarly.

4. Plot all limits on a graph, as in Figure 5-10, and then sketch the two confidence bands (which should be parabolic in shape).

5-10 Prediction of a New Value of Y at X_0

We have just dealt with estimating the mean $\mu_{Y|X_0}$ at $X = X_0$. In practice, we may wish instead to estimate the response Y of a single individual based on the fitted regression line; that is, we may want to predict an individual's $\hat{Y}$ given his $X = X_0$. Note that the obvious point estimate to use in this case is $\hat{Y}_{X_0} = \hat{\beta}_0 + \hat{\beta}_1 X_0$. Thus, $\hat{Y}_{X_0}$ is used to estimate both the mean $\mu_{Y|X_0}$ and an individual's response Y at X_0.

It is, of course, necessary to put some bounds (i.e., limits) on this estimate to take into account its variability. Here, however, we could not say that we are constructing a confidence interval for Y, since Y is not a parameter, nor could we perform a test of hypothesis for the same reason. The term used to describe the "hybrid limits" we require is the *prediction interval* (abbreviated PI), which is given by

$$\bar{Y} + \hat{\beta}_1(X_0 - \bar{X}) \pm t_{n-2,1-\alpha/2} S_{Y|X} \sqrt{1 + \frac{1}{n} + \frac{(X_0 - \bar{X})^2}{(n-1)S_X^2}} \tag{5.17}$$

We first note that an estimate of an individual's response should naturally have more variability than an estimate of a group's mean response. This is reflected by the extra term 1 under the square root sign in (5.17), which is not found in the square root part of the confidence interval formula for $\mu_{Y|X}$ (see (5.15)). To be more specific, in predicting an actual observed Y for a given individual, there are two sources of error operating: individual error as measured by σ^2 and the error in estimating $\mu_{Y|X_0}$ using $\hat{Y}_{X_0}$. More precisely, this can be expressed by the equation

$$\underbrace{Y - \hat{Y}_{X_0}}_{\substack{\text{error in} \\ \text{predicting} \\ \text{an individual's} \\ Y \text{ at } X_0}} = \underbrace{(Y - \mu_{Y|X_0})}_{\substack{\text{deviation of} \\ \text{individual's} \\ Y \text{ from true} \\ \text{mean at } X_0}} + \underbrace{(\mu_{Y|X_0} - \hat{Y}_{X_0})}_{\substack{\text{deviation of} \\ \hat{Y}_{X_0} \text{ from true} \\ \text{mean at } X_0}}$$

This representation allows us to write the variance of an individual's predicted response at X_0 as

$$\text{Var } Y + \text{Var } \hat{Y}_{X_0} = \sigma^2 \left[1 + \frac{1}{n} + \frac{(X_0 - \bar{X})^2}{(n-1)S_X^2} \right]$$

This variance expression is estimated by replacing σ^2 by its estimate $S_{Y|X}^2$. This accounts for the expression on the right-hand side of the prediction interval in (5.17).

To describe individual predictions over the *entire* range of X-values, *prediction bands* may be determined in a manner analogous to that used for the computation of confidence bands. Figure 5-10 gives 90% prediction bands for the age–systolic blood pressure data. As expected, the 90% prediction bands in this figure are wider than the corresponding 90% confidence bands.

5-11 Assessing the Appropriateness of the Straight-Line Model

In Section 5-1, it was pointed out that the usual strategy for regression with a single independent variable is to assume that the straight-line model is appropriate. This assumption is then rejected if the data indicate that a more complex model is warranted.

Many methods may be used to assess whether the straight-line assumption is reasonable. These will be discussed separately later. The basic techniques include tests for lack of fit and are understood most easily in terms of polynomial regression models (Chapter 13). Many regression diagnostics (Chapter 12) also help in evaluating the straight-line assumption either explicitly or implicitly. With the linear model, the assumptions of linearity, homoscedasticity, and normality are so intertwined that they often are met or violated as a set.

Problems

1. The accompanying table gives the dry weights (Y) of 11 chick embryos ranging in age from 6 to 16 days (X). Also given in the table are the values of the common logarithms of the weights (Z).

Age (X) (days)	6	7	8	9	10	11
Dry weight (Y) (g)	0.029	0.052	0.079	0.125	0.181	0.261
Log$_{10}$ dry weight (Z)	-1.538	-1.284	-1.102	-0.903	-0.742	-0.583

Age (X) (days)	12	13	14	15	16
Dry weight (Y) (g)	0.425	0.738	1.130	1.882	2.812
Log$_{10}$ dry weight (Z)	-0.372	-0.132	0.053	0.275	0.449

a. Draw two scatter diagrams on graph paper, one showing the ages and dry weights of the chick embryos and the other showing the ages and log dry weights.

b. Sketch straight lines by eye on both scatter diagrams. Which line seems to give the better fit?

c. Calculate the least-squares estimates of the parameters of the regression line for each set of data, and draw the estimated lines on the respective scatter diagrams. The following summations will be useful (all summations go from $i = 1$ to $i = 11$):

$$\Sigma\ X_i = 121, \quad \Sigma\ Y_i = 7.714, \quad \Sigma\ Z_i = -5.879$$

$$\Sigma\ X_i^2 = 1{,}441, \quad \Sigma\ Y_i^2 = 13.578, \quad \Sigma\ Z_i^2 = 7.375$$

$$\Sigma\ X_i Y_i = 110.712, \quad \Sigma\ X_i Z_i = -43.12$$

   d. For each of the estimated lines, find a 95% confidence interval for the true slope and then interpret the interval in each case with regard to the null hypothesis that the true slope is 0.

   e. For each of the estimated lines, calculate and sketch the 90% confidence and prediction bands.

   f. In your opinion, which of the two regression lines has the better fit? Explain.

2. The table gives the systolic blood pressure (SBP), body size (QUET),[3] age (AGE), and smoking history (SMK = 0 if nonsmoker, SMK = 1 if a current or previous smoker) for a hypothetical sample of 32 white males over 40 years old from the town of Angina.

Person	SBP	QUET	AGE	SMK
1	135	2.876	45	0
2	122	3.251	41	0
3	130	3.100	49	0
4	148	3.768	52	0
5	146	2.979	54	1
6	129	2.790	47	1
7	162	3.668	60	1
8	160	3.612	48	1
9	144	2.368	44	1
10	180	4.637	64	1
11	166	3.877	59	1
12	138	4.032	51	1
13	152	4.116	64	0
14	138	3.673	56	0
15	140	3.562	54	1
16	134	2.998	50	1
17	145	3.360	49	1
18	142	3.024	46	1
19	135	3.171	57	0
20	142	3.401	56	0
21	150	3.628	56	1
22	144	3.751	58	0
23	137	3.296	53	0
24	132	3.210	50	0
25	149	3.301	54	1
26	132	3.017	48	1
27	120	2.789	43	0
28	126	2.956	43	1
29	161	3.800	63	0
30	170	4.132	63	1
31	152	3.962	62	0
32	164	4.010	65	0

[3] QUET stands for "quetelet index," a measure of size defined by QUET = 100 (weight/height^2).

a. Draw scatter diagrams on graph paper for each of the following variable pairs:
   (1) SBP (Y) vs. QUET (X)
   (2) SBP (Y) vs. SMK (X)
   (3) QUET (Y) vs. AGE (X)
   (4) SBP (Y) vs. AGE (X)
b. For scatter diagrams (1), (3), and (4), sketch by eye a line that fits the data reasonably well. Comment on the relationships described.
c. (1) Determine the least-squares estimates of slope (β_1) and intercept (β_0) for the straight-line regression of SBP (Y) on SMK (X). A computer printout of the essential information is provided.

Computer Printout for Problem 2[4]

```
SBP (Y) on SMK (X)

MEANS AND STANDARD DEVIATIONS
X: 0.53125 +- .50693   Y: 144.53125 +- 14.39755
CORRELATIONS⁵
PEARSON: 0.247333  SPEARMAN: 0.213763  KENDALL: 0.176227
LINEAR REGRESSION OF Y ON X
SLOPE:         7.02353 +- 5.02360
INTERCEPT:   140.80000 +- 3.66147
S(Y|X):       14.18032
ANALYSIS OF VARIANCE: F(1,30) = 1.95478

SBP (Y) on QUET (X)

MEANS AND STANDARD DEVIATIONS
X: 3.44109 +- 0.49708   Y: 144.53125 +- 14.39755
CORRELATIONS
PEARSON: 0.742004  SPEARMAN: 0.746928  KENDALL: 0.572019
LINEAR REGRESSION OF Y ON X
SLOPE:        21.49167 +- 3.54515
INTERCEPT:    70.57640 +- 12.32187
S(Y|X):        9.81160
ANALYSIS OF VARIANCE: F(1,30) = 36.75122

QUET (Y) on AGE (X)

MEANS AND STANDARD DEVIATIONS
X: 53.25000 +- 6.95608   Y: 3.44109 +- 0.49708
```

---

[4] This computer printout typifies the output obtained from a standard straight-line regression program; one major goal of these exercises is to help the reader learn how to interpret and utilize such output.

[5] The Pearson correlation coefficient is discussed in detail in Chapter 6. The Spearman and Kendall correlations are *nonparametric* measures of association (see Siegel, 1956).

```
CORRELATIONS
PEARSON: 0.802751
LINEAR REGRESSION OF Y ON X
SLOPE:       5.73642E - 02 +- 7.77991E - 03
INTERCEPT:   0.38645 +- 0.41769
S(Y|X):      0.30131
ANALYSIS OF VARIANCE: F(1,30) = 54.36664

SBP (Y) on AGE (X)
MEANS AND STANDARD DEVIATIONS
X: 53.25000 +- 6.95608   Y: 144.53125 +- 14.39755
CORRELATIONS
PEARSON: 0.775204   SPEARMAN: 0.762119   KENDALL: 0.603708
LINEAR REGRESSION OF Y ON X
SLOPE:       1.60450 +- 0.23872
INTERCEPT:   59.09163 +- 12.81626
S(Y|X):      9.24543
ANALYSIS OF VARIANCE: F(1,30) = 45.17692
```

(2) Compare the value of $\hat{\beta}_0$ with the mean SBP for nonsmokers. Compare the value of $\hat{\beta}_0 + \hat{\beta}_1$ with the mean SBP for smokers. Explain the results of these comparisons.

(3) Test the hypothesis that the true slope (β_1) is 0. (Make sure that you compute the P-value for your test.)

(4) Is the test in part (3) equivalent to the usual two-sample t test for the equality of two population means assuming equal but unknown variances? Demonstrate your answer numerically.

d. (1) Determine the least-squares estimates of slope and intercept for the straight-line regression of SBP (Y) on QUET (X). Refer to the computer printout if you wish.

(2) Sketch the estimated regression line on the scatter diagram involving SBP and QUET.

(3) Test the hypothesis of zero slope (again, make sure to compute the P-value).

(4) Find a 95% confidence interval for $\mu_{Y|\bar{X}}$.

(5) Calculate 95% prediction bands.

(6) Based on the above, would you conclude that blood pressure increases as body size increases?

(7) Are any of the assumptions for straight-line regression clearly not satisfied in this example?

e. Answer questions (1) to (3) in part (d) of Problem 2 for the regression of QUET (Y) on AGE (X).

f. Answer questions (1) to (3) in part (d) of Problem 2 for the regression of SBP (Y) on AGE (X).

3. For married couples with one or more offspring, a demographic study was conducted to determine the effect of the husband's annual income (at marriage) on the

time (in months) between marriage and the birth of the first child. The table gives the husband's annual income (INC) and time between marriage and birth of the first child (TIME) for a hypothetical sample of 20 couples.

INC	TIME	INC	TIME
$ 5,775	16.2	$ 4,608	9.7
9,800	35.0	24,210	20.0
13,795	37.2	19,625	38.2
4,120	9.0	18,000	41.25
25,015	24.4	13,000	44.0
12,200	36.75	5,400	9.2
7,400	31.75	6,440	20.0
9,340	30.0	9,000	40.2
20,170	36.0	18,180	32.0
22,400	30.8	15,385	39.2

a. Draw a scatter diagram on graph paper for the variables TIME (Y) and INC (X).

b. Attempt to sketch by eye a line that fits the data reasonably well.

c. What does this line tell you about the relationship described?

d. Use the computer printout to determine the least-squares estimates of slope (β_1) and intercept (β_0) for the straight-line regression of TIME (Y) on INC (X).

Computer Printout for Problem 3

```
MEANS AND STANDARD DEVIATIONS
X: 13193.15000 +- 6824.87806   Y: 29.04250 +- 11.31785
CORRELATIONS
PEARSON: 0.430411   SPEARMAN: 0.428733   KENDALL: 0.311347
LINEAR REGRESSION OF Y ON X
SLOPE:        7.13761E - 04 +- 3.52813E - 04
INTERCEPT:   19.62575 +- 5.21291
S(Y|X):      10.49580
ANALYSIS OF VARIANCE: F(1,18) = 4.09277
```

e. Draw the estimated regression line on the scatter diagram for TIME on INC.

f. Are any of the assumptions for straight-line regression clearly not satisfied in this example?

g. Test the null hypothesis that the true slope β_1 is 0 at the $\alpha = .05$ level. Interpret the results of this test.

h. Can you suggest a model other than a straight line that would better describe the TIME–INC relationship?

4. A sociologist assigned to a correctional institution was interested in studying the relationship between intelligence and delinquency. A delinquency index (ranging from 0 to 50) was formulated to account for both the severity and frequency of crimes committed, while intelligence was measured by IQ. The table displays the delinquency index (DI) and IQ of a sample of 18 convicted minors.

DI (Y)	IQ (X)	DI (Y)	IQ (X)
26.2	110	22.1	92
33.0	89	18.6	116
17.5	102	35.5	85
25.25	98	38.0	73
20.3	110	30.0	90
31.9	98	19.7	104
21.1	122	41.1	82
22.7	119	39.6	134
10.7	120	25.15	114

a. Draw a scatter diagram on graph paper for the variable pair (DI, IQ).

b. Given that $\hat{\beta}_1 = -0.249$ and $\hat{\beta}_0 = 52.273$, draw the estimated regression line on the scatter diagram.

c. How do you account for the fact that when IQ $= 0$, $\hat{Y} = 52.273$, even though the delinquency index goes no higher than 50?

d. Find a 95% confidence interval for the true slope β_1 using the fact that $S_{Y|X} = 7.704$ and $S_X = 16.192$.

e. Interpret this confidence interval with regard to testing the null hypothesis of zero slope at the $\alpha = .05$ level.

f. Notice that the convicted minor with an IQ $= 134$ and DI $= 39.6$ appears to be quite out of place in the data. Decide whether this outlier has any effect on estimating the IQ–DI relationship by looking at the graph for the fitted line obtained when the outlier is omitted. (Note that $\hat{\beta}_0 = 70.846$ and $\hat{\beta}_1 = -0.444$ when the outlier is removed.)

g. Test the null hypothesis of zero slope when the outlier is removed given that $S_{Y|X} = 4.933$, $S_X = 14.693$, and $n = 17$. (Use $\alpha = .05$.)

h. For these data would you conclude that the delinquency index decreases as IQ increases?

5. Following the last congressional election, a political scientist attempted to investigate the relationship between campaign expenditures on television advertisements and subsequent voter turnout. The table presents the percent of total campaign expenditures relegated to television advertisements (TVEXP) and the percent of registered voter turnout (VOTE) for a hypothetical sample of 20 congressional districts.

VOTE (Y)	TVEXP (X)	VOTE (Y)	TVEXP (X)
35.4	28.5	40.8	31.3
58.2	48.3	61.9	50.1
46.1	40.2	36.5	31.3
45.5	34.8	32.7	24.8
64.8	50.1	53.8	42.2
52.0	44.0	24.6	23.0
37.9	27.2	31.2	30.1
48.2	37.8	42.6	36.5
41.8	27.2	49.6	40.2
54.0	46.1	56.6	46.1

Computer Printout for Problem 5

```
MEANS AND STANDARD DEVIATIONS
X: 36.99000 +- 8.76758   Y: 45.71000 +- 10.81670
CORRELATIONS
PEARSON: 0.954000   SPEARMAN: 0.955181   KENDALL: 0.858743
LINEAR REGRESSION OF Y ON X
SLOPE:        1.17696 +- 8.71805E - 02
INTERCEPT:    2.17407 +- 3.30974
S(Y|X):       3.33177
ANALYSIS OF VARIANCE: F(1,18) = 182.25892
```

    a. Draw a scatter diagram on graph paper for the observation pairs on VOTE (Y) and TVEXP (X).

    b. Use the printout to determine the least-squares line of VOTE on TVEXP.

    c. Draw the estimated line on the scatter diagram constructed in part (a).

    d. Are any of the assumptions for straight-line regression clearly *not* satisfied in this example?

    e. Test the hypothesis that the slope β_1 is 0 using $\alpha = .05$, and interpret your result.

    f. Test the hypothesis $H_0: \mu_{Y|X_0} = 45$ for $X_0 = \bar{X} = 36.99$ at $\alpha = .05$.

    g. Calculate the corresponding 95% confidence interval for $\mu_{Y|36.99}$.

6. A group of 13 children and adolescents (considered healthy) participated in a psychological study designed to analyze the relationship between age and average total sleep time (ATST). To obtain a measure for ATST (in minutes), recordings were taken on each subject on three consecutive nights and then averaged. The results obtained are displayed in the table.

ATST (min/24 h)	AGE (yr)	ATST (min/24 h)	AGE (yr)
586.0	4.4	515.2	8.9
461.75	14.0	493.0	11.1
491.1	10.1	528.3	7.75
565.0	6.7	575.9	5.5
462.0	11.5	532.5	8.6
532.1	9.6	530.5	7.2
477.6	12.4		

    a. Draw a scatter diagram on graph paper for the variable pair ATST (Y) and AGE (X).

    b. Calculate the least-squares estimates of slope and intercept for the straight-line regression of ATST (Y) on AGE (X). Check your results with the computer printout.

    c. Draw the estimated regression line on the scatter diagram.

    d. Are any of the assumptions for straight-line regression obviously violated?

Computer Printout for Problem 6

```
MEANS AND STANDARD DEVIATIONS
X: 9.05769 +- 2.77518   Y: 519.30385 +- 40.95056
CORRELATIONS
PEARSON: -0.951547 SPEARMAN: -0.928571 KENDALL: -0.820513
LINEAR REGRESSION OF Y ON X
SLOPE:        -14.04105 +- 1.36812
INTERCEPT:    646.48334 +- 12.91773
S(Y|X):        13.15238
ANALYSIS OF VARIANCE: F(1,11) = 105.33021
```

    e. Obtain a 95% confidence interval for β_1.
    f. Would you reject the null hypothesis H_0: $\beta_1 = 0$ based on the confidence interval obtained in part (e)?
    g. Calculate and graph 95% confidence bands based on the estimated line.

7. Several research workers associated with the office of highway safety were evaluating the relationship between driving speed (MPH) and the distance a vehicle travels once brakes are applied (DIST). The results of 19 experimental tests are displayed in the table.

MPH (X)	DIST (Y_1)	$\sqrt{\text{DIST}}$ (Y_2)	MPH (X)	DIST (Y_1)	$\sqrt{\text{DIST}}$ (Y_2)
25	37.4	6.12	50	170.0	13.04
35	57.7	7.60	20	20.0	4.47
60	337.6	18.37	15	13.5	3.67
45	142.5	11.94	27.5	40.8	6.39
50	182.4	13.51	55	207.8	14.42
37.5	67.5	8.22	40	105.0	10.25
30	37.5	6.12	45	132.6	11.52
55	225.0	15.00	17.5	19.1	4.37
60	258.1	16.07	22.5	25.0	5.00
65	297.4	17.25			

    a. Draw two scatter diagrams, one for the variable pair DIST (Y_1) and MPH (X) and the other for the pair $\sqrt{\text{DIST}}$ (Y_2) and MPH (X).
    b. Use the printout to determine and graph the least-squares lines for each variable pair given in part (a).
    c. Which of the two variable pairs seems to be better suited for straight-line regression?
For the variable pair $\sqrt{\text{DIST}}$ (Y_2) and MPH (X):
    d. Test the hypothesis that the true slope β_1 is equal to 1. (Use $\alpha = .01$.)
    e. Construct a 99% confidence interval for β_1.
    f. Calculate and graph 95% confidence and prediction bands.

Computer Printout for Problem 7

```
MEANS AND STANDARD DEVIATIONS
X: 39.73684 +- 15.76509   Y₁: 125.10000 +- 102.87605
CORRELATIONS
PEARSON: 0.954260   SPEARMAN: 0.993850
LINEAR REGRESSION OF Y₁ ON X
SLOPE:          6.22708 +-  0.47319
INTERCEPT:   -122.34459 +- 20.15624
S(Y₁|X):        31.64950
ANALYSIS OF VARIANCE: F(1,17) = 173.18128
MEANS AND STANDARD DEVIATIONS
X: 39.73684 +- 15.76509   Y₂: 10.17526 +- 4.77577
CORRELATIONS
PEARSON: 0.986289   SPEARMAN: 0.992528
LINEAR REGRESSION OF Y₂ ON X
SLOPE:          0.29878 +- 1.21247E - 02
INTERCEPT:     -1.59712 +- 0.51647
S(Y₂|X):        0.81097
ANALYSIS OF VARIANCE: F(1,17) = 607.2
```

8. The table presents the starting annual salaries (SAL) of a group of 30 college graduates who have recently entered the job market, along with their cumulative-grade-point averages (CGPA).

SAL (Y)	CGPA (X)	SAL (Y)	CGPA (X)	SAL (Y)	CGPA (X)
$10,455	2.58	$12,500	3.55	$13,255	3.55
9,680	2.31	13,310	3.64	13,004	3.55
7,300	2.47	12,105	3.72	8,000	2.47
9,388	2.52	6,200	2.24	8,224	2.47
12,496	3.22	11,522	2.70	10,750	2.78
11,812	3.37	8,000	2.30	11,669	2.78
9,224	2.43	12,548	2.83	12,322	2.98
11,725	3.08	7,700	2.37	11,002	2.58
11,320	2.78	10,028	2.52	10,666	2.58
12,000	2.98	13,176	3.22	10,839	2.58

a–c. Answer the same questions as in Problem 6(a)–(c) for the variable pair SAL (Y) and CGPA (X).

Refer to the printout when necessary to answer the following:

d. Obtain a 95% confidence interval for β_1.

e. Would you reject the null hypothesis H_0: $\beta_1 = 4{,}000$ at the $\alpha = .05$ level?

Computer Printout for Problem 8

```
MEANS AND STANDARD DEVIATIONS
X: 2.83833 +- 0.44841   Y: 10740.66667 +- 1967.65248
CORRELATIONS
PEARSON: 0.827368   SPEARMAN: 0.933515   KENDALL: 0.799219
LINEAR REGRESSION OF Y ON X
SLOPE:      3630.56128 +- 465.76874
INTERCEPT:   435.92357 +- 1337.85967
S(Y|X):     1124.71499
ANALYSIS OF VARIANCE: F(1,28) = 60.75847
```

    f. Calculate 95% confidence and prediction bands.

    g. Graph the respective 95% confidence and prediction bands.

    h. Would you reject the hypothesis H_0: $\mu_{Y|X_0} = 11,500$ at $X_0 = 2.75$? (Use $\alpha = .05$.)

9. In an experiment designed to describe the dose–response curve for vitamin K, individual rats were depleted of their vitamin K reserves and then fed dried liver for 4 days at different dosage levels.[6] The response of each rat was measured as the concentration of a clotting agent needed to clot a sample of its blood in 3 minutes. The results of the experiment on 12 rats are given in the table; values are expressed in common logarithms for both dose and response.

Rat	Log_{10}(concentration) (Y)	Log_{10}(dose) (X)
1	2.65	0.18
2	2.25	0.33
3	2.26	0.42
4	1.95	0.54
5	1.72	0.65
6	1.60	0.75
7	1.55	0.83
8	1.32	0.92
9	1.13	1.01
10	1.07	1.04
11	0.95	1.09
12	0.88	1.15

    a. Draw a scatter diagram of these data.

    b. Determine the least-squares estimates of slope and intercept for the straight-line regression of Y on X.

    c. Plot the estimated regression line on the scatter diagram.

    d. Determine and sketch 99% confidence bands based on the estimated regression line.

[6] Adapted from a study of Schønheyder (1936).

Computer Printout for Problem 9

```
MEANS AND STANDARD DEVIATIONS
X: 0.7425 +- 0.3200   Y: 1.6108 +- 0.5736
CORRELATION
PEARSON: -.99568
LINEAR REGRESSION OF Y ON X
SLOPE:      -1.78501 +- 0.05267
INTERCEPT:   2.93620 +- 0.04230
S(Y|X):      0.05589
ANALYSIS OF VARIANCE: F(1,10) = 1148.762
```

    e. Convert the fitted straight line to an equation in the original units $Y' = 10^Y$ and $X' = 10^X$.

    f. For the converted equation obtained in part (e), determine 99% confidence intervals for the true mean responses at the maximum and minimum doses used in the experiment. (*Note:* $\hat{Y}(\max) = 0.8834$, $\hat{Y}(\min) = 2.6149$.)

    g. If each of the values for X and Y on each rat are converted to their original units X' and Y', the following fitted straight line is obtained: $\hat{Y}' = 237.16095 - 21.32117X'$. How would you evaluate whether using the variables X' and Y' was better or worse than using X and Y for the regression analysis?

10. The susceptibility of catfish to a certain chemical pollutant was determined by immersing individual fish in 2 liters of an emulsion containing the pollutant and measuring the survival time in minutes.[7] The data in the table give the common log of survival time (Y) and the common log concentration (X) of the pollutant in parts per million for 18 fish.

Computer Printout for Problem 10

```
MEANS AND STANDARD DEVIATIONS
X: 4.500 +- 0.3515   Y: 2.997 +- 0.3550
CORRELATION
PEARSON: -.98823
LINEAR REGRESSION OF Y ON X
SLOPE:      -0.99810 +- 0.03863
INTERCEPT:   7.49110 +- 0.17429
S(Y|X):       .05597
ANALYSIS OF VARIANCE: F(1,16) = 667.728
```

---

[7] Adapted from a study by Nagasawa, Osano, and Kondo (1964).

Fish	Log$_{10}$(survival time) (Y)	Log$_{10}$(concentration) (X)
1	2.516	5.0
2	2.572	5.0
3	2.438	5.0
4	2.621	4.8
5	2.742	4.8
6	2.689	4.8
7	2.830	4.6
8	2.910	4.6
9	2.983	4.6
10	3.175	4.4
11	3.056	4.4
12	3.095	4.4
13	3.332	4.2
14	3.221	4.2
15	3.293	4.2
16	3.447	4.0
17	3.523	4.0
18	3.551	4.0

    a. Plot the scatter diagram for these data and draw the estimated straight line of Y on X on your scatter diagram.

    b. Test for the significance of the straight-line regression.

    c. Determine 95% confidence intervals for the true mean survival time $\mu_{Y|X}$ (where $Y = 10^Y$) at values of $X = 5$, 4.5, and 4.

11. An experiment was conducted to determine the extent to which the growth rate of a certain fungus could be affected by filling test tubes containing the same medium at the same temperature with different inert gases.[8] Three such experiments were performed for each of six gases, and the average growth rate over these three tests was used as the response. The table gives the molecular weight (X) of each gas used and the average growth rate (Y) in milliliters per hour for the three tests.

Gas	Average Growth Rate (Y)	Molecular Weight (X)
A	3.85	4.0
B	3.48	20.2
C	3.27	28.2
D	3.08	39.9
E	2.56	83.8
F	2.21	131.3

    a. Find the least-squares estimates of slope and intercept for the straight-line regression of Y on X and draw the estimated straight line on the scatter diagram for this data set.

---

[8] Adapted from a study by Schreiner, Gregoine, and Lawrie (1962).

b. Test for significant slope of the fitted straight line.

c. What information has not been used that might improve the sensitivity of the analysis?

d. What is the 90% confidence interval for the true average growth rate when the gas used has a molecular weight of 100?

e. Why would it be inappropriate to use the fitted line to estimate the growth rate for a gas with a molecular weight of 200?

f. Based on the choice of X-values used in this study, how would you criticize the accuracy of prediction obtained in this experiment using the fitted straight line?

12. Consider the data in the table below.[9]

Years	Months	Vocabulary Size
\multicolumn{2}{c}{Age}	
0	8	0
0	10	1
1	0	3
1	3	19
1	6	22
1	9	118
2	0	272
2	6	446
3	0	896
3	6	1,222
4	0	1,540
4	6	1,870
5	0	2,072
5	6	2,289
6	0	2,562

* Data from M. E. Smith, "An Investigation of the Development of the Sentence and the Extent of Vocabulary in Young Children," *Studies in Child Welfare* (University of Iowa) 3, no. 5 (1926).

a. Draw a scatter diagram for the variable pair vocabulary (Y) and age (X).

b. First convert ages to decimal years (e.g., 1 year 6 months gives 1.5 years). Then calculate the least-squares estimates of the parameters of the regression line. Sketch this regression line on the scatter diagram.

c. Add a 16th observation to the vocabulary data, with values 0.00 years and 0 words. Plot this new point. Logically this value should be on the line of vocabulary growth. Is it near the fitted line from (b)?

d. Recompute the least-squares estimates of the regression line including the (0, 0) observation ($n = 16$). Sketch the new line in a distinct way on the scatter diagram.

---

[9] Taken from Bourne, Ekstrand, and Dominowski (1971, table 14.3).

   e. Is either line fitted acceptably? If not, sketch your idea of the true relationship. If the data were to include observations through age 30 years, what should the curve look like?

   f. The data appear to be from one child. If this is true, what assumption of the least-squares approach is most likely violated and why?

13. The table gives rat body weights (grams) and latency to seizure (minutes), following injection of 40 mg/kg of body weight of metrazol.

Latency	Weight
2.30	348
1.95	372
2.90	378
2.30	390
1.10	392
2.50	395
1.30	400
2.00	409
1.70	413
2.00	415
2.95	423
1.25	428
2.05	464
3.70	468

   a. Draw a scatter diagram on graph paper with latency as a function of weight.

   b. Determine the least-squares estimates of slope and intercept for the straight-line regression of latency on weight.

   c. Test whether the slope is equal to 0. Use $\alpha = .10$.

   d. Test whether the intercept is equal to 0. Use $\alpha = .10$.

   e. Sketch the estimated regression line.

   f. Distinctly sketch your choice for a regression line if it differs from that from (e). Explain why you agree or disagree with (e), noting whether any assumptions appear to be violated.

14. Stevens (1966), citing Dimmick and Hubbard (1939), reported data from 20 studies of the color perception of unitary hues. The wavelength (millimeters) of light called green by subjects in each experiment is given below, along with the year the study was conducted.

   a. These data stimulate the question "Is there any linear trend over time in the wavelength of light called green?" Evaluate this question by finding the least-squares estimates of the straight-line regression functions for predicting wavelength from year. (*Hint:* Subtract 500 from the wavelengths and 1850 from the year.)

   b. Find a 95% confidence interval for the true slope.

   c. Draw a scatter diagram on graph paper.

Study	Wavelength	Date
1	532	1874
2	535	1884
3	495	1888
4	527	1890
5	505	1898
6	505	1898
7	503	1907
8	506	1909
9	509	1911
10	520	1912
11	514	1920
12	504	1922
13	515	1926
14	498	1927
15	500	1928
16	506	1931
17	528	1934
18	530	1935
19	512	1935
20	515	1939

15. The following data are from a study by Benignus and others (1981). Blood and brain levels of toluene (a commonly used solvent) were measured following a 3-hour inhalation exposure to either 50, 100, 500, or 1000 parts per million (ppm) toluene (PPM_TOLU). Blood toluene (BLOODTOL) and brain toluene (BRAIN-TOL) are expressed in parts per million, weight in grams, and age in days.

a. Provide a scatter diagram with BLOODTOL as the response and PPM_TOLU as the predictor.

b. Compute least-squares estimates of the straight-line regression coefficients. Plot the line on the scatter diagram.

Rat	BLOODTOL	BRAINTOL	PPM_TOLU	WEIGHT	AGE	LN_BLDTL	LN_BRNTL	LN_PPMTL
1	0.553	0.481	50	393	95	−0.593	−0.732	3.912
2	0.494	0.584	50	378	95	−0.706	−0.538	3.912
3	0.609	0.585	50	450	95	−0.495	−0.536	3.912
4	0.763	0.628	50	439	95	−0.270	−0.465	3.912
5	0.420	0.533	50	397	95	−0.868	−0.629	3.912
6	0.397	0.490	50	301	84	−0.923	−0.713	3.912
7	0.503	0.719	50	406	84	−0.687	−0.330	3.912
8	0.534	0.585	50	302	84	−0.628	−0.536	3.912
9	0.531	0.675	50	382	84	−0.633	−0.393	3.912
10	0.384	0.442	50	355	84	−0.957	−0.816	3.912
11	0.215	0.492	50	405	85	−1.536	−0.709	3.912
12	0.552	0.859	50	405	85	−0.595	−0.152	3.912
13	0.420	0.650	50	387	85	−0.868	−0.431	3.912
14	0.324	0.528	50	358	85	−1.127	−0.639	3.912

Rat	BLOODTOL	BRAINTOL	PPM_TOLU	WEIGHT	AGE	LN_BLDTL	LN_BRNTL	LN_PPMTL
15	0.387	0.546	50	311	85	−0.949	−0.605	3.912
16	1.036	1.262	100	355	86	0.035	0.233	4.605
17	1.065	1.584	100	440	86	0.063	0.460	4.605
18	1.084	1.773	100	421	86	0.081	0.573	4.605
19	0.944	1.307	100	370	86	−0.058	0.268	4.605
20	0.994	1.338	100	375	86	−0.006	0.291	4.605
21	1.146	1.180	100	368	83	0.136	0.166	4.605
22	1.167	1.108	100	321	83	0.154	0.103	4.605
23	0.833	0.939	100	359	83	−0.183	−0.063	4.605
24	0.630	0.909	100	367	83	−0.462	−0.095	4.605
25	0.955	1.078	100	363	83	−0.046	0.075	4.605
26	0.687	1.152	100	388	86	−0.376	0.141	4.605
27	0.723	1.796	100	404	86	−0.324	0.586	4.605
28	0.705	1.262	100	454	86	−0.349	0.233	4.605
29	0.696	1.865	100	389	86	−0.363	0.623	4.605
30	0.868	1.892	100	352	86	−0.142	0.638	4.605
31	8.223	19.843	500	367	83	2.107	2.988	6.215
32	10.604	24.450	500	406	83	2.361	3.197	6.215
33	12.085	29.297	500	371	83	2.492	3.377	6.215
34	7.936	18.098	500	408	83	2.071	2.896	6.215
35	11.164	25.196	500	305	83	2.413	3.227	6.215
36	10.289	18.266	500	391	84	2.331	2.905	6.215
37	11.140	19.486	500	396	84	2.411	2.970	6.215
38	9.647	18.479	500	347	84	2.267	2.917	6.215
39	13.343	21.920	500	372	84	2.591	3.087	6.215
40	11.292	20.861	500	331	84	2.424	3.038	6.215
41	7.524	22.130	500	365	85	2.018	3.097	6.215
42	10.783	18.301	500	348	85	2.378	2.907	6.215
43	8.595	17.038	500	416	85	2.151	2.835	6.215
44	9.616	22.423	500	344	85	2.263	3.110	6.215
45	11.956	15.452	500	398	85	2.481	2.738	6.215
46	30.274	44.900	1000	417	93	3.410	3.804	6.908
47	32.923	35.500	1000	351	93	3.494	3.570	6.908
48	28.619	30.800	1000	378	93	3.354	3.428	6.908
49	28.761	38.500	1000	338	93	3.359	3.651	6.908
50	25.402	31.500	1000	433	93	3.235	3.450	6.908
51	35.464	42.330	1000	342	85	3.569	3.745	6.908
52	32.706	34.030	1000	319	85	3.488	3.527	6.908
53	29.347	30.760	1000	440	85	3.379	3.426	6.908
54	26.481	32.360	1000	363	85	3.276	3.477	6.908
55	33.401	41.830	1000	336	85	3.509	3.734	6.908
56	39.541	54.930	1000	378	86	3.677	4.006	6.908
57	28.155	39.780	1000	420	86	3.338	3.683	6.908
58	25.629	49.290	1000	346	86	3.244	3.898	6.908
59	33.188	47.490	1000	413	86	3.502	3.861	6.908
60	33.505	42.660	1000	432	86	3.512	3.753	6.908

c. Repeat (a) but for the natural logarithms LN_BLDTL and LN_PPMTL.
d. Repeat (b) for the natural logarithms.
e. Which transformation leads to the best representation of the data? Note in your comments the validity of the regression assumptions.

References

Benignus, V. A., Muller, K. E., Barton, C. N., and Bittekofer, J. A. (1981). "Toluene Levels in Blood and Brain of Rats during and after Respiratory Exposure." *Toxicology and Appl. Pharmacology*, *61*: 326–334.

Bourne, L. E., Ekstrand, B. E., and Dominowski, R. L. (1971). *The Psychology of Thinking*. Englewood Cliffs, N.J.: Prentice-Hall.

Dimmick, F. L., and Hubbard, M. R. (1939). "The Spectral Location of Psychologically Unique Yellow, Green, and Blue." *Amer. J. Psychol.*, *52*: 242.

Morrison, D. F. (1976). *Multivariate Statistical Methods*. New York: McGraw-Hill Book Company.

Nagasawa, S., Osano, S., and Kondo, K. (1964). "An Analytical Method for Evaluating the Susceptibility of Fish Species to an Agricultural Chemical." *Jap. J. Appl. Ent. Zool.*, *8*: 118–122.

Schønheyder, F. (1936). "The Quantitative Determination of Vitamin K." *Biochem. J.*, *30*: 890–896.

Schreiner, H. R., Gregoine, R. C., and Lawrie, J. A. (1962). "New Biological Effects of the Gases of the Helium Group." *Science*, *136*: 653–654.

Siegel, S. (1956). *Nonparametric Statistics for the Behavioral Sciences*. New York: McGraw-Hill Book Company.

Stevens, S. S. (1966). *Handbook of Experimental Psychology*. New York: John Wiley & Sons, Inc.

Timm, N. H. (1975). *Multivariate Analysis with Applications in Education and Psychology*. Monterey, Calif.: Brooks/Cole Publishing Company.

6

The Correlation Coefficient and Straight-Line Regression Analysis

6-1 Definition of r

The correlation coefficient is an often-used statistic that not only provides a measure of how two random variables are associated in a sample but has properties that closely relate it to straight-line regression. We define the *sample correlation coefficient r* for two variables X and Y by the formula

$$r = \frac{\sum_{i=1}^{n} (X_i - \bar{X})(Y_i - \bar{Y})}{\left[\sum_{i=1}^{n} (X_i - \bar{X})^2 \sum_{i=1}^{n} (Y_i - \bar{Y})^2 \right]^{1/2}} \tag{6.1}$$

For hand-calculating purposes, another version of this formula is given by

$$r = \frac{\sum_{i=1}^{n} X_i Y_i - \left(\sum_{i=1}^{n} X_i \right)\left(\sum_{i=1}^{n} Y_i \right) / n}{\left\{ \left[\sum_{i=1}^{n} X_i^2 - \left(\sum_{i=1}^{n} X_i \right)^2 / n \right]\left[\sum_{i=1}^{n} Y_i^2 - \left(\sum_{i=1}^{n} Y_i \right)^2 / n \right] \right\}^{1/2}}$$

An equivalent formula for r that illustrates its mathematical relationship to the least-squares estimate of the slope of a fitted regression line is

$$r = \frac{S_X}{S_Y} \hat{\beta}_1 \tag{6.2}$$

Example For the age–systolic blood pressure data in Table 5-1, r is calculated as follows:

$$r = \frac{199,576 - (1,354)(4,276)/30}{\{[67,894 - (1,354)^2/30][624,260 - (4,276)^2/30]\}^{1/2}} = .66$$

Alternatively, from (6.2) we have

$$r = \frac{15.29}{22.58}(0.97) = .66$$

Three important mathematical properties of *r* are:

1. The possible values of *r* range from -1 to 1.

2. *r* is a dimensionless quantity; that is, *r* is independent of the units of measurement of X and Y.

3. *r* is positive, negative, or zero as $\hat{\beta}_1$ is positive, negative, or zero, and vice versa. This property follows directly, of course, from (6.2).

6-2 *r* as a Measure of Association

In the statistical assumptions for straight-line regression analysis discussed earlier, we did not consider the variable X to be random. Nevertheless, it often makes sense to view the regression problem as one where *both* X and Y are random variables. The measure *r* can then be interpreted as an *index of association* between X and Y in the following sense:

1. The more positive *r* is, the more positive the association is. This means that when *r* is close to 1, an individual with a high value for one variable will likely have a high value for the other, and an individual with a low value for one variable will likely have a low value for the other (Figure 6-1a).

2. The more negative *r* is, the more negative the association is; that is, an individual with a high value for one variable will likely have a low value for the other when *r* is close to -1, and conversely (Figure 6-1b).

3. If *r* is close to 0, there is little, if any, *linear* association between X and Y (Figure 6-1c).[1]

(By "association," we mean the lack of statistical independence between X and Y. More loosely, the lack of an association means that the value of one variable cannot be reasonably anticipated by knowing the value of the other variable.)

Since *r* is an index obtained from a *sample* of *n* observations, it follows that it can be considered as an estimate of an unknown population parameter. This unknown parameter is called the *population correlation coefficient* and is generally denoted by the symbol ρ_{XY}, or simply ρ if it is clearly understood which two variables are being considered. We shall agree to use ρ unless confusion is possible.

(The parameter ρ_{XY} is defined as $\rho_{XY} = \sigma_{XY}/\sigma_X\sigma_Y$, where σ_X and σ_Y denote the population standard deviations of the random variables X and Y and where σ_{XY} is called the *covariance* between X and Y. The covariance σ_{XY} is a population parameter describing the average amount that two variables "covary." In actuality, it is the population mean of the random variable $(X - \bar{X})(Y - \bar{Y})$.)

---

[1] Later we will see that a value of *r* close to 0 does not rule out a possible *nonlinear* association.

FIGURE 6-1   Correlation coefficient as a measure of association

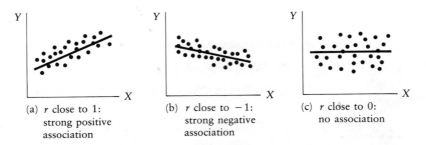

(a) r close to 1: strong positive association

(b) r close to -1: strong negative association

(c) r close to 0: no association

Figure 6-2 provides informative examples of scatter diagrams. Data were generated (via computer simulation) to have means and variances similar to the age–systolic blood pressure data of Chapter 5. The six samples observed here were produced by selecting 30 paired observations at random from each of six populations for which the population correlation coefficient ρ varied in value.

In Figure 6-2, the sample correlations range in value from .037 to .894. It should be clear that an "eyeball" analysis of the relative strengths of association is difficult, even though $n = 30$. For example, the difference between $r = .037$ in Figure 6-2a and $r = .220$ in Figure 6-2b is apparently due to the influence of just a few points. The study of so-called influential data points will be described in Chapter 12 on regression diagnostics.

In evaluating a scatter diagram, it is helpful to include reference lines at the X and Y means, as in Figure 6-2f. Roughly speaking, the proportions of observations in each quadrant reflect the strength of association. Notice that most of the observations in this figure are located in quadrants B and C, which are often referred to as the *positive quadrants*. Quadrants A and D are called the *negative quadrants*. When more observations are in positive quadrants than in negative quadrants, the sample correlation coefficient r will usually be positive. On the other hand, if more observations are in negative quadrants, then r will probably be negative.

To understand why this is so, we need to examine the numerator part of equation (6.1), namely,

$$\sum_{i=1}^{n} (X_i - \bar{X})(Y_i - \bar{Y})$$

(Note that the denominator of (6.1) is simply a positive scaling factor that makes r be dimensionless and satisfy the inequality $-1 \leq r \leq 1$.) This numerator describes how X and Y covary in terms of the n cross-products $(X_i - \bar{X})(Y_i - \bar{Y})$, $i = 1, 2, \ldots, n$. For a given i, such a cross-product term will be either positive or negative (or zero) depending on how X_i compares with $\bar{X}$ and how Y_i compares with $\bar{Y}$. In particular, if the ith observation (X_i, Y_i) is in quadrant B, then both $X_i > \bar{X}$ and $Y_i > \bar{Y}$; hence, the product of $(X_i - \bar{X})$ and $(Y_i - \bar{Y})$ must be positive. Similarly, if (X_i, Y_i) is in quadrant C, $X_i < \bar{X}$ and $Y_i < \bar{Y}$, so that $(X_i - \bar{X})(Y_i - \bar{Y})$ is again positive. Thus, observations in the positive quadrants B and C contribute positive values to the numerator of (6.1). Similarly, it follows that observations in the negative quadrants A and D contribute negative values to this numerator. So, roughly

FIGURE 6-2 Examples of a range of observed correlations between age and systolic blood pressure (SBP) for simulated data

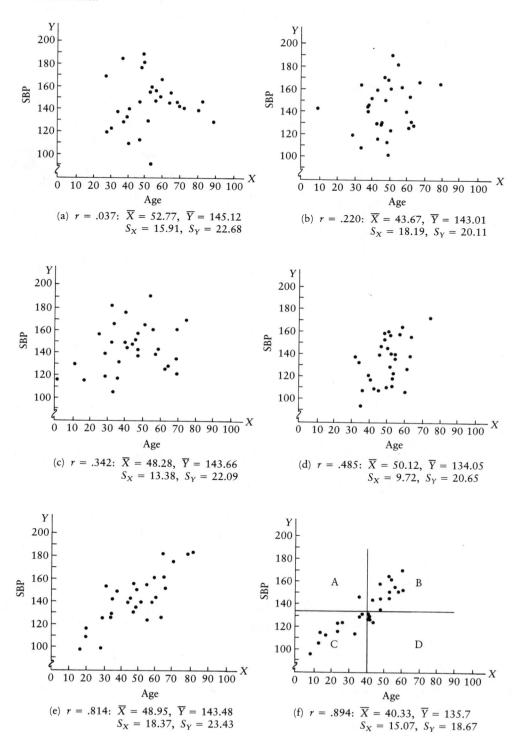

(a) $r = .037$: $\overline{X} = 52.77$, $\overline{Y} = 145.12$
$S_X = 15.91$, $S_Y = 22.68$

(b) $r = .220$: $\overline{X} = 43.67$, $\overline{Y} = 143.01$
$S_X = 18.19$, $S_Y = 20.11$

(c) $r = .342$: $\overline{X} = 48.28$, $\overline{Y} = 143.66$
$S_X = 13.38$, $S_Y = 22.09$

(d) $r = .485$: $\overline{X} = 50.12$, $\overline{Y} = 134.05$
$S_X = 9.72$, $S_Y = 20.65$

(e) $r = .814$: $\overline{X} = 48.95$, $\overline{Y} = 143.48$
$S_X = 18.37$, $S_Y = 23.43$

(f) $r = .894$: $\overline{X} = 40.33$, $\overline{Y} = 135.7$
$S_X = 15.07$, $S_Y = 18.67$

speaking, the sign of the correlation coefficient reflects the distribution of observations in these positive and negative quadrants.

6-3 The Bivariate Normal Distribution[2]

Another way of looking at straight-line regression is to consider X and Y as random variables having the bivariate normal distribution, which is a generalization of the *univariate normal distribution*. Just as the univariate normal distribution is described by a density function that is a bell-shaped curve when plotted in two dimensions, so the bivariate normal distribution is described by a *joint density function* whose plot looks like a bell-shaped surface in three dimensions (Figure 6-3).

One property of the bivariate normal distribution that has implications for straight-line regression analysis is the following: If the bell-shaped surface is cut by a plane *parallel* to the YZ-plane and passing through a specific X-value, the curve, or trace, that results is a normal distribution. In other words, the distribution of Y for fixed X is univariate-normal. We call such a distribution the *conditional distribution* of Y at X, and we denote the corresponding random variable as Y_X. Let us also denote the mean of this distribution as $\mu_{Y|X}$ and the variance as $\sigma^2_{Y|X}$. Then it follows from statistical theory that the mean and variance of Y_X can be written in terms of μ_X, μ_Y, σ^2_X, σ^2_Y, and ρ_{XY} as follows:

$$\mu_{Y|X} = \mu_Y + \rho_{XY} \frac{\sigma_Y}{\sigma_X} (X - \mu_X) \tag{6.3}$$

and

$$\sigma^2_{Y|X} = \sigma^2_Y (1 - \rho^2_{XY}) \tag{6.4}$$

Now suppose that we let $\beta_1 = \rho_{XY}(\sigma_Y/\sigma_X)$ and $\beta_0 = \mu_Y - \beta_1\mu_X$. We can then see that (6.3) has been transformed into the familiar expression for a straight-line model; that is, $\mu_{Y|X} = \beta_0 + \beta_1 X$. Furthermore, if we substitute the estimators $\bar{X}$, $\bar{Y}$, S_X, S_Y, and r for their respective parameters μ_X, μ_Y, σ_X, σ_Y, and ρ_{XY} in (6.3), we obtain the formula

$$\hat{\mu}_{Y|X} = \bar{Y} + r \frac{S_Y}{S_X} (X - \bar{X})$$

The right-hand side is exactly equivalent to the expression for the least-squares straight line given by (5.8), since

$$\hat{\beta}_1 = r \frac{S_Y}{S_X}$$

Thus, *the least-squares formulas for $\hat{\beta}_0$ and $\hat{\beta}_1$ can be developed by assuming that X and Y are random variables having the bivariate normal distribution and by substituting the usual estimates of μ_X, μ_Y, σ_X, σ_Y, and ρ_{XY} into the expression for $\mu_{Y|X}$, the conditional mean of Y given X.*

---

[2] This section is not essential for understanding the correlation coefficient as it relates to regression analysis.

FIGURE 6-3   The bivariate normal distribution

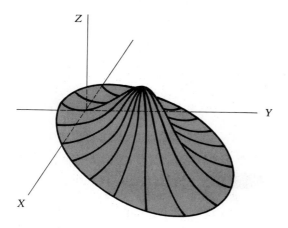

Note also that our estimate $S^2_{Y|X}$ of $\sigma^2_{Y|X}$ can be obtained by substituting the estimates S^2_Y and r for σ^2_Y and ρ_{XY} in (6.4). This is so because, from (5.11) and (6.2),

$$S^2_{Y|X} = \frac{n-1}{n-2}(S^2_Y - \hat{\beta}^2_1 S^2_X) = \frac{n-1}{n-2}S^2_Y(1 - r^2)$$

which approximates $S^2_Y(1 - r^2)$ even when n is just moderately large. Finally, (6.4) can be algebraically manipulated into the form

$$\rho^2_{XY} = \frac{\sigma^2_Y - \sigma^2_{Y|X}}{\sigma^2_Y} \tag{6.5}$$

This equation describes the square of the population correlation coefficient as the proportionate reduction in the variance of Y due to conditioning on X. The importance of (6.5) in describing the strength of the straight-line relationship will be discussed in the next section.

6-4   *r* and the Strength of the Straight-Line Relationship

In order to quantify what we mean by the *strength* of the linear relationship between X and Y, we should first consider what our predictor of Y would be if we did not use X at all. The best predictor in this case would simply be $\bar{Y}$, the sample mean of the Y's. The sum of the squares of deviations associated with the naive predictor $\bar{Y}$ would then be given by the formula

$$SSY = \sum_{i=1}^{n} (Y_i - \bar{Y})^2$$

Now, if the variable X is of any value in predicting the variable Y, the residual sum of squares given by

$$SSE = \sum_{i=1}^{n} (Y_i - \hat{Y}_i)^2$$

should be considerably less than SSY. If so, the least-squares model $\hat{Y} = \hat{\beta}_0 + \hat{\beta}_1 X$ fits the data better than the horizontal line $\hat{Y} = \bar{Y}$ (Figure 6-4). A quantitative measure of the improvement in the fit obtained by using X is given by the *square of the sample correlation coefficient r*, which can be written in the suggestive form

$$r^2 = \frac{SSY - SSE}{SSY}, \tag{6.6}$$

This quantity naturally varies between 0 and 1 since r itself varies between -1 and 1.

What interpretation can be given to the quantity r^2? To answer this question, we first note that the difference, or reduction, in SSY due to using X to predict Y may be measured by (SSY − SSE), which is always nonnegative. Furthermore, the *proportionate reduction* in SSY due to using X to predict Y is this difference divided by SSY. Thus, r^2 measures the strength of the linear relationship between X and Y in the sense that it gives the proportionate reduction in the sum of the squares of vertical deviations obtained using the least-squares line $\hat{Y} = \hat{\beta}_0 + \hat{\beta}_1 X$ relative to the naive model $\hat{Y} = \bar{Y}$, the predictor of Y if X is ignored. The larger the value of r^2, the greater the reduction in SSE relative to $\sum_{i=1}^{n}(Y_i - \bar{Y})^2$, and the stronger the linear relationship between X and Y.

The largest value that r^2 can attain is 1, which occurs when $\hat{\beta}_1$ is nonzero and when SSE = 0 (i.e., when there is a perfect positive or negative straight-line relationship between X and Y). By "perfect" we mean that *all* the data points lie on the fitted straight line. In other words, when $Y_i = \hat{Y}_i$ for all i, we must have

$$SSE = \sum_{i=1}^{n} (Y_i - \hat{Y}_i)^2 = 0$$

so that

$$r^2 = \frac{SSY - SSE}{SSY} = \frac{SSY}{SSY} = 1$$

FIGURE 6-4   Predictions of Y using and not using X

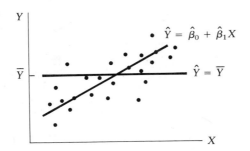

FIGURE 6-5   Examples of perfect linear association

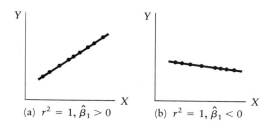

(a) $r^2 = 1, \hat{\beta}_1 > 0$ (b) $r^2 = 1, \hat{\beta}_1 < 0$

Figure 6-5 illustrates examples of perfect positive and perfect negative linear association.

The smallest value that r^2 may take, of course, is 0. This value means that there is no improvement in predictive power using X; that is, SSE = SSY. Furthermore, appealing to (6.2), we see that a correlation coefficient of 0 implies a zero estimated slope and, consequently, the absence of any linear relationship (although a nonlinear relationship is still possible).

Finally, one should *not* be led to a false sense of security by considering the magnitude of r, rather than of r^2, when assessing the strength of the linear association between X and Y. For example, when r is .5, r^2 is only .25, and it takes $r > .7$ to make $r^2 > .5$. Also, when r is .3, r^2 is .09, which indicates that only 9% of the variation in Y is explained with the help of X. For the age–systolic blood pressure data, r^2 is .44, compared with an r of .66.

6-5 What r Does Not Measure

There are two common misconceptions about r (or, equivalently, about r^2) that occasionally lead a researcher to make spurious interpretations about the relationship between X and Y:

1. *r^2 is not a measure of the magnitude of the slope of the regression line.* That is, if the value of r^2 is high (i.e., close to 1), this does not necessarily mean that the magnitude of the slope $\hat{\beta}_1$ is large. This phenomenon is illustrated in Figure 6-5. Notice that r^2 equals 1 in both parts, despite the fact that the slopes are different. Another way to look at this follows from the fact that, using (6.2),

$$\hat{\beta}_1^2 = \frac{S_Y^2}{S_X^2} \quad \text{when } r^2 = 1$$

Thus, if two different sets of data have the same amount of X variation but the first set has less Y variation than the second set, the magnitude of the slope for the first set will be smaller than that for the second.

2. *r^2 is not a measure of the appropriateness of the straight-line model.* Notice that $r^2 = 0$ in parts (a) and (b) of Figure 6-6, even though there is no evidence of any association between X and Y in (a) and strong evidence of a nonlinear association

FIGURE 6-6   Examples showing that r^2 is not a measure of the
appropriateness of the straight-line model

Examples when $r^2 = 0$

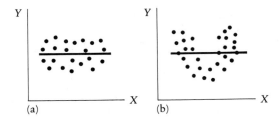

Examples when r^2 is high

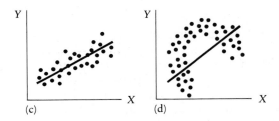

in (b). Also, notice that r^2 is high in parts (c) and (d), even though a straight-line model is quite appropriate in (c) but not completely appropriate in (d).

6-6 Tests of Hypotheses and Confidence Intervals for the Correlation Coefficient

Researchers interested in the association between two interval variables X and Y often desire a test of the null hypothesis $H_0: \rho = 0$.

6-6-1 Test of $H_0: \rho = 0$

A test of $H_0: \rho = 0$ turns out to be mathematically equivalent to the test of the hypothesis $H_0: \beta_1 = 0$ described in Section 5-8. This equivalence is suggested by the formulas $\beta_1 = \rho\sigma_Y/\sigma_X$ and $\hat{\beta}_1 = rS_Y/S_X$, which tell us, for example, that β_1 is positive, negative, or zero as ρ is positive, negative, or zero, and that an analogous relationship exists between $\hat{\beta}_1$ and r. The test statistic for the hypothesis $H_0: \rho = 0$ can be written entirely in terms of r and n, so that one can perform the test without having to fit the straight line. This test statistic is given by the formula

$$T = \frac{r\sqrt{n-2}}{\sqrt{1-r^2}}$$ (6.7)

which has the t distribution with $n - 2$ degrees of freedom when the null hypothesis $H_0: \rho = 0$ (or, equivalently, $H_0: \beta_1 = 0$) is true. Formula (6.7) yields exactly the same numerical answer as does (5.12), given by

$$T = \frac{\hat{\beta}_1 S_X \sqrt{n - 1}}{S_{Y|X}}$$

Example   For the age–systolic blood pressure data of Table 5-1, for which $r = .66$, the statistic in (6.7) is calculated as follows:

$$T = \frac{.66\sqrt{30 - 2}}{\sqrt{1 - (.66)^2}} = 4.62$$

which is the same value as that obtained for the test for slope in Table 5-3.

6-6-2   Test of $H_0: \rho = \rho_0, \rho_0 \neq 0$

A test concerning the null hypothesis $H_0: \rho = \rho_0$ $(\rho_0 \neq 0)$ cannot be directly related to a test concerning β_1; also, the hypothesis $H_0: \rho = \rho_0$ $(\rho_0 \neq 0)$ is not equivalent to the hypothesis $H_0: \beta_1 = \beta_1^{(0)}$ for some value $\beta_1^{(0)}$. Nevertheless, a test of $H_0: \rho = \rho_0$ $(\rho_0 \neq 0)$ is meaningful when previous experience or theory suggests a particular value to use for ρ_0.

The test statistic in this case can be obtained by considering the distribution of the sample correlation coefficient r. This distribution happens to be symmetric, like the normal distribution, *only* when ρ is 0. When ρ is nonzero, the distribution of r is skewed. This lack of normality prevents us from using the usual form of test statistic, which has a normally distributed estimator in the numerator and an estimate of its standard deviation in the denominator. However, it turns out that through an appropriate transformation, r can be changed to a statistic that is approximately normal. This transformation is called *Fisher's Z transformation*.[3] The formula for this transformation is

$$\frac{1}{2} \ln \frac{1 + r}{1 - r} \tag{6.8}$$

This quantity has approximately the normal distribution with mean $\frac{1}{2} \ln [(1 + \rho)/(1 - \rho)]$ and variance $1/(n - 3)$ when n is not too small (e.g., $n \geq 20$). In testing the hypothesis $H_0: \rho = \rho_0$ $(\rho_0 \neq 0)$, we can then use the test statistic

$$Z = \frac{\frac{1}{2} \ln [(1 + r)/(1 - r)] - \frac{1}{2} \ln [(1 + \rho_0)/(1 - \rho_0)]}{1/\sqrt{n - 3}} \tag{6.9}$$

This test statistic has approximately the standard normal distribution (i.e., $Z \sim N(0, 1)$) under H_0. To test $H_0: \rho = \rho_0$ $(\rho_0 \neq 0)$, we therefore use one of the following critical regions for significance level α:

$$Z \geq z_{1-\alpha} \quad \text{(upper one-tailed alternative } H_A: \rho > \rho_0\text{)}$$
$$Z \leq -z_{1-\alpha} \quad \text{(lower one-tailed alternative } H_A: \rho < \rho_0\text{)}$$
$$|Z| \geq z_{1-\alpha/2} \quad \text{(two-tailed alternative } H_A: \rho \neq \rho_0\text{)}$$

---

[3] Named after R. A. Fisher, who introduced it in 1925.

where $z_{1-\alpha}$ denotes the $100(1 - \alpha)\%$ point of the standard normal distribution. Computation of Z can be aided by using Appendix Table A-5, which gives values of $\frac{1}{2} \ln [(1 + r)/(1 - r)]$ for given values of r.

Example Suppose that from previous experience it can be hypothesized that the true correlation between age and systolic blood pressure is $\rho_0 = .85$. To test the hypothesis $H_0: \rho = .85$ against the two-sided alternative $H_A: \rho \neq .85$, we perform the following calculations using $r = .66$, $\rho_0 = .85$, and $n = 30$:

$$\frac{1}{2} \ln \frac{1 + \rho_0}{1 - \rho_0} = \frac{1}{2} \ln \frac{1 + .85}{1 - .85} = 1.2561 \qquad \text{(from Table A-5)}$$

$$\frac{1}{2} \ln \frac{1 + r}{1 - r} = \frac{1}{2} \ln \frac{1 + .66}{1 - .66} = .7928 \qquad \text{(from Table A-5)}$$

$$Z = \frac{.7928 - 1.2561}{1/\sqrt{30 - 3}} = -2.41$$

For $\alpha = .05$, the critical region is

$$|Z| \geq z_{.975} = 1.96$$

Since $|Z| = 2.41$ exceeds 1.96, the hypothesis $H_0: \rho_0 = .85$ is rejected at the .05 significance level. Further calculations show that the P-value for this test is $P = .0151$, which tells us that the result is not significant at $\alpha = .01$.

6-6-3 Confidence Interval for ρ

A $100(1 - \alpha)\%$ confidence interval for ρ can be obtained using Fisher's Z transformation (6.8) as follows. First, compute a $100(1 - \alpha)\%$ confidence interval for the parameter $\frac{1}{2} \ln[(1 + \rho)/(1 - \rho)]$ using the formula

$$\frac{1}{2} \ln \frac{1 + r}{1 - r} \pm \frac{z_{1-\alpha/2}}{\sqrt{n - 3}} \qquad (6.10)$$

where $z_{1-\alpha/2}$ is as defined previously.

Denote the lower limit of the confidence interval (6.10) by L_Z and the upper limit by U_Z; then use Appendix Table A-5 (in reverse) to determine the lower and upper confidence limits L_ρ and U_ρ for the confidence interval for ρ. That is, determine L_ρ and U_ρ from the following formulas:

$$L_Z = \frac{1}{2} \ln \frac{1 + L_\rho}{1 - L_\rho} \quad \text{and} \quad U_Z = \frac{1}{2} \ln \frac{1 + U_\rho}{1 - U_\rho}$$

The $100(1 - \alpha)\%$ confidence interval is then of the form

$$L_\rho < \rho < U_\rho$$

Example Suppose that we seek a 95% confidence interval for ρ using the age–systolic blood pressure data for which $r = .66$ and $n = 30$. A 95% confidence interval for $\frac{1}{2} \ln[(1 + \rho)/(1 - \rho)]$ is then given by

$$\frac{1}{2} \ln \frac{1 + .66}{1 - .66} \pm \frac{1.96}{\sqrt{30 - 3}}$$

which is equal to

$$.793 \pm .377$$

providing a lower limit of $L_Z = 0.416$ and an upper limit of $U_Z = 1.170$; that is,

$$0.416 < \frac{1}{2} \ln \frac{1 + \rho}{1 - \rho} < 1.170$$

To transform this to a confidence interval for ρ, we determine those values of L_ρ and U_ρ that satisfy

$$0.416 = \frac{1}{2} \ln \frac{1 + L_\rho}{1 - L_\rho} \quad \text{and} \quad 1.170 = \frac{1}{2} \ln \frac{1 + U_\rho}{1 - U_\rho}$$

Using Table A-5 we see that a value of 0.416 corresponds to an r of about .394, so that $L_\rho = .394$. Similarly, a value of 1.170 corresponds to an r of about .824, so that $U_\rho = .824$. The 95% confidence interval for ρ is then given by

$$.394 < \rho < .824$$

Notice that this confidence interval does not contain the value .85, which agrees with the conclusion of the previous section that $H_0 : \rho = .85$ is to be rejected at the 5% level (two-tailed test).

6-7 Testing for the Equality of Two Correlations

Suppose that independent random samples of sizes n_1 and n_2 are selected from two populations. Further, suppose that it is of interest to test $H_0: \rho_1 = \rho_2$ versus, say, $H_A: \rho_1 \neq \rho_2$. An appropriate test statistic can be developed based on the results given in Section 6-6. In this section, we will also consider the situation where the sample correlations to be compared are calculated using the same data set; in this case, these sample correlations are themselves "correlated."

6-7-1 Test of H_0: $\rho_1 = \rho_2$ Using Independent Random Samples

Let us assume that independent random samples of sizes n_1 and n_2 have been selected from two populations. For each population, the straight-line regression analysis assumptions given in Chapter 5, including that of normality, will hold.

An approximate test of $H_0: \rho_1 = \rho_2$ can be based on the use of Fisher's Z transformation. Let r_1 be the sample correlation calculated using the n_1 observations from the first population, and let r_2 be defined similarly. Using (6.8), let

$$Z_1 = \frac{1}{2} \ln \frac{1 + r_1}{1 - r_1} \tag{6.11}$$

and

$$Z_2 = \frac{1}{2} \ln \frac{1 + r_2}{1 - r_2} \tag{6.12}$$

Appendix Table A-5 can be used to determine Z_1 and Z_2.

To test $H_0: \rho_1 = \rho_2$, one can compute the test statistic

$$Z = \frac{Z_1 - Z_2}{\sqrt{1/(n_1 - 3) + 1/(n_2 - 3)}} \tag{6.13}$$

For large n_1 and n_2, this test statistic will have (approximately) the standard normal distribution when H_0 is true. Hence, the following critical regions for significance level α should be used:

$$Z \geq z_{1-\alpha} \qquad \text{(upper one-tailed alternative } H_A: \rho_1 > \rho_2\text{)}$$
$$Z \leq -z_{1-\alpha} \qquad \text{(lower one-tailed alternative } H_A: \rho_1 < \rho_2\text{)}$$
$$|Z| \geq z_{1-\alpha/2} \qquad \text{(two-tailed alternative } H_A: \rho_1 \neq \rho_2\text{)}$$

To illustrate this procedure, let us test whether the data sets plotted in Figures 6-2b and 6-2c reflect populations with different correlations. In other words, we wish to test $H_0: \rho_1 = \rho_2$ versus the two-sided alternative $H_A: \rho_1 \neq \rho_2$.

For the data in Figure 6-2b, $r_1 = .220$; for the Figure 6-2c data, $r_2 = .342$. Using Fisher's Z transformation and Table A-5, we can calculate Z_1 and Z_2 as

$$Z_1 = \frac{1}{2} \ln \frac{1 + r_1}{1 - r_1} = \frac{1}{2} \ln \frac{1 + .220}{1 - .220} = 0.2237$$

and

$$Z_2 = \frac{1}{2} \ln \frac{1 + r_2}{1 - r_2} = \frac{1}{2} \ln \frac{1 + .342}{1 - .342} = 0.3564$$

Then the test statistic (6.13) takes the value

$$Z = \frac{0.2237 - 0.3564}{\sqrt{1/(30 - 3) + 1/(30 - 3)}} = \frac{-0.1327}{0.2722} = -0.488$$

For $\alpha = .01$, the critical region is

$$|Z| \geq z_{.005} = 2.576$$

Since $|Z| = 0.488$ is less than 2.576, we cannot reject $H_0: \rho_1 = \rho_2$.

6-7-2 Single Sample Test of $H_0: \rho_{12} = \rho_{13}$

Consider testing the null hypothesis that the correlation ρ_{12} of variable 1 with variable 2 is the same as the correlation ρ_{13} of variable 1 with variable 3. Let us assume that a single random sample of n subjects is selected and the three sample correlations, r_{12}, r_{13}, and r_{23}, are calculated. Clearly these sample correlations are not independent, since they all are computed using the same data set. Under the usual straight-line regression analysis assumptions, it can be shown (we omit the details) that an appropriate large-sample test statistic for testing $H_0: \rho_{12} = \rho_{13}$ is

$$Z = \frac{(r_{12} - r_{13}) \sqrt{n}}{\sqrt{(1 - r_{12}^2)^2 + (1 - r_{13}^2)^2 - 2r_{23}^3 - 2(r_{23} - r_{12}r_{13})(1 - r_{12}^2 - r_{13}^2 - r_{23}^2)}} \tag{6.14}$$

For large n, this test statistic has approximately the standard normal distribution under H_0: $\rho_{12} = \rho_{13}$ (Olkin and Siotani, 1964; Olkin, 1967).

Let us now consider the following example. Assume that the weight, height, and age have been measured on each member of a sample of 12 nutritionally deficient children. Such a small sample brings into question the normal approximation involved in the use of (6.14). The data to be analyzed appear in Table 8-1 in Chapter 8. For these data, the three sample correlations are:

$$r_{12} = r_{(weight, height)} = .814$$

$$r_{13} = r_{(weight, age)} = .767$$

$$r_{23} = r_{(height, age)} = .614$$

We wish to test whether height and age are equally correlated with weight (i.e., $H_0: \rho_{12} = \rho_{13}$) versus the two-tailed alternative that they are not (i.e., $H_A: \rho_{12} \neq \rho_{13}$). Using (6.14), the test statistic takes the value

$$Z = \frac{(.814 - .767)\sqrt{12}}{\sqrt{\begin{array}{c}[1 - (.814)^2]^2 + [1 - (.767)^2]^2 - 2(.614)^3 \\ - 2[.614 - (.814)(.767)][1 - (.814)^2 - (.767)^2 - (.614)^2]\end{array}}}$$

$$= \frac{.1628}{\sqrt{.3374 + .4117 - .4630 - 2(-.0103)(-.6279)}} = \frac{0.1628}{\sqrt{0.2732}} = 0.3115$$

It is clear for these data that we cannot reject the null hypothesis of equal correlation of weight with height and age.

Problems

1. a. For the data set of Problem 1 of Chapter 5, compute the correlation coefficients of (1) dry weight with age and (2) log dry weight with age.
   b. Using Fisher's Z transformation, obtain a 95% confidence interval for ρ based on each of the correlations obtained in part (a).
   c. For each straight-line regression, calculate r^2 directly by squaring the r obtained in part (a), and also calculate r^2 using the formula $r^2 = (SSY - SSE)/SSY$.
   d. Based on the results above, which of the two regression lines do you believe provides the better fit? Explain. Does this agree with your earlier conclusion?

2. Examine the five pairs of data points given in the table.

i	1	2	3	4	5
X_i	-2	-1	0	1	2
Y_1	4	1	0	1	4

   a. What is the mathematical relationship between X and Y?
   b. Show by computation that for the straight-line regression of Y on X, $\hat{\beta}_1 = 0$.
   c. Show by computation that $r = 0$.
   d. Why is there apparently no relationship between X and Y as indicated by the estimates of β_1 and ρ?

3. Consider the data in the table.

i	1	2	3	4	5	6	7	8	9	10
X_i	1	1	1	2	2	2	3	3	3	20
Y_i	1	2	3	1	2	3	1	2	3	20

   a. Find the sample correlation coefficient r.
   b. Show that the test statistic $T = \hat{\beta}_1 / (S_{Y|X}/S_X \sqrt{n-1})$ for testing $H_0: \beta_1 = 0$ (based on a straight-line regression relationship between Y and X) is exactly equivalent to the test statistic $T' = r\sqrt{n-2}/\sqrt{1-r^2}$ for testing H_0: $\rho = 0$. (*Hint:* Use $\hat{\beta}_1 = rS_Y/S_X$ and $S_{Y|X}^2 = [(n-1)/(n-2)](S_Y^2 - \hat{\beta}_1^2 S_X^2)$.)
   c. Using T', test $H_0: \rho = 0$ versus $H_A: \rho \neq 0$.
   d. Despite the conclusion obtained in part (c), why should it bother you to conclude that the two variables are linearly related? ("A graph is worth a thousand words.")

4–6. Answer the following questions concerning the straight-line regressions of Y on X referred to in parts (d), (e), and (f) of Problem 2 in Chapter 5.
   a. Compute r and r^2 and interpret your results.
   b. Find a 99% confidence interval for ρ and interpret your result with regard to the test of $H_0: \rho = 0$ versus $H_A: \rho \neq 0$ at $\alpha = .01$.

7–12. Answer the following questions for each of the data sets of Problems 3–8 in Chapter 5.
   a. Compute r and r^2 for each variable pair and interpret your results.
   b. Test $H_0: \rho = 0$ versus $H_A: \rho \neq 0$ and interpret your findings.
   c. Find a 95% confidence interval for ρ.

13. Suppose that in a study on geographic variation in a certain species of beetle,[4] the mean tibia length (U) and the mean tarsus length (V) were obtained for samples of size 50 from each of 10 different regions spanning five southern states. Suppose further that the results were as given in the table.

Region	1	2	3	4	5	6	7	8	9	10
U	7.500	7.164	7.512	8.544	7.380	7.860	7.836	8.100	7.584	7.344
V	1.680	1.596	1.680	1.908	1.632	1.752	1.776	1.860	1.692	1.680

   a. Compute the correlation coefficient between tarsus length and tibia length.
   b. Find a 99% confidence interval for ρ.

14. In a sample of 23 young adult men, the correlation between total hemoglobin (THb) measured from venipuncture and measured from a finger needle puncture was .82. For a sample of 32 women of similar age, the correlation was .74. The two samples from each person were collected within 1 hour of each other. Assume that the straight-line regression assumptions hold.
   a. Test the hypothesis that the two population correlations are equal. Use a two-tailed test. What do you conclude?

---

[4] Adapted from a study by Sokal and Thomas (1965).

    b. If the experimenter planned to do so before collecting the data, a valid one-tailed test could be conducted. With this assumption, repeat (a) but use a one-tailed test to test whether the correlation for women is lower than that for men. What do you conclude?

    c. Assume that the researcher had planned to conduct a one-tailed test of the hypothesis that the correlation for women is higher than that for men. What test should be conducted? What do you conclude?

15. A university admissions officer regularly administers a test to all entering freshmen. A new version of the test is marketed by the testing company. In order to evaluate the new form, the admissions officer has 121 freshmen take both the old and the new versions. After the end of the school year, the admissions officer correlates the two scores with each other and with the freshman grade point average (GPA). With 1 indicating the old version, 2 the new version, and G the GPA,

$$r_{12} = .6969, \qquad r_{1G} = .5514, \qquad r_{2G} = .4188$$

Test the hypothesis that the two forms of the test are equally correlated with GPA. Use a two-tailed test with $\alpha = .05$. Assume that the straight-line regression assumptions hold.

16–21. Answer the following questions for each of the data sets of Problems 12(a), 12(c), 13, 14, 15(a), and 15(c) in Chapter 5.

    a. Compute r and r^2 for each variable pair and interpret your results.

    b. Test H_0: $\rho = 0$ versus H_A: $\rho \neq 0$ and interpret your findings.

    c. Find a 95% confidence interval for ρ.

References

Olkin, I. (1967). "Correlation Revisited." In *Improving Experimental Design and Statistical Analysis*, ed. Julian C. Stanley. Chicago: Rand McNally.

Olkin, I., and Siotani, M. (1964). "Asymptotic Distribution Functions of a Correlation Matrix." Stanford University Laboratory for Quantitative Research in Education, Report No. 6, Stanford, Calif.

Sokal, R. R., and Thomas, P. A. (1965). "Geographic Variation of *Pemphigus populitransversus* in Eastern North America: Stem Mothers and New Data on Alates." *Univ. Kansas Sci. Bull.*, 46: 201–252.

7

The Analysis-of-Variance
Table

7-1 Preview

An overall summary of the results of any regression analysis, whether straight line or not, can be provided by a table called an *analysis-of-variance (ANOVA) table*. This name derives primarily from the fact that the basic information in an ANOVA table consists of several estimates of variance. These estimates, in turn, can be used to answer the principal inferential questions of regression analysis. In the straight-line case, these questions are: (1) Is the true slope β_1 zero? (2) What is the strength of the straight-line relationship? (3) Is the straight-line model appropriate?

It should also be pointed out that, historically, the name "analysis-of-variance table" was coined to describe the overall summary table for the statistical procedure known as *analysis of variance*. As we mentioned in Chapter 2 and will see later when discussing the ANOVA method, regression analysis and analysis of variance are very closely related. More precisely, analysis-of-variance problems can be expressed in a regression framework. Thus, it should not be surprising to find that such a table can be used to summarize the results obtained from either method.

7-2 The ANOVA Table for Straight-Line Regression

Various textbooks, researchers, and computer program printouts give several slightly different ways of presenting the ANOVA table associated with straight-line regression analysis. This section will describe the most common form.

The simplest version of the ANOVA table for straight-line regression is given in Table 7-1, as applied to the age–systolic blood pressure data of Table 5-1. The mean-square term is obtained by dividing the sum of squares by its degrees of freedom. The variance ratio, or F statistic, is obtained by dividing the regression mean square by the residual mean square.

TABLE 7-1 ANOVA table for age–systolic blood pressure data of Table 5-1

Source	Degrees of Freedom (df)	Sum of Squares (SS)	Mean Square (MS)	Variance Ratio (F)
Regression (X)	1	SSY − SSE = 6,394.02	6,394.02	21.33 (P < .001)
Residual	28	SSE = 8,393.44	299.77	
Total*	29	SSY = 14,787.46		

* Corrected for the mean; $r^2 = .43$.

Finally, r^2 is obtained by dividing the regression sum of squares by the total (corrected) sum of squares.

To explain how this table summarizes the regression results, let us recall that in describing the correlation coefficient, we observed in (6.6) that

$$r^2 = \frac{SSY - SSE}{SSY}$$

where $SSY = \sum_{i=1}^{n}(Y_i - \bar{Y})^2$ is the sum of the squares of deviations of the observed Y's from the mean $\bar{Y}$, and $SSE = \sum_{i=1}^{n}(Y_i - \hat{Y}_i)^2$ is the sum of squares of deviations of observed Y's from the fitted regression line. Since SSY represents the total variation of Y *before* accounting for the linear effect of the variable X, we usually call SSY the *total unexplained variation* or the *total sum of squares about* (or *corrected for*) *the mean*. Because SSE measures the amount of variation in the observed Y's that is left *after* accounting for the linear effect of the variable X, we usually call (SSY − SSE) the *sum of squares due to* (or *explained by*) *regression*. It turns out, in addition, that (SSY − SSE) is mathematically equivalent to the expression

$$\sum_{i=1}^{n}(\hat{Y}_i - \bar{Y})^2$$

which represents the sum of squares of deviations of the predicted values from the mean $\bar{Y}$. We thus have the following mathematical result:

total unexplained variation = variation due to regression
+ unexplained residual variation

or

$$\sum_{i=1}^{n}(Y_i - \bar{Y})^2 = \sum_{i=1}^{n}(\hat{Y}_i - \bar{Y})^2 + \sum_{i=1}^{n}(Y_i - \hat{Y}_i)^2 \tag{7.1}$$

Equation (7.1) is often called the *fundamental equation of regression analysis*, and it holds for any general regression situation. Figure 7-1 illustrates this equation.

It should be pointed out that the mean-square residual term is simply the estimate $S_{Y|X}^2$ presented earlier. If the true regression model is a straight line, then, as mentioned in Section 5.6, $S_{Y|X}^2$ is an estimate of σ^2. On the other hand, the mean-square regression term (SSY − SSE) provides an estimate of σ^2 *only* if the variable X does not help to predict the

FIGURE 7-1   Variation explained and unexplained by (straight-line) regression

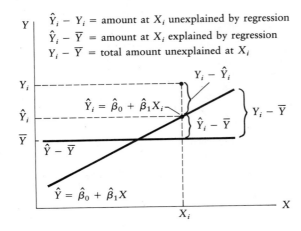

$\hat{Y}_i - Y_i$ = amount at X_i unexplained by regression
$\hat{Y}_i - \overline{Y}$ = amount at X_i explained by regression
$Y_i - \overline{Y}$ = total amount unexplained at X_i

dependent variable Y, that is, if the hypothesis $H_0: \beta_1 = 0$ is true. If in fact $\beta_1 \neq 0$, the mean-square regression term will be inflated in proportion to the magnitude of β_1 and will correspondingly overestimate σ^2.

It can be shown that the mean-square residual and mean-square regression terms are *statistically* independent of one another. Thus, if $H_0: \beta_1 = 0$ is true, the ratio of these terms represents the ratio of two independent estimates of the same variance σ^2. Under the normality assumption on the Y's, such a ratio has the F distribution, and this F statistic (with the value 21.33 in Table 7-1) can be used to test the hypothesis H_0: "no significant straight-line relationship of Y on X" (i.e., $H_0: \beta_1 = 0$ or $H_0: \rho = 0$).

Fortunately, this way of testing H_0 is *equivalent* to the use of the two-sided t test previously discussed. This is so because of the mathematical result that, for ν degrees of freedom,

$$F_{1,\nu} = T_\nu^2 \tag{7.2}$$

so that

$$F_{1,\nu,1-\alpha} = t_{\nu,1-\alpha/2}^2 \tag{7.3}$$

The expression in (7.3) states that the $100(1 - \alpha)\%$ point of the F distribution with 1 and ν degrees of freedom is exactly the same as the square of the $100(1 - \alpha/2)\%$ point of the t distribution with ν degrees of freedom.

To illustrate the equivalence of the F and t tests, we can see from our age–systolic blood pressure example that $F = 21.33$ and $T^2 = 4.62^2 = 21.33$, where 4.62 is the figure obtained for T at the end of Section 6-6-1. Also, it can be seen that $F_{1,28,0.95} = 4.20$ and that $t_{28,0.975}^2 = (2.05)^2 = 4.20$.

As a result of these equalities, the critical region

$$|T| > t_{28,0.975} = 2.05$$

TABLE 7-2   Alternative ANOVA table for age–systolic blood pressure data of
Table 5-1

Source	Degrees of Freedom (df)	Sum of Squares (SS)	Mean Square (MS)	Variance Ratio (F)
Regression $\begin{cases} \bar{Y} \\ X\|\bar{Y} \end{cases}$	1	$\dfrac{\left(\sum\limits_{i=1}^{n} Y_i\right)^2}{n} = 609{,}472.53$		
	1	6,394.02	6,394.02	21.33 $(P < .001)$
Residual	28	8,393.45	299.77	
Total*	30	$\sum\limits_{i=1}^{n} Y_i^2 = 624{,}260.00$		

* Uncorrected for the mean; $r^2 = .432$.

for testing $H_0: \beta_1 = 0$ against the two-sided alternative $H_A: \beta_1 \neq 0$ is exactly the same as the critical region

$$F > F_{1,28,0.95} = 4.20$$

That is, if $|T|$ exceeds 2.05, so then will F exceed 4.20. Similarly, if F exceeds 4.20, so will $|T|$ exceed 2.05. Thus, the null hypothesis $H_0: \beta_1 = 0$, or, equivalently, $H_0:$ "no significant straight-line relationship of Y on X," is rejected at the $\alpha = .05$ level of significance.

An alternative but less common representation of the ANOVA table is given in Table 7-2. This table differs from Table 7-1 only in that it splits up the total sum of squares corrected for the mean, SSY, into its two components, the *total uncorrected sum of squares*, $\sum_{i=1}^{n} Y_i^2$, and the *correction factor*, $(\sum_{i=1}^{n} Y_i)^2/n$. The relationship between these components is given by the equation

$$\sum_{i=1}^{n} (Y_i - \bar{Y})^2 = \sum_{i=1}^{n} Y_i^2 - \frac{\left(\sum\limits_{i=1}^{n} Y_i\right)^2}{n}$$

In the total (uncorrected) sum of squares $\sum_{i=1}^{n} Y_i^2$, the n observations on Y are considered before any estimation of the population mean of Y. The regression $(\bar{Y})$ listed in Table 7-2 refers to the variability explained by using a model involving only β_0 (which is estimated by $\bar{Y}$). This is necessarily the same amount of variability as is explained by using only $\bar{Y}$ to predict Y, without attempting to account for the linear contribution of X to the prediction of Y. The regression $(X|\bar{Y})$ describes the contribution of the variable X to predicting Y *over and above* that contributed by $\bar{Y}$ alone. Usually regression $(X|\bar{Y})$ is written simply as "regression (X)," the "given $\bar{Y}$" part being suppressed for notational simplicity. We will see more of this notation when we discuss multiple regression in subsequent chapters.

Problems

1. a. Determine the ANOVA table for the regression of dry weight (Y) on age (X) for the data of Problem 1 in Chapter 5.

    b. Perform the F test for the significance of the straight-line regression of Y on X. Verify that this test is equivalent to the t test for zero slope (and for zero correlation).

    c. Determine the ANOVA table for the regression of log dry weight (Z) on age (X) for the data of Problem 1 in Chapter 5.

    d. Perform the F test for the significance of the straight-line regression of Z on X.

2–4. Answer the same questions as in parts (a) and (b) of Problem 1 for each of the regressions of Y on X using the data in parts (d), (e), and (f) of Problem 2 in Chapter 5.

    5. a. Determine the ANOVA table for the regression of TIME (Y) on INC (X) in Problem 3 of Chapter 5.

    b. Test the hypothesis H_0: "no significant straight-line regression" using an F test.

    c. Compare the value of the test statistic F obtained in part (b) with the value of T^2, the square of the test statistic for testing H_0: $\beta_1 = 0$ required in Problem 3 of Chapter 5.

    d. Determine the value of r^2 using the ANOVA table obtained in part (a); check your answer with the r^2 value obtained in Problem 7 of Chapter 6.

    6. a. Determine the ANOVA table for the straight-line regression of DI (Y) on IQ (X) in Problem 4 of Chapter 5 (excluding the outlier).

    b. Determine the value of r^2 using the ANOVA table.

7–11. Answer the same questions as in parts (a) and (b) of Problem 6 for each of the data sets in Problems 5 through 8 and Problem 10 of Chapter 5.

    12. A biologist wished to study the effects of the temperature of a certain medium on the growth of human amniotic cells in a tissue culture. Using the same parent batch, she conducted an experiment in which five cell lines were cultured at each of four temperatures. The procedure involved initially inoculating a fixed number (0.25 million) of cells into a fresh culture flask and then, after 7 days, removing a small sample from the growing surface to be used for estimating the total number of cells in the flask. The results are given in the table, together with a computer printout for straight-line regression.

No. of Cells ($\times 10^{-6}$) after 7 Days (Y)	Temperature (X)	No. of Cells ($\times 10^{-6}$) after 7 Days (Y)	Temperature (X)
1.13	40	2.30	80
1.20	40	2.15	80
1.00	40	2.25	80
0.91	40	2.40	80
1.05	40	2.49	80
1.75	60	3.18	100
1.45	60	3.10	100
1.55	60	3.28	100
1.64	60	3.35	100
1.60	60	3.12	100

Computer Printout for Problem 12

```
MEANS AND STANDARD DEVIATIONS
X: 70.0000 +- 22.9416 Y: 2.0450 +- 0.8334
CORRELATIONS
PEARSON: .9860
LINEAR REGRESSION OF Y ON X
SLOPE:       0.03582 +- 0.00143
INTERCEPT: -0.46240 +- 0.10481
S(Y|X):      0.14263
ANALYSIS OF VARIANCE: F(1,18) = 630.716
```

    a. Construct the scatter diagram for this data set and comment on whether you think a straight line gives an adequate fit.

    b. Draw the least-squares straight line on the scatter diagram.

    c. Determine the ANOVA table for the straight-line regression of Y on X.

    d. Test for significance of the straight-line regression using an F test.

13–18. Answer the same questions as in parts (a) and (b) of Problem 6 for each of the data sets in Problems 12(a), 12(c), 13, 14, 15(a) and 15(c) in Chapter 5.

8

Multiple Regression Analysis: General Considerations

8-1 Preview

Multiple regression analysis can be looked upon as an extension of straight-line regression analysis (which involves only one independent variable) to the situation where there is more than one independent variable to be considered. Several general applications of multiple regression analysis[1] were described in Chapter 4, and specific examples given in Chapter 1. In this chapter we shall describe the multiple regression method in detail, stating the required assumptions, describing the procedures for estimating important parameters, explaining how to make and interpret inferences about these parameters, and providing examples illustrating the use of the techniques of multiple regression analysis. Before proceeding, however, it is important to mention that dealing with several independent variables simultaneously in a regression analysis is considerably more difficult than dealing with a single independent variable, for the following reasons:

1. It is more difficult to choose the best model, since there are sometimes several reasonable candidates.

2. It is more difficult to visualize what the fitted model looks like (especially if there are more than two independent variables), since it is not possible to plot directly in more than three dimensions either the data or the fitted model.

3. It is sometimes more difficult to interpret what the best-fitting model means in real-life terms.

4. Computations are virtually impossible without access to a high-speed computer and a reliable packaged computer program.

---

[1] We shall generally refer to multiple regression analysis simply as "regression analysis" throughout the remainder of the text.

8-2 Multiple Regression Models

One example of a multiple regression model is given by any second- or higher-order polynomial. The addition of higher-order terms (e.g., an X^2 or X^3 term) to the model can be considered as equivalent to the addition of new independent variables. Thus, if we rename X as X_1 and X^2 as X_2, the second-order model

$$Y = \beta_0 + \beta_1 X + \beta_2 X^2 + E$$

can be rewritten as

$$Y = \beta_0 + \beta_1 X_1 + \beta_2 X_2 + E$$

Of course, in polynomial regression we really have only one basic independent variable, the others being simple mathematical functions of this basic variable. In more general multiple regression problems, however, the number of basic independent variables may be greater than one. The general form of a regression model for k independent variables is given by

$$Y = \beta_0 + \beta_1 X_1 + \beta_2 X_2 + \cdots + \beta_k X_k + E$$

where $\beta_0, \beta_1, \beta_2, \ldots, \beta_k$ are the *regression coefficients* that need to be estimated. The *independent variables* $X_1, X_2, \ldots, X_k$ may all be separate basic variables, or some may be functions of a few basic variables.

Example   Suppose that we want to investigate how weight (WGT) varies with height (HGT) and age (AGE) for children with a particular kind of nutritional deficiency. The dependent variable here is $Y = $ WGT, and our two basic independent variables are $X_1 = $ HGT and $X_2 = $ AGE.

(Perhaps the main question associated with this type of study is whether the relationship for nutritionally deficient children is the same as that for "normal" children. To answer this question would require additional data on normal children and some kind of comparison of the models obtained for each group. Although we will learn how to deal with this kind of question in Chapter 14, we focus here on the methods needed to describe the relationship of weight to height and age for this single group of nutritionally deficient children.)

To continue, then, suppose that a random sample consists of 12 children who attend a certain clinic. The WGT, HGT, and AGE data obtained for each child are given in Table 8-1.
In describing the relationship of WGT to HGT and AGE, we may want to consider the model

$$Y = \beta_0 + \beta_1 X_1 + \beta_2 X_2 + E$$

TABLE 8-1   WGT, HGT, and AGE of a random sample of 12 nutritionally deficient children

Child	1	2	3	4	5	6	7	8	9	10	11	12
WGT (Y)	64	71	53	67	55	58	77	57	56	51	76	68
HGT (X_1)	57	59	49	62	51	50	55	48	42	42	61	57
AGE (X_2)	8	10	6	11	8	7	10	9	10	6	12	9

if we are interested only in first-order terms. If we want to consider, in addition, the higher-order term X_1^2, our model would be given by

$$Y = \beta_0 + \beta_1 X_1 + \beta_2 X_2 + \beta_3 X_3 + E$$

where $X_3 = X_1^2$. If we want to consider all possible first- and second-order terms, we would look at the model

$$Y = \beta_0 + \beta_1 X_1 + \beta_2 X_2 + \beta_3 X_3 + \beta_4 X_4 + \beta_5 X_5 + E$$

where $X_3 = X_1^2$, $X_4 = X_2^2$, and $X_5 = X_1 X_2$, or, equivalently,

$$Y = \beta_0 + \beta_1 X_1 + \beta_2 X_2 + \beta_3 X_1^2 + \beta_4 X_2^2 + \beta_5 X_1 X_2 + E$$

If we want to find the best predictive model, we might consider all of the models above (as well as some others) and then choose the best model according to some reasonable criterion.

We shall discuss the question of model selection in Chapter 16; the interpretation of product terms such as $X_1 X_2$ as interaction effects is explained in Chapter 11. For now we shall focus on the methods used and the interpretations that can be made when the choice of independent variables to be used in the model is not at issue.

8-3 Graphical Look at the Problem

When we are dealing with only one independent variable, our problem can easily be described graphically as that of finding the curve that best fits the scatter of points (X_1, Y_1), $(X_2, Y_2), \ldots, (X_n, Y_n)$ obtained on n individuals. Thus, we have a *two-dimensional* representation involving a plot of the form shown in Figure 8-1. Furthermore, as we know, the *regression equation* for this problem is defined as the path described by the mean values of the distribution of Y when X is allowed to vary.

When the number k of (basic) independent variables is two or more, the (graphical) dimension of the problem increases. The regression equation will no longer be a curve in two-dimensional space but a *hypersurface in $(k + 1)$-dimensional space*. Obviously, we will not be able to represent in a single plot either the scatter of data points or the regression

FIGURE 8-1   Scatter plot for a single independent
variable

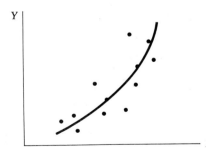

equation if there are more than two basic independent variables. In the special case $k = 2$, as in the example just given where X_1 = HGT, X_2 = AGE, and Y = WGT, the problem is to find the *surface* in three-dimensional space that best fits the scatter of points (X_{11}, X_{21}, Y_1), (X_{12}, X_{22}, Y_2), . . . , (X_{1n}, X_{2n}, Y_n), where (X_{1i}, X_{2i}, Y_i) denotes the X_1-, X_2-, and Y-values for the ith individual in the sample. The *regression equation* in this case is therefore the surface described by the mean values of Y at various combinations of values of X_1 and X_2; that is, corresponding to *each* distinct pair of values of X_1 and X_2 is a distribution of Y values with mean $\mu_{Y|X_1,X_2}$ and variance $\sigma^2_{Y|X_1,X_2}$.

Just as the simplest curve in two-dimensional space is a straight line, the simplest surface in three-dimensional space is a *plane*, which has the statistical model form $Y = \beta_0 + \beta_1 X_1 + \beta_2 X_2 + E$. Thus, finding the best-fitting plane is frequently the first step in determining the best-fitting surface in three-dimensional space when there are two independent variables, just as fitting the best straight line is the first step when one independent variable is involved. A graphical representation of a planar fit to data in the three-dimensional situation is given in Figure 8-2.

For the three-dimensional case, the least-squares solution giving the best-fitting plane is determined by minimizing the sum of squares of the distances between the observed values Y_i and the corresponding predicted values $\hat{Y}_i = \hat{\beta}_0 + \hat{\beta}_1 X_{1i} + \hat{\beta}_2 X_{2i}$ based on the fitted plane; that is, the quantity

$$\sum_{i=1}^{n} (Y_i - \hat{Y}_i)^2 = \sum_{i=1}^{n} (Y_i - \hat{\beta}_0 - \hat{\beta}_1 X_{1i} - \hat{\beta}_2 X_{2i})^2$$

is minimized to find the least-squares estimates $\hat{\beta}_0$ of β_0, $\hat{\beta}_1$ of β_1, and $\hat{\beta}_2$ of β_2.

It is natural at this point to wonder how much can be learned by considering the independent variables in the multivariable problem separately. Probably the best answer is that you can learn something about what is going on, but you will have too many separate

FIGURE 8-2   Best-fitting plane for three-dimensional data

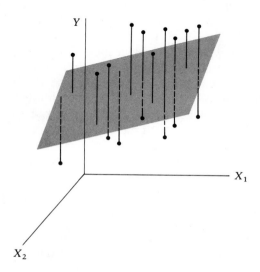

FIGURE 8-3   Separate scatter diagrams of WGT versus HGT, WGT versus AGE, and AGE versus HGT

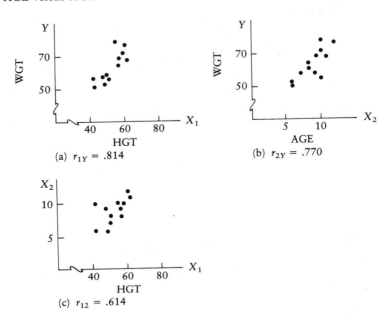

(a) $r_{1Y} = .814$

(b) $r_{2Y} = .770$

(c) $r_{12} = .614$

(univariable) pieces of information to be able to complete the (multivariable) puzzle. For example, consider the data previously given for Y = WGT, X_1 = HGT, and X_2 = AGE. If we plot separate scatter diagrams of WGT on HGT, WGT on AGE, and AGE on HGT, we get the results given in Figure 8-3.

First, we note that HGT is highly positively correlated with WGT ($r_{1Y} = .814$), as is AGE ($r_{2Y} = .770$). Thus, if we used each of these independent variables separately, we would likely find two separate, significant straight-line regressions. Does this mean that the best-fitting plane with both variables in the model together will also have significant predictive ability? The answer is probably yes. But what will the plane look like? This is difficult to say. We can get some idea of the difficulty if we consider the plot of HGT versus AGE, which reflects a positive correlation ($r_{12} = .614$). If, instead, these two variables were negatively correlated, we would expect a different orientation of the plane, although we could not clearly quantify either orientation. Thus, treating each independent variable separately does not help very much because the relationships between the independent variables themselves are not taken directly into account. The techniques of multiple regression, however, account for all these intercorrelations with regard to both estimation and inference making.

8-4 Assumptions of Multiple Regression

In the previous section we described the multiple regression problem in some generality and also hinted at some of the assumptions involved. We will now state these assumptions somewhat more formally.

8-4-1 Statement of Assumptions

Assumption 1: *Existence*   For each specific combination of values of the (basic) independent variables $X_1, X_2, \ldots, X_k$ (e.g., $X_1 = 57$, $X_2 = 8$ for the data given above), Y is a (univariate) random variable with a certain probability distribution having finite mean and variance.

Assumption 2: *Independence*   The Y observations are statistically independent of one another.

Assumption 3: *Linearity*   The mean value of Y for each specific combination of X_1, $X_2, \ldots, X_k$ is a linear function of $X_1, X_2, \ldots, X_k$; that is,

$$\mu_{Y|X_1, X_2, \ldots, X_k} = \beta_0 + \beta_1 X_1 + \beta_2 X_2 + \cdots + \beta_k X_k \tag{8.1}$$

or

$$Y = \beta_0 + \beta_1 X_1 + \beta_2 X_2 + \cdots + \beta_k X_k + E \tag{8.2}$$

where E is the error component reflecting the difference between an individual's observed response Y and the true average response $\mu_{Y|X_1, X_2, \ldots, X_k}$. Some comments are in order regarding Assumption 3.

1.   The surface described by (8.1) is called the *regression equation* (or *response surface* or *regression surface*).

2.   If some of the independent variables are higher-order functions of a few basic independent variables (e.g., $X_3 = X_1^2$, $X_5 = X_1 X_2$), the expression $\beta_0 + \beta_1 X_1 + \beta_2 X_2 + \cdots + \beta_k X_k$ is really nonlinear in the basic variables (hence the term *surface* rather than *plane*).

    (The techniques of multiple regression that we will be describing are applicable as long as the model under consideration is *inherently linear* in the regression coefficients (regardless of how the independent variables are defined). For example, a model of the form $\mu_{Y|X} = \beta_0 e^{\beta_1 X}$ is inherently linear because it can be transformed to the equivalent form $\mu_{Y|X}^* = \beta_0^* + \beta_1 X$, where $\mu_{Y|X}^* = \ln \mu_{Y|X}$ and $\beta_0^* = \ln \beta_0$. However, the model $\mu_{Y|X_1, X_2} = e^{\beta_1 X_1} + e^{\beta_2 X_2}$ cannot be transformed directly to a form that is linear in β_1 and β_2, and so estimating β_1 and β_2 requires the use of *nonlinear regression* procedures (e.g., see Gallant, 1975). A discussion of these procedures is beyond the scope of this text.)

3.   As with straight-line regression, E is the amount by which any individual's observed response deviates from the response surface. Thus, E is the *error component* in the model.

Assumption 4: *Homoscedasticity*   The variance of Y is the same for any fixed combination of $X_1, X_2, \ldots, X_k$; that is,

$$\sigma^2_{Y|X_1, X_2, \ldots, X_k} = \text{Var}\,(Y|X_1, X_2, \ldots, X_k) \equiv \sigma^2 \tag{8.3}$$

As before, this is called the assumption of homoscedasticity.

This assumption may seem very restrictive. However, as mentioned earlier in the discussion of the assumptions for straight-line regression analysis, variance heteroscedasticity must be considered only when the data show very obvious and significant departures from homogeneity. In general, mild departures will not have too adverse an effect on the results.

Assumption 5: Normality   For any fixed combination of $X_1, X_2, \ldots, X_k$, the variable Y is normally distributed. In other words,

$$Y \sim N(\mu_{Y|X_1, X_2, \ldots, X_k}, \sigma^2) \tag{8.4}$$

This assumption is not necessary for the least-squares fitting of the regression model but is required, in general, for inference making. In this regard, the usual parametric tests of hypotheses and confidence intervals used in a regression analysis are robust in the sense that only extreme departures of the distribution of Y from normality can yield spurious results. (This statement is based on both theoretical and experimental evidence.)

8-4-2 Summary and Comments

Our assumptions for simple linear (i.e., straight-line) regression analysis can be generalized to multiple linear regression analysis. Here, homoscedasticity and normality apply to $Y|X_1, X_2, \ldots, X_k$ rather than Y (i.e., to the conditional distribution of Y given $X_1, X_2, \ldots, X_k$ rather than the so-called unconditional or marginal distribution of Y).

Regarding the distribution of the random error component E, the assumptions for multiple linear regression analysis dictate that E has a normal distribution with mean 0 and variance σ^2. Of course, the linearity, existence, and independence assumptions must also hold.

Again, it is important to remember that Y is an observable random variable, while $X_1, X_2, \ldots, X_k$ are fixed (nonrandom) known quantities. The constants $\beta_0, \beta_1, \ldots, \beta_k$ are unknown population parameters, and E is a random variable that is unobservable. If one estimates $\beta_0, \beta_1, \ldots, \beta_k$ with $\hat{\beta}_0, \hat{\beta}_1, \ldots, \hat{\beta}_k$, then an acceptable estimate of E is

$$\hat{E} = Y - \hat{Y} = Y - (\hat{\beta}_0 + \hat{\beta}_1 X_1 + \cdots + \hat{\beta}_k X_k)$$

The estimated error $\hat{E}$ is usually called a *residual*.

The *HEIL GAUSS* mnemonic described in Chapter 5 applies here also. The assumption of a Gaussian distribution is required to justify the use of procedures of statistical inference involving the t and F distributions.

8-5 Determining the Best Estimate of the Multiple Regression Equation

As with straight-line regression, there are two basic approaches to estimating a multiple regression equation: the least-squares approach and the minimum-variance approach. In the straight-line case, both approaches yield the same solution.

(We are assuming, as previously noted, that we already know the best form of regression model to use; that is, we have already settled on a fixed set of k independent variables X_1,

$X_2, \ldots, X_k$. The problem of determining the best model form via algorithms for choosing the most important independent variables will be discussed in detail in Chapter 16.)

(The multiple regression model may also be fitted by using other statistical methodology, such as maximum likelihood (see Chapter 21). Under the assumption of a Gaussian distribution, the least-squares estimates of the regression coefficients are identical to the maximum-likelihood estimates.)

8-5-1 Least-Squares Approach

In general, the least-squares method chooses as the best-fitting model that model which minimizes the sum of squares of the distances between the observed responses and those predicted by the fitted model. Again, the idea is that the better the fit, the smaller the deviations of observed from predicted values. Thus, if we let

$$\hat{Y} = \hat{\beta}_0 + \hat{\beta}_1 X_1 + \hat{\beta}_2 X_2 + \cdots + \hat{\beta}_k X_k$$

denote the fitted regression model, the sum of squares of deviations of observed Y-values from corresponding predicted values using the fitted regression model is given by

$$\sum_{i=1}^{n} (Y_i - \hat{Y}_i)^2 = \sum_{i=1}^{n} (Y_i - \hat{\beta}_0 - \hat{\beta}_1 X_{1i} - \cdots - \hat{\beta}_k X_{ki})^2 \qquad (8.5)$$

The least-squares solution then consists of those values $\hat{\beta}_0, \hat{\beta}_1, \ldots, \hat{\beta}_k$ (called the "least-squares estimates") for which the sum in (8.5) is a minimum. This minimum sum of squares is generally called the *residual sum of squares* (or, equivalently, the *error sum of squares* or the *sum of squares about regression*) and, as with polynomial regression, is referred to as the SSE.

8-5-2 Minimum-Variance Approach

As with the straight-line case, the minimum-variance approach to estimating the multiple regression equation determines the best-fitting surface to be the one utilizing the minimum-variance (linear) unbiased estimates $\hat{\beta}_0, \hat{\beta}_1, \ldots, \hat{\beta}_k$ of $\beta_0, \beta_1, \ldots, \beta_k$, respectively.

8-5-3 Comments on the Least-Squares Solutions

It is not worthwhile, in our opinion, to present in this text matrix formulas for calculating the least-squares estimates $\hat{\beta}_0, \hat{\beta}_1, \ldots, \hat{\beta}_k$. Computer programs are readily available to perform the necessary calculations. Even so, we have provided in Appendix B a discussion of matrices and their use in regression analysis; by using matrix mathematics, the general regression model and the associated least-squares methodology can be represented in compact form. Also, an understanding of the matrix formulation for regression analysis carries over to more complex modeling problems, such as those concerning multivariate data (i.e., data involving two or more dependent variables).

It is of value, however, to mention some important properties of the least-squares solutions:

1. Each of the estimates $\hat{\beta}_0, \hat{\beta}_1, \ldots, \hat{\beta}_k$ is a linear function of the Y-values. This linearity property makes determining the statistical properties of these estimates

fairly straightforward. In particular, since the Y-values are assumed to be normally distributed, each of the estimates $\hat{\beta}_0, \hat{\beta}_1, \ldots, \hat{\beta}_k$ will be normally distributed, with easily computable standard deviations.

2. The least-squares regression equation $\hat{Y} = \hat{\beta}_0 + \hat{\beta}_1 X_1 + \hat{\beta}_2 X_2 + \cdots + \hat{\beta}_k X_k$ is the unique linear combination of the independent variables $X_1, X_2, \ldots, X_k$ that has maximum possible correlation with the dependent variable. In other words, of all possible linear combinations of the form $b_0 + b_1 X_1 + b_2 X_2 + \cdots + b_k X_k$, the linear combination $\hat{Y}$ is such that the correlation

$$r_{Y,\hat{Y}} = \frac{\sum_{i=1}^{n} (Y_i - \bar{Y})(\hat{Y}_i - \bar{\hat{Y}})}{\sqrt{\sum_{i=1}^{n} (Y_i - \bar{Y})^2 \sum_{i=1}^{n} (\hat{Y}_i - \bar{\hat{Y}})^2}} \tag{8.6}$$

is a maximum, where $\hat{Y}_i$ is the predicted value of Y for the ith individual and $\bar{\hat{Y}}$ is the mean of the $\hat{Y}_i$'s. Incidentally, it is always true that $\bar{\hat{Y}} = \bar{Y}$; that is, the mean of the predicted values is equal to the mean of the observed values.

3. Just as straight-line regression is related to the bivariate normal distribution, multiple regression can be related to the multivariate normal distribution. We shall return to this point later.

   Example   For the data given in Table 8-1 on the variables Y = WGT, X_1 = HGT, and X_2 = AGE, the least-squares algorithm applied to the model

   $$WGT = \beta_0 + \beta_1 HGT + \beta_2 AGE + \beta_3 (AGE)^2 + E$$

   produces the estimated equation

   $$\widehat{WGT} = 3.438 + 0.724 HGT + 2.777 AGE - 0.042 (AGE)^2$$

   so that

   $$\hat{\beta}_0 = 3.438, \quad \hat{\beta}_1 = 0.724, \quad \hat{\beta}_2 = 2.777, \quad \hat{\beta}_3 = -0.042$$

8-6 The ANOVA Table for Multiple Regression

As with straight-line regression, an ANOVA table can be used to provide an overall summary of a multiple regression analysis. The particular form of an ANOVA table can vary, depending on how the contributions of the independent variables are to be considered (e.g., individually or collectively in some fashion). A simple form reflects the contribution that all independent variables considered collectively make to prediction. For example, consider Table 8-2, an ANOVA table based on the use of HGT, AGE, and $(AGE)^2$ as independent variables for the data of Table 8-1.

As before, the term $SSY = \sum_{i=1}^{n} (Y_i - \bar{Y})^2 = 888.25$ is called the *total sum of squares*, and this figure represents the total variability in the Y observations before accounting for the joint effect of using the independent variables HGT, AGE, and $(AGE)^2$. The term $SSE = \sum_{i=1}^{n} (Y_i - \hat{Y}_i)^2 = 195.19$ is the *residual sum of squares* (or the *sum of squares due to error*), and this represents the amount of Y variation left unexplained after the independent

TABLE 8-2   ANOVA table for WGT regressed on HGT, AGE, and $(AGE)^2$

Source	df	SS	MS	F	R^2
Regression	$k = 3$	$SSY - SSE = 693.06$	231.02	9.47**	.7802
Residual	$n - k - 1 = 8$	$SSE = 195.19$	24.40		
Total	$n - 1 = 11$	$SSY = 888.25$			

Note: See Section 9-2 for an explanation of **.

variables have been used in the regression equation to predict Y. Finally, $SSY - SSE = \sum_{i=1}^{n} (\hat{Y}_i - \bar{Y})^2 = 693.06$ is called the *regression sum of squares* and measures the reduction in variation (or the variation explained) due to the independent variables in the regression equation. We thus have the familiar partition

total sum of squares = regression sum of squares + residual sum of squares

or

$$\sum_{i=1}^{n} (Y_i - \bar{Y})^2 \quad = \quad \sum_{i=1}^{n} (\hat{Y}_i - \bar{Y})^2 \quad + \quad \sum_{i=1}^{n} (Y_i - \hat{Y}_i)^2$$

In Table 8-2 the "SS" column, as in ANOVA tables for straight-line regression, identifies the various sums of squares. The "df" column gives the corresponding degrees of freedom: The regression degrees of freedom is k (the number of independent variables in the model), the residual degrees of freedom is $n - k - 1$, and the total degrees of freedom is $n - 1$. The "MS" column contains the mean-square terms, obtained by dividing the sum-of-squares terms by their corresponding degrees-of-freedom values. The F ratio is obtained by dividing the mean-square regression by the mean-square residual; the interpretation of this F ratio will be discussed in the next chapter on hypothesis testing.

The R^2 in Table 8-2 (with the value .7802) provides a quantitative measure of how well the fitted model containing the variables HGT, AGE, and $(AGE)^2$ predicts the dependent variable WGT; the computational formula for R^2 is

$$R^2 = \frac{SSY - SSE}{SSY} \tag{8.7}$$

The quantity R^2 lies between 0 and 1. If the value is 1, we say that the fit of the model is perfect. It is always true that R^2 increases as more variables are added to the model. Note, however, that a very small increase in R^2 may be neither practically nor statistically important. Additional properties of R^2 are discussed in Chapter 10.

8-7 Numerical Examples

We conclude this chapter with some examples of the type of output to be expected from a typical regression computer algorithm. This output generally consists of the values of the estimated regression coefficients, their estimated standard errors, the associated partial F (or T^2) statistics,[2] and an ANOVA table. For the data of Table 8-1, the six models we have

---

[2] Partial F statistics will be discussed in the next chapter on hypothesis testing.

chosen to consider are by no means the only possible models (e.g., no interaction terms were included). The results are provided below.

Model 1 $WGT = \beta_0 + \beta_1 HGT + E$

Coefficient $(\hat{\beta})$	Standard Error $(S_{\hat{\beta}})$	Partial F $(\hat{\beta}^2/S_{\hat{\beta}}^2)$
$\hat{\beta}_0 = 6.190$		
$\hat{\beta}_1 = 1.073$	$S_{\hat{\beta}_1} = 0.242$	19.66**

Estimated model: $\widehat{WGT} = 6.190 + 1.073 HGT$.

ANOVA TABLE

Source	df	SS	MS	F	R^2
Regression	1	588.92	588.92	19.67**	.6630
Residual	10	299.33	29.93		
Total	11	888.25			

Model 2 $WGT = \beta_0 + \beta_2 AGE + E$

Coefficient $(\hat{\beta})$	Standard Error $(S_{\hat{\beta}})$	Partial F $(\hat{\beta}^2/S_{\hat{\beta}}^2)$
$\hat{\beta}_0 = 30.571$		
$\hat{\beta}_2 = 3.643$	$S_{\hat{\beta}_2} = 0.955$	14.55**

Estimated model: $\widehat{WGT} = 30.571 + 3.643 AGE$.

ANOVA TABLE

Source	df	SS	MS	F	R^2
Regression	1	526.39	526.39	14.55**	.5926
Residual	10	361.86	36.19		
Total	11	888.25			

Model 3 $WGT = \beta_0 + \beta_3 (AGE)^2 + E$

Coefficient $(\hat{\beta})$	Standard Error $(S_{\hat{\beta}})$	Partial F $(\hat{\beta}^2/S_{\hat{\beta}}^2)$
$\hat{\beta}_0 = 45.998$		
$\hat{\beta}_3 = 0.206$	$S_{\hat{\beta}_3} = 0.055$	14.03**

Estimated model: $\widehat{WGT} = 45.998 + 0.206 (AGE)^2$.

ANOVA TABLE

Source	df	SS	MS	F	R^2
Regression	1	521.93	521.93	14.25**	.5876
Residual	10	366.32	36.63		
Total	11	888.25			

(The F-values disagree numerically due to rounding off the values of $\hat{\beta}_3$ and $S_{\hat{\beta}_3}$.)

Model 4 $WGT = \beta_0 + \beta_1 HGT + \beta_2 AGE + E$

Coefficient ($\hat{\beta}$)	Standard Error ($S_{\hat{\beta}}$)	Partial F ($\hat{\beta}^2/S_{\hat{\beta}}^2$)
$\hat{\beta}_0 = 6.553$		
$\hat{\beta}_1 = 0.722$	$S_{\hat{\beta}_1} = 0.261$	7.65*
$\hat{\beta}_2 = 2.050$	$S_{\hat{\beta}_2} = 0.937$	4.79 (.05 < P < .1)

Note: See Section 9-2 for an explanation of *.

Estimated model: $\widehat{WGT} = 6.553 + 0.722HGT + 2.050AGE$.

ANOVA TABLE

Source	df	SS	MS	F	R^2
Regression	2	692.82	346.41	15.95**	.7800
Residual	9	195.43	21.71		
Total	11	888.25			

Model 5 $WGT = \beta_0 + \beta_1 HGT + \beta_3 (AGE)^2 + E$

Coefficient ($\hat{\beta}$)	Standard Error ($S_{\hat{\beta}}$)	Partial F ($\hat{\beta}^2/S_{\hat{\beta}}^2$)
$\hat{\beta}_0 = 15.118$		
$\hat{\beta}_1 = 0.726$	$S_{\hat{\beta}_1} = 0.263$	7.62*
$\hat{\beta}_3 = 0.115$	$S_{\hat{\beta}_3} = 0.054$	4.54 (.05 < P < .1)

Estimated model: $\widehat{WGT} = 15.118 + 0.726HGT + 0.115(AGE)^2$.

ANOVA TABLE

Source	df	SS	MS	F	R^2
Regression	2	689.65	344.82	15.63**	.7764
Residual	9	198.60	22.07		
Total	11	888.25			

Model 6 $WGT = \beta_0 + \beta_1 HGT + \beta_2 AGE + \beta_3 (AGE)^2 + E$

Coefficient $(\hat{\beta})$	Standard Error $(S_{\hat{\beta}})$	Partial F $(\hat{\beta}^2/S_{\hat{\beta}}^2)$
$\hat{\beta}_0 = 3.438$		
$\hat{\beta}_1 = 0.724$	$S_{\hat{\beta}_1} = 0.277$	6.83^*
$\hat{\beta}_2 = 2.777$	$S_{\hat{\beta}_2} = 7.427$	0.14
$\hat{\beta}_3 = -0.042$	$S_{\hat{\beta}_3} = 0.422$	0.01

Estimated model: $\widehat{WGT} = 3.438 + 0.724 HGT + 2.777 AGE - 0.042 (AGE)^2$.

ANOVA TABLE

Source	df	SS	MS	F	R^2
Regression	3	693.06	231.02	9.47^{**}	.7802
Residual	8	195.19	24.40		
Total	11	888.25			

Although we will discuss model selection more fully in Chapter 16, the reader may appreciate from these results that model 4, involving HGT and AGE, is the best of the lot if we use R^2 and model simplicity as our criteria for selecting a model. The R^2-value of .7800 achieved using this model is, for all practical purposes, the same as the maximum R^2-value obtained using all three variables.

Problems

1. The multiple regression relationship of SBP (Y) to AGE (X_1), SMK (X_2), and QUET (X_3) was studied using the data in Problem 2 of Chapter 5. Three regression models were considered, yielding least-squares estimates and ANOVA tables as shown.

Model	Independent Variables Used	$\hat{\beta}_0$	$\hat{\beta}_1$	$\hat{\beta}_2$	$\hat{\beta}_3$	$S_{\hat{\beta}_1}$	$S_{\hat{\beta}_2}$	$S_{\hat{\beta}_3}$
1	AGE (X_1)	59.092	1.605			0.2387		
2	AGE (X_1), SMK (X_2)	48.050	1.709	10.294		0.2018	2.7681	
3	AGE (X_1), SMK (X_2), QUET (X_3)	45.103	1.213	9.946	8.592	0.3238	2.6561	4.4987

ANOVA RESULTS

| | Model 1 | | | |
Source	df	SS		
Regression (X_1)	1	3,861.630		
Residual	30	2,564.338		

| | Model 2 | | |
Source	df	SS
Regression (X_1, X_2)	2	4,689.684
Residual	29	1,736.285

| | Model 3 | |
Source	df	SS
Regression (X_1, X_2, X_3)	3	4,889.826
Residual	28	1,536.143

a. Using model 3: (1) What is the predicted SBP for a 50-year-old smoker with a quetelet index of 3.5? (2) What is the predicted SBP for a 50-year-old non-smoker with a quetelet index of 3.5? (3) For 50-year-old smokers, give an estimate of the change in SBP corresponding to an increase in quetelet index from 3.0 to 3.5.

b. Using the ANOVA tables, compute and compare the R^2-values for models 1, 2, and 3.

c. Conduct (separately) the overall F tests for significant regression under models 1, 2, and 3. Be sure to state your null hypothesis for each model in terms of regression coefficients.

2. A psychiatrist wanted to know whether the level of pathology (Y) in psychotic patients 6 months after treatment could be predicted with reasonable accuracy from knowledge of pretreatment symptom ratings of thinking disturbance (X_1) and hostile suspiciousness (X_2). The table gives the data that were collected on 53 patients.

Patient	Y	X_1	X_2	Patient	Y	X_1	X_2	Patient	Y	X_1	X_2	Patient	Y	X_1	X_2
1	44	2.80	6.1	15	26	3.24	6.0	28	8	2.63	6.9	41	23	2.18	6.1
2	25	3.10	5.1	16	27	2.65	6.0	29	11	2.65	5.8	42	31	2.88	5.8
3	10	2.59	6.0	17	4	3.41	7.6	30	7	3.26	7.2	43	20	3.04	6.8
4	28	3.36	6.9	18	14	2.58	6.2	31	23	3.15	6.5	44	65	3.32	7.3
5	25	2.80	7.0	19	21	2.81	6.0	32	16	2.60	6.3	45	9	2.80	5.9
6	72	3.35	5.6	20	22	2.80	6.4	33	26	2.74	6.8	46	12	3.29	6.8
7	45	2.99	6.3	21	60	3.62	6.8	34	8	2.72	5.9	47	21	3.56	8.8
8	25	2.99	7.2	22	10	2.74	8.4	35	11	3.11	6.8	48	13	2.74	7.1
9	12	2.92	6.9	23	60	3.27	6.7	36	12	2.79	6.7	49	10	3.06	6.9
10	24	3.23	6.5	24	12	3.78	8.3	37	50	2.90	6.7	50	4	2.54	6.7
11	46	3.37	6.8	25	28	2.90	5.6	38	9	2.74	5.5	51	18	2.78	7.2
12	8	2.72	6.6	26	39	3.70	7.3	39	13	2.70	6.9	52	10	2.81	5.2
13	15	3.47	8.4	27	14	3.40	7.0	40	22	3.08	6.3	53	7	3.26	6.6
14	28	2.70	5.9												

a. The least-squares equation involving both independent variables is given by $\hat{Y} = -0.628 + 23.639X_1 - 7.147X_2$. Using this equation, determine the predicted level of pathology $(\hat{Y})$ for a patient with pretreatment scores of 2.80 on thinking disturbance and 7.0 on hostile suspiciousness. How does the predicted value obtained compare with the value actually obtained for patient 5?

b. Using the ANOVA tables below, carry out the overall regression F tests for models containing both X_1 and X_2, X_1 alone, and X_2 alone.

Source	df	SS
Regression (X_1)	1	1,546
Residual	51	12,246

Source	df	SS
Regression (X_2)	1	160
Residual	51	13,632

Source	df	SS
Regression (X_1, X_2)	2	2,784
Residual	50	11,008

c. Based on your results in part (b), how would you rate the importance of the two variables in predicting Y?

d. What are the R^2-values for the three regressions referred to in part (b)?

e. What is the best model involving either one or both of the independent variables?

3. The accompanying table presents the weight (X_1), age (X_2), and plasma lipid levels of total cholesterol (Y) for a hypothetical sample of 25 patients with hyperlipoproteinemia before drug therapy.

Patient	Total Cholesterol (Y) (mg/100 ml)	Weight (X_1) (kg)	Age (X_2) (yr)
1	354	84	46
2	190	73	20
3	405	65	52
4	263	70	30
5	451	76	57
6	302	69	25
7	288	63	28
8	385	72	36
9	402	79	57
10	365	75	44
11	209	27	24
12	290	89	31
13	346	65	52
14	254	57	23
15	395	59	60
16	434	69	48
17	220	60	34
18	374	79	51
19	308	75	50
20	220	82	34
21	311	59	46
22	181	67	23
23	274	85	37
24	303	55	40
25	244	63	30

a. Given the accompanying ANOVA tables for the separate straight-line regressions of Y on X_1 (model 1) and Y on X_2 (model 2), which of the two independent variables would you say is the more important predictor of Y?

Source	df	SS		Source	df	SS
Regression (X_1)	1	10,231.7		Regression (X_2)	1	101,932.7
Residual	23	135,144.3		Residual	23	43,444.3

b. The estimated regression models resulting from the separate fits of Y on both X_1 and X_2, on X_1 alone, and on X_2 alone are given as follows:

$$\hat{Y} = 77.983 + 0.417X_1 + 5.217X_2$$
$$\hat{Y} = 199.2975 + 1.622X_1$$
$$\hat{Y} = 102.5751 + 5.321X_2$$

For each of these models, determine the predicted cholesterol level (Y) for patient 4 (with $Y = 263$, $X_1 = 70$, and $X_2 = 30$) and compare these predicted cholesterol levels with the observed value. Comment on your findings.

c. Given the ANOVA table below based on the regression involving both X_1 and X_2, carry out the overall F test for this two-variable model and the partial F test for the addition of X_1 to the model, given that X_2 is already in the model.

Source	df	SS
Regression (X_1, X_2)	2	102,570.8
Residual	22	42,806.2

d. Compute and compare the R^2-values for each of the three models considered in part (b).

e. Based on the results obtained in parts (a)–(d), what do you consider to be the best predictive model involving either one or both of the independent variables considered?

4. A sociologist investigating the recent increase in the incidence of homicide throughout the United States studied the extent to which the homicide rate per 100,000 population (Y) is associated with population size (X_1), the percent of families with yearly income less than $5,000 ($X_2$), and the rate of unemployment (X_3). Data are provided in the table for a hypothetical sample of 20 cities.

City	Y	X_1 (thousands)	X_2	X_3	City	Y	X_1 (thousands)	X_2	X_3
1	11.2	587	16.5	6.2	11	14.5	7,895	18.1	6.0
2	13.4	643	20.5	6.4	12	26.9	762	23.1	7.4
3	40.7	635	26.3	9.3	13	15.7	2,793	19.1	5.8
4	5.3	692	16.5	5.3	14	36.2	741	24.7	8.6
5	24.8	1,248	19.2	7.3	15	18.1	625	18.6	6.5
6	12.7	643	16.5	5.9	16	28.9	854	24.9	8.3
7	20.9	1,964	20.2	6.4	17	14.9	716	17.9	6.7
8	35.7	1,531	21.3	7.6	18	25.8	921	22.4	8.6
9	8.7	713	17.2	4.9	19	21.7	595	20.2	8.4
10	9.6	749	14.3	6.4	20	25.7	3,353	16.9	6.7

a. Given the ANOVA tables below, based on regressions involving each two-variable combination of the three independent variables above, conduct the overall regression F test for each two-variable regression model.

Source	df	SS
Regression (X_1, X_2)	2	1,317.80
Residual	17	537.40

Source	df	SS
Regression (X_1, X_3)	2	1,395.49
Residual	17	459.71

Source	df	SS
Regression (X_2, X_3)	2	1,477.37
Residual	17	377.82

b. Based on the results obtained in part (a), which two-variable model would you recommend?

c. Compute the R^2-values for each two-variable model above and relate the results to the conclusion reached in part (b).

d. Given the ANOVA table below, involving all three independent variables, conduct the test of overall regression.

Source	df	SS
Regression (X_1, X_2, X_3)	3	1,507.18
Residual	16	348.03

e. Determine and comment on the increase in R^2 in going from a model with just X_2 and X_3 to a model that includes all three independent variables.

f. The ANOVA table resulting from fitting a model with independent variables X_2, X_3, and $X_4 = X_2 X_3$ is as follows.[3] Use this table to provide a test of overall regression.

---

[3] The coefficient of the product term X_4 measures what is generally called an *interaction effect* associated with the variables X_2 and X_3, which concerns whether the relationship between Y and one of these two variables depends upon the levels of the other variable. A more detailed discussion of the concept of interaction is given in Chapter 11.

Source	df	SS
Regression (X_2, X_3, X_4)	3	1,480.46
Residual	16	377.73

5. A panel of educators in a large urban community was interested in evaluating the effects of educational resources on student performance. They examined the relationship between twelfth-grade mean verbal SAT scores (Y) and the following independent variables for a random sample of 25 high schools: X_1 = per pupil expenditure (in dollars), X_2 = percent of teachers with a master's degree or higher, and X_3 = pupil–teacher ratio. The ANOVA table given next summarizes the key results obtained from the regression of Y on X_1, X_2, and X_3.

Source	df	SS
Regression (X_1, X_2, X_3)	3	25,974.00
Residual	21	2,248.23
Total	24	28,222.23

  a. Conduct the overall F test for the model.
  b. Compute R^2.

6. A team of environmental epidemiologists used data from 23 counties to investigate the relationship between respiratory cancer mortality rates (Y) for a given year and the following three independent variables: X_1 = air pollution index for the county, X_2 = mean age (over 21) for the county, and X_3 = percent of working force in the county employed in a certain industry. The ANOVA table given next summarizes the key results of the regression of Y on X_1, X_2, and X_3.

Source	df	SS
Regression (X_1, X_2, X_3)	3	1,835.93
Residual	19	551.723
Total	22	2,387.653

  a. Conduct the overall F test for the regression model that includes all three independent variables.
  b. Determine R^2 for the model containing all three independent variables.

7. In an experiment to describe the toxic action of a certain chemical on silkworm larvae,[4] the relationship of $\log_{10}(\text{dose})$ and $\log_{10}(\text{larva weight})$ to $\log_{10}(\text{survival time})$ was sought. The data, obtained by feeding each larva a precisely measured dose of the chemical in an aqueous solution and then recording the survival time (i.e., time until death), are given in the table. Also given are relevant computer results and the ANOVA table.

---

[4] Adapted from a study by Bliss (1936).

Larva	1	2	3	4	5	6	7	8
Log_{10}(survival time) (Y)	2.836	2.966	2.687	2.679	2.827	2.442	2.421	2.602
Log_{10}(dose) (X_1)	0.150	0.214	0.487	0.509	0.570	0.593	0.640	0.781
Log_{10}(weight) (X_2)	0.425	0.439	0.301	0.325	0.371	0.093	0.140	0.406

Larva	9	10	11	12	13	14	15
Log_{10}(survival time) (Y)	2.556	2.441	2.420	2.439	2.385	2.452	2.351
Log_{10}(dose) (X_1)	0.739	0.832	0.865	0.904	0.942	1.090	1.194
Log_{10}(weight) (X_2)	0.364	0.156	0.247	0.278	0.141	0.289	0.193

$$\hat{Y} = 2.952 - 0.550X_1$$
$$\hat{Y} = 2.187 + 1.370X_2$$
$$\hat{Y} = 2.593 - 0.381X_1 + 0.871X_2$$

Source	df	SS		Source	df	SS
Regression (X_1, X_2)	2	0.4637		Regression (X_1)	1	0.3633
Residual	12	0.0476		Residual	13	0.1480

Source	df	SS
Regression (X_2)	1	0.3332
Residual	13	0.1780

a. Test for the significance of overall regression involving both independent variables X_1 and X_2.

b. Test to see whether using X_1 alone significantly helps in predicting survival time.

c. Test to see whether using X_2 alone significantly aids in predicting survival time.

d. Compute R^2 for each of the three models.

e. Which independent variable do you consider to be the best single predictor of survival time?

f. Which model involving one or both of the independent variables do you prefer and why?

g. Using the fitted model containing both X_1 and X_2, what log_{10}(dose) (X_1) is required to kill a larva for which $X_2 = 0.200$ in the same time that a larva for which $X_2 = 0.400$ and $X_1 = 0.500$ is killed? (*Hint*: For the heavier larva, the estimated survival time $\hat{Y}_H$ is given by $\hat{Y}_H = \hat{\beta}_0 + \hat{\beta}_1(0.500) + \hat{\beta}_2(0.400)$, where $\hat{\beta}_0$, $\hat{\beta}_1$, and $\hat{\beta}_2$ are the least-squares coefficients for the model containing both X_1 and X_2. The estimated survival time for the lighter larva weighing $X_2 = 0.200$ at a log_{10}(dose) of X_1 is $\hat{Y}_L = \hat{\beta}_0 + \hat{\beta}_1 X_1 + \hat{\beta}_2(0.200)$. The problem is to find that value of X_1 which makes $\hat{Y}_L = \hat{Y}_H$.)

h. Based on the hint in part (g), give a general formula for the log_{10}(dose) (X_1) necessary to kill a larva weighing (in log_{10} units) X_2^0 in the same time as a larva weighing twice this log_{10}(weight) and subject to a log_{10}(dose) equal to X_1^0; that is, what is X_1 in terms of X_2^0 and X_1^0?

8. An experiment to evaluate the effects of certain variables on soil erosion was performed on 10-foot-square plots of sloped farmland subjected to 2 inches of artificial rain applied for a 20-minute period.[5] The data and related computer results are as follows.

Plot	1	2	3	4	5	6	7	8	9	10	11
SL (Y)	27.1	35.6	31.4	37.8	40.2	39.8	55.5	43.6	52.1	43.8	35.7
SG (X_1)	0.43	0.47	0.44	0.48	0.48	0.49	0.53	0.50	0.55	0.51	0.48
LOBS (X_2)	1.95	5.13	3.98	6.25	7.12	6.50	10.67	7.08	9.88	8.72	4.96
PGC (X_3)	0.34	0.32	0.29	0.30	0.25	0.26	0.10	0.16	0.19	0.18	0.28

Note: SL denotes soil lost (in pounds/acre), SG denotes slope gradient of plot, LOBS denotes length (in inches) of the largest opening of bare soil on any boundary, and PGC denotes percent of ground cover.

Source	df	SS
Regression	3	680.4913
Residual	7	16.0943
Total		696.5856

a. The fitted model involving all three independent variables is given by $\hat{Y} = -1.879 + 77.326X_1 + 1.559X_2 - 23.904X_3$. Compute and compare observed and predicted values of Y for plots 1, 5, and 7.

b. Test for significant overall regression of the model containing all three independent variables.

9. In a study by Yoshida (1961), the oxygen consumption of wireworm larva groups was measured at five temperatures. The dependent variable, the rate of oxygen consumption per larva group in milliliters per hour, was transformed to 0.5 less than the common logarithm. Another independent variable (other than temperature) of importance was larva group weight, which was also transformed to common logarithms. The data and the ANOVA table are given below.

a. The fitted multiple regression model containing both X_1 and X_2 is given by

$$\hat{Y} = -0.6835 + 0.5917X_1 + 0.0393X_2$$

Using this fitted model, how much of a change in oxygen consumption would be predicted for a larva group with fixed weight X_1 if the temperature was increased from $X_2 = 20$ to $X_2 = 25$?

b. For a temperature of 20°C, compute and compare the predicted values of $\hat{Y}$ for weights of 0.250 and 0.500.

c. What is R^2 for each of the three models?

---

[5] Adapted from a study by Packer (1951).

Oxygen Consumption (Y) (log ml/hr −0.5)	Larva Group Weight (X_1) (log cg)	Temperature (X_2) (°C)
0.054	0.130	15.5
0.154	0.215	15.5
0.073	0.250	15.5
0.182	0.267	15.5
0.241	0.389	15.5
0.316	0.490	15.5
0.290	0.491	15.5
0.061	0.004	20.0
0.143	0.164	20.0
0.188	0.225	20.0
0.176	0.314	20.0
0.248	0.447	20.0
0.357	0.477	20.0
0.403	0.505	20.0
0.342	0.537	20.0
0.335	−0.046	25.0
0.408	0.176	25.0
0.366	0.199	25.0
0.482	0.292	25.0
0.545	0.380	25.0
0.596	0.483	25.0
0.590	0.491	25.0
0.631	0.491	25.0
0.610	0.519	25.0
0.482	0.053	30.0
0.477	0.114	30.0
0.551	0.137	30.0
0.516	0.190	30.0
0.561	0.210	30.0
0.588	0.230	30.0
0.561	0.240	30.0
0.580	0.260	30.0
0.674	0.389	30.0
0.718	0.470	30.0
0.754	0.521	30.0
0.800	0.544	30.0
0.654	−0.004	35.0
0.744	0.033	35.0
0.711	0.049	35.0
0.855	0.140	35.0
0.932	0.204	35.0
0.927	0.210	35.0
0.914	0.215	35.0
0.914	0.265	35.0
0.973	0.346	35.0
1.000	0.462	35.0
0.998	0.468	35.0

Source	df	SS	Source	df	SS
Regression (X_1)	1	0.0661	Regression (X_2)	1	2.7742
Residual	45	3.3399	Residual	45	0.6318

Source	df	SS
Regression (X_1, X_2)	2	3.2112
Residual	44	0.1948

References

Bliss, C. I. (1936). "The Size Factor in Action of Arsenic upon Silkworms' Larvae." *J. Exptl. Biol.*, *13*: 95–110.

Gallant, A. R. (1975). "Non-linear Regression." *Amer. Statistician*, *29*: 73-81.

Packer, P. E. (1951). "An Approach to Watershed Protection Criteria." *J. Forestry*, *49*: 638–644.

Yoshida, M. (1961). "Ecological and Physiological Researches on the Wireworm, *Melanotus caudex* Lewis. Iwata." *Shizuoka Pref.*, Japan.

9

Testing Hypotheses
in Multiple Regression

9-1 Preview

Once we have fit a multiple regression model and obtained estimates for the various parameters of interest, we want to answer questions about the contributions of various independent variables to the prediction of Y. There are three basic types of such questions. These are:

1. *An overall test.* Taken collectively, does the *entire set* of independent variables (or, equivalently, the fitted model itself) contribute significantly to the prediction of Y?

2. *Test for addition of a single variable.* Does the addition of *one* particular independent variable of interest add significantly to the prediction of Y over and above that achieved by other independent variables already present in the model?

3. *Test for addition of a group of variables.* Does the addition of some *group* of independent variables of interest add significantly to the prediction of Y obtained through other independent variables already present in the model?

These questions are typically answered by performing statistical tests of hypotheses. The null hypotheses for these tests can be stated in terms of the unknown parameters (the regression coefficients) in the model. The form of these hypotheses will differ depending on the question being asked. (We will see in the next chapter that there are alternative, but equivalent, ways to state such null hypotheses in terms of population correlation coefficients).

In the sections below, we shall describe the statistical test appropriate for each of the above questions. Each of these tests can be expressed as an F test; that is, the test statistic will have an F distribution when the stated null hypothesis is true. In some cases, the test may be equivalently expressed as a t test. (At this time, the reader may wish to review the material concerning the F and t distributions given in Chapter 3.)

124

A key feature of F tests used in regression analyses is that they all involve a ratio of two independent estimates of variance, say, $F = \hat{\sigma}_0^2/\hat{\sigma}^2$. Under the assumptions for the standard multiple linear regression analysis given earlier, the term $\hat{\sigma}_0^2$ estimates σ^2 if H_0 is true; the term $\hat{\sigma}^2$ estimates σ^2 whether H_0 is true or not. The specific forms that these variance estimates take will be described in the sections below. In general, each will be a mean-square term that can be found in an appropriate ANOVA table. If H_0 is not true, then $\hat{\sigma}_0^2$ estimates some quantity larger than σ^2. Thus, we would expect a value of F close to 1 ($= \sigma^2/\sigma^2$) if H_0 is true, but larger than 1 if H_0 is not true. The larger the value of F, then, the likelier it is that H_0 is not true.

Another general characteristic of the tests to be discussed below is that *each test can be interpreted as a comparison of two models*. One of these models will be referred to as the *full* or *complete* model; the other will be called the *reduced* model (i.e., the model to which the complete model reduces under the null hypothesis).

As a simple example, consider the two models:

$$Y = \beta_0 + \beta_1 X_1 + \beta_2 X_2 + E$$

and

$$Y = \beta_0 + \beta_1 X_1 + E$$

Under $H_0: \beta_2 = 0$, the larger (full) model reduces to the smaller (reduced) model. A test of $H_0: \beta_2 = 0$ is then essentially equivalent to determining which of these two models is more appropriate.

The reader can also notice from the above example that the set of independent variables in the reduced model (namely, X_1) is a subset of the independent variables in the full model (namely, X_1 and X_2). This is a characteristic common to all the basic types of tests to be described in this chapter. (More generally, this subset characteristic need not always be present. Suppose, for example, that we have $H_0: \beta_1 = \beta_2$. Then, the reduced model may be written as $Y = \beta_0 + \beta X + E$ with $\beta = \beta_1 = \beta_2$ and $X = X_1 + X_2$.)

9-2 Test for Significant Overall Regression

We now consider the first question stated above concerning an overall test for a model containing k independent variables, say,

$$Y = \beta_0 + \beta_1 X_1 + \beta_2 X_2 + \cdots + \beta_k X_k + E$$

The null hypothesis for this test may be generally stated as H_0: "all k independent variables considered together do not explain a significant amount of the variation in Y." Equivalently, we may state the null hypothesis as H_0: "there is no significant overall regression using all k independent variables in the model" or as $H_0: \beta_1 = \beta_2 = \cdots = \beta_k = 0$. Under this last version of H_0, the full model is reduced to a model that contains only the intercept term β_0.

To perform the test, we make use of the mean-square quantities provided in our ANOVA tables (see Table 8-2 of Chapter 8). We calculate the F statistic

$$F = \frac{\text{MS regression}}{\text{MS residual}} = \frac{(\text{SSY} - \text{SSE})/k}{\text{SSE}/(n - k - 1)} \tag{9.1}$$

where $SSY = \sum_{i=1}^{n}(Y_i - \bar{Y})^2$ and $SSE = \sum_{i=1}^{n}(Y_i - \hat{Y}_i)^2$ are the total and error sums of squares, respectively. The computed value of F can then be compared with the critical point $F_{k,n-k-1,1-\alpha}$, with α being the preselected significance level. We would reject H_0 if the computed F exceeded the critical point. Alternatively, we could compute the P-value for this test as the area under the curve of the $F_{k,n-k-1}$ distribution to the right of the computed F statistic. It can be shown that an equivalent expression for (9.1) in terms of R^2 is

$$F = \frac{R^2/k}{(1 - R^2)/(n - k - 1)} \tag{9.2}$$

For the example summarized in Table 8-2, which concerns the regression of WGT on HGT, AGE, and $(AGE)^2$ for a sample of $n = 12$ children, we have $k = 3$, MS regression $= 231.02$, MS residual $= 24.40$, and $R^2 = .7802$, so that

$$F = \frac{231.02}{24.40} = \frac{.7802/3}{(1 - .7802)/(12 - 3 - 1)} = 9.47$$

The critical point for $\alpha = .01$ is $F_{3,8,0.99} = 7.59$. Thus, we would reject H_0 at $\alpha = .01$; that is, the P-value is less than .01. We usually denote $P < .01$ by putting a ** next to the computed F, as in Table 8-2. When $.01 < P < .05$, we usually use only one *.)

In interpreting the results of this test, we can conclude that, taken together, the variables HGT, AGE, and $(AGE)^2$ significantly help to predict WGT based on the observed data. Note that the above conclusion does not mean that *all three* variables are needed for significant prediction of Y; perhaps only one or two of them are sufficient. In other words, a more parsimonious model than the one involving all three variables may be adequate. To determine this requires further tests, to be described in the next section.

(For the interested reader, we will comment on some statistical characteristics of the overall F test. The mean-square residual term, which is the denominator of the F in (9.1), is given by the formula

$$\frac{1}{n - k - 1} SSE = \frac{1}{n - k - 1} \sum_{i=1}^{n} (Y_i - \hat{Y}_i)^2$$

This quantity provides an estimate of σ^2 under the assumed model. The mean-square regression term $\sum_{i=1}^{n}(\hat{Y}_i - \bar{Y})^2/k$, which is the numerator of the F in (9.1), provides an independent estimate of σ^2 only if the null hypothesis of no significant overall regression is true. Otherwise, the numerator overestimates σ^2 directly in proportion to the absolute values of the regression coefficients $\beta_1, \beta_2, \ldots, \beta_k$; this is why an F-value that is "too large" favors rejection of H_0. Thus, the F statistic (9.1) is the ratio of two independent estimates of the same variance only if the null hypothesis $H_0: \beta_1 = \beta_2 = \cdots = \beta_k = 0$ is true.)

9-3 Partial F Test

Some important additional information concerning the fitted regression model can be obtained by presenting the ANOVA table as shown in Table 9-1. What we have done in this representation is to partition the regression sum of squares into three components:

1. $SS(X_1)$: the sum of squares explained by using only $X_1 = $ HGT to predict Y

2. $SS(X_2|X_1)$: the extra sum of squares explained by using X_2 = AGE in addition to X_1 to predict Y

3. $SS(X_3|X_1, X_2)$: the extra sum of squares explained by using X_3 = (AGE)^2 in addition to X_1 and X_2 to predict Y.

We can use the extra information in the table to answer the following questions:

1. Does X_1 = HGT alone significantly aid in predicting Y?

2. Does the addition of X_2 = AGE significantly contribute to the prediction of Y after accounting (or controlling) for the contribution of X_1?

3. Does the addition of X_3 = (AGE)^2 significantly contribute to the prediction of Y after accounting for the contribution of X_1 and X_2?

We already know how to answer question 1. This simply involves fitting the straight-line regression model using X_1 = HGT as the single independent variable. The value 588.92 therefore is the regression sum of squares for this straight-line regression model. The SSE for this model can be obtained from Table 9-1 by adding 195.19, 103.90, and 0.24 together, which yields the sum of squares 299.33, having 10 degrees of freedom (i.e., 10 = 8 + 1 + 1). The F statistic for testing whether there is significant straight-line regression when using only X_1 = HGT is then given by $F = (588.92/1)/(299.33/10) = 19.67$, which has a P-value less than .01 (i.e., X_1 contributes significantly to the linear prediction of Y).

To answer questions 2 and 3, we must use what is called a *partial F test*. This test assesses whether the addition of any specific independent variable, given others already in the model, significantly contributes to the prediction of Y. The test, therefore, allows for the deletion of variables that are of no help in predicting Y and thus enables one to reduce the set of possible independent variables to an economical set of "important" predictors.

9-3-1 The Null Hypothesis

Let us assume that we wish to test whether adding a variable X^* significantly improves the prediction of Y once variables $X_1, X_2, \ldots, X_p$ are already in the model. The null hypothesis may then be stated as H_0: "X^* does not significantly add to the prediction of Y given that $X_1, X_2, \ldots, X_p$ are already in the model" or, equivalently, as H_0: $\beta^* = 0$ in the model $Y = \beta_0 + \beta_1 X_1 + \beta_2 X_2 + \cdots + \beta_p X_p + \beta^* X^* + E$.

TABLE 9-1 ANOVA table for WGT regressed on HGT, AGE, and (AGE)^2 containing components of the regression sum of squares

Source		df	SS	MS	F	R^2
Regression	X_1	1	588.92	588.92	19.67**	.7802
	$X_2\|X_1$	1	103.90	103.90	4.78 (.05 < P < .10)	
	$X_3\|X_1, X_2$	1	0.24	0.24	0.01	
Residual		8	195.19	24.40		
Total		11	888.25			

As can be inferred from the second statement, the test procedure essentially compares two models: The *full* model contains $X_1, X_2, \ldots, X_p$ and X^* as independent variables; the *reduced* model contains $X_1, X_2, \ldots, X_p$, but not X^* (since $\beta^* = 0$ under the null hypothesis). The goal is to determine which model is more appropriate based on how much additional information X^* provides about Y over that already provided by $X_1, X_2, \ldots, X_p$. In the next chapter, we shall see that an equivalent statement of H_0 can be given in terms of a partial correlation coefficient.

9-3-2 The Procedure

To perform a partial F test concerning a variable X^*, say, given that variables $X_1, X_2, \ldots, X_p$ are already in the model, we must first compute the extra sum of squares from adding X^*, given $X_1, X_2, \ldots, X_p$, which we place in our ANOVA table under the source heading "Regression $X^* | X_1, X_2, \ldots, X_p$." This sum of squares is computed by the formula

$$\begin{matrix} \text{extra sum of squares} \\ \text{from adding } X^*, \text{ given} \\ X_1, X_2, \ldots, X_p \end{matrix} = \begin{matrix} \text{regression sum of squares} \\ \text{when } X_1, X_2, \ldots, X_p \\ \text{and } X^* \text{ are } all \\ \text{in the model} \end{matrix} - \begin{matrix} \text{regression sum of squares} \\ \text{when } X_1, X_2, \ldots, X_p \\ \text{(and } not \ X^*) \text{ are} \\ \text{in the model} \end{matrix} \qquad (9.3)$$

or, more compactly,

$$\boxed{\begin{aligned} SS(X^* | X_1, X_2, \ldots, X_p) &= \text{regression } SS(X_1, X_2, \ldots, X_p, X^*) \\ &\quad - \text{regression } SS(X_1, X_2, \ldots, X_p) \end{aligned}}$$

(Since, for any model, $\sum_{i=1}^{n} (Y_i - \bar{Y})^2$ can be split into two components, the regression sum of squares and the residual sum of squares, it follows that

$$\begin{aligned} SS(X^* | X_1, X_2, \ldots, X_p) &= \text{residual } SS(X_1, X_2, \ldots, X_p) \\ &\quad - \text{residual } SS(X_1, X_2, \ldots, X_p, X^*) \end{aligned}$$

is an equivalent expression.)

Thus, for our example,

$$\begin{aligned} SS(X_2 | X_1) &= \text{regression } SS(X_1, X_1) - \text{regression } SS(X_1) \\ &= 692.82 - 588.92 \\ &= 103.90 \end{aligned}$$

and

$$\begin{aligned} SS(X_3 | X_1, X_2) &= \text{regression } SS(X_1, X_2, X_3) - \text{regression } SS(X_1, X_2) \\ &= 693.06 - 692.82 \\ &= 0.24 \end{aligned}$$

To test the null hypothesis H_0: "the addition of X^* to a model already containing $X_1, X_2, \ldots, X_p$ does not significantly improve the prediction of Y," we compute

$$\begin{aligned} &F(X^* | X_1, X_2, \ldots, X_p) \\ &= \frac{\text{extra sum of squares from adding } X^*, \text{ given } X_1, X_2, \ldots, X_p}{\text{mean-square residual for the model containing all the variables } X_1, X_2, \ldots, X_p, X^*} \end{aligned}$$

or, more compactly,

$$F(X^*|X_1, X_2, \ldots, X_p) = \frac{SS(X^*|X_1, X_2, \ldots, X_p)}{MS \text{ residual } (X_1, X_2, \ldots, X_p, X^*)} \tag{9.4}$$

This F statistic has an F distribution with 1 and $n - p - 2$ degrees of freedom under H_0, so we would reject H_0 if the computed F exceeds $F_{1,n-p-2,1-\alpha}$. For our example, the partial F statistics are (from Table 9-1)

$$F(X_2|X_1) = \frac{SS(X_2|X_1)}{MS \text{ residual } (X_1, X_2)} = \frac{103.90}{(195.19 + 0.24)/9} = 4.78$$

and

$$F(X_3|X_1, X_2) = \frac{SS(X_3|X_1, X_2)}{MS \text{ residual } (X_1, X_2, X_3)} = \frac{0.24}{24.40} = 0.01$$

The quantity MS residual (X_1, X_2) can either be obtained directly from the ANOVA table for only X_1 and X_2 or indirectly from the partitioned ANOVA table for X_1, X_2, and X_3 by using the formula

$$MS \text{ residual } (X_1, X_2) = \frac{\text{residual } SS(X_1, X_2, X_3) + SS(X_3|X_1, X_2)}{8 + 1}$$

The statistic $F(X_2|X_1) = 4.78$ has a P-value satisfying $.05 < P < .10$, since $F_{1,9,0.90} = 3.36$ and $F_{1,9,0.95} = 5.12$. Thus, we would reject H_0 at $\alpha = .10$ and conclude that the addition of X_2 after accounting for X_1 significantly adds to the prediction of Y at the $\alpha = .10$ level. At $\alpha = .05$, however, we would not reject H_0.

The statistic $F(X_3|X_1, X_2)$ equals 0.01, and so obviously H_0 would not be rejected regardless of the significance level; we would therefore conclude that once $X_1 = $ HGT and $X_2 = $ AGE are in the model, the addition of $X_3 = (AGE)^2$ is superfluous.

9-3-3 The t Test Alternative

An equivalent way to perform the partial F test for the variable added last is to use a t test. (The reader may recall that an F statistic with 1 and $n - k - 1$ degrees of freedom is the square of a t statistic with $n - k - 1$ degrees of freedom.) The t test alternative focuses on a test of the null hypothesis $H_0: \beta^* = 0$, where β^* is the coefficient of X^* in the regression equation $Y = \beta_0 + \beta_1 X_1 + \beta_2 X_2 + \cdots + \beta_p X_p + \beta^* X^* + E$. The equivalent statistic for testing this null hypothesis is

$$T = \frac{\hat{\beta}^*}{S_{\hat{\beta}^*}} \tag{9.5}$$

where $\hat{\beta}^*$ is the corresponding estimated coefficient and $S_{\hat{\beta}^*}$ is the estimate of the standard error of $\hat{\beta}^*$, both of which are printed by standard regression programs.

In performing this test, we reject $H_0: \beta^* = 0$ if

$$\begin{cases} |T| > t_{n-p-2,1-\alpha/2} & \text{(two-sided test; } H_A: \beta^* \neq 0) \\ T > t_{n-p-2,1-\alpha} & \text{(upper one-sided test; } H_A: \beta^* > 0) \\ T < -t_{n-p-2,1-\alpha} & \text{(lower one-sided test; } H_A: \beta^* < 0) \end{cases}$$

It can be shown that a two-sided t test is equivalent to the partial F test described above. For example, in testing $H_0: \beta_3 = 0$ in the model $Y = \beta_0 + \beta_1 X_1 + \beta_2 X_2 + \beta_3 X_3 + E$ fit to the data in Table 8-1, we compute

$$T = \frac{\hat{\beta}_3}{S_{\hat{\beta}_3}} = \frac{-0.0417}{0.4224} = -0.10$$

Squaring, we get

$$T^2 = 0.01 = \text{partial } F(X_3 | X_1, X_2) \quad \text{from Table 9-1}$$

9-3-4 Comments

An important general application of the partial F test concerns the control of extraneous variables (e.g., confounders, which will be discussed in Chapter 11). Consider, for example, a situation with one main study variable of interest, S, and p control variables C_1, $C_2, \ldots, C_p$. The effect of S on the outcome variable Y, controlling for $C_1, C_2, \ldots, C_p$, may be assessed by considering the model

$$Y = \beta_0 + \beta_1 C_1 + \beta_2 C_2 + \cdots + \beta_p C_p + \beta_{p+1} S + E$$

The appropriate null hypothesis would be $H_0: \beta_{p+1} = 0$. The partial F statistic in this situation is given by $F(S | C_1, C_2, \ldots, C_p)$ using (9.4) with $X^* = S$ and $C_i = X_i$, $i = 1, 2, \ldots, p$.

With several study variables (i.e., several S's), the task involves determining which of the S's are important and perhaps even rank-ordering them by their relative importance. Such a task concerns finding a best model, a topic we will address in Chapter 16 (where the term *best* will be carefully defined). For now, it is worthwhile to note that one strategy (detailed in Chapter 16) is to work backwards by *deleting* S variables one at a time until a best model is obtained. This requires performing several partial F tests (as described in Chapter 16). If the starting model of interest is

$$Y = \beta_0 + \beta_1 C_1 + \cdots + \beta_p C_p + \beta_{p+1} S_1 + \cdots + \beta_{p+k} S_k + E$$

then the first backward step involves the consideration of k partial F tests, $F(S_i | C_1, C_2, \ldots, C_p, S_1, S_2, \ldots, S_k$ except $S_i)$, where $i = 1, 2, \ldots, k$. The corresponding (separate) null hypotheses are $H_0: \beta_i = 0$, where $i = p + 1, p + 2, \ldots, p + k$. The usual backward procedure identifies that variable S_l associated with the smallest partial F value. This variable becomes the first to be deleted from the model, *provided* its partial F is not significant. Then the above elimination process starts all over again for the reduced model with S_l removed. Of course, if the smallest partial F value is significant, then no S variables are deleted.

Each partial F test made at the first backward step concerns the contribution of a specific S variable given that it is the last S variable to enter the model. It would therefore be inappropriate to delete more than one S variable at this first step. For example, it would be inappropriate to delete simultaneously *all* S variables from the model if all partial F's were nonsignificant at this first step. This is because, given that one particular S variable (say, S_l) is deleted, the remaining S variables may now become important (based on consideration of their partial F's under the reduced model).

For example, suppose we fit the model

$$Y = \beta_0 + \beta_1 C_1 + \beta_2 S_1 + \beta_3 S_2 + E$$

and obtain the following partial F results:

$$F(S_1 | C_1, S_2) = 0.01 \qquad (P = .90)$$
$$F(S_2 | C_1, S_1) = 0.85 \qquad (P = .25)$$

Then, S_1 is "less significant" than S_2, controlling for C_1 and the other S variable. Under the strategy of backward elimination then, S_1 should be deleted before the elimination of S_2 is considered. However, to delete both S_1 and S_2 at this point would be incorrect. In fact, when considering the reduced model $Y = \beta_0 + \beta_1 C_1 + \beta_2 S_2 + E$, $F(S_2 | C_1)$ may be highly significant. In other words, if S_1 is not significant given S_2 and C_1, and S_2 is not significant given S_1 and C_1, then S_2 is not necessarily unimportant in a reduced model containing S_2 and C_1 but not S_1.

9-4 Multiple-Partial *F* Test

This testing procedure addresses the more general problem of assessing the additional contribution of two or more independent variables over and above that made by other variables already in the model. For the example involving $Y = \text{WGT}$, $X_1 = \text{HGT}$, $X_2 = \text{AGE}$, and $X_3 = (\text{AGE})^2$, we may be interested in testing whether the AGE variables, taken collectively, significantly improve the prediction of WGT given that HGT is already in the model. In contrast to the partial F test discussed in Section 9-3, the multiple-partial F test concerns the simultaneous addition of two or more variables to a model. Nevertheless, the test procedure is a straightforward extension of the partial F test.

9-4-1 The Null Hypothesis

We wish to test whether the addition of the k variables $X_1^*, X_2^*, \ldots, X_k^*$ significantly improves the prediction of Y once the p variables $X_1, X_2, \ldots, X_p$ are already in the model. The (full) model of interest is thus

$$Y + \beta_0 + \beta_1 X_1 + \beta_p X_p + \beta_1^* X_1^* + \cdots + \beta_k^* X_k^* + E$$

Then, the null hypothesis of interest may be stated as H_0: "$X_1^*, X_2^*, \ldots, X_k^*$ do not significantly add to the prediction of Y given that $X_1, X_2, \ldots, X_p$ are already in the model", or, equivalently, $H_0: \beta_1^* = \beta_2^* = \cdots = \beta_k^* = 0$ in the (full) model.[1]

From the second version of H_0, it follows that the *reduced* model is of the form

$$Y = \beta_0 + \beta_1 X_1 + \beta_2 X_2 + \cdots + \beta_p X_p + E$$

(i.e., the X_i^* terms are dropped from the full model).

Using the above example, the (full) model is

$$\text{WGT} = \beta_0 + \beta_1 \text{HGT} + \beta_1^* \text{AGE} + \beta_2^* (\text{AGE})^2 + E$$

The null hypothesis here is $H_0: \beta_1^* = \beta_2^* = 0$.

---

[1] In Chapter 10, an equivalent expression for this null hypothesis will be given in terms of a multiple-partial correlation coefficient.

9-4-2 The Procedure

As with the partial F test, we must compute the extra sum of squares due to the addition of the X_i^* terms to the model. In particular, we have

$$SS(X_1^*, X_2^*, \ldots, X_k^* | X_1, X_2, \ldots, X_p)$$
$$= \text{regression } SS(X_1, X_2, \ldots, X_p, X_1^*, X_2^*, \ldots, X_k^*) - \text{regression } SS(X_1, X_2, \ldots, X_p)$$
$$= \text{residual } SS(X_1, X_2, \ldots, X_p) - \text{residual } SS(X_1, X_2, \ldots, X_p, X_1^*, X_2^*, \ldots, X_k^*)$$

Using this extra sum of squares, we then obtain the following F statistic:

$$F(X_1^*, X_2^*, \ldots, X_k^* | X_1, X_2, \ldots, X_p) = \frac{SS(X_1^*, X_2^*, \ldots, X_k^* | X_1, X_2, \ldots, X_p)/k}{\text{MS residual } (X_1, X_2, \ldots, X_p, X_1^*, X_2^*, \ldots, X_k^*)}$$

(9.6)

This F statistic has an F distribution with k and $n - p - k - 1$ degrees of freedom under H_0: $\beta_1^* = \beta_2^* = \cdots = \beta_k^* = 0$.

Note that in (9.6) we must divide the extra sum of squares by k, the number of regression coefficients specified to be zero under the null hypothesis of interest. This number k is also the numerator degrees of freedom for the F statistic. The denominator of the F is the mean-square residual for the full model; its degrees of freedom is $n - (p + k + 1)$, which is $n - 1$ minus the number of variables in this model (namely, $p + k$).

An alternative way to write this F statistic is

$$F(X_1^*, X_2^*, \ldots, X_k^* | X_1, X_2, \ldots, X_p) = \frac{[\text{regression } SS(\text{full}) - \text{regression } SS(\text{reduced})]/k}{\text{MS residual (full)}}$$

$$= \frac{[\text{residual } SS(\text{reduced}) - \text{residual } SS(\text{full})]/k}{\text{MS residual (full)}}$$

Using the information in Table 9-1 involving WGT, HGT, AGE, and $(AGE)^2$, we can test $H_0: \beta_1^* = \beta_2^* = 0$ in the model $WGT = \beta_0 + \beta_1 HGT + \beta_1^* AGE + \beta_2^* (AGE)^2 + E$ as follows:

$$F(AGE, (AGE)^2 | HGT) = \frac{\{\text{regression } SS[HGT, AGE, (AGE)^2] - \text{regression } SS(HGT)\}/2}{\text{MS residual } [HGT, AGE, (AGE)^2]}$$

$$= \frac{[(588.92 + 103.90 + 0.24) - 588.92]/2}{24.40}$$

$$= 2.13$$

For $\alpha = .05$, the critical point is

$$F_{k, n-p-k-1, 0.95} = F_{2, 12-1-2-1, 0.95} = 4.46,$$

so that H_0 would not be rejected at $\alpha = .05$.

In the above calculation, note that we used the relationship

regression $SS[HGT, AGE, (AGE)^2]$
$$= \text{regression } SS(HGT) + \text{regression } SS(AGE|HGT) + \text{regression } SS[(AGE)^2 | HGT, AGE]$$
$$= 588.92 + 103.90 + 0.24$$

Alternatively, we could form two ANOVA tables (Table 9-2), one for the full and one for the reduced model, and then extract the appropriate regression and/or residual sum-of-

TABLE 9-2   ANOVA tables for WGT regressed on HGT, AGE, and $(AGE)^2$

Full Model					Reduced Model			
Source	df	SS	MS		Source	df	SS	MS
Regression (X_1, X_2, X_3)	3	693.06	231.02		Regression (X_1)	1	588.92	588.92
Residual	8	195.19	24.40		Residual	10	299.33	29.93
Total	11	888.25			Total	11	888.25	

square terms from these tables. More examples of partial *F* calculations will be given at the end of this chapter.

9-4-3 Comments

As with the partial *F* test, the multiple-partial *F* test is very useful for assessing the importance of extraneous variables. In particular, it is often used to test whether a "chunk" (i.e., a group) of variables having some trait in common is important when considered together. An example of a chunk would be a collection of variables all of a certain order (e.g., $(AGE)^2$, $HGT \times AGE$, and $(HGT)^2$ are all of order 2).

Another example would be a collection of two-way product terms (e.g., X_1X_2, X_1X_3, X_2X_3); this latter group is sometimes referred to as a set of interaction variables (see Chapter 11). It is often of interest to assess the importance of interaction effects collectively before trying to consider individual interaction terms in a model. In fact, the initial use of such a chunk test may reduce the total number of tests to be performed, since variables may be dropped from the model as a group. This, in turn, may help to provide better control of overall Type I error rates, which may be inflated due to multiple testing (Abt, 1981).

9-5 Strategies for Using Partial *F* Tests

In applying the ideas presented in this chapter, the reader will typically use a computer program to carry out the numerical calculations required. Therefore, we will briefly describe the computer output for typical regression programs. In order to understand and use such output, we must discuss two strategies for using partial *F* tests: *variables-added-in-order tests* and *variables-added-last tests*.

Table 9-3 shows the output from a typical regression computer program[2] for the model

$$WGT = \beta_0 + \beta_1 HGT + \beta_2 AGE + \beta_3(AGE)^2 + E$$

The results here were computed with *centered* predictors (Section 12-5-2), so $(HGT - 52.75)$, $(AGE - 8.833)$, and $(AGE - 8.833)^2$ were used, with mean HGT = 52.75 and mean AGE = 8.833. Table 9-3 consists of five sections, labeled A through E. Section A provides the overall ANOVA table for the regression model. Note that computer output typically presents numbers with far more significant digits than can be justified. Section B provides a test for significant overall regression, the multiple R^2-value, the mean $(\bar{Y})$ of the dependent variable (WGT), the WGT residual standard deviation or "root-mean-square error" (*s*), and the coefficient of variation $(100s/\bar{Y})$.

---

[2] This particular output was produced by the SAS program GLM. We have tried to avoid referring to particular programs, since many good ones exist.

TABLE 9-3   Typical computer output for data from Table 8-1 (HGT and AGE centered)

```
              GENERAL LINEAR MODELS PROCEDURE
DEPENDENT VARIABLE: WEIGHT     BODY WEIGHT IN POUNDS
SOURCE               DF   SUM OF SQUARES        MEAN SQUARE⎫
MODEL                 3     693.06046340        231.02015447 ⎬A
ERROR                 8     195.18953660         24.39869208⎭
CORRECTED TOTAL      11     888.25000000


MODEL F =         9.47                    PR > F = 0.0052⎫
R-SQUARE          C.V.        ROOT MSE      WEIGHT MEAN ⎬B
0.780254          7.8717      4.93950322    62.75000000⎭


SOURCE               DF      TYPE I SS     F VALUE   PR > F⎫
HEIGHT                1     588.92252318     24.14   0.0012 ⎬C
AGE                   1     103.90008336      4.26   0.0730
AGE*AGE               1       0.23785686      0.01   0.9238⎭


SOURCE               DF      TYPE III SS   F VALUE   PR > F⎫
HEIGHT                1     166.58195495      6.83   0.0310 ⎬D
AGE                   1     101.80889273      4.17   0.0754
AGE*AGE               1       0.23785686      0.01   0.9238⎭

                            T FOR HO:   PR > T  STD ERROR OF⎫
PARAMETER      ESTIMATE    PARAMETER=0            ESTIMATE ⎬E
INTERCEPT    62.88786380        31.51   0.0001   1.99570812
HEIGHT        0.72369024         2.61   0.0310   0.27696316
AGE           2.04005621         2.04   0.0754   0.99869425
AGE*AGE      -0.04170670        -0.10   0.9238   0.42240715⎭
```

Section C provides certain tests concerning the importance of each predictor in the model. Section D provides a different set of tests regarding these predictors, and Section E provides yet a third set.

9-5-1 Basic Principles

Two methods (or strategies) are widely used for evaluating whether a variable should be included in a model: partial F tests for variables-added-in-order and partial F tests for variables-added-last.[3] For the first method, the following procedure is employed: (1) An

---

[3] Searle (1971), among others, refers to these methods as "ignoring" and "eliminating" tests, respectively.

Instead of considering variables-added-in-order, it may be of interest to consider *variables-deleted-in-order*. This latter strategy would apply, for example, in polynomial regression where a backward selection algorithm is used to determine the proper degree of the polynomial. As can be seen from the computer output in Table 9-3, section C, the same set of sums of squares would be produced whether the variables are considered to be added in one order or deleted in the reverse order.

order for adding variables one at a time is specified; (2) the significance of the (straight-line) model involving only the variable ordered first is assessed; (3) the significance of adding the second variable to the model involving only the first variable is assessed; (4) the significance of adding the third variable to the model containing the first and second variables is assessed; and so on.

For the second (variables-added-last) method, the following procedure is used: (1) An initial model containing two or more variables is specified; (2) the significance of each variable in the initial model is assessed separately as if it were the last variable to enter the model (i.e., if there are k variables in the initial model, then there will be k variables-added-last tests). In either method, each test is conducted using a partial F test for the addition of a single variable.

Variables-added-in-order tests can be illustrated with the weight example. One possible ordering is HGT first, followed by AGE and then $(AGE)^2$. For this ordering, the smallest model considered is

$$WGT = \beta_0 + \beta_1 HGT + E$$

The overall regression F test of $H_0: \beta_1 = 0$ is used to assess the contribution of HGT. Next, the model

$$WGT = \beta_0 + \beta_1 HGT + \beta_2 AGE + E$$

is fit. The significance of adding AGE to a model already containing HGT is then assessed using the partial $F(AGE|HGT)$. Finally, the full model is fit using HGT, AGE, and $(AGE)^2$. The importance of the last variable is tested with the partial $F[(AGE)^2|HGT, AGE]$. The tests used are those discussed in this chapter and summarized in Table 9-1. Also, these are the tests that are provided in section C of Table 9-3 (using Type I sums of squares). Note that each test in Table 9-1 involves a different residual sum of squares, while those in Table 9-3 use a common residual sum of squares. More will be said about this issue later.

In order to describe variables-added-last tests, consider again the full model

$$WGT = \beta_0 + \beta_1 HGT + \beta_1 AGE + \beta_2 (AGE)^2 + E$$

The contribution of HGT when added last is assessed by comparing the full model to the model with HGT deleted, namely,

$$WGT = \beta_0 + \beta_2 AGE + \beta_3 (AGE)^2 + E$$

The partial F statistic, using (9.4), has the form $F[(HGT|AGE, (AGE)^2]$. The sum of squares for HGT added last is then the difference in the error sum of squares (or the regression sum of squares) for the two models above. Similarly, the reduced model with AGE deleted is

$$WGT = \beta_0 + \beta_1 HGT + \beta_3 (AGE)^2 + E$$

with the corresponding partial F statistic being $F[AGE|HGT, (AGE)^2]$, and the reduced model with $(AGE)^2$ omitted is

$$WGT = \beta_0 + \beta_1 HGT + \beta_2 AGE + E$$

with partial F statistic $F[(AGE)^2|HGT, AGE]$. The three F statistics just described are provided in section D of Table 9-3 (using Type III sums of squares).

An important characteristic of variables-added-in-order sums of squares is that they decompose the regression sum of squares into a set of mutually exclusive and exhaustive pieces. For example, the sums of squares provided in section C of Table 9-3 (588.922, 103.900, and 0.238) add to 693.060, which is the regression sum of squares given in section A. The variables-added-last sums of squares do not generally have this property (e.g., the sums of squares given in section D of Table 9-3 do not add to 693.060).

Each of these two testing strategies has its own advantages, and the situation being considered will determine which to employ. For example, if all variables are considered of equal importance, the variables-added-last tests are usually preferable. Such tests treat all variables equally; because the importance of each variable is assessed as if it were the last variable to enter the model, the order of entry is not a consideration.

In contrast, if the order in which the predictors enter the model is an important consideration, then the variables-added-in-order testing approach may be better. An example where the entry order is important is the situation where main effects (e.g., X_1, X_2, and X_3) are forced into the model, followed by their cross-products (X_1X_2, X_1X_3, and X_2X_3,) or so-called interaction terms (see Chapter 11). Such tests evaluate the contribution of a variable and adjust *only* for those variables just preceding it into the model.

9-5-2 Commentary

As discussed above, section C of Table 9-3 provides variables-added-in-order tests, which are also given for the same data in Table 9-1. Section D of Table 9-3 provides variables-added-last tests. Finally, section E provides t tests (which are equivalent to the variables-added-last F tests in section D), as well as regression coefficient estimates and their standard errors.

Table 9-4 gives an ANOVA table for the variables-added-last tests for the weight example. (We recommend that the reader consider how this table was extracted from the computer output in Table 9-3.) The variables-added-last tests usually give a different ANOVA table than one based on the variables-added-in-order tests. Note that a different residual sum of squares is used for each variables-added-in-order test in Table 9-1, while the same residual sum of squares (based on the three-variable model involving HGT, AGE, and (AGE)^2) is used for all the variables-added-last tests in Table 9-4.

An argument can be made that it is preferable to use the residual sum of squares for the three-variable model (i.e., that "largest" model containing all candidate predictors) for all tests. This is because the error variance σ^2 will *not* be correctly estimated by a model ignoring important predictors but will be correctly estimated (under the usual regression assumptions)

TABLE 9-4   ANOVA table for WGT regressed on HGT, AGE, and (AGE)^2 using variables-added-last tests

Source	df	SS	MS	F	R^2
$X_1\|X_2, X_3$	1	166.58	166.58	6.83*	.7802
$X_2\|X_1, X_3$	1	101.81	101.81	4.17	
$X_3\|X_1, X_2$	1	.24	.24	.01	
Residual	8	195.19	24.40		
Total	11	888.25			

* Exceeds .05 critical value of 5.32 for F with 1 and 8 degrees of freedom.

by a model containing all candidate predictors (even if some are not important). In other words, overfitting a model in estimating σ^2 is safer than underfitting. Of course, extreme overfitting will lose precision, but it will still provide a valid estimate of residual variation. We generally prefer using the residual sum of squares based on fitting the "largest" model, although some statisticians would disagree.

9-5-3   Models Underlying the Source Tables

Tables 9-5 and 9-6 present the models being compared based on the computer output in Table 9-3 and the associated ANOVA tables. Table 9-5 summarizes the models and residual sums of squares needed to conduct variables-added-last tests for the full model containing HGT, AGE, and $(AGE)^2$. Table 9-6 lists the models that must be fitted to provide variables-added-in-order tests for the order of entry HGT, then AGE, and then $(AGE)^2$.

Table 9-7 details computations of regression sums of squares for both types of tests. For example, the first line, where $24{,}227.3650 = 24{,}422.5545 - 195.1895$, is the difference in the error sums of squares for models 1 and 5 given in Table 9-5. These results can then be used to produce any of the F tests given in Tables 9-1, 9-2, and 9-3.

TABLE 9-5   Variables-added-last regression models and residual sums of squares for data from Table 8-1

Model No.	Model	SSE
1	$WGT = \qquad\quad \beta_1 HGT + \beta_2 AGE + \beta_3 (AGE)^2 + E$	24,422.5545
2	$WGT = \beta_0 \qquad\quad + \beta_2 AGE + \beta_3 (AGE)^2 + E$	361.7715
3	$WGT = \beta_0 + \beta_1 HGT \qquad\quad + \beta_3 (AGE)^2 + E$	296.9984
4	$WGT = \beta_0 + \beta_1 HGT + \beta_2 AGE \qquad\quad + E$	195.4274
5	$WGT = \beta_0 + \beta_1 HGT + \beta_2 AGE + \beta_3 (AGE)^2 + E$	195.1895

TABLE 9-6   Variables-added-in-order regression models and residual sums of squares for data from Table 8-1

Model No.	Model	SSE
6	$WGT = \qquad\qquad\qquad\qquad\qquad + E$	48,139.0000
7	$WGT = \beta_0 \qquad\qquad\qquad\qquad + E$	888.2500
8	$WGT = \beta_0 + \beta_1 HGT \qquad\qquad\quad + E$	299.3275
4	$WGT = \beta_0 + \beta_1 HGT + \beta_2 AGE \qquad + E$	195.4274
5	$WGT = \beta_0 + \beta_1 HGT + \beta_2 AGE + \beta_3 (AGE)^2 + E$	195.1895

TABLE 9-7   Computations for regression sum of squares for data from Table 8-1

Parameter	Variable	Regression SS Added-Last	Regression SS In-Order
β_0	Intercept	SSE(1) − SSE(5) = 24,227.3650	SSE(6) − SSE(7) = 47,250.7500
β_1	HGT	SSE(2) − SSE(5) = 166.5820	SSE(7) − SSE(8) = 588.9225
β_2	AGE	SSE(3) − SSE(5) = 101.8089	SSE(8) − SSE(4) = 103.9001
β_3	$(AGE)^2$	SSE(4) − SSE(5) = 0.2379	SSE(4) − SSE(5) = 0.2379

9-6 Tests Involving the Intercept

Inferences about the intercept β_0 are occasionally of interest in multiple regression analysis. A test of $H_0: \beta_0 = 0$ is usually carried out with an intercept-added-last test, although an intercept-added-in-order test is also feasible (where the intercept is the first term added to the model). Many computer programs provide only a t test involving the intercept. The t-test statistic for the intercept in Table 9-3 corresponds exactly to a partial F test for adding the intercept last. The two models being compared are

$$Y = \beta_0 + \beta_1 X_1 + \beta_2 X_2 + \cdots + \beta_k X_k + E$$

and

$$Y = \beta_1 X_1 + \beta_2 X_2 + \cdots + \beta_k X_k + E$$

The null hypothesis of interest is $H_0: \beta_0 = 0$ versus $H_A: \beta_0 \neq 0$. The test is computed as

$$F = \frac{(\text{SSE without } \beta_0 - \text{SSE with } \beta_0)/1}{\text{SSE with } \beta_0/(n - k - 1)}$$

This F statistic has 1 and $n - k - 1$ degrees of freedom and is equal to the square of the t statistic used for testing $H_0: \beta_0 = 0$. For the weight example, an intercept-added-last test is reported in Table 9-3 as a t test. The corresponding partial F equals $(31.51)^2 = 992.88$ and has 1 and 8 degrees of freedom.

An intercept-added-in-order test can also be conducted. In this case, the two models being compared are

$$Y = E$$

and

$$Y = \beta_0 + E$$

Again, the null hypothesis is $H_0: \beta_0 = 0$ versus the alternative $H_A: \beta_0 \neq 0$. The special nature of this test leads to the simple expression

$$F = \frac{n\bar{Y}^2/1}{\sum_{i=1}^{n} (Y_i - \bar{Y})^2/(n - 1)}$$

an F statistic with 1 and $n - 1$ degrees of freedom. This statistic involves the residual sum of squares from a model with just an intercept (such as model 7 in Table 9-6). Alternatively, the residual sum of squares from the "largest" model may be used. Using this latter approach, the F statistic for the weight data becomes (see Table 9-6)

$$F = \frac{[\text{SSE}(6) - \text{SSE}(7)]/1}{\text{SSE}(5)/8}$$
$$= \frac{(48,139.00 - 888.25)/1}{195.1895/8}$$
$$= 193.61$$

with 1 and 8 degrees of freedom. In general, using the residual from the largest model (with k predictors) gives $n - k - 1$ error degrees of freedom, so that the F statistic is compared to a critical value with 1 and $n - k - 1$ degrees of freedom.

Problems

1. The following ANOVA tables are for the data considered in Problem 2 of Chapter 8.

Source	df	SS
X_1	1	1,546
$X_2 \vert X_1$	1	1,238
Residual	50	11,008

Source	df	SS
X_2	1	160
$X_1 \vert X_2$	1	2,623
Residual	50	11,008

Y is the level of pathology, X_1 is the thinking disturbance rating, and X_2 is the hostile suspiciousness rating.

a. Provide variable-added-in-order tests for both variables, with thinking disturbance added first. Use $\alpha = .05$.
b. Provide variable-added-in-order tests for both variables, with hostile suspiciousness added first.
c. Provide a table of variable-added-last tests.
d. What, if any, differences are present in the three approaches?
e. Which predictors appear to be necessary, and why?

2. A psychologist examined the regression relationship between anxiety level (Y), measured on a scale from 1 to 50 as the average of an index determined at three points in a 2-week period, and the following three independent variables: X_1 = systolic blood pressure, X_2 = IQ, and X_3 = job satisfaction (measured on a scale of 1 to 25). The ANOVA table below summarizes results obtained from a variable-added-in-order regression analysis on data on 22 outpatients undergoing therapy at a certain clinic.

Source		df	SS
Regression	(X_1)	1	981.326
	$(X_2 \vert X_1)$	1	190.232
	$(X_3 \vert X_1, X_2)$	1	129.431
Residual		18	442.292

a. Test for the significance of each independent variable as it enters the model. State the null hypothesis for each test in terms of regression coefficients.
b. Test for the significance of the addition of both X_2 and X_3 to a model already containing X_1. State the null hypothesis in terms of regression coefficients and in terms of (multiple-partial) correlation coefficients.

   c. In terms of regression sums of squares, what test corresponds to comparing the two models

$$Y = \beta_0 + \beta_1 X_1 + \beta_2 X_2 + \beta_3 X_3 + E$$

and

$$Y = \beta_0 + \beta_3 X_3 + E$$

Why can't this test be done using the ANOVA table? Describe the appropriate test procedure.

   d. Based on the tests made, what would you recommend as the most appropriate statistical model? (Use $\alpha = .05$.)

3. An educator examined the relationship between the number of hours devoted to reading each week (Y) and the independent variables social class (X_1), number of years of school completed (X_2), and reading speed measured by pages read per hour (X_3). The ANOVA table obtained from a stepwise regression analysis on data for a sample of 19 women over 60 is shown.

Source		df	SS
Regression	(X_3)	1	1,058.628
	$(X_2\|X_3)$	1	183.743
	$(X_1\|X_2, X_3)$	1	37.982
Residual		15	363.300

   a. Test for the significance of each variable as it enters the model.
   b. Test $H_0: \beta_1 = \beta_2 = 0$ in the model $Y = \beta_0 + \beta_1 X_1 + \beta_2 X_2 + \beta_3 X_3 + E$.
   c. Why can't we test $H_0: \beta_1 = \beta_3 = 0$ using the ANOVA table given? What formula would you use for this test?
   d. What is your overall evaluation concerning the appropriate model to use given the results in parts (a) and (b)?

4. An experiment was conducted regarding a quantitative analysis of factors found in high-density lipoprotein (HDL) in a sample of human blood serum. Three variables though to be predictive or associated with HDL measurement (Y) were the total cholesterol (X_1) and total triglyceride (X_2) concentrations in the sample, plus the presence or absence of a certain sticky component found in the serum called sinking pre-beta, or SPB (X_3), coded as 0 if absent and 1 if present. The data obtained are shown in the table, and the ANOVA results follow.

Y	X_1	X_2	X_3	Y	X_1	X_2	X_3
47	287	111	0	57	192	115	1
38	236	135	0	42	349	408	1
47	255	98	0	54	263	103	1
39	135	63	0	60	223	102	1
44	121	46	0	33	316	274	0
64	171	103	0	55	288	130	0
58	260	227	0	36	256	149	0
49	237	157	0	36	318	180	0
55	261	266	0	42	270	134	0
52	397	167	0	41	262	154	0
49	295	164	0	42	264	86	0
47	261	119	1	39	325	148	0
40	258	145	1	27	388	191	0
42	280	247	1	31	260	123	0
63	339	168	1	39	284	135	0
40	161	68	1	56	326	236	1
59	324	92	1	40	248	92	1
56	171	56	1	58	285	153	1
76	265	240	1	43	361	126	1
67	280	306	1	40	248	226	1
57	248	93	1	46	280	176	1

Source	df	SS
Regression (X_1)	1	46.2356
Residual	40	4,567.3835

Source	df	SS
Regression (X_2)	1	21.3397
Residual	40	4,592.2793

Source	df	SS
Regression (X_3)	1	735.2054
Residual	40	3,878.4136

Source	df	SS
Regression (X_1, X_2)	2	135.3820
Residual	39	4,478.2369

Source	df	SS
Regression (X_1, X_3)	2	783.1691
Residual	39	3,830.4500

Source	df	SS
Regression (X_2, X_3)	2	737.8069
Residual	39	3,875.8122

Source	df	SS
Regression (X_1, X_2, X_3)	3	819.7473
$X_1X_3, X_2X_3 \mid X_1, X_2, X_3$	2	74.7443
Residual	36	3,719.0517

Source	df	SS
Regression (X_1, X_3)	2	783.1691
$X_1X_3 \mid X_1, X_3$	1	62.4247
Residual	38	3,768.0252

Source	df	SS
Regression (X_2, X_3)	2	737.8069
$X_2X_3 \mid X_2, X_3$	1	1.5539
Residual	38	3,874.2583

a. Test whether X_1, X_2, or X_3 alone significantly helps in predicting Y.
b. Test whether X_1, X_2, and X_3 taken together significantly help to predict Y.
c. Test whether the true coefficients of the product terms X_1X_3 and X_2X_3 are simultaneously zero in the model containing X_1, X_2, and X_3 plus these product terms. Specify the two models being compared, and state the null hypothesis in terms of regression coefficients. If this test is not rejected, what can you conclude about the relationship of Y to X_1 and X_2 when X_3 equals 1 compared with when X_3 equals 0.
d. Test (at $\alpha = .05$) whether X_3 is associated with Y after taking into account the combined contribution of X_1 and X_2. What does your result together with your answer to part (c) tell you about the relationship of Y with X_1 and X_2 when SPB is present as compared with when it is absent?
e. What overall conclusion can you draw about the association of Y with the three independent variables for this data set? Specify the two models being compared, and state the appropriate null hypothesis in terms of regression coefficients.

5. Consider the ANOVA tables given in Chapter 8, Problem 3. Use $\alpha = .05$.
   a. Provide variable-added-in-order tests for both variables with weight entered first.
   b. Repeat (a), but have age entered first.
   c. Provide variable-added-last tests for both weight and age.
   d. What, if any, differences are present in parts (a) through (c)?
   e. Which predictors appear to be necessary, and why?

6. The following tables provide additional information about the data from Chapter 8, Problem 4.

Source	df	SS		Source	df	SS
X_2	1	1,308.34		X_3	1	1,360.14
$X_1\|X_2$	1	9.46		$X_1\|X_3$	1	35.35
Residual	17	537.40		Residual	17	459.71

Source	df	SS
X_3	1	1,360.14
$X_2\|X_3$	1	117.23
Residual	17	377.82

Source	df	SS
Regression (X_1, X_2, X_3)	3	1,507.18
Residual	16	348.03

Answer the following questions using the ANOVA tables above and in Chapter 8, Problem 4. Use $\alpha = .05$.
a. Provide variable-added-in-order tests for the order X_2, X_1, and X_3.
b. Provide variable-added-in-order tests for the order X_3, X_1, and X_2.
c. List all orders that can be tested using the ANOVA tables here and in Chapter 8, Problem 4. List all orders that *cannot* be so computed.

d. Provide variable-added-last tests for X_1, X_2, and X_3.

e. The ANOVA table resulting from fitting a model with predictors X_2, X_3, and $X_4 = X_2X_3$ is as follows.

Source	df	SS	MS
X_3	1	1,360.14	1,360.14
$X_2\|X_3$	1	117.23	117.23
$X_4\|X_2, X_3$	1	0.09	0.09
Residual	16	377.73	23.61

Use this table to test whether X_4 significantly improves the prediction of Y, given that X_2 and X_3 are already in the model.

7. The following ANOVA table is based on the data discussed in Chapter 8, Problem 5. Use $\alpha = .05$.

Source		df	SS
Regression	(X_1)	1	18,953.04
	$(X_3\|X_1)$	1	7,010.03
	$(X_2\|X_1, X_3)$	1	10.93
Residual		21	2,248.23
Total		24	28,222.23

a. Provide a test to compare the following two models:

$$Y = \beta_0 + \beta_1X_1 + \beta_2X_2 + \beta_3X_3 + E$$

and

$$Y = \beta_0 + \beta_1X_1 + E$$

b. Provide a test to compare the following two models:

$$Y = \beta_0 + \beta_1X_1 + \beta_3X_3 + E$$

and

$$Y = \beta_0 + E$$

c. State which two models are being compared in computing

$$F = \frac{(18,953.04 + 7,010.03 + 10.93)/3}{2248.23/21}$$

References

Abt, K. (1981). "Problems of Repeated Significance Testing." *Controlled Clinical Trials*, 1: 377–381.

Searle, S. R. (1971). *Linear Models*. New York: John Wiley & Sons, Inc.

Correlations:
Multiple, Partial,
and Multiple-Partial

10-1 Preview

We saw in Chapter 5 that the essential features of straight-line regression, except for the quantitative prediction formula provided by the fitted regression equation, can also be described in terms of the correlation coefficient r. These features are summarized as follows:

1. The squared correlation coefficient, r^2, measures the strength of the linear relationship between the dependent variable Y and the independent variable X. The closer r^2 is to 1, the stronger the linear relationship; the closer r^2 is to 0, the weaker the linear relationship.

2. $r^2 = (\text{SSY} - \text{SSE})/\text{SSY}$ is the proportionate reduction in the total sum of squares achieved by using a straight-line model in X to predict Y.

3. $r = \hat{\beta}_1(S_X/S_Y)$, where $\hat{\beta}_1$ is the estimated slope of the regression line.

4. r (or r_{XY}) is an estimate of the population parameter ρ (or ρ_{XY}), which describes the correlation between X and Y, both considered as random variables.

5. Assuming that X and Y have a bivariate normal distribution with parameters μ_X, μ_Y, σ_X^2, σ_Y^2, and ρ_{XY}, the conditional distribution of Y given X is $N(\mu_{Y|X}, \sigma_{Y|X}^2)$, where

$$\mu_{Y|X} = \mu_Y + \rho \frac{\sigma_Y}{\sigma_X}(X - \mu_X) \quad \text{and} \quad \sigma_{Y|X}^2 = \sigma_Y^2(1 - \rho^2)$$

Here r^2 estimates ρ^2, which can be expressed as

$$\rho^2 = \frac{\sigma_Y^2 - \sigma_{Y|X}^2}{\sigma_Y^2}$$

6. r can be used as a general index of linear association between two random variables in the following sense:

   a. The more highly positive r is, the more "positive" the linear association; that is, an individual with a high value of one variable will likely have a high value of the other, and an individual with a low value of one variable will probably have a low value of the other.

   b. The more highly negative r is, the more "negative" is the linear association; that is, an individual with a high value of one variable will likely have a low value of the other, and conversely.

   c. If r is close to 0, there is little evidence of linear association, which indicates that there is nonlinear association or no association at all.

This connection between regression and correlation can also be extended to the multiple regression case, as we will discuss in this chapter. However, when several independent variables are involved, the essential features of regression are described not by a single correlation coefficient as in the straight-line case, but by several correlations. These include a set of (zero-order) correlations such as r plus a whole group of additional (higher-order) indices called multiple correlations, partial correlations, and multiple-partial correlations.[1] These higher-order correlations allow us to answer many of the same questions that can be answered by fitting a multiple regression model. In addition, the correlation analog has been found particularly useful in uncovering spurious relationships among variables, identifying intervening variables, and making certain types of causal inferences.[2]

10-2 Correlation Matrix

When dealing with more than one independent variable, the collection of all zero-order correlation coefficients (i.e., the r's between all possible pairs of variables) can be represented most compactly in *correlation matrix form*. For example, when there are $k = 3$ independent variables X_1, X_2, and X_3, and one dependent variable Y, there are $C_2^4 = 6$ zero-order correlations, and the correlation matrix has the general form

$$
\begin{array}{c}
Y \\ X_1 \\ X_2 \\ X_3
\end{array}
\begin{array}{cccc}
Y & X_1 & X_2 & X_3 \\
\begin{bmatrix}
1 & r_{Y1} & r_{Y2} & r_{Y3} \\
 & 1 & r_{12} & r_{13} \\
 & & 1 & r_{23} \\
 & & & 1
\end{bmatrix}
\end{array}
$$

Here r_{Yj} $(j = 1, 2, 3)$ is the correlation between Y and X_j, and r_{ij} $(i, j = 1, 2, 3)$ is the correlation between X_i and X_j.

---

[1] The *order* of a correlation coefficient, as the term is used here, is the number of variables being controlled or adjusted for (see Section 10-5).

[2] See Blalock (1971) for a discussion of techniques for inferring causes.

For the data in Table 8-1, this matrix takes the form

	WGT	HGT	AGE	$(AGE)^2$
WGT	1	.814	.770	.767
HGT		1	.614	.615
AGE			1	.994
$(AGE)^2$				1

Each of these correlations taken separately describes the strength of the linear relationship between the two variables involved. In particular, the correlations $r_{Y1} = .814$, $r_{Y2} = .770$, and $r_{Y3} = .767$ measure the strength of the linear association with the dependent variable WGT for each of the independent variables taken separately. As we can see, HGT ($r_{Y1} = .814$) is the independent variable with the strongest linear relationship to WGT, followed by AGE and then $(AGE)^2$.

Nevertheless, these zero-order correlations describe neither (1) the overall relationship of the dependent variable WGT to the independent variables HGT, AGE, and $(AGE)^2$ considered together, nor (2) the relationship between WGT and AGE after controlling for HGT,[3] nor (3) the relationship between WGT and the combined effects of AGE and $(AGE)^2$ after controlling for HGT. The measure that describes (1) is called the multiple correlation coefficient of WGT on HGT, AGE, and $(AGE)^2$. The measure that describes (2) is called the partial correlation coefficient between WGT and AGE controlling for HGT. Finally, the measure that describes (3) is called the multiple-partial correlation coefficient between WGT and the combined effects of AGE and $(AGE)^2$ controlling for HGT.

It should be noted that even though AGE and $(AGE)^2$ are very highly correlated in our example, it is possible, if the general relationship of AGE to WGT is nonlinear, that $(AGE)^2$ will be significantly correlated with WGT even after AGE has been controlled for. In fact, this is what happens in general when the addition of a second-order term in polynomial regression significantly improves the prediction of the dependent variable.

10-3 Multiple Correlation Coefficient

The *multiple correlation coefficient,* denoted as $R_{Y|X_1, X_2, \ldots, X_k}$, is a measure of the overall *linear association* of one (dependent) variable Y with several other (independent) variables $X_1, X_2, \ldots, X_k$.

(By "linear association" we mean that $R_{Y|X_1, X_2, \ldots, X_k}$ measures the strength of the association between Y and the best-fitting linear combination of the X's, which is the least-squares solution $\hat{Y} = \hat{\beta}_0 + \hat{\beta}_1 X_1 + \hat{\beta}_2 X_2 + \cdots + \hat{\beta}_k X_k$. In fact, no other linear combination of the X's will have as great a correlation with Y. It can also be shown that $R_{Y|X_1, X_2, \ldots, X_k}$ is always nonnegative.)

Thus, the multiple correlation coefficient is a direct generalization of the simple correlation coefficient r to the case of several independent variables. We have actually dealt with this

---

[3] In this case the phrase "controlling for" pertains to determining the extent to which the variables WGT and AGE are related after removing the effect of HGT on WGT and AGE.

measure up to now under the name R^2, which is the square of the multiple correlation coefficient.

Two computational formulas provide useful interpretations of the multiple correlation coefficient $R_{Y|X_1,X_2,\dots,X_k}$ and its square. These are

$$(1) \quad R_{Y|X_1,X_2,\dots,X_k} = \frac{\displaystyle\sum_{i=1}^{n}(Y_i - \bar{Y})(\hat{Y}_i - \bar{\hat{Y}})}{\sqrt{\displaystyle\sum_{i=1}^{n}(Y_i - \bar{Y})^2 \sum_{i=1}^{n}(\hat{Y}_i - \bar{\hat{Y}})^2}}$$

and

$$(2) \quad R^2_{Y|X_1,X_2,\dots,X_k} = \frac{\displaystyle\sum_{i=1}^{n}(Y_i - \bar{Y})^2 - \sum_{i=1}^{n}(Y_i - \hat{Y}_i)^2}{\displaystyle\sum_{i=1}^{n}(Y_i - \bar{Y})^2} = \frac{\text{SSY} - \text{SSE}}{\text{SSY}}$$

where $\hat{Y}_i = \hat{\beta}_0 + \hat{\beta}_1 X_{1i} + \hat{\beta}_2 X_{2i} + \cdots + \hat{\beta}_k X_{ki}$ (the predicted value for the ith individual) and $\bar{\hat{Y}} = \sum_{i=1}^{n}\hat{Y}_i/n$. Formula (2), which we have seen several times before, is most useful for assessing the fit of the regression model. Also, it can be seen from formula (1) that $R_{Y|X_1,X_2,\dots,X_k} = r_{Y,\hat{Y}}$, the *simple linear correlation between the observed values Y and the predicted values Ŷ.*

We shall illustrate the computation of $R_{Y|X_1,X_2,\dots,X_k}$ by applying formula (1) to the data of Table 8-1, where $X_1 = \text{HGT}$, $X_2 = \text{AGE}$, and $Y = \text{WGT}$. Using only X_1 and X_2 in the model, the fitted regression equation is $\hat{Y} = 6.553 + 0.722X_1 + 2.050X_2$, and the observed and predicted values are given in Table 10-1.

TABLE 10-1   Observed and predicted values for the regression of WGT on HGT and AGE

Child	1	2	3	4	5	6	7	8	9	10	11	12
Y (observed)	64	71	53	67	55	58	77	57	56	51	76	68
Ŷ (predicted)	64.11	69.65	54.23	73.87	59.78	57.01	66.77	59.66	57.38	49.18	75.20	66.16

One can check that $\bar{Y} = \bar{\hat{Y}} = 62.75$. As we mentioned in Chapter 8, this is no coincidence, since it is a mathematical fact that $\bar{Y} = \bar{\hat{Y}}$. With this result in mind, the best computational formula for $R_{Y|X_1,X_2,\dots,X_k}$ takes the form

$$R_{Y|X_1,X_2,\dots,X_k} = \frac{\displaystyle\sum_{i=1}^{n} Y_i\hat{Y}_i - n\bar{Y}^2}{\sqrt{\left(\displaystyle\sum_{i=1}^{n} Y_i^2 - n\bar{Y}^2\right)\left(\sum_{i=1}^{n} \hat{Y}_i^2 - n\bar{Y}^2\right)}} \tag{10.1}$$

so that, for example,

$$R_{\text{WGT|HGT,AGE}} = \frac{47{,}943.60 - 12(62.75)^2}{\sqrt{[48{,}139 - 12(62.75)^2][47{,}943.544 - 12(62.75)^2]}} = .8832$$

Two other multiple correlations calculated for this data set (which correspond to different regression models) are

$$R_{\text{WGT}|\text{HGT},\text{AGE},(\text{AGE})^2} = .8833$$

and

$$R_{\text{WGT}|\text{HGT},(\text{AGE})^2} = .8811$$

The three values above support our earlier finding that the use of HGT and AGE or HGT and $(\text{AGE})^2$ does as well in predicting WGT as does the use of all three independent variables.

10-4 Relationship of $R_{Y|X_1,X_2,...,X_k}$ to the Multivariate Normal Distribution[4]

An informative way of looking at the sample multiple correlation coefficient $R_{Y|X_1,X_2,...,X_k}$ is to consider it as an estimator of a population parameter characterizing the joint distribution of all the variables $Y, X_1, X_2, \ldots, X_k$ taken together. When we had two variables X and Y and had assumed that their joint distribution was bivariate normal $N_2(\mu_Y, \mu_X, \sigma_Y^2, \sigma_X^2, \rho_{XY})$, we saw that r_{XY} estimated ρ_{XY}, which satisfied the formula $\rho_{XY}^2 = (\sigma_Y^2 - \sigma_{Y|X}^2)/\sigma_Y^2$, where $\sigma_{Y|X}^2$ was the variance of the conditional distribution of Y given X. Now, when we have k independent variables and one dependent variable, we get an analogous result if we assume that their joint distribution is *multivariate normal*. Let us now consider what happens with just *two* independent variables. In this case the *trivariate normal distribution* of Y, X_1, and X_2 can be described as

$$N_3(\mu_Y, \mu_{X_1}, \mu_{X_2}, \sigma_Y^2, \sigma_{X_1}^2, \sigma_{X_2}^2, \rho_{Y1}, \rho_{Y2}, \rho_{12})$$

where μ_Y, μ_{X_1}, and μ_{X_2} are the three (unconditional) means; σ_Y^2, $\sigma_{X_1}^2$, and $\sigma_{X_2}^2$ are the three (unconditional) variances; and ρ_{Y1}, ρ_{Y2}, and ρ_{12} are the three correlation coefficients. The *conditional distribution of Y given X_1 and X_2* is then a univariate normal distribution with a (conditional) mean denoted by $\mu_{Y|X_1,X_2}$ and a (conditional) variance denoted by $\sigma_{Y|X_1,X_2}^2$; we usually write this compactly as

$$Y|X_1, X_2 \frown N(\mu_{Y|X_1,X_2}, \sigma_{Y|X_1,X_2}^2)$$

In fact, it turns out that

$$\mu_{Y|X_1,X_2} = \mu_Y + \rho_{Y1}\frac{\sigma_{Y|X_2}}{\sigma_{X_1|X_2}}(X_1 - \mu_{X_1}) + \rho_{Y2}\frac{\sigma_{Y|X_1}}{\sigma_{X_2|X_1}}(X_2 - \mu_{X_2})$$

and

$$\sigma_{Y|X_1,X_2}^2 = (1 - \rho_{Y,\mu_{Y|X_1,X_2}}^2)\sigma_Y^2$$

where $\rho_{Y,\mu_{Y|X_1,X_2}}$ is the population correlation coefficient between the random variables Y and $\mu_{Y|X_1,X_2} = \beta_0 + \beta_1 X_1 + \beta_2 X_2$ (where we are considering X_1 and X_2 as random variables). Also, $\sigma_{Y|X_2}^2$, $\sigma_{Y|X_1}^2$, $\sigma_{X_1|X_2}^2$, and $\sigma_{X_2|X_1}^2$ are the conditional variances, respectively, of Y given X_2, Y given X_1, X_1 given X_2, and X_2 given X_1.

---

[4] This section is not essential for the application-oriented reader.

This parameter $\rho_{Y,\mu_{Y|X_1,X_2}}$ is the *population analog of the sample multiple correlation coefficient* $R_{Y|X_1,X_2}$, and we write $\rho_{Y,\mu_{Y|X_1,X_2}}$ simply as $\rho_{Y|X_1,X_2}$. Furthermore, from the formula for $\sigma^2_{Y|X_1,X_2}$, it can be seen (with a little algebra) that

$$\rho^2_{Y|X_1,X_2} = \frac{\sigma^2_Y - \sigma^2_{Y|X_1,X_2}}{\sigma^2_Y}$$

which is the *proportionate reduction in the unconditional variance of Y due to conditioning on* X_1 *and* X_2.

Generalizing these findings to the case of k independent variables, we may summarize the characteristics of the multiple correlation coefficient $R_{Y|X_1,X_2,\ldots,X_k}$ as follows:

1. $R^2_{Y|X_1,X_2,\ldots,X_k}$ measures the proportionate reduction in the total sum of squares $\sum(Y_i - \bar{Y})^2$ to $\sum(Y_i - \hat{Y}_i)^2$ due to the multiple linear regression of Y on X_1, X_2, $\ldots$, X_k.

2. $R_{Y|X_1,X_2,\ldots,X_k}$ is the correlation $r_{Y,\hat{Y}}$ of the observed values (Y) with the predicted values $(\hat{Y})$.

3. $R_{Y|X_1,X_2,\ldots,X_k}$ is an estimate of $\rho_{Y|X_1,X_2,\ldots,X_k}$, which is the correlation of Y with the mean of the conditional distribution of Y given X_1, X_2, $\ldots$, X_k (i.e., the correlation of Y with the true regression equation $\beta_0 + \beta_1 X_1 + \beta_2 X_2 + \cdots + \beta_k X_k$, where the X's are considered to be random).

4. $R^2_{Y|X_1,X_2,\ldots,X_k}$ is an estimate of the proportionate reduction in the unconditional variance of Y due to conditioning on X_1, X_2, $\ldots$, X_k; that is, it estimates

$$\rho^2_{Y|X_1,X_2,\ldots,X_k} = \frac{\sigma^2_Y - \sigma^2_{Y|X_1,X_2,\ldots,X_k}}{\sigma^2_Y}$$

10-5 Partial Correlation Coefficient

The *partial correlation coefficient* is a measure of the strength of the linear relationship between two variables after controlling for the effects of other variables. If the two variables of interest are Y and X, and the control variables are Z_1, Z_2, $\ldots$, Z_p, then we denote the corresponding partial correlation coefficient by $r_{YX|Z_1,Z_2,\ldots,Z_p}$. The order of the partial correlation depends on the number of variables that are being controlled for. Thus, *first-order* partials have the form $r_{YX|Z}$, *second-order* partials have the form $r_{YX|Z_1,Z_2}$, and, in general, pth *order* partials have the form $r_{YX|Z_1,Z_2,\ldots,Z_p}$.

For the three independent variables HGT, AGE, and (AGE)^2 in our example, the highest-order partial possible is second order. The values of most of the partial correlations that can be computed from this data set are given in Table 10-2.

The easiest way to obtain a partial correlation is to use a standard computer program. A computing formula for use with a desk calculator, which helps to highlight the structure of the partial correlation coefficient, will be given a little later. First, however, let us see how we can use the information in Table 10-2 to describe our data.

If we look back at our (zero-order) correlation matrix, we see that the variable most highly correlated with WGT is HGT ($r_{Y1} = .814$). Thus, of the three independent variables

TABLE 10-2 Partial correlations for the WGT, HGT, and
AGE data of Table 8-1

Order	Controlling Variables	Form of Correlation	Computed Value
1	HGT	$r_{WGT,AGE|HGT}$	.589
1	HGT	$r_{WGT,(AGE)^2|HGT}$	.580
1	HGT	$r_{AGE,(AGE)^2|HGT}$	.988
1	AGE	$r_{WGT,HGT|AGE}$	.678
1	AGE	$r_{WGT,(AGE)^2|AGE}$	.015
1	AGE	$r_{HGT,(AGE)^2|AGE}$	.060
1	$(AGE)^2$	$r_{WGT,HGT|(AGE)^2}$	.677
1	$(AGE)^2$	$r_{WGT,AGE|(AGE)^2}$	.111
1	$(AGE)^2$	$r_{HGT,AGE|(AGE)^2}$	.022
2	HGT, AGE	$r_{WGT,(AGE)^2|HGT,AGE}$	−.015
2	HGT, $(AGE)^2$	$r_{WGT,AGE|HGT,(AGE)^2}$	.131
2	AGE, $(AGE)^2$	$r_{WGT,HGT|AGE,(AGE)^2}$	.679

we are considering, HGT is the most important according to the strength of its linear relationship with WGT.

We might now ask the following question: After HGT, what is the next most important variable to the linear prediction of WGT? Since the first-order partial $r_{WGT,AGE|HGT} = .589$ is larger than $r_{WGT,(AGE)^2|HGT} = .580$, it makes sense to conclude that AGE is next in importance, after we have accounted for HGT. (If we want to test the significance of this partial correlation coefficient, we would use a partial F test as described in Chapter 9. We shall return to this point shortly.)

The only variable left to consider is $(AGE)^2$. We might now ask: Once we have accounted for HGT and AGE, does $(AGE)^2$ add anything to our knowledge of WGT? To answer this, we can look at the second-order partial correlation $r_{WGT,(AGE)^2|HGT,AGE} = -.015$. Notice that the magnitude of this partial correlation is very small. Thus, we would be inclined to conclude that $(AGE)^2$ provides essentially no additional information about WGT once HGT and AGE have been used together as predictors.

The procedure for selecting variables just described, which starts with the most important variable and continues step by step to add variables in order of importance while controlling for variables already selected, is called a *forward selection* procedure. Alternatively, we could have handled the variable selection problem by working backward. By this, we mean that we start with all the variables and delete (step by step) variables that do not contribute much to the description of the dependent variable. We shall discuss procedures for selecting variables further in Chapter 16.

10-5-1 Tests of Significance for Partial Correlations

Regardless of the procedure used to select variables, it will be necessary to decide at each step whether a particular partial correlation coefficient is significantly different from zero or not. Actually, we already described how to test for such significance in a slightly different context when we were considering the various uses of the ANOVA table in regression analysis. When we wanted to test whether adding a variable to the regression model was worthwhile, given that certain other variables were already in the model, we used a partial F test. It can be shown that this partial F test is exactly equivalent to a test of significance for the corresponding partial correlation coefficient. Thus, to test whether $r_{YX|Z_1,Z_2,...,Z_p}$ is significantly different from zero, we compute the corresponding partial $F(X|Z_1, Z_2, ..., Z_p)$ and reject the null hypothesis if this F statistic exceeds an appropriate critical value of the $F_{1,n-p-2}$ distribution. For example, in testing whether $r_{WGT,(AGE)^2|HGT,AGE}$ is significant, we find that the partial $F[(AGE)^2|HGT, AGE] = 0.010$ does not exceed $F_{1,12-2-2,0.90} = F_{1,8,0.90} = 3.46$. Therefore, we conclude that this partial correlation is not significantly different from zero, and so $(AGE)^2$ does not contribute to the prediction of WGT once we have accounted for HGT and AGE.

The null hypothesis for this test can be stated more formally by considering the population analog of the sample partial correlation coefficient $r_{YX|Z_1,Z_2,...,Z_p}$. This corresponding population parameter, usually denoted by $\rho_{YX|Z_1,Z_2,...,Z_p}$, is called the *population partial correlation coefficient*. The null hypothesis can then be stated as $H_0: \rho_{YX|Z_1,Z_2,...,Z_p} = 0$ and the associated alternative hypothesis as $H_A: \rho_{YX|Z_1,Z_2,...,Z_p} \neq 0$.

10-5-2 Relating the Test for Partial Correlation to the Partial F Test

The structure of the population partial correlation helps us to relate this form of higher-order correlation to regression. Let us, for simplicity, consider this relationship for the special case of two independent variables. The formula for the square of $\rho_{YX_1|X_2}$ can be written

$$\rho_{YX_1|X_2}^2 = \frac{\sigma_{Y|X_2}^2 - \sigma_{Y|X_1,X_2}^2}{\sigma_{Y|X_2}^2}$$

Thus, the square of the sample partial correlation $r_{YX_1|X_2}$ is an estimate of the proportionate reduction in the conditional variance of Y given X_2 due to conditioning on both X_1 and X_2.

(The partial correlation $\rho_{YX_1|X_2}$ can also be described as a zero-order correlation for a conditional bivariate distribution. If the joint distribution of Y, X_1, and X_2 is trivariate normal, the conditional joint distribution of Y and X_1 given X_2 is bivariate normal. The zero-order correlation between Y and X_1 for this conditional distribution is what we call $\rho_{YX_1|X_2}$; this is exactly the partial correlation between X_1 and Y controlling for X_2.)

It then follows that an analogous formula for the squared sample partial correlation coefficient is

$$r_{YX_1|X_2}^2 = \frac{\text{residual SS (using only } X_2 \text{ in the model)} - \text{residual SS (using } X_1 \text{ and } X_2 \text{ in the model)}}{\text{residual SS (using only } X_2 \text{ in the model)}}$$

$$= \frac{\text{extra sum of squares due to adding } X_1 \text{ to the model, given } X_2 \text{ is already in the model}}{\text{residual SS (using only } X_2 \text{ in the model)}} \tag{10.2}$$

It should be clear from the structure of (10.2) and from the discussion of partial F statistics given in Chapters 8 and 9 why the test of $H_0: \rho_{YX_1|X_2} = 0$ is performed using $F(X_1|X_2)$ as the test statistic.

10-5-3 Another Way of Describing Partial Correlations

Another way to compute a first-order partial correlation is to use the formula

$$r_{YX|Z} = \frac{r_{YX} - r_{YZ}r_{XZ}}{\sqrt{(1 - r_{YZ}^2)(1 - r_{XZ}^2)}} \qquad (10.3)$$

For example, to compute $r_{WGT,AGE|HGT}$, we calculate

$$\frac{r_{WGT,AGE} - r_{WGT,HGT}r_{AGE,HGT}}{\sqrt{(1 - r_{WGT,HGT}^2)(1 - r_{AGE,HGT}^2)}} = \frac{.770 - .814(.614)}{\sqrt{[1 - (.814)^2][1 - (.614)^2]}}$$

$$= \frac{.770 - .500}{\sqrt{.338(.623)}} = .589$$

Notice that the first correlation in the numerator is the simple zero-order correlation between WGT and AGE. The *control variable* HGT appears in the second expression in the numerator (where it is correlated separately with each of the variables WGT and AGE) and in both terms in the denominator. *It is by use of (10.3)* that we can interpret the partial correlation coefficient as an adjustment of the simple correlation coefficient to take into account the effect of the control variable. In particular, if r_{YZ} and r_{XZ} have the same sign, then controlling for Z *reduces* (i.e., makes less positive or more negative, as the case may be) the zero-order correlation r_{YX} between Y and X. On the other hand, if r_{YZ} and r_{XZ} have opposite signs, controlling for Z *increases* r_{YX}.

To compute higher-order partial correlations, we simply reapply this formula using the appropriate next-lower-order partials. For example, the second-order partial correlation is an adjustment of the first-order partial, which, in turn, is an adjustment of the simple zero-order correlation. In particular, we have the following general formula for a second-order partial correlation:

$$r_{YX|Z,W} = \frac{r_{YX|Z} - r_{YW|Z}r_{XW|Z}}{\sqrt{(1 - r_{YW|Z}^2)(1 - r_{XW|Z}^2)}} = \frac{r_{YX|W} - r_{YZ|W}r_{XZ|W}}{\sqrt{(1 - r_{YZ|W}^2)(1 - r_{XZ|W}^2)}} \qquad (10.4)$$

To compute $r_{WGT,(AGE)^2|HGT,AGE}$, for example we have

$$\frac{r_{WGT,(AGE)^2|HGT} - r_{WGT,AGE|HGT}r_{(AGE)^2,AGE|HGT}}{\sqrt{(1 - r_{WGT,AGE|HGT}^2)(1 - r_{(AGE)^2,AGE|HGT}^2)}} = \frac{.580 - (.589)(.988)}{\sqrt{[1 - (.589)^2][1 - (.988)^2]}}$$

$$= -.015$$

10-5-4 Partial Correlation as a Correlation of Residuals of Regression

There is still another important interpretation concerning partial correlations. For the variables Y, X, and Z, suppose that we fit the two straight-line regression equations $Y = \beta_0 + \beta_1 Z + E$ and $X = \beta_0^* + \beta_1^* Z + E$. Let $\hat{Y} = \hat{\beta}_0 + \hat{\beta}_1 Z$ be the fitted line of Y on Z, and let $\hat{X} = \hat{\beta}_0^* + \hat{\beta}_1^* Z$ be the fitted line of X on Z. Then the deviations, or *residuals*, $(\hat{Y} - Y)$ and $(\hat{X} - X)$ represent what remains after the variable Z has explained all the variation it can in the variables Y and X separately.

If we now correlate these residuals (i.e., find $r_{\hat{Y}-Y,\hat{X}-X}$), we obtain a measure that is independent of the effects of Z. It can be shown that *the partial correlation between Y and X controlling for Z can be defined as the correlation of the residuals of the straight-line regressions of Y on Z and X on Z*; that is, $r_{YX|Z} = r_{\hat{Y}-Y,\hat{X}-X}$.

10-5-5 Semipartial Correlations

An alternative form of partial correlation is sometimes considered. The partial correlation was just characterized as a correlation between Y adjusted for Z and X adjusted for Z. Some statisticians refer to this as a "full" partial, since both variables being correlated have been adjusted.

The *semipartial correlation* (or "part" correlation) may be characterized as the correlation between two variables when only one of the two has been adjusted for a third variable. For example, one may consider the semipartial correlation between Y and X with only X adjusted for Z or with only Y adjusted for Z. The first will be denoted by $r_{Y(X|Z)}$, and the second by $r_{X(Y|Z)}$. Thus, we have $r_{Y(X|Z)} = r_{Y,X-\hat{X}}$ and $r_{X(Y|Z)} = r_{Y-\hat{Y},X}$, where $\hat{X}$ and $\hat{Y}$ are obtained from straight-line regressions on Z.

Another way of describing these semipartials is in terms of zero-order correlations, as follows:

$$r_{Y(X|Z)} = \frac{r_{YX} - r_{YZ}r_{XZ}}{\sqrt{1 - r_{XZ}^2}} \tag{10.5}$$

and

$$r_{X(Y|Z)} = \frac{r_{YX} - r_{YZ}r_{XZ}}{\sqrt{1 - r_{YZ}^2}} \tag{10.6}$$

It is instructive to compare these formulae with formula (10.3) for the full partial. The numerator is the same in all three expressions: the partial covariance between Y and X adjusted for Z (with all three variables standardized to have variance 1). Clearly, then, these correlations all have the same sign; and if any one equals 0, they all do. For significance testing, it is appropriate to use the extra-sum-of-squares test described earlier.

These three correlations have different interpretations. They each describe the relationship between Y and X but with adjustment for different quantities. The semipartial $r_{Y(X|Z)}$ is the correlation between Y and X with X adjusted for Z; the semipartial $r_{X(Y|Z)}$ is the correlation between Y and X with Y adjusted for Z. Finally, the full partial $r_{YX|Z}$ is the correlation between Y and X with both Y and X adjusted for Z.

TABLE 10-3   Possible relationships among variables X and Y and nuisance variable Z

Case	Nuisance Relationship	Diagram	Preferred Correlation
1	Neither X nor Y affected by Z	$X \longleftrightarrow Y$	r_{XY}
2	Both X and Y affected by Z	$X \longleftrightarrow Y$ Z	$r_{YX|Z}$
3	Only X affected by Z	$X \longleftrightarrow Y$ Z	$r_{Y(X|Z)}$
4	Only Y affected by Z	$X \longleftrightarrow Y$ Z	$r_{X(Y|Z)}$

Choosing the proper correlation coefficient depends on the relationship among the three variables X, Y, and Z (the nuisance variable). Table 10-3 shows the four possible types of relationship. Case 1 involves assessing the relationship between X and Y without a nuisance variable present; here, the simple correlation, r_{XY}, should be used. Case 2 illustrates the situation where the nuisance variable Z is related to both X and Y, so that the use of $r_{YX|Z}$ is appropriate. In cases 3 and 4, the nuisance variable affects just one of the two variables X and Y. In these cases, semipartial correlations permit just one of two primary variables to be adjusted for the effects of a nuisance variable.

10-5-6 Summary of the Features of the Partial Correlation Coefficient

1. The partial correlation $r_{YX|Z_1,Z_2,...,Z_p}$ measures the strength of the linear relationship between two variables X and Y while controlling for variables $Z_1, Z_2, \ldots, Z_p$.

2. The square of the partial correlation $r_{YX|Z_1,Z_2,...,Z_p}$ measures the proportion of the residual sum of squares that is accounted for by the addition of X to a regression model already involving $Z_1, Z_2, \ldots, Z_p$; that is,

$$r^2_{YX|Z_1,Z_2,...,Z_p} = \frac{\text{extra sum of squares due to adding } X \text{ to a model already containing } Z_1, Z_2, \ldots, Z_p}{\text{residual SS (using only } Z_1, Z_2, \ldots, Z_p \text{ in the model)}}$$

3. The partial correlation coefficient $r_{YX|Z_1,Z_2,...,Z_p}$ is an estimate of the population parameter $\rho_{YX|Z_1,Z_2,...,Z_p}$, which is the correlation between Y and X in the conditional joint distribution of Y and X given $Z_1, Z_2, \ldots, Z_p$. Also, the square of this population partial correlation coefficient is given by the equivalent formula

$$\rho^2_{YX|Z_1,Z_2,...,Z_p} = \frac{\sigma^2_{Y|Z_1,Z_2,...,Z_p} - \sigma^2_{Y|X,Z_1,Z_2,...,Z_p}}{\sigma^2_{Y|Z_1,Z_2,...,Z_p}}$$

where $\sigma^2_{Y|Z_1,Z_2,...,Z_p}$ is the variance of the conditional distribution of Y given $Z_1, Z_2, \ldots, Z_p$ (and where $\sigma^2_{Y|X,Z_1,Z_2,...,Z_p}$ is similarly defined).

4. The partial F statistic $F(X|Z_1, Z_2, \ldots, Z_p)$ is used to test H_0: $\rho_{YX|Z_1,Z_2,...,Z_p} = 0$.

5. The (first-order) partial correlation coefficient $r_{YX|Z}$ is an adjustment of the (zero-order) correlation r_{YX} that takes into account the effect of the control variable Z. This is seen from the formula

$$r_{YX|Z} = \frac{r_{YX} - r_{YZ}r_{XZ}}{\sqrt{(1 - r^2_{YZ})(1 - r^2_{XZ})}}$$

Higher-order partial correlations are computed by reapplying this formula using the next-lower-order partials.

6. The partial correlation $r_{YX|Z}$ can be defined as the correlation of the residuals of the straight-line regressions of Y on Z and of X on Z; that is, $r_{YX|Z} = r_{\hat{Y}-Y, \hat{X}-X}$.

10-6 Alternative Representation of the Regression Model

With the correlation analog to multiple regression, we can express the regression model $\mu_{Y|X_1, X_2, \ldots, X_k} = \beta_0 + \beta_1 X_1 + \beta_2 X_2 + \cdots + \beta_k X_k$ in terms of partial correlation coefficients and conditional variances. When $k = 3$, this representation takes the form

$$\mu_{Y|X_1, X_2, X_3} = \mu_Y + \rho_{Y1|23} \frac{\sigma_{Y|23}}{\sigma_{1|23}} (X_1 - \mu_{X_1}) + \rho_{Y2|13} \frac{\sigma_{Y|13}}{\sigma_{2|13}} (X_2 - \mu_{X_2})$$

$$+ \rho_{Y3|12} \frac{\sigma_{Y|12}}{\sigma_{3|12}} (X_3 - \mu_{X_3}) \tag{10.7}$$

where, for example, $\rho_{Y1|23} = \rho_{YX_1|X_2, X_3}$, and where we define

$$\beta_1 = \rho_{Y1|23} \frac{\sigma_{Y|23}}{\sigma_{1|23}}, \qquad \beta_2 = \rho_{Y2|13} \frac{\sigma_{Y|13}}{\sigma_{2|13}}, \qquad \beta_3 = \rho_{Y3|12} \frac{\sigma_{Y|12}}{\sigma_{3|12}}$$

Note the similarity between this representation and that for the straight-line case, where β_1 is equal to $\rho(\sigma_Y/\sigma_X)$. Also, we can see here that

$$\beta_0 = \mu_Y - \beta_1 \mu_{X_1} - \beta_2 \mu_{X_2} - \beta_3 \mu_{X_3}$$

An equivalent method to that of least squares for estimating the coefficients β_0, β_1, β_2, and β_3 is to substitute for the population parameters in the above formulae the corresponding estimates

$$\hat{\mu}_Y = \bar{Y}, \qquad \hat{\mu}_{X_1} = \bar{X}_1, \qquad \hat{\mu}_{X_2} = \bar{X}_2, \qquad \hat{\mu}_{X_3} = \bar{X}_3,$$

$$\hat{\beta}_1 = r_{Y1|23} \frac{S_{Y|23}}{S_{1|23}}, \qquad \hat{\beta}_2 = r_{Y2|13} \frac{S_{Y|13}}{S_{2|13}}, \qquad \hat{\beta}_3 = r_{Y3|12} \frac{S_{Y|12}}{S_{3|12}}$$

10-7 Multiple-Partial Correlation

10-7-1 The Coefficient and Its Associated F Test

The *multiple-partial correlation coefficient* is used to describe the overall relationship between a dependent variable and *two or more* independent variables while controlling for still other variables. For example, suppose, in addition to the independent variables HGT(X_1), AGE(X_2), and (AGE)^2 (X_2^2), we also consider the variable $X_1^2 = $ (HGT)^2 and the product term $X_1 X_2 = $ HGT $\times$ AGE. Our complete regression model is then of the form

$$Y = \beta_0 + \beta_1 X_1 + \beta_2 X_2 + \beta_{11} X_1^2 + \beta_{22} X_2^2 + \beta_{12} X_1 X_2 + E$$

We call such a model a *complete second-order model* since it includes all possible variables up through second-order terms. For such a complete model, we frequently want to ask whether any of the second-order terms are important or, in other words, whether a first-order model involving only X_1 and X_2 (i.e., $Y = \beta_0 + \beta_1 X_1 + \beta_2 X_2 + E$) is adequate. There are two equivalent ways to represent this question as a hypothesis-testing problem. One way is to test $H_0: \beta_{11} = \beta_{22} = \beta_{12} = 0$ (i.e., all second-order coefficients are zero). The other is to test the hypothesis $H_0: \rho_{Y(X_1^2,X_2^2,X_1X_2)|X_1,X_2} = 0$, where $\rho_{Y(X_1^2,X_2^2,X_1X_2)|X_1,X_2}$ is the population multiple-partial correlation of Y with the second-order variables, controlling for the effects of the first-order variables. (In general, we write the multiple-partial as $\rho_{Y(X_1,X_2,\ldots,X_k)|Z_1,Z_2,\ldots,Z_p}$.) This parameter is estimated by the sample multiple-partial correlation $r_{Y(X_1^2,X_2^2,X_1X_2)|X_1,X_2}$. This measure describes the overall multiple contribution of adding the second-order terms to the model after the effects of the first-order terms are partialed out or controlled for (hence the term *multiple-partial*). Two equivalent formulae for calculating $r^2_{Y(X_1^2,X_2^2,X_1X_2)|X_1,X_2}$ are:

$$r^2_{Y(X_1^2,X_2^2,X_1X_2)|X_1,X_2} = \frac{\begin{array}{c}\text{residual SS (only } X_1 \text{ and } X_2 \text{ in the model)}\\ -\text{ residual SS (all first- and second-order terms in the model)}\end{array}}{\text{residual SS (only } X_1 \text{ and } X_2 \text{ in the model)}}$$

$$= \frac{\begin{array}{c}\text{extra sum of squares due to the addition of the}\\ \text{second-order terms } X_1^2, X_2^2, \text{ and } X_1X_2 \text{ to a model}\\ \text{containing only the first-order terms } X_1 \text{ and } X_2\end{array}}{\text{residual SS (only } X_1 \text{ and } X_2 \text{ in the model)}} \qquad (10.8)$$

and

$$r^2_{Y(X_1^2,X_2^2,X_1X_2)|X_1,X_2} = \frac{R^2_{Y|X_1,X_2,X_1^2,X_2^2,X_1X_2} - R^2_{Y|X_1,X_2}}{1 - R^2_{Y|X_1,X_2}}$$

To test either $H_0: \rho_{Y(X_1^2,X_2^2,X_1X_2)|X_1,X_2} = 0$ or, equivalently, $H_0: \beta_{11} = \beta_{22} = \beta_{12} = 0$, we calculate the multiple-partial F statistic given in general form by expression (9.6) in Chapter 9. For this example, then, the formula becomes

$$F(X_1^2, X_2^2, X_1X_2|X_1, X_2) = \frac{\begin{array}{c}[\text{regression SS}(X_1, X_2, X_1^2, X_2^2, X_1X_2)\\ -\text{ regression SS}(X_1, X_2)]/3\end{array}}{\text{MS residual }(X_1, X_2, X_1^2, X_2^2, X_1X_2)}$$

$$= \frac{\begin{array}{c}[\text{residual SS}(X_1, X_2)\\ -\text{ residual SS}(X_1, X_2, X_1^2, X_2^2, X_1X_2)]/3\end{array}}{\text{MS residual }(X_1, X_2, X_1^2, X_2^2, X_1X_2)}$$

We would reject H_0 at the α level of significance if the calculated value of $F(X_1^2, X_2^2, X_1X_2|X_1, X_2)$ exceeded the critical value $F_{3,n-6,1-\alpha}$.

In general, then, the null hypothesis $H_0: \rho_{Y(X_1,X_2,\ldots,X_k)|Z_1,Z_2,\ldots,Z_p} = 0$ is equivalent to the hypothesis $H_0: \beta_1^* = \beta_2^* = \cdots = \beta_k^* = 0$ in the model $Y = \beta_0 + \beta_1 Z_1 + \beta_2 Z_2 + \cdots + \beta_p Z_p + \beta_1^* X_1 + \beta_2^* X_2 + \cdots + \beta_k^* X_k + E$. The appropriate test statistic is the multiple-partial F given by

$$F(X_1, X_2, \ldots, X_k|Z_1, Z_2, \ldots, Z_p)$$

which has the F distribution with k and $n - p - k - 1$ degrees of freedom under H_0.

The general F statistic given by (9.6) may be expressed in terms of squared multiple correlations (R^2 terms) involving the two models being compared. The general form for this alternative expression is

$$F = \frac{[R^2(\text{larger model}) - R^2(\text{smaller model})]/[\text{regression df (larger model)} - \text{regression df (smaller model)}]}{[1 - R^2(\text{larger model})]/\text{residual df (larger model)}} \qquad (10.9)$$

10-8 Concluding Remarks

As have seen throughout this chapter, a regression F test is associated with many equivalent null hypotheses. As an example, the following null hypotheses all make the same statement but in different forms:

1. H_0: "adding variables to the smaller model to form the larger model does not improve the prediction of Y"

2. H_0: "the population regression coefficients for the variables in the larger model but not in the smaller model are all equal to 0"

3. H_0: "the extra variability explained by the variables added to form the larger model is not larger than random variability as measured by σ^2"

4. H_0: "the population multiple-partial correlation between Y and variables added to produce the larger model, controlling for the variables in the smaller model, is 0"

5. H_0: "the value of the population squared multiple correlation coefficient for the larger model is not greater than the value of that parameter for the smaller model"

The first three null hypotheses involve prediction, while the latter two involve association. The investigator, for interpretative purposes, can choose either group depending on whether his or her focus is on the predictive ability of the regression model or on a particular association of interest.

Problems

1. The correlation matrix obtained for the variables SBP (Y), AGE (X_1), SMK (X_2), and QUET (X_3), using the data in Problem 2 of Chapter 5, is given by

	SBP	AGE	SMK	QUET
SBP	1	.7752	.2473	.7420
AGE	.7752	1	−.1395	.8028
SMK	.2473	−.1395	1	−.0714
QUET	.7420	.8028	−.0714	1

   a. Based on this matrix, which of the independent variables AGE, SMK, and QUET explains the largest proportion of the total variation in the dependent variable SBP?

   b. Using either an available computer program or an appropriate computational formula, determine the partial correlations $r_{SBP,SMK|AGE}$ and $r_{SBP,QUET|AGE}$.

   c. Test for the significance of $r_{SBP,SMK|AGE}$ using the ANOVA results given in Problem 1 of Chapter 8. Express the appropriate null hypothesis in terms of a population partial correlation coefficient.

   d. Determine the second-order partial $r_{SBP,QUET|AGE,SMK}$ and test for the significance of this partial correlation (again, using the results of Problem 1 in Chapter 8).

   e. Based on the results in parts (a) through (d), how would you rank the independent variables by their importance in predicting Y? Which of these variables are relatively unimportant?

f. Compute the squared multiple-partial correlation $r^2_{SBP(QUET,SMK)|AGE}$ using the ANOVA tables in Problem 1 of Chapter 8. Test for the significance of this correlation. Does this test result alter your previous choice in (e) of those variables to be included in the regression model?

2. An equivalent way of performing a partial F test for the significance of the addition of a new variable to a model while controlling for variables already in the model is to perform a t test using the appropriate partial correlation coefficient. If the dependent variable is Y, the independent variable of interest is X, and the controlling variables are $Z_1, Z_2, \ldots, Z_p$, then the t test for $H_0: \rho_{YX|Z_1,Z_2,\ldots,Z_p} = 0$ against $H_A: \rho_{YX|Z_1,Z_2,\ldots,Z_p} \neq 0$ is given by the test statistic

$$T = r_{YX|Z_1,Z_2,\ldots,Z_p} \frac{\sqrt{n - p - 2}}{\sqrt{1 - r^2_{YX|Z_1,Z_2,\ldots,Z_p}}}$$

which has a t distribution under H_0 with $n - p - 2$ degrees of freedom. The critical region for this test is therefore given by

$$|T| \geq t_{n-p-2,1-\alpha/2}$$

a. In a study of the relationship of water hardness to sudden-death rates in $n = 88$ North Carolina counties, the following partial correlations were obtained:

$$r_{YX_3|X_1} = .124 \quad \text{and} \quad r_{YX_3|X_1,X_2} = .121$$

where

Y = sudden-death rate in county
X_1 = distance from county seat to main hospital center
X_2 = population per physician
X_3 = water hardness index for county

Test separately the following hypotheses:

$$\rho_{YX_3|X_1} = 0 \quad \text{and} \quad \rho_{YX_3|X_1,X_2} = 0$$

b. An alternative way of forming the ANOVA table associated with a regression analysis is to use partial correlation coefficients. For example, if three independent variables X_1, X_2, and X_3 are used, the ANOVA table looks as follows:

Source		df	SS
Regression	X_1	1	$r^2_{YX_1}SSY$
	$X_2\|X_1$	1	$r^2_{YX_2|X_1}(1 - r^2_{YX_1})SSY$
	$X_3\|X_1, X_2$	1	$r^2_{YX_3|X_1,X_2}(1 - r^2_{YX_2|X_1})(1 - r^2_{YX_1})SSY$
Residual		$n - 4$	$(1 - r^2_{YX_3|X_1,X_2})(1 - r^2_{YX_2|X_1})(1 - r^2_{YX_1})SSY$
Total		$n - 1$	$\sum_{i=1}^{n} (Y_i - \bar{Y})^2$

Determine the ANOVA table for the three independent variables in the water hardness study using the following information:

$$r_{YX_1} = -.196, \qquad r_{YX_2} = .033, \qquad r_{X_1 X_2} = .038,$$
$$r_{YX_3 | X_1, X_2} = .121, \qquad SSY = 21.05, \qquad n = 88$$

c. In addition to the independent variables X_1, X_2, and X_3 considered above, the predictive ability of the following independent variables was also studied:

$X_4 =$ per capita income

$X_5 =$ coroner habit

$Z_1 = 1$ if county is in Piedmont area and 0 otherwise

$Z_2 = 1$ if county is in Coastal Plains area and 0 otherwise

$Z_3 = 1$ if county is in Tidewater area and 0 otherwise

Furthermore, 25 first-order product (i.e., interaction) terms of the form $X_1 X_2$, $X_1 Z_1$, and so on, were also considered (excluding $Z_1 Z_2$, $Z_1 Z_3$, and $Z_2 Z_3$). Three ANOVA tables that were obtained are given here.

ONLY X_1 USED

Source	df	SS	MS
Regression	1	0.3846	0.3846
Residual	86	20.6703	0.2404

ONLY "MAIN EFFECTS" USED

Source	df	SS	MS
Regression	8	2.6853	0.3357
Residual	79	18.3696	0.2325

MAIN EFFECTS PLUS FIRST-ORDER INTERACTIONS USED

Source	df	SS	MS
Regression	33	7.4143	0.2247
Residual	54	13.6406	0.2528

Test whether the addition of all the interaction terms to the model significantly aids in the prediction of the dependent variable after controlling for the main effects. State the null hypothesis for this test in terms of the appropriate multiple-partial correlation coefficient.

d. Test whether there is significant overall prediction based on each of the three ANOVA tables. Determine the multiple R^2-values for each of the three tables.

e. What can you conclude from these results about the relationship of water hardness to sudden death?

3. Two variables X and Y are said to have a *spurious correlation* if their correlation solely reflects each variable's relationship to a third (antecedent) variable Z (and

possibly to other variables). For example, recently the correlation between the total annual income from all sources for members of the U.S. Congress (Y) and the number of persons owning color television sets (X) has been quite high. Yet, during this time, there has been a general upward trend in buying power (Z_1) and in wages of all types (Z_2), which, in turn, would be reflected in increased purchases of color TVs as well as in increased income of members of Congress. Thus, the high correlation between X and Y more than likely merely reflects the influence of inflation on each of these two variables. It is therefore a spurious correlation, since, without controlling for Z_1 and Z_2, say, this correlation misleadingly suggests a relationship between the color TV sales and the income of members of Congress.

  a. How would you attempt to detect statistically whether a correlation between X and Y such as described in spurious?

  b. In a hypothetical study concerning socioecological determinants of respiratory morbidity for a sample of 25 communities, the accompanying correlation matrix was obtained for four variables.

	Unemployment Level (X_1)	Average Temperature (X_2)	Air Pollution Level (X_3)	Respiratory Morbidity Rate (Y)
Unemployment level (X_1)	1	.51	.41	.35
Average temperature (X_2)	—	1	.29	.65
Air pollution level (X_3)	—	—	1	.50
Respiratory morbidity rate (Y)	—	—	—	1

    (1) Determine the partial correlations $r_{YX_1|X_2}$, $r_{YX_1|X_3}$, and $r_{YX_1|X_2,X_3}$.

    (2) Use the results in (1) to determine whether the correlation of .35 between unemployment level (X_1) and respiratory morbidity rate (Y) is spurious. (Use the testing formula given in Problem 2 to make the appropriate tests.)

  c. Describe a relevant example concerning spurious correlation in your field of interest. (Use only interval variables and define them carefully.)

4. a. Using the information provided in Problem 1 of Chapter 9, determine the proportion of residual variation that is explained by the addition of X_2 to a model already containing X_1; that is, compute

$$Q = \frac{\text{regression SS}(X_1, X_2) - \text{regression SS}(X_1)}{\text{residual SS}(X_1)}$$

  b. How is the formula given in part (a) related to the partial correlation $r_{YX_2|X_1}$?

  c. Test the hypothesis H_0: $\rho_{YX_2|X_1} = 0$ using both an F test and a two-sided t test. Check to see that these tests are equivalent.

5. Refer to Problem 6 of Chapter 9 to answer the following questions concerning the relationship of homicide rate (Y) to population size (X_1), percentage of families with yearly incomes less than \$5,000 (X_2), and unemployment rate (X_3):

  a. Determine the squared partial correlations $r^2_{YX_1|X_3}$ and $r^2_{YX_2|X_3}$ using the ANOVA tables. Check the computation of $r^2_{YX_2|X_3}$ by means of an alternative formula using the information that $r_{YX_2} = .8398$, $r_{YX_3} = .8562$, and $r_{X_2X_3} = .8074$.

  b. Based on the results in part (a), which variable (if any) should next be considered for entry into the model given that X_3 is already in the model?

  c. Test H_0: $\rho_{YX_2|X_3} = 0$ using the t test described in Problem 2.

d. Determine the squared partial correlation $r^2_{YX_1|X_2,X_3}$ from the ANOVA tables presented in Problem 6 and then test $H_0: \rho_{YX_1|X_2,X_3} = 0$.

e. Determine the squared multiple-partial correlation $r^2_{Y(X_1,X_2)|X_3}$ and test $H_0: \rho_{Y(X_1,X_2)|X_3} = 0$.

f. Based on the results in parts (a) through (e), what variables would you include in your final regression model? (Use $\alpha = .05$.)

6. Using the ANOVA table given in Problem 7 of Chapter 9, which considers the regression relationship of twelfth-grade mean verbal SAT scores (Y) to per pupil expenditures (X_1), percent of teachers with advanced degrees (X_2), and pupil–teacher ratio (X_3), test the following null hypotheses:

a. $H_0: \rho_{YX_3|X_1} = 0$

b. $H_0: \rho_{YX_2|X_1,X_3} = 0$

c. $H_0: \rho_{Y(X_2,X_3)|X_1} = 0$

d. Based on these results, and assuming that X_1 is an important predictor of Y, what additional variables would you include in your regression model?

7. Using the ANOVA table given below from data in Problem 6 of Chapter 8, which considers the regression relationship of respiratory cancer mortality rates (Y) to the air pollution index (X_1), mean age (X_2), and percent of working force employed in a certain industry (X_3), test the following hypotheses:

a. $H_0: \rho_{YX_2|X_1} = 0$

b. $H_0: \rho_{YX_3|X_1,X_2} = 0$

c. $H_0:$ "the addition of X_2 and X_3 to a model already containing X_1 does not significantly improve the prediction of Y"

d. Based on these results, which variables would you consider to be important predictors of Y? (Use $\alpha = .05$.)

e. State (a) in terms of equivalent tests of semipartial correlations.

Source	df	SS
X_1	1	1,523.658
$X_2|X_1$	1	181.743
$X_3|X_1, X_2$	1	130.529
Residual	19	551.723
Total	22	2,387.653

8. Refer to the computer results given below from data in Problem 8 of Chapter 8 to answer the following questions dealing with factors related to soil erosion.

Stepwise*			Fitting X_1 First, Then Letting X_2 and X_3 Enter Stepwise		
Source	df	SS	Source	df	SS
X_2	1	667.7280	X_1	1	640.4250
$X_3|X_2$	1	5.8228	$X_2|X_1$	1	32.7819
$X_1|X_3, X_2$	1	6.9405	$X_3|X_1, X_2$	1	7.2844
Residual	7	16.0943	Residual	7	16.0943

* That is, in order of relative importance.

    a. Using the correlation matrix

$$R = \begin{array}{c} \\ Y \\ X_1 \\ X_2 \\ X_3 \end{array} \begin{array}{cccc} Y & X_1 & X_2 & X_3 \\ \left[\begin{array}{cccc} 1 & .959 & .979 & -.904 \\ & 1 & .951 & -.819 \\ & & 1 & -.879 \\ & & & 1 \end{array}\right] \end{array}$$

    compute $r_{YX_2|X_1}$ and $r_{YX_3|X_1}$.

    b. Based on your results in part (a), which variable (if any) should next be entered into a regression model that already contains X_1?

    c. Test $H_0: \rho_{YX_2|X_1} = 0$ using the t test described in Problem 2.

    d. Determine the squared multiple-partial correlation $r^2_{Y(X_2, X_3)|X_1}$ and test H_0: $\rho_{Y(X_2, X_3)|X_1} = 0$.

9. Using the computer results in Problem 9 of Chapter 8:

    a. Test $H_0: \rho_{YX_1} = 0$ and $H_0: \rho_{YX_2} = 0$.

    b. Test $H_0: \rho_{YX_1|X_2} = 0$ and $H_0: \rho_{YX_2|X_1} = 0$.

    c. Based on your results in parts (a) and (b), which variables (if any) should be included in the regression model, and what is their relative order of importance?

10. Use the correlation matrix from Problem 1 to answer the following questions.

    a. Compute the semipartial $r_{SBP(SMK|AGE)}$.

    b. Compute the semipartial $r_{SMK(SBP|AGE)}$.

    c. Compare these correlations to the (full) partial $r_{SBP,SMK|AGE}$, computed in Problem 1.

11. For the data discussed in Problem 7, provide estimates of the following:

    a. $r^2_{YX_1}$

    b. $R^2_{Y|X_1, X_2}$

    c. $R^2_{Y|X_1, X_2, X_3}$

    d. $r^2_{YX_3|X_1, X_2}$

    e. $r^2_{YX_2|X_1}$

12. Using the results in Problem 8, answer the following questions:

    a. Provide an estimate of $r^2_{YX_1|X_3, X_2}$.

    b. What two models would be compared in testing whether the correlation in (a) is zero in the population?

    c. Provide an estimate of $r^2_{YX_1}$.

    d. What does the difference between (a) and (c) say about the relationships among the three predictor variables?

    e. What two models would be compared in testing whether the correlations in (a) and (c) differ in the population?

Reference

Blalock, H. M., Jr., ed. (1971). *Causal Models in the Social Sciences*. Chicago: Aldine Publishing Company.

Confounding and Interaction
in Regression

11-1 Preview

Two different goals of a regression analysis are (1) to predict the dependent variable using a set of independent variables and (2) to quantify the relationship of one or more independent variables to a dependent variable. These goals differ because the first focuses on finding a model that fits the observed data and predicts future data as well as possible, whereas the second pertains to producing accurate estimates of one or more regression coefficients in the model. The second goal, moreover, is of particular interest when the research question concerns disease etiology, such as trying to identify one or more determinants of a disease or other health-related outcome.

Confounding and *interaction* are two methodological concepts relevant to attaining the second goal. In this chapter, we describe these concepts using regression terminology. More general discussion of this subject can be found elsewhere (e.g., Kleinbaum, Kupper, and Morgenstern, 1982) within the context of epidemiological research, which typically addresses etiologic questions involving the second goal above. We begin here with a general overview of these concepts, after which we discuss the regression formulation of each concept separately. In Chapter 15 we shall describe a popular regression procedure, analysis of covariance (ANACOVA), which may be used to adjust or correct for problems of confounding. Subsequently, in Chapter 16, we shall briefly describe a strategy for obtaining a "best" regression model that incorporates the assessment of both confounding and interaction.

11-2 An Overview

Confounding and interaction, though different concepts, both involve the assessment of an association between two or more variables so that additional variables that may affect this association are accounted for. The measure of association that is chosen usually depends on the characteristics of the variables of interest. For example, if both variables are continuous, as in the classic regression context, the measure of association will typically be a regression coefficient. The additional variables to be considered are synonymously referred to as *extraneous variables*, *control variables*, or *covariates*. The essential question concerning these variables is whether and how they should be incorporated into a model with which the association of interest can be estimated.

In more practical terms, suppose we consider a study to assess whether physical activity level (PAL) is associated with systolic blood pressure (SBP), accounting (i.e., controlling) for AGE. The extraneous variable here is AGE. We need to determine whether we can ignore AGE in our analysis and still correctly assess the PAL–SBP association. In particular, we need to address the following two questions: (1) Is the estimate of the association between PAL and SBP meaningfully different depending on whether we ignore AGE? (2) Is the estimate of the association between PAL and SBP meaningfully different for different values of AGE? The first question is concerned with confounding, the second question with interaction.

In general, *confounding exists if meaningfully different interpretations of the relationship of interest result when an extraneous variable is ignored or included in the data analysis.* In practice, the assessment of confounding requires a comparison between a *crude* estimate of an association (which ignores the extraneous variable(s) of interest) and an *adjusted* estimate of association (which accounts in some way for the extraneous variables). If the crude and adjusted estimates are meaningfully different, then we say that confounding is present and one or more extraneous variables must be included in our data analysis. Note that this definition does not require a statistical test but rather a comparison of estimates obtained from the data (see Kleinbaum, Kupper, and Morgenstern, 1982, chap. 13, for further discussion of this point).

For example, using the above illustration, a crude estimate of the relationship between PAL and SBP (ignoring AGE) is given by the regression coefficient, say $\hat{\beta}_1$, of the variable PAL in the straight-line model that predicts SBP using just PAL. In contrast, an adjusted estimate is given by the regression coefficient, $\hat{\beta}_1^*$, of the same variable, PAL, in the multiple regression model that predicts SBP using both PAL and AGE. In particular, if PAL is defined dichotomously (e.g., PAL = 1 or 0 for high or low physical activity, respectively), then the crude estimate is simply the crude difference between the mean systolic blood pressures in each physical activity group, and the adjusted estimate represents an adjusted difference in these two mean systolic blood pressures that controls for AGE. In general, confounding is present if there is any meaningful difference between the crude and adjusted estimates.

Interaction is the condition where the relationship of interest is different at different levels (i.e., values) of the extraneous variable(s). In contrast to confounding, the assessment of interaction does not consider either a crude estimate or an (overall) adjusted estimate, but instead focuses on describing the relationship of interest at different values of the extraneous variables. For example, in assessing interaction due to AGE in describing the PAL–SBP relationship, the issue is whether some description (i.e., estimate) of this relationship varies with different values of AGE (e.g., whether the relationship is strong at older ages and weak at younger ages). If the PAL–SBP relationship does vary with AGE, then we say that there is an AGE × (read "by") PAL interaction. To assess interaction a statistical test may be employed in addition to subjective evaluation of the meaningfulness (e.g., clinical importance) of an estimated interaction effect. Again, for further discussion, see Kleinbaum, Kupper, and Morgenstern (1982).

When both confounding and interaction are considered for the same data set, the use of an overall (adjusted) estimate as a summary index of the relationship of interest would tend to mask any (strong) interaction effects that may be present. For example, if the PAL–SBP association differs meaningfully at different values of AGE, the use of a single overall estimate, such as the regression coefficient of PAL in a multiple regression model containing both AGE and PAL, would hide this interaction finding. This illustrates the following important principle: *Interaction should be assessed before confounding is; the use of a summary (adjusted) estimate that controls for confounding is recommended only when there is no meaningful interaction* (Kleinbaum, Kupper, and Morgenstern, 1982, chap. 13).

Thus, in general, confounding and interaction are different phenomena. A variable may manifest both confounding and interaction, neither, or only one of the two. Nevertheless, if strong interaction is found, an adjustment for confounding is inappropriate.

We are now ready to address how these concepts can be employed using regression terminology, assuming a linear model and a continuous dependent variable. A regression analog for a dichotomous outcome variable could, for example, involve a logistic rather than a linear model. Logistic modeling is discussed briefly in Chapter 21; a more detailed discussion in which confounding and interaction are considered can be found in Kleinbaum, Kupper, and Morgenstern (1982, chaps. 20–24).

11-3 Interaction in Regression

In this section, we shall describe how two independent variables can interact to affect a dependent variable and how such an interaction can be represented by an appropriate regression model.

11-3-1 An Example

To illustrate the concept of interaction, we shall consider the following simple example. Suppose it is of interest to determine how two independent variables, temperature (T) and catalyst concentration (C), jointly affect the growth rate (Y) of organisms in a certain biological system. Further, suppose that two particular temperature levels (T_0 and T_1) and two particular levels of catalyst concentration (C_0 and C_1) are to be examined, and that an experiment is performed in which an observation on Y is obtained for each of the four combinations of temperature–catalyst concentration level, (T_0, C_0), (T_0, C_1), (T_1, C_0), and (T_1, C_1).

(In statistical parlance, this experiment is called a *complete factorial experiment*, because observations on Y are obtained for all combinations of settings for the independent variables (or factors). The advantage of a factorial experiment is that any existing interaction effects can be detected and measured efficiently.)

Now, let us consider two graphs based on two hypothetical data sets for the experiment scheme described above. Figure 11-1a suggests that the rate of change in the growth rate as a function of temperature is the same regardless of the level of catalyst concentration; in other words, the relationship between Y and T does not in any way depend on C.

(For those readers familiar with calculus, the phrase "rate of change" is related to the notion of a derivative of a function. In particular, Figure 11-1a portrays a situation where the partial derivative with respect to T of the response function relating the mean of Y to T and C is independent of C.)

It is <u>important to point out</u> that we are not saying that Y and C are unrelated, but that the relationship between Y and T does not vary as a function of C. When this is the case, we say that T and C do not interact or, equivalently, that there is no $T \times C$ interaction effect. Practically speaking, this means that we can investigate the effects of T and C on Y independently of one another and that we can legitimately talk about the separate effects (sometimes called the main effects) of T and C on Y.

One way to quantify the relationship depicted in Figure 11-1a is with a regression model of the form

$$\mu_{Y|T,C} = \beta_0 + \beta_1 T + \beta_2 C \qquad (11.1)$$

Here, the change in the mean of Y for a 1-unit change in T is equal to β_1, regardless of the level of C. In fact, changing the level of C in (11.1) has only the effect of shifting the straight line relating $\mu_{Y|T,C}$ and T either up or down without affecting the value of the slope β_1, as seen in Figure 11-1a. In particular, $\mu_{Y|T,C_0} = (\beta_0 + \beta_2 C_0) + \beta_1 T$ and $\mu_{Y|T,C_1} = (\beta_0 + \beta_2 C_1) + \beta_1 T$.

In general, then, one might say that no interaction is synonymous with parallelism in the sense that the response curves of Y versus T for fixed values of C are parallel; in other words, these response curves (which may be linear or nonlinear) all have the same general shape, differing from one another only by additive constants independent of T (e.g., see Figure 11-2).

In contrast, Figure 11-1b depicts the situation where the relationship between Y and T depends on C; in particular, Y appears to increase with increasing T when $C = C_0$ but to decrease with increasing T when $C = C_1$. In other words, the behavior of Y as a function of temperature cannot be considered independently of catalyst concentration. When this is the case, we say that T and C interact or, equivalently, that there is a $T \times C$ interaction effect. Practically speaking, this means that it really does not make much sense to talk about the separate (or main) effects of T and C on Y, since T and C do not operate independently of one another in their effects on Y.

FIGURE 11-1   (a) No interaction versus (b) interaction

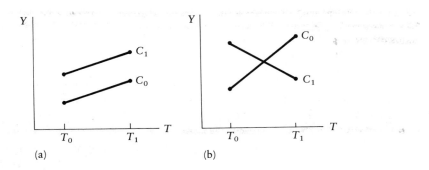

(a)   (b)

FIGURE 11-2   Illustration of no interaction

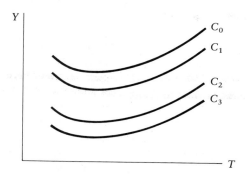

One way to represent such an interaction effect mathematically is to consider a regression model of the form

$$\mu_{Y|T,C} = \beta_0 + \beta_1 T + \beta_2 C + \beta_{12} TC \qquad (11.2)$$

Here the change in the mean value of Y for a 1-unit change in T is equal to $\beta_1 + \beta_{12}C$, which clearly depends on the level of C. In other words, introducing a product term such as $\beta_{12}TC$ in a regression model of the type (11.2) is one way to account for the fact that two such factors as T and C do not operate independently of one another. For our particular example, when $C = C_0$, model (11.2) can be written as

$$\mu_{Y|T,C} = (\beta_0 + \beta_2 C_0) + (\beta_1 + \beta_{12} C_0) T$$

and when $C = C_1$, model (11.2) becomes

$$\mu_{Y|T,C} = (\beta_0 + \beta_2 C_1) + (\beta_1 + \beta_{12} C_1) T$$

In particular, Figure 11-1b suggests that the interaction effect β_{12} is negative, with the linear effect $(\beta_1 + \beta_{12} C_0)$ of T at C_0 being positive and the linear effect $(\beta_1 + \beta_{12} C_1)$ of T at C_1 being negative. A negative interaction effect is to be expected here, since Figure 11-1b suggests that the slope of the linear relationship between Y and T decreases (i.e., goes from positive to negative in sign) as C changes from C_0 to C_1. Of course, it is possible for β_{12} to be positive, in which case the interaction effect would manifest itself as a larger positive value for the slope when $C = C_1$ than when $C = C_0$.

11-3-2   Interaction Modeling in General

As the preceding illustration suggests, interaction among independent variables can generally be described in terms of a regression model that involves product terms. Unfortunately, there are no precise rules for specifying such terms. For example, if interaction involving three variables X_1, X_2, and X_3 is of interest, one model to consider is:

$$Y = \beta_0 + \beta_1 X_1 + \beta_2 X_2 + \beta_2 X_3 + \beta_4 X_1 X_2 + \beta_5 X_1 X_3 + \beta_6 X_2 X_3 + \beta_7 X_1 X_2 X_3 + E \quad (11.3)$$

In this model, the two-factor products of the form $X_i X_j$ are often referred to as *first-order interactions*, whereas three-factor products like $X_1 X_2 X_3$ are called *second-order interactions*, and so on for higher-order products. The higher the order of interaction, the more difficult it becomes to interpret its meaning.

Model (11.3) is not the most general model possible when considering the three variables X_1, X_2, and X_3. Additional product terms such as $X_i X_j^2$, $X_i X_j^3$, $X_i^2 X_j^2$, and so on can also be included. Nevertheless, there is a limit on the total number of such terms: The model cannot contain more than $n - 1$ independent variables when n is the total number of observations in the data. Moreover, it may not even be possible to fit reliably a model with fewer than $n - 1$ variables if some of the variables (e.g., higher-order products) are highly correlated with other variables in the model, as would be the case when the model contains several interaction terms. This problem, called *collinearity*, is discussed in Chapter 12.

Model (11.3) may, on the other hand, be considered too general if one is focusing on particular interactions of interest. For example, if the purpose of one's study is to describe the relationship between X_1 and Y controlling for the possible confounding and/or interaction effects of X_2 and X_3, the following simpler model may be of more interest than (11.3):

$$Y = \beta_0 + \beta_1 X_1 + \beta_2 X_2 + \beta_3 X_3 + \beta_4 X_1 X_2 + \beta_5 X_1 X_3 + E \qquad (11.4)$$

The terms $X_1 X_2$ and $X_1 X_3$ describe the interactions of X_2 and X_3, respectively, with X_1. In contrast, the term $X_2 X_3$, which is not contained in model (11.4), does not concern interaction involving X_1.

In using statistical testing to evaluate interaction for a given regression model, a number of options are available. (A more detailed discussion of how to select variables is given in Chapter 16.) One approach is to test globally for the presence of any kind of interaction and then, if significant interaction is found, to identify particular interaction terms of importance by using other tests. For example, in considering model (11.3), one could first test $H_0: \beta_4 = \beta_5 = \beta_6 = \beta_7 = 0$ using the multiple-partial F statistic

$$F(X_1 X_2, X_1 X_3, X_2 X_3, X_1 X_2 X_3 | X_1, X_2, X_3)$$

which has an $F_{4, n-8}$ distribution when H_0 is true. If this F statistic is found to be significant, individually important interaction terms might then be identified by using selected partial F tests.

A second way to assess interaction is to test for interaction in a hierarchical sequence, beginning with the highest-order terms and then proceeding sequentially to lower-order terms if higher-order terms are not significant. Using model (11.3), for example, one might first test $H_0: \beta_7 = 0$, which considers the second-order interaction, and then test $H_0: \beta_4 = \beta_5 = \beta_6 = 0$ in a reduced model (excluding the three-way product term $X_1 X_2 X_3$) if the first test is nonsignificant.

11-3-3 A Second Example

We now consider a study to assess physical activity level (PAL) as a predictor of systolic blood pressure (SBP), controlling for AGE and SEX. A model that allows for possible interactions of both AGE with PAL and SEX with PAL is given by

$$SBP = \beta_0 + \beta_1 (PAL) + \beta_2 (AGE) + \beta_3 (SEX) + \beta_4 (PAL \times AGE) + \beta_5 (PAL \times SEX) + E$$

Note the absence of a term involving AGE × SEX; such a term does not indicate interaction associated with the study variable of interest (PAL).

To assess interaction for this model, one might first perform a multiple-partial F test of $H_0: \beta_4 = \beta_5 = 0$; if the test was significant, then partial F tests could be conducted to determine whether one or more of these product terms should be kept in the model. If the first test was found nonsignificant, one would then simplify the full model by removing these two product terms entirely, giving the reduced model SBP $= \beta_0 + \beta_1(\text{PAL}) + \beta_2(\text{AGE}) + \beta_3(\text{SEX}) + E$. At this point the interaction phase of model building would be complete. The next step would involve the assessment of confounding, which we discuss in the next section.

11-4 Confounding in Regression

We have emphasized earlier (Section 11-1) that the assessment of confounding is questionable in the presence of interaction. Thus, in our discussion of confounding here, we shall assume throughout that there is no interaction.[1]

11-4-1 Controlling for One Extraneous Variable

Let us suppose that we are interested in describing the relationship between an independent variable T and a continuous dependent variable Y, taking into account the possible confounding effect of a third variable C. As described in the previous section, the assessment of confounding requires the comparison of a crude estimate of the $T-Y$ relationship, which ignores the effect of the control variable (C), with an estimate of the relationship that accounts (or controls) for this variable. This comparison can be expressed in terms of the following two regression models:

$$Y = \beta_0 + \beta_1 T + \beta_2 C + E \tag{11.5}$$

and

$$Y = \beta_0 + \beta_1 T + E \tag{11.6}$$

The assumption of no $T \times C$ interaction precludes the need to consider a product term of the form TC in these models.

From model (11.5), the relationship between T and Y adjusted for the variable C can be expressed in terms of the (partial) regression coefficient (β_1) of the T variable. The estimate of β_1, which we will denote by $\hat{\beta}_{1|C}$, obtained from least-squares fitting of model (11.5), is an adjusted-effect measure in the sense that it gives the estimated change in Y per unit change in T after accounting for C (i.e., with C in the model).

A crude estimate of the $T-Y$ relationship is the estimated coefficient of T (namely, $\hat{\beta}_1$) based on model (11.6), a model that does not involve the variable C.

---

[1] It is possible, however, to assess confounding for variables that are not components of interaction terms. For example, if one is considering the model $Y = \beta_0 + \beta_1 X_1 + \beta_2 X_2 + \beta_3 X_3 + \beta_4 X_1 X_3 + E$, where X_1 is the study variable of interest, one might wish to consider whether X_2 is a confounder, since it is not a component of $X_1 X_3$, the only interaction term in the model. For more realistic examples, see Kleinbaum, Kupper, and Morgenstern (1982, chap. 23).

Thus, we have the following general rule for assessing the presence of confounding when only one independent variable is to be controlled: *Confounding is present if the estimate of the coefficient (β_1) of the study variable T meaningfully changes when the variable C is removed from model (11.5)*, that is, if

$$\hat{\beta}_{1|C} \neq \hat{\beta}_1 \qquad\qquad (11.7)$$

where $\hat{\beta}_{1|C}$ denotes the (adjusted) estimate of β_1 using model (11.5) and $\hat{\beta}_1$ denotes the (crude) estimate of β_1 using model (11.6).

The $\neq$ sign in expression (11.7) indicates that a subjective decision is required as to whether the two estimates are meaningfully different; that is, one needs to determine subjectively whether the two estimates each describe a different interpretation of the $T-Y$ association in question. A statistical test is neither required nor appropriate (Kleinbaum, Kupper, and Morgenstern, 1982, chap. 13).

As an example, suppose Y denotes SBP, T denotes PAL, and C denotes AGE. For some set of data, suppose it was found that

$$\hat{\beta}_{1|AGE} = 4.1 \quad \text{and} \quad \hat{\beta}_1 = 15.9$$

Then, it can be concluded that a 1-unit change in PAL yields a 16-unit change in SBP when AGE is ignored, whereas, when AGE is controlled, a 1-unit change in PAL yields only a 4.1-unit change in SBP: that is, the association between PAL and SBP is much weaker after controlling for AGE. (As a special case, if PAL is a 0–1 variable, then $\hat{\beta}_1$ gives the crude difference in mean systolic blood pressures between the two PAL groups, and $\hat{\beta}_{1|AGE}$ gives an adjusted [for AGE] difference in mean blood pressures.) Thus, AGE would be labeled a confounder and should be controlled in the analysis.

As another example, suppose that

$$\hat{\beta}_{1|AGE} = 6.2 \quad \text{and} \quad \hat{\beta}_1 = 6.1$$

Here, we would be inclined to say that AGE is not a confounder because there is no meaningful difference between the estimates 6.2 and 6.1. Unfortunately, an investigator may have to deal with much more difficult comparisons, such as $\hat{\beta}_{1|AGE} = 4.1$ versus $\hat{\beta}_1 = 5.5$. When comparing such estimates numerically, one must also consider the clinical importance of the numerical difference between estimates based on (a priori) knowledge of the variable(s) involved. For instance, since the coefficients 4.1 and 5.5 estimate, respectively, adjusted and crude differences in mean blood pressures between high and low PAL groups, it is important to decide whether a mean difference of 5.5 is clinically more important than a mean difference of 4.1. One approach to this problem is to control for any variable (as a confounder) that changes the crude effect estimate by some prespecified amount determined by clinical judgment.

(One approach sometimes used to assess confounding is, for example, to conduct a statistical test of $H_0: \beta_2 = 0$ in model (11.5). Such a test does not address confounding, but rather *precision*; that is, such a test evaluates whether significant additional variation in Y is explained by adding C to a model already containing T. An almost equivalent approach is to determine whether a confidence interval for β_1, the coefficient of T, is considerably narrower when C is in the model than when it is not. Precision is often an important issue when considering extraneous factors, but it is a different issue from confounding. In fact, for etiologic questions, confounding, which concerns validity (i.e., do you have the right answer?), usually takes precedence over precision. Another reason for not focusing on β_2 is that if $\hat{\beta}_2 \neq 0$, it does not follow that $\hat{\beta}_{1|C} \neq \hat{\beta}_1$. That is, $\hat{\beta}_2 \neq 0$ is not a sufficient condition for confounding.)[2]

Before turning to criteria for confounding involving several covariates, we comment on the practical problem of deciding what type of variables (i.e., covariates) should be considered for control as potential confounders. Although the answer here is somewhat debatable, we take the position that a list of eligible variables should be constructed based on prior knowledge and/or research about the relationship of the dependent variable to each covariate under consideration. In particular, we recommend that only variables known to be reasonably predictive of (i.e., associated with) the dependent variable should be considered as potential confounders and/or effect modifiers. In epidemiological terms, such variables are generally referred to as *risk factors* (Kleinbaum, Kupper, and Morgenstern, 1982). The idea here is to restrict attention to the control of only those (previously studied) extraneous variables that the investigator anticipates may account for the hypothesized relationship between T and Y presently being studied. To develop such a list, the investigators will have to make a subjective decision.[3]

11-4-2 Controlling for Several Extraneous Variables

Suppose that we wish to describe the association between T and Y, taking into account several covariates $C_1, C_2, \ldots, C_p$. Analogous to the procedure described for one covariate, we can assess confounding by comparing a crude estimate of the $T-Y$ relationship to some adjusted estimate. As before, the crude estimate can be defined in terms of a regression model like (11.6), which describes the relationship between T and Y ignoring all covariates. To obtain the adjusted estimate, however, we must now consider an extended model defined as follows:

$$Y = \beta_0 + \beta_1 T + \beta_2 C_1 + \beta_3 C_2 + \cdots + \beta_{p+1} C_p + E \qquad (11.8)$$

(Like model (11.5), model (11.8) assumes no interaction involving T since no product terms of the form TC_i are included.)

Using this model, we can define confounding involving several variables as follows: *Confounding is present if the estimate of the regression coefficient (β_1) of T in a regression model like (11.6), which ignores the variables $C_1, C_2, \ldots, C_p$, is meaningfully different from the corresponding estimate of β_1 based on a model like (11.8), which controls for $C_1, C_2, \ldots, C_p$,* that is, if

$$\hat{\beta}_{1|C_1, C_2, \ldots, C_p} \neq \hat{\beta}_1 \qquad (11.9)$$

---

[2] Suppose $n = 6$ and we have the following data for (T, C, Y): $(1, 0, 4)$, $(1, 1, 5)$, $(1, 2, 6)$, $(0, 0, 1)$, $(0, 1, 2)$, and $(0, 2, 3)$. Then unweighted least-squares fitting gives $\hat{Y} = 1 + 3T + C$ when T and C are predictors, whereas $\hat{Y} = 2 + 3T$ when C is ignored. Thus, $\hat{\beta}_2 = 1$ ($\neq 0$), yet there is no confounding, since $\hat{\beta}_1 = 3 = \hat{\beta}_{1|C}$.

[3] As a caveat to the above recommendations, certain variables usually referred to as *intervening variables* should not be considered as potential confounders (Kleinbaum, Kupper, and Morgenstern, 1982). A variable C is called intervening between T and Y if T causes C and then C causes Y. Controlling intervening variables may spuriously reduce or eliminate any manifestation in the data of a true association between T and Y.

where $\hat{\beta}_{1|C_1, C_2, \ldots, C_p}$ denotes the (adjusted) estimate of β_1 using (11.8) and $\hat{\beta}_1$ is the (crude) estimate of β_1 using (11.6).

One problem with applying the above definition, however, is that it addresses the question of whether confounding is present without directly identifying specific variables to be controlled.[4] In other words, when confounding is deemed to be present based on (11.9), it may still be the case that only a subset of $C_1, C_2, \ldots, C_p$ is required for adequate control. How does one identify such a subset? More specifically, why bother to identify such a subset rather than simply to control for all variables $C_1, C_2, \ldots, C_p$?

The answer to the latter question is that, when addressing the control of covariates, the possible gains in precision must be considered in addition to the control of confounding. In particular, a subset of C_i variables might be preferred to the entire set because the subset may provide equivalent control of confounding (i.e., may give the same adjusted estimate) while providing greater precision in estimating the adjusted association of interest. However, there is no guarantee that precision will be increased by using a subset; in fact, precision may be reduced. In any case, confounding should take precedence over precision in the sense that no subset should be considered unless it gives the same adjusted-effect estimate as that obtained when controlling for all C_i's.

To illustrate, suppose $p = 5$; that is, we consider controlling for $C_1, C_2, \ldots, C_5$ using model (11.8). Suppose also that the estimate of β_1 takes on the following values depending on which sets of $C_1, C_2, \ldots, C_5$ are controlled.

$$\hat{\beta}_{1|C_1, C_2, \ldots, C_5} = 4.0, \qquad \hat{\beta}_{1|C_1, C_2} = 4.3, \qquad \hat{\beta}_1 = 16.0$$

Then, because 16.0 is much different from 4.0, one can argue that confounding is present. Yet since 4.0 is not meaningfully different from 4.3, it can also be argued that C_3, C_4, and C_5 do not need to be controlled, since essentially the same (adjusted) estimate is obtained when controlling only for C_1 and C_2 as when adjusting for all C_i's.

Thus, for this example, we have identified two sets of C_i variables that we can use for control. Which set do we choose? The answer depends on an evaluation of precision. One approach is to compare interval estimates for some parameter of interest, one interval being derived from a model that controls for C_1 and C_2 only, and the other interval from a model that controls for C_1 through C_5. The logical parameter for this example is the population regression coefficient, β_1, of the variable T when controlling for a particular set of C_i's. That is, we may compare an interval estimate for β_1 when only C_1 and C_2 are controlled to a corresponding interval estimate for β_1 when C_1 through C_5 are controlled. The narrower interval of the two is then the interval reflecting the most precision. For example, if the two 95% interval estimates are (2.6, 7.4) for $\beta_{1|C_1, C_2}$ and (1.7, 7.6) for $\beta_{1|C_1, C_2, \ldots, C_5}$, then the former interval is narrower; in this case, some precision is gained by dropping C_3, C_4, and C_5 from the model.

---

[4] Another problem concerns how to assess confounding when there are two or more study variables, say, T_1 and T_2, of interest. For this general situation, confounding may be defined to be present if (11.9) is satisfied for the coefficient of *any* study variable of interest, given a model containing all such study variables and all control variables. Unfortunately, this definition has the practical drawback of requiring several subjective decisions, one for each study variable of interest.

(An alternative, but not exactly equivalent, approach to evaluating precision is to perform a statistical test for the significance of the addition of C_3, C_4, and C_5 to a model containing T, C_1, and C_2. The null hypothesis for this test may be stated as $H_0: \beta_4 = \beta_5 = \beta_6 = 0$ in the model (11.8) with $p = 5$. If this test is not significant, then it may be argued that retaining C_3, C_4, and C_5 does not provide additional precision (i.e., explanation of variance). This would indicate that only C_1 and C_2 should be controlled for greater precision.

Because this testing approach will not always lead to the same conclusion as the approach of estimating intervals, the investigator may need to choose between them. In most situations, however, both approaches will usually lead to similar results.)

Now we shall address the question on identifying which set to control. We have seen, by example, that we must first identify a baseline-adjusted estimate (i.e., a "gold standard") with which we can make comparisons. The ideal gold standard is the regression coefficient estimate that controls for all C_i's being considered. Then, any subset of C_i's that gives essentially the same adjusted estimate (i.e., an estimate that is not meaningfully different from the gold standard when only the C_i's in that subset are controlled) is a candidate set for control. It is even conceivable that several such candidates are possible (Kleinbaum, Kupper, and Morgenstern, 1982, chap. 14).

Which set does one finally use? The answer, again, is based on precision: Use that set which gives the most precision (e.g., the tightest confidence interval for the adjusted effect under study). (For "political" reasons, that is, to convince people that all variables have been controlled, it might be better to control for $C_1, C_2, \ldots, C_p$ unless some subset of C_i's leads to a large increase in precision.)

To illustrate, suppose that the candidate sets in Table 11-1 can be identified when $p = 5$ in model (11.8). All three proper subsets of C_1, C_2, C_3, C_4, and C_5 may be considered candidates for control since they all give adjusted estimates roughly equal to the gold standard $\hat{\beta}_{1|C_1, C_2, \ldots, C_5} = 4.0$. Of these candidates, the subset involving C_1, C_2, and C_4 gives the best precision (narrowest confidence interval); therefore this subset can be used both to control confounding and to enhance precision.

TABLE 11-1 An example of candidate sets for control

Candidate Set	$\hat{\beta}_{1|\text{Candidate Set}}$	95% Confidence Interval for $\beta_{1|\text{Candidate Set}}$
C_1, C_2, C_3, C_4, C_5 (baseline)	4.0	(2.3, 7.2)
C_1, C_2	4.3	(2.6, 7.6)
C_1, C_4	4.2	(2.1, 7.0)
C_1, C_2, C_4	3.8	(1.9, 6.2)

How can we get this output

11-4-3 An Example Revisited

In Section 11-3-3 we considered a hypothetical study to assess the relationship between physical activity level (PAL) and systolic blood pressure (SBP) while controlling for both AGE and SEX. A model that allows for possible interactions of AGE and SEX with PAL was considered, and methods for testing for such interactions were described. Assuming that no significant interaction effects were found, the resulting reduced model is as follows:

$$\text{SBP} = \beta_0 + \beta_1(\text{PAL}) + \beta_2(\text{AGE}) + \beta_3(\text{SEX}) + E$$

Given this no-interaction model, the next step is to assess confounding; that is, does the coefficient of PAL change when AGE and/or SEX are dropped from the model? To answer this, we can examine the estimate of the coefficient of PAL in four models, namely, one including both AGE and SEX, one involving either AGE or SEX but not both, and one involving neither. The gold standard model for comparison is the one (given above) that contains both control variables and PAL. Then, for example, if the estimate of β_1 changes considerably when at least one control variable is dropped from this gold standard model, we need to control for both AGE and SEX. However, if we obtain essentially the same estimate of β_1 (as obtained using the gold standard model) when only AGE is in the model, we then do not need to retain SEX in the model to control for confounding. However, inclusion of the sex variable in addition to AGE may increase or decrease precision. Thus, the decision as to whether to control for just AGE or for both AGE and SEX would depend, for example, on a comparison of confidence intervals for β_1. If the confidence interval is considerably narrower when only AGE is controlled, then we wouldn't retain SEX in the model.

Finally, once a decision is made about which variables are to be controlled (i.e., which is the best model for providing a valid and precise estimate of the coefficient of PAL), we then make statistical inferences about the true PAL–SBP relationship. Given the no-interaction model, this involves testing $H_0: \beta_1 = 0$ in the best model and then obtaining an interval estimate of β_1.

11-5 Summary and Conclusions

Confounding and interaction are two methodological concepts pertaining to the assessment of a relationship between independent and dependent variables.

Interaction, which takes precedence over confounding, exists when the relationship of interest is different at different levels of extraneous (control) variables. In linear regression, interaction is evaluated using statistical tests about product terms involving basic independent variables in the model.

Confounding, which is not evaluated with statistical testing, is present when the effect of interest differs depending on whether an extraneous variable is ignored or retained in the analysis. In regression terms, confounding is assessed by comparing crude versus adjusted regression coefficients from different models.

When several potential confounders are being considered, it may be worthwhile to identify nonconfounders that can be dropped from the model to gain precision; this may not be possible (i.e., precision may be lost by dropping variables) in some situations.

When there is strong interaction involving a certain extraneous variable, the assessment of confounding for that extraneous variable is irrelevant. Moreover, in such a situation the assessment of confounding involving other extraneous variables, though possible, is quite complex and extremely subjective. Consequently, the assessment of confounding is not usually recommended when important interaction effects have been identified.

Problems

1. Consider the numerical examples given in Section 8-7 of Chapter 8. These concern the assessment of the relationship of the independent variables HGT, AGE, and $(AGE)^2$ to the dependent variable WGT. Suppose that HGT is considered to be the independent variable of primary concern, so that interest lies in evaluating the relationship of HGT to WGT controlling for the possible confounding effects of AGE and $(AGE)^2$.

   a. Assuming no interaction of any kind, state an appropriate regression model that should be used as the baseline (i.e., gold standard) for decisions about confounding.

   b. Using an appropriate (partial) regression coefficient as your measure of association, determine if there is confounding due to AGE and/or $(AGE)^2$.

   c. Can $(AGE)^2$ be dropped from your initial model in part (a) because it is not needed to control adequately for confounding? Explain your answer (using a regression coefficient as your measure of association).

   d. Should $(AGE)^2$ be retained in the final model for the sake of precision? Explain.

   e. Based on considerations of both confounding and precision, what should be your final model? Why?

   f. How would you modify your initial model in part (a) to allow for the assessment of interactions?

   g. Regarding your answer to part (f), how would you test for interaction?

2. Consider the computer results provided below that describe regression analyses involving two independent variables X_1 and X_2 and the dependent variable Y. Assume it is of interest to assess the relationship of X_1 with Y controlling for the possible confounding effects of X_2.

DATA

Observation	Y	X_1	X_2	Observation	Y	X_1	X_2
1	4	0	0	9	6	1	0
2	4	0	0	10	8	1	0
3	4	0	0	11	13	1	1
4	6	0	0	12	13	1	1
5	6	0	0	13	13	1	1
6	6	0	0	14	15	1	1
7	11	0	1	15	15	1	1
8	13	0	1	16	15	1	1

CORRELATION COEFFICIENTS

$$\begin{array}{c c c c} & Y & X_1 & X_2 \\ Y & 1 & .65273 & .94943 \\ X_1 & & 1 & .50000 \\ X_2 & & & 1 \end{array}$$

Note: $r_{YX_1|X_2} = .6547$.

ANOVA TABLES AND PARAMETER ESTIMATES

Source	df	MS	Parameter	Estimate	T for H_0: Parameter = 0
Regression (X_1)	1	121.000	Intercept	6.750	5.60 (P = .0001)
Residual	14	11.643	β_1	5.500	3.22 (P = .0061)

Source	df	MS		Parameter	Estimate	T for H_0: Parameter $= 0$
Regression (X_1, X_2)	2	134.000		Intercept	5.000	11.80 ($P = .0001$)
Residual	13	1.231		β_1	2.000	3.12 ($P = .0081$)
				β_2	7.000	10.93 ($P = .0001$)

a. Using an appropriate regression coefficient as your measure of association, determine if there is confounding. Explain.

b. Suppose confounding had been defined to require a comparison of crude versus adjusted (partial) correlation coefficients. What conclusion would be drawn? Explain.

c. What is the moral of this example?

3. a–c. Consider the computer results provided below that describe regression analyses involving two independent variables X_1 and X_2 and the dependent variable Y (using a different data set than that used in Problem 2). Answer the same questions as in Problem 2 for this new printout.

d. What does this example illustrate regarding the use of a test of the hypothesis $H_0: \beta_2 = 0$ for assessing confounding?

DATA

Observation	Y	X_1	X_2
1	4	0	0
2	6	0	0
3	11	0	1
4	13	0	1
5	6	1	0
6	8	1	0
7	13	1	1
8	15	1	1

CORRELATION COEFFICIENTS

	Y	X_1	X_2
Y	1	.26491	.92717
X_1		1	.00000
X_2			1

Note: $r_{YX_1|X_2} = .707$.

ANOVA TABLES AND PARAMETER ESTIMATES

Source	df	MS	R^2		Parameter	Estimate	T for H_0: Parameter $= 0$
Regression (X_1)	1	8.000	.0702		Intercept	8.500	4.04 ($P = .0068$)
Residual	6	17.667	—		β_1	2.000	0.67 ($P = .5260$)

Source	df	MS	R^2		Parameter	Estimate	T for H_0: Parameter $= 0$
Regression (X_1, X_2)	2	53.000	.9298		Intercept	5.000	6.45 ($P = .0013$)
Residual	5	1.600	—		β_1	2.000	2.24 ($P = .0756$)
					β_2	7.000	7.83 ($P = .0005$)

Computer Printout (SPSS) for Problem 4

```
DEPENDENT VARIABLE.. SBPSL

VARIABLE(S) ENTERED ON STEP NUMBER 1.. SBP 1

MULTIPLE R       0.45834      ANALYSIS OF VARIANCE    DF    SUM OF SQUARES    MEAN SQUARE        F
R SQUARE         0.21007      REGRESSION              1.       14.79083        14.79083    13.56308
STANDARD ERROR   1.04428      RESIDUAL               51.       55.61661         1.09052

----------VARIABLES IN THE EQUATION----------         ----------VARIABLES NOT IN THE EQUATION----------

VARIABLE        B        BETA    STD ERROR B     F     VARIABLE  BETA IN  PARTIAL  TOLERANCE      F
SBP 1      -0.04660  -0.45834    0.01265     13.563    RW        0.23166  0.26007  0.99553     3.627
(CONSTANT)  5.10797                                    SR        0.23074  0.25953  0.99933     3.611


VARIABLE(S) ENTERED ON STEP NUMBER 2.. RW

MULTIPLE R       0.51332      ANALYSIS OF VARIANCE    DF    SUM OF SQUARES    MEAN SQUARE        F
R SQUARE         0.26350      REGRESSION              2.       18.55240         9.27620     8.94445
STANDARD ERROR   1.01838      RESIDUAL               50.       51.85504         1.03710

----------VARIABLES IN THE EQUATION----------         ----------VARIABLES NOT IN THE EQUATION----------

VARIABLE        B        BETA    STD ERROR B     F     VARIABLE  BETA IN  PARTIAL  TOLERANCE      F
SBP 1      -0.04817  -0.47382    0.01237     15.174    SR        0.04646  0.00450  0.00690     0.001
RW          0.02252   0.23166    0.01182      3.627
(CONSTANT)  5.38484


VARIABLE(S) ENTERED ON STEP NUMBER 3.. SR

MULTIPLE R       0.51334      ANALYSIS OF VARIANCE    DF    SUM OF SQUARES    MEAN SQUARE        F
R SQUARE         0.26352      REGRESSION              3.       18.55345         6.18448     5.84409
STANDARD ERROR   1.02871      RESIDUAL               49.       51.85399         1.05824

----------VARIABLES IN THE EQUATION----------         ----------VARIABLES NOT IN THE EQUATION----------

VARIABLE        B        BETA    STD ERROR B     F     VARIABLE  BETA IN  PARTIAL  TOLERANCE      F
SBP 1      -0.04798  -0.47193    0.01391     11.899
RW          0.01801   0.18527    0.14372      0.016
SR          0.00004   0.04646    0.00122      0.001
(CONSTANT)  5.36183
```

From *Statistical Package for the Social Sciences* by Nie et al. Copyright © 1975 by McGraw-Hill Book Company and Dr. Norman Nie, President, SPSS, Inc.

4. A regression analysis using data on $n = 53$ males considered the following variables:

Y = SBPSL (estimated slope based on the straight-line regression of an individual's blood pressure over time)

X_1 = SBP1 (initial blood pressure)

X_2 = RW (relative weight)

$X_3 = X_1 X_2$ = SR (product of SBP1 and RW)

The computer printout on page 177 was obtained using a standard stepwise regression program (SPSS). Using this output, complete the following exercises:

a. Fill in the ANOVA table for the fit of the model $Y = \beta_0 + \beta_1 X_1 + \beta_2 X_2 + \beta_3 X_3 + E$.

Source		df	SS	MS
Regression $\begin{cases} X_1 \\ X_2\vert X_1 \\ X_3\vert X_1, X_2 \end{cases}$				
Residual				
Total		52		

b. Test $H_0: \rho_{YX_2 \vert X_1} = 0$.

c. Test H_0: "the addition of X_3 to the model given that X_1 and X_2 are already in the model is not significant."

d. Test $H_0: \rho_{Y(X_2, X_3)\vert X_1} = 0$.

e. Based on the tests above, what would you consider to be the most appropriate regression model? (Use $\alpha = .05$.)

f. Based on the information provided, can you assess whether X_1 is a confounder of the X_2–Y relationship? Explain.

5. An experiment was conducted regarding a quantitative analysis of factors found in high-density lipoprotein (HDL) in a sample of human blood serum. Three variables thought to be predictive or associated with HDL measurement (Y) were the total cholesterol (X_1) and total triglyceride (X_2) concentrations in the sample, plus the presence or absence of a certain sticky component found in the serum called sinking pre-beta, or SPB (X_3), which has coded as 0 if absent and 1 if present. The data obtained are shown in the table and the computer results below.

DATA

Y	X_1	X_2	X_3	Y	X_1	X_2	X_3
47	287	111	0	57	192	115	1
38	236	135	0	42	349	408	1
47	255	98	0	54	263	103	1
39	135	63	0	60	223	102	1
44	121	46	0	33	316	274	0
64	171	103	0	55	288	130	0
58	260	227	0	36	256	149	0
49	237	157	0	36	318	180	0

DATA (Cont.)

Y	X_1	X_2	X_3	Y	X_1	X_2	X_3
55	261	266	0	42	270	134	0
52	397	167	0	41	262	154	0
49	295	164	0	42	264	86	0
47	261	119	1	39	325	148	0
40	258	145	1	27	388	191	0
42	280	247	1	31	260	123	0
63	339	168	1	39	284	135	0
40	161	68	1	56	326	236	1
59	324	92	1	40	248	92	1
56	171	56	1	58	285	153	1
76	265	240	1	43	361	126	1
67	280	306	1	40	248	226	1
57	248	93	1	46	280	176	1

ANOVA TABLES

Source	df	SS
Regression (X_1)	1	46.2356
Residual	40	4,567.3835

Source	df	SS
Regression (X_2)	1	21.3397
Residual	40	4,592.2793

Source	df	SS
Regression (X_3)	1	735.2054
Residual	40	3,878.4136

Source	df	SS
Regression (X_1, X_2)	2	135.3820
Residual	39	4,478.2369

Source	df	SS
Regression (X_1, X_3)	2	783.1691
Residual	39	3,830.4500

Source	df	SS
Regression (X_2, X_3)	2	737.8069
Residual	39	3,875.8122

Source	df	SS
Regression (X_1, X_2, X_3)	3	819.7473
X_1X_3, $X_2X_3 \mid X_1, X_2, X_3$	2	74.7443
Residual	36	3,719.0517

Source	df	SS
Regression (X_1, X_3)	2	783.1691
$X_1X_3 \mid X_1, X_3$	1	62.4247
Residual	38	3,768.0252

Source	df	SS
Regression (X_2, X_3)	2	737.8069
$X_2X_3 \mid X_2, X_3$	1	1.5539
Residual	38	3,874.2583

a. Test whether X_1, X_2, or X_3 alone significantly helps in predicting Y.
b. Test whether X_1, X_2, and X_3 taken together significantly help to predict Y.
c. Test whether the true coefficients of the product terms X_1X_3 and X_2X_3 are simultaneously zero in the model containing X_1, X_2, and X_3 plus these product terms. State the null hypothesis in terms of a multiple-partial correlation coeffi-

cient. If this test is not rejected, what can you conclude about the relationship of Y to X_1 and X_2 when X_3 equals 1 as compared with when X_3 equals 0?

d. Test (at $\alpha = .05$) whether X_3 is associated with Y after taking into account the combined contribution of X_1 and X_2. State the appropriate null hypothesis in terms of a partial correlation coefficient. What does your result together with your answer to part (c) tell you about the relationship of Y with X_1 and X_2 when SPB is present as compared with when it is absent?

e. Describe how you would determine whether X_1, X_2, or both X_1 and X_2 need to be retained in the model in order to control for confounding and possibly enhance precision. Assume that there is no interaction and that the study variable of interest is X_3.

f. Based on the information provided, can confounding of X_1 and/or X_2 be assessed in evaluating the relationship of X_3 to Y? Explain.

Reference

Kleinbaum, D. G., Kupper, L. L., and Morgenstern, H. (1982). *Epidemiologic Research*. Belmont, Calif.: Lifetime Learning Publications.

12

Regression Diagnostics

12-1 Preview

The purpose of this chapter is to provide a general overview of statistical techniques known as *regression diagnostics*.[1] Such techniques are employed to check the assumptions and to assess the accuracy of computations for a multiple regression analysis. We shall focus primarily on methods for analyzing residuals, assessing the influence of outliers, and assessing collinearity. All these methods are essentially diagnostic tools; in addition, procedures for analyzing the data that solve, avoid, or help correct diagnosed problems are briefly discussed.

12-2 Simple Approaches to Diagnosing Problems in Data

In analyzing data, it is important to be familiar with their basic characteristics. Such familiarity helps avoid many errors. For example, it is essential to know the following:

1. the type of subject or experimental unit (e.g., small pine tree needles, elderly male humans),

2. the procedure for data collecting,

3. the unit of measurement for each variable (e.g., kilograms, meters, square inches, cubic centimeters), and

4. a plausible range of values and a typical value for each variable.

---

[1] Our use of the term *regression diagnostics* is much less restrictive than is common in the statistical literature. Our usage is intended to encourage the reader to examine the validity of all aspects of a regression analysis.

This knowledge can then be combined with descriptive statistics computed for a set of data in order to detect errors in the data and pinpoint potential violations of the assumptions of a planned analysis. For regression analysis in particular, simple descriptive statistics computed on the response variable and the predictor variables can be extremely helpful in detecting potential violations of the assumptions. With the use of a comprehensive statistical computer package, it is easy to generate descriptive statistics for all variables in an analysis.

As a first step in detecting data base problems, we recommend listing the five largest and five smallest values for every variable. This is a very simple but extremely powerful technique for examining data. It must be combined with knowledge of the type of subject, data collection procedure, measurement units, and plausible ranges. One can then immediately detect many data-recording errors, format errors in computer input, and some outliers. In addition, the extremes are a good indication of whether the data are in the expected range. Note that a listing of the data may be used to check and correct errors. This can reveal some blatant errors, such as format errors in computer input. However, experience indicates that detecting individual data point problems is very unlikely, particularly as the number of observations increases beyond 50.

A useful second step in detecting problems and assessing analysis assumptions is to calculate descriptive statistics. The choice of statistics depends on the scale of measurement (nominal, ordinal, or interval; see Chapter 3); many good alternatives are available. One simple approach is to consider frequency tables for variables with few distinct values, and the mean, standard deviation, and minimum and maximum values for continuous variables. For continuous variables, more detailed information can usually be presented conveniently in some form of frequency histogram (discussed in detail below). It is often helpful to compute such descriptive statistics separately for important groups within the sample (such as males and females). Such descriptive statistics should be compared with what is expected from the study design and from scientific knowledge about the variables. Suspicious and implausible data should be investigated more carefully.

More elaborate descriptive approaches can also be made part of a regression analysis, including correlations among pairs of variables and plots of the response as a function of each predictor. As will be discussed below, very high between-predictor correlations may signal collinearity problems. The plots may indicate the presence of nonlinear relationships or troublesome unevenness in the distribution of the data.

We shall illustrate many of the techniques described in this chapter using two specific examples. The data for the first are taken from Lewis and Taylor (1967). WEIGHT, HEIGHT, and AGE were all recorded for a sample of boys and girls. In this chapter, we shall consider only the data from the 127 boys and we shall evaluate a model in which the response is the subject's WEIGHT and the predictors are HEIGHT, (HEIGHT)^2, AGE, and (AGE)^2. The data for the second example arise from a hypothetical calibration experiment in environmental engineering. The 17 values of the "concentration of a certain pollutant" (X) and the "instrument reading" (Y) are given in Table 12-1, along with other information to be described below.

In our first example, the first step in assessing the appropriateness of the regression assumptions would be to consider descriptive statistics for the three variables WEIGHT, HEIGHT, and AGE. Frequency histograms and their close relatives, the schematic plot and the stem-and-leaf diagram (all to be described later in this chapter), are the most useful tools for this task. Analysis of such raw data can be helpful in detecting very serious errors in

TABLE 12-1   Residuals for the calibration experiment

Predictor (X_i)	Response (Y_i)	Predicted Value $(\hat{Y}_i)$	Unstandardized $(e_i = Y_i - \hat{Y}_i)$	Standardized $(z_i = e_i/S)$	Studentized $\left(r_i = e_i/\sqrt{S^2(1 - h_i)}\right)$	Jackknife $\left(r_{(-i)} = e_i/\sqrt{S^2_{(-i)}(1 - h_i)}\right)$
0.0	10.7	15.39	−4.69	−1.14	−1.29	−1.32
0.5	14.2	17.29	−3.09	−0.75	−0.83	−0.82
1.0	16.7	19.18	−2.48	−0.60	−0.65	−0.64
1.5	19.1	21.07	−1.97	−0.48	−0.51	−0.50
2.0	24.9	22.97	1.93	0.47	0.49	0.48
2.5	25.4	24.86	0.54	0.13	0.14	0.13
3.0	32.3	26.75	5.55	1.35	1.40	1.45
3.5	30.8	28.65	2.15	0.52	0.54	0.53
4.0	39.6	30.54	9.06	2.20	2.27	2.71
4.5	30.3	32.43	−2.13	−0.52	−0.54	−0.52
5.0	37.2	34.33	2.87	0.70	0.72	0.71
5.5	37.8	36.22	1.58	0.38	0.40	0.39
6.0	37.5	38.11	−0.61	−0.15	−0.16	−0.15
6.5	38.6	40.01	−1.41	−0.34	−0.37	−0.35
7.0	42.6	41.90	0.70	0.17	0.18	0.18
7.5	44.3	43.79	0.51	0.12	0.14	0.13
8.0	37.2	45.69	−8.49	−2.06	−2.33	−2.82

The table has grouped headers: "Variables" spans Predictor and Response; "Residuals" spans Unstandardized, Standardized, Studentized, and Jackknife.

individual observations or in suggesting an appropriate model. Analysis of residuals and other regression diagnostic procedures provide the most refined and accurate evaluation of model assumptions.

As mentioned earlier, plots are informative prior to regression analysis. The data from Lewis and Taylor (1967) are plotted in scatter diagrams with the response variable, WEIGHT, on the vertical axis and the two predictors, HEIGHT and AGE, on the horizontal (Figures 12-1 and 12-2, respectively). In both cases, a single observation (which is circled) appears isolated at the upper edge of the plot, making it highly suspect. In this case, a listing of the five largest and smallest values for each variable would also reveal the same point. Possible explanations for such an unusual observation include (1) the values for the observation were measured, recorded, or entered in the computer incorrectly and (2) the values are correct and the observation must be evaluated for its effect on the analysis. Consequently, the fact that an observation appears to be unusual when compared with the rest of the data does not automatically mean that it should be dropped. The observation appears below the main data cluster (but not nearly as suspiciously) in Figure 12-3, which plots HEIGHT against AGE. We shall see how regression diagnostic procedures permit easier detection of such observations and indicate their influence on a regression analysis. Furthermore, most of the regression diagnostics to be described are as easy to use with several predictors as with one or two.

FIGURE 12-1   Children's body weight as a function of height; data from Lewis and Taylor (1967; *n* = 127)

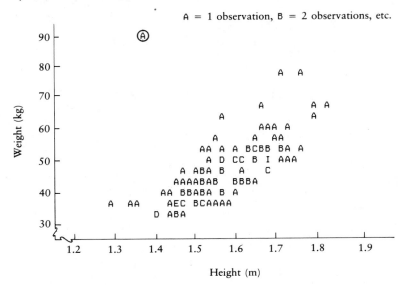

FIGURE 12-2   Children's body weight as a function of age; data from Lewis and Taylor (1967; *n* = 127)

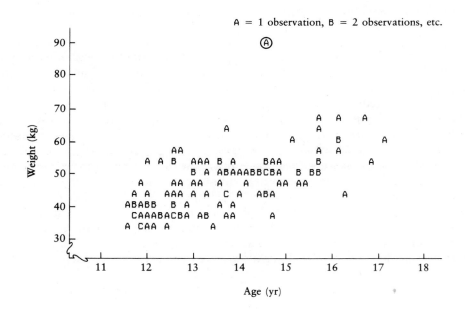

FIGURE 12-3   Children's height as a function of age; data from Lewis and Taylor (1967; $n = 127$)

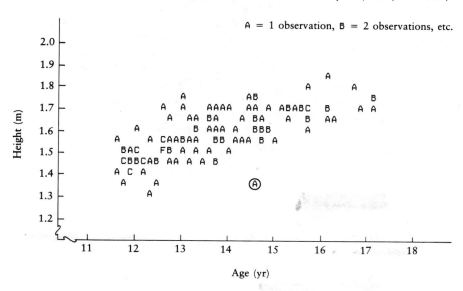

12-3 Residual Analysis

12-3-1 Some Properties of Residuals

Analytic Properties

Given n observations $(Y_i, X_{i1}, X_{i2}, \ldots, X_{ik})$, $i = 1, 2, \ldots, n$, recall that the methodology of regression analysis is concerned with the least-squares fitting of a model describing the observed response Y_i as

$$Y_i = \beta_0 + \beta_1 X_{i1} + \beta_2 X_{i2} + \cdots + \beta_k X_{ik} + E_i \qquad i = 1, 2, \ldots, n$$

in which E_i denotes the (unobserved) error term for the ith response. We write the fitted model as

$$\hat{Y} = \hat{\beta}_0 + \hat{\beta}_1 X_1 + \hat{\beta}_2 X_2 + \cdots + \hat{\beta}_k X_k$$

and then the predicted response at the ith data point is

$$\hat{Y}_i = \hat{\beta}_0 + \hat{\beta}_1 X_{i1} + \hat{\beta}_2 X_{i2} + \cdots + \hat{\beta}_k X_{ik}$$

With this framework in mind, we define the ith residual, e_i, to be the difference between the observed value, Y_i, and the predicted value, $\hat{Y}_i$, namely,

$$e_i = Y_i - \hat{Y}_i \qquad i = 1, 2, \ldots, n$$

In words, then, the $\{e_i\}$ reflect the amount of discrepancy between observed and predicted values that is still present after having fitted the least-squares model. Also, each e_i represents an estimate of the corresponding unobserved error E_i. Now, the usual assumptions (described in Chapter 8) made about the unobserved error terms $\{E_i\}$ for regression analysis are that they are independent, have zero mean, have a common variance σ^2, and follow a normal distribution (the normality assumption is required for performing parametric tests of significance). If the model is indeed appropriate for the data under analysis, it is reasonable to expect that the *observed* residuals, $\{e_i\}$, should exhibit properties not at odds with the stated assumptions. The basic strategy underlying the statistical procedure generally referred to as *residual analysis* is to assess the appropriateness of a model according to the behavior of the set of observed residuals. Our purpose here is to discuss methods for making such an assessment. The methodology discussed will be applicable to any context in which a model is fitted and a set of residuals is produced (e.g., as in analysis of variance and nonlinear regression, as well as multiple linear regression).

It is important to keep in mind the following characteristics of the n residuals e_1, $e_2, \ldots, e_n$ and functions thereof:

1. The mean of the $\{e_i\}$ is 0:

$$\bar{e} = \frac{1}{n} \sum_{i=1}^{n} e_i = 0$$

2. The estimate of population variance computed from the sample of the n residuals is defined to be

$$S^2 = \frac{1}{n - k - 1} \sum_{i=1}^{n} e_i^2$$

    which is exactly the residual mean square, $\text{SSE}/(n - k - 1)$; S^2 is an unbiased estimator of σ^2 if the model involving $k + 1$ parameters is correct.

3. The $\{e_i\}$ are not independent random variables. This is obvious from the fact that the $\{e_i\}$ sum to zero. In general, if the number of residuals (n) is large relative to the number of independent variables (k), this dependency effect can, for all practical purposes, be ignored in any analysis of the residuals (see Anscombe and Tukey, 1963, for a discussion of the effect of this dependency on graphical procedures involving residuals).

4. The quantity

$$z_i = \frac{e_i}{S}$$

    is called a *standardized residual*; it is often examined, rather than e_i, in a residual analysis. Note that, as with the $\{e_i\}$, the standardized residuals sum to 0 and hence are not independent. The standardized residuals have unit variance in the sense that

$$\frac{1}{n - k - 1} \sum_{i=1}^{n} z_i^2 = \frac{1}{n - k - 1} \sum_{i=1}^{n} \left(\frac{e_i}{S}\right)^2 = \frac{1}{S^2}\left(\frac{1}{n - k - 1} \sum_{i=1}^{n} e_i^2\right) = 1$$

5. The quantity

$$r_i = \frac{e_i}{S\sqrt{1 - h_i}} = \frac{z_i}{\sqrt{1 - h_i}}$$

    is called a *studentized residual*, so named because it approximately follows a

Student's t distribution with $n - k - 1$ degrees of freedom if the data follow the *HEIL GAUSS* assumptions (Chapter 8). The standard error of e_i is $S\sqrt{1 - h_i}$. The quantity h_i, the *leverage*,[2] is a measure of the importance of the ith observation in determining the model fit. Leverage values are such that $0 \le h_i \le 1$. Their role in helping diagnose regression problems will be treated in detail in Section 12-4. Studentized residuals have a mean near 0 (but not exactly 0), and a variance

$$\frac{1}{n - k - 1} \sum_{i=1}^{n} r_i^2$$

slightly larger than 1.

6. The quantity

$$r_{(-i)} = r_i \sqrt{\frac{S^2}{S_{(-i)}^2}} = \frac{e_i}{\sqrt{S_{(-i)}^2(1 - h_i)}} = r_i \sqrt{\frac{(n - k - 1) - 1}{(n - k - 1) - r_i^2}}$$

is called a *jackknife residual*. The quantity $S_{(-i)}^2$ is the residual variance computed with the ith observation deleted. Hence, the ith jackknife residual is standardized by a function of h_i and by a standard deviation based upon $n - 1$ of the observations (deleting the ith observation). In contrast, a standardized residual involves S, the standard deviation based on all n observations. Jackknife residuals have a mean near 0 and a variance

$$\frac{1}{(n - k - 1) - 1} \sum_{i=1}^{n} r_{(-i)}^2$$

slightly greater than 1. If the usual *HEIL GAUSS* assumptions are met, each jackknife residual exactly follows a t distribution with $(n - k - 1) - 1$ error degrees of freedom. If the standard regression assumptions are satisfied and approximately the same number of observations are made at all predictor values, then patterns in standardized, studentized, and jackknife residuals will look very similar. As potential problems arise, however, studentized residuals and especially jackknife residuals will make suspicious values more obvious to the data analyst. For example, if the ith observation lies far from the rest of the data, $S_{(-i)}$ will tend to be much smaller than S, which in turn will make $r_{(-i)}$ large in comparison to r_i. Thus $r_{(-i)}$ will tend to stand out more than r_i, thereby further highlighting the outlier. Also, large h_i-values for high leverage observations lead correspondingly to larger $r_{(-i)}$-values than r_i-values.

7. Studentized and jackknife residuals are distributed like t random variables under the usual *HEIL GAUSS* assumptions. Furthermore, as error degrees of freedom ($n - k - 1$ for studentized and $n - k - 2$ for jackknife) increase much above 30, the distributions of residuals are more and more closely approximated by a standard normal (mean 0, variance 1) distribution. This information is helpful in evaluating the size of observed residuals by appealing to properties of a standard

---

[2] The quantity h_i is the ith diagonal element of the $(n \times n)$ matrix $\mathbf{X(X'X)}^{-1}\mathbf{X'} = \mathbf{H}$, called the hat matrix, since $\hat{\mathbf{y}} = \mathbf{Hy}$, in which $\hat{\mathbf{y}} = (\hat{y}_1, \hat{y}_2, \ldots, \hat{y}_n)'$ and $\mathbf{y} = (y_1, y_2, \ldots, y_n)'$.

normal distribution. For example, no more than 5% of the residuals would be expected to exceed 1.96 in absolute value, if the residuals approximately represented a random sample from a $N(0, 1)$ distribution.

Residual Properties for the Weight Example

Using the WEIGHT, HEIGHT, and AGE data introduced above, let us assume that a model has been fitted with WEIGHT as the response and HEIGHT, (HEIGHT)^2, AGE and (AGE)^2 as predictors. We shall now analyze the residuals from that fitted model. Because the sample included 127 boys, there are 127 residuals. The 5 smallest and 5 largest residuals, $\{e_i\}$, are $\{-11.5, -10.8, -8.3, -8.1, -7.8\}$ and $\{13.0, 14.8, 15.6, 18.4, 45.2\}$. The corresponding studentized residuals, $\{r_i\}$, are $\{-1.74, -1.60, -1.22, -1.20, -1.19\}$ and $\{1.92, 2.30, 2.33, 2.70, 7.21\}$. The corresponding jackknife residuals, $\{r_{(-i)}\}$, are $\{-1.76, -1.61, -1.22, -1.21, -1.19\}$ and $\{1.94, 2.34, 2.38, 2.78, 9.48\}$. Obviously, the largest value seems suspiciously detached from the others.

With WEIGHT measured in kilograms, the residuals have variance 45.6, skewness 2.81, and kurtosis 15.33. We shall now describe these latter two characteristics. *Skewness* indicates the degree of asymmetry of a distribution. Just as the variance is the average squared deviation of observations about the mean, skewness is the average cubed deviation about the mean. To simplify comparisons between samples and to help account for estimation in small samples, skewness is usually computed as

$$sk(X) = \left(\frac{n}{n-2}\right)\left(\frac{1}{n-1}\right) \sum_{i=1}^{n} \left(\frac{X_i - \bar{X}}{S_X}\right)^3$$

The 2.81 value of skewness suggests that the normality assumption is questionable, since skewness is 0 for any symmetric distribution (such as a normal distribution). In addition, the positive value ($+2.81$) indicates that relatively more values are above the mean than below it; the sample values are thus said to be "positively skewed." A negative value for skewness indicates that relatively more values are below the mean than above it.

Kurtosis indicates the heaviness of the tails relative to the middle of the distribution. Kurtosis is the average of the fourth power of the deviations about the mean and is therefore always nonnegative. Standardized kurtosis may be computed as

$$Kur(X) = \left[\frac{n(n+1)}{(n-2)(n-3)}\right]\left(\frac{1}{n-1}\right) \sum_{i=1}^{n} \left(\frac{X_i - \bar{X}}{S_X}\right)^4$$

The term in brackets approaches 1.00 as n increases and helps to account for estimation based on a small sample. Since standardized kurtosis for a standard normal distribution is 3.0, this value is often subtracted from Kur(X). The resulting statistic can be as small as -3 for flat distributions with short tails, is 0 for a normal distribution, and is positive for heavy-tailed distributions. Thus, the positive kurtosis value in our example suggests a distribution with tails heavier than in a normal distribution. Finally, the reader should be cautioned that skewness and kurtosis statistics are highly variable in small samples and hence are often difficult to interpret. However, since we have over 100 observations in this example, these measures should be reasonably stable.

We have described four types of residuals for regression analysis. Which type is to be preferred? Since unstandardized and standardized residuals differ only by a multiplicative constant, they contain exactly the same information. For example, plots involving these two types of residuals would have exactly the same shape. Standardized residuals (as do studentized and jackknife residuals) have the advantage of being measured in a scale very similar to z scores. Consequently, the values of standardized, studentized, and jackknife residuals do not change if the Y variable is measured, for example, in inches rather than centimeters. Furthermore, one gains experience in interpretation that is easily transferred across sets of data. If no problems with outliers are present, $n - k - 1$ is not too small, and all X-values are roughly equally represented (approximately follow a uniform distribution), then standardized, studentized, and jackknife residuals are all essentially the same. As one strays from these conditions, then studentized residuals are preferred to standardized residuals, and jackknife residuals are in turn preferred over studentized residuals. Consequently, we shall emphasize the use of jackknife residuals in further discussions.

12-3-2 Graphical Analysis of Residuals

Often the most direct and revealing way to examine a set of residuals is to make a series of plots of the residuals. Two basic kinds of plots are useful: one-dimensional and two-dimensional displays. The former kind employs only the properties and relationships of the observed residuals among themselves. The latter kind considers the relationships of the residuals to other variables (such as the response and the predictors). In such a graphical analysis, a violation of a specific assumption (e.g., independence, model correctness, normality, or homogeneity of variance) will sometimes be much more evident from one type of plot than another. Here we discuss available types of plots and the interpretations that can be drawn from them.

One-Dimensional Displays

The simplest plots are one-dimensional displays. Three kinds of one-dimensional displays of residuals are most useful: (1) histograms, especially stem-and-leaf versions, (2) schematic plots, and (3) normal probability plots. Each will be described separately below.

We shall first consider our second example, an instrument calibration experiment in environmental engineering. The 17 values of the "concentration of a certain pollutant" (X) and the "instrument reading" (Y) are given in Table 12-1, along with the predicted values $\{\hat{Y}_i\}$, the residuals $\{e_i\}$, the standardized residuals $\{e_i/S\}$, the studentized residuals $\{r_i\}$, and the jackknife residuals $\{r_{(-i)}\}$. All are based on the least-squares straight line $\hat{Y} = 15.39 + 3.79X_i$, for which $S = 4.11$.

First we shall examine a histogram of the jackknife residuals, shown in Figure 12-4a. What should we look for in a figure of this type? Since we usually assume that $E_i N \sim (0, \sigma^2)$, we would expect (if the standard regression assumptions hold) the histogram of the $\{r_{(-i)}\}$ to look like one based on a random sample of 17 observations from a t distribution (with mean 0). Of course, as $n - k - 2$ increases much beyond 30, the distribution should reflect sampling from a standard normal distribution. For this example, with $n - k - 2 = 12$, one should see only slightly heavier tails than for a $N(0, 1)$ distribution. Otherwise, the picture should approximate the familiar bell-shaped curve.

For the calibration experiment, the residuals present a completely acceptable picture. As is customary for a histogram, the endpoints of the grouping intervals are indicated on the plot. For example, one observation, the largest, falls between 2.0 and 3.0, and six observations fall between -1.0 and 0.0. A frequency histogram can convey even more information if

FIGURE 12-4   Frequency histogram and stem-and-leaf
diagram of jackknife residuals for calibration experiment ($n = 17$)

```
  3 | *              2 | 7
  2 | *              1 | 4
  1 | *******        0 | 1124557
  0 | ******        -0 | 865542
 -1 | *             -1 | 3
 -2 | *             -2 | 8
 -3 |
```

(a) Histogram   (b) Stem-and-leaf diagram

it is converted to a stem-and-leaf diagram, as in Figure 12-4b. The lower (upper) endpoints of the grouping interval serve as *stems* for positive (negative) values. For example, the largest jackknife residual, rounded to two significant digits, is 2.7. Its stem is 2, the first significant digit. The *leaf* on the stem for the largest jackknife residual is 7, the second significant digit. The next largest value is read as 1.4. The third largest value is 0.7, then 0.5, 0.5, 0.4, and so on, down to the smallest value of -2.8. The figure may be compared with Table 12-1, which lists all residuals.

The second kind of useful one-dimensional plot is a *schematic* plot. Figure 12-5 presents a schematic plot of the jackknife residuals for the calibration data ($n = 17$). A schematic plot is based entirely on the order of the values in the sample. The quartiles are the most important order-based statistics for the schematic plot. The *first quartile*, or 25th percentile, is that value at or below which 25% of the data values lie; the *second quartile*, or

FIGURE 12-5   Schematic plot of jackknife residuals for
calibration experiment ($n = 17$)

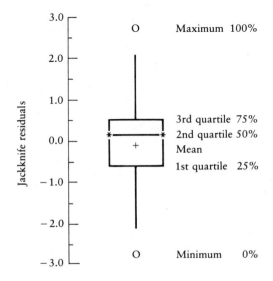

50th percentile (or median), is that value at or below which 50% of the data values lie; the *third quartile* is that value at or below which 75% of the data values lie. The *interquartile range* (IQR) is equal to the value of the third quartile minus the value of the first quartile. It is a measure of the spread of a distribution, like the variance. However, an important difference between the IQR and the variance is illustrated by the fact that although doubling the largest value in the sample would in general increase the variance dramatically, it would not change the IQR. For the jackknife residuals, the quartiles are approximately $-0.6, 0.1$, and 0.5, with an interquartile range of 1.1. For comparison, from Appendix Table A-1, it is easy to verify that for a $N(0, 1)$ distribution, the corresponding values are $-0.675, 0.0, +0.675$, and 1.35, which compare quite closely (even though $n = 17$).

A schematic plot is also sometimes known as a *box-and-whisker plot,* or simply *box plot,* due to its appearance. The box is outlined by three horizontal lines, which mark the values of the first quartile, the second quartile (the median, indicated by asterisks), and the third quartile (refer to Figure 12-5). The scale is determined by the units and range of the data. The mean is indicated by a $+$ on the backbone of the plot. Since symmetric distributions have equal mean and median, their schematic plots have the mean $+$ in the middle of the median line, indicated by $*---*$, giving $*-+-*$. Symmetry also manifests itself in equal distances from the median to the other quartiles. The whiskers (vertical lines) extend from the box as far as the data extend up and down, to a limit of 1.5 IQRs (in the vertical direction). An "O" at the end of a whisker indicates an outside value, which occurs in about 1 observation out of 20 for a Gaussian sample (i.e., a sample from a normal distribution). An * at the end of a whisker indicates a detached value, which occurs in about 1 observation out of 200 in a Gaussian sample. A value is *outside* if it is more than 1.5 IQRs beyond the box and *detached* if it is more than 3.0 IQRs beyond.

Referring to Figure 12-5, we see one positive outside value and one negative outside value. Recalling that we should expect slightly heavy tails and that jackknife residuals are sensitive to heavy tails, these two outsiders do not appear overly bothersome. Even so, it is still appropriate to examine such values more closely. With a little practice, one can extract a great deal of information about residuals from examining schematic plots. For more details about schematic plotting, see Tukey (1977).

Another graphical approach is to construct a normal or half-normal plot of the residuals (or their standardized, studentized, or jackknife counterparts) using normal probability paper. Figure 12-6 represents such a plot for the 17 jackknife residuals from the instrument calibration study. The plotting procedure first entails ordering the residuals from smallest to largest, followed by marking off the horizontal axis to include all the $r_{(-i)}$-values. The cumulative relative frequencies (i.e., i/n, with i indicating the ith ordered residual) up to each value are then plotted as ordinate values versus the $r_{(-i)}$-values as abscissae.[3] For example, the jackknife residual 1.45 is the second from largest ($i = 16$) of the 17 residuals, so we plot $16/17$ (or 0.94) as the ordinate value corresponding to 1.45.

An important property associated with the use of this particular type of graph is that the cumulative relative frequencies for a normal distribution plot as a straight line. The line for the $N(0, 1)$ distribution is drawn on the graph and can be used as a yardstick in assessing

---

[3] To avoid treating a cumulative frequency of 1 (see Figure 12-6) and to provide better statistical estimation, it is often recommended that $(i - 1/2)/n, i/(n + 1)$, or $(i - 3/8)/(n + 1/4)$ be plotted versus residual values, rather than i/n. Here, i denotes the ith ordered residual.

FIGURE 12-6   Normal probability plot of jackknife residuals for linear
model fitted to data from calibration experiment ($n = 17$)

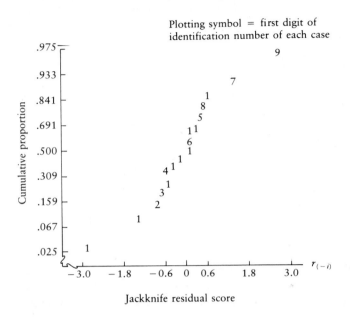

whether the scatter of points reflects any obvious deviation from normality. In our case, neither the stem-and-leaf diagram, the schematic plot, nor the normal probability plot suggests any blatant departures from the normality assumption, although such a statement is necessarily qualitative. More quantitative criteria will be described below in the section on significance tests about residuals.

Two-Dimensional Displays

As mentioned earlier, plotting the observed response values against the predictor values is a good way to check the validity of regression assumptions. With a single predictor, one can plot Y-values or residual values against X-values, for example, instrument readings versus pollutant levels for the calibration experiment. With many predictors, the situation is more complex. For example, assume that air temperature is known to be important in determining the response of the instrument. Depending on the observed pattern of combinations of pollutant and temperature level, a plot of instrument reading against pollutant level might erroneously suggest, for example, heterogeneity of variance. In general, it is advisable to plot residuals not only versus each of the predictor variables but also versus the predicted responses, as well as the observed responses versus the predicted responses.

The nature of predicted values helps explain the potential for misleading plots. For regression involving a single predictor,

$$\hat{Y} = \hat{\beta}_0 + \hat{\beta}_1 X$$

and for multiple regression,

$$\hat{Y} = \hat{\beta}_0 + \hat{\beta}_1 X_1 + \hat{\beta}_2 X_2 + \cdots + \hat{\beta}_k X_k$$

The predicted value, $\hat{Y}$, represents that linear combination of the X-variables that is most highly correlated with Y. For univariate regression,

$$r^2(Y, \hat{Y}) = r^2(Y, X)$$

This tells us that the strength of the (linear) relationship between Y and $\hat{Y}$ is the same as that between Y and X. For multiple regression, on the other hand,

$$r^2(Y, \hat{Y}) = R^2(Y|X_1, X_2, \ldots, X_k)$$

In general, however,

$$r^2(Y, \hat{Y}) \geq r^2(Y, X_j)$$

As a special case, for uncorrelated X-variables,

$$R^2(Y|X_1, X_2, \ldots, X_k) = r^2(Y, \hat{Y}) = r^2(Y, X_1) + r^2(Y, X_2) + \cdots + r^2(Y, X_k)$$

This helps demonstrate why the relationship between Y and any single X must be considered in light of all other X-variables. Unfortunately, even if all of the X-variables are mutually uncorrelated, consideration cannot be confined only to single-predictor plots, since observations may be *multivariate* outliers (i.e., outliers if the X-variables are considered together).

One valid way to plot observed responses against predictors is to use a partial regression plot for each predictor. One plots the response variable adjusted for $k - 1$ predictors against the predictor adjusted for the same $k - 1$ predictors. In particular, assume that the overall model of primary interest is

$$Y_i = \beta_0 + \beta_1 X_{i1} + \beta_2 X_{i2} + \cdots + \beta_k X_{ik} + E_i$$

To produce the partial regression plot for the kth predictor, first fit two models:

$$Y_i = \beta_0 + \beta_1 X_{i1} + \beta_2 X_{i2} + \cdots + \beta_{k-1} X_{i(k-1)} + E_i$$
$$X_{ik} = \alpha_0 + \alpha_1 X_{i1} + \alpha_2 X_{i2} + \cdots + \alpha_{k-1} X_{i(k-1)} + E_i$$

Then plot the residuals from the two models against each other, $Y_i - \hat{Y}_i$ versus $X_{ik} - \hat{X}_{ik}$, $i = 1, 2, \ldots, n$. The best-fitting (i.e., least-squares) line for these paired residual data will have an intercept of 0 and a slope equal to $\hat{\beta}_k$, the estimated regression coefficient based on fitting the original model. The reader should note the similarity of this process to the method of computing partial correlations presented in Chapter 10. The simple correlation between $Y_i - \hat{Y}_i$ and $X_{ik} - \hat{X}_{ik}$ is the sample multiple-partial correlation between Y and X_k, controlling for X_1 through X_{k-1}.

Of the possible two-dimensional plots, some of the more useful graphs for checking assumptions in multiple regression will involve plotting residuals (especially studentized or jackknife) versus predicted or predictor values. A few of the general patterns that can emerge from a plot of residuals versus predicted values are portrayed in Figure 12-7. Figure 12-7a illustrates the type of pattern to be expected if all basic assumptions hold. A horizontal band of points should be obtained with no hint of any systematic trends.

Figure 12-7b illustrates a systematic pattern to be expected from a departure from linearity, indicating a need for curvilinear terms in the regression model. Naturally, different types of model inappropriateness result in different residual patterns.

Figure 12-7c represents the pattern to be expected when the error variance increases directly with $\hat{Y}$. It is possible, of course, to encounter situations in which the error variance

FIGURE 12-7   Typical jackknife residual plots as a function of predicted value $\hat{Y}$ or time of data collection for hypothetical data

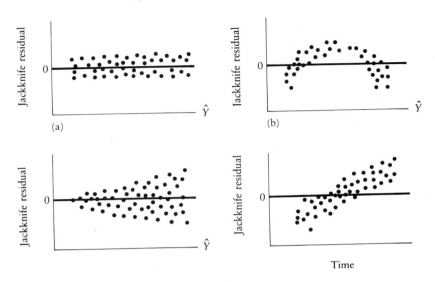

appears to be an even more complex function of the predicted value. In any case, transformations of the data are very often helpful in eliminating such heteroscedasticity of the variance (see Section 12-8-3). Also, it is helpful to take replicate observations at as many values of X as possible. This provides homogeneous (with respect to X) observation groups for which only variability due to "pure" error distinguishes the observations in a group. Observations associated with distinct predictor values vary, and their residuals vary, as a function of the *true* model. Without groups of replicates, inappropriateness of the fitted model (sometimes called *misspecification*) is difficult to distinguish from heterogeneity of error variance. Such confusion could lead to the wrong choice of corrective actions.

Figure 12-7d is a plot of the jackknife residuals versus time; a linear time-related effect is clearly present. Whenever variables not included in the regression model (e.g., the variable "time" when the data are collected in a time sequence) may have a significant effect (such as that reflected in Figure 12-7d by a strong positive correlation with the residuals), it is often informative to construct plots involving such variables.

For the calibration data, a plot of the jackknife residuals versus $\{\hat{Y}\}$ for the model $\hat{Y} = 15.39 + 3.79X$ is presented in Figure 12-8. One can immediately recognize a systematic pattern similar to that of Figure 12-7b. This suggests introducing a quadratic term into the model. If we do this, the resulting model is $\hat{Y} = 10.00 + 8.10X - 0.54X^2$, with $S = 2.84$. A plot of jackknife residuals versus $\{\hat{Y}\}$ for this model is given in Figure 12-9. Although Figure 12-9 reflects considerable variation (due in part to the small sample size), the pattern of points roughly reflects the horizontal band in Figure 12-7a.

Another way to examine the distribution of residuals is the following. Since the studentized and jackknife residuals are assumed to represent a sample from a distribution that is approximately standard normal, we would expect roughly 68% of these standardized residuals to lie in the interval $(-1.65, +1.65)$, about 95% to be contained in the interval $(-1.96, +1.96)$, and so on. (If $n - k - 1$ is small, with k = number of predictors, then 68% and 95%

FIGURE 12-8   Jackknife residuals as a function of the predicted values for calibration experiment: linear model

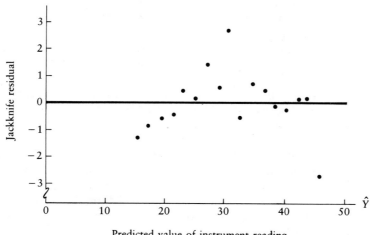

Predicted value of instrument reading

limits of the *t* distribution can be used, with $n - k - 1$ degrees of freedom for standardized and studentized, but $n - k - 2$ for jackknife.) For these data, only 2 (i.e., 12%) of the 17 jackknife residuals exceed 1.65 in absolute value. The same two jackknife residuals exceed the $n - k - 2 = 14$ degrees of freedom, two-tailed *t* critical value of 2.14 when $\alpha = .05$. Also, because the values of these two particular residuals are large (-2.82 and 2.71), the validity of the normality assumption might be questioned. Recall that the jackknife approach tends to highlight extreme residual values.

FIGURE 12-9   Jackknife residuals as a function of the predicted values for calibration experiment: quadratic model

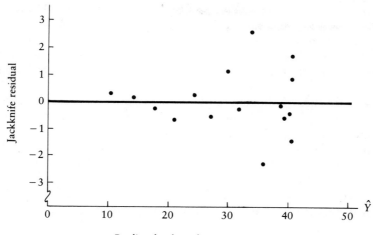

Predicted value of instrument reading

12-3-3 Significance Tests

More quantitative criteria for assessing the validity of the normality assumption can, of course, be based on the use of standard statistical testing procedures (e.g., the chi-square and Kolmogorov–Smirnov tests; see Stephens, 1974). An additional significance test of interest is the Shapiro–Wilks's (1965) test for normality, which is appropriate for small sample sizes, say, those less than 50. Since such tests are often discussed in a basic applied statistics course and are now easily conducted using standard computer programs, such discussion and computation will not be given here.

As a word of caution, the analysis of residuals with respect to departures from normality is generally difficult because the distribution of a set of residuals is affected by several factors. For example, the residuals may exhibit a nonnormal pattern because of an inappropriate regression model, nonhomogeneity of the variance, or even simply too few residuals. In this last case, there will not be a pattern of sufficient stability to permit one to make a valid statistical inference about the nature of the underlying probability distribution. Hence, it is good practice to gather evidence concerning each of the possible types of departures from the assumptions and then to examine all evidence before making specific indictments. Following this strategy, we shall defer judgment about violation of the normality assumption in our calibration example until we have examined the possibility of other violations.

As mentioned earlier, a graphical analysis of a set of residuals is necessarily somewhat subjective. However, as demonstrated in the previous section, a careful graphical approach involving the simultaneous evaluation of several different types of residual plots often reveals whatever anomalies exist in the data. Nevertheless, there are occasions when it is desirable to utilize statistical testing procedures to answer specific questions. We have already alluded to the fact that standard goodness-of-fit tests are available for examining the normality assumption. In fact, there are statistical tests available to examine each of the specific regression assumptions in question. For example, when a set of residuals is gathered in a time sequence, a nonparametric "runs test" (e.g., see Siegel, 1956, pp. 52–60) is frequently used to determine whether the time sequence of positive and negative residuals is unusual enough to be considered more than just a random occurrence.

An additional method for assessing the validity of the independence assumption is to use Durbin–Watson (1951) statistics, which test the null hypothesis of independence (no autocorrelation) over time. An *autocorrelation,* $r(Y_t, Y_{t-1})$, is the correlation between measurements taken at times t and $t - 1$. Assume, for example, that the data for our hypothetical instrument calibration experiment were collected in a particular order over time, such as one measurement per day. If the readings were slightly higher on each successive day (perhaps due to instrument drift), then one would expect a positive autocorrelation among the residuals.

Tests for variance nonhomogeneity could be based on a comparison, by means of F statistics, of sample variances calculated from replicate observations obtained at each of a series of values of an independent variable (e.g., see Bartlett, 1947). However, Bartlett's test has been criticized for being overly sensitive to deviations from normality. As an alternative, to assess the amount of variance nonhomogeneity in terms of a suspected monotonic relationship between the error variance and an independent variable, one can test the significance of the Spearman rank correlation (again, see Siegel, 1956, pp. 202–213) between the absolute value of the residual and the value of the independent variable.

12-4 Treating Outliers

An outlier among a set of residuals is one that is much larger than the rest in absolute value, perhaps lying as many as three or more standard deviations from the mean of the residuals. Clearly, the presence of such an extreme value can significantly affect the least-squares fitting of a model, and so it is important to determine if the analysis should be modified in some way (such as by deleting the observation in question). An outlier in the data may indicate special circumstances warranting further investigation (e.g., such as the presence of an unanticipated interaction effect). Therefore, we do not recommend immediately discarding the observation unless there is strong evidence that it resulted from a mistake (e.g., an error in data recording or some other cause independent of the process under study, like an obvious instrument malfunction). Statistical procedures for evaluating outliers are presented below. Further discussion of some of these methods can be found in papers by Anscombe (1960), Anscombe and Tukey (1963), and Stevens (1984). See Tukey (1977) for a thorough discussion of outliers and their detection.

12-4-1 Definitions

It is important to recognize differences among possible types of extreme values. As described above, an outlier is any rare or unusual observation appearing at one of the extremes of the data range. All regression observations, and hence outliers in particular, may be evaluated as to (1) reasonableness given knowledge of the variable, (2) response extremeness, and (3) predictor extremeness. The goal is to identify observations that are important in affecting either the choice of variables in the model or the accuracy of estimations of the regression coefficients and associated standard errors. We shall discuss these concepts briefly, then describe procedures for quantifying them.

If an observation has been identified as an outlier, it should be checked for *plausibility*. As discussed earlier, it is important that the data analyst be familiar with the basic characteristics of the data. Consider, as an example, body temperature as a response. The number 38.1 is plausible if the units are degrees Celsius and the subjects are humans. In contrast, if the units are degrees Fahrenheit, then 38.1 is an implausible value. More generally, one may classify any observation as being impossible, highly implausible, or plausible. It is then necessary to consider the importance of an observation in determining the choice of variables in the model, coefficient estimates, and associated statistics before deciding what, if any, action to take. Importance includes *leverage* and *influence*, concepts discussed in detail below.

Traditionally, outliers among observations (the set of values $Y, X_1, X_2, \ldots, X_k$) were detected by considering the residuals. It may also be helpful to consider the location of a particular response value (Y) relative to the values of the other responses. An extreme response value may deserve attention in order to determine its plausibility and importance.

A more difficult task is to evaluate the extremeness of predictor values. The set $\{X_{i1}, X_{i2}, \ldots, X_{ik}\}$ represents a point in a k-dimensional space. With two predictors, for example, the space is a plane and the points (X_{i1}, X_{i2}), $i = 1, 2, \ldots, n$, are easily plotted. With more than two predictors, two-dimensional plots are not sufficient to locate outliers in the predictor set. Fortunately, quantitative methods described below can help solve this problem.

12-4-2 Detecting Outliers

Methods for detecting outliers have received a great deal of attention recently (see, for example, Belsley, Kuh, and Welsch, 1980; Cook and Weisberg, 1982; and Stevens, 1984). We shall describe three regression diagnostic statistics for evaluating outliers: (1) jackknife residuals, (2) leverages, and (3) Cook's distance (a measure of influence). Other closely related statistics have been suggested. Our opinion is that *some* diagnostic statistics should be employed as part of any regression analysis. Most good regression computer programs available now provide a selection of diagnostics but require the user to request these diagnostics via options.

Jackknife residuals and their utility relative to standardized residuals were discussed above. To understand their use in marking outliers, consider the components of the formula for the ith jackknife residual, namely,

$$r_{(-i)} = \frac{e_i}{S_{(-i)} \sqrt{1 - h_i}}$$

Three quantities merit discussion: e_i, $S_{(-i)}$, and h_i. Since

$$e_i = Y_i - \hat{Y}_i$$

the numerator of $r_{(-i)}$ reflects the extremeness of the ith observed response, Y_i, relative to the predicted value, $\hat{Y}_i$.

Recall that the variance of the residuals is S^2, that is,

$$S^2 = \frac{1}{n - k - 1} \sum_{i=1}^{n} e_i^2$$

The S in the denominator of standardized residuals reflects the goodness-of-fit of the model and scales these residuals to have unit variance. In turn, $S_{(-i)}^2$ is a jackknifed estimate of the residual variance. Its use helps prevent an outlier from masking its own effect (since its contribution to S^2 is ignored). To complete the dissection of $r_{(-i)}$, note that the best estimates of the variances of $\hat{Y}_i$ and $e_i = Y_i - \hat{Y}_i$ are, respectively,

$$S_{\hat{Y}_i}^2 = S^2 h_i \tag{12.1}$$

and

$$S_{(Y_i - \hat{Y}_i)}^2 = S^2 (1 - h_i) \tag{12.2}$$

The leverage, h_i, is a measure of the geometric distance of the ith predictor point, $(X_{i1}, X_{i2}, \ldots, X_{ik})$, from the center point $(\bar{X}_1, \bar{X}_2, \ldots, \bar{X}_k)$ of the predictor (X) space. Consequently, an observation may be associated with an outlier jackknife residual if the observation is an outlier in the response variable (Y) or in the predictor space (of $X_1, X_2, \ldots, X_k$), or if it strongly affects the fit of the model (as reflected in the difference between S^2 and $S_{(-i)}^2$). Naturally, a combination of two or all three effects could yield a large outlier jackknife residual.

It is easy to test whether a particular jackknife residual is significantly different from 0 (i.e., has an extreme value not due to chance alone). Recall that a single jackknife residual exactly follows a t distribution with $n - k - 2$ degrees of freedom if the usual *HEIL GAUSS* assumptions are met. A corrected significance level must be used to account for the fact that n

tests, one for each observation, will be conducted. For example, if 50 subjects are observed and one chooses the .05 significance level, it is appropriate to require a P-value of $.05/50 = .001$. Since usually either extreme positive or negative values are of concern, one may use $.025/50 = .0005$ for a two-tailed test. Using .025 rather than the corrected value would lead, on average, to falsely declaring two or three observations as outliers. Using this corrected significance level is an application of the Bonferroni (α-splitting) correction.

Consider, for example, the instrument calibration data in Table 12-1. Table A-8 in Appendix A provides two-tailed critical values for a useful range of n (number of observations) and k (number of predictors). For $\alpha = .05$, $n = 15$, and $k = 1$, the critical value for studentized residuals is 3.58. Note that a different column is used for jackknife residuals, giving a critical value of 3.65 for $n = 15$, $k = 1$. Since $n = 17$ is not tabled, interpolation should be used. For jackknife residuals, this interpolation gives

$$3.65 + \frac{(3.54 - 3.65)(20 - 17)}{20 - 15} = 3.584$$

Since the largest jackknife residual (in absolute value) is 2.82, this value gives no cause for concern. Even the liberal $\alpha = .10$ $n = 15$ value of 3.27 is much larger than 2.82.

Another regression diagnostic is the set of *leverage values,* $\{h_i\}$, introduced in our discussion of studentized residuals. From equation (12.1), note that

$$h_i = \frac{S_{\hat{Y}_i}^2}{S^2}$$

For the univariate model

$$Y_i = \beta_0 + \beta_1 X_i + E_i$$

the leverage value for the ith observation takes the special form

$$h_i = \frac{1}{n} + \frac{(X_i - \bar{X})^2}{(n-1)S_X^2}$$

in which

$$S_X^2 = \frac{1}{n-1} \sum_{i=1}^{n} (X_i - \bar{X})^2$$

The main component of the formula for leverage is the squared standardized distance of the X_i-value from the center (mean) of the set of X-values, namely, the squared z score

$$z_{X_i}^2 = \left(\frac{X_i - \bar{X}}{S_X}\right)^2$$

For simple linear regression involving a single predictor, then, leverage indicates the extremeness of an observation in the range of X-values. More generally, for multiple regression, a leverage value measures the extremeness of an observation in the k-dimensional space of X_1, $X_2, \ldots, X_k$. For the special case in which all predictor variables $X_1, X_2, \ldots, X_k$ have mean 0 and are uncorrelated,

$$h_i = \frac{1}{n} + \sum_{j=1}^{k} \frac{X_{ij}^2}{(n-1)S_j^2}$$

in which

$$S_j^2 = \frac{1}{n-1} \sum_{i=1}^{n} X_{ij}^2$$

Leverages are related to an alternate regression diagnostic, *Mahalanobis distance* (see Stevens, 1984). Leverage measures the distance of an observation from the set of X-variable means, namely, from $\{\bar{X}_1, \bar{X}_2, \ldots, \bar{X}_k\}$.

Interpretation of the size and extremeness of leverage values is simplified by noting the following properties. First, in general,

$$0 \leq h_i \leq 1$$

However, $h_i \geq 1/n$ if the regression model under consideration contains an intercept (i.e., a β_0 term). A leverage h_i of 1 indicates that $\hat{Y}_i = Y_i$, which guarantees that the model is forced (levered) to fit the ith observed response exactly. Furthermore, with k being the number of predictors in the model $Y = \beta_0 + \beta_1 X_1 + \cdots + \beta_k X_k + E$,

$$\sum_{i=1}^{n} h_i = k + 1$$

Consequently, the average leverage value is

$$\bar{h} = \frac{k+1}{n}$$

Hoaglin and Welsch (1978) recommended examining more closely any observation for which $h_i > 2(k+1)/n$. If the model is correct, then the set of leverages for a sample has a frequency histogram that looks like a chi-square density. If the predictors follow a Gaussian distribution (i.e., each predictor has a normal distribution), then for any single h_i,

$$F_i = \frac{[h_i - (1/n)]/k}{(1 - h_i)/(n - k - 1)}$$

follows an F distribution with k and $n - k - 1$ degrees of freedom under the null hypothesis that the ith observation is a random sample of size 1 from the Gaussian predictor population. Thus, a test of extreme leverage can be conducted by comparing F_i to an F critical value of $F_{k,(n-k-1),1-\alpha/n}$. For example, with $\alpha = .05$ for a sample of 100 subjects, use a P-value of $.05/100 = .0005$ (using the Bonferroni correction) to avoid spuriously declaring too many leverages as outliers. Appendix Table A-9 provides leverage critical values corresponding to $F_{k,(n-k-1),1-\alpha/n}$ for a useful range of k, the number of predictors, and n, the sample size. Recall that the predictor values are assumed to be fixed values.[4] Even so, the F statistic just presented can be a rough indication of troublesome observations.

The regression diagnostic usually referred to as *Cook's distance* is a measure of the *influence* of an observation. Cook's distance measures how much the regression coefficients are changed by deleting the particular observation in question. For the special case of mean

---

[4] The reader should note that we have assumed throughout our treatment of multiple regression that the predictor values are all fixed, known constants. Violation of this assumption can seriously bias estimates of variability and hence distort tests of hypotheses.

0, equal variance, and uncorrelated predictors, Cook's distance for the ith observation d_i is proportional to

$$\sum_{j=0}^{k} [\hat{\beta}_j - \hat{\beta}_{j(-i)}]^2 = [\hat{\beta}_0 - \hat{\beta}_{0(-i)}]^2 + [\hat{\beta}_1 - \hat{\beta}_{1(-i)}]^2 + \cdots + [\hat{\beta}_k - \hat{\beta}_{k(-i)}]^2$$

Here $\hat{\beta}_j$ is the estimated regression coefficient using all the data, and $\hat{\beta}_{j(-i)}$ is the corresponding estimated regression coefficient with the ith observation deleted. If the predictors do not have mean 0 and equal variance and are not uncorrelated, then Cook's distance is proportional to a weighted sum of the terms $[\hat{\beta}_j - \hat{\beta}_{j(-i)}]^2$. For any set of data, Cook's distance for the ith observation, d_i, can be expressed in terms of leverages and studentized residuals as

$$d_i = \left(\frac{1}{k+1}\right) r_i^2 \left(\frac{h_i}{1-h_i}\right) = \frac{e_i^2 h_i}{(k+1)S^2(1-h_i)^2}$$

These formulae show the close relationship of d_i to the leverage h_i and the studentized residual r_i. Clearly, d_i may be large either because the observation is extreme in the predictor space (i.e., h_i near 1) or because it has a large studentized residual r_i.

Obviously, $d_i \geq 0$, and, in general, d_i may be arbitrarily large. Cook and Weisberg (1982) suggest that a $d_i > 1$ may deserve closer scrutiny. If the model is correct, d_i can be expected to be less than about 1.0. This is a simple approximation to a d_i-value that keeps the regression coefficient estimates based on deleting the ith observation within a 50% confidence region of the original estimates. The value of 1 is based on the fact that d_i is roughly like an F random variable with k and $n - k - 1$ degrees of freedom. The question being addressed is whether the set of regression coefficients is different with the ith subject deleted than it is with all subjects included. Appendix Table A-10 provides F values for 50% critical values for a range of k and n. If one wishes to consider the fact that n tests are being done (this occurs implicitly by considering the maximum d_i), the 50%/n critical value should be used. These are available from most F tables, such as Appendix Table A-4.

It is clear that some observation must be the most extreme in every sample. It would be silly to delete automatically this most extreme observation, or some cluster of extreme observations, based on just statistical testing procedures. The goal of regression diagnostics in evaluating outliers is to warn the data analyst to examine more closely such extreme observations. Scientific judgment is more important here than statistical tests, once influential observations have been flagged. One should be cautioned that deleting the most deviant observations will in all cases slightly improve, and sometimes substantially improve, the fit of the model. One must be careful not to data snoop simply in order to polish the fit of the model by discarding troublesome data points.

12-4-3 Diagnostics for the Weight Example

One-dimensional plots are provided for jackknife residuals (Figure 12-10), leverages (Figure 12-11), and Cook's distance statistics (Figure 12-12) for the WEIGHT regression example. If the usual assumptions are correct, residuals should mimic a bell-shaped curve, while stem-and-leaf diagrams for both leverage and Cook's distance should look roughly like chi-square densities. Hence the latter two diagrams will typically have a long tail to the right and many observations bunched near 0. Figure 12-10 (the jackknife residual plot) looks acceptably bell shaped, the important exception being the detached extreme value of 9.5. No significance test is needed to encourage examining this particular observation more closely.

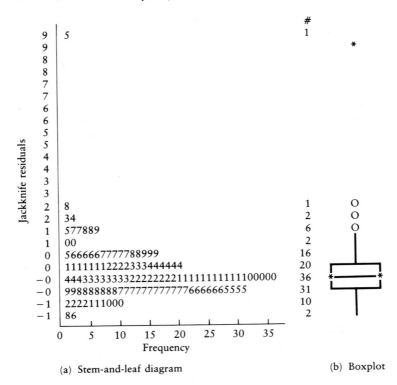

(a) Stem-and-leaf diagram   (b) Boxplot

FIGURE 12-11 Leverages for four-predictor model of children's weights; data from
Lewis and Taylor (1967; $n = 127$)

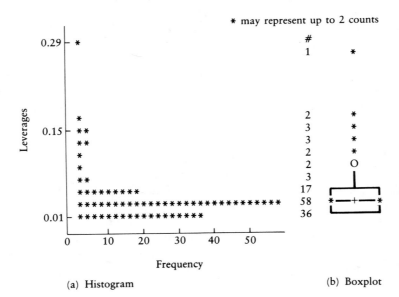

(a) Histogram   (b) Boxplot

FIGURE 12-12   Cook's d_i statistics for four-predictor model of children's weights; data from Lewis and Taylor (1967; $n = 127$)

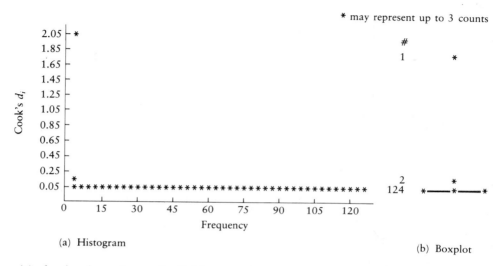

(a) Histogram   (b) Boxplot

The critical value from Appendix Table A-8 is between 4.06 and 4.15 for $\alpha = .01$, $n = 127$, $k = 4$. The leverage histogram also closely resembles the shape expected if all assumptions hold, an important exception being the detached value of .29. Since leverages are between 0 and 1, with a mean here of $(k + 1)/n = 5/127 = .039$, even the values in the neighborhood of .15 may deserve attention. (In fact, linear interpolation in Appendix Table A-9 gives a critical value of .151, for $\alpha = .01$, $n = 127$, $k = 4$.) Certainly the detached value of .29 is troublesome; as expected, it corresponds to the detached jackknife residual value of 9.5. Not surprisingly, the maximum influence statistic of 2.05 highlights the same observation. Appendix Table A-10 provides a comparison value of between .88 and .87, which is consistent with the rule of thumb that any $d_i > 1$ is potentially troublesome.

All regression diagnostics for the example indicate the need to consider one particular observation. In addition, a Kolmogorov-Smirnov test indicates that the residuals are not normally distributed. The particularly troublesome observation has variable values WEIGHT = 88.9 kg (195.6 lb), HEIGHT = 1.37 m (54 in.), and AGE = 14.5 yr. Although not impossible, this is an extremely unusual weight-and-height combination for a teenage boy. Consequently, for all subsequent analyses, this particular observation will be deleted.

It is informative to consider how the regression analysis changes as a result of deleting the observation in question. The R^2-value jumps from .52 (for $n = 127$ with the outlier) to .68 (for $n = 126$ without the outlier). The regression coefficients also change dramatically, as would be expected from the large Cook's distance statistic value of 2.05.

For an analysis excluding the outlier, one-dimensional plots are provided for jackknife residuals (Figure 12-13), leverage values (Figure 12-14), and influence values (Figure 12-15). The critical value ($n = 126$, $k = 4$) for jackknife residuals is approximately 3.64, with $\alpha = .05$. Hence, the maximum jackknife residual of 3.9 exceeds the .05 critical value but fails to exceed the .01 critical value of 4.06 for $n = 100$. Note that the stem-and-leaf diagram of the jackknife residuals is approximately bell shaped and that no obviously detached points are present.

FIGURE 12-13   Jackknife residuals of children's weights without outlier; data from Lewis and Taylor (1967; $n = 126$)

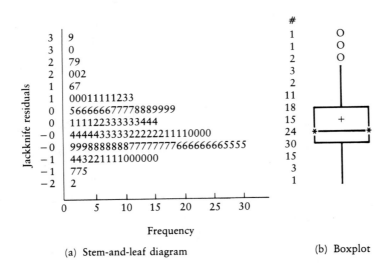

(a) Stem-and-leaf diagram   (b) Boxplot

FIGURE 12-14   Leverage values of children's weights without outlier; data from Lewis and Taylor (1967; $n = 126$)

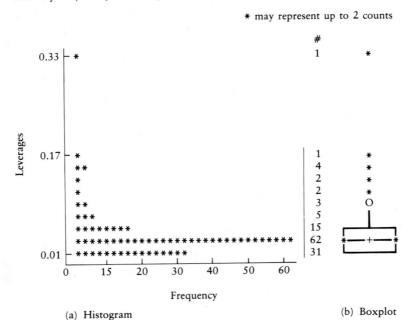

(a) Histogram   (b) Boxplot

FIGURE 12-15   Cook's d_i statistics of children's weights without outlier; data from Lewis and Taylor (1967; $n = 126$)

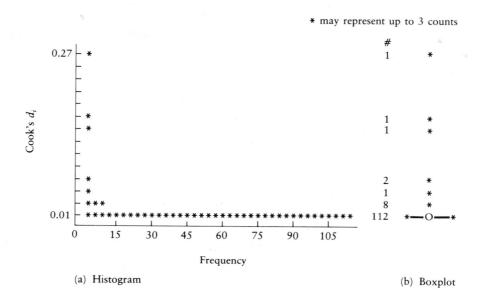

(a) Histogram   (b) Boxplot

The jackknife residual, leverage, and influence analyses each nominate different observations for attention. The jackknife residual of 3.9 corresponds to the observation WEIGHT = 43.1 kg, HEIGHT = 1.54 m, and AGE = 13.67 yr; the leverage of .33 corresponds to WEIGHT = 35.8 kg, HEIGHT = 1.28 m, and AGE = 12.25 yr; the influence statistic of 0.27 (Figure 12-15) corresponds to WEIGHT = 77.8 kg, HEIGHT = 1.71 m, and AGE = 17.1 yr. These WEIGHT–HEIGHT–AGE combinations are all plausible for teenage boys. The maximum leverage of .33 is much larger than a cutoff of about .17 (from Appendix Table A-9, $\alpha = .05$, $n = 126$, $k = 4$). The maximum influence statistic of 0.27 is below the suggested cutoff of 0.88 (from Appendix Table A-10) or the rule-of-thumb value 1. The small value of .27 for the maximum influence statistic reflects the fact that any single observation must be very unusual in order to change regression estimates noticeably if the assumptions are met and the error degrees-of-freedom value $n - k - 1$ is reasonably large. This is a very good reason to avoid small samples! In addition, these numerical findings support *not* deleting plausible observations, such as the one with a 3.9 jackknife residual or the one with .33 leverage, since the deletion would not influence one's conclusions.

In the next stage of analysis, the residuals of the model for predictor values surrounding these subjects would be considered. It is debatable whether the leverage and influence diagrams are sufficiently like chi-square distributions and whether detached values are present. In any case, $R^2 = .68$, and the distribution of residuals (Figure 12-13) indicates that the fitted model is a reasonable one. For more details concerning the interpretation of regression diagnostic statistics, see Weisberg (1980), Cook and Weisberg (1982), Belsley, Kuh, and Welsch (1980), Stevens (1984), and Hoaglin and Welsch (1978).

12-5 Collinearity

In this section, we describe certain features of a regression analysis that can result in numerical problems which in turn lead to inaccurate estimates of (1) regression coefficients, (2) variability, and (3) P-values. These problems can be loosely grouped into one of two types: collinearity and scaling (including centering). Collinearity concerns the relationship of the independent variables (predictors) to one another. Scaling pertains to the units in which the variables under study are measured and their means. Certain kinds of collinearity problems (e.g., those involving polynomial regression terms) can be expressed as scaling problems and therefore can be easily resolved. These concepts will be expanded below, and methods for combating these problems will also be outlined.

12-5-1 Collinearity with Two Predictors

The problems emanating from collinearity can be illustrated with simple two-variable regression examples. Consider fitting the model

$$Y_i = \beta_0 + \beta_1 X_{i1} + \beta_2 X_{i2} + E_i$$

to produce $\hat{\beta}_0$, $\hat{\beta}_1$, and $\hat{\beta}_2$. In general, it can be shown that

$$\hat{\beta}_j = c_j \left[\frac{1}{1 - r^2(X_1, X_2)} \right]$$

for $j = 1$ or $j = 2$. Here c_j is a value that depends on the data, and $r^2(X_1, X_2)$ is the squared correlation between X_1 and X_2. In turn,

$$\hat{\beta}_0 = \bar{Y} - \hat{\beta}_1 \bar{X}_1 - \hat{\beta}_2 \bar{X}_2$$

so that

$$\bar{Y} - \hat{\beta}_0 = \hat{\beta}_1 \bar{X}_1 + \hat{\beta}_2 \bar{X}_2 = \left[\frac{1}{1 - r^2(X_1, X_2)} \right] (c_1 \bar{X}_1 + c_2 \bar{X}_2)$$

Here $\bar{X}_1$ and $\bar{X}_2$ are the means of X_1 and X_2, respectively. These expressions tell us that $\hat{\beta}_1$, $\hat{\beta}_2$, and $\bar{Y} - \hat{\beta}_0$ are all proportional to $1/[1 - r^2(X_1, X_2)]$. More specifically, it is informative to consider fitting the model

$$Y_i = \theta_0 + \theta_1 X_{i1} + \theta_2 X_{i1} + E_i$$

In this case, a single variable, X_1, is included in the model twice. What are the estimates of the regression coefficients? Since $r^2(X_1, X_2) = r^2(X_1, X_1) = 1$, it follows that $1 - r^2(X_1, X_2) = 0$, and

$$\hat{\theta}_j = c_j' \left(\frac{1}{0} \right) = ?$$

From this we conclude that the estimates of the regression coefficients are indeterminate. Since the estimates of the variances of the regression coefficients are also proportional to the "inflation factor" $1/[1 - r^2(X_1, X_2)]$, they are also indeterminate. In turn, the P-values for tests about the coefficients are also indeterminate, since they involve the estimates just discussed. Note that the above model can be rewritten in the form

$$Y_i = \theta_0 + (\theta_1 + \theta_2) X_{i1} + E_i$$

From this, one can see that an infinite number of values of θ_1 and θ_2 add to the same coefficient value; thus an estimate of the coefficient of X_1, say $\overline{\theta_1 + \theta_2}$, does not permit a unique determination of the individual estimates $\hat{\theta}_1$ and $\hat{\theta}_2$. Consequently, θ_1 and θ_2 cannot be estimated separately. For example, if the actual estimate for the slope of X_1 is 12.5, then $\hat{\theta}_1 = 12.5$, $\hat{\theta}_2 = 0$ are possible components, as are $\hat{\theta}_1 = 0$, $\hat{\theta}_2 = 12.5$ and $\hat{\theta}_1 = 6$, $\hat{\theta}_2 = 6.5$, and so on. In this extreme example, one of the predictor variables is a *perfect linear combination* of the other, namely

$$X_1 = \alpha + \beta X_2 = 0 + (1)X_2 = X_2$$

Geometrically the points (X_{i1}, X_{i2}) all fall on the straight line $X_1 = X_2$, hence the term *collinear*.

The following data provide a slightly more general example:

X_1	X_2
0	3
1	5
3	9
4	11
7	17

Plotting X_2 against X_1 gives a very simple picture, as shown in Figure 12-16. From the plot, it is obvious that all of the data points (X_{i1}, X_{i2}) fall exactly on the straight line $X_2 = 3 + 2X_1$. This demonstrates that X_1 and X_2 are exactly collinear. Also, $r^2(X_1, X_2) = 1$ directly measures this perfect collinearity. Note that the collinearity issue involves only the predictor variables and does not depend on the relationship between the response and any of the predictors. As we shall see, as $r^2(X_1, X_2)$ decreases, the collinearity problem between X_1 and X_2 becomes less severe, the ideal situation occurring if X_1 and X_2 are uncorrelated.

FIGURE 12-16   Perfectly collinear set of pairs of variable values

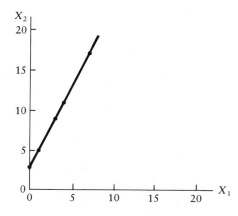

As another example, consider fitting the quadratic model

$$Y_i = \beta_0 + \beta_1 X_i + \beta_2 X_i^2 + E_i$$

to the engineering calibration data. In this case, the size of $r^2(X_1, X_2) = r^2(X, X^2)$ is a measure of the degree of collinearity between X and X^2. The mean of the X-values (listed in Table 12-1) is 4.0. Table 12-2 displays the effect of subtracting various constants from all of the X-values. The new values are computed as $X_i - a$ for various values of a between -4 and 4.1, including $a = 0$ (the original data). The mean of $X - a$ is shifted from the mean of X by a, and the variance of $X - a$ is the same as that of X. Table 12-2 also gives the mean and standard deviation of $(X - a)^2$ and the covariance (Cov) and correlation (r) between $X - a$ and $(X - a)^2$. The final column shows the associated value of the inflation factor $1/(1 - r^2)$. The squared correlation (r^2) ranges from .00 to .98, and the inflation factor from 1 to approximately 50. The only change to the data is a constant shift in the X-values. Situations depicted across the rows range from near collinearity to no collinearity, depending on the value of a. The process of subtracting a constant from a variable is a form of scaling. In the example, if $a = \bar{X} = 4.0$, it is called *centering*. As can be seen, centering in polynomial regression can reduce collinearity substantially. More will be said about this later in this chapter and more so in Chapter 13.

In general, for a two-variable model, if $r^2(X_1, X_2)$ is nearly 1.0, then a near collinearity is present. Although the coefficient estimates can be computed, they are very unstable. In particular, this instability is reflected in large estimates of the coefficient variances, since such variance estimates are proportional to

$$\frac{1}{1 - r^2(X_1, X_2)}$$

As $r^2(X_1, X_2)$ gets closer to 1.0, this factor becomes large, thereby inflating the estimated variances of the regression coefficients. Later we shall see that this factor is a special case of a variance inflation factor.

TABLE 12-2   Effects of subtracting a constant on $r[X^*, (X^*)^2]$ for calibration example ($n = 17$), $X^* = X - a$

	Mean		Standard Deviation				
a	X^*	$(X^*)^2$	X^*	$(X^*)^2$	$\mathrm{Cov}[X^*, (X^*)^2]$	$r[X^*, (X^*)^2]$	$\{1 - r^2[X^*, (X^*)^2]\}^{-1}$
-4.0	8.0	70.00	2.52	40.77	102.00	.99	50.25
0.0	4.0	22.00	2.52	20.93	51.00	.96	12.76
1.0	3.0	15.00	2.52	16.12	38.25	.94	8.59
2.0	2.0	10.00	2.52	11.50	25.50	.88	4.43
3.0	1.0	7.00	2.52	7.47	12.75	.68	1.86
3.5	0.5	6.25	2.52	6.05	6.375	.42	1.21
3.7	0.3	6.09	2.52	5.71	3.825	.26	1.07
3.9	0.1	6.01	2.52	5.52	1.275	.09	1.01
4.0	0.0	6.00	2.52	5.50	0	.00	1.00
4.1	-0.1	6.01	2.52	5.52	-1.275	$-.09$	1.01

12-5-2 Collinearity Concepts

In this section, we shall generalize the discussion of the last section to treat any number of predictors. In addition, we shall describe methods for quantifying the degree of collinearity in fitting a regression model to a particular set of data.

As has been seen in the examples just discussed, collinearity concerns relationships among the predictor variables and does not directly involve the response variable. As such, one informative way to examine collinearity is to consider what happens if each predictor variable is the response variable in a multiple regression model in which the independent variables are all of the remaining predictors. For k predictors, then, there would be k such models. For example, with the four-predictor model

$$Y_i = \beta_0 + \beta_1 X_{i1} + \beta_2 X_{i2} + \beta_3 X_{i3} + \beta_4 X_{i4} + E_i$$

four models would be fitted as follows:

$$X_{i1} = \alpha_{01} \qquad\quad + \alpha_{21} X_{i2} + \alpha_{31} X_{i3} + \alpha_{41} X_{i4} + E_i$$
$$X_{i2} = \alpha_{02} + \alpha_{12} X_{i1} \qquad\quad + \alpha_{32} X_{i3} + \alpha_{42} X_{i4} + E_i$$
$$X_{i3} = \alpha_{03} + \alpha_{13} X_{i1} + \alpha_{23} X_{i2} \qquad\quad + \alpha_{43} X_{i4} + E_i$$
$$X_{i4} = \alpha_{04} + \alpha_{14} X_{i1} + \alpha_{24} X_{i2} + \alpha_{34} X_{i3} \qquad\quad + E_i$$

To assess collinearity, the associated R^2-values based on fitting these four models are needed, namely, $R^2(X_1|X_2, X_3, X_4)$, $R^2(X_2|X_1, X_3, X_4)$, $R^2(X_3|X_1, X_2, X_4)$, and $R^2(X_4|X_1, X_2, X_3)$. If any of these multiple R^2-values equals 1.0, then a perfect collinearity is said to exist among the set of predictors. The term *collinearity* is used to indicate that one of the predictors is an *exact* linear combination of the others. As described in earlier examples, in the special case $k = 2$, perfect collinearity means that X_1 is a straight-line function of X_2, say $X_1 = \alpha + \beta X_2$, so that the points (X_{i1}, X_{i2}) all fall on the line (hence the term *collinearity*).

Consider, for example, predicting college grade point average (CGPA) using high school grade point average (HGPA) and College Board scores on the mathematics (MATH) and verbal (VERB) tests. The model is

$$\text{CGPA}_i = \beta_0 + \beta_1 \text{HGPA}_i + \beta_2 \text{MATH}_i + \beta_3 \text{VERB}_i + E_i$$

Now imagine trying to improve prediction by adding the total (combined) board scores (TOT = MATH + VERB) to create a new model:

$$\text{CGPA}_i = \alpha_0 + \alpha_1 \text{HGPA}_i + \alpha_2 \text{MATH}_i + \alpha_3 \text{VERB}_i + \alpha_4 \text{TOT}_i + E_i$$

This model has a perfect collinearity, which means that the parameters in the model cannot be estimated uniquely. To see this, begin by rewriting the model as

$$\text{CGPA}_i = \alpha_0 + \alpha_1 \text{HGPA}_i + \alpha_2 \text{MATH}_i + \alpha_3 \text{VERB}_i + \alpha_4 (\text{MATH}_i + \text{VERB}_i) + E_i$$

It then follows that

$$\text{CGPA}_i = \alpha_0 + \alpha_1 \text{HGPA}_i + (\alpha_2 + \alpha_4)\text{MATH}_i + (\alpha_3 + \alpha_4)\text{VERB}_i + E_i$$

With this version of the model, consider choosing $\alpha_4 = 0$. Then $\alpha_2 = \beta_2$ and $\alpha_3 = \beta_3$ give the correct original model. Next, choose $\alpha_4 = 3$. Then $\alpha_2 = \beta_2 - 3$ and $\alpha_3 = \beta_3 - 3$ also gives the correct original model. In fact, this demonstrates that, for any choice of α_4, we can choose α_2 and α_3 to provide the correct model. Since the α_4 parameter is actually not needed (it could best be set equal to 0), a model with a perfect collinearity is sometimes said to be *overparameterized*.

An important related concept is that of *near* collinearity, which arises if the multiple R^2 of one predictor with the remaining predictors is nearly 1. We saw an example of this in considering the squared correlation, $r^2(X, X^2)$, between X and X^2 in the preceding section (see Table 12-2 when $a = 0$). For a more general model involving k predictors, say,

$$Y_i = \beta_0 + \beta_1 X_{i1} + \beta_2 X_{i2} + \cdots + \beta_k X_{ik} + E_i$$

the multiple R^2-value of interest for the first predictor is $R^2(X_1|X_2, X_3, \ldots, X_k)$, the multiple R^2-value of interest for the second predictor is $R^2(X_2|X_1, X_3, \ldots, X_k)$, and so on. These quantities are generalizations of the statistic $r^2(X_1, X_2)$ for a $k = 2$ variable model. For convenience, we shall denote by R_j^2 the squared multiple correlation based on regressing X_j on the remaining $k - 1$ predictors.

The *variance inflation factor* (VIF) is often used to measure collinearity in a multiple regression analysis. It may be computed as

$$\text{VIF}_j = \frac{1}{1 - R_j^2} \qquad j = 1, 2, \ldots, k$$

The quantity VIF_j is clearly a generalization of the variance inflation factor for a two-predictor model, $1/[1 - r^2(X_1, X_2)]$. Clearly, $\text{VIF}_j \geq 0$. As for the two-variable case, the estimates of the variances for the regression coefficients are proportional to the VIFs, namely,

$$S_{\hat{\beta}_j}^2 = c_j^* (\text{VIF}_j) \qquad j = 1, 2, \ldots, k$$

Clearly, this expression suggests that the larger the value of VIF_j, the more troublesome is the variable X_j. A rule of thumb for evaluating VIFs is to be concerned with any value larger than 10.0. For VIF_j, this corresponds to $R_j^2 > .90$ or, equivalently, $R_j > .95$. Some people prefer to consider

$$\text{tolerance}_j = \frac{1}{\text{VIF}_j} = 1 - R_j^2$$

The choice among R_j^2, $1 - R_j^2$, or VIF_j is a matter of personal preference since they all contain exactly the same information. As R_j^2 goes to 1.0, $1 - R_j^2$ (the tolerance) goes to 0 and VIF_j goes to infinity.

Because of its special nature, the intercept requires separate treatment in evaluating collinearity. For the general model involving k predictors,

$$Y_i = \beta_0 + \beta_1 X_{i1} + \beta_2 X_{i2} + \cdots + \beta_k X_{ik} + E_i$$

we find regression coefficient estimates $\hat{\beta}_0, \hat{\beta}_1, \ldots, \hat{\beta}_k$. The intercept estimate can be expressed simply as

$$\hat{\beta}_0 = \bar{Y} - (\hat{\beta}_1 \bar{X}_1 + \hat{\beta}_2 \bar{X}_2 + \cdots + \hat{\beta}_k \bar{X}_k) = \bar{Y} - \sum_{j=1}^{k} \hat{\beta}_j \bar{X}_j$$

Here $\bar{Y} = \sum_{i=1}^{n} Y_i/n$ is the mean of the response values, and $\bar{X}_j = \sum_{i=1}^{n} X_{ij}/n$ is the mean of the values for predictor j. From this, it can be deduced that the estimated intercept is affected by the VIF$_j$'s, $j = 1, 2, \ldots, k$, since it is a function of the $\hat{\beta}_j$'s. The problem disappears if the means of all X_j's are 0 (e.g., if the predictor data are centered). In that case, $\bar{Y}$ is the estimated intercept.

In general, even if the predictor data are not centered, a variance inflation factor for $\hat{\beta}_0$, VIF_0, can be defined and is interpreted in the same way as VIF_j. First, define

$$\text{VIF}_0 = \frac{1}{1 - R_0^2},$$

Here R_0^2 is the generalized squared multiple correlation for the regression model

$$I_i = \alpha_1 X_{i1} + \alpha_2 X_{i2} + \cdots + \alpha_k X_{ik} + E_i$$

in which I_i is identically 1 (which may be thought of as the score for the intercept variable). Note that no intercept is included in this model (as a predictor), which is why R_0^2 is called a *generalized* squared correlation. As with the VIF_j's, $\text{VIF}_0 \geq 0$, and

$$S_{\hat{\beta}_0}^2 = c_0^*(\text{VIF}_0)$$

Hence, the interpretation for VIF_0 is the same as for VIF_j, $j = 1, 2, \ldots, k$.

Controversy surrounds the treatment of the intercept in regression diagnostics. For some, it is simply another predictor; for others, it should be eliminated from discussion. We take a middle position, arguing that the model and data at hand determine the role of the intercept. (See, for example, the discussions following Belsley, 1984.) This leads us to discuss diagnostics both with and without the intercept included, which corresponds to the cases with and without centering of the predictors and response variables.

The reader should be cautioned that the presence of collinearity or (more typically) near collinearity presents computational difficulties in calculating numerically reliable estimates of the R_j^2-values, tolerances, and the VIF_j's using standard regression procedures. This apparent impasse can be solved in at least three ways. The first way is to use computational algorithms that detect collinearity problems as they arise in the midst of the calculations. A discussion of such algorithms is beyond the scope of this text.

A second way by which the impasse may be avoided is to scale the data appropriately. By *scaling*, we mean the choice of measurement unit (e.g., degrees Celsius versus degrees Fahrenheit) and the choice of measurement origin (e.g., degrees Celsius versus kelvin). We shall consider only linear changes in scale, such as

$$X_1 = \alpha + \beta X_2$$
$$C = \frac{5}{9}(F - 32) = -32\left(\frac{5}{9}\right) + \left(\frac{5}{9}\right)(F)$$

or

$$K = (-273) + (1)(C)$$

where C, F, and K are temperatures in degrees Celsius, degrees Fahrenheit, and kelvin, respectively.

Often scaling refers just to multiplying by a constant rather than also adding a constant. An example of scaling is converting from feet to inches. One important case of adding a constant, a form of scaling, is *centering*. A set of scores $\{X_{ij}\}$ is centered by subtracting the mean $\bar{X}_j$ of the scores from each individual score, giving

$$X_{ij}^* = X_{ij} - \bar{X}_j$$

in which $\bar{X}_j = \sum_{i=1}^n X_{ij}/n$.

Computing standardized scores (z scores) is a closely related method of scaling; in particular, the standardized score corresponding to X_{ij} is

$$z_{ij} = \frac{X_{ij} - \bar{X}_j}{S_j}$$

in which $S_j^2 = \sum_{i=1}^n (X_{ij} - \bar{X}_j)^2/(n-1)$. Centered and standardized scores have mean 0 since $\sum_{i=1}^n X_{ij}^* = \sum_{i=1}^n z_{ij} = 0$. Also note that $\sum_{i=1}^n z_{ij}^2/(n-1) = 1$, so that the set $\{z_{ij}\}$ of standardized scores has a variance equal to 1. We shall delay the discussion of detecting and fixing scaling problems to later in this chapter. For now, we need the concept of scaling in order to understand the following discussion of other diagnostic statistics.

A third way to avoid the impasse created by collinearity and near collinearity is to use alternate computational methods to diagnose collinearity. In particular, a very popular method for characterizing near and/or exact collinearities among the predictors involves computing the *eigenvalues* of the predictor variable *correlation matrix*. The eigenvalues are connected with the *principal component analysis* of the predictors. This technique is discussed in more detail in Chapter 24. The principal components of the predictors are a set of new variables that are linear combinations of the original predictors. These components have the very special properties that (1) they are not correlated with each other and (2) each, in turn, has maximum variance, given that all are mutually uncorrelated. The principal components provide idealized predictor variables that still retain all of the same information as the original variables. The variances of these components (the new variables) are called *eigenvalues*. The larger the eigenvalue, the more important is the associated principal component in representing the information in the predictors. *As an eigenvalue approaches zero, the presence of a near collinearity among the original predictors is indicated*. The presence of an eigenvalue that is exactly 0 means that a perfect linear dependency (i.e., an exact collinearity) exists among the predictors.

If a set of k predictor variables does *not* involve an *exact* collinearity, then k principal components are needed to reproduce exactly all of the information contained in the original variables. If one of the predictors is a perfect linear combination of the others, then only $k - 1$ principal components are needed to provide all of the information in the original variables. *The number of zero (or near-zero) eigenvalues is the number of collinearities (or near collinearities) among the predictors*. Even a single eigenvalue near 0 presents a serious problem that must be resolved.

In using the eigenvalues to determine the presence of near collinearity, three kinds of statistics are usually employed: the *condition index* (CI), the *condition number* (CN), and the *variance proportions*.

Consider again the general linear model

$$Y_i = \beta_0 + \beta_1 X_{i1} + \beta_2 X_{i2} + \cdots + \beta_k X_{ik} + E_i$$

Often collinearity is assessed by considering only the k parameters β_1 through β_k and ignoring the intercept β_0. This is accomplished by centering the response and predictor variables. This corresponds to fitting the model

$$Y_i - \bar{Y} = \beta_1(X_{i1} - \bar{X}_1) + \beta_2(X_{i2} - \bar{X}_2) + \cdots + \beta_k(X_{ik} - \bar{X}_k) + E_i$$

Noting that $\hat{\beta}_0 = \bar{Y} - \sum_{j=1}^k \hat{\beta}_j \bar{X}_j$ in the original model, we can see that centering the predictors and the response forces the estimated intercept in that centered model to be 0. Hence, it can be dropped from that model.

It is also common to assess collinearity after centering and standardizing both the predictors and the response. This leads to the standardized model

$$\frac{Y_i - \bar{Y}}{S_Y} = \beta_1^* \frac{(X_{i1} - \bar{X}_1)}{S_1} + \beta_2^* \frac{(X_{i2} - \bar{X}_2)}{S_2} + \cdots + \beta_k^* \frac{(X_{ik} - \bar{X}_k)}{S_k} + E_i^*$$

The coefficients for this standardized model are often called the *standardized regression coefficients*; in particular,

$$\beta_j^* = \beta_j \left(\frac{S_j}{S_Y}\right) \qquad j = 1, 2, \ldots, k$$

Table 12-3 summarizes an eigenanalysis of the predictor correlation matrix $\mathbf{R}_{xx}$ for a hypothetical four-predictor ($k = 4$) standardized regression model.[5] With $k = 4$ predictors, four eigenvalues (denoted by λ's) exist. Later we shall include the intercept in the analysis and have $k + 1$ eigenvalues. In the present case, $\lambda_1 = 2.0$, $\lambda_2 = 1.0$, $\lambda_3 = 0.6$, and $\lambda_4 = 0.4$. (Note that the sum of the eigenvalues for a correlation matrix involving k predictors is *always* equal to k.) It is customary to list the eigenvalues from largest (λ_1) to smallest (λ_k). A condition index can be computed for each eigenvalue as

$$\text{CI}_j = \sqrt{\lambda_1/\lambda_j}$$

In particular, CI_3 for this example is $\sqrt{2.0/0.6} = 1.87$. The first (largest) eigenvalue always has an associated condition index (CI_1) of 1.0. The largest CI_j always involves the largest (λ_1) and smallest (λ_k) eigenvalues. It is called the *condition number* and is given by the formula

$$\text{CN} = \sqrt{\lambda_1/\lambda_k}$$

For this example, $\text{CN} = \sqrt{2.0/0.4} = 2.24$. Since eigenvalues are variances, CI_j and CN are ratios of standard deviations (of principal components, the idealized predictors). As with VIFs, the CI_j's and CN are nonnegative, and larger values suggest potential near collinearity. Belsley, Kuh, and Welsch (1980) recommended interpreting a CN of 30 or more as reflecting moderate to severe collinearity, worthy of further investigation. Of course, such a CN may be associated with two or more CI_j's greater than or equal to 30.

A variance proportion indicates, for each predictor, the proportion of total variance of its estimated regression coefficient associated with a particular principal component. The variance proportions suggest collinearity problems if more than one predictor has a high

TABLE 12-3   Eigenanalysis of the predictor correlation matrix for hypothetical four-variable model

Variable	Eigenvalue	Condition Index	Variance Proportions X_1	X_2	X_3	X_4
1	2.0	1.00	.09	.11	.08	.15
2	1.0	1.41	.32	.10	.25	.07
3	0.6	1.87	.40	.52	.12	.13
4	0.4	2.24	.19	.27	.55	.65

[5] The predictor correlation matrix $\mathbf{R}_{xx}$ for this standardized model satisfies the relationship $(n - 1)\mathbf{R}_{xx} = \mathbf{X}_s'\mathbf{X}_s$, where $\mathbf{X}_s'\mathbf{X}_s$ is the cross-products matrix. Hence, each eigenvalue of $\mathbf{X}_s'\mathbf{X}_s$ is $(n - 1)$ times the corresponding eigenvalue of $\mathbf{R}_{xx}$.

variance proportion (loads highly) on a principal component having a high condition index. Two or more proportions of at least .5 for such a component suggest a problem. One should definitely be concerned with two or more loadings greater than .9 on a component with a large condition index.

Table 12-3 includes variance proportions for the hypothetical four-predictor model under consideration. The entries represent a typical pattern of condition indices and proportions for a regression analysis with no major collinearity problems. Notice that each column sums to a total proportion of 1.00 since each estimated regression coefficient has its own (total) variance partitioned among the four components. The last row involves two loadings greater than .5, corresponding to the smallest eigenvalue. Since the smallest eigenvalue has a CI of only 2.24 (far from the suggested warning level of 10.0 for mild to moderate collinearity), the proportion pattern does not indicate any major problem.

The intercept can play an important role in collinearity. In defining regression models, it is standard practice to include the intercept term β_0. The intercept is a regression coefficient for a variable having the constant value of 1. Consequently, any variable with near-zero variance will be nearly a constant multiple of the intercept and hence nearly collinear with it. This problem can arise spuriously when variables are improperly scaled (discussed below). To evaluate this possibility, the eigenanalysis discussed above can be modified to include the intercept by basing it on the scaled cross-products matrix.[6] The eigenanalysis including the intercept may suggest that this constant is nearly collinear with one or more predictors.

Centering may help to decrease collinearity. From a purely theoretical perspective, statisticians disagree as to when centering helps regression calculations (for example, see Belsley, 1984; Smith and Campbell, 1980; also see the related comments in the same issues). For actual computations, centering can increase numerical accuracy in many situations. Polynomial regression (Chapter 13) is one situation in which centering is recommended. In the example to follow, we shall numerically illustrate the effects of centering on procedures for diagnosing collinearity.

12-5-3 An Example of Near Collinearity

The boys' WEIGHT data will be used to illustrate the use of collinearity diagnostics. First, consider the correlations among the predictor variables as presented in Table 12-4. Since two correlations among predictors are greater than .99, one should expect trouble. These high correlations arise because $(AGE)^2$ is almost perfectly collinear with AGE, and

TABLE 12-4   Correlations for weight example without centering
($n = 126$)

	HEIGHT	$(HEIGHT)^2$	AGE	$(AGE)^2$	WEIGHT
HEIGHT	1	.999	.774	.766	.790
$(HEIGHT)^2$		1	.777	.770	.794
AGE			1	.998	.699
$(AGE)^2$				1	.704

[6] The eigenanalysis would then be based on the $(k + 1) \times (k + 1)$ cross-products matrix $\mathbf{X'X}$ (see footnote 2), suitably scaled to have 1's on the diagonal, rather than on the $k \times k$ correlation matrix $\mathbf{R}_{xx}$. The sum of the eigenvalues for this scaled cross-products matrix is equal to $(k + 1)$.

TABLE 12-5   Eigenanalysis of the scaled cross-products matrix for the model where WEIGHT is predicted from HEIGHT, (HEIGHT)^2, AGE, and (AGE)^2 without centering ($n = 126$)

Variable	Eigenvalue	Condition Index	Variance Proportions				
			Intercept	HEIGHT	(HEIGHT)^2	AGE	(AGE)^2
1	4.969	1.0	.00	.00	.00	.00	.00
2	0.025	14.1	.00	.00	.00	.00	.00
3	0.006	29.2	.00	.00	.00	.00	.00
4	2×10^{-5}	490.2	.27	.00	.00	.84	.83
5	5×10^{-6}	1,028.0	.73	1.00	1.00	.16	.17

(HEIGHT)^2 is essentially collinear with HEIGHT. In general, any between-predictor correlation above .9 (in absolute value) merits further attention.

Table 12-5 indicates that serious collinearity problems exist. This can be seen in several ways. First, Belsley, Kuh, and Welsch (1980) suggested that any CI greater than or equal to 30 suggests the presence of moderate to severe collinearity. In this example, two condition indices, CI_4 and CI_5, are much greater than 30. Second, the smallest eigenvalue (5×10^{-6}) is associated with maximum variance proportions (i.e., loadings) for HEIGHT and (HEIGHT)^2. The fact that the intercept loads highly on a problem component (i.e., one with an eigenvalue near 0) would suggest to an experienced analyst that centering may help to reduce the problem.

To evaluate the utility of centering for this example, the predictor variables were transformed in the following way:

HEIGHT$_*$ = HEIGHT − 1.577

(HEIGHT$_*$)^2 = (HEIGHT − 1.577)^2

AGE$_*$ = AGE − 13.67

(AGE$_*$)^2 = (AGE − 13.67)^2

Because the mean HEIGHT (for $n = 126$) was 1.577 m and the mean AGE was 13.67 yr, the variables HEIGHT$_*$ and AGE$_*$ are centered. This also means that the definitions of the parameters have changed, leading to different tests and interpretations. Next, the same regression was recomputed using the transformed variables. As expected, the R^2-value was still .68. However, predictor correlations (Table 12-6) and collinearity diagnostics for the scaled cross-products matrix (Table 12-7) show radical improvement. The maximum between-predictor correlation in Table 12-6 has dropped from .999 to .774. The maximum condition index (the condition number) is now 3.7 in Table 12-7, compared with the value of

TABLE 12-6   Correlations for weight example with centering of predictors ($n = 126$)

	HEIGHT	(HEIGHT)^2	AGE	(AGE)^2	WEIGHT
HEIGHT	1	−.067	.774	.221	.790
(HEIGHT)^2		1	.032	.372	.053
AGE			1	.430	.699
(AGE)^2				1	.396

TABLE 12-7   Eigenanalysis of the scaled cross-products matrix for the model where WEIGHT is predicted from HEIGHT, (HEIGHT)^2, AGE, and (AGE)^2 with centering ($n = 126$)

Variable	Eigenvalue	Condition Index	Variance Proportions				
			Intercept	HEIGHT	(HEIGHT)^2	AGE	(AGE)^2
1	2.353	1.0	.05	.01	.06	.01	.06
2	1.764	1.1	.02	.09	.02	.08	.00
3	0.376	2.5	.65	.05	.48	.03	.01
4	0.334	2.6	.03	.21	.44	.03	.53
5	0.172	3.7	.25	.64	.00	.85	.40

1,028 in Table 12-5. The only hint of any problem arises in the loadings for the principal component with the smallest eigenvalue (the last row), although these variance proportions of .85 and .64 should not be bothersome because the associated condition index is low (namely, 3.7).

The data and polynomial-type model just considered clearly illustrate a collinearity problem that could seriously affect the accuracy of regression calculations. In this example, simple centering of the original variables solved the problem, since the collinearity was due to high correlations among different powers of the same variable. A general treatment of numerical problems with polynomial regression models will be described in Chapter 13.

12-5-4 Avoidable Collinearities

Problems with regression calculations due to collinearity problems may not be easy to detect. It is therefore important when doing regression analyses to be aware of potential dangers. The following examples are chosen to highlight common occurrences in regression analyses that may lead to near collinearities (or even to exact linear dependencies among the predictor variables).

In many situations, one wishes to consider as predictors the powers of a continuous variable. An example would be the use of HEIGHT, (HEIGHT)^2, and (HEIGHT)^3 as predictors for the weight example considered above. These predictors are often called the natural polynomials, and careless use of them can often lead to near collinearities. One solution to this problem, in addition to centering, is to use orthogonal polynomials. This topic is covered in detail in Chapter 13.

Collinearity problems can arise if particularly extreme data values are incorrectly included in the data set via errors in data collection. This data-handling problem can be detected by using the methods to analyze outliers described earlier. Again, the reader is cautioned about blithely discarding troublesome observations.

The use of dummy variables may inadvertently introduce exact collinearities. The discussion of how to avoid this problem will be deferred until Chapter 14.

Interaction terms generally create the atmosphere for collinearity problems, especially if such terms are overused. For example, if the predictors are AGE, HEIGHT, and AGE × HEIGHT, then a near collinearity may show up due to the close functional relationship between the product term and the two basic predictors AGE and HEIGHT. The amount of collinearity introduced will depend upon the range of heights and weights in the data and the number of repeated values of each variable. In general, fitting a model containing several interaction terms will almost guarantee that some collinearity problems will arise. An important special case in which interactions are not generally troublesome in this way is in ANOVA designs with equal or nearly equal numbers of subjects in each cell (Chapter 19).

A subtler form of near collinearity occurs with the following set of predictors: head-of-household income, education, number of years in work force, and age. Since these four variables tend to be very positively correlated with one another, it is likely that one of the four will be nearly perfectly predicted from (some linear combination of) the remainder. This illustrates one of the most troublesome types of collinearity. Ideally one can avoid the problem by eliminating one or more of the variables. If this solution is unpalatable, then certain esoteric methods of analysis (such as ridge regression) can be employed.

12-6 Scaling Problems

A general class of problems in regression analysis may arise due to improper scaling of the predictors and/or the response variable. Specifically, such problems concern the loss of computational accuracy. The resulting inaccuracy can be so great as to give coefficient estimates with the wrong sign. A scaling problem may occur if a predictor has too wide a range of values. For example, recording adult human body weight in grams might be problematic. As another example, a problem might occur with data for people who use vastly greater quantities of, say, vitamin C than the average person. Such properly measured values of vitamin C intake would still be outliers, and their use could lead to modeling problems. Similarly, if the mean of a variable is large with little variability, computational problems may arise. A simple example would be recording human body temperature in kelvin (degrees Celsius minus 273).

Most scaling problems can be avoided by proper data validation and rescaling before performing regression analyses. Some problems, such as the vitamin C intake example, may require changes in some part of the modeling strategy. Alternate analysis methods are discussed below.

12-7 Treating Collinearity and Scaling Problems

Both near collinearity and bad scaling create numerical problems, including the inaccurate computation of (1) estimates of the regression coefficients, (2) estimates of standard errors, and (3) hypothesis test statistics. In the worst cases, the analysis may change substantially if the data are input into a program in a different order, or if a few observations are deleted. All reduce our confidence in the analysis.

The first step in treating numerical problems is to validate the data adequately before attempting any regression modeling. Procedures for such validation can detect most scaling problems and suggest solutions. For continuous variables, the numerical ranges of the predictor variables should be as similar as possible. For example, do not measure one set of weights in grams and another in tons. In general, numbers that are convenient to write and plot are also preferred for data analysis. Typically this gives ranges such as 1 to 10 or 10 to 100.

The second step in treating numerical problems is to utilize regression diagnostics and collinearity diagnostics (e.g., eigenvalues and condition indices). These methods allow one to detect numerical problems but do not necessarily indicate a good solution.

The third step in treating numerical problems is to attempt to eliminate redundant variables. An ordering of the importance of the variables is central to this task. More sophisticated techniques involve trying to formalize the selection of variables (Chapter 16; also see McCabe, 1984). Also, analysis of the principal components may be applied to the process of reducing variables (see the brief discussion to follow and Chapters 16 and 24).

12-8 Alternate Strategies of Analysis

When one or more of the basic underlying assumptions of regression analysis are clearly not satisfied and/or when numerical problems are identified, the analyst may want to turn to other strategies of analysis. In the subsections to follow we shall (1) list some alternate analysis methods, (2) briefly mention generalizations to linear regression that may be adequate in some applications, and (3) briefly describe transformations to the data that may allow the use of multiple linear regression.

12-8-1 Alternate Approaches

If the analyst decides that the regression model used does not fit the data and cannot be made to fit via simple generalizations of linear regression (such as weighted least squares), then other methods must be used. If the response variable cannot be modeled as a linear function of the parameters, then analysis methods do exist for nonlinear functions (Gallant, 1975). If the problem concerns nonnormal distributions, then methods of rank analysis (Conover and Iman, 1981) or categorical data analysis (see Chapter 22) may be appropriate.

As an aside, some rank analysis methods may be thought of as essentially two-step processes. The first step is to replace the original data with appropriate ranks. The second is to conduct linear regression analysis on the ranks. The reader should be cautioned not to presume that this approach will work in any particular case, but the brief discussion is still helpful in understanding ranking methods.

18-8-2 Generalizations of Linear Regression

Survey of Methods

Generalizations of linear regression may be loosely grouped into exact and approximate methods. Exact methods have procedures of estimation and hypothesis testing with known properties for finite samples, while approximate methods have only asymptotic results available. Therefore the approximate methods must be used with caution in small samples. The generalized approximate methods include regression on principal components, ridge regression, and robust regression. Exact generalizations include multivariate techniques and exact weighted least squares (i.e., weights are known without error).

Regression on principal components and ridge regression are often recommended for treating collinearity problems. In the analysis of principal components, the original predictor variables are replaced by a set of mutually uncorrelated variables, the principal component scores. If necessary, components associated with eigenvalues near 0 are dropped from the analysis, thus eliminating the attendant collinearity problem. Ridge regression involves perturbing the eigenvalues of the original predictor variable cross-products matrix to "push" them away from 0, thus reducing the amount of collinearity. A detailed discussion of these methods is beyond the scope of this book. See, for example, Gunst and Mason (1980). The reader should note that both methods lead to biased regression estimates of the parameters under the assumed model. In addition, P-values for statistical tests may be optimistically small when using such biased estimation methods.

Robust regression involves weighting or transforming the data so as to minimize the effects of extreme observations. The goal is to make the analysis more robust (i.e., less sensitive) to any particular observation and also less sensitive to the basic assumptions of regression analysis. (See Huber, 1981, for a discussion of such procedures.)

Multivariate methods include multivariate multiple regression, multivariate analysis of variance, discriminant analysis, canonical correlation, and growth curve analysis, among others. In all these procedures, one can account for nonindependence among observations, such nonindependence being explicitly modeled. The types of predictors and response variables that may be used in these methods differ, as do what hypotheses are testable and how the nonindependence is modeled. Many multivariate books are available. See, for example, Timm (1975) and Morrison (1976).

Weighted-Least-Squares Analysis

The *weighted-least-squares* method of analysis is a modification of standard regression analysis procedures and is used when a regression model is to be fit to a set of data for which the assumptions of variance homogeneity and/or independence do not hold. We shall briefly describe here the weighted-least-squares approach for dealing with variance heterogeneity. We refer the reader to other sources, such as Draper and Smith (1981) and Neter, Wasserman, and Kutner (1983), for a discussion of the general method of weighted regression, which incorporates a discussion of treating nonindependence.

When the variance of Y varies for different values of the independent variable(s), weighted-least-squares analysis can be used provided these variances (i.e., σ_i^2 for the ith observation on Y) are known or can be assumed to be of the form $\sigma_i^2 = \sigma^2/W_i$, where the weights $\{W_i\}$ are known. The methodology then involves determining the regression coefficients $\hat{\beta}_0', \hat{\beta}_1', \ldots, \hat{\beta}_k'$ that minimize the expression

$$\sum_{i=1}^{n} W_i(Y_i - \hat{\beta}_0' - \hat{\beta}_1' X_{1i} - \hat{\beta}_2' X_{2i} - \cdots - \hat{\beta}_k' X_{ki})^2$$

where the weight W_i is given by $1/\sigma_i^2$ when the $\{\sigma_i^2\}$ are known or is exactly the W_i in the expression σ^2/W_i when this latter form applies.

The specific weighted-least-squares solution for the straight-line regression case (i.e., $Y = \beta_0 + \beta_1 X + E$) is given by the formulas

$$\hat{\beta}_1' = \frac{\sum_{i=1}^{n} W_i(X_i - \bar{X}')(Y_i - \bar{Y}')}{\sum_{i=1}^{n} W_i(X_i - \bar{X}')^2} = \frac{\sum_{i=1}^{n} W_i X_i Y_i - \left(\sum_{i=1}^{n} W_i X_i\right)\left(\sum_{i=1}^{n} W_i Y_i\right)\bigg/\sum_{i=1}^{n} W_i}{\sum_{i=1}^{n} W_i X_i^2 - \left(\sum_{i=1}^{n} W_i X_i\right)^2\bigg/\sum_{i=1}^{n} W_i}$$

and

$$\hat{\beta}_0' = \bar{Y}' - \hat{\beta}_1' \bar{X}'$$

in which

$$\bar{Y}' = \frac{\sum_{i=1}^{n} W_i Y_i}{\sum_{i=1}^{n} W_i} \quad \text{and} \quad \bar{X}' = \frac{\sum_{i=1}^{n} W_i X_i}{\sum_{i=1}^{n} W_i}$$

Under the usual normality assumption on the Y-variable, the same general procedures are applicable as are used in the unweighted case regarding t tests, F tests, and confidence intervals about the various regression parameters; for example, to test $H_0: \beta_1 = 0$, the following test statistic may be used:

$$T = \frac{\hat{\beta}_1' - 0}{S'_{Y|X}/S'_X\sqrt{n-1}} \frown t_{n-2} \quad \text{under } H_0$$

in which

$$S_{Y|X}'^2 = \frac{1}{n-2} \sum_{i=1}^{n} W_i(Y_i - \hat{\beta}_0' - \hat{\beta}_1'X_i)^2$$

and

$$S_X'^2 = \frac{1}{n-1} \sum_{i=1}^{n} W_i(X_i - \bar{X}')^2$$

12-8-3 Transformations

The three primary reasons for using data transformations are (1) *to stabilize* the variance of the dependent variable if the homoscedasticity assumption is violated, (2) *to normalize* (i.e., to transform to the normal distribution) the dependent variable if the normality assumption is noticeably violated, and (3) *to linearize* the regression model if the original data suggest a model that is nonlinear in either the regression coefficients and/or the original variables (dependent or independent). It is fortunate that the same transformation often helps to accomplish the first two goals and sometimes even the third, rather than achieving one goal at the expense of either of the other two.

A more thorough discussion of the properties of various transformations can be found in several sources, notably Armitage (1971), Draper and Smith (1981), and Neter, Wasserman, and Kutner (1983). In addition, Box and Cox (1964) described an approach to making an exploratory search for one of a family of transformations (see also Box and Cox, 1984, and Carroll and Ruppert, 1984). Nevertheless, we consider it useful to describe a few of the more commonly used transformations:

1. The *log transformation* ($Y' = \log Y$). Used (provided Y takes on only positive values) to stabilize the variance of Y if it increases markedly with increasing Y, to normalize the dependent variable if the distribution of the residuals for Y is positively skewed, and to linearize the regression model if the relationship of Y to some independent variable suggests a model with consistently increasing slope.

2. The *square root transformation* ($Y' = \sqrt{Y}$). Used to stabilize the variance if it is proportional to the mean of Y. This is particularly appropriate if the dependent variable has the Poisson distribution.

3. The *reciprocal transformation* ($Y' = 1/Y$). Used to stabilize the variance if it is proportional to the fourth power of the mean of Y, which indicates that there is a huge increase in variance above some threshold value of Y. This transformation minimizes the effect of large values of Y, since for these values the transformed Y'-values will be close to 0, and large increases in Y will cause only trivial decreases in Y'.

4. The *square transformation* $(Y' = Y^2)$. Used to stabilize the variance if it decreases with the mean of Y, to normalize the dependent variable if the distribution of the residuals for Y is negatively skewed, and to linearize the model if the original relationship with some independent variable is curvilinear downward (i.e., the slope consistently decreases as the independent variable increases).

5. The *arcsin transformation* $(Y' = \arcsin \sqrt{Y} = \sin^{-1}\sqrt{Y})$. Used to stabilize the variance if Y is a proportion or rate.

12-9 An Important Caution

All of the techniques in this chapter involve checking the validity of the assumptions and estimates for a regression analysis. Toward that end, outlier analysis and collinearity reduction both suggest deleting either observations or predictor variables in order to improve the quality of the model. Naturally, the observations and variables that are most at odds with the fitted model are deleted. All of the techniques lead to underestimation of the variability and give P-values that are optimistically small. The potential penalty associated with such improvements in fitting a model is biased results. The methods discussed in this chapter are safest to use with large samples and are most unreliable with small samples. Unfortunately, of course, the potential influence of a single observation is largest in small samples. This is a strong argument against using small samples in regression analysis.

In some cases a problem can be resolved only by evaluating a second sample of data. Picard and Cook (1984) discussed the issue of optimism in selecting a regression model. They recommended considering split samples. In a split-sample design, part of the data is used for exploratory data analysis, and the remainder is used for confirming the validity and reliability of the exploratory results. This technique will be described in Chapter 16 in the context of selecting the best regression model.

Problems

Note to the student: Some of the sets of data used in the following problems have hidden problems: outliers, collinearities, and variables in need of transformation. At some requests some computer programs may either balk or blithely produce garbage. This state of affairs is realistic. Having "nice" data, as we provide in most other problem sets, is not realistic but allows us to focus on the particular chapter topics.

1. Consider the data in Problem 1 of Chapter 5.
   a. Fit a univariate linear model with dry weight as the response and age as the predictor.
   b. Provide a plot of studentized or jackknife residuals versus the predictor.
   c. Provide a frequency histogram and schematic plot of the residuals.
   d. Report and interpret a test of whether any residuals have extreme values not due to chance alone.
   e. Do these analyses highlight any potentially troublesome observations? Explain why or why not.

f. Why is it only approximately correct to test studentized residuals for normality? What role does sample size play?

2. a–e. Repeat parts (a) through (e) of Problem 1 using $\log_{10}$(dry weight) as the response.

f. Does using the original data or using logarithms lead to the best-behaved residuals? Explain.

3–14. a–e. Repeat parts (a) through (e) of Problem 1 using the data from Problems 3 through 14 of Chapter 5.

15. a–e. Consider the data from Problem 15 of Chapter 5. Repeat parts (a) through (e) of Problem 1 for BLOODTOL as predicted by PPM_TOLU.

16. a–e. Repeat Problem 15 but for LN_BLDTL as predicted by LN_PPMTL.

f. Does using the original data or logarithms lead to the best-behaved residuals? Explain.

17. Repeat Problem 15 replacing BLOODTOL by BRAINTOL.

18. Repeat Problem 16 replacing LN_BLDTL by LN_BRNTL.

19. The following data are from an article by Bethel and others (1985). All subjects are asthmatics.

Subject	AGE (yr)	SEX	HEIGHT (cm)	WEIGHT (kg)	FEV$_1$* (L)
1	24	M	175	78.0	4.7
2	36	M	172	67.6	4.3
3	28	F	171	98.0	3.5
4	25	M	166	65.5	4.0
5	26	F	166	65.0	3.2
6	22	M	176	65.5	4.7
7	27	M	185	85.5	4.3
8	27	M	171	76.3	4.7
9	36	M	185	79.0	5.2
10	24	M	182	88.2	4.2
11	26	M	180	70.5	3.5
12	29	M	163	75.0	3.2
13	33	F	180	68.0	2.6
14	31	M	180	65.0	2.0
15	30	M	180	70.4	4.0
16	22	M	168	63.0	3.9
17	27	M	168	91.2	3.0
18	46	M	178	67.0	4.5
19	36	M	173	62.0	2.4

* Forced expiratory volume in 1 second.

a. Fit a model of FEV$_1$ as predicted by HEIGHT, WEIGHT, and AGE.

b. Provide variable-added-last tests for all predictors and a test of the intercept.

c. Provide variance inflation factors for each predictor.

    d. Provide a correlation matrix including all predictors and the response.

    e. Provide eigenvalues, condition indexes, and condition numbers for the correlation matrix (excluding the intercept).

    f. Provide eigenvalues, condition indexes, and condition numbers for the scaled cross-products matrix (including the intercept).

    g. Provide residuals (preferably studentized) and leverage values. Do any observations seem bothersome? Explain.

    h. Does there appear to be any problem with collinearity? Explain.

20. Create a new variable, FEMALE, for the data in Problem 19, where FEMALE = 1 if sex is F and FEMALE = 0 if sex is M.

    a–h. Repeat parts (a) through (h) of Problem 19 adding FEMALE as a predictor.

21. Create three new interaction variables for the data in Problem 20:

$$FEMAGE = FEMALE*AGE$$

$$FEMHT = FEMALE*HEIGHT$$

$$FEMWT = FEMALE*WEIGHT$$

    a–h. Repeat parts (a) through (h) of Problem 19 adding FEMALE and the three interaction variables as predictors.

    i. Explain how the proportion of females leads to difficulties in Problems 20 and 21. Suggest a simple solution for analyzing these data. Suggest a solution for future research.

22. Freund (1979) reported the following analysis of daily soil evaporation, EVAP, as a function of the following predictor variables:

    MAXAT = maximum daily air temperature

    MINAT = minimum daily air temperature

    AVAT = integrated area under the daily air temperature curve, i.e., a measure of average air temperature

    MAXST = maximum daily soil temperature

    MINST = minimum daily soil temperature

    AVST = integrated area under the soil temperature curve

    MAXH = maximum daily relative humidity

    MINH = minimum daily relative humidity

    AVH = integrated area under the daily humidity curve

    WIND = total wind, measured in miles per day

In addition, he provided the overall ANOVA table, tests of coefficients, and the raw data, as follows.

Source	df	SS	MS	F
Regression	10	8159.83	815.98	19.27
Residual	35	1482.27	42.35	
Total	45	9642.11		

Variable	$\hat{\beta}$	t	VIF
MAXAT	0.5011	0.88	8.828
MINAT	0.3041	0.39	8.887
AVAT	0.09219	0.42	22.21
MAXST	2.232	2.22	39.29
MINST	0.2049	0.19	14.08
AVST	0.7426	−2.12	52.36
MAXH	1.110	0.98	1.981
MINH	0.7514	1.54	25.38
AVH	−0.5563	−3.44	24.12
WIND	0.00892	0.97	1.985

a. Compute the R^2 and provide a test of significance (use $\alpha = .01$).
b. Compute R_j^2 for each predictor.
c. Which, if any, variables are implicated as possibly inducing collinearity?
d. Freund (1979) noted that "some of the coefficients have 'wrong' signs." Using your knowledge of evaporation, explain which coefficients have wrong signs, paying particular attention to those with extreme t-values.

23. The data for Problem 22, from Freund (1979), are given below.
   a. Fit the model considered in Problem 22.
   b. Provide a test of the intercept.
   c. Note any numerical discrepancies between the output you have and that of Problem 22.
   d–h. Repeat parts (d) through (h) of Problem 19 for these data.

Observation	Month	Day	MAXST	MINST	AVST	MAXAT	MINAT	AVAT	MAXH	MINH	AVH	WIND	EVAP
1	6	6	84	65	147	85	59	151	95	40	398	273	30
2	6	7	84	65	149	86	61	159	94	28	345	140	34
3	6	8	79	66	142	83	64	152	94	41	388	318	33
4	6	9	81	67	147	83	65	158	94	50	406	282	26
5	6	10	84	68	167	88	69	180	93	46	379	311	41
6	6	11	74	66	131	77	67	147	96	73	478	446	4
7	6	12	73	66	131	78	69	159	96	72	462	294	5
8	6	13	75	67	134	84	68	159	95	70	464	313	20
9	6	14	84	68	161	89	71	195	95	63	430	455	31
10	6	15	86	72	169	91	76	206	93	56	406	604	38
11	6	16	88	73	178	91	76	208	94	55	393	610	43
12	6	17	90	74	187	94	76	211	94	51	385	520	47
13	6	18	88	72	171	94	75	211	96	54	405	663	45
14	6	19	88	72	171	92	70	201	95	51	392	467	45
15	6	20	81	69	154	87	68	167	95	61	448	184	11
16	6	21	79	68	149	83	68	162	95	59	436	177	10
17	6	22	84	69	160	87	66	173	95	42	392	173	30
18	6	23	84	70	160	87	68	177	94	44	392	76	29
19	6	24	84	70	168	88	70	169	95	48	398	72	23

Observation	Month	Day	MAXST	MINST	AVST	MAXAT	MINAT	AVAT	MAXH	MINH	AVH	WIND	EVAP
20	6	25	77	67	147	83	66	170	97	60	431	183	16
21	6	26	87	67	166	92	67	196	96	44	379	76	37
22	6	27	89	69	171	92	72	199	94	48	393	230	50
23	6	28	89	72	180	94	72	204	95	48	394	193	36
24	6	29	93	72	186	92	73	201	94	47	386	400	54
25	6	30	93	74	188	93	72	206	95	47	389	339	44
26	7	1	94	75	199	94	72	208	96	45	370	172	41
27	7	2	93	74	193	95	73	214	95	50	396	238	45
28	7	3	93	74	196	95	70	210	96	45	380	118	42
29	7	4	96	75	198	95	71	207	93	40	365	93	50
30	7	5	95	76	202	95	69	202	93	39	357	269	48
31	7	6	84	73	173	96	69	173	94	58	418	128	17
32	7	7	91	71	170	91	69	168	94	44	420	423	20
33	7	8	88	72	179	89	70	189	93	50	399	415	15
34	7	9	89	72	179	95	71	210	98	46	389	300	42
35	7	10	91	72	182	96	73	208	95	43	384	193	44
36	7	11	92	74	196	97	75	215	96	46	389	195	41
37	7	12	94	75	192	96	69	198	95	36	380	215	49
38	7	13	96	75	195	95	67	196	97	24	354	185	53
39	7	14	93	76	198	94	75	211	93	43	364	466	53
40	7	15	88	74	188	92	73	198	95	52	405	399	21
41	7	16	88	74	178	90	74	197	95	61	447	232	1
42	7	17	91	72	175	94	70	205	94	42	380	275	44
43	7	18	92	72	190	95	71	209	96	44	379	166	44
44	7	19	92	73	189	96	72	208	93	42	372	189	46
45	7	20	94	75	194	95	71	208	93	43	373	164	47
46	7	21	96	76	202	96	71	208	94	40	368	139	50

References

The following list includes sources that are not cited in Chapter 12 but that may be helpful to the reader seeking discussions of topics discussed in this chapter.

Anscombe, F. J. (1960). "Rejection of Outliers." *Technometrics*, 2: 123–147.

Anscombe, F. J., and Tukey, J. W. (1963). "The Examination and Analysis of Residuals." *Technometrics*, 5: 141–160.

Armitage, P. (1971). *Statistical Methods in Medical Research*. Oxford: Blackwell Scientific Publications.

Barnett, V., and Lewis, T. (1978). *Outliers in Statistical Data*. New York: John Wiley & Sons, Inc.

Bartlett, M. S. (1947). "The Use of Transformations." *Biometrics*, 3: 39–52.

Belsley, D. A. (1984). "Demeaning Conditioning Diagnostics Through Centering." *Amer. Statistician*, 38: 73–77. (Also see comments in same issue.)

Belsley, D. A., Kuh, E., and Welsch, R. E. (1980). *Regression Diagnostics: Identifying Influential Data and Sources of Collinearity*. New York: John Wiley & Sons, Inc.

Bethel, R. A., Sheppard, D., Geffroy, B., Tam, E., Nadel, J. A., and Boushey, J. A. (1985). "Effect of 0.25 ppm Sulfur Dioxide on Airway Resistance in Freely Breathing, Heavily Exercising, Asthmatic Subjects." *Amer. Rev. Resp. Dis.*, *131*: 659–661.

Box, G. E. P., and Cox, D. R. (1984). "An Analysis of Transformations Revisited, Rebuttal." *J. Amer. Statist. Assoc.*, *17*: 209–210.

Box, G. E. P., and Cox, D. R. (1984). "An Analysis of Transformations Revisited, Rebuttal." *J. Amer. Statist. Assoc.*, *17*: 209–210.

Carroll, R. J., and Ruppert, D. (1984). "Power Transformations When Fitting Theoretical Models to Data." *J. Amer. Statist. Assoc.*, *79*: 321–328.

Conover, W. J., and Iman, R. L. (1981). "Rank Transformations as a Bridge Between Parametric and Nonparametric Statistics." *Amer. Statistician*, *35*: 124–128.

Cook, R. D., and Weisberg, S. (1982). *Residuals and Influence in Regression*. New York: Chapman and Hall.

Draper, N. R., and Smith, H. (1981). *Applied Regression Analysis*. New York: John Wiley & Sons, Inc.

Durbin, J., and Watson, G. S. (1951). "Testing for Serial Correlation in Least Squares Regression." *Biometrika*, *37*: 409–428.

Freund, R. J. (1979). "Multicollinearity etc., Some 'New' Examples." *Proceedings of the Statistical Computing Section*, Amer. Statist. Assoc., pp. 111–112.

Gallant, A. R. (1975). "Nonlinear Regression." *Amer. Statistician*, *29*: 73–81.

Gunst, R. F., and Mason, R. L. (1980). *Regression Analysis and Its Application*. New York: Marcel Dekker, Inc.

Hackney, O. J., and Hocking, R. R. (1979). "Diagnostic Techniques for Identifying Data Problems in Multiple Linear Regression." *Proceedings of the Statistical Computing Section*, Amer. Statist. Assoc., pp. 94–98.

Hinkley, D. V., and Runger, G. (1984). "The Analysis of Transformed Data." *J. Amer. Statist. Assoc.*, *79*: 302–320.

Hoaglin, D. C., and Welsch, R. E. (1978). "The Hat Matrix in Regression and ANOVA." *Amer. Statistician*, *32*: 17–22.

Hocking, R. R. (1983). "Developments in Linear Regression Methodology: 1959–1982." *Technometrics*, *25*: 219–248.

Huber, P. J. (1981). *Robust Statistics*. New York: John Wiley & Sons, Inc.

Lewis, T., and Taylor, L. R. (1967). *Introduction to Experimental Ecology*. New York: Academic Press.

McCabe, G. P. (1984). "Principal Variables." *Technometrics*, *26*: 137–144.

Morrison, D. F. (1976). *Multivariate Statistical Methods*. New York: McGraw-Hill Book Company.

Mosteller, F., and Tukey, J. W. (1977). *Data Analysis and Regression*. Reading, Mass.: Addison-Wesley Publishing Company, Inc.

Neter, J., Wasserman, W., and Kutner, M. H. (1983). *Applied Linear Regression Models*. Homewood, Ill.: Richard D. Irwin.

Picard, R. R., and Cook, R. D. (1984). "Cross-validation of Regression Models." *J. Amer. Statist. Assoc.*, *79*: 575–583.

Shapiro, S. S., and Wilks, M. B. (1965). "An Analysis of Variance Test for Normality (Complete Samples)." *Biometrika*, *52*: 591–611.

Siegel, S. (1956). *Nonparametric Statistics for the Behavioral Sciences*. New York: McGraw-Hill Book Company.

Smith, G., and Campbell, F. (1980). "A Critique of Some Ridge Regression Methods." *J. Amer. Statist. Assoc., 75*: 74–81.

Stephens, M. A. (1974). "EDF Statistics for Goodness of Fit and Some Comparisons." *J. Amer. Statist. Assoc., 69*: 730–737.

Stevens, J. P. (1984). "Outliers and Influential Data Points in Regression Analysis." *Psych. Bull., 95*: 334–344.

Timm, N. H. (1975). *Multivariate Analysis with Applications in Education and Psychology*. Belmont, Calif.: Wadsworth Publishing Company, Inc.

Tukey, J. W. (1977). *Exploratory Data Analysis*. Reading, Mass.: Addison-Wesley Publishing Company, Inc.

Weisberg, S. (1980). *Applied Linear Regressions*. New York: John Wiley & Sons, Inc.

I3

Polynomial Regression

13-1 Preview

In this chapter, we focus on a special case of the multiple regression model, the *polynomial model*, which is often of interest whenever there is only *one basic* independent variable (say X) to be considered. We initially considered a straight-line model (Chapter 5) for this situation; however, we might wish to determine whether we can significantly improve prediction by increasing the complexity of the fitted straight-line model. The simplest extension of the straight-line model is the second-order polynomial, or parabola, which involves a second term, X^2, in addition to X. The addition of high-order terms like X^2, X^3, and so on, which are simple functions of a single basic variable, can be considered equivalent to adding new independent variables. Thus, if we renamed X as X_1 and X^2 as X_2, the second-order model

$$Y = \beta_0 + \beta_1 X + \beta_2 X^2 + E$$

would become

$$Y = \beta_0 + \beta_1 X_1 + \beta_2 X_2 + E$$

In general, polynomial models are special cases of the general multiple regression model. However, since only one basic independent variable is being considered, any polynomial model can be represented by a curvilinear plot on a two-dimensional graph (rather than as a surface in higher-dimensional space). As mentioned in Chapter 5, when there is only one basic independent variable X, the fundamental goal is to find that curve which best fits the data so that the relationship between X and Y is appropriately described. Because a higher-order curve may be more appropriate than a straight line, it usually is important to consider fitting such (polynomial) curves.

We first consider methods for fitting and evaluating the second-order (parabolic) model, after which we consider higher-order polynomial models. Since these models are special cases of the general multiple regression model, the fitting of these models and the methods for inference are essentially the same as described more generally in Chapters 8 and

228

9. Since the independent variables in a polynomial model are functions of the same basic variable (X), they are inherently correlated. This, in turn, can lead to computational difficulties due to collinearity (Chapter 12). Fortunately, techniques are available, such as centering and the use of orthogonal polynomials, which help to remedy such problems, and these procedures will be discussed later in this chapter. We shall also see that the use of orthogonal polynomials helps to simplify hypothesis testing.

13-2 Polynomial Models

The most general kind of curve usually considered for describing the relationship between a single independent variable X and a response Y is called a *polynomial*. Mathematically, a polynomial of order k in x is an expression of the form

$$y = c_0 + c_1 x + c_2 x^2 + \cdots + c_k x^k$$

in which the c's and k (which must be a nonnegative whole number) are constants. We have already considered the simple polynomial corresponding to $k = 1$ (i.e., the straight line having the form $y = c_0 + c_1 x$). The second-order polynomial corresponding to $k = 2$ (i.e., the parabola) has the general form $y = c_0 + c_1 x + c_2 x^2$.

In going from a *mathematical* model to a *statistical* model, as we did in the straight-line case, we may write a parabolic model in either of the following forms:

$$\mu_{Y|X} = \beta_0 + \beta_1 X + \beta_2 X^2 \tag{13.1}$$

or

$$Y = \beta_0 + \beta_1 X + \beta_2 X^2 + E \tag{13.2}$$

In these equations, capital Y's and X's denote statistical variables; β_0, β_1, and β_2 denote the unknown parameters called regression coefficients; $\mu_{Y|X}$ denotes the mean of Y at a given X; and E denotes the error component, which represents the difference between the observed response Y at X and the true average response $\mu_{Y|X}$ at X.

If we tentatively assume that a parabolic model as given by either (13.1) or (13.2) is appropriate for describing the relationship between X and Y, we must then determine a specific estimated parabola that best fits the data. As in the straight-line case, this best-fitting parabola may be determined by employing the least-squares method as described in the next section.

13-3 Least-Squares Procedure for Fitting a Parabola

The least-squares estimates of the parameters β_0, β_1, and β_2 in a parabolic model are chosen so as to minimize the sum of squares of deviations of observed points from corresponding points on the fitted parabola (Figure 13-1). Letting $\hat{\beta}_0$, $\hat{\beta}_1$, and $\hat{\beta}_2$ denote the least-squares estimates of the unknown regression coefficients in the parabolic model (13.1), and letting $\hat{Y}$ denote the value of the predicted response at X, we can write the estimated parabola as follows:

$$\hat{Y} = \hat{\beta}_0 + \hat{\beta}_1 X + \hat{\beta}_2 X^2 \tag{13.3}$$

FIGURE 13-1   Deviations of observed points from the least-squares parabola

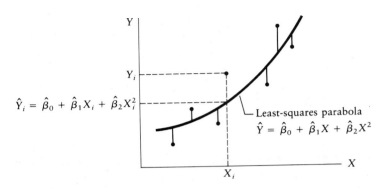

The minimum sum of squares obtained using this least-squares parabola is

$$\text{SSE} = \sum_{i=1}^{n} (Y_i - \hat{Y}_i)^2 = \sum_{i=1}^{n} (Y_i - \hat{\beta}_0 - \hat{\beta}_1 X_i - \hat{\beta}_2 X_i^2)^2 \qquad (13.4)$$

As with the general regression model, we do not find it necessary to present the precise formulae for calculating the least-squares estimates $\hat{\beta}_0$, $\hat{\beta}_1$, and $\hat{\beta}_2$. These formulae are quite complex and become even more so for polynomials of order higher than two. The researcher is not likely to employ such polynomial regression methods without using a packaged computer program, which can perform the necessary calculations and print the numerical results. (We have provided in Appendix B a discussion of matrices and their relationship to regression analysis; by using matrix mathematics, the general regression model and the associated least-squares methodology can be compactly represented.)

Example   For the age–systolic blood pressure data of Table 5-1 with the outlier removed,[1] the least-squares estimates for the parabolic regression coefficients are computed to be:

$$\hat{\beta}_0 = 113.41, \qquad \hat{\beta}_1 = 0.088, \qquad \hat{\beta}_2 = 0.010$$

The fitted model given by (13.3) then becomes

$$\hat{Y} = 113.41 + 0.088X + 0.010X^2 \qquad (13.5)$$

This equation can be compared with the straight-line equation obtained in Section 5-5 for these data with the outlier removed, namely,

$$\hat{Y} = 97.08 + 0.95X \qquad (13.6)$$

When comparing (13.5) to (13.6), it is important to notice that the estimates of β_0 and β_1 are different in the two models, indicating that the estimation of β_2 affects the estimation of β_0 and β_1 in the quadratic model.

---

[1] As described in Section 5-5-3, the outlier corresponds to the data point ($X = 47$, $Y = 220$) for the second individual listed in Table 5-1.

13-4 ANOVA Table for Second-Order Polynomial Regression

As with the straight-line case, the essential results based on a second- or higher-order polynomial model can be summarized in an ANOVA table. The ANOVA table for a parabolic fit to the age–systolic blood pressure data of Table 5-1 (with the outlier removed) is given in Table 13-1.

The contents of Table 13-1 deserve comment. First, only variables-added-in-order tests are described. Natural variable orderings suggest themselves, either from the largest to the smallest power of the predictor or vice versa. Consequently, a variables-added-last test for each term should be avoided with polynomial models. Using variables-added-in-order tests will aid in choosing the most parsimonious yet relevant model possible. Also, all such tests should utilize the residual from the largest model considered; this notion will be discussed more fully in Section 13-10.

TABLE 13-1   ANOVA table for a parabola fit to age–systolic blood pressure data of Table 5-1 with outlier removed

Source		df	SS	MS	F
Regression	X	1	6,110.10	6,110.10	68.89
	$X^2 \mid X$	1	163.30	163.30	1.84
Residual		26	2,306.05	88.69	
Total (corrected)*		28	8,579.45		

Note: The residual from the largest model was used for all tests.

* $R^2 = .731$.

13-5 Inferences Associated with Second-Order Polynomial Regression

There are three basic inferential questions associated with second-order polynomial regression. These are as follows:

1. Is the overall regression significant; that is, is more of the variation in Y explained by the second-order model than by ignoring X completely (and just using $\bar{Y}$)?

2. Does the second-order model provide significantly more predictive power than that provided by the straight-line model?

3. Given that a second-order model is more appropriate than a straight-line model, should we add higher-order terms (e.g., X^3, X^4, etc.) to the second-order model?

13-5-1 Test for Overall Regression and Strength of the Overall Parabolic Relationship

To determine whether the overall regression is significant involves testing the null hypothesis H_0: "there is no significant overall regression using X and X^2" (i.e., $\beta_1 = \beta_2 = 0$). The testing procedure used for this null hypothesis involves the overall F test described in Chapter 9, namely, computing

$$F = \frac{\text{MS regression}}{\text{MS residual}}$$

Then compare the value of this F statistic with an appropriate critical point of the F distribution, which (in our example) has 2 and 26 degrees of freedom in the numerator and denominator, respectively. For $\alpha = .001$, we find that $F = 35.37 > F_{2,26,0.999} = 9.12$, and so we reject the null hypothesis of nonsignificant overall regression ($P < .001$).

To obtain a quantitative measure of how well the second-order model predicts the dependent variable, we can use the squared multiple correlation coefficient (the multiple R^2). As with r^2 in straight-line regression, R^2 represents the proportionate reduction in the error sum of squares obtained by using X and X^2 instead of the naive predictor $\bar{Y}$. The formula for calculating R^2 is given by

$$R^2 \text{ (second-order model)} = \frac{\text{SSY} - \text{SSE (second-order model)}}{\text{SSY}} , \qquad (13.7)$$

For this example, $R^2 = .731$. The F test above (with $P < .001$) tells us that this R^2 is significantly different from 0.

> (It is possible, although not likely, that the overall F test for the second-order model will not lead to the rejection of H_0 even if the t test (or the equivalent F test) for significant regression of the straight-line model leads to rejection. This possibility arises because the loss of 1 degree of freedom in SSE in going from the straight-line model to the second-order model may result in a smaller computed F, coupled with an altered critical point of the F distribution. In our example, the computed F is reduced from 66.81 (68.89 for the variables-added-in-order test in Table 13-1) for the straight-line model to 35.37 for the second-order model, and the critical point of the F distribution for $\alpha = .001$ is reduced from $F_{1,27,0.999} = 13.6$ to $F_{2,26,0.999} = 9.12$.)

13-5-2 Test for the Addition of the X^2 Term to the Model

To answer the second question about increased predictive power, we must perform a partial F test of the null hypothesis H_0: "the addition of the X^2 term to the straight-line model does not significantly improve the prediction of Y over and above that achieved by the straight-line model itself" (i.e., $\beta_2 = 0$). To test this null hypothesis, we compute the partial F statistic

$$F(X^2|X) = \frac{(\text{extra SS due to adding } X^2)/1}{\text{MS residual (second-order model)}} \qquad (13.8)$$

and then compare this F value to an appropriate F percentage point (which is an $F_{1,26}$ value in our example).

(Since X^2 is the last variable added, this is a variables-added-last test. Alternatively, we can divide the estimated coefficient $\hat{\beta}_2$ by its estimated standard deviation to form a statistic that has a t distribution under H_0 with 26 degrees of freedom.)

The ANOVA information needed to compute this F test for our example is given in Table 13-1. The extra sum of squares for $X^2|X$ is 163.30 and is computed as the difference between the sum-of-squares regression values for the first- and second-order models. The partial F statistic (13.8) is then computed to be

$$F = \frac{163.30}{88.69} = 1.84$$

Since $F_{1,26,0.90} = 2.91$, we would not reject H_0 at the $\alpha = .10$ level. Furthermore, the P-value for this test satisfies the inequality $.10 < P < .25$. Thus, for this example, we conclude that adding a quadratic term to the straight-line model does not significantly improve prediction. In addition to the results of this partial F test, this conclusion is supported by a scatter diagram of the data (Figure 5-8 in Chapter 5), which offers no evidence of a parabolic relationship between X and Y. Finally, the conclusion is also supported by the small increase in R^2 when the X^2 term is added to the straight-line model. Since $r^2 = R^2 = .712$ for the straight-line model, and $R^2 = .731$ for the parabolic model, the increase in R^2 is $.731 - .712 = .019$.

13-5-3 Testing for Adequacy of the Second-Order Model

The previous analysis of the Table 5-1 data has shown that a straight-line model fits the data adequately, is significantly predictive of the response, and is preferable to a parabolic model. Consequently, it would be superfluous in this case to evaluate whether a model of order higher than two would be significantly better than a straight-line model. Nevertheless, we note that, in general, any question of model adequacy can be addressed (with a *lack-of-fit* test) for any model (of any order) being considered at a given stage of an analysis. Any such lack-of-fit test can be characterized by a partial or multiple-partial F test for the addition of one or more terms to the model under study. More detailed discussion of lack-of-fit tests will follow in the sections below.

13-6 An Example Requiring a Second-Order Model

We now turn to another hypothetical example to illustrate the methods of polynomial regression; this example will lead us to a different conclusion regarding the appropriateness of a second-order model.

Suppose that a laboratory study is undertaken to determine the relationship between the dosage (X) of a certain drug and the weight gain (Y). Eight laboratory animals of the same sex, age, and size are selected and randomly assigned to one of eight dosage levels.

(The study design here can certainly be criticized for not having more than one animal receiving the same dosage, as well as for involving such a small total sample size. Replication at each dosage would provide a reliable estimate of animal-to-animal variation in the data. However, for some laboratory studies, sufficient numbers of animals are not easily

TABLE 13-2   Weight gain after 2 weeks as a function of
dosage level

Dosage level (X)	1	2	3	4	5	6	7	8
Weight gain (Y) (dag)	1	1.2	1.8	2.5	3.6	4.7	6.6	9.1

available; also, cost and time are often limiting factors. It should also be noted that the data for this example have been contrived to simplify the analysis and to present a relationship that is clearly second-order in nature.)

The gain in weight (in dekagrams) is measured for each animal after a two-week time period, during which all animals are subject to the same dietary regimen and general laboratory conditions. The data are given in Table 13-2, and a scatter diagram for these data in Figure 13-2. By simply eyeballing this diagram, it is apparent that a parabolic curve is a more appropriate model than a straight line. We shall now proceed to quantify this visual impression.

The complete ANOVA table based on fitting a parabola to the data in Table 13-2 is given in Table 13-3. The equation for the least-squares parabola on which the ANOVA table is based has the form

$$Y = 1.13 - 0.41X + 0.17X^2$$

Let us investigate the information contained in this ANOVA table. First, by combining the regression $X^2|X$ sum of squares with the residual sum of squares, we are able to test whether there is a significant straight-line regression effect before we add the X^2 term to the model; in particular, the ANOVA table for straight-line regression derived from Table 13-3 is given in Table 13-4. The least-squares line is $\hat{Y} = 1.20 + 1.11X$. The null hypothesis of no significant linear regression is clearly rejected, since an F of 61.95 exceeds $F_{1,6,0.999} = 35.51$ ($P < .001$). Our next step is to examine the complete ANOVA table and decide whether the

FIGURE 13-2   Scatter diagram of hypothetical data
for animal weight gain study

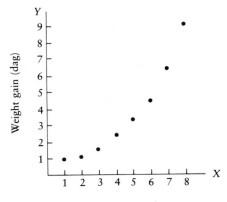

TABLE 13-3   Regression ANOVA table for quadratic model fit to weight gain data

Source		df	SS	MS	F
Regression $\begin{cases} X \\ X^2 \vert X \end{cases}$	X	1	52.04	52.04	61.95
	$X^2 \vert X$	1	4.83	4.83	120.75
Residual		5	0.20	0.04	
Total (corrected)*		7	57.07		

* $R^2 = .997$.

addition of the X^2 term significantly improves the prediction of Y over and above that obtained from a simple straight-line model. In doing so, we are asking whether the increase in R^2 of .085 (.997 − .912) obtained by including the X^2 term in the model significantly improves the fit. The appropriate test statistic to be used in answering this question is the partial F statistic

$$F = \frac{\text{(extra sum of squares due to adding } X^2)/1}{\text{MS residual (second-order model)}}$$

$$= \frac{4.83}{0.04} = 120.75$$

which exceeds $F_{1,5,0.999} = 47.18$ ($P < .001$). Therefore, the addition of the X^2 term to the model significantly improves prediction. As might be expected, a test for overall significant regression of the second-order model yields a highly significant F, namely,

$$F = \frac{\text{MS regression (second-order model)}}{\text{MS residual (second-order model)}}$$

$$= \frac{(52.04 + 4.83)/2}{0.04} = 710.88$$

What have we shown up to this point? We have concluded so far that a first-order (straight-line) model is not as good as a second-order model. We now need to determine whether adding higher-order terms to the second-order model is warranted. For example, we can add an X^3 term to the second-order model and then test whether prediction is significantly improved. Fitting this third-order model by least squares results in the ANOVA table given in Table 13-5. To test whether the addition of the third-order term significantly

TABLE 13-4   Regression ANOVA table for linear model fit to weight gain data

Source	df	SS	MS	F
Regression (X)	1	52.04	52.04	61.95
Residual	6	5.03	0.84	
Total (corrected)*	7	57.07		

* $R^2 = .912$.

TABLE 13-5   Regression ANOVA table for the
third-order model fit to weight gain data

Source		df	SS	MS	F
	X	1	52.040	52.04	
Regression	$X^2\vert X$	1	4.830	4.83	
	$X^3\vert X, X^2$	1	0.140	0.14	10.00[a]
Residual		4	0.056	0.014	
Total (corrected)[b]		7	57.066		

[a] $.025 < P < .05$.

[b] $R^2 = .999$.

improves the fit, the following statistic is calculated:

$$F = \frac{(\text{extra sum of squares due to adding } X^3)/1}{\text{MS residual (third-order model)}}$$

$$= \frac{0.14}{0.014} = 10.00$$

This F statistic has an F distribution with 1 and 4 degrees of freedom under H_0: "the addition of the X^3 term is not worthwhile" (i.e., $\beta_3 = 0$). Since $F_{1,4,0.95} = 7.71$ and $F_{1,4,0.975} = 12.22$, we have $.025 < P < .05$. This P-value would thus reject H_0 for $\alpha = .05$ but not for $\alpha = .025$. This makes the decision of whether to include the X^3 term in the model a little difficult. However, some other factors should be taken into consideration: (1) The R^2-value for the parabolic fit is very high, namely, .997; (2) the R^2-value only increases from .997 to .999 in going from a second-order model to a third-order model; (3) the scatter diagram clearly suggests a second-order curve; and (4) when in doubt, the simplest model is preferable because it is the easiest to interpret. All things considered, then, the most sensible conclusion is that the second-order model is the most appropriate.

In summary, for the data in Table 13-2, the best-fitting model is

$$\hat{Y} = 1.13 - 0.41X + 0.17X^2$$

with an R^2 of .997.

Finally, it is also valuable to have the standard deviations (or standard errors) of the estimated regression coefficients. These are difficult to compute by hand for models involving two or more predictors. However, all commonly used regression programs print the numerical values of the estimated coefficients and their estimated standard errors. For the second-order model fit to the data in Table 13-2, we obtain $S_{\hat{\beta}_1} = 0.141$ and $S_{\hat{\beta}_2} = 0.015$. For example, then, a $100(1 - \alpha)\%$ confidence interval for β_2 would be

$$\hat{\beta}_2 \pm t_{5,1-\alpha/2} S_{\hat{\beta}_2}$$

where the degrees of freedom for the appropriate critical t value are the degrees of freedom associated with the residual sum of squares in Table 13-3. In particular, a 95% confidence interval for β_2 in our example is

$$0.17 \pm (2.571)(0.015)$$

or (0.13, 0.21). Note that this interval does not include 0, which agrees with our ANOVA table conclusion concerning the importance of the X^2 term in the quadratic model.

13-7 Fitting and Testing Higher-Order Models

So far we have seen how the basic ideas of multiple regression may be applied to fitting and testing quadratic and cubic polynomial models. These same methods generalize to all higher-order polynomial models. Nevertheless, several related issues need discussion: the use of orthogonal polynomials and strategies for choosing a polynomial model.

How large an order of polynomial model to consider depends on the problem being studied and the amount and type of data being collected. One consideration, typically of interest for studies in the biological and social sciences, is whether the regression relationship can be described by a monotonic function (i.e., one that is always increasing or decreasing). If only monotonic functions are of interest, usually a second- or third-order model will suffice (although monotonicity is not guaranteed since, for example, some parabolas increase and then decrease). A large number of well-placed predictor values and a small error variance are needed to fit reliably models of higher order than cubic.

A more general consideration is the number of bends (more technically, *points of inflection*) in the polynomial curve that one wishes to fit. For example, a first-order model has no bends; a second-order model has no more than one bend, and each higher-order term adds another potential bend. In practice, fitting polynomial models higher than order three usually leads to models which are neither always decreasing nor always increasing. Substantial theoretical and/or empirical evidence should exist to support the employment of such complicated nonmonotonic models.

The amount of data directly limits the maximum order of a polynomial that may be fit. Consider the weight gain data (Table 13-2). With eight distinct values, a polynomial of order seven would fit the eight points perfectly, giving an SSE value of 0 and an R^2-value of 1. (Nevertheless, because the fitted equation would have eight estimated parameters, no gain in parsimony is made over simply listing the eight data points.) *Generally, the maximum order of polynomial that may be fit is one less than the number of distinct X-values.* For example, consider the age–systolic blood pressure data in Table 5.1 (with the outlier removed). Of the 29 observations, 5 are replicates, a fact which implies that 24 distinct X-values exist. Hence, a polynomial of order 23 could be fit to these data, although it would be absurd to consider fitting such a model.

13-8 Lack-of-Fit Tests

Given that a polynomial model has been fitted and the estimated regression coefficients tested for their significance, how can one be confident that a model of order higher than the highest order tested is not likely needed? A lack-of-fit (abbreviated LOF) test can be used to address this question. Conceptually, a lack-of-fit test concerns the evaluation of a model more complex than the one under primary consideration. Historically, the term was sometimes used to describe the classical procedure discussed in the next paragraph.

TABLE 13-6   Replicates and pure error
estimates from repeated
observations for age–systolic
blood pressure data of
Table 5-1

X	Y	SS*	df
39	144, 120	288.0	1
42	124, 128	8.0	1
45	138, 135	4.5	1
56	154, 150	8.0	1
67	170, 158	72.0	1
		380.5 (SS_{PE})	5

* The sum of squares for a given X is calculated
using the formula $\sum_m (Y_{mx} - \bar{Y}_x)^2$, where Y_{mx} is
the mth observation of Y at $X = x$ and $\bar{Y}_x$ is the
mean of all replicates at $X = x$.

The classical lack-of-fit test can be applied only if there are replicate observations. By
the term *replicate,* we mean that an experimental unit (subject) has the same X-value as
another experimental unit. With n total observations, if d X-values are distinct, then the
number of replicates is $r = n - d$. Recall that a polynomial curve of order $d - 1$ can pass
through exactly d distinct points. A classical lack-of-fit test compares the fit of a polynomial
of order $d - 1$ with the fit of the polynomial model currently under consideration.

For the age–systolic blood pressure example (with the outlier removed), 5 X-values out
of the total of 29 involve replicates (i.e., $r = 5$). These are listed in Table 13-6. In a classical
lack-of-fit test of these data, a polynomial is considered of order $d - 1 = n - r - 1 = 29 -
5 - 1 = 23$, which is the higher order of the two polynomial models being compared. The
lower-order polynomial is the model of primary interest, such as the second-order model fit
to the age–systolic blood pressure data. For our example, the two models would be

$$Y = \beta_0 + \beta_1 X + \beta_2 X^2 + E$$

and

$$Y = \beta_0 + \beta_1 X + \beta_2 X^2 + \beta_3 X^3 + \cdots + \beta_{23} X^{23} + E$$

Table 13-7 contains the ANOVA information needed to compare these two models.
The F test for such a comparison is a multiple-partial F test of the null hypothesis $H_0: \beta_j =
0, j = 3, 4, \ldots, 23$. (It should be emphasized that trying to fit the larger 23rd-degree
polynomial directly will lead to serious collinearity problems; a possible alternative method
of fitting, described in Section 13-9, is the use of orthogonal polynomials.)

The ANOVA framework for the classical lack-of-fit test (see Tables 13-7 and 13-8)
partitions the residual sum of squares (SSE) for the model whose fit is being questioned into
two components: a pure error sum of squares SS_{PE} (with degrees of freedom df_{PE}) and a lack-
of-fit sum of squares SS_{LOF}. The test statistic, which is equivalent to the multiple-partial F test

TABLE 13-7 Regression ANOVA table for classical lack-of-fit test of second-order polynomial fit to age–systolic blood pressure data of Table 5-1

Source		df
X		1
$X^2\|X$		1
Residual { Lack of fit		21
Pure error		5
Total		28

described above, can be written as $F = \text{MS}_{\text{LOF}}/\text{MS}_{\text{PE}}$. In actuality, SS_{PE} is the error sum of squares for the higher-degree polynomial model being compared, of order $d - 1 = n - r - 1$. The quantity SS_{LOF} is the extra sum of squares due to the addition to the lower-order model of all higher-order terms needed to construct the higher-order model. In the example being

TABLE 13-8 Regression ANOVA table for classical lack-of-fit test of second-order polynomial fit to age–systolic blood pressure data of Table 5-1

Source		df
X		1
$X^2\|X$		1
Residual { Lack of fit $(X^3, X^4, \ldots, X^{23}\|X, X^2)$		21
Pure error		5
Total		28

Note: The lack-of-fit test statistic is given by

$$F = F(X^3, X^4, \ldots, X^{23}|X, X^2)$$
$$= \frac{[\text{regression SS}(X, X^2, \ldots, X^{23}) - \text{regression SS}(X, X^2)]/21}{\text{residual SS}(X, X^2, \ldots, X^{23})/5}$$
$$= \frac{\text{MS}_{\text{LOF}}}{\text{MS}_{\text{PE}}}$$

Under H_0: "no lack of fit of second-order model," this F statistic should have an F distribution with 21 and 5 degrees of freedom.

considered,

$$SS_{LOF} = \text{regression } SS(X, X^2, \ldots, X^{23}) - \text{regression } SS(X, X^2)$$

and

$$SS_{PE} = \text{residual } SS(X, X^2, \ldots, X^{23})$$

With the availability of standard computer regression packages, the use of a multiple-partial F test for lack of fit can be less cumbersome to compute than identifying replicate observations in order to compute SS_{PE} directly. Note that orthogonal polynomials (described below) *must* be used to avoid serious inaccuracy in the multiple regression computations leading to SS_{LOF}.

13-9 Orthogonal Polynomials

In Chapter 12, we illustrated collinearity problems when using polynomial models, and we also demonstrated that centering the predictor helped to remedy such problems for a second-order polynomial model. A more sophisticated approach is needed for higher-order models. The polynomials we have discussed so far have all been *natural polynomials*. This terminology derives from the fact that each of the independent variables (X, X^2, X^3, etc.) in a polynomial of the form

$$Y = \beta_0 + \beta_1 X + \beta_2 X^2 + \cdots + \beta_k X^k + E$$

is a simple polynomial by itself. Another method of fitting polynomial models involves orthogonal polynomials. As we shall show, orthogonal polynomials constitute a new set of independent variables that are defined in terms of the simple polynomials but have more complicated structures. In this section we shall explain the method and describe its advantages and disadvantages. The basic motivation behind the use of orthogonal polynomials is to avoid the serious collinearity inherent in the use of natural polynomials.

In Section 13-2, we alluded to the natural polynomial model

$$Y = \beta_0 + \beta_1 X + \beta_2 X^2 + \cdots + \beta_k X^k + E$$

The simple polynomials $X, X^2, \ldots, X^k$ are the predictors. Orthogonal polynomial variables are new predictor variables that are linear combinations of these simple polynomials. Denoting these orthogonal polynomials as $X_1^*, X_2^*, \ldots, X_k^*$, we can write them as linear combinations of the form

$$X_1^* = a_{01} + a_{11}X$$
$$X_2^* = a_{02} + a_{12}X + a_{22}X^2$$
$$\vdots$$
$$X_k^* = a_{0k} + a_{1k}X + a_{2k}X^2 + \cdots + a_{kk}X^k$$

where the a's are constants that relate the X^*'s to the original predictors. Note that each linear combination has the form of a polynomial. It can be shown that each of the simple polynomials can be written as a linear combination of the X^*'s as follows:

$$X = b_{01} + b_{11}X_1^*$$
$$X^2 = b_{02} + b_{12}X_1^* + b_{22}X_2^*$$
$$\vdots$$
$$X^k = b_{0k} + b_kX_1^* + b_{2k}X_2^* + \cdots + b_{kk}X_k^*$$

in which the b's are constants. With no loss of information, we can write either

$$Y = \beta_0 + \beta_1 X + \beta_2 X^2 + \cdots + \beta_k X^k + E$$

or

$$Y = \beta_0^* + \beta_1^* X_1^* + \beta_2^* X_2^* + \cdots + \beta_k^* X_k^* + E$$

The parameters $\{\beta_i^*\}$ in the latter model will be numerically different from those in the former. Moreover, the two sets of variables differ in that the simple polynomials are highly correlated with one another, while the orthogonal polynomials (via a judicious choice of the a's) are pairwise uncorrelated.[2] However, although the parameters and predictors in the two models have very different interpretations, it can be shown that the multiple R^2-values and the overall regression F tests obtained by fitting these two models are exactly the same.

In summary, the orthogonal polynomial values represent a recoding of the original predictors with two desirable basic properties: (1) *The orthogonal polynomial variables contain exactly the same information as the simple polynomial variables*, and (2) *the orthogonal polynomial variables are uncorrelated with each other*. The first property means that all questions about the natural polynomial model can be answered using the orthogonal polynomial model. For example, one can assess the overall strength of the regression relationship, conduct the overall regression F test, and even compute partial F tests using the orthogonal polynomial model. The second property, zero pairwise correlation, completely eliminates any collinearity.

With regard to partial F tests, it can be shown that the partial F test of H_0: $\beta_i^* = 0$ for the orthogonal polynomial model of order k is equivalent to the partial F test of H_0: $\beta_i = 0$ in the reduced natural polynomial model

$$Y = \beta_0 + \beta_1 X + \beta_2 X^2 + \cdots + \beta_i X^i + E$$

for any $i \le k$. Thus, only one orthogonal polynomial model has to be fit (of the highest order, say k, of interest) if one is interested in testing sequentially for the significance of each simple polynomial term in the original natural polynomial model. This may be summarized by saying that the k tests of H_0: $\beta_i^* = 0$, $i = 1, 2, \ldots, k$, using the orthogonal polynomial model are equivalent to the k variables-added-in-order tests of H_0: $\beta_i = 0$, $i = 1, 2, \ldots, k$, using an appropriate sequence of k natural polynomial models.

As an illustration, suppose one is considering a third-order natural polynomial model

$$Y = \beta_0 + \beta_1 X + \beta_2 X^2 + \beta_3 X^3 + E$$

---

[2] $r_{X_i X_j} = 0$ for all $i \ne j$.

Suppose, further, that it is of interest to determine a best model by proceeding backward, starting with a test of $H_0: \beta_3 = 0$ in the above cubic model. If this test is nonsignificant, one would proceed to a test of $H_0: \beta_2 = 0$ in a (reduced) parabolic model; in turn, if $H_0: \beta_2 = 0$ is not rejected, then one would perform a test of $H_0: \beta_1 = 0$ in a (reduced) straight-line model. One way to conduct these three tests would be to fit three separate natural polynomial models (cubic, parabolic, and straight-line models) and then to perform partial F (or t) tests, starting with the cubic model, to assess the significance of the highest-order term in each model, stopping at any point if significance is obtained. Unfortunately, collinearity will generally compromise reliable fitting of the cubic and (without centering) the quadratic model. To avoid this problem, one can fit a *single* third-order orthogonal polynomial model, then test sequentially $H_0: \beta_3^* = 0$, $H_0: \beta_2^* = 0$, and finally $H_0: \beta_1^* = 0$, stopping at any point if significance is obtained. This approach generally insures computational accuracy.

13-9-1 Transforming from Natural to Orthogonal Polynomials

In this section we describe one method for expressing orthogonal polynomials in terms of the original simple polynomials. This procedure involves specifying the a's in the above

TABLE 13-9   Observed weight gains and simple polynomial values

Obser-vation	WGTGAIN (Y)	DOSE1 (X)	DOSE2 (X^2)	DOSE3 (X^3)	DOSE4 (X^4)	DOSE5 (X^5)	DOSE6 (X^6)	DOSE7 (X^7)
1	0.9	1	1	1	1	1	1	1
2	1.1	2	4	8	16	32	64	128
3	1.6	3	9	27	81	243	729	2187
4	2.3	4	16	64	256	1024	4096	16384
5	3.5	5	25	125	625	3125	15625	78125
6	5.0	6	36	216	1296	7776	46656	279936
7	6.6	7	49	343	2401	16807	117649	823543
8	8.7	8	64	512	4096	32768	262144	2097152
9	0.9	1	1	1	1	1	1	1
10	1.1	2	4	8	16	32	64	128
11	1.6	3	9	27	81	243	729	2187
12	2.1	4	16	64	256	1024	4096	16384
13	3.4	5	25	125	625	3125	15625	78125
14	4.5	6	36	216	1296	7776	46656	279936
15	6.7	7	49	343	2401	16807	117649	823543
16	8.6	8	64	512	4096	32768	262144	2097152
17	0.8	1	1	1	1	1	1	1
18	1.2	2	4	8	16	32	64	128
19	1.4	3	9	27	81	243	729	2187
20	2.2	4	16	64	256	1024	4096	16384
21	3.2	5	25	125	625	3125	15625	78125
22	4.8	6	36	216	1296	7776	46656	279936
23	6.7	7	49	343	2401	16807	117649	823543
24	8.8	8	64	512	4096	32768	262144	2097152

Note: The linear predictor variable X is called DOSE1. The quadratic simple polynomial X^2 is called DOSE2, etc.; in general, the term X^i is called DOSEi.

equations relating the X^*'s to powers of X. We shall illustrate the method with a specific example. Assume that the scientist who collected the data in Table 13-2 decided to try to confirm the conclusions made in the first study by collecting new data. The same eight dosage values, from 1 to 8, were used. Three new observations were taken at each dosage, giving a total of 24 observations. Since eight distinct X-values are available, the highest-order polynomial that can be fitted is of the seventh order. We shall consider fitting third- and seventh-order polynomials to these data to demonstrate the use of orthogonal polynomials and an associated lack-of-fit test.

Table 13-9 provides the data from the new study and also gives the values of the X^2, $X^3, \ldots, X^7$ terms (which are labeled DOSE2, DOSE3, $\ldots$, DOSE7 in the table). These are the natural polynomial values. Table 13-10 includes the same response (Y) values, but the predictor (X) values have been replaced by orthogonal polynomial values (labeled ODOSE1, ODOSE2, $\ldots$, ODOSE7). These values were obtained using Table A-7 in Appendix A, as we shall now describe.

We first note that *Table A-7 can be used only if the predictor (X) values are equally spaced and if the same number of observations (i.e., replicates) occur at each value*. If either of these conditions is not satisfied, then Table A-7 cannot be used. In that case, a reasonable

TABLE 13-10   Observed weight gains and orthogonal polynomial values

Obser-vation	WGTGAIN	ODOSE1	ODOSE2	ODOSE3	ODOSE4	ODOSE5	ODOSE6	ODOSE7
1	0.9	−7	7	−7	7	−7	1	−1
2	1.1	−5	1	5	−13	23	−5	7
3	1.6	−3	−3	7	−3	−17	9	−21
4	2.3	−1	−5	3	9	−15	−5	35
5	3.5	1	−5	−3	9	15	−5	−35
6	5.0	3	−3	−7	−3	17	9	21
7	6.6	5	1	−5	−13	−23	−5	−7
8	8.7	7	7	7	7	7	1	1
9	0.9	−7	7	−7	7	−7	1	−1
10	1.1	−5	1	5	−13	23	−5	7
11	1.6	−3	−3	7	−3	−17	9	−21
12	2.1	−1	−5	3	9	−15	−5	35
13	3.4	1	−5	−3	9	15	−5	−35
14	4.5	3	−3	−7	−3	17	9	21
15	6.7	5	1	−5	−13	−23	−5	−7
16	8.6	7	7	7	7	7	1	1
17	0.8	−7	7	−7	7	−7	1	−1
18	1.2	−5	1	5	−13	23	−5	7
19	1.4	−3	−3	7	−3	−17	9	−21
20	2.2	−1	−5	3	9	−15	−5	35
21	3.2	1	−5	−3	9	15	−5	−35
22	4.8	3	−3	−7	−3	17	9	21
23	6.7	5	1	−5	−13	−23	−5	−7
24	8.8	7	7	7	7	7	1	1

Note: The linear orthogonal polynomial predictor variable is called ODOSE1, the quadratic orthogonal polynomial variable is called ODOSE2, etc.; in general, the ith-order variable is called ODOSEi.

alternative is to use a computer program (such as the ORPOL function in SAS[3] PROC MATRIX or PROC IML) to calculate the appropriate orthogonal polynomial values.

To go from Table 13-9 to 13-10, begin by noting that the variable "dosage" has eight distinct, equally spaced values and that three replicates are taken at each of these values. Hence, Table A-7 can be used. In Table A-7, k indicates the number of distinct values of X, in our case 8. The row in Table A-7 labeled "Linear" for $k = 8$ consists of the entries $(-7, -5, -3, -1, 1, 3, 5, 7)$. These eight entries will be used to replace the eight original values for the linear term X. Thus, a subject who received a dosage of 1 (as did the first subject) is given a linear orthogonal polynomial score of -7. Similarly, a subject who received a dosage of 2 is given a linear orthogonal polynomial score of -5, and so on, with a dosage of 8 corresponding to a score of 7. This pattern is repeated in Table 13-10 twice more for the replicate observations. Analogous operations yield the remaining columns appearing in Table 13-10. For example, the eight quadratic entries for $k = 8$ $(7, 1, -3, -5, -5, -3, 1, 7)$ give the orthogonal polynomial values for the variable ODOSE2 in Table 13-10, which replace the original eight values for the quadratic term X^2 (given in the corresponding column of Table 13-9).

Finally, note that the rightmost column of Table A-7 gives the sum of squared values $(\sum p_i^2)$ for each row in the table. For $k = 8$, dividing each linear orthogonal polynomial score by $\sqrt{168}$, each quadratic score by $\sqrt{168}$, each cubic score by $\sqrt{264}$, and so on, will make the variance of each set of orthogonal polynomial scores (i.e., each set of row entries) equal to 1. This standardization helps in two ways. First, numerical accuracy is improved by avoiding scaling problems. Second, we shall see later that the estimated standard errors of the resulting estimated regression coefficients are all equal. This fact makes it simpler to compare and interpret such regression coefficients.

13-9-2 Regression Analysis with Orthogonal Polynomials

It is instructive to consider the regression ANOVA tables based on fitting third- and seventh-order polynomial models to the weight gain data in Tables 13-9 and 13-10. Table 13-11 summarizes the results for the third-order natural polynomial model. First, note that the model fits extremely well ($R^2 = .998$). Furthermore, the variable-added-in-order tests give P-values of .0001, .0001, and .3995 for linear, quadratic, and cubic tests, respectively. These results clearly argue for a second-order model. Since the simple polynomials X, X^2, and X^3 are correlated, the variable-added-last tests are more difficult to interpret. Nevertheless, the cubic test again gives $P = .3995$ since it is the last variable to enter; the linear effect is not significant ($P = .1359$), but the quadratic effect is ($P = .0046$). When using variable-added-last tests, proceeding backwards is recommended, starting with the cubic model and including in the final model all lower-order terms below the highest term deemed significant. Then, the series of variable-added-last tests also supports the choice of the second-order model.

This preference for a second-order model is further supported by comparing R^2-values for the second- and third-order models. It can be shown, using the variable-added-in-order SS entries in Table 13-11, that the R^2-value for the second-order model is

$$R^2 = \frac{155.6666 + 14.0333}{170.039} = .998$$

which is the same (within round-off error) as the R^2-value for the third-order model.

---

[3] SAS is a registered trademark of SAS Institute, Cary, N.C.

TABLE 13-11   Third-order natural polynomial model for weight gain data ($N = 24$)

Source	df	SS	MS
Model[a]	3	162.7121	56.5707
Error	20	0.3274	0.0163
Total (corrected)[b]	23	170.0395	

Source	df	Variables-Added-in-Order SS	F
DOSE1 (X)	1	155.6666	9508.98 $(P = .0001)$
DOSE2 $(X^2\|X)$	1	14.0333	857.23 $(P = .0001)$
DOSE3 $(X^3\|X, X^2)$	1	0.0121	0.74 $(P = .3995)$

Source	df	Variables-Added-Last SS	F
DOSE1 $(X\|X^2, X^3)$	1	0.0395	2.41 $(P = 0.1359)$
DOSE2 $(X^2\|X, X^3)$	1	0.1662	10.15 $(P = 0.0046)$
DOSE3 $(X^3\|X, X^2)$	1	0.0121	0.74 $(P = 0.3995)$

Parameter	Estimate	T for H_0: Parameter = 0	$S_{\hat{\beta}_j}$
Intercept	1.0261	5.64 $(P = .0001)$	0.182
DOSE1	−0.2558	−1.55 $(P = .1359)$	0.164
DOSE2	0.1316	3.19 $(P = .0046)$	0.041
DOSE3	0.0026	0.86 $(P = .3995)$	0.003

[a] $F = 3,455.65 \ (P = .0001)$.

[b] $R^2 = .998075$.

Note that the t test P-values are the same as the P-values for the variable-added-last F test (.1359, .0046, and .3995 for linear, quadratic, and cubic); this is because these t tests are equivalent to the variable-added-last F tests (note that $T_{20}^2 = F_{1,20}$). All such tests use the residual from the largest model, which has 20 degrees of freedom.

For comparison, consider the results of fitting the third-order orthogonal polynomial model, as summarized in Table 13-12. The most important difference between Tables 13-11 and 13-12 is that the two types of sum of squares (namely, variable-added-in-order and variable-added-last) are equal in the latter. Consequently, the corresponding partial F test results are the same and coincide with the t test results. Furthermore, the standard errors of the estimated linear, quadratic, and cubic regression coefficients are identical. As a result, the estimated regression coefficients can be compared directly.

The above results, whether from analysis using natural or orthogonal polynomials, indicate that it is unnecessary to consider any polynomial model higher than second order. Nevertheless, it is instructive to use these data to illustrate particular problems that arise if the model order exceeds two. Specifically, we address the (hidden) problem of collinearity. The existence of this collinearity problem can be demonstrated by considering the predictor correlations given in Table 13-13 for both the simple polynomials and their centered counterparts, that is, $(X - \bar{X})$, $(X - \bar{X})^2$, ..., as well as X, X^2.... We first focus on the

TABLE 13-12   Third-order orthogonal polynomial
model for weight gain data ($N = 24$)

Source	df	SS	MS
Model[a]	3	162.7121	56.5707
Error	20	0.3274	0.0163
Total (corrected)[b]	23	170.0395	

Source	df	Variables-Added-in-Order SS	F
ODOSE1	1	155.6666	9508.98 ($P = .0001$)
ODOSE2	1	14.0333	857.23 ($P = .0001$)
ODOSE3	1	0.0121	0.74 ($P = .3995$)

Source	df	Variables-Added-Last SS	F
ODOSE1	1	155.6666	9508.98 ($P = .0001$)
ODOSE2	1	14.0333	857.23 ($P = .0001$)
ODOSE3	1	0.0121	0.74 ($P = .3995$)

Parameter	Estimate	T for H_0: Parameter = 0	$S_{\hat{\beta}_j^*}$
Intercept	3.6541	139.91 ($P = .0001$)	0.0261
ODOSE1	7.2033	97.51 ($P = .0001$)	0.0738
ODOSE2	2.1628	29.28 ($P = .0001$)	0.0738
ODOSE3	0.0635	0.86 ($P = .3995$)	0.0738

[a] $F = 3,455.65$ ($P = .0001$).

[b] $R^2 = .998075$.

correlations among the linear, quadratic, and cubic terms in a third-order model; this information is contained in the upper left corner of each array (above and to the left of the dashed lines). The existence of a collinearity problem is suggested by the presence of three very high correlations (all above .93) for the noncentered predictor data. Notice that centering X helps reduce collinearity: Two of the three correlations become 0 after centering. Turning to the complete array, we again see, for the noncentered polynomials, that the smallest off-diagonal correlation is .779, and that four are greater than .990. As before, centering X helps to reduce collinearity substantially since correlations between odd and even powers are then all 0. Nevertheless, the remaining correlations are high, with two correlations greater than .99. Note that the analogous correlation array using orthogonal polynomials would have all zeros off the diagonal.

In Table 13-14 the collinearity problem suggested by Table 13-13 can be examined further. Predictor correlation matrix eigenvalues (see Chapter 12) are reported here for third- and seventh-order natural, centered, and orthogonal polynomial models. For the third-order natural polynomial model, the condition number (CN; see Chapter 12) is 70.8, suggesting a collinearity problem. If the X-variable is centered (i.e., the dosage mean $\bar{X} = 4.5$ is subtracted from the dosage X before the second- and third-order terms are computed), the condition number is reduced to 5.1, indicating no severe collinearity problem. Unfortunately, as shown

TABLE 13-13 Predictor correlations for weight gain data ($N = 24$)

Simple Polynomials

	X	X^2	X^3	X^4	X^5	X^6	X^7
X	1	.976	.932	.887	.846	.810	.779
X^2		1	.988	.963	.935	.908	.882
X^3			1	.993	.978	.960	.941
X^4				1	.996	.986	.973
X^5					1	.997	.990
X^6						1	.998
X^7							1

Centered Simple Polynomials

	$(X - \bar{X})$	$(X - \bar{X})^2$	$(X - \bar{X})^3$	$(X - \bar{X})^4$	$(X - \bar{X})^5$	$(X - \bar{X})^6$	$(X - \bar{X})^7$
$(X - \bar{X})$	1	0	.926	0	.855	0	.813
$(X - \bar{X})^2$		1	0	.969	0	.932	0
$(X - \bar{X})^3$			1	0	.985	0	.965
$(X - \bar{X})^4$				1	0	.992	0
$(X - \bar{X})^5$					1	0	.996
$(X - \bar{X})^6$						1	0
$(X - \bar{X})^7$							1

TABLE 13-14   Eigenvalues of predictor correlation
matrices for polynomial models for
weight gain data ($N = 24$)

Third-Order Model

	Natural Polynomial		Orthogonal
Eigenvalue	Uncentered	Centered	Polynomial
1	2.931	1.926	1.000
2	0.069	1.000	1.000
3	6×10^{-4}	0.074	1.000
CN	70.8	5.1	1.0

Seventh-Order Model

	Natural Polynomial		Orthogonal
Eigenvalue	Uncentered	Centered	Polynomial
1	6.638	3.773	1.000
2	0.348	2.929	1.000
3	0.013	0.221	1.000
4	3×10^{-4}	0.071	1.000
5	4×10^{-6}	0.006	1.000
6	2×10^{-8}	5×10^{-4}	1.000
7	2×10^{-11}	2×10^{-5}	1.000
CN	569,664.0	430.0	1.0

TABLE 13-15 Seventh-order orthogonal polynomial
model for weight gain data ($N = 24$)

Source	df	SS	MS
Model[a]	7	169.7795	24.2542
Error	16	0.2600	0.0162
Total (corrected)[b]	23	170.0395	

Source	df	Variables-Added-in-Order SS	F
ODOSE1	1	155.6666	9579.98 ($P = .0001$)
ODOSE2	1	14.0333	863.59 ($P = .0001$)
ODOSE3	1	0.0121	0.75 ($P = .4003$)
ODOSE4	1	0.0509	3.13 ($P = .0958$)
ODOSE5	1	0.0000	0.00 ($P = .9924$)
ODOSE6	1	0.0036	0.22 ($P = .6420$)
ODOSE7	1	0.0128	0.79 ($P = .3871$)

Source	df	Variables-Added-Last SS	F
ODOSE1	1	155.6666	9579.48 ($P = .0001$)
ODOSE2	1	14.0333	863.59 ($P = .0001$)
ODOSE3	1	0.0121	0.75 ($P = .4003$)
ODOSE4	1	0.0509	3.13 ($P = .0958$)
ODOSE5	1	0.0000	0.00 ($P = .9924$)
ODOSE6	1	0.0036	0.22 ($P = .6420$)
ODOSE7	1	0.0128	0.79 ($P = .3871$)

Parameter	Estimate	T for H_0: Parameter $= 0$	$S_{\hat{\beta}_j}$
Intercept	3.6541	140.43 ($P = .0001$)	0.0260
ODOSE1	7.2033	97.87 ($P = .0001$)	0.0735
ODOSE2	2.1628	29.39 ($P = .0001$)	0.0735
ODOSE3	0.0635	0.86 ($P = .4003$)	0.0735
ODOSE4	−0.1302	−1.77 ($P = .0958$)	0.0735
ODOSE5	0.0007	0.01 ($P = .9924$)	0.0735
ODOSE6	−0.0348	0.47 ($P = .6420$)	0.0735
ODOSE7	−0.0654	−0.89 ($P = .3871$)	0.0735

[a] $F = 1,492.57$ ($P = .0001$).

[b] $R^2 = .998075$.

by the correlation arrays in Table 13-3, such centering will *not* solve the collinearity problem for higher-order models. For a seventh-order model based on centered dosage data, the condition number is a very disturbing 430.0. Note also that the condition number for an uncentered, seventh-order natural polynomial model is an alarming 569,664. In contrast, the condition number for both the third-order and the seventh-order orthogonal polynomial model is 1.0. This will be true for any fitted orthogonal polynomial model.

The condition numbers above lead us to recommend only the use of orthogonal polynomials (i.e., we rule out using natural polynomials) for conducting lack-of-fit tests.[4] Table 13-15 summarizes the ANOVA results based on fitting a seventh-order orthogonal polynomial model to the $N = 24$ dosage–weight gain observations. Notice that the lower-order polynomial results (up to, say, order three) remain virtually unchanged. This follows from the fact that the quadratic model fits so well and from the properties of orthogonal polynomials. After the .0001 P-values for the linear and quadratic terms, the next smallest P-value is .0958 for the fourth-order term. As before, a second-order polynomial model seems most appropriate.

As discussed in Section 13-8, the lack-of-fit test for the second-order model may be performed using a multiple-partial F statistic of the form

$$F = \frac{[\text{regression SS (seventh order)} - \text{regression SS (second order)}]/(7 - 2)}{\text{SSE (seventh order)}/(24 - 1 - 7)}$$

The actual value of the statistic in our example is

$$F = \frac{(169.7795 - 169.6999)/5}{.2600/16} = 0.98$$

With 5 and 16 degrees of freedom for $\alpha = .25$, the critical F value is 1.48. Hence, we fail to reject the null hypothesis of no lack of fit; i.e., a second-order model provides a good description of these data.

13-10 Strategies for Choosing a Polynomial Model

In our discussion of polynomial models, we have sometimes started with the smallest model, involving only a linear term, and sequentially added increasing powers of X. This is a forward selection model-building strategy. Although this is a natural approach, it can give misleading results for inference-making procedures.

With a forward selection strategy, one usually tests for the importance of a candidate predictor by comparing the extra regression sum of squares for the addition of that predictor to the residual mean square. This residual mean square is based on fitting a model containing the candidate (predictor) variable and those variables already in the model. The corresponding partial F statistic is of the form

$$F(X^i | X, X^2, \ldots, X^{i-1}) = \frac{\text{SS } (X^i | X, X^2, \ldots, X^{i-1})/1}{\text{MS residual } (X, X^2, \ldots, X^i)}$$

when one is testing for the importance of X^i in a polynomial model already containing terms up through X^{i-1}. Notice that the mean-square residual in the above expression is not based on terms of order higher than X^i, even though such terms may actually belong in the final model.

---

[4] In addition, we note that centering X may be required when using a computer program to calculate orthogonal polynomial values. Although it is usually not the case, orthogonal polynomial calculations can be inaccurate for large samples and/or for very high order models (without centering).

The forward selection testing approach described above can lead to underfitting the data (i.e., the forward selection algorithm is likely to quit too soon, thereby choosing a model with an order lower than is actually required). The reason for this problem is that, when proceeding forward, the residual mean-square error estimate of σ^2 at any step will be biased upward if the polynomial model at that step is of too low a degree. Since the denominator in the above partial F expression then tends to be too large, the F statistic itself may be too small and hence nonsignificant, thus stopping the forward selection algorithm prematurely.

This underfitting bias can be avoided by using a backward elimination strategy (see Chapter 16), where the partial F test at each backward step always involves the residual mean error for the full (or largest) model fitted. When using this backward elimination approach, however, it is possible to overfit the data (i.e., to choose a final model of order slightly higher than required). Fortunately, the residual mean-square error estimate from the full model is still a valid (unbiased) estimate of σ^2. Consequently, using this estimate in the denominator of the partial F test at any backward step will be a statistically valid procedure. What is lost by slightly overfitting the data is some statistical power, but usually this loss is negligible.

Thus, for fitting polynomial models, we generally recommend a backward elimination strategy for selecting variables, using in all partial F tests the mean-square error estimate based on the largest-order polynomial model fitted. When implementing this strategy, we recommend first choosing this full model to be of third order or lower to simplify interpretation and to enhance computational accuracy. Second, proceeding backward in a stepwise fashion starting with the largest power term, one should sequentially delete terms that are nonsignificant, stopping at the first power term that is significant; this term and all lower-order terms should be retained in the final model. Third, conduct a multiple-partial F test for lack of fit. Fourth, the residual analysis methods of Chapter 12 should be applied, as with all regression modeling. Of particular use for polynomial regression is a plot of jackknife residuals against X. Any need for a higher-order model will often appear as a nonlinear trend in the residuals.

Problems

1. In an environmental engineering study of a certain chemical reaction, the concentrations of 18 separately prepared solutions were recorded at different times (three measurements at each of six times). The natural logarithms of the concentrations were also computed. The data are as given in the table.

Solution Number (i)	TIME (X_i) (hr)	Concentration (Y_i) (mg/ml)	Ln Concentration (ln Y_i)
1	6	0.029	−3.540
2	6	0.032	−3.442
3	6	0.027	−3.612
4	8	0.079	−2.538
5	8	0.072	−2.631
6	8	0.088	−2.430

Solution Number (i)	TIME (X_i) (hr)	Concentration (Y_i) (mg/ml)	Ln Concentration (ln Y_i)
7	10	0.181	−1.709
8	10	0.165	−1.802
9	10	0.201	−1.604
10	12	0.425	−0.856
11	12	0.384	−0.957
12	12	0.472	−0.751
13	14	1.130	0.122
14	14	1.020	0.020
15	14	1.249	0.222
16	16	2.812	1.034
17	16	2.465	0.902
18	16	3.099	1.131

a. Plot on separate sheets of graph paper:
   (1) Concentration (Y) versus time (X).
   (2) Natural logarithm of concentration (ln Y) versus time (X).

b. Using the computer printout, obtain the following:
   (1) The estimated equation of the straight-line (degree 1) regression of Y on X.
   (2) The estimated equation of the quadratic (degree 2) regression of Y on X.
   (3) The estimated equation of the straight-line (degree 1) regression of ln Y on X.
   (4) Plots of each of these fitted equations on their respective scatter diagrams.

c. Based on the computer printout, complete the table for the straight-line regression of Y on X.

Source	df	SS	MS	F
Regression	1			
Residual { Lack of fit	4			
{ Pure error	12			
Total	17			

d. Based on the computer printout, complete the ANOVA table below.

Source	df	SS	MS	F
Regression { Degree 1 (X)	1			
{ Degree 2 ($X^2|X$)	1			
Residual { Lack of fit	3			
{ Pure error	12			
Total	17			

e. Determine and compare the proportions of the total variation in Y explained by the straight-line regression on X and by the quadratic regression on X.

Computer Printout for Problem 1

Concentration (Y) on Time (X):

FITTING DEGREE 1
MULTIPLE R-SQUARED = .732

REGRESSION COEFFICIENTS

0   -1.9318
1   .24597

FITTING DEGREE 2
MULTIPLE R-SQUARED = .957

REGRESSION COEFFICIENTS

0   3.1721
1   -.78102
2   .46682E-01

ANALYSIS OF VARIANCE

SOURCE	DF	SS	MS
DEGREE 1	1	12.705	12.705
DEGREE 2	1	3.9051	3.9051
LACK OF FIT	3	.51446	.17149
PURE ERROR	12	.23248	.19374E-01
TOTAL	17	17.357	1.0210

Log$_e$ Concentration (ln Y) on Time (X):

FITTING DEGREE 1
MULTIPLE R-SQUARED = .996

REGRESSION COEFFICIENTS

0   -6.2096
1   .45117

ANALYSIS OF VARIANCE

SOURCE	DF	SS	MS
DEGREE 1	1	42.746	42.746
LACK OF FIT	4	.27836E-01	.69591E-02
PURE ERROR	12	.12247	.10206E-01
TOTAL	17	42.896	2.5253

    f. Carry out F tests for the significance of the straight-line regression of Y on X and for the adequacy of fit of the estimated regression line.

    g. Carry out an overall F test for the significance of the quadratic regression of Y on X, a test for the significance of the addition of X^2 to the model, and an F test for the adequacy of fit of the estimated quadratic model.

    h. For the straight-line regression of $\ln Y$ on X, carry out F tests for the significance of the overall regression and for the adequacy of fit of the straight-line model.

    i. What proportion of the variation in $\ln Y$ is explained by the straight-line regression of $\ln Y$ on X? Compare this result with that obtained in part (e) for the quadratic regression of Y on X.

    j. A fundamental assumption in regression analysis is variance homoscedasticity.

        (1) By examining the scatter diagrams constructed in part (a), state why taking natural logarithms of the concentrations helps with regard to the assumption of variance homogeneity.

        (2) Do you think the straight-line regression of $\ln Y$ on X is better for describing this set of data than the quadratic regression of Y on X? Explain.

    k. What key assumption about the data would be in question if, instead of 18 different solutions, there were only 3 different solutions, each of which was analyzed at the six different time points?

2. With the addition of five pairs of observations, (18,000, 39.2), (22,400, 27.9), (24,210, 22.3), (5,400, 11.7), and (9,340, 32.5), to the data in Problem 3 of Chapter 5, the accompanying printout is obtained for the regression of TIME (Y) on INC (X).

    a. Using the computer output, complete the ANOVA table for the straight-line regression of TIME (Y) on INC (X).

Source		df	SS	MS	F
Regression		1			
Residual {	Lack of fit	18			
	Pure error	5			
Total		24			

    b. Using the computer results, complete the ANOVA table for the quadratic regression of TIME (Y) on INC (X).

Source		df	SS	MS	F
Regression {	Degree 1 (X)	1			
	Degree 2 ($X^2\|X$)	1			
Residual {	Lack of fit	17			
	Pure error	5			
Total		24			

    c. Plot on the scatter diagram of the data for this problem the fitted straight-line (degree 1) equation and the fitted quadratic (degree 2) equation.

Computer Printout for Problem 2

FITTING DEGREE 1
MULTIPLE R-SQUARED = 0.153

REGRESSION COEFFICIENTS

```
0    20.17655
1     0.00061
```

FITTING DEGREE 2
MULTIPLE R-SQUARED = 0.880

REGRESSION COEFFICIENTS

```
0   -19.86602
1     0.00787
2    -0.00000025
```

FITTING DEGREE 3
MULTIPLE R-SQUARED = 0.901

REGRESSION COEFFICIENTS

```
0   -35.29278
1     0.01223
2    -0.0000006
3     0.0000000
```

ANALYSIS OF VARIANCE

SOURCE	DF	SS	MS
DEGREE 1	1	442.91410	442.91410
DEGREE 2	1	2100.08113	2100.08113
DEGREE 3	1	61.10018	61.10018
LACK OF FIT	16	271.74872	16.98430
PURE ERROR	5	15.20121	3.04024
TOTAL	24	2891.04534	120.46022

d. Calculate and compare the R^2-values obtained for the straight-line, quadratic, and cubic fits.

e. Carry out F tests for the significance of the straight-line regression and for the adequacy of fit of the straight-line model.

f. Carry out F tests for the significance of the quadratic regression, of the addition of the quadratic term to the model, and of the adequacy of fit of the quadratic model.

g. Which model is most appropriate: straight line, quadratic, or cubic?

3. a–g. For the data on DIST (Y) and MPH (X) in Problem 7 of Chapter 5, use the information provided below to answer the same questions as in parts (a) through (g) of Problem 2 above.

$$\text{Degree 1 fit:} \quad \hat{Y} = -122.345 + 6.227X$$

$$\text{Degree 2 fit:} \quad \hat{Y} = 32.901 - 3.051X + 0.1176X^2$$

$$\text{Degree 3 fit:} \quad \hat{Y} = 114.621 - 10.620X + 0.3247X^2 - 0.00173X^3$$

Source		df	SS	MS
Regression	X	1	173,473.96	173,473.96
	$X^2 \mid X$	1	10,515.44	10,515.44
	$X^3 \mid X, X^2$	1	415.19	415.19
Residual	Lack of fit	11	2,664.15	242.20
	Pure error	4	3,433.93	858.48
Total		18	190,502.67	

4. For the date on VOTE (Y) and TVEXP (X) in Problem 5 of Chapter 5, it was found that the straight-line model was adequate. Using the computer printout for quadratic regression, do the following:

a. Plot the fitted straight-line model and the fitted quadratic model on the scatter diagram for the data of this problem.

b. Determine the change in R^2 in going from a degree 1 to a degree 2 model.

c. Test for the significance of the addition of the X^2 term to the model.

d. Do the results in parts (a) through (c) contradict your earlier conclusion about the adequacy of fit of the straight-line model?

Computer Printout for Problem 4

```
FITTING DEGREE 1                    FITTING DEGREE 2
MULTIPLE R-SQUARED = 0.910          MULTIPLE R-SQUARED = 0.911

REGRESSION COEFFICIENTS             REGRESSION COEFFICIENTS

0          2.17407                  0          9.65678
1          1.17697                  1          0.75077
                                    2          0.00575
```

Source	df	SS	MS
Degree 1	1	2,023.20500	2,023.20500
Degree 2	1	2.45015	2.45015
Lack of fit	12	166.80279	13.90023
Pure error	5	30.55988	6.11198
Total	19	2,223.01782	117.00094

5. For the regression of PCI (Y) on YNG (X) for African countries considered in Problem 2 of Chapter 14, use the information provided to do the following:
   a. Plot the estimated straight-line and quadratic models on the scatter diagram for the data of the African countries.
   b. Test for the significance of the straight-line regression and for the adequacy of fit of the straight-line model.
   c. Test for the significance of the addition of the X^2 term to the model.
   d. Which model is more appropriate, the straight-line model or the quadratic model?

$$\text{Degree 1 fit:}\quad \hat{Y} = 893.57 - 17.276X$$
$$\text{Degree 2 fit:}\quad \hat{Y} = 732.05 - 9.203X - 0.0996X^2$$

Source	df	SS	MS
Regression $\begin{cases} X \\ X^2\mid X \end{cases}$	1	153,784.8	153,784.8
	1	88.3	88.3
Residual $\begin{cases} \text{Lack of fit} \\ \text{Pure error} \end{cases}$	15	2,773.9	184.9
	8	911.5	113.9
Total	25	157,558.5	

6. For the data given in Problem 12 of Chapter 7, which concerns the relationship between the temperature (X) of a certain medium and the growth (Y) of human amniotic cells in a tissue culture, it is of interest to evaluate whether a parabolic model is more appropriate than a straight-line model. Use the computer results given to answer the following questions:
   a. Plot the fitted straight-line model and the fitted quadratic model on the same scatter diagram.
   b. Test for the significance of the addition of the X^2 term to the model.
   c. Determine the change in R^2 in going from a straight line to a parabolic model.
   d. How do the results in parts (b) and (c) compare with your results in Problem 12 of Chapter 7 for the earlier test of adequacy of fit of the straight-line model?
   e. Which model is more appropriate, straight line or parabolic?
7. The skin response in rats to different concentrations of a newly developed vaccine was measured in an experiment, resulting in the data, models, and ANOVA table that follow.

Computer Printout for Problem 6 (from SPSS package)

```
DEPENDENT VARIABLE.. Y

VARIABLE(S) ENTERED ON STEP NUMBER 1.. X

MULTIPLE R         0.98603      ANALYSIS OF VARIANCE    DF    SUM OF SQUARES   MEAN SQUARE        F
R SQUARE           0.97225      REGRESSION              1.       12.83072       12.83072      630.71608
ADJUSTED R SQUARE  0.97071      RESIDUAL               18.        0.36618        0.02034
STANDARD ERROR     0.14263

           ———— VARIABLES IN THE EQUATION ————            ———— VARIABLES NOT IN THE EQUATION ————
VARIABLE       B        BETA    STD ERROR B      F       VARIABLE    BETA IN    PARTIAL   TOLERANCE     F
X          0.03582    0.98603    0.00143      630.716       XX       0.84502    0.64297    0.01606    11.981
(CONSTANT) -0.46240


VARIABLE(S) ENTERED ON STEP NUMBER 2.. XX

MULTIPLE R         0.99183      ANALYSIS OF VARIANCE    DF    SUM OF SQUARES   MEAN SQUARE        F
R SQUARE           0.98372      REGRESSION              2.       12.98210        6.49105      513.73362
ADJUSTED R SQUARE  0.98181      RESIDUAL               17.        0.21480        0.01264
STANDARD ERROR     0.11241

           ———— VARIABLES IN THE EQUATION ————            ———— VARIABLES NOT IN THE EQUATION ————
VARIABLE       B        BETA    STD ERROR B      F       VARIABLE    BETA IN    PARTIAL   TOLERANCE     F
X          0.00537    0.14782    0.00887      0.367
XX         0.00022    0.84502    0.00006     11.981
(CONSTANT) 0.49460

MAXIMUM STEP REACHED
```

Note: B stands for the regression coefficient $\hat{\beta}$, XX stands for X^2, and you can ignore for now the terms "BETA," "PARTIAL," and "TOLERANCE." Also,

$$\text{adjusted } R^2 = R^2 - \left(\frac{k}{n - k - 1}\right)(1 - R^2).$$

From *Statistical Package for the Social Sciences* by Nie et al. Copyright © 1975 by McGraw-Hill Book Company and Dr. Norman Nie, President, SPSS Inc.

Concentration (X) (ml/l)	0.5	0.5	1.0	1.0	1.5	1.5	2.0	2.0	2.5	2.5	3.0	3.0
Skin response (Y) (mm)	13.90	13.81	14.08	13.99	13.75	13.60	13.32	13.39	13.45	13.53	13.59	13.64

Degree 1: $\hat{Y} = 13.986 - 0.1802X$

Degree 2: $\hat{Y} = 14.270 - 0.6065X + 0.1218X^2$

Degree 3: $\hat{Y} = 13.362 + 1.6800X - 1.3929X^2 + 0.2885X^3$

Degree 1				Degree 2				Degree 3		
Source	df	SS		Source	df	SS		Source	df	SS
Regression	1	0.2844		Regression	2	0.3536		Regression	3	0.5222
Residual	10	0.3461		Residual	9	0.2769		Residual	8	0.1083

a. Plot the straight-line, quadratic, and cubic equations on the scatter diagram for this data set.

b. Test sequentially for significant straight-line fit, for significant addition of X^2, and for significant addition of X^3 to the model.

c. Which of the three models do you recommend and why? (*Note:* You might also wish to consider R^2 for each model.)

8. This problem uses the data presented in the table in Problem 12 of Chapter 5. Use $\alpha = .05$.

a. Use a computer program to fit a natural polynomial cubic model for predicting vocabulary size as a function of age in years. Provide estimated regression coefficients.

b. Using variables-added-in-order tests, determine the best model.

c. Report variables-added-last tests. Explain any differences between results here and in part (b).

d. Report appropriate collinearity diagnostics for the model and evaluate them. Include predictor correlations.

e. Plot jackknife (or studentized) residuals against predicted values for the best model based on results in (b). Provide a frequency histogram or schematic plot of the residuals, and comment on it.

f. Compare these results to those from Chapter 5, Problem 12.

9. a–e. Repeat Problem 8, parts (a) through (e), after centering the predictor (age).

f. Compare the results to those in Problem 8.

10. a–e. Repeat Problem 8, (a) through (e), but use orthogonal polynomials.

f. Compare the results to those in Problems 8 and 9. *Hint:* Table A-7 cannot be used.

11. This problem uses the data from Problem 13 in Chapter 5. Use $\alpha = .10$.

a. Use a computer program to fit a quadratic natural polynomial model for predicting latency as a function of weight minus average weight.

b–e. Repeat parts (b) through (e) from Problem 8 for this analysis.

f. Compare these results to those in Chapter 5.

12. This problem uses the data from Problem 15 of Chapter 5.

  a. Using a computer program, fit a cubic polynomial model with BLOODTOL as the response and PPM_TOLU as the predictor. Center PPM_TOLU. Provide the prediction equation.

  b–e. Repeat parts (b) through (e) from Problem 8 for this analysis.

  f. For each predictor value, compute an estimate of variance of the response variable. Are these approximately equal?

  g. Compare these results to those from Chapter 5.

13. a. For the data from Problem 15 of Chapter 5, specify the orthogonal polynomial codings needed for coding PPM_TOLU linear, quadratic, and cubic terms. Repeat part (a) of Problem 12, but use the orthogonal coding. *Hint*: Table A-7 cannot be used.

  b–e. Repeat parts (b) through (e) from Problem 8 for this analysis.

Dummy Variables
in Regression

14-1 Preview

To this point we have considered only continuous variables as predictors. The methods of regression analysis can be generalized to treat categorical predictors as well. The generalization is based entirely on the use of dummy variables, the central idea of this chapter.

It is by using dummy variables that the application of regression analysis can be broadened. In particular, dummy variables allow one to employ regression analysis to produce the same information obtained by such seemingly distinct analytical procedures as analysis of covariance (Chapter 15), analysis of variance (Chapters 17 through 20), and discriminant analysis (Chapter 23).

In this chapter we shall focus on an important application of dummy variables: comparing several regression equations by use of a single multiple regression model. We shall also describe an alternative method that may be used if only two equations are being compared.

14-2 Definitions

A *dummy*, or *indicator*, *variable* is any variable in a regression equation that takes on a finite number of values so that different categories of a nominal variable can be identified. The term *dummy* simply relates to the fact that the values taken on by such variables (usually values like 0, 1, and -1) indicate no meaningful measurement but rather the categories of interest.

Examples of dummy variables include the following:

$$X_1 = \begin{cases} 1 & \text{if treatment A is used} \\ 0 & \text{otherwise} \end{cases}$$

$$X_2 = \begin{cases} 1 & \text{if subject is female} \\ -1 & \text{if subject is male} \end{cases}$$

$$Z_1 = \begin{cases} 1 & \text{if residence is in western United States} \\ 0 & \text{if residence is in central United States} \\ -1 & \text{if residence is in eastern United States} \end{cases}$$

$$Z_2 = \begin{cases} 0 & \text{if residence is in western United States} \\ 1 & \text{if residence is in central United States} \\ -1 & \text{if residence is in eastern United States} \end{cases}$$

The variable X_1 indicates a nominal variable describing "treatment group" (either treatment A or not treatment A), the variable X_2 indexes the levels of the nominal variable "sex," and variables Z_1 and Z_2 work in tandem to describe the nominal variable "geographical residence." Note that in the latter case, the three categories of geographical residence are described by the following combination of the two variables Z_1 and Z_2:

residence in western United States: $Z_1 = 1, Z_2 = 0$

residence in central United States: $Z_1 = 0, Z_2 = 1$

residence in eastern United States: $Z_1 = -1, Z_2 = -1$

14-3 Rule for Defining Dummy Variables

We recommend that the following simple rule always be applied to avoid collinearity in defining a dummy variable for regression analysis: *If the nominal independent variable of interest has k categories, then one must define exactly k − 1 dummy variables to index these categories, provided that the regression model contains a constant term (i.e., an intercept β_0). If the regression model does not contain an intercept, then k dummy variables are needed to index the k categories of interest.* For example, if there are $k = 3$ categories, the number of dummy variables should be $k - 1 = 2$ for a model containing an intercept.

(One exception needs to be pointed out. If an intercept is not included in an overall regression model designed to compare several regression equations, the dummy variables can be defined so that each of the regression equations derived from the overall model has its own intercept. Thus, the use of an intercept in the overall model generally depends upon how the investigator prefers to code the dummy variables.)

In applying this rule, the following should be noted:

1. If an intercept is used in the regression equation, proper definition of the $k - 1$ dummy variables automatically indexes all the k categories.

2. If k dummy variables are used to describe a nominal variable with k categories in a model containing an intercept, all the coefficients in the model cannot be uniquely estimated because collinearity is present.

3. There are many different ways to properly define the $k - 1$ dummy variables to index the k categories of a given nominal variable. For example, two equivalent

ways to describe the nominal variable "geographical residence" (represented earlier by Z_1 and Z_2) are

$$Z_1^* = \begin{cases} 1 & \text{if residence is in western United States} \\ 0 & \text{otherwise} \end{cases}$$

$$Z_2^* = \begin{cases} 1 & \text{if residence is in central United States} \\ 0 & \text{otherwise} \end{cases}$$

and

$$Z_1' = \begin{cases} 1 & \text{if residence is in western United States} \\ 0 & \text{otherwise} \end{cases}$$

$$Z_2' = \begin{cases} 1 & \text{if residence is in eastern United States} \\ 0 & \text{otherwise} \end{cases}$$

Among the coding schemes available for regression, we recommend the method often referred to as *reference cell coding*, which uses $k - 1$ dummy variables as suggested above. Each variable takes on only values of 1 and 0, and each variable indicates group membership (1 for a specific group, 0 otherwise). The reader is urged to adopt this method unless another method is advantageous for a particular application. Note that some computer programs use 1 and -1 for coding. Since the choice of coding scheme affects analysis and interpretation, it is important to specify which coding method is being used.

We will now illustrate using dummy variables to compare two or more regression models. We begin by considering two straight-line models and then extend the discussion to comparing more than two multiple regression models.

14-4 Comparing Two Straight-Line Regression Equations: An Example

In Chapters 5 through 7, the age–systolic blood pressure example was used to illustrate the main principles and methods of straight-line regression analysis. These data, although hypothetical, were shown upon analysis to support the commonly found epidemiological observation that blood pressure increases with age.

Another common observation is that males tend to have higher blood pressure than females of similar age. This observation can also be supported by using the principles of regression analysis. However, the problem here is one of comparison: The straight-line regression of systolic blood pressure versus age for females must be compared with the corresponding regression for males (see Figure 14-1).

To illustrate the problem of comparing straight lines, it is convenient to continue with our age–systolic blood pressure example. However, since it was pointed out in Chapter 5 that the data point (47, 220) is an outlier quite distinct from the rest of the data, we will discard this data point in all further analyses. We shall therefore assume that the 29 observations on age and systolic blood pressure considered previously were made on females and that a second sample of observations on age and systolic blood pressure was collected on 40 males. The entire data set is presented in Table 14-1. In addition, Table 14-1 also provides

FIGURE 14-1   Comparison by sex of straight-line
regressions of systolic blood pressure
on age

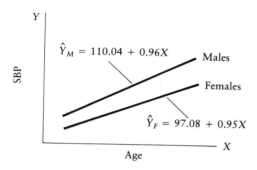

the information needed to compare the two fitted straight lines. For each data set, this information consists of the sample size (n), the intercept ($\hat{\beta}_0$), the slope ($\hat{\beta}_1$), the sample mean $\bar{X}$ of the X's, the sample mean $\bar{Y}$ of the Y's, the sample variance S_X^2 of the X's, and the residual mean-square error ($S_{Y|X}^2$). To distinguish between male and female data, we have used the subscripts M and F, respectively. Thus, n_M, $\hat{\beta}_{0M}$, $\hat{\beta}_{1M}$, and $S_{Y|X_M}^2$ denote the sample size, intercept, slope, and mean-square error for the male data, whereas n_F, $\hat{\beta}_{0F}$, $\hat{\beta}_{1F}$, and $S_{Y|X_F}^2$ denote the corresponding information for the female data.

The least-squares lines are then given as follows:[1]

Males: $\hat{Y}_M = 110.04 + 0.96X$
Females: $\hat{Y}_F = 97.08 + 0.95X$

These lines are sketched in Figure 14-1. It can be seen from the figure that the male line lies completely above the female line. This fact alone supports the contention that males have higher blood pressure than females over the age range being considered. Nevertheless, it is necessary to explore statistically whether the observed differences between the regression lines could have occurred by chance. In other words, to be statistically precise when comparing two regression lines, it is necessary to consider the sampling variability of the data by using statistical test(s) and/or confidence interval(s). In the sections below, a number of statistical procedures for dealing with this comparison problem are described.

14-5 Questions for Comparing Two Straight Lines

There are three basic questions to consider when comparing two straight-line regression equations. These are:

1. Are the two slopes the same or different (regardless of whether the intercepts are different)?[2]

---

[1] It can be shown that the straight-line model for males, like that for females, is appropriate based on a lack-of-fit test.

[2] If the two slopes are not different, we say that the two lines are *parallel*.

TABLE 14-1   Data on systolic blood pressure (SBP) and age for 40 males and 29 females and
associated data for comparing two straight-line regression equations

Male (i)	SBP (Y)	Age (X)	Male (i)	SBP (Y)	Age (X)	Male (i)	SBP (Y)	Age (X)
1	158	41	15	142	44	28	144	33
2	185	60	16	144	50	29	139	23
3	152	41	17	149	47	30	180	70
4	159	47	18	128	19	31	165	56
5	176	66	19	130	22	32	172	62
6	156	47	20	138	21	33	160	51
7	184	68	21	150	38	34	157	48
8	138	43	22	156	52	35	170	59
9	172	68	23	134	41	36	153	40
10	168	57	24	134	18	37	148	35
11	176	65	25	174	51	38	140	33
12	164	57	26	174	55	39	132	26
13	154	61	27	158	65	40	169	61
14	124	36						

Female (i)	SBP (Y)	Age (X)	Female (i)	SBP (Y)	Age (X)
1	144	39	16	135	45
2	138	45	17	114	17
3	145	47	18	116	20
4	162	65	19	124	19
5	142	46	20	136	36
6	170	67	21	142	50
7	124	42	22	120	39
8	158	67	23	120	21
9	154	56	24	160	44
10	162	64	25	158	53
11	150	56	26	144	63
12	140	59	27	130	29
13	110	34	28	125	25
14	128	42	29	175	69
15	130	48			

Group	n	$\hat{\beta}_0$	$\hat{\beta}_1$	$\bar{X}$	$\bar{Y}$	S_X^2	$S_{Y|X}^2$
Males	40	110.04	0.96	46.93	155.15	221.15	71.90
Females	29	97.08	0.95	45.07	139.86	242.14	91.46

2. Are the two intercepts the same or different (regardless of whether the slopes are different)?

3. Are the two lines coincident (that is, the same), or do they differ in slope and/or intercept?

Situations pertaining to these three questions are illustrated in Figure 14-2.

FIGURE 14-2   Possible conclusions from comparison of two straight-line
regressions

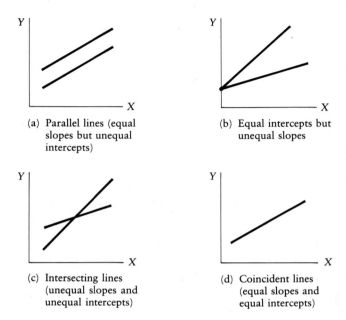

   (a) Parallel lines (equal
       slopes but unequal
       intercepts)

   (b) Equal intercepts but
       unequal slopes

   (c) Intersecting lines
       (unequal slopes and
       unequal intercepts)

   (d) Coincident lines
       (equal slopes and
       equal intercepts)

For our particular age–systolic blood pressure example, a conclusion that the lines are parallel (Figure 14-2a) can be interpreted to mean that one sex has a consistently higher systolic blood pressure than the other at all ages, but that the rate of change with respect to age is the same for both sexes. If it is concluded that the two lines have a common intercept but different slopes (Figure 14-2b), this means that the two sexes begin at an early age with the same average blood pressure but that the blood pressure changes with respect to age at a different rate for one sex than for the other. If the two lines have different slopes and different intercepts[3] (Figure 14-2c), this means that the relationship between age and systolic blood pressure is different for the two sexes in that both the origins and the rates of change are different for the sexes. Furthermore, if the lines intersect in the range of X-values of interest, this indicates that at early ages one sex has a higher average systolic blood pressure than the other, whereas at later ages the other sex does.

14-6 Methods of Comparing Two Straight Lines

There are two general approaches to answering the three questions above on comparing two straight lines:

Method I   Treat the male and female data separately by fitting the two separate regression equations

---

[3] It is possible for two lines to have unequal slopes and unequal intercepts and yet not intersect within the range of X-values of interest. This is illustrated by our example in Figure 14-1.

$$Y_M = \beta_{0M} + \beta_{1M}X + E \tag{14.1}$$

and

$$Y_F = \beta_{0F} + \beta_{1F}X + E \tag{14.2}$$

and then make appropriate two-sample t tests.

Method II   Define the dummy variable Z to be 0 if the subject is male and 1 if female. Thus, for the n_M observations on males, $Z = 0$; for the n_F observations on females, $Z = 1$. Our data will then be of the form:

Males: $(X_{1M}, Y_{1M}, 0), (X_{2M}, Y_{2M}, 0), \ldots, (X_{n_M M}, Y_{n_M M}, 0)$
Females: $(X_{1F}, Y_{1F}, 1), (X_{2F}, Y_{2F}, 1), \ldots, (X_{n_F F}, Y_{n_F F}, 1)$

Then, for the combined data above, the single multiple regression model

$$Y = \beta_0 + \beta_1 X + \beta_2 Z + \beta_3 XZ + E \tag{14.3}$$

yields the following two models for the two values of Z:

$$\begin{cases} Z = 0: & Y_M = \beta_0 + \beta_1 X + E \\ Z = 1: & Y_F = (\beta_0 + \beta_2) + (\beta_1 + \beta_3)X + E \end{cases}$$

This allows us to write the regression coefficients for the separate models for method I in terms of the coefficients of model (14.3) as follows:

$$\beta_{0M} = \beta_0, \qquad \beta_{0F} = \beta_0 + \beta_2, \qquad \beta_{1M} = \beta_1, \qquad \beta_{1F} = \beta_1 + \beta_3$$

Thus, *model (14.3) incorporates the two separate regression equations within a single model* and allows for different slopes (β_1 for males and $\beta_1 + \beta_3$ for females) and different intercepts (β_0 for males and $\beta_0 + \beta_2$ for females). We shall now describe the details involved in making statistical inferences for these two methods.

14-7 Method I: Comparing Two Straight Lines Using Separate Regression Fits

14-7-1 Testing for Parallelism

Using (14.1) and (14.2), the appropriate null hypothesis for comparing the slopes (i.e., for a test of parallelism) is given by

$$\boxed{H_0: \beta_{1M} = \beta_{1F}}$$

(When the null hypothesis $H_0: \beta_{1M} = \beta_{1F}$ is true, then the two regression lines simplify to $Y_M = \beta_{0M} + \beta_1 X + E$ for males and $Y_F = \beta_{0F} + \beta_1 X + E$ for females, where β_1 ($= \beta_{1M} = \beta_{1F}$) is the common slope. An estimate of this common slope β_1 is given by the following formula, which is a weighted average of the two separate slope estimates:

$$\hat{\beta}_1 = \frac{(n_M - 1)S^2_{X_M}\hat{\beta}_{1M} + (n_F - 1)S^2_{X_F}\hat{\beta}_{1F}}{(n_M - 1)S^2_{X_M} + (n_F - 1)S^2_{X_F}}$$

Note that $\hat{\beta}_1$ equals the slope computed from the pooled data.)

Any of the following three alternative hypotheses can be used:

$$H_A: \begin{cases} \beta_{1M} > \beta_{1F} & \text{(one sided)} \\ \beta_{1M} < \beta_{1F} & \text{(one sided)} \\ \beta_{1M} \neq \beta_{1F} & \text{(two sided)} \end{cases}$$

The test statistic for evaluating parallelism is then given by

$$T = \frac{\hat{\beta}_{1M} - \hat{\beta}_{1F}}{S_{\hat{\beta}_{1M} - \hat{\beta}_{1F}}} \qquad (14.4)$$

where

$\hat{\beta}_{1M}$ = least-squares estimate of the slope β_{1M} using the n_M observations (on males)

$\hat{\beta}_{1F}$ = least-squares estimate of the slope β_{1F} using the n_F observations (on females)

$S_{\hat{\beta}_{1M} - \hat{\beta}_{1F}}$ = estimate of the standard deviation of the estimated difference between slopes $(\hat{\beta}_{1M} - \hat{\beta}_{1F})$

This standard deviation involves pooling and summing the estimated variances of the slopes of the fitted regression lines.[4] It is equal to the square root of the following variance:

$$S^2_{\hat{\beta}_{1M} - \hat{\beta}_{1F}} = S^2_{P,Y|X} \left[\frac{1}{(n_M - 1)S^2_{X_M}} + \frac{1}{(n_F - 1)S^2_{X_F}} \right] \qquad (14.5)$$

where

$$S^2_{P,Y|X} = \frac{(n_M - 2)S^2_{Y|X_M} + (n_F - 2)S^2_{Y|X_F}}{n_M + n_F - 4} \qquad (14.6)$$

is a pooled estimate of σ^2 (see footnote 4) based on combining residual mean-square errors for males and females and where

$S^2_{Y|X_M}$ = residual mean-square error for the male data

$S^2_{Y|X_F}$ = residual mean-square error for the female data

$S^2_{X_M}$ = variance of the X's for the male data

$S^2_{X_F}$ = variance of the X's for the female data

The test statistic given by (14.4) will, under the usual regression assumptions, be distributed as a Student's t with $n_M + n_F - 4$ degrees of freedom when H_0 is true. We then have the following critical regions for different hypotheses and significance level α:

$$\begin{cases} T \geq t_{n_M + n_F - 4, 1 - \alpha} & \text{for } H_A: \beta_{1M} > \beta_{1F} \\ T \leq -t_{n_M + n_F - 4, 1 - \alpha} & \text{for } H_A: \beta_{1M} < \beta_{1F} \\ |T| > t_{n_M + n_F - 4, 1 - \alpha/2} & \text{for } H_A: \beta_{1M} \neq \beta_{1F} \end{cases}$$

---

[4] This pooling is valid only if the variance of Y_M, say σ^2_M, is equal to the variance of Y_F, say σ^2_F (i.e., only if the assumption of homogeneity of error variance holds).

The associated $100(1 - \alpha)\%$ confidence interval for $\beta_{1M} - \beta_{1F}$ is of the form

$$(\hat{\beta}_{1M} - \hat{\beta}_{1F}) \pm t_{n_M + n_F - 4, 1 - \alpha/2} S_{\hat{\beta}_{1M} - \hat{\beta}_{1F}}$$

Example Using the data given in Table 14-1, the estimates $S^2_{P,Y|X}$, $S^2_{\hat{\beta}_{1M} - \hat{\beta}_{1F}}$, $S^2_{\hat{\beta}_{1M}}$, and $S^2_{\hat{\beta}_{1F}}$ are computed as follows:

$$S^2_{P,Y|X} = \frac{(n_M - 2) S^2_{Y|X_M} + (n_F - 2) S^2_{Y|X_F}}{n_M + n_F - 4} = \frac{38(71.90) + 27(91.46)}{40 + 29 - 4}$$

$$= \frac{5,201.62}{65} = 80.02$$

$$S^2_{\hat{\beta}_{1M} - \hat{\beta}_{1F}} = S^2_{P,Y|X} \left[\frac{1}{(n_M - 1) S^2_{X_M}} + \frac{1}{(n_F - 1) S^2_{X_F}} \right]$$

$$= 80.02 \left[\frac{1}{39(221.15)} + \frac{1}{28(242.14)} \right]$$

$$= 0.021$$

$$S^2_{\hat{\beta}_{1M}} = \frac{S^2_{Y|X_M}}{(n_M - 1) S^2_{X_M}} = \frac{71.90}{39(221.15)} = 0.0083$$

$$S^2_{\hat{\beta}_{1F}} = \frac{S^2_{Y|X_F}}{(n_F - 1) S^2_{X_F}} = \frac{91.46}{28(242.14)} = 0.0135$$

The test statistic (14.4) is then computed as

$$T = \frac{\hat{\beta}_{1M} - \hat{\beta}_{1F}}{S_{\hat{\beta}_{1M} - \hat{\beta}_{1F}}} = \frac{0.96 - 0.95}{\sqrt{0.021}} = \frac{0.01}{0.145} = 0.069$$

For this test statistic, the critical value for a two-sided test (i.e., $H_A: \beta_{1M} \neq \beta_{1F}$) with $\alpha = .05$ is given by

$$t_{65, 0.975} = 1.9964$$

Since $|T| = 0.069$ does not exceed 1.9964, we do not reject H_0. Thus, we conclude that there is not sufficient evidence to reject the hypothesis of parallelism (i.e., the lines for males and females have the same slope).

14-7-2 Comparing Two Intercepts

We now describe how to use separate regression fits to determine whether both straight lines have the same intercept, regardless of the two slopes. The null hypothesis in this case is given by

$$H_0: \beta_{0M} = \beta_{0F}$$

(If $H_0: \beta_{0M} = \beta_{0F}$ is true, the two regression lines simplify to $Y_M = \beta_0 + \beta_{1M} X + E$ and $Y_F = \beta_0 + \beta_{1F} X + E$, where $\beta_0 \, (= \beta_{0M} = \beta_{0F})$ is the common intercept. An estimate of this common intercept β_0 is given by the equation

$$\hat{\beta}_0 = \frac{n_M \hat{\beta}_{0M} + n_F \hat{\beta}_{0F}}{n_M + n_F}$$

which is the weighted average of the two separate intercept estimates.)

The test statistic in this case is given by:

$$T = \frac{\hat{\beta}_{0M} - \hat{\beta}_{0F}}{S_{\hat{\beta}_{0M}-\hat{\beta}_{0F}}} \tag{14.7}$$

where $\hat{\beta}_{0M}$ and $\hat{\beta}_{0F}$ are the intercept estimates for males and females, respectively, and where $S^2_{\hat{\beta}_{0M}-\hat{\beta}_{0F}}$ estimates the variance of the estimated difference between the intercepts by means of the formula

$$S^2_{\hat{\beta}_{0M}-\hat{\beta}_{0F}} = S^2_{P,Y|X}\left[\frac{1}{n_M} + \frac{1}{n_F} + \frac{\bar{X}_M^2}{(n_M-1)S_{X_M}^2} + \frac{\bar{X}_F^2}{(n_F-1)S_{X_F}^2}\right] \tag{14.8}$$

The statistic T given in (14.7) will have the t distribution with $n_M + n_F - 4$ degrees of freedom when $H_0: \beta_{0M} = \beta_{0F}$ is true. We therefore have the following critical regions for different hypotheses and significance level α:

$$\begin{cases} T \geq t_{n_M+n_F-4,1-\alpha} & \text{for } H_A: \beta_{0M} > \beta_{0F} \\ T \leq -t_{n_M+n_F-4,1-\alpha} & \text{for } H_A: \beta_{0M} < \beta_{0F} \\ |T| \geq t_{n_M+n_F-4,1-\alpha/2} & \text{for } H_A: \beta_{0M} \neq \beta_{0F} \end{cases}$$

The associated $100(1-\alpha)\%$ confidence interval for $\beta_{0M} - \beta_{0F}$ is

$$(\hat{\beta}_{0M} - \hat{\beta}_{0F}) \pm t_{n_M+n_F-4,1-\alpha/2}S_{\hat{\beta}_{0M}-\hat{\beta}_{0F}}$$

Example For the data of Table 14-1 and the value of $S^2_{P,Y|X}$ obtained in Section 14-7-1, the estimates $S^2_{\hat{\beta}_{0M}-\hat{\beta}_{0F}}$, $S^2_{\hat{\beta}_{0M}}$, and $S^2_{\hat{\beta}_{0F}}$ are computed as follows:

$$\begin{aligned} S^2_{\hat{\beta}_{0M}-\hat{\beta}_{0F}} &= S^2_{P,Y|X}\left[\frac{1}{n_M} + \frac{1}{n_F} + \frac{\bar{X}_M^2}{(n_M-1)S_{X_M}^2} + \frac{\bar{X}_F^2}{(n_F-1)S_{X_F}^2}\right] \\ &= 80.02\left[\frac{1}{40} + \frac{1}{29} + \frac{(46.93)^2}{39(221.15)} + \frac{(45.07)^2}{28(242.14)}\right] \\ &= 80.02(0.0250 + 0.0345 + 0.2554 + 0.2996) \\ &= 49.17 \end{aligned}$$

$$\begin{aligned} S^2_{\hat{\beta}_{0M}} &= S^2_{Y|X_M}\left[\frac{1}{n_M} + \frac{\bar{X}_M^2}{(n_M-1)S_{X_M}^2}\right] = 71.90\left[\frac{1}{40} + \frac{(46.93)^2}{39(221.15)}\right] \\ &= 71.40(0.0250 + 0.2554) = 20.16 \end{aligned}$$

$$\begin{aligned} S^2_{\hat{\beta}_{0F}} &= S^2_{Y|X_F}\left[\frac{1}{n_F} + \frac{\bar{X}_F^2}{(n_F-1)S_{X_F}^2}\right] = 91.46\left[\frac{1}{29} + \frac{(45.07)^2}{28(242.14)}\right] \\ &= 91.46(0.0345 + 0.2996) = 30.56 \end{aligned}$$

From these results, the T statistic of (14.7) is computed as follows:

$$T = \frac{\hat{\beta}_{0M} - \hat{\beta}_{0F}}{S_{\hat{\beta}_{0M}-\hat{\beta}_{0F}}} = \frac{110.04 - 97.08}{\sqrt{49.17}} = \frac{12.96}{7.01} = 1.85$$

For a two-sided test ($H_A: \beta_{0M} \neq \beta_{0F}$) with $\alpha = .05$, we find that $|T| = 1.85$ does not exceed $t_{65,0.975} = 1.9964$ (i.e., $.05 < P < .1$). Thus, the null hypothesis of common intercepts is not rejected at $\alpha = .05$ but is rejected at $\alpha = .1$.

14-7-3 Testing for Coincidence from Separate Straight-Line Fits

Two straight lines are coincident if their slopes and their intercepts are equal. In considering the male–female regression equations given by (14.1) and (14.2), the null hypothesis of coincidence is therefore equivalent to testing H_0: $\beta_{0M} = \beta_{0F}$ and $\beta_{1M} = \beta_{1F}$ *simultaneously*. If so, the two regression models both reduce to the general form

$$Y = \beta_0 + \beta_1 X + E$$

where β_0 ($= \beta_{0M} = \beta_{0F}$) and β_1 ($= \beta_{1M} = \beta_{1F}$) are the common intercept and slope, respectively.

(The estimates of the common slope β_1 and common intercept β_0 are obtained by simply pooling all observations on males and females together and determining the usual least-squares slope and intercept estimates using the pooled data set.)

As mentioned earlier, a preferred way to test the null hypothesis of coincident lines is to employ a multiple regression model involving dummy variables. Nevertheless, another procedure that is generally not quite as efficient (e.g., not as powerful) is often convenient when separate models are fit. In practice, this procedure often yields the same conclusion as that obtained from using dummy variables.

Using separate regression fits, the procedure is to perform *both* the test of H_0: $\beta_{0M} = \beta_{0F}$ of equal intercepts and the test of H_0: $\beta_{1M} = \beta_{1F}$ of equal slopes. If *either one or both* of these null hypotheses are rejected, one can conclude that there is statistical evidence that the two lines are not coincident. If neither is rejected, then there is no evidence in the data of noncoincidence.

A valid criticism of this testing procedure, which reflects on its power, is that it involves two separate tests rather than a single test. This fact results in two difficulties:

1. The procedure does not precisely test for coincidence.

2. If α is the significance level of each separate test, the overall significance level for the two tests combined is greater than α; that is, there is more chance of rejecting a true H_0 (i.e., of making a Type I error).

One reasonable (but fairly conservative) way to get around this second difficulty is to use $\alpha/2$ for each separate test to guarantee an overall significance level of no more than α (the Bonferroni correction). Nevertheless, using $\alpha/2$ for each test is conservative (i.e., makes it harder to reject either H_0), thus making it difficult to detect a real difference between the two lines.

With regard to the first difficulty, it is important to point out that even if both tests are not rejected, it is still possible (although unlikely) for the two lines not to coincide. This is so because each separate test (e.g., the test of H_0: $\beta_{1M} = \beta_{1F}$) allows the remaining parameters (β_{0M} and β_{0F}) to be unequal. In other words, the test for equal slopes does not assume equal intercepts, nor does the test for equal intercepts assume equal slopes. The multiple regression procedure, which involves the use of a single model containing a dummy variable for group status, avoids this drawback and permits testing for common slope and common intercept *simultaneously*.

Example   We saw in Sections 14-7-1 and 14-7-2 that the null hypothesis of equal slopes (regardless of the intercepts) was not rejected ($P > .40$) and that the null hypothesis of equal intercepts (regardless of the slopes) was associated with a P-value of between .05 and .1. Putting these two facts together, we would be inclined to support the position that there is no *strong* evidence for noncoincidence. As we shall see shortly, the more appropriate test procedure involving a single model yields a different conclusion.

14-8   Method II: Comparing Two Straight Lines Using a Single Regression Equation

Another approach for comparing regression equations uses a single multiple regression model that contains one or more dummy variables to distinguish the groups being compared. When comparing two straight lines, the model is given by (14.3), which we restate for the reader's convenience:

$$Y = \beta_0 + \beta_1 X + \beta_2 Z + \beta_3 XZ + E$$

where Y = SBP, X = AGE, and Z is a dummy variable indicating gender (1 if female, 0 if male). For the data of Table 14-1 ($n_M = 40$, $n_F = 29$), the fitted model is

$$\hat{Y} = 110.04 + 0.96X - 12.96Z - 0.012XZ$$

which yields the following separate straight-line equations:

$$Z = 0: \quad \hat{Y}_M = 110.04 + 0.96X$$
$$Z = 1: \quad \hat{Y}_F = 97.08 + 0.95X$$

These two straight-line equations are exactly the same as obtained in Section 14-4 by fitting separate regressions.

Table 14-2 provides ANOVA results needed to answer statistical inference questions about this model. This table provides variables-added-in-order tests for the fitted regression equation and allows us to perform appropriate tests for parallelism, for equal intercepts, and for coincidence.

14-8-1   Test of Parallelism: Single-Model Approach

Referring again to the above dummy variable model (14.3), the null hypothesis that the two regression lines are parallel is equivalent to $H_0: \beta_3 = 0$. If $\beta_3 = 0$, the slope for females, $\beta_{1F} = \beta_1 + \beta_3$, simplifies to β_1, which is the slope for males (i.e., the two lines are parallel). The test statistic for testing $H_0: \beta_3 = 0$ is the partial F statistic (or equivalent t test) for the significance of the addition of the variable XZ to a model already containing X and Z.[5]

---

[5] If H_0 is not rejected by the test of $H_0: \beta_3 = 0$, model (14.3) can be revised by eliminating the β_3 term. This revised (or reduced) model becomes $Y = \beta_0 + \beta_1 X + \beta_2 Z + E$, which is in the form of an analysis-of-covariance model (see Chapter 15).

TABLE 14-2   Three models for age–systolic blood pressure
              example

Source	df	SS	MS	F
Regression (X)	1	14,951.25	14,951.25	121.27
Residual	67	8,260.51	123.29	
Regression (X, Z)	2	18,009.78	9,004.89	114.25
Residual	66	5,201.99	78.82	
Regression (X, Z, XZ)	3	18,010.33	6,003.44	75.02
Residual	65	5,201.44	80.02	

In our example, this test statistic is computed as follows:

$$F(XZ|X, Z) = \frac{\text{regression SS}(X, Z, XZ) - \text{regression SS}(X, Z)}{\text{MS residual }(X, Z, XZ)}$$

$$= \frac{18,010.33 - 18,009.78}{80.02}$$

$$= 0.007 \quad (P = .9342)$$

This F statistic, with 1 and 65 degrees of freedom, is extremely small (P is very large); so we do not reject H_0 and therefore have no statistical basis for believing that the two lines are not parallel. This was the same decision made on the basis of separate regression fits. In fact, the F computed here is (theoretically) the square of the corresponding T computed when using separate straight-line fits in Section 14-7-1, although the numerical answers may not exactly agree due to round-off errors.

14-8-2 Test of Equal Intercepts: Single-Model Approach

The hypothesis that the two intercepts are equal, allowing for unequal slopes, is equivalent to H_0: $\beta_2 = 0$ for the overall model (14.3). The test compares the overall model

$$Y = \beta_0 + \beta_1 X + \beta_2 Z + \beta_3 XZ + E$$

to the reduced model

$$Y = \beta_0 + \beta_1 X + \beta_3 XZ + E$$

This is a variables-added-last test considering Z, the sex group dummy variable.[6] Another approach involves a variables-added-in-order test comparing

$$Y = \beta_0 + \beta_1 X + \beta_2 Z + E$$

to the reduced model

$$Y = \beta_0 + \beta_1 X + E$$

Note that this latter test presumes equal slopes, so it is essentially a test for coincidence, assuming parallelism. Not surprisingly, neither test is uniformly preferred (see Section

---

[6] The partial F statistic for the variables-added-last test for equal intercepts, $F(Z|X, XZ)$, is the square of the statistic given by (14.7) when fitting separate straight-line models.

TABLE 14-3 ANOVA table for method II for the age–systolic blood pressure example

Source	df	SS	MS	F	P
X (AGE)	1	14,951.25	14,951.25	186.84	.0001
Z (SEX)$\mid X$	1	3,058.52	3,058.52	38.22	.0001
$X \times Z \mid X, Z$	1	0.55	0.55	0.01	.9342
Residual	65	5,201.44	80.02		
Total (corrected)	68	23,211.76			

14-10). As discussed in Chapters 9 and 13, we recommend using the residual from the full model, (14.3), for either test.

For the example under consideration, we shall opt for the second approach, since the slope test was not significant. The variables-added-in-order test for H_0: $\beta_2 = 0$ is then computed as follows:

$$F(Z\mid X) = \frac{\text{regression SS}(Z, X) - \text{regression SS}(X)}{\text{MS residual }(X, Z, XZ)}$$

$$= \frac{18,009.78 - 14,951.25}{80.02}$$

$$= 38.22$$

(Note that the test statistic above is modified from the usual partial F statistic in that we are now using for the denominator the mean-square residual for the full model, which contains X, Z, and XZ; the usual partial F statistic would use the mean-square residual for the model containing only X and Z.)

With 1 and 65 degrees of freedom, we see from Table 14-3 that $P < .0001$. Hence the intercepts are judged to be different for the male and female straight-line models.

14-8-3 Test of Coincidence: Single-Model Approach

The hypothesis that the two regression lines are coincident is H_0: $\beta_2 = \beta_3 = 0$. When both β_2 and β_3 are 0, the model for females, $Y_F = (\beta_0 + \beta_2) + (\beta_1 + \beta_3)X + E$, reduces to $Y_M = \beta_0 + \beta_1 X + E$ for males (i.e., the two lines are coincident). The test of H_0: $\beta_2 = \beta_3 = 0$ is thus a multiple-partial F test, since it concerns a subset of regression coefficients.[7] The two models being compared are therefore

$$Y = \beta_0 + \beta_1 X + \beta_2 Z + \beta_3 XZ + E$$

and

$$Y = \beta_0 + \beta_1 X + E$$

For our example, the information in either Table 14-2 or 14-3 leads to the following computation:

---

[7] If the test for coincidence is not rejected, model (14.3) can be reduced to the form $Y = \beta_0 + \beta_1 X + E$.

$$F(XZ, Z|X) = \frac{[\text{regression SS}(X, Z, XZ) - \text{regression SS}(X)]/2}{\text{MS residual } (X, Z, XZ)}$$

$$= \frac{(18,010.33 - 14,951.25)/2}{80.02}$$

$$= 19.1$$

Comparing this F with $F_{2,65,0.999} = 7.72$, we reject H_0 with $P < .001$ and conclude that there is very strong evidence that the two lines are *not* coincident. This conclusion contradicts our earlier conclusion (Section 14-7-3) using the results from separate tests for equal slopes and equal intercepts.

14-9 Comparison of Methods I and II

The question naturally arises at this point whether the method that uses dummy variables differs from the method that fits two separate regression equations. And if there is a difference, is one of the methods preferable to the other?

(The t tests described earlier for separate regression fits assume that the estimates of the residual variances for males and females can be pooled. The dummy variable method described in this chapter also assumes that such pooling is appropriate. However, when pooling variance estimates is found to be invalid (i.e., when variance homogeneity does not hold), the dummy variable approach should be avoided. Armitage (1971, p. 122) discusses some alternative procedures to pooling when there are unequal variances.)

In deciding whether one method is preferable, we first point out that the two methods will yield exactly the same estimated regression coefficients for the two straight-line models. That is, if we fit the model $Y = \beta_0 + \beta_1 X + \beta_2 Z + \beta_3 XZ + E$ by the least-squares method to obtain estimated coefficients $\hat{\beta}_0$, $\hat{\beta}_1$, $\hat{\beta}_2$, and $\hat{\beta}_3$, the straight-line equations obtained by setting Z equal to 0 and to 1 in this estimated model will be the same as would be obtained if the two straight lines were fit separately. In particular, if $\hat{\beta}_{0M}$, $\hat{\beta}_{0F}$, $\hat{\beta}_{1M}$, and $\hat{\beta}_{1F}$ denote the estimated regression coefficients based on separate regression fits, then $\hat{\beta}_{0M} = \hat{\beta}_0$, $\hat{\beta}_{0F} = \hat{\beta}_0 + \hat{\beta}_2$, $\hat{\beta}_{1M} = \hat{\beta}_1$, and $\hat{\beta}_{1F} = \hat{\beta}_1 + \hat{\beta}_3$. As for statistical tests involving regression coefficients estimated by the two methods, the following may be said:

1. *The tests for parallel lines are exactly equivalent*; that is, the T statistic with $n_1 + n_2 - 4$ degrees of freedom computed for testing H_0: $\beta_3 = 0$ in the dummy variable model is exactly the same as the T statistic given by (14.4) for testing H_0: $\beta_{1M} = \beta_{1F}$ based on fitting two separate models.

2. *The tests for coincident lines are different*, and the one using the dummy variable model is generally preferable. The approach using separate regressions tests H_0: $\beta_{1M} = \beta_{1F}$ and H_0: $\beta_{0M} = \beta_{0F}$ separately and then rejects the null hypothesis of coincident lines if either or both null hypotheses are rejected. This is exactly equivalent to performing two separate tests of H_0: $\beta_2 = 0$ and H_0: $\beta_3 = 0$ and using the same decision rule for the dummy variable approach, but it is not equivalent to testing the single null hypothesis H_0: $\beta_2 = \beta_3 = 0$ (i.e., testing whether β_2 and β_3 are both simultaneously 0).

14-10 Testing Strategies and Interpretation: Comparing Two Straight Lines

Several strategies can be used to identify a best model for comparing two straight lines. Strategies for more general situations are described in Chapter 16. We prefer a backward strategy for most situations, that is, starting with the largest model of interest and then trying to reduce the model through a sequence of hypothesis tests. A flow diagram of this strategy for comparing two straight lines is given in Figure 14-3. In this case, the largest model to be

FIGURE 14-3   Comparing two straight lines: Backward testing strategy

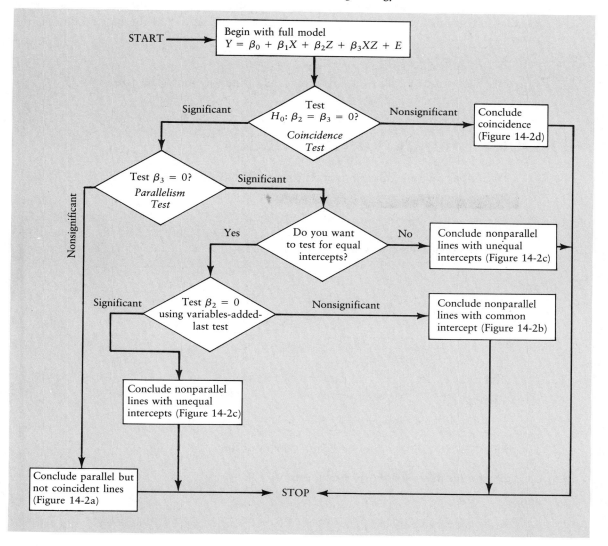

considered is model (14.3), which contains X, Z, and XZ as independent variables. To reduce the model, we perform tests for coincidence, then for parallelism, and then for equal intercepts as follows: (1) If the test for coincidence is nonsignificant, we stop further testing and conclude that the best model is $Y = \beta_0 + \beta_1 X + E$ (i.e., coincident lines, Figure 14-2d). (2) If the test for coincidence is significant and the test for parallelism is nonsignificant, then the data argue for parallel but noncoincident lines (Figure 14-2a). (3) If the test for coincidence is significant and the test for parallelism is significant, then we might not even be interested in a test for equal intercepts; if we are, however, the appropriate test procedure would involve the variables-added-last statistic $F(Z|X, XZ)$, which does not assume parallel lines. If this latter test is significant, we would argue for Figure 14-2c; if not significant, we would tend to support Figure 14-2b.

Applying this strategy to the age–systolic blood pressure data, we would conclude, based on tests reported above, that the test for coincidence is significant and the test for parallelism is nonsignificant, so that our overall conclusion is that the best model has the form

$$Y = \beta_0 + \beta_1 X + \beta_2 Z + E$$

In words, we assume parallel (and noncoincident) lines, as shown in Figure 14-2a.

14-11 Other Dummy Variable Models

Two other dummy variable models that could have been used instead of (14.3) are

$$Y = \beta_0^* + \beta_1^* X + \beta_2^* Z^* + \beta_3^* XZ^* + E \qquad (14.9)$$

for which

$$Z^* = \begin{cases} 1 & \text{if subject is male} \\ -1 & \text{if subject is female} \end{cases}$$

and

$$Y = \beta_{0M} Z_1' + \beta_{0F} Z_2' + \beta_{1M} XZ_1' + \beta_{1F} XZ_2' + E \qquad (14.10)$$

for which

$$Z_1' = \begin{cases} 1 & \text{if subject is male} \\ 0 & \text{if subject is female} \end{cases}$$

$$Z_2' = \begin{cases} 0 & \text{if subject is male} \\ 1 & \text{if subject is female} \end{cases}$$

For model (14.9), the separate regression equations are

$$Z^* = 1: \quad Y_M = (\beta_0^* + \beta_2^*) + (\beta_1^* + \beta_3^*) X + E$$

and

$$Z^* = -1: \quad Y_F = (\beta_0^* - \beta_2^*) + (\beta_1^* - \beta_3^*) X + E$$

The test for parallel lines is equivalent to testing $H_0: \beta_3^* = 0$, the test for equal intercepts is equivalent to testing $H_0: \beta_2^* = 0$, and the test for coincident lines is equivalent to testing $H_0: \beta_2^* = \beta_3^* = 0$.[8]

For model (14.10), the separate regression equations are

$$Z_1' = 1, Z_2' = 0: \quad Y_M = \beta_{0M} + \beta_{1M}X + E$$

and

$$Z_1' = 0, Z_2' = 1: \quad Y_F = \beta_{0F} + \beta_{1F}X + E$$

The test for parallel lines is then equivalent to testing $H_0: \beta_{1M} = \beta_{1F}$ (not necessarily equal to 0), the test for equal intercepts is then equivalent to testing $H_0: \beta_{0M} = \beta_{0F}$ (not necessarily equal to 0), and the test for coincident lines is equivalent to testing $H_0: \beta_{1M} = \beta_{1F}$ and $\beta_{0M} = \beta_{0F}$ simultaneously.[9]

(The test of $H_0: \beta_{1M} = \beta_{1F}$ and $\beta_{0M} = \beta_{0F}$ differs from previously discussed multiple-partial F tests because the coefficients under this H_0 are not equal to 0. The testing procedure in such a case is given as follows:

1. Reduce the model according to the specifications under the null hypothesis; for example, for the test of coincidence, the full model (14.10) becomes

$$Y = \beta_{0M}(Z_1' + Z_2') + \beta_{1M}(XZ_1' + XZ_2') + E$$
$$= \beta_{0M} + \beta_{1M}X + E \quad \text{since } Z_1' + Z_2' = 1$$

2. Find the residual sum of squares for this reduced model.

3. Compute the following F statistic:

$$F = \frac{[\text{residual SS (reduced model)} - \text{residual SS (full model)}]/\nu^*}{\text{MS residual (full model)}}$$

where $\nu^* =$ number of independent linear parametric functions specified to be 0 under H_0 (in our case $\nu^* = 2$ since the null hypothesis of coincidence specifies that $\beta_{1M} - \beta_{1F} = 0$ and that $\beta_{0M} - \beta_{0F} = 0$).

4. Test H_0 using F tables with ν^* and $n - 4$ degrees of freedom, where 4 is the number of parameters in the full model.)

It is important to note that model (14.10) does not include an overall intercept. Care must be taken in using this particular model with most computer programs (which generally include an overall intercept by default). Collinearity often labeled "LESS THAN FULL RANK MODEL" may occur.

------

[8] We can express the coefficients of model (14.9) in terms of the regression coefficients for the separate male and female models:

$$\beta_0^* = \frac{\beta_{0M} + \beta_{0F}}{2}, \qquad \beta_1^* = \frac{\beta_{1M} + \beta_{1F}}{2}, \qquad \beta_2^* = \frac{\beta_{0M} - \beta_{0F}}{2}, \qquad \beta_3^* = \frac{\beta_{1M} - \beta_{1F}}{2}$$

[9] The coefficients of model (14.10) are identical to the correspondingly labeled coefficients for the separate male and female models.

14-12 Comparing Four Regression Equations

Suppose that we wish to compare the separate multiple regressions of systolic blood pressure on age and weight for four social class groups. For each individual in each social class group, we observe values of the variables $Y = \text{SBP}$, $X_1 = \text{AGE}$, and $X_2 = \text{WEIGHT}$; further, we shall suppose that there are n_i individuals in the ith social class (SC) group, $i = 1$, 2, 3, 4. We begin by defining three dummy variables Z_1, Z_2, and Z_3 as follows:

$$Z_1 = \begin{cases} 1 & \text{if SC2 member} \\ 0 & \text{otherwise} \end{cases} \quad Z_2 = \begin{cases} 1 & \text{if SC3 member} \\ 0 & \text{otherwise} \end{cases} \quad Z_3 = \begin{cases} 1 & \text{if SC4 member} \\ 0 & \text{otherwise} \end{cases}$$

The complete model to be used (if no interaction between AGE and WEIGHT is considered) is then given as follows:

$$Y = \beta_0 + \beta_1 X_1 + \beta_2 X_2 + \beta_3 Z_1 + \beta_4 Z_2 + \beta_5 Z_3 + \beta_6 X_1 Z_1 + \beta_7 X_2 Z_1$$
$$+ \beta_8 X_1 Z_2 + \beta_9 X_2 Z_2 + \beta_{10} X_1 Z_3 + \beta_{11} X_2 Z_3 + E \qquad (14.11)$$

For each particular social class, model (14.11) specializes as follows:

SC1$(Z_1 = Z_2 = Z_3 = 0)$: $Y = \beta_0 + \beta_1 X_1 + \beta_2 X_2 + E$

SC2$(Z_1 = 1, Z_2 = Z_3 = 0)$: $Y = (\beta_0 + \beta_3) + (\beta_1 + \beta_6)X_1 + (\beta_2 + \beta_7)X_2 + E$

SC3$(Z_1 = Z_3 = 0, Z_2 = 1)$: $Y = (\beta_0 + \beta_4) + (\beta_1 + \beta_8)X_1 + (\beta_2 + \beta_9)X_2 + E$

SC4$(Z_1 = Z_2 = 0, Z_3 = 1)$: $Y = (\beta_0 + \beta_5) + (\beta_1 + \beta_{10})X_1 + (\beta_2 + \beta_{11})X_2 + E$

14-12-1 Tests of Hypotheses

The following hypotheses concerning the parameters in model (14.11) are of interest.

1. *Are all four regression equations parallel* (i.e., test H_0: $\beta_6 = \beta_7 = \beta_8 = \beta_9 = \beta_{10} = \beta_{11} = 0$)? When H_0 is true, the models for each social class reduce to

    SC1: $Y = \beta_0 + \beta_1 X_1 + \beta_2 X_2 + E$

    SC2: $Y = (\beta_0 + \beta_3) + \beta_1 X_1 + \beta_2 X_2 + E$

    SC3: $Y = (\beta_0 + \beta_4) + \beta_1 X_1 + \beta_2 X_2 + E$

    SC4: $Y = (\beta_0 + \beta_5) + \beta_1 X_1 + \beta_2 X_2 + E$

    Thus, the coefficients of X_1 and X_2 are the same for each social class when H_0 is true (i.e., the four regression equations are said to be parallel). The test statistic for testing H_0 is given by the multiple-partial

    $$F(X_1 Z_1, X_2 Z_1, X_1 Z_2, X_2 Z_2, X_1 Z_3, X_2 Z_3 | X_1, X_2, Z_1, Z_2, Z_3)$$

    which has 6 and $n_1 + n_2 + n_3 + n_4 - 12$ degrees of freedom.

2. *Are all four regression equations coincident* (i.e., test H_0: $\beta_3 = \beta_4 = \beta_5 = \beta_6 = \beta_7 = \beta_8 = \beta_9 = \beta_{10} = \beta_{11} = 0$)? When H_0 is true, all four social class models reduce to the form

    $$Y = \beta_0 + \beta_1 X_1 + \beta_2 X_2 + E$$

The test statistic is the multiple-partial

$$F(Z_1, Z_2, Z_3, X_1Z_1, X_1Z_2, X_2Z_1, X_2Z_2, X_1Z_3, X_2Z_3 | X_1, X_2)$$

which has 9 and $n_1 + n_2 + n_3 + n_4 - 12$ degrees of freedom.

14-2-2 An Alternate Dummy Variable Model

Another dummy variable coding scheme for comparing the four social class groups is as follows: Define

$$Z_1^* = \begin{cases} -1 & \text{if SC1 member} \\ 1 & \text{if SC2 member} \\ 0 & \text{if SC3 member} \\ 0 & \text{if SC4 member} \end{cases} \qquad Z_2^* = \begin{cases} -1 & \text{if SC1 member} \\ 0 & \text{if SC2 member} \\ 1 & \text{if SC3 member} \\ 0 & \text{if SC4 member} \end{cases}$$

$$Z_3^* = \begin{cases} -1 & \text{if SC1 member} \\ 0 & \text{if SC2 member} \\ 0 & \text{if SC3 member} \\ 1 & \text{if SC4 member} \end{cases}$$

Then use the model

$$\begin{aligned} Y = {} & \beta_0^* + \beta_1^* X_1 + \beta_2^* X_2 + \beta_3^* Z_1^* + \beta_4^* Z_2^* + \beta_5^* Z_3^* + \beta_6^* X_1 Z_1^* \\ & + \beta_7^* X_2 Z_1^* + \beta_8^* X_1 Z_2^* + \beta_9^* X_2 Z_2^* + \beta_{10}^* X_1 Z_3^* + \beta_{11}^* X_2 Z_3^* + E \end{aligned} \qquad (14.12)$$

For the above dummy variable coding, the four regression equations for the four social classes based on model (14.12) are

$$
\begin{aligned}
\text{SC1}(Z_1^* = Z_2^* = Z_3^* = -1): \quad Y = {} & (\beta_0^* - \beta_3^* - \beta_4^* - \beta_5^*) \\
& + (\beta_1^* - \beta_6^* - \beta_8^* - \beta_{10}^*) X_1 \\
& + (\beta_2^* - \beta_7^* - \beta_9^* - \beta_{11}^*) X_2 + E \\
\text{SC2}(Z_1^* = 1, Z_2^* = Z_3^* = 0): \quad Y = {} & (\beta_0^* + \beta_3^*) + (\beta_1^* + \beta_6^*) X_1 \\
& + (\beta_2^* + \beta_7^*) X_2 + E \\
\text{SC3}(Z_1^* = 0, Z_2^* = 1, Z_3^* = 0): \quad Y = {} & (\beta_0^* + \beta_4^*) + (\beta_1^* + \beta_8^*) X_1 \\
& + (\beta_2^* + \beta_9^*) X_2 + E \\
\text{SC4}(Z_1^* = Z_2^* = 0, Z_3^* = 1): \quad Y = {} & (\beta_0^* + \beta_5^*) + (\beta_1^* + \beta_{10}^*) X_1 \\
& + (\beta_2^* + \beta_{11}^*) X_2 + E
\end{aligned}
$$

So, the null hypotheses to be tested concerning parallelism and coincidence (using appropriate multiple-partial F tests) are:

Parallelism: $H_0: \beta_6^* = \beta_7^* = \beta_8^* = \beta_9^* = \beta_{10}^* = \beta_{11}^* = 0$

Coincidence: $H_0: \beta_3^* = \beta_4^* = \beta_5^* = \beta_6^* = \beta_7^* = \beta_8^* = \beta_9^* = \beta_{10}^* = \beta_{11}^* = 0$

14-13 Comparing Several Regression Equations Involving Two Nominal Variables

Suppose that we want to compare eight regression equations of SBP (Y) on AGE (X_1) and WEIGHT (X_2) corresponding to the eight combinations of SEX (Q) and social class (SC) groups. Then the following regression model can be used:

$$
\begin{aligned}
Y = {} & \beta_0 + \beta_1 X_1 + \beta_2 X_2 + \beta_3 Z_1 + \beta_4 Z_2 + \beta_5 Z_3 + \beta_6 Q + \beta_7 Z_1 Q + \beta_8 Z_2 Q \\
& + \beta_9 Z_3 Q + \beta_{10} X_1 Z_1 + \beta_{11} X_2 Z_1 + \beta_{12} X_1 Z_2 + \beta_{13} X_2 Z_2 + \beta_{14} X_1 Z_3 \\
& + \beta_{15} X_2 Z_3 + \beta_{16} X_1 Q + \beta_{17} X_2 Q + \beta_{18} X_1 Z_1 Q + \beta_{19} X_2 Z_1 Q + \beta_{20} X_1 Z_2 Q \\
& + \beta_{21} X_2 Z_2 Q + \beta_{22} X_1 Z_3 Q + \beta_{23} X_2 Z_3 Q + E
\end{aligned}
\tag{14.13}
$$

in which the dummy variables are defined as

$$
Z_1 = \begin{cases} 1 & \text{if SC2 member} \\ 0 & \text{otherwise} \end{cases} \qquad
Z_2 = \begin{cases} 1 & \text{if SC3 member} \\ 0 & \text{otherwise} \end{cases}
$$

$$
Z_3 = \begin{cases} 1 & \text{if SC4 member} \\ 0 & \text{otherwise} \end{cases} \qquad
Q = \begin{cases} 1 & \text{if subject is male} \\ 0 & \text{if subject is female} \end{cases}
$$

For each SEX–SC combination, we have:

SC1–male: $Y = (\beta_0 + \beta_6) + (\beta_1 + \beta_{16})X_1 + (\beta_2 + \beta_{17})X_2 + E$

SC2–male: $Y = (\beta_0 + \beta_3 + \beta_6 + \beta_7) + (\beta_1 + \beta_{10} + \beta_{16} + \beta_{18})X_1$
$\qquad\qquad + (\beta_2 + \beta_{11} + \beta_{17} + \beta_{19})X_2 + E$

SC3–male: $Y = (\beta_0 + \beta_4 + \beta_6 + \beta_8) + (\beta_1 + \beta_{12} + \beta_{16} + \beta_{20})X_1$
$\qquad\qquad + (\beta_2 + \beta_{13} + \beta_{17} + \beta_{21})X_2 + E$

SC4–male: $Y = (\beta_0 + \beta_5 + \beta_6 + \beta_9) + (\beta_1 + \beta_{14} + \beta_{16} + \beta_{22})X_1$
$\qquad\qquad + (\beta_2 + \beta_{15} + \beta_{17} + \beta_{23})X_2 + E$

SC1–female: $Y = \beta_0 + \beta_1 X_1 + \beta_2 X_2 + E$

SC2–female: $Y = (\beta_0 + \beta_3) + (\beta_1 + \beta_{10})X_1 + (\beta_2 + \beta_{11})X_2 + E$

SC3–female: $Y = (\beta_0 + \beta_4) + (\beta_1 + \beta_{12})X_1 + (\beta_2 + \beta_{13})X_2 + E$

SC4–female: $Y = (\beta_0 + \beta_5) + (\beta_1 + \beta_{14})X_1 + (\beta_2 + \beta_{15})X_2 + E$

The model therefore includes SEX × SC interaction, but not WEIGHT × AGE interaction. The following hypotheses concerning the parameters in model (14.13) are of interest.

1. *Male and female regression equations are parallel (controlling for SC).* This is a test of

$$H_0\colon \beta_{16} = \beta_{17} = \beta_{18} = \beta_{19} = \beta_{20} = \beta_{21} = \beta_{22} = \beta_{23} = 0$$

When this null hypothesis is true, the eight equations above reduce to

$\begin{cases} \text{SC1–male:} & Y = (\beta_0 + \beta_6) + \beta_1 X_1 + \beta_2 X_2 + E \\ \text{SC1–female:} & Y = \beta_0 + \beta_1 X_1 + \beta_2 X_2 + E \end{cases}$

$\begin{cases} \text{SC2–male:} & Y = (\beta_0 + \beta_3 + \beta_6 + \beta_7) + (\beta_1 + \beta_{10})X_1 + (\beta_2 + \beta_{11})X_2 + E \\ \text{SC2–female:} & Y = (\beta_0 + \beta_3) + (\beta_1 + \beta_{10})X_1 + (\beta_2 + \beta_{11})X_2 + E \end{cases}$

$\begin{cases} \text{SC3–male:} & Y = (\beta_0 + \beta_4 + \beta_6 + \beta_8) + (\beta_1 + \beta_{12})X_1 + (\beta_2 + \beta_{13})X_2 + E \\ \text{SC3–female:} & Y = (\beta_0 + \beta_4) + (\beta_1 + \beta_{12})X_1 + (\beta_2 + \beta_{13})X_2 + E \end{cases}$

$$\begin{cases} \text{SC4–male:} & Y = (\beta_0 + \beta_5 + \beta_6 + \beta_9) + (\beta_1 + \beta_{14})X_1 + (\beta_2 + \beta_{15})X_2 + E \\ \text{SC4–female:} & Y = (\beta_0 + \beta_5) + (\beta_1 + \beta_{14})X_1 + (\beta_2 + \beta_{15})X_2 + E \end{cases}$$

Thus, within any specific social class group, the male and female regression equations are parallel (since they have the same X_1 and X_2 coefficients).

2. *Male and female regression equations are coincident (controlling for SC).* This is a test of

$$H_0: \beta_6 = \beta_7 = \beta_8 = \beta_9 = \beta_{16} = \beta_{17} = \beta_{18} = \beta_{19} = \beta_{20} = \beta_{21} = \beta_{22} = \beta_{23} = 0$$

3. *All four social class equations are parallel (controlling for SEX).* This is a test of

$$H_0: \beta_{10} = \beta_{11} = \beta_{12} = \beta_{13} = \beta_{14} = \beta_{15} = \beta_{18} = \beta_{19} = \beta_{20} = \beta_{21} = \beta_{22} = \beta_{23} = 0$$

When this hypothesis is true, the eight equations reduce to

$$\begin{cases} \text{SC1–male:} & Y = (\beta_0 + \beta_6) + (\beta_1 + \beta_{16})X_1 + (\beta_2 + \beta_{17})X_2 + E \\ \text{SC2–male:} & Y = (\beta_0 + \beta_3 + \beta_6 + \beta_7) + (\beta_1 + \beta_{16})X_1 + (\beta_2 + \beta_{17})X_2 + E \\ \text{SC3–male:} & Y = (\beta_0 + \beta_4 + \beta_6 + \beta_8) + (\beta_1 + \beta_{16})X_1 + (\beta_2 + \beta_{17})X_2 + E \\ \text{SC4–male:} & Y = (\beta_0 + \beta_5 + \beta_6 + \beta_9) + (\beta_1 + \beta_{16})X_1 + (\beta_2 + \beta_{17})X_2 + E \end{cases}$$

$$\begin{cases} \text{SC1–female:} & Y = \beta_0 + \beta_1 X_1 + \beta_2 X_2 + E \\ \text{SC2–female:} & Y = (\beta_0 + \beta_3) + \beta_1 X_1 + \beta_2 X_2 + E \\ \text{SC3–female:} & Y = (\beta_0 + \beta_4) + \beta_1 X_1 + \beta_2 X_2 + E \\ \text{SC4–female:} & Y = (\beta_0 + \beta_5) + \beta_1 X_1 + \beta_2 X_2 + E \end{cases}$$

Thus, within any given sex group, all four regression equations have the same coefficients for X_1 and X_2.

4. *All four social class equations are coincident (controlling for SEX).* This is a test of

$$H_0: \beta_3 = \beta_4 = \beta_5 = \beta_7 = \beta_8 = \beta_9 = \beta_{10} = \beta_{11} = \beta_{12} = \beta_{13} = \beta_{14}$$
$$= \beta_{15} = \beta_{18} = \beta_{19} = \beta_{20} = \beta_{21} = \beta_{22} = \beta_{23} = 0$$

5. *All eight regression equations are parallel.* This is a test of

$$H_0: \text{``}\beta_{10} \text{ through } \beta_{23} \text{ are simultaneously } 0\text{''}$$

When this hypothesis is true, the eight equations are all of the form

$$Y = \beta_{0(i)} + \beta_1 X_1 + \beta_2 X_2 + E \qquad \text{for } i = 1, 2, \ldots, 8$$

(i.e., the eight models differ only in intercept).

6. *All eight regression equations are coincident.* This is a test of

$$H_0: \text{``}\beta_3 \text{ through } \beta_{23} \text{ are simultaneously } 0\text{''}$$

7. *There is no interaction effect between SEX and SC.* This is a test of

$$H_0: \beta_7 = \beta_8 = \beta_9 = \beta_{18} = \beta_{19} = \beta_{20} = \beta_{21} = \beta_{22} = \beta_{23} = 0$$

From the form of the complete model (14.13), each of the coefficients in H_0 corresponds to a product term of the general form $Z_i Q$ or $X_i Z_i Q$, which involves the product of a social class variable and the sex variable.

Problems

1. Using the data in Problem 2 of Chapter 5 and/or the printout given here, answer the following questions concerning the separate straight-line regressions of SBP on QUET for smokers (SMK = 1) and nonsmokers (SMK = 0), respectively.
   a. Determine the least-squares line of SBP (Y) on QUET (X) separately for smokers and nonsmokers.
   b. Test H_0: "the slopes are the same for the populations of smokers and nonsmokers being sampled" versus H_A: "nonsmokers have a more positive slope."
   c. Test H_0: "the intercepts are the same for the populations of smokers and nonsmokers being sampled" versus H_A: "the intercepts are different."
   d. Test H_0: "the straight lines are coincident for the populations of smokers and nonsmokers being sampled" versus H_A: "the straight lines are not coincident."

Computer Printout for Problem 1

```
SMOKERS

MEANS AND STANDARD DEVIATIONS
X: 3.40829 +- 0.56785    Y: 147.82353 +- 15.21198

CORRELATIONS
PEARSON: 0.750995   SPEARMAN: 0.718137   KENDALL: 0.588235

LINEAR REGRESSION OF Y ON X

SLOPE:      20.11804 +- 4.56719
INTERCEPT:  79.25533 +- 15.76837
S(Y|X):     10.37401

ANALYSIS OF VARIANCE: F(1,15) = 19.40317

NONSMOKERS

MEANS AND STANDARD DEVIATIONS
X: 3.47827 +- 0.41930    Y: 140.80000 +- 12.90183

CORRELATIONS
PEARSON: 0.854814   SPEARMAN: 0.919501   KENDALL: 0.778882

LINEAR REGRESSION OF Y ON X

SLOPE:      26.30283 +- 4.42865
INTERCEPT:  49.31176 +- 15.50814
S(Y|X):      6.94794

ANALYSIS OF VARIANCE: F(1,13) = 35.27456
```

2. A topic of major concern to demographers and economists alike is the effect of a high fertility rate on per capita income. The first two accompanying tables display values of per capita income (PCI) and population percentage under age 15 (YNG)

for a hypothetical sample of developing countries in Latin America and Africa, respectively. The third table summarizes the results of straight-line regressions of PCI (Y) on YNG (X) for each group of countries.

LATIN AMERICAN COUNTRIES

YNG (X)	PCI (Y)	YNG (X)	PCI (Y)	YNG (X)	PCI (Y)
32.2	788	44.0	292	35.0	685
47.0	202	44.0	321	47.4	220
34.0	825	43.0	300	48.0	195
36.0	675	43.0	323	37.0	605
38.7	590	40.0	484	38.4	530
40.9	408	37.0	625	40.6	480
45.0	324	39.0	525	35.8	690
45.4	235	44.6	340	36.0	685
42.2	338	33.0	765		

AFRICAN COUNTRIES

YNG (X)	PCI (Y)	YNG (X)	PCI (Y)	YNG (X)	PCI (Y)
34.0	317	41.0	188	39.0	225
36.0	270	42.0	166	39.0	232
38.2	208	45.0	132	37.0	260
43.0	150	36.0	290	37.0	250
44.0	105	42.6	160	46.0	92
44.0	128	33.0	300	45.6	110
46.0	85	33.0	320	42.0	180
48.0	75	47.0	85	38.8	235
40.0	210	47.0	75		

SUMMARY OF SEPARATE STRAIGHT-LINE FITS

Location	n	$\hat{\beta}_0$	$\hat{\beta}_1$	$\bar{X}$	$\bar{Y}$	S_X^2	$S_{Y|X}^2$	r
Latin American	26	2170.67	−42.0	40.277	478.846	21.633	1391.756	−.983
African	26	893.571	−17.28	40.931	186.462	20.611	157.238	−.988

a–d. Do the same as in parts (a) through (d) of Problem 1 for the straight-line regressions of PCI (Y) on YNG (X) for Latin American and African countries, respectively.

e. Test H_0: "the population correlation coefficients are equal for the two groups of countries under study." (Use $\alpha = .05$.) Does your conclusion here clash with your findings regarding the equality of slopes? *Hint*: See Section 6-7.

(*Note*: For each test above, assume that the alternative hypothesis is two sided.)

3. A team of anthropologists and nutrition experts were investigating the influence of protein content in diet on the relationship between AGE and height (HT) for New Guinean children. The two accompanying tables display values of HT (in centimeters) and AGE for a hypothetical sample of children with protein-rich and protein-poor diets, respectively.

PROTEIN-RICH DIET

AGE (X)	0.2	0.5	0.8	1.0	1.0	1.4	1.8	2.0	2.0	2.5	2.5	3.0	2.7
HT (Y)	54	54.3	63	66	69	73	82	83	80.3	91	93.2	94	94

PROTEIN-POOR DIET

AGE (X)	0.4	0.7	1.0	1.0	1.5	2.0	2.0	2.4	2.8	3.0	1.3	1.8	0.2	3.0
HT (Y)	52	55	61	63.4	66	68.5	67.9	72	76	74	65	69	51	77

SUMMARY OF SEPARATE STRAIGHT-LINE FITS

Diet	n	$\hat{\beta}_0$	$\hat{\beta}_1$	$\bar{X}$	$\bar{Y}$	S_X^2	$S_{Y|X}^2$	r
Protein rich	13	50.324	16.009	1.646	76.677	0.808	5.841	.937
Protein poor	14	51.225	3.686	1.650	65.557	0.873	4.598	.969

a–d. Do the same as in parts (a) through (d) of Problem 1 for the straight-line regressions of HT (Y) on AGE (X) for the two diets. (Consider a two-sided alternative in each case.)

e. Test whether the population correlation coefficient for children with a protein-rich diet is significantly different from that for children with a protein-poor diet. (Consider a two-sided alternative.)

4. For the data of DI (Y) on IQ (X) in Problem 4 of Chapter 5, assume that this sample of 17 observations (outlier removed) consisted of males only. Now suppose that another sample of observations on DI (Y) and IQ (X) has been obtained for 14 females. The information needed to compare the straight-line regression equations for males and females is given in the table.

Sex Group	n	$\hat{\beta}_0$	$\hat{\beta}_1$	$\bar{X}$	$\bar{Y}$	S_X^2	$S_{Y|X}^2$	r
Males	17	70.846	−0.444	101.411	25.812	215.882	24.335	−.807
Females	19	61.871	−0.438	101.053	17.579	175.497	16.692	−.825

a–d. Do the same as in parts (a) through (d) of Problem 1 for the straight-line regressions of DI (Y) on IQ (X) for the two sexes. (Consider two-sided alternatives.)

e. Perform a two-sided test as to whether the population correlation coefficients for males and females are equal.

5. Assume that the data of VOTE (Y) on TVEXP (X) in Problem 5 of Chapter 5 came from congressional districts in New York. Now, suppose that a second sample of 17 congressional districts in California was then selected and the same information recorded. The table provides the information needed for comparing the straight-line regression equations for New York and California.

Location	n	$\hat{\beta}_0$	$\hat{\beta}_1$	$\bar{X}$	$\bar{Y}$	S_X^2	$S_{Y|X}^2$	r
New York	20	2.174	1.177	36.99	45.71	76.870	11.101	.954
California	17	8.030	1.036	36.371	45.706	97.335	13.492	.945

a–d. Do the same as in parts (a) through (d) of Problem 1 for the straight-line regression of VOTE (Y) on TVEXP (X) for the two states. Consider the one-sided alternative H_A: $\beta_{1(CAL)} < \beta_{1(NY)}$ for the test for slope and the one-sided alternative H_A: $\beta_{0(CAL)} < \beta_{0(NY)}$ for the test for intercept.

e. Perform a two-sided test regarding whether the correlation coefficients for New York and California are equal.

6. The data in the table represent four-week growth rates for depleted chicks at different dosage levels of vitamin B, by sex.[10]

Males		Females	
Growth Rate (Y)	$\text{Log}_{10}(\text{Dose})$ (X)	Growth Rate (Y)	$\text{Log}_{10}(\text{Dose})$ (X)
17.1	0.301	18.5	0.301
14.3	0.301	22.1	0.301
21.6	0.301	15.3	0.301
24.5	0.602	23.6	0.602
20.6	0.602	26.9	0.602
23.8	0.602	20.2	0.602
27.7	0.903	24.3	0.903
31.0	0.903	27.1	0.903
29.4	0.903	30.1	0.903
30.1	1.204	28.1	0.903
28.6	1.204	30.3	1.204
34.2	1.204	33.0	1.204
37.3	1.204	35.8	1.204
33.3	1.505	32.6	1.505
31.8	1.505	36.1	1.505
40.2	1.505	30.5	1.505

Using the information provided in the next table:

a. Determine the dose–response straight lines separately for each sex, and plot them on the same graph.

b. Test whether the slopes for males and females are different.

c. Find a 99% confidence interval for the true difference between the male and female slopes.

d. Test for coincidence of the two straight lines.

Sex Group	n	$\hat{\beta}_0$	$\hat{\beta}_1$	$\bar{X}$	$\bar{Y}$	S_X^2	$S_{Y|X}^2$	r
Males	16	13.767	14.966	0.941	27.84	0.1797	11.4403	.889
Females	16	15.656	12.735	0.903	27.16	0.1812	8.4719	.888

7. The results in the table below were obtained in a study of the amount of energy metabolized by two similar species of birds under constant temperature.[11] Infor-

---

[10] Adapted from a study by Clark, Lechyeka, and Cook (1940).

[11] Adapted from a study by Davis (1955).

mation based on separate straight-line fits to each data set is summarized in the second table.

Species A		Species B	
Calories (Y)	Temperature (X) (°C)	Calories (Y)	Temperature (X) (°C)
36.9	0	41.1	0
35.8	2	40.6	2
34.6	4	38.9	4
34.3	6	37.9	6
32.8	8	37.0	8
31.7	10	36.1	10
31.0	12	36.3	12
29.8	14	34.2	14
29.1	16	33.4	16
28.2	18	32.8	18
27.4	20	32.0	20
27.8	22	31.9	22
25.5	24	30.7	24
24.9	26	29.5	26
23.7	28	28.5	28
23.1	30	27.7	30

Species	n	$\hat{\beta}_0$	$\hat{\beta}_1$	$\bar{X}$	$\bar{Y}$	S_X^2	$S_{Y\|X}^2$	r
A	16	36.579	−0.4528	15.00	29.79	90.6666	0.1662	.9959
B	16	40.839	−0.4368	15.00	34.29	90.6666	0.1757	.9953

a. Plot the least-squares straight lines for each species on the same graph.
b. Test whether the two lines are parallel.
c. Test whether the two lines have the same intercept.
d. At 15°C, give a 95% confidence interval for the true difference between the mean amounts of energy metabolized by each species. *Hint*: Use a confidence interval of the form

$$(\hat{Y}_{15}^A - \hat{Y}_{15}^B) \pm t_{n_A+n_B-4,1-\alpha/2} \sqrt{S_{\hat{Y}_{15}^A}^2 + S_{\hat{Y}_{15}^B}^2}$$

where $\hat{Y}_{15}^i$ is the predicted value at 15°C for species i (i = A, B) and $S_{\hat{Y}_{15}^i}^2$ is the estimated variance of $\hat{Y}_{15}^i$ given by the general formula (for $X = X_0$):

$$S_{\hat{Y}_{X_0}^i}^2 = S_{P,Y\|X}^2 \left[\frac{1}{n_i} + \frac{(X_0 - \bar{X})^2}{(n_i - 1)S_{X_i}^2} \right]$$

8. In Problem 1 of this chapter, separate straight-line regressions of SBP on QUET were compared for smokers (SMK = 1) and nonsmokers (SMK = 0).
   a. Define a single multiple regression model that uses the data for both smokers and nonsmokers and that defines straight-line models for each group with possibly differing intercepts and slopes. Define the intercept and slope for each straight-line model in terms of the regression coefficients of the single regression model.

b. Using the computer results provided below, determine and plot on graph paper the two fitted straight lines obtained from the fit of the regression model.

c. Test the following null hypotheses:

H_0: "the two lines are parallel"

H_0: "the two lines are coincident"

For each of these tests, state the appropriate null hypothesis in terms of the regression coefficients of the regression model.

d. Compare your answers in parts (b) and (c) with those obtained from fitting separate regressions in Problem 1.

COMPUTER RESULTS FOR PROBLEM 8

Variable Entered on Step 1: QUET

Variable	$\hat{\beta}$	$S_{\hat{\beta}}$	Partial F
QUET	21.49167	3.54515	36.751
(intercept)	70.57643		

ANOVA (Step 1)

Source	df	SS
Regression	1	3,537.94538
Residual	30	2,888.02337

Variable Entered on Step 2: SMK

Variable	$\hat{\beta}$	$S_{\hat{\beta}}$	Partial F
QUET	22.11560	3.22996	46.882
SMK	8.57101	3.16670	7.326
(intercept)	63.87606		

ANOVA (Step 2)

Source	df	SS
Regression	2	4,120.36603
Residual	29	2,305.60272

Variable Entered on Step 3: QUET × SMK

Variable	$\hat{\beta}$	$S_{\hat{\beta}}$	Partial F
QUET	26.30282	5.70349	21.268
SMK	29.94357	24.16355	1.536
QUET × SMK	−6.18479	6.93171	0.796
(intercept)	49.31178		

ANOVA (Step 3)

Source	df	SS
Regression	3	4,184.10718
Residual	28	2,241.86157

9. Use the computer results that follow to compare the separate regressions of SBP on AGE and QUET for smokers and nonsmokers (using the data in Problem 2 of Chapter 5) as follows:

a. State the appropriate regression model incorporating both equations for smokers and nonsmokers.

b. Determine the fitted regression equations for smokers and nonsmokers separately using the fitted regression model.

c. Test for parallelism of the two models, stating the null hypothesis in terms of appropriate regression coefficients.

d. Test for coincidence of the two models, stating the null hypothesis in terms of regression coefficients.

COMPUTER RESULTS FOR PROBLEM 9

Variable(s) Entered on Step 1: AGE, QUET

Variable	$\hat{\beta}$	$S_{\hat{\beta}}$	Partial F
AGE	1.04516	0.38606	7.329
QUET	9.75073	5.40245	3.258
(intercept)	55.32344		

ANOVA (Step 1)

Source	df	SS
Regression	2	4,120.59219
Residual	29	2,305.37656

Variable Entered on Step 2: SMK

Variable	$\hat{\beta}$	$S_{\hat{\beta}}$	Partial F
AGE	1.21271	0.32382	14.025
QUET	8.59245	4.49868	3.648
SMK	9.94557	2.65606	14.021
(intercept)	45.10320		

ANOVA (Step 2)

Source	df	SS
Regression	3	4,889.82563
Residual	28	1,536.14312

Variable(s) Entered on Step 3: AGE × SMK, QUET × SMK

Variable	$\hat{\beta}$	$S_{\hat{\beta}}$	Partial F
AGE	1.02892	0.50177	4.205
QUET	10.45104	9.13014	1.310
SMK	−0.53744	23.23004	0.001
AGE × SMK	0.43733	0.71279	0.376
QUET × SMK	−3.70682	10.76763	0.119
(intercept)	48.61271		

ANOVA (Step 3)

Source	df	SS
Regression	5	4,915.63033
Residual	26	1,510.33842

10. The results presented below are based on data from a study by Gruber (1970) to determine the extent and manner in which changes in blood pressure over time depend on initial blood pressure (at the beginning of the study), the sex of the individual, and the relative weight of the individual. Data were collected on $n = 104$ persons, for which the $k = 7$ independent variables were examined by multiple regression. The variables used in the study are defined below:

Y = SBPSL (estimated slope based on the straight-line regression of an individual's blood pressure over time)[12]

X_1 = SBP1 (initial systolic blood pressure)

X_2 = SEX (male = 1, female = −1)

X_3 = RW (relative weight)

$X_4 = X_1 X_2, \qquad X_5 = X_1 X_3$

$X_6 = X_2 X_3, \qquad X_7 = X_1 X_2 X_3$

---

[12] This choice of dependent variable can be criticized because observations on an individual's blood pressure taken over time are *not* independent, thus violating Assumption 2 in Chapter 8. A preferred statistic to that used by Gruber could be obtained by weighted regression (see Chapter 12) or by growth curve analysis (see Allen and Grizzle, 1969).

Variable	$\hat{\beta}$	$S_{\hat{\beta}}$	Partial F ($\hat{\beta}^2/S_{\hat{\beta}}^2$)
X_1(SBP1)	-0.045	0.00762	34.987
X_2(SEX)	0.695	0.86644	0.643
X_3(RW)	0.027	0.07049	0.149
X_4(X_1X_2)	-0.0029	0.00762	0.145
X_5(X_1X_3)	-0.00018	0.00062	0.084
X_6(X_2X_3)	-0.0092	0.07049	0.017
X_7($X_1X_2X_3$)	0.00022	0.00062	0.125
(intercept)	4.667		

Source	df	SS	MS	F
Overall regression	7	37.148	5.307	$\dfrac{5.307}{76.246/96} = 6.68^{**}$
Regression X_1	1	24.988	24.988	$\dfrac{24.988}{88.406/102} = 28.83^{**}$
$X_2\|X_1$	1	7.886	7.886	$\dfrac{7.886}{80.520/101} = 9.89^{**}$
$X_3\|X_1, X_2$	1	1.057	1.057	$\dfrac{1.057}{79.463/100} = 1.33$
$X_4\|X_1, X_2, X_3$	1	0.020	0.020	$\dfrac{0.020}{79.443/99} = 0.025$
$X_5\|X_1, X_2, X_3, X_4$	1	0.254	0.254	$\dfrac{0.254}{79.189/98} = 0.314$
$X_6\|X_1, X_2, X_3, X_4, X_5$	1	2.844	2.844	$\dfrac{2.844}{76.345/97} = 3.613$
$X_7\|X_1, X_2, X_3, X_4, X_5, X_6$	1	0.099	0.099	$\dfrac{0.099}{76.246/96} = 0.125$
Residual	96	76.246	0.794	
Total (corrected)	103	113.394		

Note: $F_{7,96,0.95} = 2.11$, $F_{1,96,0.95} = 3.95$, $F_{1,102,0.95} = 3.94$.

Using only the ANOVA table, answer the following:

a. Determine the form of the separate fitted regression models of SBPSL (Y) on SBP1 (X_1), RW (X_3), and SBP1 $\times$ RW (X_5) for each sex in terms of the regression coefficients of the fitted regression model.

b. Using this ANOVA table, why can't one test whether these two regression equations are either parallel or coincident? Describe the appropriate testing procedure in each case.

11. For the data in Problem 2 of this chapter, address the following using the information provided:

a. State the regression model that incorporates the straight-line models for each group of countries.

b. Determine and plot the separate fitted straight lines based on the fitted regression model given in part (a).

c. Test for parallelism of the two straight lines.

d. Test for coincidence of the two straight lines.

e. Compare your results in parts (b) through (d) to those obtained in Problem 2.

COMPUTER RESULTS FOR PROBLEM 11

Variable Entered on Step 1: YNG

Variable	$\hat{\beta}$	$S_{\hat{\beta}}$	Partial F
YNG	−32.1236	4.6767	47.182
(intercept)	1,636.9963		

ANOVA (Step 1)

Source	df	SS
Regression (YNG)	1	1,095,554.6
Residual	50	1,160,995.1

Variable Entered on Step 2: Z

Variable	$\hat{\beta}$	$S_{\hat{\beta}}$	Partial F
YNG	−29.9394	1.9587	233.637
$Z\begin{cases}1 & \text{if Latin American}\\0 & \text{otherwise}\end{cases}$	272.8088	17.7008	237.536
(intercept)	1,411.9048		

ANOVA (Step 2)

Source	df	SS
Regression (YNG, Z)	2	2,058,010.1
Residual	49	198,539.6

Variable Entered on Step 3: YNG × Z

Variable	$\hat{\beta}$	$S_{\hat{\beta}}$	Partial F
YNG	−17.2758	1.2260	198.561
Z	1,277.0988	70.0061	332.795
YNG × Z	−24.7291	1.7132	208.347
(intercept)	893.5713		

ANOVA (Step 3)

Source	df	SS
Regression (YNG, Z, YNG × Z)	3	2,219,373.9
Residual	48	37,175.8

12. Using the information given, answer the same questions as in Problem 11 concerning the regression of height (Y) on age (X) for children in one of two diet categories. (This problem is based on the data in Problem 3.)

COMPUTER RESULTS FOR PROBLEM 12

Variable Entered on Step 1: AGE

Variable	$\hat{\beta}$	$S_{\hat{\beta}}$	Partial F
AGE	12.0445	1.5353	61.546
(intercept)	51.0600		

ANOVA (Step 1)

Source	df	SS
Regression (AGE)	1	3,053.35
Residual	25	1,240.28

Variable Entered on Step 2: Z

Variable	$\hat{\beta}$	$S_{\hat{\beta}}$	Partial F
AGE	12.0583	0.8897	183.699
$Z\begin{cases}1 & \text{if protein rich}\\0 & \text{otherwise}\end{cases}$	11.1662	1.5721	50.449
(intercept)	45.6610		

ANOVA (Step 2)

Source	df	SS
Regression (AGE, Z)	2	3,893.80
Residual	24	399.83

Variable Entered on Step 3: AGE × Z				ANOVA (Step 3)		
Variable	$\hat{\beta}$	$S_{\hat{\beta}}$	Partial F	Source	df	SS
AGE	8.6860	0.6762	164.999	Regression (AGE, Z, AGE × Z)	3	4,174.21
Z	−0.9015	1.8619	0.234	Residual	23	119.42
AGE × Z	7.3229	0.9965	54.005			
(intercept)	51.2252					

13. In Gruber's (1970) study of $n = 104$ individuals (discussed in Problem 10), the relationship between blood pressure change (SBPSL) and relative weight (RW) controlling for initial blood pressure (SBP1) was compared for three different geographical backgrounds and for three different psychosocial orientations using the following 15 variables:

$Y = $ SBPSL

$X_1 = $ SBP1 (initial blood pressure)

$X_2 = $ R (1 if rural background, 0 if town, −1 if urban)

$X_3 = $ T (1 if town background, 0 if rural, −1 if urban)

$X_4 = $ TD (1 if traditional orientation, 0 if transitional, −1 if modern)

$X_5 = $ TN (1 if transitional orientation, 0 if traditional, −1 if modern)

$X_6 = $ RW (relative weight)

$X_7 = $ T × TD

$X_8 = $ T × TN

$X_9 = $ R × TD

$X_{10} = $ R × TN

$X_{11} = $ R × TD × RW

$X_{12} = $ R × TN × RW

$X_{13} = $ T × TD × RW

$X_{14} = $ T × TN × RW

A run of a standard stepwise-regression program using these data yielded the accompanying ANOVA table (variables were forced to enter in the order presented) based on the model

$$Y = \beta_0 + \beta_1 X_1 + \beta_2 X_2 + \cdots + \beta_{14} X_{14} + E$$

a. Using the regression model given, determine the form of the nine fitted regression equations corresponding to the nine possible combinations of background with orientation (i.e., R = 1 and TD = 1, R = 0 and TD = 1, R = −1 and TD = 1, etc.). (Note: Each of the nine equations will be of the form $\hat{Y} = \hat{\beta}_0^* + \hat{\beta}_1^*$ (SBP1) $+ \hat{\beta}_2^*$ (RW).)

b. Test the null hypothesis that the nine regression equations determined in part (a) are parallel. State the null hypothesis in terms of the regression coefficients of the original 14-variable regression model.

   c. Test the hypothesis H_0: "the three regression equations corresponding to the three backgrounds (rural, town, and urban) are parallel (but not necessarily coincident)" against the alternative H_A: "they are not parallel."

   d. Set up the formula for testing H_0: "the nine regression equations dealt with in part (a) are coincident" against H_A: "they are not coincident." State the null hypothesis in terms of the coefficients in the regression equation.

Source		df	SS	F
	X_1	1	24.9878	28.830
	$X_2 \mid X_1$	1	0.5218	0.600
	$X_3 \mid X_1, X_2$	1	0.0057	0.006
	$X_4 \mid X_1, X_2, X_3$	1	1.0520	1.199
	$X_5 \mid X_1, X_2, X_3, X_4$	1	1.1116	1.271
	$X_6 \mid X_1{-}X_5$	1	0.8321	0.951
Regression	$X_7 \mid X_1{-}X_6$	1	0.2919	0.331
	$X_8 \mid X_1{-}X_7$	1	1.6601	1.902
	$X_9 \mid X_1{-}X_8$	1	0.5843	0.667
	$X_{10} \mid X_1{-}X_9$	1	0.2266	0.257
	$X_{11} \mid X_1{-}X_{10}$	1	1.1916	1.355
	$X_{12} \mid X_1{-}X_{11}$	1	2.0853	2.407
	$X_{13} \mid X_1{-}X_{12}$	1	1.5915	1.854
	$X_{14} \mid X_1{-}X_{13}$	1	0.0208	0.024
Residual		89	77.2303	
Total		103	113.3934	

14. The Environmental Protection Agency conducted an experiment to assess the characteristics of sampling procedures designed to measure the suspended particulate concentration (X) in a particular city. At each of two distinct locations (designated location 1 and location 2), two identical sampling units were set up side by side and readings were taken on each of 10 days. The data are given here in tabular form, where X_{ij1} and X_{ij2} are, respectively, the measured concentration for samplers 1 and 2 at location i on day j ($i = 1, 2; j = 1, 2, \ldots, 10$). It was hypothesized that the inherent variation in the observations depends on the level of concentration being measured. To quantify this hypothesis, it is proposed to fit a model of the form

$$|d_{ij}| = \beta_0 + \beta_1 Z + \beta_2(X_{ij1} + X_{ij2}) + \beta_3(X_{ij1} + X_{ij2})Z + E$$

where Z is 1 if the observation pertains to location 1 and is 0 otherwise, and where $|d_{ij}| = |X_{ij1} - X_{ij2}|$.

   a. Using the results provided, determine and interpret the fitted straight-line relationship between $|d_{ij}|$ and $(X_{ij1} + X_{ij2})$ at each location.

   b. Test for parallelism of the two lines.

Day	Location 1 X_{1j1}	X_{1j2}	$d_{1j} = X_{1j1} - X_{1j2}$	Location 2 X_{2j1}	X_{2j2}	$d_{2j} = X_{2j1} - X_{2j2}$
1	4	3	1	6	5	1
2	8	6	2	3	1	2
3	12	16	−4	1	2	−1
4	1	1	0	10	12	−2
5	7	6	1	17	17	0
6	11	8	3	4	7	−3
7	14	10	4	8	6	2
8	10	12	−2	12	12	0
9	2	2	0	10	9	1
10	15	20	−5	20	19	1

c. Test for coincidence of the two lines.

d. How would one test whether the level of concentration is significantly related to the inherent variation in the observations for at least one of the two locations?

COMPUTER RESULTS FOR PROBLEM 14

Variable Entered on Step 1: $X_{ij1} + X_{ij2}$

Variable	$\hat{\beta}$	$S_{\hat{\beta}}$	Partial F
$X_{ij1} + X_{ij2}$	0.0455	0.0288	2.493
(intercept)	0.9560		

ANOVA (Step 1)

Source	df	SS
Regression ($X_{ij1} + X_{ij2}$)	1	4.8348
Residual	18	34.9152

Variable Entered on Step 2: Z

Variable	$\hat{\beta}$	$S_{\hat{\beta}}$	Partial F
$X_{ij1} + X_{ij2}$	0.0482	0.0277	3.031
$Z \begin{cases} 1 & \text{if location 1} \\ 0 & \text{otherwise} \end{cases}$	0.9626	0.5981	2.590
(intercept)	0.4279		

ANOVA (Step 2)

Source	df	SS
Regression ($X_{ij1} + X_{ij2}, Z$)	2	9.4514
Residual	17	30.2986

Variable Entered on Step 3: $(X_{ij1} + X_{ij2}) \times Z$

Variable	$\hat{\beta}$	$S_{\hat{\beta}}$	Partial F
$X_{ij1} + X_{ij2}$	−0.0392	0.0202	3.746
Z	−2.4279	0.6177	15.448
$(X_{ij1} + X_{ij2}) \times Z$	0.1951	0.0302	41.640
(intercept)	2.0086		

ANOVA (Step 3)

Source	df	SS
Regression (all three variables)	3	31.3395
Residual	16	8.4105

15. A biologist compared the effect of temperature of each of two mediums on the growth of human amniotic cells in a tissue culture. The data obtained are as shown.

a. Assuming that a parabolic model is appropriate for describing the relationship between Y and X for each medium, provide a single regression model that will incorporate two separate parabolic models, one corresponding to each medium.

   b. Use the computer printout provided to determine and plot the separate fitted parabolas for each medium. (*Note*: $Z = 0$ for medium A and $Z = 1$ for medium B.)
   c. Test for "parallelism" of the two parabolas.
   d. Test for coincidence of the two parabolas.
   e. Using only the computer results given, is it possible to test whether a quadratic term should be included in the model for each medium? Explain.

Medium A		Medium B	
No. of Cells $\times 10^{-6}$ (Y)	Temperature (°F) (X)	No. of Cells $\times 10^{-6}$ (Y)	Temperature (°F) (X)
1.13	40	0.98	40
1.20	40	1.05	40
1.00	40	0.92	40
0.91	40	0.90	40
1.05	40	0.89	40
1.75	60	1.60	60
1.45	60	1.45	60
1.55	60	1.40	60
1.64	60	1.50	60
1.60	60	1.56	60
2.30	80	2.20	80
2.15	80	2.10	80
2.25	80	2.20	80
2.40	80	2.30	80
2.49	80	2.26	80
3.18	100	3.10	100
3.10	100	3.00	100
3.28	100	3.13	100
3.35	100	3.20	100
3.12	100	3.07	100

COMPUTER RESULTS FOR PROBLEM 15

Source	df	SS	Source	df	SS
Regression (X)	1	25.6686	Regression (X, X^2)	2	25.9593
Residual	38	0.7036	Residual	37	0.4129

Source	df	SS	Source	df	SS
Regression (X, X^2, Z)	3	26.0685	Regression (X, X^2, Z, ZX)	4	26.0685
Residual	36	0.3037	Residual	35	0.3037

Source	df	SS
Regression (X, X^2, Z, ZX, ZX^2)	5	26.0686
Residual	34	0.3036

Computer Printout (SPSS) for Problem 16

```
VARIABLE(S) ENTERED ON STEP NUMBER 1.. X

MULTIPLE R         0.88583       ANALYSIS OF VARIANCE    DF    SUM OF SQUARES    MEAN SQUARE         F
R SQUARE           0.78470       REGRESSION              1.    1041.33728        1041.33728    109.33735
ADJUSTED R SQUARE  0.77752       RESIDUAL               30.     285.72230          9.52408
STANDARD ERROR     3.08611

        ——— VARIABLES IN THE EQUATION ———                        ——— VARIABLES NOT IN THE EQUATION ———
VARIABLE       B        BETA     STD ERROR B        F       VARIABLE   BETA IN   PARTIAL   TOLERANCE      F
X         13.85501    0.88583     1.32502       109.337     Z          0.01293   0.02784   0.99791     0.022
(CONSTANT) 14.72828                                         XZ         0.04475   0.08801   0.83267     0.226


VARIABLE(S) ENTERED ON STEP NUMBER 2.. Z (= 1 if MALE, = 0 if FEMALE)

MULTIPLE R         0.88592       ANALYSIS OF VARIANCE    DF    SUM OF SQUARES    MEAN SQUARE         F
R SQUARE           0.78486       REGRESSION              2.    1041.55874         520.77937     52.89862
ADJUSTED R SQUARE  0.77002       RESIDUAL               29.     285.50084           9.84486
STANDARD ERROR     3.13765

        ——— VARIABLES IN THE EQUATION ———                        ——— VARIABLES NOT IN THE EQUATION ———
VARIABLE       B        BETA     STD ERROR B        F       VARIABLE   BETA IN   PARTIAL   TOLERANCE      F
X         13.84577    0.88524     1.34656       105.413     XZ         0.19139   0.15355   0.13847     0.676
Z          0.16655    0.01293     1.11049         0.022
(CONSTANT) 14.65352


VARIABLE(S) ENTERED ON STEP NUMBER 3.. XZ

MULTIPLE R         0.88678       ANALYSIS OF VARIANCE    DF    SUM OF SQUARES    MEAN SQUARE         F
R SQUARE           0.78639       REGRESSION              3.    1043.28988         349.42996     35.09721
ADJUSTED R SQUARE  0.76743       RESIDUAL               28.     278.76970           9.95606
STANDARD ERROR     3.15532

        ——— VARIABLES IN THE EQUATION ———                        ——— VARIABLES NOT IN THE EQUATION ———
VARIABLE       B        BETA     STD ERROR B        F       VARIABLE   BETA IN   PARTIAL   TOLERANCE      F
X         12.73533    0.81424     1.91389        44.278
Z         -1.88944   -0.14670     2.73852         0.476
XZ         2.23020    0.19139     2.71233         0.676
(CONSTANT) 15.65624
```

From *Statistical Package for the Social Sciences* by Nie et al. Copyright © 1975 by McGraw-Hill Book Company and Dr. Norman Nie, President, SPSS Inc. Used with permission of McGraw-Hill Book Company and Dr. Norman Nie, President, SPSS Inc.

Regression coefficients for the fitted model containing all five independent variables are as follows: intercept, 0.4946; X, 0.00537; X^2, 0.00022; Z, -0.1437; ZX, 0.00123; ZX^2, -0.00001.

16. Answer the following questions using the computer results given based on the data of Problem 6 concerning the growth rates (Y) of depleted chicks at different (log) dosage levels (X) of vitamin B for males and females:

   a. Define a single multiple regression model that incorporates different straight-line models for males and females.

   b. Plot the fitted straight lines for each sex on graph paper.

   c. Test for parallelism.

   d. Test for coincidence.

   e. Compare your results regarding parts (b), (c), and (d) to those obtained for Problem 6.

References

Allen, D. M., and Grizzle, J. E. (1969). "Analysis of Growth and Dose Response Curves." *Biometrics*, 25: 357–382.

Armitage, P. (1971). *Statistical Methods in Medical Research*. Oxford: Blackwell Scientific Publications.

Clark, M. F., Lechyeka, M., and Cook, C. A. (1940). "The Biological Assay of Riboflavia." *J. Nutr.*, 20: 133–144.

Davis, E. A., Jr. (1955). "Seasonal Changes in the Energy Balance of the English Sparrow." *Auk.*, 72(4): 385–411.

Gruber, F. J. (1970). "Industrialization and Health." Ph.D. dissertation, Department of Epidemiology, University of North Carolina, Chapel Hill, N.C.

Analysis of Covariance and Other Methods for Adjusting Continuous Data

15-1 Preview

In Chapter 11 we discussed the issue of controlling for extraneous variables when assessing an association of interest. Three reasons for considering control are to assess *interaction,* to correct for *confounding,* and to increase the *precision* in estimating the association of interest. In regression the usual approach for carrying out such control is to fit a regression model that contains as independent variables not only the study factors (exposure variables) of interest, but also those extraneous variables considered to be important (and perhaps even product terms involving these variables). The focus then becomes one of determining the effects of the study factors on the response variable *adjusted* for the presence of the control variables in the model.

In this chapter we describe in some detail how to carry out this process of adjustment by using a popular procedure for regression modeling called analysis of covariance (ANACOVA). This technique involves a multiple regression model in which the study factors of interest are all treated as nominal variables, whereas the variables being controlled, that is, the *covariates,* may be measurements on any measurement scale. As discussed in the previous chapter, nominal variables are incorporated into regression models by means of dummy variables. Thus, the general ANACOVA model usually contains a mixture of dummy variables and other types of variables, and the dependent variable is considered continuous. In using the ANACOVA model, it is also assumed that there is no interaction of covariates with study variables, although, as discussed below, this assumption should be assessed in the analysis.

In addition to considering ANACOVA, we shall also briefly review other regression-type methods for controlling for extraneous factors.

15-2 Adjustment Problem

Suppose, as in the example in Section 14-4, that we are considering a sample of observations on the dependent variable systolic blood pressure and the independent variable age for $n_F = 29$ women and $n_M = 40$ men.

Two questions often of interest when analyzing such data are as follows:

1. Is the true straight-line relationship between blood pressure and age (given that a straight-line model is adequate) the same for the male and female populations?

2. Do the mean blood pressure levels for the male and female groups differ significantly after taking into account (i.e., after adjusting for or controlling for) the possible confounding effect due to differing age distributions in the two groups?

Although the statistical techniques required to answer these questions are not completely unrelated, these questions nevertheless differ in emphasis: The first focuses on a comparison of straight-line regression equations, whereas the second focuses on a comparison of the mean blood pressure levels in the two groups.

We have already considered question 1 (in Chapter 14) through use of the model

$$Y = \beta_0 + \beta_1 X + \beta_2 Z + \beta_3 XZ + E \tag{15.1}$$

where Z is a dummy variable identifying the sex group ($Z = 1$ if female and $Z = 0$ if male). By using appropriate tests of hypotheses concerning the parameters in this model, one of three important conclusions could be reached regarding question 1. These are:

1. The lines are coincident (i.e., $\beta_2 = \beta_3 = 0$).

2. The lines are parallel but not coincident (i.e., $\beta_3 = 0$ but $\beta_2 \neq 0$).

3. The lines are not parallel ($\beta_3 \neq 0$).

These conclusions have a great deal to do with the answer to question 2. If conclusion 1 is appropriate, we would say that the two sex groups do not differ in mean blood pressure level after controlling for the effect of age. If conclusion 2 holds, we would say that the sex group associated with the higher straight line has a higher mean blood pressure level at all ages. If conclusion 3 is valid, we would have to look closer at the orientation of the two straight lines: If they do not intersect in the age range of interest, we would say that the sex group associated with the higher curve has a higher mean blood pressure level at each age; if they do intersect in the age range of interest, we could say that one sex group has a higher mean blood pressure level than the other group at lower ages and a lower mean blood pressure level at higher ages (i.e., there is an age–sex group interaction effect).

Thus, by considering question 1 as outlined above, we may draw reasonable inferences about the relationship between the true average blood pressure levels in the two groups as a function of age. Nevertheless, additional statistical considerations are necessary for estimating the true adjusted mean difference as well as for estimating the adjusted means for each group. In this regard, question 2 is concerned with the problem of determining an appropriate method for adjusting the sample mean blood pressure levels to take into account the effect of age and the problem of providing a statistical test to compare these adjusted mean scores.

In the example we are considering, the need for adjustment stems from the fact that age is a factor known to be strongly associated with blood pressure and that the two groups, as sampled, may have widely different age distributions. Without adjusting the sample mean values to reflect any difference in the age distributions in the two groups, it would not be possible to determine (e.g., through the use of a two-sample t test based on the unadjusted sample means) whether a significant difference was due solely to the effect of age or to the effects of other factors. With adjustment, however, it could be determined whether any findings were solely attributable, for example, to the fact that the females in the sample were older than the males (or vice versa).

15-3 Analysis of Covariance

The usual statistical technique for handling the adjustment problem described above is called the *analysis of covariance*. In this approach, a regression model is fitted of the form

$$Y = \beta_0 + \beta_1 X + \beta_2 Z + E \tag{15.2}$$

where X (age) is referred to as the *covariate* and Z is a dummy variable that indexes the two groups to be compared ($Z = 1$ if female, $Z = 0$ if male). This model assumes, in contrast to model (15.1), which contains a $\beta_3 XZ$ term, that the regression lines for males and females are parallel. Under this model, *the adjusted mean scores for males and females are defined to be the predicted values obtained by evaluating the model at $Z = 0$ and $Z = 1$ when X is set equal to the overall mean age for the two groups. Furthermore, a partial F test of the hypothesis $H_0: \beta_2 = 0$ is used to determine whether these adjusted mean scores are signifi- cantly different.*

In computing these adjusted scores, we need to consider the two straight lines obtained by fitting model (15.2):

$$\begin{aligned} \text{Males } (Z = 0): \quad & \hat{Y}_M = \hat{\beta}_0 + \hat{\beta}_1 X \\ \text{Females } (Z = 1): \quad & \hat{Y}_F = (\hat{\beta}_0 + \hat{\beta}_2) + \hat{\beta}_1 X \end{aligned} \tag{15.3}$$

Explicit formulas for the estimated coefficients in (15.3) are as follows:

$$\begin{aligned} \hat{\beta}_1 &= \frac{(n_M - 1)S_{X_M}^2 \hat{\beta}_{1M} + (n_F - 1)S_{X_F}^2 \hat{\beta}_{1F}}{(n_M - 1)S_{X_M}^2 + (n_F - 1)S_{X_F}^2} \\ \hat{\beta}_0 &= \bar{Y}_M - \hat{\beta}_1 \bar{X}_M \\ \hat{\beta}_0 + \hat{\beta}_2 &= \bar{Y}_F - \hat{\beta}_1 \bar{X}_F \end{aligned} \tag{15.4}$$

where $\hat{\beta}_{1M}$ and $\hat{\beta}_{1F}$ are the estimated slopes based on separate straight-line fits for males and females, $S_{X_M}^2$ and $S_{X_F}^2$ are the sample variances (on X) for males and females, $\bar{Y}_M$ and $\bar{Y}_F$ are the mean blood pressures for the male and female samples, and $\bar{X}_M$ and $\bar{X}_F$ are the mean ages for the male and female samples, respectively. Note that $\hat{\beta}_1$ is a weighted average of the slopes $\hat{\beta}_{1M}$ and $\hat{\beta}_{1F}$, which are estimated separately from the male and female data sets.

Based on (15.3) and (15.4), two alternative formulae for computing adjusted mean scores are as follows:

Sex	Z	Adjusted Score	Formula 1	Formula 2	
Male	0	$\bar{Y}_M$ (adj)	$\hat{\beta}_0 + \hat{\beta}_1\bar{X}$	$\bar{Y}_M - \hat{\beta}_1(\bar{X}_M - \bar{X})$	(15.5)
Female	1	$\bar{Y}_F$ (adj)	$(\hat{\beta}_0 + \hat{\beta}_2) + \hat{\beta}_1\bar{X}$	$\bar{Y}_F - \hat{\beta}_1(\bar{X}_F - \bar{X})$	

In this table, $\bar{X}$ is the overall mean age for the combined data on males and females,

$$\bar{X} = \frac{n_M\bar{X}_M + n_F\bar{X}_F}{n_M + n_F}$$

Formula 1 is useful when model (15.2) has been estimated directly using a standard multiple regression program; formula 2, on the other hand, can be used without resorting to multiple regression procedures, although separate straight lines must be fitted to the male and female data sets.

Given that the parallel straight-line assumption of model (15.2) is appropriate (we shall discuss this assumption in Section 15-4), the two formulae provide a comparison of the mean blood pressure levels for the two sex groups as if they both had the same age distribution. In this regard, the covariance approach described above attempts to artificially equate the age distributions by treating both sex groups as if they had the same mean age, the best estimate of which is $\bar{X}$. The adjusted scores, then, represent the predicted $\hat{Y}$-values for each fitted line at $\bar{X}$, the assumed common mean age. This is depicted graphically in Figure 15-1.

That the partial F test of $H_0: \beta_2 = 0$ addresses the question of whether there is a significant difference between the adjusted means follows because, from formula 1 in (15.5), the difference in the two adjusted mean scores is exactly equal to $\hat{\beta}_2$; that is,

$$\hat{\beta}_2 = \bar{Y}_F(adj) - \bar{Y}_M(adj) = [(\hat{\beta}_0 + \hat{\beta}_2) + \hat{\beta}_1\bar{X}] - (\hat{\beta}_0 + \hat{\beta}_1\bar{X})$$

Example   The least-squares fitting of the model (15.2) for the male–female data we have been discussing yields the following estimated model:

$$\hat{Y} = 110.29 + 0.96X - 13.51Z$$

The separate fitted equations for males and females, respectively, are

Males ($Z = 0$): $\hat{Y}_M = 110.29 + 0.96X$

Females ($Z = 1$): $\hat{Y}_F = 96.78 + 0.96X$

The adjusted mean scores obtained from these equations using formula 1 of (15.5) with $\bar{X} = 46.14$ are

$$\bar{Y}_M(adj) = 110.29 + 0.96(46.14) = 154.40$$
$$\bar{Y}_F(adj) = 96.78 + 0.96(46.14) = 140.89$$

A comparison of these adjusted mean scores with the unadjusted means gives the following:

Sex	Unadjusted $\bar{Y}$	Adjusted
Male	155.15	154.40
Female	139.86	140.89

FIGURE 15-1   Adjusted mean systolic blood pressure (SBP) scores for
males and females, controlling for age using analysis
of covariance

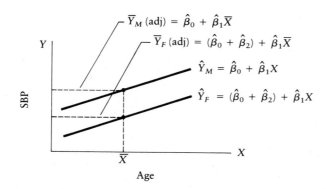

Notice that the adjusted mean for males is slightly lower than the unadjusted mean for males, whereas the female adjusted mean is slightly higher than its unadjusted counterpart. The direction of these changes accurately reflects the fact that, in this sample, the males are somewhat older on the average ($\bar{X}_M = 46.93$) than the females ($\bar{X}_F = 45.07$). Use of the adjusted mean scores, in effect, removes the influence of age on the comparison of mean blood pressures in the two groups by considering what the mean blood pressures in the two groups would be if both groups had the same mean age ($\bar{X} = 46.14$).

Nevertheless, whether adjusted or not, the mean blood pressure for males in this example appears to be considerably higher than that for females. In fact, the covariance adjustment has done little to change this impression, since the discrepancy between the male and female groups is 15.29 using unadjusted mean scores and 13.51 using adjusted mean scores. To test whether this difference in adjusted mean scores is significant, we use the partial F test of the hypothesis $H_0: \beta_2 = 0$ based on model (15.2), which can be computed from the following analysis-of-variance presentation:

Source		SS	df	MS
Reduced model ($\beta_2 = 0$) {	Regression (X)	14,951.25	1	14,951.25
	Residual	8,260.51	67	123.29
Complete model (15.2) {	Regression (X, Z)	18,009.78	2	9,004.89
	Residual	5,201.99	66	78.82

From this presentation, the appropriate partial F statistic is obtained as follows:

$$F(Z|X) = \frac{\text{regression SS}(X, Z) - \text{regression SS}(X)}{\text{MS residual }(X, Z)}$$

$$= \frac{18,009.78 - 14,951.25}{78.82}$$

$$= 38.80$$

which has 1 and 66 degrees of freedom. The P-value for this test satisfies $P < .001$, so we would reject H_0 and conclude that the two adjusted scores are significantly different.

15-4 Assumption of Parallelism: A Potential Drawback

A potential problem in using the ANACOVA pertains to the assumption of parallelism of the regression lines. It is certainly conceivable that, in certain applications, these regression lines may have different slopes. In such a case, the parallelism assumption would be invalid and the covariance method of adjustment just described should be avoided. To guard against applying the covariance method of adjustment incorrectly, we recommend that a test for parallelism be made before proceeding with the ANACOVA. This amounts to a test of $H_0: \beta_3 = 0$ for the complete model (15.1).

(We saw in Chapter 14 that the parallelism hypothesis ($H_0: \beta_3 = 0$) is not rejected for the age–systolic blood pressure data. This result supports the use of the ANACOVA model for these data (given, of course, that the assumption of variance homogeneity also holds).)

If this test supports the conclusion that the regression lines are not truly parallel, what should be done, if anything, about adjustment? It is usually recommended in this case that no adjustment at all be made to the sample means, since any such adjustment would be misleading; that is, a direct comparison of means is not appropriate when the true difference between the mean blood pressure levels in the two groups varies with age (i.e., when there is an age–sex interaction). In this case, since the main feature of the data would then be that the two regression lines describe very different relationships between age and blood pressure, an analysis that allows two separate regression lines to be fitted (without assuming parallelism) and that quantifies how the lines differ would be sufficient.

15-5 Analysis of Covariance: Several Groups and Several Covariates

In the example discussed in the previous sections, we compared two groups and adjusted for the single covariate, age. In general, ANACOVA may be used to provide adjusted scores when there are several (say, k) groups and when it is necessary to adjust simultaneously for several (say, p) covariates. The regression model describing this general situation is written

$$Y = \beta_0 + \beta_1 X_1 + \beta_2 X_2 + \cdots + \beta_p X_p + \beta_{p+1} Z_1 + \beta_{p+2} Z_2$$
$$+ \cdots + \beta_{p+k-1} Z_{k-1} + E \tag{15.6}$$

For this model, there are p covariates $X_1, X_2, \ldots, X_p$ and k groups, which are represented by the $k - 1$ dummy variables $Z_1, Z_2, \ldots, Z_{k-1}$. As discussed in Chapter 14, there is flexibility in defining these dummy variables, but for our purposes we shall assume that the Z's are defined as follows:

$$Z_j = \begin{cases} 1 & \text{if group } j \\ 0 & \text{otherwise} \end{cases} \qquad j = 1, 2, \ldots, k - 1$$

The fitted regression equations for the k different groups are then determined by specifying the appropriate combinations of values for the Z's. These are

group 1 ($Z_1 = 1$, other $Z_j = 0$): $\hat{Y}_1 = (\hat{\beta}_0 + \hat{\beta}_{p+1}) + \hat{\beta}_1 X_1 + \hat{\beta}_2 X_2$
$+ \cdots + \hat{\beta}_p X_p$

group 2 ($Z_2 = 1$, other $Z_j = 0$): $\hat{Y}_2 = (\hat{\beta}_0 + \hat{\beta}_{p+2}) + \hat{\beta}_1 X_1 + \hat{\beta}_2 X_2$
$+ \cdots + \hat{\beta}_p X_p$

$$\vdots \qquad\qquad\qquad\qquad\qquad \vdots \qquad\qquad\qquad (15.7)$$

group $k - 1$ ($Z_{k-1} = 1$, other $Z_j = 0$): $\hat{Y}_{k-1} = (\hat{\beta}_0 + \hat{\beta}_{p+k-1}) + \hat{\beta}_1 X_1 + \hat{\beta}_2 X_2$
$+ \cdots + \hat{\beta}_p X_p$

group k (all $Z_j = 0$): $\hat{Y}_k = \hat{\beta}_0 + \hat{\beta}_1 X_1 + \hat{\beta}_2 X_2 + \cdots + \hat{\beta}_p X_p$

Notice from (15.7) that the corresponding coefficients of the covariates $X_1, X_2, \ldots, X_p$ in each of the k equations are exactly the same; thus, these regression equations represent "parallel" hypersurfaces in $(p + 1)$-space, which is a natural generalization of the situation for the single covariate case. The adjusted mean score for a particular group is then computed as the predicted Y-value obtained by evaluating the fitted equation for that group at the mean values $\bar{X}_1, \bar{X}_2, \ldots, \bar{X}_p$ of the p covariates based on the combined data for all k groups:

$$\bar{Y}_j(\text{adj}) = (\hat{\beta}_0 + \hat{\beta}_{p+j}) + \hat{\beta}_1 \bar{X}_1 + \cdots + \hat{\beta}_p \bar{X}_p \qquad j = 1, 2, \ldots, k - 1$$
$$\bar{Y}_k(\text{adj}) = \hat{\beta}_0 + \hat{\beta}_1 \bar{X}_1 + \cdots + \hat{\beta}_p \bar{X}_p \qquad\qquad\qquad\qquad (15.8)$$

To determine whether the k adjusted mean scores are significantly different from one another, we test the null hypothesis

$$H_0: \beta_{p+1} = \beta_{p+2} = \cdots = \beta_{p+k-1} = 0$$

using a multiple-partial F test with $k - 1$ and $n - p - k$ degrees of freedom based on model (15.6). If H_0 is rejected, we conclude that there are significant differences among the adjusted means (although we cannot, without further inspection, determine where the major differences are).

Example   In the Ponape study (Patrick et al., 1974) of the effect of rapid cultural change on health status, one research goal was to determine whether blood pressure was associated with a measure of the strength of a Ponapean's prestige in the modern (i.e., western) part of his culture relative to the traditional culture. A measure of prestige discrepancy (PD) was developed and then measured by questionnaire for each of 550 Ponapean males. Of particular interest was whether a higher prestige discrepancy score corresponded to higher blood pressure. It was also considered necessary to adjust or control for the effects of the covariates age and body size in considering these questions.

To perform this analysis, prestige discrepancy was categorized into three groups:

group 1:   modern prestige much higher than traditional prestige
group 2:   modern prestige not much different from traditional prestige
group 3:   traditional prestige much higher than modern prestige

Then an ANACOVA was carried out with diastolic blood pressure (DBP) as the dependent variable and with AGE and quetelet index (QUET; a measure of body size) as the two covariates. A covariance model in this case is given by

TABLE 15-1   ANACOVA example using Ponape study data

$\overline{\text{DBP}}$	PD Group 1 $(n_1 = 87)$	PD Group 2 $(n_2 = 383)$	PD Group 3 $(n_3 = 80)$	P Value for F Test of $H_0: \beta_3 = \beta_4 = 0^*$
Unadjusted	71.68	68.16	68.55	.001 $< P <$ .005
Adjusted	72.07	68.02	68.80	

* Using model (15.9) with 2 and 545 df.

$$DBP = \beta_0 + \beta_1 AGE + \beta_2 QUET + \beta_3 Z_1 + \beta_4 Z_2 + E \qquad (15.9)$$

where

$$Z_1 = \begin{cases} 1 & \text{if group 1} \\ 0 & \text{otherwise} \end{cases} \quad \text{and} \quad Z_2 = \begin{cases} 1 & \text{if group 2} \\ 0 & \text{otherwise} \end{cases}$$

(A test of the parallelism assumption implicit in model (15.9) considers the null hypothesis $H_0: \beta_5 = \beta_6 = \beta_7 = \beta_8 = 0$ in the full model

$$DBP = \beta_0 + \beta_1 AGE + \beta_2 QUET + \beta_3 Z_1 + \beta_4 Z_2 + \beta_5 Z_1 AGE$$
$$+ \beta_6 Z_2 AGE + \beta_7 Z_1 QUET + \beta_8 Z_2 QUET + E$$

This multiple-partial F test (with 4 and 541 degrees of freedom) was not rejected using the Ponape data, thus supporting the use of model (15.9).)

The adjusted DBP mean scores for the three groups were then determined by substituting the values of the overall means $\overline{AGE}$ and $\overline{QUET}$ into the fitted equations for the three groups as follows:

$$\overline{DBP}_1(\text{adj}) = (\hat{\beta}_0 + \hat{\beta}_3) + \hat{\beta}_1 \overline{AGE} + \hat{\beta}_2 \overline{QUET}$$
$$\overline{DBP}_2(\text{adj}) = (\hat{\beta}_0 + \hat{\beta}_4) + \hat{\beta}_1 \overline{AGE} + \hat{\beta}_2 \overline{QUET}$$
$$\overline{DBP}_3(\text{adj}) = \hat{\beta}_0 + \hat{\beta}_1 \overline{AGE} + \hat{\beta}_2 \overline{QUET}$$

The test for the equality of these adjusted means was based on the use of a multiple-partial F test of the null hypothesis $H_0: \beta_3 = \beta_4 = 0$ under model (15.9). Table 15-1 summarizes these calculations.

The results indicate the presence of highly significant differences among the adjusted mean blood pressure scores, with group 1 having a somewhat higher adjusted mean blood pressure than the other two groups. It should be noted, however, that despite the statistical significance found, the adjusted mean blood pressure of 72.07 for group 1 is close enough (clinically speaking) to the adjusted means for the other groups to cast doubt on the clinical significance of these results.

15-6 Comments and Cautions

15-6-1 Rationale for Adjustment

ANACOVA adjusts for disparities in covariate distributions over groups by artificially assuming that all groups have the same set of *mean* covariate values. For example, if age and weight are the covariates and there are two groups being compared, the ANACOVA adjustment procedure treats both groups as if they had the same mean age and the same mean weight.

The ANACOVA adjustment procedure is equivalent to artificially assuming a common covariate *distribution* based on the combined sample over all groups. That is, not only are the means assumed to be equal, but the entire distribution of the covariates in the combined sample is assumed to be the same as the distribution of the covariates in each group. To see this, it can be shown that the adjusted score for any group can be expressed as the average over the combined sample of the predicted scores for that group; that is,

$$\hat{Y}_j(\text{adj}) = \frac{1}{n} \sum_{k=1}^{n} \hat{Y}_{jk} \qquad j = 1, 2, \ldots, k$$

where $\hat{Y}_j(\text{adj})$ is the adjusted score for group j defined by (15.8), n is the number of subjects in the combined sample, and $\hat{Y}_{jk}$ is the predicted response in the jth group using the set of equations (15.7) and based on the covariate values of the kth individual in the combined sample. For example,

$$\hat{Y}_{1k} = (\hat{\beta}_0 + \hat{\beta}_{p+1}) + \hat{\beta}_1 X_{1k} + \hat{\beta}_2 X_{2k} + \cdots + \hat{\beta}_p X_{pk}$$

is the predicted group 1 response for the kth individual in the combined sample, where $(X_{1k}, X_{2k}, \ldots, X_{pk})$ are the covariate values for that individual. Thus, the adjusted score for a given group can be obtained by artificially assuming that all n persons in the combined sample comprise the given group; it then follows that the covariate distribution of the group is assumed to be that of the combined sample.

Thus the method of adjustment using ANACOVA does what one hopes it will do: It corrects the disparity in covariate distributions over groups by assuming a common distribution (not just a common set of means).

15-6-2   The Parallelism Assumption

As stated earlier, the ANACOVA method is inappropriate when the relationship between the covariates and the response is not the same in each group. Such "nonparallelism" or "interaction" might be reflected, for example, by males having higher blood pressures than females at older ages and females having higher blood pressures than males at younger ages. Consequently, the use of a standard ANACOVA could lead to adjusted (for age) mean scores for each (gender) group that are roughly equal; this would create the misleading impression that there was little difference between male and female blood pressures, when, in fact, there were large differences in certain age categories. This example illustrates why we recommend that no method of adjustment be used in the presence of interaction; instead, the nature of the interaction itself should be quantified. The method for assessing interaction in this context, as previously indicated, requires the ANACOVA model (15.6) to be expanded to include product terms between covariates and group variables and then the coefficients of these product terms to be tested for significance. The extended ANACOVA model that allows for such interaction terms has the form

$$Y = \beta_0 + \sum_{i=1}^{p} \beta_i X_i + \sum_{j=1}^{k-1} \beta_{p+j} Z_j + \sum_{i=1}^{p} \sum_{j=1}^{k-1} \gamma_{ij} X_i Z_j + E \qquad (15.10)$$

Using this model, a chunk test for parallelism would test H_0: all $\gamma_{ij} = 0$ and would involve a multiple-partial F statistic of the form

$$F(\text{all } X_i Z_j \text{ product terms} | \text{all } X_i \text{ and } Z_j \text{ terms})$$

which would have an F distribution under H_0 with $p(k-1)$ and $n-1-p-(k-1)-p(k-1)$ degrees of freedom. The use of ANACOVA adjusted scores is appropriate only when the above test for interaction is not significant.

15-6-3 Validity and Precision

As discussed in Chapter 11, validity and precision are two reasons for considering the control of covariates. In using ANACOVA, validity is achieved by adjusting for confounding, thereby giving an estimate of association that would otherwise be distorted (biased) if the covariate(s) of interest was (were) ignored in the analysis.

Although validity should be the first consideration, it is nevertheless possible to find no confounding in one's data and still control for one or more covariates in order to gain precision. To assess precision, we can consider either the variances of the estimators of the association(s) of interest (the smaller these variances, the greater the precision) or confidence intervals for the association(s) of interest (the narrower the confidence intervals, the greater the precision). For example, consider an ANACOVA model involving two groups, say males and females, with systolic blood pressure as the dependent variable and age as the only covariate of interest. If the age distribution for males was exactly the same as that for females in the data, then age would not be a confounder. Nevertheless, because age is strongly positively associated with systolic blood pressure, precision will likely be increased by adjusting for age, even though confounding (i.e., validity) is not at issue.

15-6-4 Alternatives to ANACOVA

When adjusting for covariates, a model can be fitted that contains the covariates and the study variables of primary interest (and perhaps even product terms so that interaction can be assessed) without requiring the study variables to be categorical. A best model would then need to be derived using criteria for selecting variables (see Chapter 16). If, for such a model, significant interaction is found, it would *not* be appropriate (as noted earlier for categorical study variables) to derive adjusted scores. Moreover, even if there is no interaction, it is impossible to obtain adjusted scores for groups, simply because the study variables are not defined categorically.

Nevertheless, predicted values based on the best regression model can be considered to be adjusted values in the broad sense that the covariates are being taken into account in the modeling process. Furthermore, adjusted scores for distinct values of the study variables can be obtained by computing predicted values using the overall mean covariate values in the best model, as was done for ANACOVA using categorical study variables. For example, if the best model relating systolic blood pressure to age and weight is determined to be

$$\widehat{SBP} = \hat{\beta}_0 + \hat{\beta}_1 \text{AGE} + \hat{\beta}_2 \text{WEIGHT}$$

then adjusted blood pressure scores for persons weighing 150 and 175 pounds (controlling for age) could be computed as

$$\widehat{SBP}_{150}(\text{adj}) = \hat{\beta}_0 + \hat{\beta}_1 \overline{\text{AGE}} + 150\hat{\beta}_2$$

and

$$\widehat{SBP}_{175}(\text{adj}) = \hat{\beta}_0 + \hat{\beta}_1\overline{AGE} + 175\hat{\beta}_2$$

These adjusted values are similar to group adjusted scores in the sense that the value of 150 can be thought of as representing a group of people whose weights are all close to 150; a similar interpretation can be given to the 175-pound value.

In a more restrictive alternative to ANACOVA, *all* variables, even covariates, are treated as categorical. If the set of dummy variables defining the covariates is distinguished from the set of dummy variables defining the study variables, then the regression model being considered may be treated as a two-way analysis-of-variance model with unequal cell numbers (see Chapter 20). This alternative, however, might be inappropriate if the underlying scales of measurement of some of the covariates are actually noncategorical (e.g., are continuous). If, for example, the inherently continuous variable age is categorized into three age groups, then a completely categorical model may lead to different results than might be obtained by using a model that treats age continuously.

15-7 Summary

In this chapter, we have described the most popular approach to controlling for covariates, ANACOVA. To use ANACOVA, the study variables of interest must be treated as categorical variables, whereas the covariates are not so restricted. ANACOVA also requires the assumption that there is no interaction between covariates and study variables. This assumption can be checked by testing for the significance of appropriate product terms in an extended ANACOVA model. If the test for interaction is nonsignificant, then adjusted scores for different groups can be obtained by substituting into the group-specific ANACOVA model the mean covariate values for the combined sample. If the test for interaction is significant, then adjusted scores should not be used; instead, the nature of the interaction should be described.

Problems

1. In Problem 8 of Chapter 14, straight-line regression fits of SBP on QUET were compared for smokers and nonsmokers, and it was demonstrated that these straight lines could be considered parallel. Use the results based on that problem (and the fact that the overall sample mean value $\overline{QUET} = 3.441$) to address the following issues relating to an ANACOVA for these data:

   a. State the appropriate ANACOVA regression model to use for comparing the mean blood pressures in the two smoking categories controlling for QUET.

   b. Determine the adjusted SBP means for smokers and nonsmokers. Compare these values to the unadjusted mean values

   $$\overline{SBP}(\text{smokers}) = 147.823 \quad \text{and} \quad \overline{SBP}(\text{nonsmokers}) = 140.800$$

    c. Test whether the true adjusted mean blood pressures in the two groups are equal. State the null hypothesis in terms of the regression coefficients in the ANACOVA model given in part (a).

    d. Obtain a 95% confidence interval for the true difference in adjusted SBP means.

2. **a–d.** Answer the same questions as in parts (a) through (d) of Problem 1 regarding an analysis of covariance designed to control for *both* AGE and QUET. (*Note*: $\overline{AGE} = 53.250$.) Use the results from Problem 9 of Chapter 14.

    e. Is it necessary to control for *both* AGE and QUET as opposed to, say, just one of the two covariates?

3. In an experiment conducted at the National Institute of Environmental Health Sciences, the absorption (or uptake) of a chemical by a rat on one of two different diets, I and II, was known to be affected by the weight (or size) of the rat. A completely randomized design utilizing four rats for each diet was employed in the experiment, and the initial weight of each rat was recorded so that the diets could be compared after adjusting for the effect of initial weight. The data for the experiment are given in the table.

Initial Weight (X)	3	1	4	4	5	2	3	2
Diet (Z)	I	I	I	I	II	II	II	II
Response (Y)	14	13	14	15	16	15	15	14

    a. Using the initial weight as a covariate, state the ANACOVA regression model for comparing the two diets (set $Z = -1$ if diet I is used and $Z = 1$ if diet II is used).

    b. Use the results in the table to determine the adjusted mean responses for each diet, controlling for initial weight.

    c. Test whether the two diets are significantly different using the ANACOVA regression model defined in part (a).

    d. Test whether the two diets are significantly different, completely ignoring the covariate. How do the two testing procedures compare?

    e. Determine a 95% confidence interval for the true difference in the adjusted mean responses.

Fitted Covariance Model			ANOVA		
Variable	$\hat{\beta}$	$S_{\hat{\beta}}$	Source	df	SS
X	0.5	0.1291	X	1	3.0
Z	0.5	0.1581	$Z\mid X$	1	2.0
(intercept)	13.0		Residual	5	1.0
			Total	7	6.0

4. A political scientist developed a questionnaire to determine political tolerance scores (Y) for a random sample of faculty members at her university. She wanted to compare mean scores adjusted for age (X) for each of three categories: full

professors, associate professors, and assistant professors. The results are given in the tables (the higher the score, the more tolerant the individual).

GROUP 1: FULL PROFESSORS

Age (X)	65	61	47	52	49	45	41	41	40	39
Tolerance (Y)	3.03	2.70	4.31	2.70	5.09	4.02	3.71	5.52	5.29	4.62

GROUP 2: ASSOCIATE PROFESSORS

Age (X)	34	31	30	35	49	31	42	43	39	49
Tolerance (Y)	4.62	5.22	4.85	4.51	5.12	4.47	4.50	4.88	5.17	5.21

GROUP 3: ASSISTANT PROFESSORS

Age (X)	26	33	48	32	25	33	42	30	31	27
Tolerance (Y)	5.20	5.86	4.61	4.55	4.47	5.71	4.77	5.82	3.67	5.29

a. State an ANACOVA regression model that can be used to compare the three groups, controlling for age.

b. What model should be used to check whether the ANACOVA model in part (a) is appropriate? Carry out the appropriate test. (Use $\alpha = .01$.)

c. Using ANACOVA, determine adjusted mean tolerance scores for each group, and test whether these are significantly different from one another. Also, compare the adjusted means with the unadjusted means. (*Note*: $\bar{X}$(overall) = 39.667, $\bar{Y}$(group 1) = 4.10, $\bar{Y}$(group 2) = 4.86, $\bar{Y}$(group 3) = 5.00.)

Fitted Covariance Model			ANOVA		
Variable	$\hat{\beta}$	$S_{\hat{\beta}}$	Source	df	SS
X	-0.0364	0.0174	X	1	6.2050
$Z_1 \begin{cases} 1 & \text{if group 1} \\ 0 & \text{otherwise} \end{cases}$	-0.3398	0.4146	$Z_1, Z_2\|X$	2	0.6434
$Z_2 \begin{cases} 1 & \text{if group 2} \\ 0 & \text{otherwise} \end{cases}$	0.0636	0.3323	$XZ_1, XZ_2\|X, Z_1, Z_2$	2	3.3610
(intercept)	6.1837		Residual	24	9.7628

5. A psychological experiment was performed to determine whether in problem-solving dyads containing one male and one female, "influencing" behavior depended on the sex of the experimenter. The problem for each dyad was a strategy game called "Rope a Steer," which required 20 separate decisions about which way to proceed toward a defined goal on a game board. For each subject in the dyad, a verbal-influence activity score was derived as a function of the number of statements made by the subject to influence the dyad to move in a particular direction. The difference in verbal-influence activity scores within a dyad was denoted as the variable VIAD, which was then used as the dependent variable in an ANACOVA designed to control for the effects of differing IQ scores of the male and female in each dyad. The data are given in the following tables.

GROUP 1: MALE EXPERIMENTER

VIAD	-10	-4	9	-15	-15	5	-8	-4	-1	13
IQ_M	115	112	106	123	125	105	115	122	138	110
IQ_F	100	110	108	135	115	112	121	132	135	126

GROUP 2: FEMALE EXPERIMENTER

VIAD	8	-5	2	-7	15	-10	-3	10	2	4
IQ_M	120	130	110	113	102	141	120	113	114	102
IQ_F	141	128	104	98	106	130	128	105	107	111

a. State an ANACOVA model appropriate for these data.

b. Determine adjusted mean VIAD scores for each group and compare these to the unadjusted means for each group.

c. Test whether the adjusted mean scores are significantly different.

d. Find a 95% confidence interval for the true difference in adjusted mean scores.

Fitted Covariance Model				ANOVA		
Variable	$\hat{\beta}$	$S_{\hat{\beta}}$		Source	df	SS
IQ_M	-0.6930	0.1939		IQ_M, IQ_F	2	612.05
IQ_F	0.2811	0.1597		$Z\|IQ_M, IQ_F$	1	131.47
$Z\begin{cases}1 \text{ if female experimenter} \\ 0 \text{ if male experimenter}\end{cases}$	5.1961	3.1329		$(IQ_M \times Z, IQ_F \times Z)\|IQ_M, IQ_F, Z$	2	3.22
(intercept)	44.5900			Residual	14	761.46

6. An experiment was conducted to compare the effects of four different drugs (A, B, C, and D) in delaying the atrophy of denervated muscles. A certain leg muscle in each of 48 rats was deprived of its nerve supply by severing the appropriate nerves. The rats were then put randomly into four groups, and each group was assigned treatment by one of the drugs. After 12 days, four rats from each group were killed and the weight W (in grams) of the denervated muscle was obtained.

Theoretically, atrophy should be measured by the loss in weight of the muscle, but the initial weight of the muscle could not have been obtained without killing the rat. Instead, the initial total body weight X (in grams) of the rat was measured. It was assumed that this figure is closely related to the initial weight of the muscle.

Drugs A and C were large and small dosages, respectively, of atropine sulfate. Drug B was quinidine sulfate. Drug D acted as a control; it was simply a saline solution, which could not have had any effect on atrophy. Compare the effects of the four drugs using ANACOVA, controlling for initial total body weight (X). Use the results in the accompanying tables to perform your analysis. (*Note:* $\bar{X} = 226.125$.)

Drug A		Drug B		Drug C		Drug D	
X	W	X	W	X	W	X	W
198	0.34	233	0.41	204	0.57	186	0.81
175	0.43	250	0.87	234	0.80	286	1.01
199	0.41	289	0.91	211	0.69	245	0.97
224	0.48	255	0.87	214	0.84	215	0.87

Fitted Covariance Model			
Variable		$\hat{\beta}$	$S_{\hat{\beta}}$
X		0.0032	0.0013
Z_1	$\begin{cases} 1 & \text{if drug A} \\ 0 & \text{otherwise} \end{cases}$	−0.3917	0.0954
Z_2	$\begin{cases} 1 & \text{if drug B} \\ 0 & \text{otherwise} \end{cases}$	−0.2257	0.0902
Z_3	$\begin{cases} 1 & \text{if drug C} \\ 0 & \text{otherwise} \end{cases}$	−0.1351	0.0878
(intercept)		0.1728	

ANOVA		
Source	df	SS
Regression (X)	1	0.3202
Residual	14	0.4571
Regression (X, Z_1, Z_2, Z_3)	4	0.6185
Residual	11	0.1587

7. Trough urine samples were analyzed for sodium content for each of two collection periods, one before and one after administration of Mercuhydrin, for each of 30 dogs. The experimenter used 7 dogs as a control group for the study.

Experimental Group ($Z = 1$)			Control Group ($Z = 0$)		
Animal	First Collection Period (X) ([Na], mM/l)	Second Collection Period (Y) ([Na], mM/l)	Animal	First Collection Period (X) ([Na], mM/l)	Second Collection Period (Y) ([Na], mM/l)
1	17.5	22.1	18	6.3	12.7
2	9.4	12.0	19	9.7	17.1
3	10.0	15.2	20	7.1	9.5
4	7.4	23.1	21	7.2	11.0
5	8.8	9.8	22	5.3	8.2
6	18.9	26.9	23	14.3	15.8
7	10.8	11.1	24	7.9	9.7
8	8.8	13.6	25	14.1	14.7
9	8.8	12.8	26	12.8	17.0
10	9.2	7.5	27	12.8	20.2
11	8.1	8.1	28	10.7	13.9
12	10.3	27.5	29	5.9	11.8
13	10.1	11.2	30	3.8	9.0
14	7.3	11.0			
15	11.1	15.3	Mean	9.7	13.9
16	9.4	11.5	S.D.	3.3	5.3
17	8.2	8.4			

Animal	Control Group ($Z = 0$)	
	First Collection Period (X) ([Na], mM/l)	Second Collection Period (Y) ([Na], mM/l)
1	11.1	9.4
2	5.1	5.9
3	6.5	14.8
4	17.2	15.5
5	11.8	23.4
6	6.6	7.3
7	4.1	8.2
Mean	8.9	12.1
S.D.	4.7	6.4

COMPUTER RESULTS FOR PROBLEM 7

Multiple R^2 = .406

Variable	$\hat{\beta}$	S.D. of $\hat{\beta}$	$\hat{\beta}$/S.D.
X	0.96155	0.20418	4.709
Z	1.06435	1.83729	0.5793
(intercept)	3.49992		

ANOVA

Source	df	MS
X	1	434.4861
Added by Z	1	6.3764
Lack of fit	30	21.0913
Pure error	4	3.3179

These dogs were not administered the drug, but their urine samples were collected for two similar time periods.

a. Using ANACOVA (computer results are given above) with the before measure as the covariate, find the adjusted mean sodium contents for both experimental and control groups.

b. Test whether the two adjusted means are significantly different.

c. What alternative testing approach (involving a t test) could be used for these data? Carry out this test. (_Hint_: Variances of before–after differences for each group are 16.78517 [experimental] and 25.915 [control].)

d. Are the two testing methods (t test versus covariance analysis) equivalent in this problem? Explain by comparing regression models appropriate for each method.

e. What do the results of performing a lack-of-fit test using the computer output given indicate about the appropriateness of the covariance model used in parts (a) and (b)?

8. Consider again the data in Problem 4.

a. Indicate how to compute appropriate cross-product variables to allow testing of whether an interaction is present between age and faculty rank.

b. State the associated regression model.

c. Using a computer, fit the model. Provide estimates of the coefficients.

d. Provide a multiple-partial F test of whether interaction is present, controlling for age and faculty rank. Use $\alpha = .05$.

e. Does this indicate that the ANACOVA was valid?

9. This problem treats data from Problem 15 of Chapter 5. Consider LN_BRNTL as the dependent variable, and dosage level of toluene (PPM_TOLU) as a categorical predictor (four levels). The experimenter wished to explore the possibility of using WEIGHT as a control variable.

   a. State the appropriate ANACOVA model. Use the 50-ppm exposure group as the reference group in coding dummy variables.

   b. Use a computer program to fit the model and provide estimated regression coefficients.

   c. Provide adjusted mean estimates. Compare these with the unadjusted means. (Do not perform any tests.)

   d. Test whether the adjusted means are all equal. Use $\alpha = .05$. State the null hypothesis in terms of the regression coefficients.

10. Repeat Problem 9 using LN_BLDTL as the dependent variable.

11. **a.** Indicate how to compute appropriate cross-product terms so that the interaction of WEIGHT and dosage for Problem 9 can be tested.

    b. State the appropriate regression model.

    c. State the null hypothesis of no interaction in terms of regression coefficients.

12. **a.** For the model fitted in Problem 9, compute adjusted predicted scores and residuals.

    b. Plot residuals against predicted scores.

    c. Compute estimates of the residual variances separately for each dosage group. Are they approximately equal?

    d. Do these diagnostics suggest any problems?

Reference

Patrick, R., Cassel, J. C., Tyroler, H. A., Stanley, L., and Wild, J. (1974). "The Ponape Study of the Health Effects of Cultural Change." Paper presented to Annual Meeting, Society for Epidemiologic Research, Berkeley, Calif.

Selecting the Best Regression Equation

16-1 Preview

The general problem to be discussed in this chapter is as follows: We have one response variable Y and a set of k predictor variables $X_1, X_2, \ldots, X_k$. We want to determine the *best* (*most important* or *most valid*) subset of the k predictors and the corresponding *best-fitting* regression model for describing the relationship between Y and the X's.[1] What exactly we mean by "best" depends in part on our overall goal for modeling. Two such goals were described in Chapter 11 and are briefly reviewed now.

One goal is to find a model that gives the best prediction of Y, given $X_1, X_2, \ldots, X_k$, for some new observation or for a batch of new observations. In practice, we emphasize estimating the regression of Y on the X's (see Chapter 8), $E(Y|X_1, X_2, \ldots, X_k)$, which expresses the mean of Y as a function of the predictors. Using this goal, we may say that our best model is *reliable* if it predicts well in a new sample. The details of the model may be of little or no consequence, such as the inclusion of any particular variable or the magnitude or sign of its regression coefficient. For example, in considering a sample of systolic blood pressures, the goal may be simply to predict blood pressure (SBP) as a function of demographic variables like AGE, RACE, and GENDER. We may not necessarily care which variables are in the model or how they are defined, as long as the final model obtained gives the best prediction possible.

In addition to the question of prediction is the question of validity, that is, obtaining valid (i.e., accurate) estimates for one or more regression coefficients in a model and then making inferences about the corresponding parameters of interest. The goal here is to quantify the relationship between one or more independent variables of interest and the dependent variable, controlling for the other variables. As an example, we might wish to describe the relationship of SBP to AGE, controlling for RACE and GENDER. In this case, we are

---

[1] Hocking (1976) provided an exceptionally thorough and well-written review of this topic. His presentation is at a technical level somewhat higher than that of this text.

focusing on regression coefficients involving AGE (including functions of AGE); even though RACE and GENDER may remain in the model for control purposes, their regression coefficients are not of primary interest.

In this chapter we shall focus on strategies for selecting the best model when the primary goal of analysis is prediction. In the last section we shall briefly mention a strategy for modeling in situations where the validity of the estimates of one or more regression coefficients is of primary importance (see also Chapter 11).

16-2 Steps in Selecting the Best Regression Equation

In selecting the best regression equation, we recommend the following steps:

1. Specify the maximum model (defined in Section 16-3) to be considered.

2. Specify a criterion for selecting a model.

3. Specify a strategy for applying the criterion.

4. Conduct the specified analysis.

5. Evaluate the reliability of the model chosen.

By following the above steps, the fuzzy idea of finding the best predictors of Y can be converted into simple, concrete actions. Each step helps to insure reliability and to reduce the work required. Specifying the maximum model forces the analyst to (1) state the analysis goal clearly, (2) recognize the limitations of the data at hand, and (3) describe the range of plausible models explicitly. All computer programs for selecting models require that a maximum model be specified. In doing so, the analyst should consider all available scientific knowledge. In turn, specifying the criterion for selecting a model and the strategy for applying that criterion simplifies and speeds the analysis process. Finally, whether the goal is prediction or validity, the reliability of the chosen model must be demonstrated. Each step will be considered separately below.

We shall illustrate the above process of analysis via two examples. The first example will be one considered earlier in Chapter 8. The data (given in Table 8-1) are the hypothetical results of measuring a response variable weight (WGT) and predictor variables height (HGT) and age (AGE) on 12 children. Once the methods have been illustrated with this simple example, similar real data for over 200 subjects (Lewis and Taylor, 1967; discussed in Chapter 12) will be used for a case study of the methods.

16-3 Step 1: Specifying the Maximum Model

The *maximum model* is defined to be the largest model (having the most predictor variables) considered at any point in the process of model selection. All other possible models can be created by deleting predictor variables from the maximum model. A model created by deleting predictors from the maximum model is called a *restriction* of the maximum model.

Throughout this chapter we shall make the important assumption that the maximum model with k variables or some restriction of it with $p \leq k$ variables is the *correct model* for the population. An important implication of this assumption is that the *population* squared multiple correlation for the maximum model, namely $\rho^2(Y|X_1, X_2, \ldots, X_k)$, is no larger than that for the correct model (which may have fewer variables); as a result, adding more predictors to the correct model does not increase the population squared multiple correlation for the correct model. In turn, for the sample at hand, $R^2(Y|X_1, X_2, \ldots, X_k)$ for the maximum model is always at least as large as that for any subset model. However, other sample-based criteria (discussed below) may not necessarily suggest that the largest model is best.

To illustrate, we consider the data from Table 8-1, reproduced below:

Child	1	2	3	4	5	6	7	8	9	10	11	12
Y (WGT)	64	71	53	67	55	58	77	57	56	51	76	68
X_1 (HGT)	57	59	49	62	51	50	55	48	42	42	61	57
X_2 (AGE)	8	10	6	11	8	7	10	9	10	6	12	9

One possible model (not necessarily the maximum one) to consider is given by

$$Y = \beta_0 + \beta_1 X_1 + \beta_2 X_2 + E$$

This model allows for only a linear (planar) relationship between the response (WGT) and the two predictors (HGT and AGE). Nevertheless, the nature of growth suggests that the relationship, although monotonic, in both HGT and AGE, may well be nonlinear. This implies that at least one quadratic (squared) term may need to be included in the model. Since HGT and AGE are highly correlated, and since the sample size is very small,[2] only $(AGE)^2$ will be considered. Should the interaction between AGE and HGT be considered? Should transformations of the predictors and/or the response be considered (e.g., replacing AGE and $(AGE)^2$ with log(AGE))? The limitations of the data lead us to define the maximum model as

$$Y = \beta_0 + \beta_1 X_1 + \beta_2 X_2 + \beta_3 X_3 + E$$

with $X_1 = $ HGT, $X_2 = $ AGE, and $X_3 = (AGE)^2$.

There are many reasons for choosing a large maximum model. The most important one is the desire to avoid making a Type II (false negative) error. In a regression analysis, a Type II error corresponds to omitting a predictor that has a truly nonzero regression coefficient in the population. Other reasons for considering a large maximum model are the wishes to include:

1. all conceivable basic predictors,

2. high-order powers of basic predictors $((AGE)^2, (HGT)^2)$,

3. other transformations of predictors (log(AGE), 1/HGT),

4. interactions among predictors (e.g., AGE × HGT) including two-way and higher-order interactions, and/or

5. all possible "control" variables, as well as their powers and interactions.

---

[2] Because the sample size is so small, it almost precludes making reliable conclusions from any regression analysis. The small sample size is used to simplify our discussion.

Recall that overfitting a model (including variables in the model with truly zero regression coefficients in the population) will not introduce bias when estimating population regression coefficients if the usual regression assumptions are met. However, we must be careful that overfitting does not introduce harmful collinearity (Chapter 12). Also, underfitting (i.e., leaving important predictors out of the final model) will introduce bias in the estimated regression coefficients.

There are also important reasons for working with a small maximum model. With a prediction goal, the need for reliability (to be discussed later) strongly argues for a small maximum model, and with a validity goal, we want to focus on a few important variables. In either case, we wish to avoid a Type I (false positive) error. In a regression analysis, a Type I error corresponds to including a predictor that has a zero population regression coefficient. The desire for parsimony is another important reason for choosing a small maximum model. Practically unimportant but statistically significant predictors can greatly confuse the interpretation of regression results; complex interaction terms are particularly troublesome in this regard.

The particular sample of data to be analyzed imposes certain constraints on the choice of the maximum model. In general, the smaller the sample size, the smaller the maximum model should be. The general idea is that a larger number of independent observations are needed to estimate reliably a larger number of regression coefficients. This notion has led to various guidelines about the size of a maximum model. The most basic constraint is that the error degrees of freedom be positive. Symbolically, we require

$$\text{df error} = n - k - 1 > 0$$

which is equivalent to the constraint

$$n > k + 1$$

As always, n is the number of observations and k is the number of predictors, giving $k + 1$ regression coefficients (including the intercept). With negative error degrees of freedom, the model has at least one perfect collinearity; consequently, unique estimates of coefficients and variances cannot be computed. With zero error degrees of freedom ($n = k + 1$), unique estimates of the coefficients can be computed, but not of the variances (recall that $\hat{\sigma}^2 = \text{SSE/df error}$). Furthermore, even if the population correlation is 0, $R^2 = 1.00$ when df error = 0, which reflects the fact that the model exactly fits the observed data (i.e., SSE = 0). This is the most extreme example of the positive bias in R^2. In such a situation, we have exchanged n values of Y for n estimated regression coefficients. Hence we have gained nothing since the dimensionality of the problem remains the same.

The question then arises as to how many error degrees of freedom are needed. Simple rules can be suggested to insure that the estimates for a *single* model are reliable, but these rules are inadequate when considering a series of models. Later in this chapter split-sample approaches will be offered as possible ways to assess reliability.

The weakest requirement is for a minimum of approximately 10 error degrees of freedom, namely,

$$n - k - 1 \geq 10$$

or

$$n \geq 10 + k + 1$$

Another rule of thumb for regression that has been suggested is to have at least 5 (or 10) observations per predictor, namely, require that

$$n \geq 5k \qquad (\text{or } n \geq 10k)$$

Assume, for example, that we wish to consider a maximum model involving 30 predictors. To have 10 error degrees of freedom requires a sample of size 41, while $n > 5k$ demands a sample of size 150. Split-sample approaches, discussed later in this chapter, may reduce the required sample size substantially.

Another constraint on the maximum model concerns the amount of variability present in the predictor values, considered either individually or jointly. If a predictor has the same value for all subjects, then it obviously cannot be used in any model. For example, consider the variable GENDER of child (male = 1, female = 0) as a candidate predictor for the weight example. If all of the subjects are male, then GENDER = 1 for all subjects, the sample variance for the variable is 0.00, and it is perfectly collinear with the intercept variable. Clearly, if all subjects are of one gender, then comparisons of gender cannot be made.

Similarly, consider the variables GENDER, RACE (white = 0, black = 1), and their interaction (GENRACE = GENDER × RACE). If no black females are represented in the data, then the 1 degree of freedom interaction effect cannot be estimated. If a race–sex cell is nearly empty, then the estimated interaction coefficient may be very unstable (i.e., have a large variance).

Polynomial terms (e.g., X^2 and X^3) and other transformations merit particular consideration when specifying the maximum model. For the weight example, one might wish to consider AGE, $(AGE)^2$, and exp(AGE) as possible predictors. Near collinearities (Chapter 12) can lead to very unstable and often uninterpretable results when trying to find the best model. (See Marquardt and Snee, 1975, for further discussion of this topic.) Consequently, one should attempt to reduce such collinearity if possible. Centering, if applicable, almost always helps to increase numerical accuracy, as can multiplying variables by various constants (a special form of scaling; see Chapter 12) to produce nearly equal variances for all predictors (say, variances around 1).

If the collinearity problem cannot be overcome, then one should probably either (1) conduct separate analyses for each form of X (for example, one analysis using X and X^2 and then a separate analysis using exp(X)), (2) eliminate some variables, or (3) impose structure (e.g., a fixed order of tests) on the testing procedure. If nothing is done about severe collinearity, then the estimated regression coefficients in the best model may be highly unstable (i.e., have high variances) and possibly be quite far from the true parameter values.

16-4 Step 2: Specifying Criteria for Selecting a Model

The second step in selecting the best model is to specify the selection criterion. A *selection criterion* is an index that can be computed for each candidate model and used to compare models. Thus, given one particular selection criterion, candidate models can be ordered from best to worst. This helps to automate the process of choosing the "best" model. As we shall see, this selection-criterion-specific process may not find the *best* model in a global sense. However, the use of a specific selection criterion can substantially reduce the work involved in finding a *good* model.

Obviously, the selection criterion should be related to the goal of the analysis. For example, if the goal is reliable prediction of future observations, the selection criterion should be somewhat liberal to avoid missing useful predictors.

(This leads us to note the distinctions among (1) numerical differences, (2) statistically significant differences, and (3) scientifically important differences. Numerical differences may or may not correspond to significant or important differences. In turn, with sensitive tests (typical of analyses of large samples), significant differences may or may not correspond to important differences. Finally, with insensitive tests (typical of analyses of very small samples), important differences may not be significant.)

Many selection criteria for choosing the best model have been suggested. Hocking (1976), for example, reviewed eight candidates. We shall consider four criteria: R_p^2, F_p, $MSE(p)$, and C_p. Each will be defined and discussed separately below.

Before discussing these criteria, we must define the notation needed to understand them. All four criteria attempt to compare two model equations: the maximum model with k predictors and a restricted model with p predictors ($p \leq k$). The maximum model is

$$Y = \beta_0 + \beta_1 X_1 + \beta_2 X_2 + \cdots + \beta_p X_p + \beta_{p+1} X_{p+1} + \cdots + \beta_k X_k + E \qquad (16.1)$$

and the reduced model (a restriction of the maximum model) is

$$Y = \beta_0 + \beta_1 X_1 + \beta_2 X_2 + \cdots + \beta_p X_p + E \qquad (16.2)$$

Let $SSE(k)$ be the error sum of squares for the k-variable model, $SSE(p)$ the error sum of squares for the p-variable model, and so on. Also, $SSY = \sum_{i=1}^{n}(Y_i - \bar{Y})^2$ is the total (corrected) sum of squares for the response Y. Often $p = k - 1$, which is the case when evaluating the addition or deletion of a single variable. We shall assume that the $k - p$ variables under consideration for addition or deletion are denoted $X_{p+1}, X_{p+2}, \ldots, X_k$ for notational convenience.

The sample squared multiple correlation R^2 is a natural candidate for deciding which model is best and is therefore the first criterion discussed. The multiple R^2 for the p-variable model is

$$R_p^2 = R^2(Y|X_1, X_2, \ldots, X_p) = 1 - \frac{SSE(p)}{SSY} \qquad (16.3)$$

Unfortunately, R_p^2 has three potentially misleading characteristics. First, it tends to *overestimate* ρ_p^2, the corresponding population value. Second, adding predictors, even useless ones, can never decrease R_p^2. In fact, adding variables will invariably increase R_p^2, at least slightly. Finally, R_p^2 is always largest for the maximum model, even though a better model may be obtained by deleting some (or even many) variables. Such a reduced (or restricted) model may be better because it may sacrifice only a negligible amount of predictive strength while substantially simplifying the model.

Another reasonable criterion for selecting the best model is the F test statistic for comparing the full and restricted models. This statistic, F_p, can be expressed in terms of sums of squares for error (SSE's) as

$$F_p = \frac{[SSE(p) - SSE(k)]/(k - p)}{SSE(k)/(n - k - 1)} = \frac{[SSE(p) - SSE(k)]/(k - p)}{MSE(k)} \qquad (16.4)$$

This statistic may be compared to an F distribution with $k - p$ and $n - k - 1$ degrees of freedom. The criterion F_p tests whether $SSE(p) - SSE(k)$, the difference between the residual sum of squares for the p-variable model and the residual sum of squares for the maximum model, is significantly different from 0. If F_p is *not* significant, we can use the smaller (p-variable) model and achieve roughly the same predictive ability as that yielded by the full model. Hence, a reasonable rule for selecting variables might be to retain p variables if both F_p is not significant and p is as small as possible. An often-used special case of F_p occurs if $p = k - 1$, in which F_p is a test of H_0: $\beta_k = 0$ in the full model.

The third criterion to be considered for selecting the best model is the estimated error variance for the p-variable model, namely,

$$MSE(p) = \frac{SSE(p)}{n - p - 1} \qquad (16.5)$$

In considering one particular model, we have earlier symbolized this estimate as S^2. The quantity $MSE(p)$ is an inviting choice for a selection criterion since we wish to find a model with small residual variance.

A less obvious candidate for a selection criterion involving $SSE(p)$ is Mallow's C_p, namely,

$$C_p = \frac{SSE(p)}{MSE(k)} - [n - 2(p + 1)] \qquad (16.6)$$

The C_p criterion helps us to decide how many variables to put in the best model, since it achieves a value of approximately $p + 1$ if $MSE(p)$ is roughly equal to $MSE(k)$ (i.e., if the correct model is of size p). Knowing the correct model size greatly aids in choosing the best model.

The criteria F_p, R_p^2, $MSE(p)$, and C_p are intimately related. For example, the F_p test can be expressed in terms of multiple squared correlations as follows:

$$F_p = \frac{(R_k^2 - R_p^2)/(k - p)}{(1 - R_k^2)/(n - k - 1)} \qquad (16.7)$$

and the C_p statistic is the following simple function of the F_p statistic:

$$C_p = (k - p)F_p + (2p - k + 1) \qquad (16.8)$$

Why consider more than one criterion? The reason is that no single criterion is always best. In practice, the alternatives can lead to different model choices. In the remainder of the chapter we shall consider all of the criteria mentioned, at least to some extent. An important aspect of our discussion will be a demonstration of the limitations of R_p^2 as the sole criterion for selecting a model. We shall favor C_p since it tends to simplify the decision about how many variables to retain in the final model.

16-5 Step 3: Specifying the Strategy for Selecting Variables

The third step in choosing the best model is to specify the strategy for selecting variables. Such a strategy is concerned with determining how many variables and also which particular variables should be in the final model. Traditionally, such strategies have focused on deciding whether a single variable should be added to a model (a forward selection method) or whether a single variable should be deleted from a model (a backward elimination method). As computers became more powerful, methods for considering more than one variable per step became practical (by generalizing single-variable methods to deal with sets, or *chunks*, of variables). Before discussing these strategies in detail, we shall first consider an algorithm for evaluating models that, if feasible to conduct, should be the method of choice.

16-5-1   All-Possible-Regressions Procedure

Whenever practical, the *all-possible-regressions procedure* is to be preferred over any other variable selection strategy. It is the only method guaranteed to find the model having the largest R_p^2, the smallest $\mathrm{MSE}(p)$, and so on. This strategy is not always used because the amount of calculation necessary becomes impractical if the number of variables k in the maximum model is large. The all-possible-regressions procedure requires that we fit each possible regression equation associated with each possible combination of the k independent variables. In particular, for our example we would need to fit the seven models corresponding to the following seven sets of independent variables: (1) HGT, (2) AGE, (3) $(AGE)^2$, (4) HGT and AGE, (5) HGT and $(AGE)^2$, (6) AGE and $(AGE)^2$, and (7) HGT, AGE, and $(AGE)^2$. For k independent variables, the number of models to be fitted would be $2^k - 1$; for example, if $k = 10$, then $2^{10} - 1 = 1,023$.

Once all $2^k - 1$ models have been fitted, one would then assemble the fitted models into sets involving from 1 to k variables and then order the models within each set according to some criterion (e.g., R_p^2, F_p, $\mathrm{MSE}(p)$, or C_p).

For our data, a summary of the results of this all-possible-regressions procedure is given in Table 16-1. From the table, the leaders (in terms of R_p^2-values) in each of the sets involving one, two, and three variables are given as follows:

One-variable set: HGT with $R_1^2 = .6630$

Two-variable set: HGT, AGE with $R_2^2 = .7800$

Three-variable set: HGT, AGE, $(AGE)^2$ with $R_3^2 = .7802$

Of the three models (models 1, 4, and 7, respectively, in Table 16-1), clearly model 4, involving HGT and AGE, should be our choice since its R^2-value is essentially the same as that for model 7 and is much higher than the value for model 1. Thus, our choice of the best regression equation based on the all-possible-regressions procedure with R_p^2 as the criterion is given by

$$\mathrm{WGT} = 6.553 + 0.722\mathrm{HGT} + 2.050\mathrm{AGE}$$

Other aspects of Table 16-1 deserve comment. Consider the partial F statistics. For a given variable in a given model, the associated partial F statistic assesses the contribution made by that variable to the prediction of Y (WGT) over and above the contributions made by other variables already in the given model. For example, for model 4, which involves only X_1 and X_2, the partial F for X_2 is $F(X_2|X_1) = 4.785$; but for model 7, which includes X_1, X_2, and X_3, the partial F for X_2 is $F(X_2|X_1, X_3) = 0.140$.

These partial F's are variables-added-last tests, each based on the MSE for the corresponding model being fit for that row. Such tests must be treated with some caution. Any test based on a model with fewer terms than the correct model will be biased, perhaps substan-

TABLE 16-1 Summary of results of all-possible-regressions procedure

Model	No. of Variables	Variables Used	Estimated Coefficients				Partial F Statistics			Overall F Statistic	R_p^2	MSE(p)	C_p
			$\hat{\beta}_0$	$\hat{\beta}_1$	$\hat{\beta}_2$	$\hat{\beta}_3$	X_1	X_2	X_3				
1	1	HGT (X_1)	6.190	1.073			19.67**			19.67**	.6630	29.93	4.27
2	1	AGE (X_2)	30.571		3.643			14.55**		14.55**	.5926	36.18	6.83
3	1	(AGE)^2 (X_3)	45.998			0.206			14.25**	14.25**	.5876	36.63	7.01
4	2	HGT, AGE	6.553	0.722	2.050		7.665*	4.785		15.95**	.7800	21.71	2.01
5	2	HGT, (AGE)^2	15.118	0.726		0.115	7.601*		4.565	15.63**	.7764	22.07	2.14
6	2	AGE, (AGE)^2	32.404		3.205	0.025		0.113	0.002	6.55**	.5927	40.20	8.83
7	3	HGT, AGE, (AGE)^2	3.438	0.724	2.777	−0.042	6.827*	0.140	0.010	9.47**	.7802	24.40	4.00

tially, because the test involves the use of biased error terms. Furthermore, so many tests are computed that the Type I error rate should be higher than the nominal rate α.

Overall F tests are also affected by the estimate of the error variance used. In variable selection, this estimate is important because biased tests conducted early in a stepwise algorithm may stop the procedure prematurely and miss important predictors. As calculated in Table 16-1, each overall F statistic is based on the corresponding MSE listed in the same row. In contrast, if the MSE for the largest model, number 7, was used in the denominator of each overall F test, each resulting test statistic would be an F_p statistic. Such a statistic would involve a comparison of the R_p^2-value for a reduced model (models 1 through 6) and the R_k^2-value for the largest model (model 7). For example, for model 4,

$$F_p = \frac{(R_k^2 - R_p^2)/(k - p)}{(1 - R_k^2)/(n - k - 1)} = \frac{(.7802 - .7800)/(3 - 2)}{(1 - .7802)/(12 - 3 - 1)} = 0.007$$

In general, such a test is a multiple-partial F test. The small F-value here indicates that the predictive abilities of the maximum model and model 4 do not differ significantly.

Table 16-1 also provides C_p-values. The value of C_p is expected to be close to $p + 1$ if the correct model is considered or if a larger model that contains the correct model is considered. If important predictors are omitted, C_p should be larger than $p + 1$. Also, if $F_p < 1$, then $C_p < (p + 1)$; this can occur when R_p^2 is close enough to R_k^2 in value. One prefers models with C_p not too far from $p + 1$. Therefore, for a model with one variable, C_p is compared with the value 2.0, for two variables with the value 3.0, and so on. For this example, no one-variable model has a C_p-value near 2.0. For the only three-variable model, C_p is exactly 4.00. The full model with k predictors is guaranteed to have C_p exactly equal to $k + 1$; to see this, examine equation (16.6), the formula for C_p. The two-variable model with AGE and $(AGE)^2$ has a C_p-value much greater than 3, while models 4 and 5 have C_p-values near the minimum possible C_p-value of

$$\frac{(n - k - 1)MSE(k)}{MSE(k)} - [n - 2(p + 1)] = (2p - k + 1)$$

which equals 2 when $k = 3$ and $p = 2$. (Note from (16.8) that this lower bound for C_p is attained when $F_p = 0$, and that it can be negative.) Such a value is better than the value of $p + 1 = 3$.

The all-possible-regressions procedure was presented first since it is preferred whenever practical. It has the distinction of being the only method guaranteed to find the best model, in the sense that any selection criterion will be numerically optimized for the particular sample under study. Naturally, this does not guarantee that one has found the correct (or population) model. In fact, in many situations, several reasonable candidates for the best model can be found, with different selection criteria suggesting different best models. Furthermore, such findings may vary from sample to sample, even though all the samples are chosen from the same population. Consequently, the choice of the best model can vary from sample to sample. These considerations motivate the discussion in Section 16-7 on evaluating the reliability of the chosen regression equation.

As mentioned earlier, the all-possible-regressions algorithm is often not practical since $2^k - 1$ models must be fitted when k candidate predictors are being evaluated. As computationally feasible alternatives, many methods have been suggested to approximate the all-possible-regressions procedure. These methods are not guaranteed to find the best model.

Nevertheless, these methods, to be discussed below, can (with careful use) glean essentially all the information in the data needed to choose the best model.

16-5-2 Backward Elimination Procedure

In the *backward elimination procedure*, we proceed as follows:

Step 1. Determine the fitted regression equation containing all independent variables. For our example we obtain

$$\widehat{WGT} = 3.438 + 0.724HGT + 2.777AGE - 0.042(AGE)^2$$

with ANOVA table

Source	df	SS	MS	F	R^2
Regression	3	693.06	231.02	9.47**	.7802
Residual	8	195.19	24.40		
Total	11	888.25			

Step 2. Calculate the partial F statistic for every variable in the model as though it were the last variable to enter (see Table 16-1).

Variable	Partial F^*
HGT	6.827**
AGE	0.140
$(AGE)^2$	0.010

* Based on 1 and 8 df.

(Recall that the partial F statistics above test whether the addition of the last variable to the model significantly helps in predicting the dependent variable given that the other variables are already in the model.)

Step 3. Focus on the lowest observed partial F test value (F_L, say). In our example, $F_L = 0.010$ for the variable $(AGE)^2$.

Step 4. Compare this value F_L with a preselected critical value (F_C, say) *of the F distribution* (i.e., test for the significance of the partial F_L). (a) *If $F_L < F_C$, remove from the model the variable under consideration, recompute the regression equation for the remaining variables, and repeat steps 2, 3, and 4.* (b) *If $F_L > F_C$, adopt the complete regression equation as calculated.* In our example, if we work at the 10% level, $F_L = 0.010 < F_{1,8,0.90} = F_C = 3.46$. Therefore, we remove $(AGE)^2$ from the model and recompute the equation using only HGT and AGE. We then obtain

$$\widehat{WGT} = 6.553 + 0.722HGT + 2.050AGE$$

with ANOVA table

Source	df	SS	MS	F	R^2
Regression	2	692.82	346.41	15.95**	.7800
Residual	9	195.43	21.71		
Total	11	888.25			

With $(AGE)^2$ out of the picture, the partial F's become 7.665 for HGT and 4.785 for AGE (see Table 16-1). Our new $F_C = F_{1,9,0.90} = 3.36$, which is less than 4.785. Therefore, the partial F for AGE is significant and we stop here with this model, which is the same model that we arrived at using the all-possible-regressions procedure.

16-5-3 Forward Selection Procedure

In the *forward selection procedure,* we proceed as follows:

Step 1. Select as the first variable to enter the model that variable most highly correlated with the dependent variable, and then fit the associated straight-line regression equation. For our example we have the following correlations: $r_{YX_1} = .814$, $r_{YX_2} = .770$, and $r_{YX_3} = .767$. Thus, the first variable to enter is $X_1 = $ HGT.

The straight-line regression equation relating WGT and HGT is

$$\widehat{WGT} = 6.190 + 1.073HGT$$

with ANOVA table

Source	df	SS	MS	F	R^2
Regression	1	588.92	588.92	19.67**	.6630
Residual	10	299.33	29.93		
Total	11	888.25			

If the F statistic in the table is not significant, we stop and conclude that no independent variables are important predictors. If the F statistic is significant (as it is), we include this variable (in our case, HGT) in the model and proceed to step 2.

Step 2. Calculate the partial F statistic associated with each remaining variable based on a regression equation containing that variable and the variable initially selected. For our example,

partial $F(X_2|X_1) = 4.785$

partial $F(X_3|X_1) = 4.565$

Step 3. Focus on the variable with the largest partial F statistic. For our data, the variable AGE has the largest partial F (4.785).

Step 4. Test for the significance of the partial F statistic associated with the variable selected in step 3. (a) If this test is significant, add the new variable to the regression equation. (b) If this test is not significant, use in the model only the variable added in step 1. For our example, since the partial F for AGE is significant at the 10% level ($F_{1,9,0.90} = 3.36$), we add AGE to get the two-variable model

$$\widehat{WGT} = 6.553 + 0.722HGT + 2.050AGE$$

Step 5. At each subsequent step, determine the partial F statistics for those variables not yet in the model and then add to the model that variable which has the largest partial F value if it is statistically significant. At any step, if the largest partial F is not significant, no more variables are included in the model and the process is terminated. For our example we have already added HGT and AGE to the model. We now see if we should add $(AGE)^2$. The partial F for $(AGE)^2$, controlling for HGT and AGE, is given by

partial $F(X_3|X_1, X_2) = 0.010$

This value is not statistically significant, since $F_{1,8,0.90} = 3.46$. Again, we have arrived at the same two-variable model chosen via the previously discussed methods.

16-5-4 Stepwise Regression Procedure

Stepwise regression is a modified version of forward regression that permits reexamination, at every step, of the variables incorporated in the model in previous steps. A variable that entered at an early stage may become superfluous at a later stage because of its relationship with other variables now in the model. To check on this possibility, at each step a partial F test for each variable presently in the model is made as though it were the most recent variable entered, irrespective of its actual entry point into the model. That variable with the smallest nonsignificant partial F statistic (if there is such a variable) is removed, the model is refitted with the remaining variables, the partial F's are obtained and similarly examined, and so on. The whole process continues until no more variables can be entered or removed.

For our example the first step, as in the forward selection procedure, would be to add the variable HGT to the model, since it has the highest significant correlation with WGT. Next, as before, we would add AGE to the model, since it has a higher significant partial correlation with WGT than does $(AGE)^2$, controlling for HGT. Now, before testing to see if $(AGE)^2$ should also be added to the model, we look at the partial F of HGT given that AGE is already in the model to see if we should now remove HGT. This partial is given by $F(X_1|X_2) = 7.665$ (see Table 16-1), which exceeds $F_{1,9,0.90} = 3.36$. Thus we do not remove HGT from the model. Next we check to see if we need to add $(AGE)^2$; of course, the answer is no, since we have dealt with this situation before.

The analysis-of-variance table that best summarizes the results obtained for our example is thus as follows:

Source		df	SS	MS	F	R^2
Regression	$\begin{cases} X_1 \\ X_2|X_1 \end{cases}$	1	588.92	588.92	19.67**	.7800
		1	103.90	103.90	4.79 ($P < .10$)	
Residual		9	195.43	21.71		
Total		11	888.25			

The ANOVA table that considers all variables is

Source		df	SS	MS	F	R^2
Regression	$\begin{cases} X_1 \\ X_2|X_1 \\ X_3|X_1, X_2 \end{cases}$	1	588.92	588.92	19.67**	.7802
		1	103.90	103.90	4.79 ($P < .10$)	
		1	0.24	0.24	0.01	
Residual		8	195.19	24.40		
Total		11	888.25			

16-5-5 Using Computer Programs

So far we have discussed decisions in stepwise variable selection in terms of F_p and its P-value, R_p^2, C_p, and $MSE(p)$. In backward elimination, the F statistic is often called an "F-to-leave," while in forward selection the F statistic is often called an "F-to-enter." Stepwise computer programs require specifying comparison or critical values for these F's or for their associated significance levels. These values must not be interpreted as they are in a single-hypothesis test. Their limitation derives from the fact that the probability of finding at least one significant independent variable when there are actually none increases rapidly as the number k of candidate independent variables increases, an approximate upper bound on this overall significance level being $1 - (1 - \alpha)^k$, where α is the significance level of any one test.

To prevent this overall significance level from exceeding α in value, a conservative but easily implemented approach is to conduct any one test at level α/k (see Pope and Webster, 1972, and Kupper, Stewart, and Williams, 1976, for further discussion). In Section 16-7 we shall discuss other techniques for helping to insure the reliability of our conclusions.

Since almost all model selection methods involve using the data to generate hypotheses for further study (often called "exploratory data analysis"), P-values and other variable selection criteria must be utilized cleverly. For example, model selection programs often allow specifying a P-value to help decide whether a variable is to be considered significant enough to be included in a model. It is very helpful to specify the "P-to-enter" to be 1.00 for forward selection. This guarantees that the process will go as far as possible, thereby providing the maximum information for choosing the best model. Similarly, specifying a value of 0.00 for the P-to-leave guarantees that a backward elimination process will remove all variables.

16-5-6 Chunkwise Methods

The methods for selecting single variables just described can be generalized in a very useful way. The basic idea is that any selection method in which a single variable is added or deleted can be generalized to adding or deleting a group of variables. Consider, for example, using backward elimination to build a model for which the response variable is blood cholesterol level. Assume that three groups of predictors are available: demographic (gender, race, age, and their pairwise interactions), anthropometric (height, weight, and their interaction), and diet recall (amounts of five food types); hence, there is a total of $6 + 3 + 5 = 14$ predictors. The three groups of variables constitute *chunks*, sets of predictors that are logically related and of equal importance (within a chunk) as candidate predictors.

Several possible chunkwise testing methods are available. Choosing among them depends on (1) the analyst's preference for backward, forward, or other selection strategies and (2) the extent to which the analyst can logically group variables (i.e., form chunks) and then order the groups in importance. In many applications, an a priori order exists among the chunks. For this example, the researcher may wish to consider diet variables *only* after controlling for important demographic and anthropometric variables. Imposing order in this fashion helps simplify the analysis, typically increases reliability, and increases the chance of finding a scientifically plausible model. Hence, whenever possible, we recommend imposing an order on chunk selection.

We shall illustrate the use of chunkwise testing methods by describing a backward elimination strategy. We emphasize that other strategies may be preferred, depending upon the situation. One approach to a backward elimination method for chunkwise testing involves requiring all but one specified chunk of variables to stay in the model, the variables in that specified chunk being candidates for deletion.[4] For our example, assume that the set of diet variables is the first chunk considered for deletion.

---

[4] Stepwise regression computer packages permit this approach by letting the user specify variables to be forced into the model. The same feature can be used for subsequent chunkwise steps.

(If an a priori order among chunks exists, then that order determines which chunk is considered first for deletion. Otherwise, it is typical practice to choose first that chunk making the least important predictive contribution (e.g., having the smallest multiple-partial F statistic).)

Thus, in our example, all of the demographic and anthropometric variables are forced to remain in the model. If, for example, the chunk multiple-partial F test for the set of diet variables is not significant, then the entire chunk can be deleted. If this test is significant, then at least one of the variables in this diet chunk should be retained.

Of course, the simplest chunkwise method adds or deletes all variables in a chunk together. However, a more sensitive approach is to manipulate single variables within a significant chunk while keeping the other chunks in place. If we assume the diet chunk to be important, then we must decide which of the diet variables to retain as important predictors. A reasonable second step here would then be to require all demographic variables and the important individual diet variables to be retained, while considering the (second) chunk of anthropometric variables for deletion. The final step for this three-chunk example would require the individual variables selected from the first two chunks to remain in the model, while variables in the third (demographic) chunk are candidates for deletion. Forward and stepwise single-variable selection methods can also be generalized for use in chunkwise testing.

Chunkwise methods for selecting variables can have substantial advantages over single-variable selection methods. First, chunkwise methods effectively incorporate into the analysis prior scientific knowledge and preferences about sets of variables. Second, the number of possible models to be evaluated is reduced. If a chunk test is not significant and the entire chunk of variables is deleted, then clearly no tests on individual variables in that chunk are carried out. In many situations, such testing for group (or chunk) significance can be more effective and reliable than testing variables one at a time.

16-6 Step 4: Conducting the Analysis

Having specified the maximum model, the criterion for selecting a model, and the strategy for applying the criterion, one must then conduct the analysis as planned. Obviously this will be done with some type of computer program. The goodness of fit of the model chosen should certainly be examined. Also, the regression diagnostic methods of Chapter 12, such as residual analysis, are needed to demonstrate that the model chosen is reasonable for the data at hand. In the next section, we shall discuss methods for evaluating whether the model chosen has a good chance of fitting new samples of data from the same (or similar) populations.

16-7 Step 5: Evaluating Reliability with Split Samples

Having chosen a model that is best for a particular sample of data, one has no assurance that such a model can be reliably applied to other samples. In a sense, we are asking the question "Will our conclusions generalize?" If the chosen model predicts well for subsequent samples from the population of interest, we say that the model is *reliable*. In this section we discuss methods for evaluating the reliability of a model. Most generally accepted methods for assessing model reliability involve some form of a *split-sample* approach. We shall discuss three approaches to assessing model reliability: the follow-up study, the split-sample analysis, and the holdout sample.

The most compelling way to assess the reliability of a chosen model is to conduct a new study and test the fit of the chosen model to the new data. However, this approach is usually expensive and sometimes intractable. The question then arises as to whether a single study can achieve the two goals of finding the best model and assessing its reliability.

A split-sample analysis attempts to achieve both of the above goals in a single study. For the cholesterol example, the simplest split-sample analysis would proceed as follows. First, randomly assign all observations to one of two groups, the training group or the holdout group. This must be done before any analysis is conducted. Subjects may be grouped in strata based on one or more important categorical variables. For the cholesterol example, appropriate strata might be defined by various gender–race combinations. If such strata are important, a stratum-specific split-sample scheme can be used. With this method, all subjects within a stratum are randomly assigned to the training or the holdout group; such assignment is done separately for each stratum. The goal of stratum-specific random assignment is to insure that the two groups of observations (training and holdout) are equally representative of the parent population.

An alternative to stratified random splitting is a pair-matching assignment scheme. With pair-matching assignment, one finds pairs of subjects that are as similar as possible and then randomly assigns one member of the pair to the training sample and the other to the holdout sample. Unfortunately, the differences between the resulting training and holdout groups tend to be fewer than corresponding differences among subsequent randomly chosen samples. This tends to create an unrealistically optimistic opinion about model reliability, which we wish to avoid.

Either of two alternatives is usually recommended as the second step in a split-sample analysis. The first is to conduct model selection separately for each of the two groups of data. Typically, any difference in predictor variables chosen by the two selection processes is taken as an indication of unreliability. In practice, the two models obtained will *almost always* differ somewhat, which is the primary reason that model selection methods have a reputation for being unreliable.

The above approach for assessing reliability is far too stringent when prediction is the goal. More realistically, we suggest that a *very good* predictive model should (1) predict as well in any new sample as it does in the sample at hand and (2) pass all regression diagnostic tests for model adequacy applied to any new sample. These comments lead us to consider a second approach to a split-sample analysis for assessing reliability.

This second approach attempts to answer the question "Will the chosen model predict well in a new sample?" This approach addresses a more modest, yet very desirable, goal, as opposed to the former approach. With this second approach one begins by conducting a model-building process using the data for the training group. Suppose that the fitted prediction equation obtained is

$$\hat{Y} = \hat{\beta}_0 + \hat{\beta}_1 X_1 + \hat{\beta}_2 X_2 + \cdots + \hat{\beta}_p X_p$$

For the children's weight example, this equation took the form

$$\widehat{WGT} = 6.553 + 0.722HGT + 2.050AGE$$

Next, use the estimated prediction equation, which is based only on the training group data, to compute predicted values for this group. Denote this set of predicted values as $\{\hat{Y}_{i1}\}$, $i = 1$, $2, \ldots, n_1$, with the subscript i indexing the subject and the subscript 1 denoting the training group. For this training sample, let

$$R^2(1) = R_1^2(Y|X_1, X_2, \ldots, X_p) = r^2(Y_1, \hat{Y}_1)$$

denote the sample squared *multiple* correlation, which equals the sample squared *univariate* correlation between the observed and predicted response values.

Next, use the prediction equation for the training group to compute predicted values for the holdout (or "validation") sample. Denote this set of predicted values as

$$\{\hat{Y}_{i2}^*\} \qquad i = n_1 + 1, n_1 + 2, \ldots, n_1 + n_2$$

where n_2 is the number of observations in the holdout group and $n = n_1 + n_2$.

Finally, compute the univariate correlation between these predicted values and the observed responses in the holdout sample (group 2), namely,

$$R_*^2(2) = r^2(Y_2, \hat{Y}_2^*)$$

The quantity $R_*^2(2)$ is called the "cross-validation correlation," and the quantity

$$R^2(1) - R_*^2(2)$$

is called the "shrinkage on cross-validation." Typically, $R_*^2(2)$, the cross-validation correlation, is a less biased estimator of the population squared multiple correlation than is the (positively) biased $R^2(1)$. Hence, the shrinkage statistic is almost always positive. How large must shrinkage be to cast doubt on model reliability? No firm rules can be given. Certainly one should presume that the fitted model is unreliable if shrinkage is 0.90 or more. In contrast, shrinkage values less than 0.10 are indicative of a reliable model.

Using only half of the data to estimate the prediction equation appears rather wasteful. If the shrinkage is small enough to indicate reliability of the model, then it seems reasonable to pool the data and calculate pooled estimates of the regression coefficients.

Depending upon the situation, using only half of the data for the training sample analysis may be inadvisable. A useful rule of thumb is to increase the relative size of the training sample as the total sample size decreases and to decrease the relative size of the training sample as the total sample size increases. To illustrate this, consider two situations. In the first situation, data from over 3,000 subjects were available for regression analysis. The maximum model included fewer than 20 variables, consisting of a chunk of demographic control variables and a chunk of pollution exposure variables. The primary goal of the study was to test for the presence of a relationship between the response variable (a measure of pulmonary function) and the pollution variables, controlling for demographic effects. To choose the best set of demographic control variables, a 10% stratified random sample ($n \approx 300$) was used as a training sample.

The second example concerned a study of the relationship between amount of body fat measured by X-ray methods (CAT scans) and traditional measures such as skinfold thickness. The primary goal was to provide an equation to predict the X-ray-measured body fat from demographic information (gender and age) and simple anthropometric measures (body weight, height, and three readings of skinfold thickness). The maximum model contained fewer than 20 predictors. Data from approximately 200 subjects were available. These data comprised nearly one year's collection from the hospital X-ray laboratory and demanded substantial human effort and computer processing to extract. Consequently, approximately 75% of the data were used as a training sample. These two examples illustrate the general concept that the splitting proportion should be tailored to the problem under study.

16-8 Example Analysis of Actual Data

We are now ready to apply the methods discussed in this chapter to a set of actual data. In Chapter 12 (on regression diagnostics), we introduced data on more than 200 children reported by Lewis and Taylor (1967). Lewis and Taylor reported body weight (WGT), standing height (HGT), AGE, and GENDER for all subjects. Our goal here will be to provide a reliable equation for predicting weight.

The first step is to choose a maximum model. The number of subjects is large enough so that we can consider a fairly wide range of models. It is natural to want to include the *linear* effects of AGE, HGT, and GENDER in the model. Quadratic and cubic terms for both AGE and HGT will also be included, since WGT is expected to increase in a nonlinear way as a function of AGE and HGT. It is reasonable to expect that the same model will not hold for both males and females. Hence, the interaction of the dichotomous variable GENDER with each power of AGE and HGT will be included. Terms involving cross-products between AGE and HGT powers will not be included, since no biological fact suggests that such interactions are important. Hence, 13 variables in all will be included in the maximum model as predictors of WGT: AGE, AGE2 = $(AGE)^2$, AGE3 = $(AGE)^3$, HGT, HT2 = $(HGT)^2$, HT3 = $(HGT)^3$, MALE = 1 for a boy and 0 for a girl, MALE × AGE, MALE × AGE2, MALE × AGE3, MALE × HGT, MALE × HT2, and MALE × HT3. (In the analysis to follow, AGE and HGT will denote centered variables.)

The second step is to choose a selection criterion. We choose to emphasize C_p, while also looking at R_p^2. The former criterion is helpful in deciding the size of the best model, while the latter provides an easily interpretable measure of predictive ability.

The third step is to choose a strategy for selecting variables. We prefer backward elimination methods, with as much structure imposed on the process as possible. As mentioned earlier, one procedure is to group the variables into chunks and then to order the chunks by degree of importance. However, we shall not consider a chunkwise strategy here.

After the seemingly best model is chosen, the fourth step is to evaluate the reliability of the model. Since the goal here is to produce a reliable prediction equation, this step is especially important. To implement a split-sample approach to assessing reliability, the subjects were randomly assigned to either data set ONE (the training group) or TWO (the holdout group). Random splitting was sex specific. The table below summarizes the observed split-sample subject distribution:

Sample	Female	Male	Total
ONE	55	63	118
TWO	56	63	119
Total	111	126	237

TABLE 16-2   Descriptive statistics for sample ONE
             ($n = 118$)

Variable	Mean	S.D.	Minimum	Maximum
WGT (lb)	100.05	17.47	63.5	150.0
AGE[a] (yr)	0	1.50	−2.00	3.91
HGT[b] (in.)	0	3.7	−8.47	10.73

[a] Equals age $- \overline{AGE}_1$; $\overline{AGE}_1 \doteq 13.59$.

[b] Equals height $- \overline{HGT}_1$; $\overline{HGT}_1 \doteq 61.27$.

Sample ONE will be used to build the best model. Then sample TWO will be used to compute cross-validation correlation and shrinkage statistics.

16-8-1 Analysis of Sample ONE

Table 16-2 provides descriptive statistics for sample ONE, and Table 16-3 provides the matrix of intercorrelations. To avoid computational problems, the predictor variables were centered (see Table 16-2 and Chapter 12).

Table 16-4 summarizes the backward elimination analysis using sample ONE. For sample ONE, SSY = 35,700.6949 and MSE = 120.0027676 for the full model. Table 16-4 specifies the order of variable elimination, the variables in the best model for each possible model size, and C_p and R_p^2 for each such model. Since the maximum model includes $k = 13$ variables, the top row corresponds to $p = 13$, for which $R_p^2 = .65042$. Since the maximum model always has $C_p = k + 1$, the C_p-value of 14 provides no information regarding the best

TABLE 16-3   Correlations ($\times$ 100) for sample ONE with centered age and height variables
             ($n = 118$)

	WGT	AGE	AGE2	AGE3	HGT	HT2	HT3	MALE	MALE × AGE	MALE × AGE2	MALE × AGE3	MALE × HGT	MALE × HT2
WGT													
AGE	59												
AGE2	25	44											
AGE3	48	78	80										
HGT	77	67	16	43									
HT2	05	10	23	11	15								
HT3	63	46	14	32	79	35							
MALE	03	−01	−04	−03	18	18	24						
MALE × AGE	46	72	28	50	62	29	46	00					
MALE × AGE2	18	26	55	38	27	35	32	51	37				
MALE × AGE3	38	58	49	60	46	30	40	12	81	70			
MALE × HGT	57	55	16	34	81	41	68	10	77	27	57		
MALE × HT2	26	22	19	17	43	78	70	50	31	53	36	47	
MALE × HT3	48	38	17	27	64	65	88	16	54	32	46	78	75

TABLE 16-4   Backward elimination using sample
ONE ($n = 118$)

p	C_p	R_p^2	Variables Used
13	14.00	.65042	HT3
12	12.00	.65042	MALE × AGE
11	10.00	.65042	AGE2
10	8.00	.65040	MALE × AGE2
9	6.00	.65040	MALE × AGE3
8	4.03	.65032	HT2
7	2.08	.65015	AGE
6	0.38	.64914	MALE
5	−0.82	.64644	MALE × HT3
4	1.15	.63312	MALE × HT2
3	−0.12	.63066	MALE × HGT
2	0.52	.62177	AGE3
1	6.72	.59422	HGT

model size. The second row corresponds to a $p = 12$ variable model. The variable HT3 (listed in the preceding row) has been deleted, giving $C_p = 12$. Similarly, the bottom row tells us that the best single-variable model includes HGT ($C_p = 6.72$, $R_p^2 = .59422$), the best two-variable model includes HGT and AGE3 ($C_p = 0.52$, $R_p^2 = .62177$), and so on. In general, for row p in the table, the variable in that row and all variables listed in rows below it are included in the best p-variable model, while all variables listed in rows above row p are excluded.

An examination of Table 16-4 does not suggest any obvious choice for a best model. The maximum R^2 is about .650, and the minimum is around .594, which is a small range in which to operate. This indicates that many different-sized models predict about equally well. This is not an unusual occurrence with moderate-to-strong predictor intercorrelations, say, above .50 in absolute value (see Table 16-3). Since $R_7^2 \doteq .650$ and $R_{13}^2 \doteq .650$, it seems unreasonable to use more than seven predictors. However, when using only R_p^2, it is difficult to decide whether a model containing fewer than seven predictors is appropriate.

In contrast, the C_p statistic suggests that only two variables are needed (recall that C_p should be compared with the value $p + 1$). Observe that C_p is greater than $p + 1$ only when $p = 1$, which indicts only the one-variable model in Table 16-4. Unfortunately, the $p = 2$ model with HGT and AGE3 as predictors is not appealing since the linear and quadratic terms AGE and AGE2 are not included. As discussed in detail in Chapter 13, we strongly recommend including such lower-order terms in polynomial-type regression models.

An apparent nonlinear relationship may indicate the need for transforming the response and/or predictor variables. Since quadratic and cubic terms are included as candidate predictors, log(WGT) and (WGT)$^{1/2}$ can be tried as alternative response variables. Such transformations can sometimes linearize a relationship (see Chapter 12). Using all 13 predictors, $R^2 = .668$ for the log transformation and .659 for the square root transformation. Because neither transformation produces a substantial gain in R^2, they will not be considered further.

In additional exploratory analyses of the sample ONE data, models can be fitted in two fixed orders, an interaction ordering (Table 16-5) and a power ordering (Table 16-6). One can interpret Tables 16-5 and 16-6 similarly to Table 16-4; each row gives p, the number of

TABLE 16-5   Interaction-ordered fitting using
sample ONE ($n = 118$)

p	C_p	R_p^2	Variables Used
13	14.00	.65042	MALE × AGE3
12	12.01	.65040	MALE × HT3
11	10.39	.64909	MALE × AGE2
10	8.39	.64909	MALE × HT2
9	6.94	.64727	AGE3
8	6.97	.64045	HT3
7	6.37	.63575	AGE2
6	7.35	.62571	HT2
5	5.36	.62567	MALE × AGE
4	3.60	.62486	MALE × HGT
3	4.98	.61352	MALE
2	5.86	.60383	AGE
1	6.72	.59422	HGT

variables in the model, and the associated C_p- and R_p^2-values. All variables on or below line p are included in the fitted model, and all above line p are excluded.

In both tables, certain values of p are natural break points for model evaluation. For example, in Table 16-5, $p = 3$ corresponds to including just the linear terms of the predictors HGT, AGE, and MALE. The fact that $4.98 > 3 + 1$ leads us to consider larger models. Including the pairwise interactions MALE × HGT and MALE × AGE gives $p = 5$ and $C_5 = 5.36 < 5 + 1$, which suggests that this five-variable model is reasonable. Notice that $R_5^2 = .626$, and that each further variable addition increases R^2 very little. In fact, the best model could be of size $p = 4$, since $C_4 = 3.60$ and R^2 is not appreciably reduced. Taken together, these comments suggest choosing a four-variable model containing HGT, AGE, MALE, and MALE × HGT, with $R_4^2 = .625$ and $C_4 = 3.60$.

Table 16-6 summarizes a similar analysis in which the powers of the continuous

TABLE 16-6   Power-ordered fitting using sample
ONE ($n = 118$)

p	C_p	R_p^2	Variables Used
13	14.00	.65042	MALE × AGE3
12	12.01	.65040	MALE × HT3
11	10.39	.64909	MALE × AGE2
10	8.39	.64909	MALE × HT2
9	6.94	.64727	MALE × AGE
8	8.00	.64698	MALE × HGT
7	4.77	.64112	AGE3
6	5.05	.63343	HT3
5	4.83	.62747	AGE2
4	6.27	.61589	HT2
3	4.98	.61352	MALE
2	5.86	.60383	AGE
1	6.72	.59422	HGT

predictors are entered into the model in a logical ordering. As in Table 16-5, $p = 3$ gives $C_3 = 4.98$, indicating the need to consider larger models. The model with all three original variables and the two quadratic terms gives $p = 5$, $C_p = 4.83$, and $R_p^2 = .62747$. Adding higher-order powers or any interactions does not improve R^2 appreciably, nor does it lead to C_p-values noticeably above $p + 1$. All models smaller than $p = 5$ have C_p-values noticeably greater than $p + 1$. The results in Table 16-6 lead to choosing a five-variable model containing HGT, AGE, MALE, HT2, and AGE2, with $R_5^2 = .627$ and $C_5 = 4.83$.

In this example, then, two possible models have been suggested, both with roughly the same R^2-value. Personal preference may be exercised within the constraints of parsimony and scientific knowledge. Here, we prefer the five-variable model with HGT, AGE, MALE, HT2, and AGE2 as predictors. We choose this model as best because we expect growth to be nonlinear and because we prefer to avoid using interaction terms. The chosen model is thus of the general form

$$\widehat{WGT} = \hat{\beta}_0 + \hat{\beta}_1 MALE + \hat{\beta}_2 HGT + \hat{\beta}_3 AGE + \hat{\beta}_4 HT2 + \hat{\beta}_5 AGE2$$

and a predicted weight for the ith subject may be computed with the formula

$$\widehat{weight}_i = 100.9 - 3.153 MALE_i + 3.53(height_i - 61.27)$$
$$+ .4199(age_i - 13.59) - .0731(height_i - 61.27)^2$$
$$+ .8067(age_i - 13.59)^2$$

Recall that MALE has a value of 1 for boys and 0 for girls. For males, the estimated intercept is $100.9 - 3.153(1) = 97.747$, while for females it is $100.9 - 3.153(0) = 100.9$. For these particular data, then, the fitted model predicts that, on the average, girls outweigh boys of the same height and age by about 3 pounds. After a model has been selected, it is necessary to consider residual analysis and other regression diagnostic procedures (see Chapter 12). It can be shown that some large residuals are present, but they do not justify a more complex model. Since they turn out not to be influential, they do not need to be deleted to help estimation accuracy.

16-8-2 Analysis of Sample TWO

Since exploratory analysis of sample ONE was used to choose the best (most appealing) model, analysis of the holdout sample TWO will be used to evaluate the reliability of the chosen prediction equation. Table 16-7 provides descriptive statistics for sample TWO. These appear very similar to those of sample ONE, as one would hope. Note that the AGE and HGT variables are centered about the sample ONE means. This is necessary to maintain the definitions of these variables as used in the fitted model. Cross-validation analysis would look spuriously bad if this were not done.

TABLE 16-7 Descriptive statistics for sample TWO
($n = 119$)

Variable	Mean	S.D.	Minimum	Maximum
WGT (lb)	102.6	21.22	50.5	171.5
AGE[a] (yr)	0.20	1.46	−1.92	3.58
HGT[b] (in.)	0.18	4.16	−10.77	9.73

[a] Equals age $- \overline{AGE}_1$; $\overline{AGE}_1 \doteq 13.59$.

[b] Equals height $- \overline{HGT}_1$; $\overline{HGT}_1 \doteq 61.27$.

When compared with Table 16-4, Table 16-8 illustrates the typical instability of stepwise methods for selecting variables. Based on a backward elimination strategy, almost no variable is in the same place as in the analysis for sample ONE (Table 16-4).

Our recommended cross-validation analysis begins by using the regression equation estimated from sample ONE (with predictors HGT, AGE, MALE, HT2, and AGE2) to predict WGT values for sample TWO. The squared multiple correlation between these predicted values and the observed WGT values in sample TWO is .621, and this is the cross-validation correlation. Since $R^2(1) = .627$ for sample ONE, shrinkage is $.627 - .621 = .006$, which is quite small and indicative of excellent reliability of estimation.

An important aspect related to assessing the reliability of a model involves considering difference scores of the form

$$y_i - \hat{y}_i$$

or, in our example,

$$WGT_i - \widehat{WGT_i}$$

where only sample TWO subjects are used and where the sample ONE equation is used to compute the predicted values. These "unstandardized residuals" can be subjected to various residual analyses. The most helpful entails univariate descriptive statistics, a box-and-whisker plot, and a plot of the difference scores $(y_i - \hat{y}_i)$ versus the predicted values $(\hat{y}_i)$ (such plots are described in Chapter 12). In our example, a few large positive residuals are present, but they are neither sufficiently implausible nor influential to require further investigation. Their presence does hint at why cubic terms and interactions keep trying to creep into the model. These residuals could be reduced in size, but only by adding many more predictors.

If the model is finally deemed acceptable, then for prediction purposes one should pool all the data and report the coefficient estimates based on the combined data. If the model is not acceptable, then one must review the model-building process, paying particular attention to variable selection and large differences between training and holdout groups of data.

TABLE 16-8 Backward elimination using
           sample TWO ($n = 119$)

p	C_p	R_p^2	Variables Used
13	14.00	.72328	AGE2
12	12.00	.72328	MALE
11	10.02	.72323	AGE3
10	8.31	.72245	MALE $\times$ AGE2
9	7.22	.72006	HT2
8	6.06	.71781	MALE $\times$ HT2
7	4.15	.71763	MALE $\times$ HGT
6	3.16	.71496	HT3
5	1.52	.71402	MALE $\times$ HT3
4	2.58	.70595	MALE $\times$ AGE
3	7.99	.68642	AGE
2	9.60	.67690	MALE $\times$ AGE3
1	33.90	.60762	HGT

16-9 Issues in Selecting the Most Valid Model

In this chapter, we have considered variable selection strategies when the primary study goal is prediction. In contrast, recall that validity-oriented strategies are concerned with providing valid estimates of one or more regression coefficients in a model. Essentially, a valid estimate is one that accurately reflects the true relationship(s) in the target population being studied. As described in Chapter 11, in a validity-based strategy for selecting variables, both confounding and interaction must be considered. Moreover, the sample of data under consideration must accurately reflect the population of interest, and the assumptions attendant upon the model and analysis being used must be reasonably satisfied.

To be slightly more specific, a variable selection strategy with a validity goal would first involve a determination of important interaction effects, followed by an evaluation of confounding, which is contingent upon the results from assessing interaction in the analysis. For examples of this type of strategy, see the epidemiologic-research-oriented text of Kleinbaum, Kupper, and Morgenstern (1982, chaps. 21–24).

Problems

Note to the student: Use C_p for all of the problems in this chapter. Supplementary problems can be created simply by using one of the other selection criteria.

1. Add the variables $(HGT)^2$ and $(AGE \times HGT)$ to the data set on WGT, HGT, AGE, and $(AGE)^2$ given in Section 16-3. Then, using an available regression program, determine the best regression model relating WGT to the five independent variables using (a) the forward approach, (b) the backward approach, and (c) an approach that first determines the best model using HGT and AGE as the only candidate independent variables and then determines whether any second-order terms should be added to the model. (Use $\alpha = .10$.) Compare and discuss the models obtained for each of the three approaches.

2. Using the data given in Problem 2 of Chapter 5 (with SBP as the dependent variable), find the best regression model using $\alpha = .05$ and the independent variables AGE, QUET, and SMK by the (a) forward approach, (b) backward approach, and (c) all-possible-regressions approach. (d) Based on the results in parts (a) through (c), select a model for further analysis to determine which, *if any*, of the following interaction (i.e., product) terms should be added to the model: AQ = AGE × QUET, AS = AGE × SMK, QS = QUET × SMK, and AQS = AGE × QUET × SMK.

3. For the same data set as considered in Problem 2, use the stepwise-regression approach to find the best regression models of SBP on QUET, AGE, and QUET × AGE for smokers and nonsmokers separately. (Use $\alpha = .05$.) Compare the results for each group with the answer to Problem 2(d).

4. For the data given in Problem 4 of Chapter 8, find (using $\alpha = .10$) the best regression model relating homicide rate (Y) to population size (X_1), percent of families with income less than $5,000 ($X_2$), and unemployment rate (X_3). Use (a) the stepwise approach, (b) the backward approach, and (c) the all-possible-regressions approach.

5. The data set listed below contains information on AGE, SEX (1 = male, 2 = female), work problems index (WP), marital conflict index (MC), and depression index (DEP) for a sample of 39 new admissions to a psychiatric clinic at a large university hospital. For each sex *separately,* determine (using $\alpha = .10$) the best regression model relating DEP to MC and WP, controlling for AGE, using the following sequential procedure: (1) Force AGE into the model first; (2) use the all-possible-regressions approach on the remaining two independent variables WP and MC; and (3) determine whether the interaction term (MC $\times$ WP) should be added to the model. Compare and discuss the results obtained for each sex.

ID No.	AGE	SEX	WP	MC	DEP
1	45	2	90	70	69
2	35	1	90	75	75
3	32	2	70	32	35
4	32	2	80	30	73
5	39	2	85	55	86
6	25	2	85	6	161
7	22	1	75	20	202
8	30	2	70	63	91
9	49	2	75	4	113
10	47	1	84	12	68
11	48	1	64	11	109
12	49	2	85	7	92
13	45	2	80	8	80
14	41	2	80	15	82
15	45	2	82	6	156
16	59	2	72	5	198
17	42	2	70	17	170
18	35	1	70	29	188
19	31	2	70	80	82
20	45	1	70	126	37
21	28	1	85	30	194
22	37	1	90	9	294
23	29	1	80	14	94
24	29	1	70	24	126
25	31	1	80	21	192
26	29	1	60	11	232
27	29	1	70	10	184
28	23	2	80	10	238
29	44	2	78	19	112
30	28	1	70	22	141
31	32	2	70	21	108
32	36	2	74	77	87
33	22	2	78	67	33
34	46	2	70	25	73
35	21	1	70	14	168
36	34	1	80	17	218
37	27	2	80	18	175
38	31	2	80	42	126
39	19	2	75	36	135

6. Use the data in Problem 15 in Chapter 5. Use LN_BRNTL as the response variable and LN_PPMTL, LN_BLDTL, AGE, and WEIGHT as predictors. Use $\alpha = .05$.

  a. Indicate a plausible fixed order for testing predictors, based upon the nature of the data. Briefly defend your choice.

  b. Use a computer program to fit the full model. Provide a test of whether the multiple correlation is 0.

  c. Using the *fixed* order, choose a best model, adding variables in a forward fashion. Do not adhere to a particular α but use your judgment.

  d. Repeat (c) but delete the variables in the fixed order (a backward method).

  e. Using a computer program employing a stepwise procedure, find the best model using a backward algorithm.

  f. Repeat (e) using a forward algorithm.

  g. Compare and contrast your conclusions for (c), (d), (e), and (f). Indicate your preferred model.

  h. Using the ideas presented in Chapter 12, indicate how one could evaluate the adequacy of the best model and the validity of the underlying assumptions.

  i. Suggest a practical split-sample approach for this particular set of data. Include recommended sample sizes, any stratification variables, and variable selection strategy.

7. Use the data of Bethel and others (1985) discussed in Problem 19 of Chapter 12. Delete the three female subjects, thereby leaving 16 observations. Use FEV_1 as the response, and AGE, WEIGHT, and HEIGHT as predictors.

  a. Use the all-possible-regressions procedure to suggest a best model.

  b. Consider a model with centered AGE, WEIGHT, HEIGHT, and their squares as predictors. Suggest a plausible forward chunkwise strategy for choosing a model and implement it.

  c. Use the all-possible-regressions procedure for the expanded model to choose a best model.

  d. Compare results from (a), (b), and (c). What model seems most plausible? How do the data limit your conclusions?

8. Use the data from Freund (1979), presented in Problem 22 of Chapter 12. Taking the model discussed there as the maximum model, repeat (a) through (h) from Problem 6. In (h), take additional note of the possible role of collinearity.

References

Bethel, R. A., Sheppard, D., Geffroy, B., Tam, E., Nadel, J. A., Boushey, J. A. (1985). "Effect of 0.25 ppm Sulfur Dioxide on Airway Resistance in Freely Breathing, Heavily Exercising, Asthmatic Subjects." *Am. Rev. Resp. Dis.*, *131*: 659–661.

Freund, R. J. (1979). "Multicollinearity etc., Some 'New' Examples." *Proceedings of the Statistical Computing Section*, American Statistical Association, pp. 111–112.

Hocking, R. R. (1976). "The Analysis and Selection of Variables in Linear Regression." *Biometrics*, *32*: 1–49.

Kleinbaum, D. G., Kupper, L. L., and Morgenstern, H. (1982). *Epidemiologic Research*. Belmont, Calif.: Lifetime Learning Publications.

Kupper, L. L., Stewart, J. R., and Williams, K. A. (1976). "A Note on Controlling Significance Levels in Stepwise Regression." *Am. J. Epidemiol.*, *103*(1): 13–15.

Lewis, T., and Taylor, L. R. (1967). *Introduction to Experimental Ecology*. New York: Academic Press.

Marquardt, D. W., and Snee, R. D. (1975). "Ridge Regression in Practice." *Amer. Statistician*, *29*(1): 3–20.

Pope, P. T., and Webster, J. T. (1972). "The Use of an *F*-statistic in Stepwise Regression Procedures." *Technometrics*, *14*(2): 327–340.

17

One-Way Analysis of Variance

17-1 Preview

This chapter is the first of four chapters concerned with *analysis of variance*. Analysis of variance (ANOVA) has been mentioned previously. Earlier we described ANOVA as a technique for assessing how several *nominal* independent variables affect a *continuous* dependent variable. An example was given in Table 2-2 (the Ponape study); the problem involved describing the effects on blood pressure of two cultural incongruity indices, each dichotomized into the two categories "high" and "low" (see Example 1.5). In this case, the dependent variable (blood pressure) was continuous and the two independent variables (the cultural incongruity indices) were both nominal.

The very fact that ANOVA is generally restricted to consideration of nominal independent variables allows for an interesting interpretation of the purpose of the technique. Loosely speaking, *ANOVA is usually concerned with comparisons involving several population means*. In fact, in the simplest special case involving a comparison of two population means, the ANOVA comparison procedure is equivalent to the usual two-sample t test (which, as we know, requires the assumption of equal population variances).

The population means to be compared are generally easily specified by cross-classifying the nominal independent variables under consideration to form different combinations of categories.[1] In the example above (dealing with the Ponape study), we need only cross-

---

[1] Such specification is not really possible if the categories of any nominal variable are considered as only a sample from a much larger population of categories of interest. We will consider such situations later when discussing what are called *random-effects models*.

classify the HI and LO categories of incongruity index 1 with the HI and LO categories of index 2. This yields the four population means corresponding to the four combinations HI–HI, HI–LO, LO–HI, and LO–LO, as indicated by the following configuration:

INDEX 2

		HI	LO
INDEX 1	HI	μ_1	μ_2
	LO	μ_3	μ_4

Assessing whether the two indices have some effect on the dependent variable "blood pressure" is equivalent to determining what kind of differences, if any, exist among the four means.

17-1-1 Why the Name ANOVA?

If the technique called ANOVA usually deals with a comparison of means, it seems somewhat inappropriate to call it analysis of *variance*. Instead, why not use the acronym ANOME, where *ME* stands for "means"? Actually, the use of the designation ANOVA is quite justifiable: Although usually means are compared, the comparisons are made using estimates of variance. The ANOVA test statistics, as with regression analysis, are F statistics and are actually ratios of estimates of variance. In fact, it is even possible, and in some cases appropriate, to specify the null hypotheses of interest in terms of population variances.

17-1-2 ANOVA versus Regression

Another general distinction concerns the difference between what may be called an "ANOVA problem" and what may be called a "regression problem." For ANOVA, *all* the independent variables are treated as being nominal; for regression analysis, any mixture of measurement scales (nominal, ordinal, or interval) is allowable for the independent variables. In fact, ANOVA is often viewed as a special case of regression analysis, and almost any ANOVA model can be represented by a regression model whose parameters can be estimated and inferred about in the usual manner. The same may be said for certain other multivariable techniques, such as analysis of covariance and even discriminant analysis. In effect, then, we may view the various names given to these techniques as indicators of different (linear) models with the same general form

$$Y = \beta_0 + \beta_1 X_1 + \beta_2 X_2 + \cdots + \beta_p X_p + E$$

yet involving different types of variables and perhaps different assumptions regarding these variables. The choice of method can then be regarded simply as being equivalent to the choice of an appropriate linear model.

17-1-3 Factors and Levels

Some additional terms must be introduced at this point. When considering the use of dummy variables in regression in Chapter 14, we saw that a nominal variable with k categories is generally incorporated into a regression model by defining $k - 1$ dummy

TABLE 17-1 Examples of random and fixed factors

Random	Fixed	Random or Fixed
Subjects	Sex	Locations
Litters	Age	Treatments
Observers	Marital status	Drugs
Days	Education	Tests
Weeks		

variables. These $k - 1$ variables collectively describe the *basic* nominal variable being considered. It is usually convenient to be able to refer to such a basic variable without having to identify the specific dummy variables used to define it in the regression model. The approach generally adopted is to call the basic nominal variable a *factor*. The different categories of the factor are often referred to as its *levels*.

For example, if we wanted to compare the effects of several drugs on some human health response, we would consider the nominal variable "drugs" as a single factor and the specific drug categories as the levels. If k drugs were being compared, these would be incorporated into a regression model by defining $k - 1$ dummy variables. If, in addition to comparing the drugs, we also wanted to consider whether males and females responded differently, we would consider the nominal variable "sex" as a second factor and the specific categories (male and female) as the levels of this dichotomous factor.

17-1-4 Fixed versus Random Factors

One final point before proceeding with the methodology of one-way ANOVA is the need to distinguish between what are called random and fixed factors. A *random factor* is a factor whose levels may be regarded as a sample from some large population of levels.[2] A *fixed factor* is one whose levels are the only levels of interest. The distinction is important in any ANOVA, since different tests of significance are required for different configurations of random and fixed factors. We will see this more specifically when considering two-way ANOVA. For now it is perhaps useful to give some examples of random and fixed factors. These are described briefly below and are summarized in Table 17-1.

1. "Subjects" or "litters" is usually considered a random factor, since we ordinarily wish to infer from the subjects used to a large population of potential subjects.

2. "Observers" is a random factor often considered when examining the effect of different observers on the response variable of interest.

3. "Days," "weeks," and so on, are usually considered as random factors in investigations of the effect of time on a response variable observed during different time periods. We usually use many levels for such temporal factors to represent a large number of time periods.

4. "Sex" is always a fixed factor, since its two levels include all possible levels of interest.

---

[2] In practice, the experimental levels of a random factor need not be selected at random as long as they are reasonably representative of the large population of levels of interest.

5. "Locations" (e.g., cities, plants, or states) may be fixed or random depending on whether only specific sites are of interest or whether a larger geographical universe is to be considered.

6. "Age" is usually considered to be a fixed factor regardless of how the different age groups are defined.

7. "Treatments," "drugs," "tests," and so on, are usually considered as fixed factors, but they may be considered random if their levels are representative of a much larger group of possible levels.

8. "Marital status" is considered to be a fixed factor.

9. "Education" is considered to be a fixed factor.

17-2 One-Way ANOVA: The Problem, Assumptions, and Data Configuration

One-way ANOVA deals with the effect of a single factor on a single response variable. When that one factor is a fixed factor, one-way ANOVA (often referred to as *fixed-effects one-way ANOVA*) involves a comparison of several (two or more) population means.[3] It can be said that the different populations correspond to the different levels of the single-factor "populations."

17-2-1 The Problem

The main analysis problem in fixed-effects one-way ANOVA is to determine whether the population means are all equal. Thus, if there are k means (denoted as $\mu_1, \mu_2, \ldots, \mu_k$), the basic null hypothesis of interest is given by

$$H_0: \mu_1 = \mu_2 = \cdots = \mu_k \tag{17.1}$$

The alternative hypothesis is given by

$$H_A: \text{"the } k \text{ population means are not all equal"}$$

If the null hypothesis (17.1) is rejected, the next problem is to find out where the differences are. For example, if $k = 3$ and $H_0: \mu_1 = \mu_2 = \mu_3$ is rejected, we might wish to determine whether the main differences are between μ_1 and μ_2, between μ_1 and μ_3, between μ_1 and the average of the other two means, and so on. Such questions fall under the general statistical subject referred to as "multiple-comparison procedures," which will be discussed in Section 17-7.

17-2-2 The Assumptions

The assumptions needed for fixed-effects one-way ANOVA may be stated simply as follows:

---

[3] We shall be focusing our attention throughout this section on situations involving only fixed factors. Random factors will be discussed in Section 17-6.

1. Random samples (individuals, animals, etc.) are selected from each of k populations or groups.

2. A value of a specified dependent variable is recorded for each experimental unit (individual, animal, etc.) sampled.

3. The dependent variable is normally distributed in each population.

4. The variance of the dependent variable is the same in each population (this common variance is denoted as σ^2).

Although these assumptions provide the theoretical justification for applying this method, it is nevertheless sometimes necessary to compare several means using fixed-effects one-way ANOVA when the necessary assumptions are not clearly satisfied. Indeed, it is a rare instance when these assumptions hold exactly. It is therefore important to consider the consequences of applying fixed-effects one-way ANOVA when the preceding assumptions are in question.

In general, fixed-effects one-way ANOVA can be applied if none of the assumptions are very badly violated. This is true for more complex ANOVA situations as well as fixed-effects one-way ANOVA. The term generally used to denote this property of broad applicability is called *robustness* (i.e., a procedure is robust with respect to moderate departures from the basic assumptions).

We must nevertheless be careful to avoid using robustness as an automatic justification for blindly applying the ANOVA method. Certain facts should be kept in mind when considering the use of ANOVA in a given situation. For example, the normality assumption does not have to be exactly satisfied as long as we are dealing with relatively large samples (e.g., 20 or more observations from each population), although the consequences of large deviations from normality are somewhat more severe for random factors than for fixed factors. The assumption of variance homogeneity can also be mildly violated without serious risk, provided that the numbers of observations selected from each population are more or less the same, although, again, the consequences are more severe for random factors.

Violation of the assumption of independence of the observations, however, can lead to very serious errors in inference for both the fixed and random cases. In general, great care should be taken to ensure that the observations are independent. This concern arises primarily in studies where repeated observations are recorded on the same experimental subjects, since very often the level of response of a subject on one occasion has a decided effect on subsequent responses.

What, then, does one do when one or more of these assumptions are in serious question? For one thing, the data might be transformed (e.g., by means of a log, square root, or other transformation) so that they more closely satisfy the assumptions. For another thing, a more appropriate method of analysis (e.g., nonparametric ANOVA or growth curve analysis) might be utilized.[4]

---

[4] A description of nonparametric methods that can be used when these assumptions are clearly and strongly violated can be found in Siegel (1956), Lehmann (1975), and Hollander and Wolfe (1973). A discussion of growth curve analysis can be found in Allen and Grizzle (1969).

17-2-3 Data Configuration for One-Way ANOVA

Computations necessary for one-way ANOVA can be easily performed even with an ordinary calculator when the data are conveniently arranged. Table 17-2 illustrates a useful way of presenting the data for the general one-way situation. It can be seen from this table that the number of observations selected from each population does *not* have to be the same; that is, there are n_i observations from the ith population, and n_i need not equal n_j if $i \neq j$. Notice also that the double-subscript notation (Y_{ij}) is used to distinguish one observation from another. The first subscript for a given observation denotes the population number, and the second distinguishes that observation from the others in that sample. Thus, Y_{23} denotes the third observation from the second population, Y_{62} denotes the second observation from the sixth population, and Y_{kn_k} denotes the last observation from the kth population. The totals for each sample (from each population) are denoted alternatively by T_i or $Y_{i\cdot}$ for the ith sample, where the $\cdot$ denotes that we are summing over all values of j (i.e., we are adding together all observations within the given sample); the grand total over all samples is denoted as $G = Y_{\cdot\cdot}$. The sample means are alternatively denoted by $\bar{Y}_{i\cdot}$ or T_i/n_i for the ith sample; these statistics are particularly important because they are the estimates of the population means of interest. Finally, the grand mean over all samples is $\bar{Y} = G/n$.

Example   In a study by Daly (1973) concerning the effects of neighborhood characteristics on health, a stratified random sample of 100 households was selected, 25 from each of the four Turnkey neighborhoods included in the study. The data configuration of CMI scores for female heads of household is given in Table 17-3. Such scores are measures (derived from questionnaires) of the overall (self-perceived) health of individuals; the higher the score, the poorer the health. It is important to point out, in addition, that each of the Turnkey neighborhoods differed in the total number of residents and in the percentage of blacks in the surrounding neighborhoods. The racial composition of the Turnkey neighborhoods themselves was over 95% black. Daly's main thesis was that the health of persons living in similar federal or state housing projects varied according to the racial composition of the surrounding neighborhoods: The "friendlier" (in terms of racial composition) the surrounding neighborhood was, the better would be the health of the residents in the project. According to Daly, federal housing planners had never used information concerning overall neighborhood racial composition and other neighborhood characteristics and their relation-

TABLE 17-2   General data configuration for one-way ANOVA

Population	Sample Size	Observations	Total	Sample Mean
1	n_1	$Y_{11}, Y_{12}, Y_{13}, \ldots, Y_{1n_1}$	$T_1 = Y_{1\cdot}$	$\bar{Y}_{1\cdot} = T_1/n_1$
2	n_2	$Y_{21}, Y_{22}, Y_{23}, \ldots, Y_{2n_2}$	$T_2 = Y_{2\cdot}$	$\bar{Y}_{2\cdot} = T_2/n_2$
3	n_3	$Y_{31}, Y_{32}, Y_{33}, \ldots, Y_{3n_3}$	$T_3 = Y_{3\cdot}$	$\bar{Y}_{3\cdot} = T_3/n_3$
$\vdots$	$\vdots$	$\vdots$	$\vdots$	$\vdots$
k	n_k	$Y_{k1}, Y_{k2}, Y_{k3}, \ldots, Y_{kn_k}$	$T_k = Y_{k\cdot}$	$\bar{Y}_{k\cdot} = T_k/n_k$
	$n = \sum_{i=1}^{k} n_i$		$G = Y_{\cdot\cdot}$	$\bar{Y} = G/n$

TABLE 17-3   Cornell Medical Index scores for a sample of women from different households in four Turnkey housing neighborhoods

Neighborhood	No. of Households	% Blacks in Surrounding Neighborhoods	Sample Size (n_i)	Observations (Y_{ij})	Total (T_i)	Sample Mean $(\bar{Y}_{i \cdot})$
Cherryview	98	17	25	49, 12, 28, 24, 16, 28, 21, 48, 30, 18, 10, 10, 15, 7, 6, 11, 13, 17, 43, 18, 7, 10, 9, 12, 12	$T_1 = 473$	$\bar{Y}_{1 \cdot} = 18.92$
Morningside	211	100	25	5, 1, 44, 11, 4, 3, 14, 2, 13, 68, 34, 40, 36, 40, 22, 25, 14, 23, 26, 11, 20, 4, 16, 25, 17	$T_2 = 518$	$\bar{Y}_{2 \cdot} = 20.72$
Northhills	212	36	25	20, 31, 19, 9, 7, 16, 11, 17, 9, 14, 10, 5, 15, 19, 29, 23, 70, 25, 6, 62, 2, 14, 26, 7, 55	$T_3 = 521$	$\bar{Y}_{3 \cdot} = 20.84$
Easton	40	65	25	13, 10, 20, 20, 22, 14, 10, 8, 21, 35, 17, 23, 17, 23, 83, 21, 17, 41, 20, 25, 49, 41, 27, 37, 57	$T_4 = 671$	$\bar{Y}_{4 \cdot} = 26.84$
			$\sum_{i=1}^{4} n_i = 100$		$G = 2{,}183$	$\bar{Y} = 21.83$

ship to health as criteria for selecting areas for such projects. This study, it was hoped, could provide some concrete recommendations for improved federal planning.

On first examination of the data in Table 17-3, we see that the sample means vary. To determine whether the observed differences in the sample means can be attributed solely to chance, one may perform a one-way ANOVA. The possibility of violations of the assumptions underlying this methodology should not be of great concern for this data set, since the sample sizes are equal and reasonably large and since observations on women from different households may be considered to be independent.

17-3 Methodology for One-Way ANOVA

The null hypothesis of equal population means (H_0: $\mu_1 = \mu_2 = \cdots = \mu_k$) is tested using an F test. The test statistic is calculated as follows:

$$F = \frac{MST}{MSE} \tag{17.2}$$

where

$$MST = \frac{\sum\limits_{i=1}^{k} (T_i^2/n_i) - G^2/n}{k - 1} \tag{17.3}$$

and

$$MSE = \frac{\sum\limits_{i=1}^{k} \sum\limits_{j=1}^{n_i} Y_{ij}^2 - \sum\limits_{i=1}^{k} (T_i^2/n_i)}{n - k} \tag{17.4}$$

When H_0 is true (i.e., when the population means are all equal), the F statistic of (17.2) has the F distribution with $k - 1$ numerator and $n - k$ denominator degrees of freedom. Thus, for a given α we would reject H_0 and conclude that some (i.e., at least two) of the population means are different from one another if

$$F \geq F_{k-1,n-k,1-\alpha}$$

where $F_{k-1,n-k,1-\alpha}$ is the $100(1 - \alpha)\%$ point of the F distribution with $k - 1$ and $n - k$ degrees of freedom. It should be pointed out that the critical region for this test involves only upper percentage points of the F distribution, since only large values of the F statistic (usually values much greater than 1) will provide evidence for rejecting H_0.

17-3-1 Numerical Illustration

For the data given in Table 17-3, the calculations needed to perform the F test proceed as follows:

$$\sum_{i=1}^{4} \sum_{j=1}^{25} Y_{ij}^2 = \underbrace{\frac{49^2 + 12^2 + \cdots + 37^2 + 57^2}{\text{a sum of 100 squared observations}}}_{} = 72{,}851.40$$

$$\sum_{i=1}^{4} \frac{T_i^2}{n_i} = \frac{473^2}{25} + \frac{518^2}{25} + \frac{521^2}{25} + \frac{671^2}{25} = 48{,}549.40$$

$$\frac{G^2}{n} = \frac{2{,}183^2}{100} = 47{,}654.89$$

$$MST = \frac{\sum\limits_{i=1}^{4} (T_i^2/n_i) - G^2/n}{4 - 1}$$

$$= \frac{48{,}549.40 - 47{,}654.89}{3}$$

$$= 298.17$$

$$MSE = \frac{\sum\limits_{i=1}^{4} \sum\limits_{j=1}^{25} Y_{ij}^2 - \sum\limits_{i=1}^{4} (T_i^2/n_i)}{100 - 4}$$

$$= \frac{72{,}851.40 - 48{,}549.40}{96}$$

$$= 253.14$$

$$F = \frac{\text{MST}}{\text{MSE}}$$
$$= \frac{298.17}{253.14}$$
$$= 1.178$$

Using these calculations we may test H_0: $\mu_1 = \mu_2 = \mu_3 = \mu_4$ (i.e., there are *no* differences among the true mean CMI scores for the four neighborhoods) against H_A: "there are differences among the true mean CMI scores." For example, if $\alpha = .1$, we would find from the F tables that $F_{3,96,0.90} = 2.15$, which is greater than the computed F of 1.178. Thus, we would not reject H_0 at $\alpha = .1$.

To find the P-value for this test, we note further that $F_{3,96,0.75} = 1.41$, which also exceeds the computed F. Thus, we know that $P > .25$. Hence, we must conclude (as did Daly) that there are no significant differences among the observed mean CMI scores for the four neighborhoods.

It should be noted that if a significant difference among the sample means had been found, it would still have been up to the investigator to determine whether the actual magnitudes of these differences were meaningful in a practical sense and whether the patterns of these differences were as hypothesized. In our example, examination of the percentages of blacks in the surrounding neighborhoods (see Table 17-3) indicates that the observed differences among the sample means are clearly not in the pattern hypothesized. That is, under Daly's conjecture, it would be expected that Cherryview (with 17% black in the surrounding neighborhood) would have the highest observed mean CMI score, followed by Northhills (36%), Easton (65%), and finally Morningside (100%). This was not the order actually obtained. Daly also examined whether her conjecture was supported when controlling for other possibly relevant factors, such as "months lived in the neighborhood," "number of children," and "marital status." However, no significant results were obtained from these analyses either.

17-3-2   Rationale for the *F* Test in One-Way ANOVA

The use of the F test described above may be motivated in a number of ways, three of which are given below. Ideally these will not only provide the user of this F test with an intuitive theoretical appreciation of its purpose but also provide some insight into the rationale behind more complex ANOVA testing procedures.

1. The F test in one-way ANOVA is a generalization of the two-sample t test. It may be easily shown with a little algebra that the numerator and denominator components in the F statistic for one-way ANOVA (17.2) are simple generalizations of the corresponding components in the square of the ordinary two-sample t test statistic. In fact, when $k = 2$ it can be shown that the F statistic for one-way ANOVA is exactly equal to the square of the corresponding t statistic. Such a result is intuitively reasonable since the numerator degrees of freedom of the F when $k = 2$ is 1, and we have previously noted that the square of a t statistic with ν degrees of freedom has the F distribution with 1 and ν degrees of freedom in numerator and denominator, respectively.

In particular, recall that the two-sample t test statistic is given by the formula

$$T = \frac{(\bar{Y}_{1.} - \bar{Y}_{2.})/\sqrt{1/n_1 + 1/n_2}}{S_p}$$

where the pooled sample variance S_p^2 is given by

$$S_p^2 = \frac{1}{n_1 + n_2 - 2} \sum_{i=1}^{2} \sum_{j=1}^{n_i} (Y_{ij} - \bar{Y}_{i\cdot})^2$$

or, equivalently, by

$$S_p^2 = \frac{(n_1 - 1)S_1^2 + (n_2 - 1)S_2^2}{n_1 + n_2 - 2}$$

where S_1^2 and S_2^2 are the sample variances for groups 1 and 2, respectively. Focusing first on the denominator (MSE) of the F statistic (17.2), it can be shown with some algebra that

$$MSE = \frac{\sum_{i=1}^{k} \sum_{j=1}^{n_i} Y_{ij}^2 - \sum_{i=1}^{k} (T_i^2/n_i)}{n - k}$$

$$= \frac{(n_1 - 1)S_1^2 + (n_2 - 1)S_2^2 + \cdots + (n_k - 1)S_k^2}{n_1 + n_2 + \cdots + n_k - k}$$

Thus, MSE is a pooled estimate of the common population variance σ^2, since it is a weighted sum of the k estimates of σ^2 obtained using the k different sets of observations. Furthermore, when $k = 2$, MSE is equal to S_p^2.

Looking at the numerator (MST) of the F statistic, it can be shown that

$$MST = \frac{\sum_{i=1}^{k} (T_i^2/n_i) - G^2/n}{k - 1}$$

$$= \frac{1}{k - 1} \sum_{i=1}^{k} n_i (\bar{Y}_{i\cdot} - \bar{Y})^2$$

which simplifies to $(\bar{Y}_{1\cdot} - \bar{Y}_{2\cdot})^2/(1/n_1 + 1/n_2)$ when $k = 2$. Thus, the equivalence is established.

2. *The F statistic is the ratio of two variance estimates.* We have already seen that MSE is a pooled estimate of the common population variance σ^2; that is, the true average (or mean) value (μ_{MSE}, say) of MSE is σ^2. It turns out, however, that MST estimates σ^2 *only* when H_0 is true, that is, *only* when the population means $\mu_1, \mu_2, \ldots, \mu_k$ are all equal. In fact, the true mean value (μ_{MST}, say) of MST has the general form

$$\mu_{MST} = \sigma^2 + \frac{1}{k - 1} \sum_{i=1}^{k} n_i (\mu_i - \bar{\mu})^2 \tag{17.5}$$

where $\bar{\mu} = \sum_{i=1}^{k} n_i \mu_i / n$. It can be seen by inspection of expression (17.5) that MST estimates σ^2 *only* when all the μ_i are equal, in which case $\mu_i = \bar{\mu}$ for every i and so $\sum_{i=1}^{k} n_i (\mu_i - \bar{\mu})^2 = 0$. Otherwise, both terms on the right-hand side of (17.5) are positive and MST estimates something greater in value than σ^2. In other words,

$$\mu_{MST} = \sigma^2 \quad \text{when } H_0 \text{ is true}$$

and

$$\mu_{MST} > \sigma^2 \quad \text{when } H_0 \text{ is not true}$$

Loosely speaking, then, the F statistic MST/MSE may be viewed as approximating in some sense the ratio of population means

$$\frac{\mu_{MST}}{\mu_{MSE}} = \frac{\sigma^2 + [1/(k - 1)] \sum_{i=1}^{k} n_i (\mu_i - \bar{\mu})^2}{\sigma^2} \tag{17.6}$$

When H_0 is true, the numerator and denominator of (17.6) both equal σ^2, and the F statistic is thus the ratio of two estimates of the same variance. Furthermore, F can be expected to give different values depending on whether or not H_0 is true; that is, F should take a value close to 1 if H_0 is true (since it would be approximating $\sigma^2/\sigma^2 = 1$), whereas F should be larger than 1 if H_0 is false (since the numerator of (17.6) would be greater than the denominator).

3. *The F statistic compares the variability between groups to the variability within groups.* As with regression analysis, the total variability in the observations in a one-way ANOVA situation is measured by a total sum of squares,

$$\text{TSS} = \sum_{i=1}^{k} \sum_{j=1}^{n_i} (Y_{ij} - \bar{Y})^2 \tag{17.7}$$

Furthermore, it can be shown that

$$\text{TSS} = \text{SST} + \text{SSE} \tag{17.8}$$

where

$$\text{SST} = (k - 1)\text{MST} = \sum_{i=1}^{k} n_i (\bar{Y}_{i\cdot} - \bar{Y})^2$$

and

$$\text{SSE} = (n - k)\text{MSE} = \sum_{i=1}^{k} \sum_{j=1}^{n_i} (Y_{ij} - \bar{Y}_{i\cdot})^2$$

The term SST can be considered a measure of the variability *between* (or *across*) populations. (The designation SST is read "sum of squares due to treatments," since the populations often represent treatment groups.) It involves components of the general form $\bar{Y}_{i\cdot} - \bar{Y}$, which is the difference between the ith group mean and the overall mean.

The term SSE is a measure of the variability *within* populations and gives no information concerning variability between populations. It involves components of the general form $Y_{ij} - \bar{Y}_{i\cdot}$, which is the difference between the jth observation in the ith group and the mean for the ith group.

If SST is quite large compared with SSE, we know that most of the total variability is due to differences *between* populations rather than to differences *within* populations. Thus, it is quite natural in such a case to suspect that the population means are not all equal.

By writing the F statistic (17.2) in the form

$$F = \frac{\text{SST}}{\text{SSE}} \left(\frac{n - k}{k - 1} \right)$$

we can see that F will be large whenever SST accounts for a much larger proportion of the total sum of squares than does SSE.

17-3-3 ANOVA Table for One-Way ANOVA

As in regression analysis, the results of any ANOVA procedure can be summarized in a table appropriately named for the method. The ANOVA table for one-way ANOVA is given in general form in Table 17-4; Table 17-5 is the ANOVA table for our example involving the CMI data. The "Source" and "SS" columns in Table 17-4 display the components of the fundamental equation of one-way ANOVA,[5]

$$TSS = SST + SSE$$

The "MS" column contains the sums of squares divided by their corresponding degrees of freedom; the two mean squares are then used to form the numerator and denominator for the F test.

TABLE 17-4   General ANOVA table for one-way ANOVA (k populations)

Source	df	SS	MS	F
Between	$k - 1$	$SST = \sum_{i=1}^{k} \dfrac{T_i^2}{n_i} - \dfrac{G^2}{n}$	$MST = \dfrac{SST}{k-1}$	$\dfrac{MST}{MSE}$
Within	$n - k$	$SSE = TSS - SST$	$MSE = \dfrac{SSE}{n-k}$	
Total	$n - 1$	$TSS = \sum_{i=1}^{k} \sum_{j=1}^{n_i} Y_{ij}^2 - \dfrac{G^2}{n}$		

TABLE 17-5   ANOVA table for CMI data ($k = 4$)

Source	df	SS	MS	F
Between (neighborhoods)	3	894.51	298.17	1.178
Within (error)	96	24,302.00	253.14	
Total	99	25,196.51		

17-4 Regression Model for Fixed-Effects One-Way ANOVA

We have stated earlier that most ANOVA procedures can also be considered in a regression analysis setting; this can be done by defining appropriate dummy variables in a

---

[5] The usual method for calculating sums of squares involves computing TSS and SST separately and then computing SSE by subtraction (i.e., SSE = TSS − SST).

regression model.[6] The ANOVA F tests are then formulated in terms of hypotheses concerning the coefficients of the dummy variables in the regression model.[7]

Example   For the example involving the CMI data of Daly's (1973) study (see Table 17-3), a number of alternative regression models could be used to describe the situation, depending on the coding schemes used for the dummy variables. One such model is

$$Y = \mu + \alpha_1 X_1 + \alpha_2 X_2 + \alpha_3 X_3 + E \qquad (17.9)$$

where the regression coefficients are denoted as μ, α_1, α_2, and α_3 and the independent variables are defined as

$$X_1 = \begin{cases} 1 & \text{if neighborhood 1} \\ -1 & \text{if neighborhood 4} \\ 0 & \text{if otherwise} \end{cases} \qquad X_2 = \begin{cases} 1 & \text{if neighborhood 2} \\ -1 & \text{if neighborhood 4} \\ 0 & \text{if otherwise} \end{cases}$$

$$X_3 = \begin{cases} 1 & \text{if neighborhood 3} \\ -1 & \text{if neighborhood 4} \\ 0 & \text{if otherwise} \end{cases}$$

(Although we have previously used the Greek letter β with subscripts to denote regression coefficients, we have changed the notation for our ANOVA regression model so that these coefficients correspond to the parameters in the classical fixed-effects ANOVA model described in Section 17-5.)

The coding scheme used here to define the dummy variables is called an *effect* coding scheme. The coefficients μ, α_1, α_2, and α_3 for this (dummy variable) model can each be expressed in terms of the underlying population means (μ_1, μ_2, μ_3, and μ_4) as follows:

$$\mu = \frac{\mu_1 + \mu_2 + \mu_3 + \mu_4}{4} \qquad (= \bar{\mu}^*, \text{ say})$$

$$\alpha_1 = \mu_1 - \bar{\mu}^*$$

$$\alpha_2 = \mu_2 - \bar{\mu}^* \qquad (17.10)$$

$$\alpha_3 = \mu_3 - \bar{\mu}^*$$

(We show that the coefficients can be expressed as in (17.10) as follows. $\mu_{Y|X_1,X_2,X_3} = \mu + \alpha_1 X_1 + \alpha_2 X_2 + \alpha_3 X_3$. Thus,

$$\mu_1 = \mu_{Y|1,0,0} = \mu + \alpha_1 \qquad \text{since } X_1 = 1, X_2 = 0, X_3 = 0 \text{ for neighborhood 1}$$

---

[6] As mentioned earlier, we are restricting out attention here entirely to models with *fixed* factors. Models involving random factors will be treated in Section 17-6.

[7] We will see later that a regression formulation is often desirable, if not mandatory, when dealing with certain nonorthogonal ANOVA problems involving two or more factors. We shall discuss such problems in Chapter 20.

$$\mu_2 = \mu_{Y|0,1,0} = \mu + \alpha_2 \qquad \text{since } X_1 = 0, X_2 = 1, X_3 = 0 \\ \text{for neighborhood 2}$$

$$\mu_3 = \mu_{Y|0,0,1} = \mu + \alpha_3 \qquad \text{since } X_1 = 0, X_2 = 0, X_3 = 1 \\ \text{for neighborhood 3}$$

$$\mu_4 = \mu_{Y|-1,-1,-1} = \mu - \alpha_1 - \alpha_2 - \alpha_3 \qquad \text{since } X_1 = X_2 = X_3 = -1 \\ \text{for neighborhood 4}$$

Adding the left-hand sides and right-hand sides of these equations yields

$$\mu_1 + \mu_2 + \mu_3 + \mu_4 = 4\mu$$

or

$$\mu = \frac{1}{4} \sum_{i=1}^{4} \mu_i \qquad (= \bar{\mu}^*)$$

Solution (17.10) is then obtained by replacing μ with $\bar{\mu}^*$ in the equations above and then solving for the regression coefficients α_i in terms of μ_1, μ_2, μ_3, and μ_4.)

Model (17.9), then, involves coefficients that describe separate comparisons of the first three population means with the overall unweighted mean, $\bar{\mu}^*$.

Also, for model (17.9), $\mu_4 - \bar{\mu}^*$ can be expressed as the negative sum of α_1, α_2, and α_3. Moreover, this regression model can be fitted to provide *exactly* the same F statistic as required in one-way ANOVA for the test of H_0: $\mu_1 = \mu_2 = \mu_3 = \mu_4$. The equivalent regression null hypothesis is H_0: $\alpha_1 = \alpha_2 = \alpha_3 = 0$,[8] the regression F statistic will have the same degrees of freedom (i.e., $k - 1 = 3$ and $n - k = 96$) as given previously, and the ANOVA table will be exactly the same as given in the last section (where we pooled the dummy variable effects into one source of variation with 3 degrees of freedom).

Other coding schemes for the independent variables will yield exactly the same ANOVA table and F test as model (17.9), although the regression coefficients themselves will represent different parameters and have different least-squares estimators. One frequently used coding scheme defines the independent variables as

$$X_i = \begin{cases} 1 & \text{if neighborhood } i \\ 0 & \text{otherwise} \end{cases} \qquad i = 1, 2, 3$$

This coding scheme is an example of *reference cell* coding. The referent group in this case is group 4, and the regression coefficients will describe separate comparisons of the first three population means with μ_4:

$$\mu = \mu_4$$
$$\alpha_1 = \mu_1 - \mu_4$$
$$\alpha_2 = \mu_2 - \mu_4$$
$$\alpha_3 = \mu_3 - \mu_4$$

---

[8] When $\alpha_1 = \alpha_2 = \alpha_3 = 0$, it follows from simple algebra using (17.10) that $\mu_1 = \mu_2 = \mu_3 = \mu_4$ (e.g., $\alpha_1 = \mu_1 - \bar{\mu}^* = 0$ implies that $\mu_1 = \bar{\mu}^*$, $\alpha_2 = \mu_2 - \bar{\mu}^* = 0$ implies that $\mu_2 = \bar{\mu}^* = \mu_1$, etc.).

17-4-1 Effect Coding Model

A model using effect coding that is analogous to that of the previous section may be given for the general situation involving k populations as follows:

$$Y = \mu + \alpha_1 X_1 + \alpha_2 X_2 + \cdots + \alpha_{k-1} X_{k-1} + E \tag{17.11}$$

in which

$$X_i = \begin{cases} 1 & \text{for population } i \\ -1 & \text{for population } k \qquad i = 1, 2, \ldots, k-1 \\ 0 & \text{otherwise} \end{cases}$$

The coefficients of this model can be expressed in terms of the k population means μ_1, $\mu_2, \ldots, \mu_k$ as follows:

$$\mu = \frac{\mu_1 + \mu_2 + \cdots + \mu_k}{k} = \bar{\mu}^*$$

$$\alpha_1 = \mu_1 - \bar{\mu}^*$$

$$\alpha_2 = \mu_2 - \bar{\mu}^* \tag{17.12}$$

$$\vdots$$

$$\alpha_{k-1} = \mu_{k-1} - \bar{\mu}^*$$

$$-(\alpha_1 + \alpha_2 + \cdots + \alpha_{k-1}) = \mu_k - \bar{\mu}^*$$

for model (17.11). Also, the F statistic for one-way ANOVA with k populations can be equivalently obtained by testing the null hypothesis $H_0: \alpha_1 = \alpha_2 = \cdots = \alpha_{k-1} = 0$ in model (17.11).

17-4-2 Reference Cell Coding Model

Another coding scheme for k-population one-way ANOVA uses the following reference cell coding:

$$X_i = \begin{cases} 1 & \text{for population } i \\ 0 & \text{otherwise} \end{cases} \qquad i = 1, 2, \ldots, k-1$$

We again emphasize that only $k - 1$ such dummy variables are needed, and the population "left out" becomes the reference population (group or cell). Thus, given X_i as defined above, group k is the reference group. (In contrast, if X_1 had been left out instead of X_k, then group 1 would have been the reference group.)

Using the above reference cell coding, the responses in each group under the model $Y = \mu + \alpha_1 X_1 + \alpha_2 X_2 + \cdots + \alpha_{k-1} X_{k-1} + E$ are as follows:

group 1:	$Y = \mu + \alpha_1 \quad + E$
group 2:	$Y = \mu + \alpha_2 \quad + E$
$\vdots$	
group $k - 1$:	$Y = \mu + \alpha_{k-1} + E$
group k:	$Y = \mu \qquad + E$

In turn, the regression coefficients can be written in terms of group means as follows:

$$\alpha_1 = \mu_1 - \mu_k$$
$$\alpha_2 = \mu_2 - \mu_k$$
$$\vdots$$
$$\alpha_{k-1} = \mu_{k-1} - \mu_k$$
$$\mu = \mu_k$$

Thus, we can see that different coding schemes (e.g., an effect or some kind of reference cell scheme) will yield regression coefficients representing different parameters (for example, $\alpha_1 = \mu_1 - \bar{\mu}^*$ for the above effect coding, but $\alpha_1 = \mu_1 - \mu_k$ for the above reference cell coding). Nevertheless, regardless of the coding scheme used, the test of the hypothesis $H_0: \mu_1 = \mu_2 = \cdots = \mu_k$ can be equivalently obtained by testing $H_0: \alpha_1 = \alpha_2 = \cdots = \alpha_{k-1} = 0$ in the regression model (17.11). In other words, the correct SST, SSE, and $F_{k-1, n-k}$-values will be obtained from the regression analysis regardless of the (legitimate) coding scheme chosen.

17-5 Fixed-Effects Model for One-Way ANOVA

Many textbooks and articles dealing strictly with ANOVA procedures use a more classical type of model than the regression model given above to describe the fixed-effects one-way ANOVA situation. This type of model is often referred to as a *fixed-effects ANOVA model*; in such a model, all the factors being considered are fixed (i.e., the levels of each factor are the only levels of interest). The "effects" referred to in such a model represent measures of the influence (i.e., the effect) that different levels of the factor have on the dependent variable.[9] Such measures are often expressed in the form of differences between a given mean and an overall mean. That is, the effect of the ith population is often measured by the amount that the ith population mean differs from an overall mean.

Example For the CMI data ($k = 4$), the fixed-effects ANOVA model is given as follows:

$$Y_{ij} = \mu + \alpha_i + E_{ij} \qquad i = 1, 2, 3, 4; \quad j = 1, 2, \ldots, 25 \qquad (17.13)$$

where

$Y_{ij} = j$th observation from the ith population

$$\mu = \frac{\mu_1 + \mu_2 + \mu_3 + \mu_4}{4} \ (= \bar{\mu}^*, \text{ the overall unweighted mean,}$$
$$\text{since } n_i = 25 \text{ for all } i)$$

$\alpha_1 = \mu_1 - \mu =$ differential effect of neighborhood 1

$\alpha_2 = \mu_2 - \mu =$ differential effect of neighborhood 2

---

[9] When dealing with models with two or more factors, effects can also refer to measures of the influence of combinations of levels of the different factors on the dependent variable.

$\alpha_3 = \mu_3 - \mu =$ differential effect of neighborhood 3

$\alpha_4 = \mu_4 - \mu =$ differential effect of neighborhood 4

$E_{ij} = Y_{ij} - \mu - \alpha_i =$ error component associated with the jth observation from the ith population

One important property of this model is that the sum of the four α effects is 0; that is, $\alpha_1 + \alpha_2 + \alpha_3 + \alpha_4 = 0$. Thus, these effects represent differentials from the overall population mean μ that average out to 0. Nevertheless, the effect of one level (i.e., a neighborhood) may be considerably different from the effect of another. If this is so, we would more than likely find that our F test leads to rejection of the null hypothesis of equal population mean CMI scores for the four neighborhoods.

Another important property of this model is that the effects α_1, α_2, α_3, and α_4, which are actually population parameters defined in terms of population means, can be simply estimated from the data. This is done for any given effect by appropriately substituting the usual estimates of the means into the expression for the effect. For our example, the estimated effects are given by

$\hat{\alpha}_1 = \bar{Y}_{1.} - \bar{Y} =$ sample mean CMI score for neighborhood 1 $-$ overall sample mean CMI score for all neighborhoods

$\hat{\alpha}_2 = \bar{Y}_{2.} - \bar{Y} =$ sample mean CMI score for neighborhood 2 $-$ overall sample mean CMI score for all neighborhoods

$\hat{\alpha}_3 = \bar{Y}_{3.} - \bar{Y} =$ sample mean CMI score for neighborhood 3 $-$ overall sample mean CMI score for all neighborhoods

$\hat{\alpha}_4 = \bar{Y}_{4.} - \bar{Y} =$ sample mean CMI score for neighborhood 4 $-$ overall sample mean CMI score for all neighborhoods

The actual numerical values obtained from these formulae are as follows:

$\hat{\alpha}_1 = 18.92 - 21.83 = -2.91$

$\hat{\alpha}_2 = 20.72 - 21.83 = -1.11$

$\hat{\alpha}_3 = 20.84 - 21.83 = -0.99$

$\hat{\alpha}_4 = 26.84 - 21.83 = 5.01$

As with the population effects, it can be easily seen that the estimated effects also sum to 0, that is, $\sum_{i=1}^{4} \hat{\alpha}_i = 0$.

If we consider the general one-way ANOVA situation (i.e., there are k populations and n_i observations from the ith population), then the fixed-effects one-way ANOVA model may be written as follows:

$$Y_{ij} = \mu + \alpha_i + E_{ij} \qquad i = 1, 2, \ldots, k; \quad j = 1, 2, \ldots, n_i \qquad (17.14)$$

where

$Y_{ij} = j$th observation from the ith population

$$\mu = \frac{\mu_1 + \mu_2 + \cdots + \mu_k}{k} \qquad (= \bar{\mu}^*)$$

$\alpha_i = \mu_i - \mu$ = differential effect of population i

$E_{ij} = Y_{ij} - \mu - \alpha_i$ = error component associated with the jth observation from the ith population

(An alternative definition of μ may be $\bar{\mu} = \sum_{i=1}^{k} n_i\mu_i/n$, the overall weighted mean of the means.)

Here, it is easy to show that the sum of the α effects is 0, that is, $\sum_{i=1}^{k} \alpha_i = 0$. Similarly, the estimated effects, $\hat{\alpha}_i = \bar{Y}_{i.} - \bar{Y}^*$, where $\bar{Y}^* = \sum_{i=1}^{k} \bar{Y}_{i.}/k$, satisfy the constraint $\sum_{i=1}^{k} \hat{\alpha}_i = 0$.

(If μ is defined to be $\mu = \bar{\mu} = \sum_{i=1}^{k} n_i\mu_i/n$, it can be shown that the weighted sum $\sum_{i=1}^{k} n_i\alpha_i = 0$ and that, similarly, the weighted sum of the estimated effects $\hat{\alpha}_i = \bar{Y}_{i.} - \bar{Y}$, where $\bar{Y} = \sum_{i=1}^{k} n_i\bar{Y}_{i.}/n$, satisfies $\sum_{i=1}^{k} n_i\hat{\alpha}_i = 0$.)

Another property worth noting is that this model (17.14) corresponds in structure to the regression model given by (17.9); that is, the regression coefficients $\alpha_1, \alpha_2, \ldots, \alpha_{k-1}$ are precisely the effects $\alpha_1 = \mu_1 - \bar{\mu}^*, \alpha_2 = \mu_2 - \bar{\mu}^*, \ldots, \alpha_{k-1} = \mu_{k-1} - \bar{\mu}^*$, the regression constant μ represents the overall (unweighted) mean $\bar{\mu}^*$, and the negative sum of the regression coefficients $(-\sum_{i=1}^{k-1} \alpha_i)$ represents the effect $\alpha_k = \mu_k - \bar{\mu}^*$. This is why we have defined each of these models using the same notation for the unknown parameters:

$$Y = \mu + \sum_{i=1}^{k-1} \alpha_i X_i + E \qquad \text{(dummy variable regression model)}$$

$$Y_{ij} = \mu + \alpha_i + E_{ij} \qquad \text{(fixed-effects ANOVA model)}$$

(Note that μ represents the unweighted average of the k population means, $\bar{\mu}^*$, rather than the weighted average, $\bar{\mu}$, even though the sample sizes can be different in the different populations. Correspondingly, the least-squares estimate of μ is $\sum_{i=1}^{k} \bar{Y}_{i.}/k$, the unweighted average of the k sample means, rather than $\bar{Y}$. Nevertheless, it is possible to redefine the dummy variables in the regression model to obtain a least-squares solution yielding $\bar{Y}$ as the estimate of μ. The following dummy variable definitions

$$X_i = \begin{cases} -n_i & \text{if population } k \\ n_k & \text{if population } i \qquad i = 1, 2, \ldots, k-1 \\ 0 & \text{otherwise} \end{cases}$$

are required.)

17-6 Random-Effects Model

In Section 17-1 we distinguished between fixed and random factors. We have also stated that for ANOVA situations involving two or more factors, the F tests required for making inferences differ depending on whether all factors are fixed, all factors are random, or there are some of both types. It turns out that the null hypotheses to be tested must be stated in different terms when random factors are involved than when only fixed factors are involved.

Example   To get some insight into the structure of random-effects models, it is useful to consider Daly's (1973) study (see Table 17-3) again. It might be argued that the four different Turnkey neighborhoods are merely a representative sample of a larger population of similar types of neighborhoods (some of which might even be predominantly white with differing percentages of blacks in the surrounding neighborhoods). If so, the neighborhood factor would have to be considered as random. The appropriate ANOVA model would then be a *random-effects* one-way ANOVA model.[10] Its form would be essentially the same as that given in (17.13), except that the α components would be treated differently. That is, the random-effects model would be of the form[11]

$$Y_{ij} = \mu + A_i + E_{ij} \qquad i = 1, 2, 3, 4; \quad j = 1, 2, \ldots, 25 \qquad (17.15)$$

In this model, the A_i's can be viewed as a random sample of random variables having a common distribution. This common distribution represents the entire population of possible effects (in our example, neighborhoods).

To perform the appropriate analysis, it is necessary to assume that the distribution of A_i is normal with zero mean:

$$A_i \frown N(0, \sigma_A^2) \qquad i = 1, 2, 3, 4 \qquad (17.16)$$

where σ_A^2 denotes the variance of A_i. We must also assume that the A_i's are independent of the E_{ij}'s and of each other.

The requirement of zero mean in (17.16) is similar in philosophy to the requirement that $\sum_{i=1}^{k} \alpha_i = 0$ for the fixed-effects model. When the random model (17.15) applies, we are assuming that the average (i.e., mean) effect of neighborhoods is 0 over the entire population of neighborhoods. That is, we assume that $\mu_{A_i} = 0$, $i = 1, 2, 3, 4$.

How then do we state our null hypothesis? Because we have required that the neighborhood effects average out to 0 over the entire population of possible effects, there is only one way to assess whether there are any significant neighborhood effects at all, and this involves consideration of σ_A^2. If there is no variability (i.e., $\sigma_A^2 = 0$), all neighborhood effects must be 0. If there is variability (i.e., $\sigma_A^2 > 0$), there are some nonzero effects in the population of neighborhood effects.

Thus, our null hypothesis of no neighborhood effects should be stated as follows:

$$H_0: \sigma_A^2 = 0 \qquad (17.17)$$

This hypothesis is therefore analogous to the null hypothesis (17.1) used in the fixed-effects case, although it happens to be stated in terms of a population variance rather than in terms of population means.

We must still explain why the *F* test given by (17.2) for the fixed-effects model is exactly the same as that used for the random-effects model.[12] Such an explanation is best made by

---

[10] This type of model is also referred to as a *components-of-variance model*.

[11] Our usual convention has been to use Latin letters (*X, Y, Z,* etc.) to denote random variables and Greek letters (β, μ, σ, τ) to denote parameters. This requires using A_i's rather than α_i's to denote random effects.

[12] Again, we point out that the *F* tests are computationally equivalent for fixed-effects and random-effects models only in one-way ANOVA. When dealing with two-way or higher-way ANOVA, the testing procedures may be different.

considering the properties of the mean squares MST and MSE. Recalling our previous argument in Section 17-3-2 for the fixed-effects model, we saw that the F statistic, MST/MSE, could be considered as a rough approximation to the ratio of the means of these mean squares,[13]

$$\frac{\mu_{MST}}{\mu_{MSE}} = \frac{\sigma^2 + [1/(k-1)] \sum_{i=1}^{k} n_i(\mu_i - \bar{\mu})^2}{\sigma^2}$$

A similar argument can be made with regard to the F statistic for the random-effects model. In particular, it still holds, for the random- as well as for the fixed-effects model, that the denominator MSE estimates σ^2, that is,

$$\mu_{MSE} = \sigma^2$$

Furthermore, it can be shown for the random-effects model applied to our example ($k = 4$, $n_i = 25$) that MST estimates

$$\mu_{MST(random)} = \sigma^2 + 25\sigma_A^2 \tag{17.18}$$

Thus, for the random-effects model, F approximates the ratio

$$\frac{\mu_{MST(random)}}{\mu_{MSE}} = \frac{\sigma^2 + 25\sigma_A^2}{\sigma^2} \tag{17.19}$$

Since the null hypothesis in this case is $H_0: \sigma_A^2 = 0$, we can see that the ratio (17.19) simplifies to $\sigma^2/\sigma^2 = 1$ when H_0 is true. Thus, we again see that the F statistic under H_0 consists of the ratio of two estimates of the same variance σ^2. Furthermore, because $\sigma_A^2 > 0$ when H_0 is not true, the more the variability among neighborhood effects, the larger should be the observed value of F.

In general, the random-effects model for one-way ANOVA is given by

$$Y_{ij} = \mu + A_i + E_{ij} \qquad i = 1, 2, \ldots, k; \quad j = 1, 2, \ldots, n_i \tag{17.20}$$

where A_i and E_{ij} are independent random variables satisfying $A_i \frown N(0, \sigma_A^2)$ and $E_{ij} \frown N(0, \sigma^2)$.

For this model it can be shown that F approximates the following ratio of expected mean squares:

$$\frac{\mu_{MST(random)}}{\mu_{MSE}} = \frac{\sigma^2 + n_0\sigma_A^2}{\sigma^2}$$

where

$$n_0 = \frac{\sum_{i=1}^{k} n_i - \left(\sum_{i=1}^{k} n_i^2 \middle/ \sum_{i=1}^{k} n_i\right)}{k-1}$$

---

[13] The parameters μ_{MST} and μ_{MSE} are often called *expected mean squares*.

TABLE 17-6   Combined one-way ANOVA table for fixed- and random-effects models

Source	df	MS	F	Expected Mean Square (EMS) Fixed Effects	Random Effects
Between	$k - 1$	MST	$\dfrac{\text{MST}}{\text{MSE}}$	$\sigma^2 + \dfrac{1}{k-1}\displaystyle\sum_{i=1}^{k} n_i(\mu_i - \bar{\mu})^2$	$\sigma^2 + n_0\sigma_A^2$
Within	$n - k$	MSE		σ^2	σ^2
Total	$n - 1$				
				$H_0: \mu_1 = \mu_2 = \cdots = \mu_k$	$H_0: \sigma_A^2 = 0$

is like an average of the n_i observations selected from each population.[14] The F statistic for the random-effects model is therefore the ratio of two estimates of σ^2 when $H_0: \sigma_A^2 = 0$ is true.

Table 17-6 summarizes both the similarities and differences between the fixed- and random-effects models. Tables with similar formats will be used in subsequent chapters to highlight distinctions for ANOVA situations with more than two factors.

17-7 Multiple-Comparison Procedures

Whenever an ANOVA F test for simultaneously comparing several population means is found to be statistically significant, it is then customarily of interest to determine which *specific* differences there are among the population means. For example, if four means are being compared (fixed-effects case) and the null hypothesis $H_0: \mu_1 = \mu_2 = \mu_3 = \mu_4$ is rejected,[15] one may wish to determine which subgroups of means are different by considering any number of more-specific hypotheses like $H_{01}: \mu_1 = \mu_2$, $H_{02}: \mu_2 = \mu_3$, $H_{03}: \mu_3 = \mu_4$, or even $H_{04}: (\mu_1 + \mu_2)/2 = (\mu_3 + \mu_4)/2$, which compares the average effect of populations 1 and 2 with the average effect of populations 3 and 4. Such specific comparisons may have been of interest to the investigator before (a priori) the data were collected or may arise in completely exploratory studies only after (a posteriori) the data have been examined. In

---

[14] When all the n_i are equal as in the Daly (1973) example (i.e., $n_i = n^*$), then n_0 is equal to n^* since

$$n_0 = \frac{kn^* - kn^{*2}/kn^*}{k - 1} = n^*$$

In the Daly (1973) example, $n^* = 25$.

[15] This section deals only with fixed-effects ANOVA problems, since the random-effects model treats the observed factor levels as a sample from a larger population of levels of interest and is therefore not directed at inferences of just the sampled levels.

either event, a seemingly reasonable first approach to making inferences about differences among the population means would be to make several t tests and to focus on all those tests found significant. For example, if all pairwise comparisons among the means were desired, this would require $_4C_2 = 6$ such tests when considering four means (or, in general, $_kC_2 = k(k - 1)/2$ tests when dealing with k means). Thus, for testing H_0: $\mu_i = \mu_j$ at the α level of significance, we could reject this H_0 when

$$|T| \geq t_{n-k,1-\alpha/2}$$

where

$$T = \frac{(\bar{Y}_i - \bar{Y}_j) - 0}{\sqrt{\text{MSE}(1/n_i + 1/n_j)}}$$

where n = total number of observations, k = number of means under consideration, n_i and n_j are the sizes of the samples selected from the ith and jth populations, respectively, $\bar{Y}_i$ and $\bar{Y}_j$ are the corresponding sample means, and MSE is the mean-square-error term with $n - k$ degrees of freedom that estimates the (homoscedastic) variance σ^2. Note that MSE is used instead of just a two-sample estimate of σ^2 based only on data from groups i and j; this is because MSE is a better estimate of σ^2 (in terms of degrees of freedom) under the assumption of variance homogeneity over all k populations.

Equivalently, one could reject H_0 if the $100(1 - \alpha)\%$ confidence interval

$$(\bar{Y}_i - \bar{Y}_j) \pm t_{n-k,1-\alpha/2} \sqrt{\text{MSE}\left(\frac{1}{n_i} - \frac{1}{n_j}\right)}$$

does not include 0.

Unfortunately, there is a serious drawback to performing several such t tests, because the more null hypotheses there are to be tested, the more likely it is to reject one of them even if all null hypotheses are actually true. In other words, if several such tests are made, each at the α level, then the probability of incorrectly rejecting *at least one* H_0 will be much larger than α and will increase with the number of tests made. Moreover, if in an exploratory study the investigator decides to compare only those sample means that are most discrepant (e.g., the largest versus the smallest), then the testing procedure would be biased in favor of rejecting H_0 because only those comparisons most likely to be significant would be made. This bias would be reflected in the fact that the true probability of falsely rejecting a given null hypothesis would exceed the α level specified for the test.

17-7-1 The LSD Approach

An approximate way to circumvent the problem of distorted significance levels when making several tests involves reducing the significance level used for each individual test sufficiently so that the *overall significance level* (i.e., the probability of falsely rejecting at least one of the null hypotheses being tested) is fixed at some desired value (say, α). Now, it can be shown that if one makes l such tests, the maximum possible value for this overall significance level is $l\alpha$. Thus, one simple way to ensure an overall significance level of at most α would be to use α/l as the significance level for each separate test; this approach is often referred to as the *least-significant-difference (LSD)* method. This is an application of the Bonferroni correction for multiple testing. For example, if all $l = _kC_2 = k(k - 1)/2$ pairwise comparisons of k population means were to be made, each test could be performed at the $\alpha/_kC_2$ significance level to ensure that the overall significance level would not exceed α.

TABLE 17-7   Potencies (dosages at death) of four cardiac substances

Sub-stance	Sample Size (n_i)	Dosage at Death (Y_{ij})	Total	Sample Mean ($\bar{Y}_i$)	Sample Variance (S_i^2)
1	10	29, 28, 23, 26, 26 19, 25, 29, 26, 28	259	25.9	9.4333
2	10	17, 25, 24, 19, 28 21, 20, 25, 19, 24	222	22.2	12.1778
3	10	17, 16, 21, 22, 23 18, 20, 17, 25, 21	200	20.0	8.6667
4	10	18, 20, 25, 24, 16 20, 20, 17, 19, 17	196	19.6	8.7111

A disadvantage to the LSD method, however, is that the *true* overall significance level may be so much less than the maximum value α that none of the individual tests are very likely to be rejected (i.e., the overall power of the method is low). Consequently, several more powerful procedures have been devised to provide an overall significance level of α. All these procedures are grouped under the heading "multiple-comparison procedures." We shall focus here on two such methods, one due to Tukey and the other to Scheffé. Discussions of these and other multiple-comparison methods can be found in Miller (1966), Guenther (1964), Lindman (1974), and Neter and Wasserman (1974).

Example   As an illustration of a multiple-comparison problem, let us consider the set of data given in Table 17-7, which was collected in an experiment designed to compare the relative potencies of four cardiac substances. In the experiment, a suitable dilution of one of the substances was slowly infused into an anesthetized guinea pig, and the dosage at which the pig died was recorded. Ten guinea pigs were used for each substance, and it was assumed that the laboratory environment and the measurement procedures were the same for each guinea pig. The main research goal was to determine whether there were any differences among the potencies of the four substances and, if so, to quantify those differences. The overall ANOVA table for comparing the mean potencies of the four cardiac substances is given in Table 17-8.

The global F test strongly rejects ($P < .001$) the null hypothesis of equality of the four population means. So the multiple-comparison question now arises: What is the best way to account for the differences found? As a crude first step, we can examine the nature of the differences with the help of a schematic diagram of ordered sample means (Figure 17-1). In the diagram an overbar has been drawn over the labels for substances 3 and 4 to indicate that the sample mean potencies for these two substances are quite similar. On the other hand, no

TABLE 17-8   ANOVA table for data of Table 17-7

Source	df	SS	MS	F
Substances	3	249.875	83.292	8.545 ($P < .001$)
Error	36	350.900	9.747	
Total	39	600.775		

FIGURE 17-1   Crude comparison of sample means for potency data

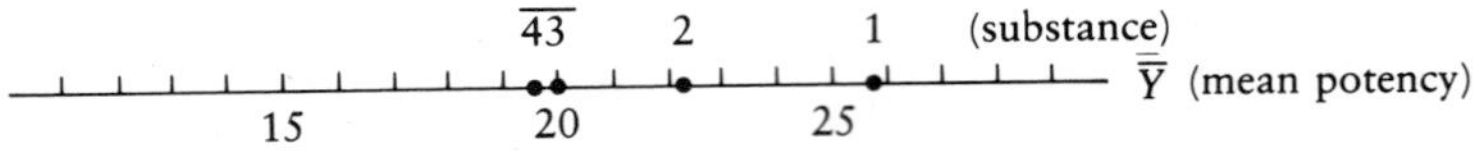

overbar has been drawn connecting 1 and 2 with each other or with 3 and 4, suggesting that substances 1 and 2 appear to be different from each other as well as from both 3 and 4.

Such an overall quantification of the differences among the population means is what is desired from a multiple-comparison analysis. Nevertheless, the purely descriptive approach above does not take into account the sampling variability associated with any estimated comparison of interest. As a result, two sample means that appear to be practically different may not, in fact, be statistically different. Since the only multiple-comparison method that we have discussed so far is the LSD method, let us now consider how we can apply this method to the data of Table 17-7, using an overall significance level of $\alpha = .05$ for all pairwise comparisons of the mean potencies of the four cardiac substances. This approach requires the computation of $_4C_2 = 6$ confidence intervals, each associated with a significance level of $\alpha/6 = .05/6 = .0083$, utilizing the formula

$$(\bar{Y}_i - \bar{Y}_j) \pm t_{36,1-0.0083/2} \sqrt{MSE \left(\frac{1}{10} + \frac{1}{10}\right)}$$

The right-hand side of this expression is calculated as follows:

$$t_{36,0.99585} \sqrt{9.747 \left(\frac{1}{5}\right)} = 2.72(1.396) = 3.798$$

Thus, the pairwise confidence intervals are given as follows:

$\mu_1 - \mu_4$: 6.3 ± 3.798; i.e., $(2.502, 10.098)^*$

$\mu_1 - \mu_3$: 5.9 ± 3.798; i.e., $(2.102, 9.698)^*$

$\mu_1 - \mu_2$: 3.7 ± 3.798; i.e., $(-0.098, 7.498)$

$\mu_2 - \mu_4$: 2.6 ± 3.798; i.e., $(-1.198, 6.398)$

$\mu_2 - \mu_3$: 2.2 ± 3.798; i.e., $(-1.598, 5.998)$

$\mu_3 - \mu_4$: 0.4 ± 3.798; i.e., $(-3.398, 4.198)$

(In this example, the term *least significant difference* refers to the fact that *any two* sample means are considered significantly different if their absolute difference exceeds 3.798. Of course, such a blanket statement cannot be made when the n_i's are not all equal.)

The intervals above indicate only two significant comparisons (starred) and translate into a diagrammatic overall ranking (Figure 17-2). These results are somewhat ambiguous since there are overlapping "sets of similarities," which indicate that substances 2, 3, and 4

FIGURE 17-2   LSD comparison of sample means for potency data

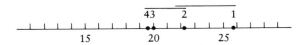

have essentially the same potency, that 1 and 2 have about the same potency, but that 1 differs from both 3 and 4. In other words, one conclusion that can be made here is that 2, 3, and 4 are to be grouped together and that 1 and 2 are to be grouped together, which is difficult to reconcile since substance 2 is common to both groups. Having to confront this ambiguity is actually quite fortuitous from a pedagogical standpoint, since such ambiguous results are not uncommon when carrying out multiple-comparison procedures. In our case the results indicate that the procedure used was not sensitive enough for evaluating substance 2. Repeating the analysis with a larger data set would help to clear up the ambiguity. Alternatively, since the LSD approach tends to be conservative (i.e., the confidence intervals tend to be wider than necessary to achieve the overall significance level desired), other multiple-comparison methods, such as those of Tukey or Scheffé, may provide more precise results (i.e., narrower confidence intervals) and so may reduce or eliminate any ambiguity.

17-7-2 Tukey's Method

Tukey's method is applicable when:

1. The sizes of the samples selected from each population are equal; that is, $n_i \equiv n^*$, say, for all $i = 1, 2, \ldots, k$, where k is the number of means being compared.

2. Pairwise comparisons of the means are of primary interest; that is, null hypotheses of the form $H_0: \mu_i = \mu_j$ are to be considered.

   (Actually, a generalized version of Tukey's procedure is available for considering more complex comparisons than just simple pairwise differences between means. However, since this procedure is most powerful when comparing simple differences between means and not when making more complex comparisons, Tukey's method should generally be used only in the former situation; the latter situation is best handled using Scheffé's method (discussed in Section 17-7-3).)

To use Tukey's method, the following confidence interval is computed for the pairwise comparison of population means μ_i and μ_j:

$$(\bar{Y}_i - \bar{Y}_j) \pm T \sqrt{\text{MSE}} \qquad (17.21)$$

where

$$T = \frac{1}{\sqrt{n^*}} q_{k, n-k, 1-\alpha}$$

and $q_{k, n-k, 1-\alpha}$ is the $100(1 - \alpha)\%$ point of the distribution of the studentized range with k and $n - k$ degrees of freedom (see Table A-6 in Appendix A), n^* is the common sample size (i.e., $n_i \equiv n^*$), k is the number of populations or groups, and $n \ (= kn^*)$ is the total number of observations.

   (The studentized range distribution with k and r degrees of freedom is the distribution of a statistic of the form R/S, where $R = \{\max_i (Y_i) - \min_i (Y_i)\}$ is the range of a set of k independent observations $Y_1, Y_2, \ldots, Y_k$ from a normal distribution with mean μ and variance σ^2, and S^2 is an estimate of σ^2 based on r degrees of freedom, which is independent of the Y's. In particular, when k means are being compared in fixed-effects one-way ANOVA, the statistic $\{\max_i (\bar{Y}_i) - \min_i (\bar{Y}_i)\}/\sqrt{\text{MSE}/n^*}$ has the studentized range distribution with k and $n - k$ degrees of freedom under H_0: $\mu_1 = \mu_2 = \cdots = \mu_k$, where $n_i = n^*$ for each i and $n = kn^*$.

An alternative to Tukey's procedure, which uses the studentized range distribution but with a modified numerator degrees of freedom, is called the *Student–Newman–Keuls* (*SNK*) method. The SNK procedure replaces the first k in $q_{k,n-k,1-\alpha}$ by $k^* =$ number of means in the range of means being tested. Thus, for comparing the second largest with the smallest of four means, we would have $k^* = 3$, whereas $k^* = 2$ when comparing the second largest with the third largest.

For unequal sample sizes, Steele and Torrie (1960) recommended a slight modification to the Tukey procedure. For each comparison involving unequal sample sizes, the term $T \sqrt{\text{MSE}}$ in (17.21) is replaced by $q_{k,n-k,1-\alpha} \sqrt{(\text{MSE}/2)(1/n_i + 1/n_j)}$, where n_i and n_j are the sample sizes associated with the ith and jth groups, respectively.)

In the set of all $_kC_2$ Tukey pairwise confidence intervals of the form (17.21), the probability is $1 - \alpha$ that these intervals simultaneously contain the associated population mean differences being estimated; that is, $1 - \alpha$ is the *overall confidence coefficient* for all pairwise comparisons taken together. In particular, if each confidence interval is used to test the appropriate pairwise hypothesis of the general form H_0: $\mu_i = \mu_j$ by determining whether the value 0 is contained in the calculated interval, the probability of falsely rejecting the null hypothesis for *at least one* of the $_kC_2$ tests is equal to α.

In applying Tukey's method to a set of data, the procedure is usually carried out stepwise as follows:

1. Rank-order the sample means $\bar{Y}_i$ from the largest to smallest (e.g., in our example, the order is $\bar{Y}_1 > \bar{Y}_2 > \bar{Y}_3 > \bar{Y}_4$).

2. Compare the largest sample mean with the smallest using (17.21), then the largest with the next smallest, and so on, until either the largest has been compared with the second largest or a nonsignificant result has been obtained, whichever comes first (e.g., in our example, we would first look at 1 versus 4, then 1 versus 3, and finally 1 versus 2).

3. Continue by comparing the second largest mean with the smallest, the second largest with the next smallest, and so on, but make no further comparisons with the second largest mean once any nonsignificant result is obtained.

4. Continue making such comparisons involving the third largest mean, then the fourth largest, and so on. At each stage, once a nonsignificant comparison is obtained, conclude that no difference exists between any means enclosed by the first nonsignificant pair.

5. Represent the overall conclusions about similarities and differences among the population means using a schematic diagram of the ordered sample means, with overbars drawn to connect those means that are not significantly different.

To illustrate Tukey's method using an overall significance level of $\alpha = .05$ for the potency data of Table 17-7, the ordering of sample means from largest to smallest indicates that the following sequence of pairwise comparisons should be made:

1 vs. 4,   1 vs. 3,   1 vs. 2,   2 vs. 4,   2 vs. 3,   3 vs. 4

Since the value of $T\sqrt{\text{MSE}}$ in (17.21) is needed for any such comparison, we compute this first as follows: For $n^* = 10$, $k = 4$, $n = 40$, and MSE $= 9.747$.

$$T = \frac{1}{\sqrt{n^*}}\, q_{k,n-k,1-\alpha} = \frac{1}{\sqrt{10}}\, q_{4,36,0.95} = \frac{1}{\sqrt{10}}\,(3.84) = 1.206$$

(the value $q_{4,36,0.95} = 3.84$ was obtained from Table A-6 by interpolation), and so

$$T\sqrt{\text{MSE}} = 1.206\sqrt{9.747} = 3.765$$

Now, using (17.21), we compare 1 and 4 as follows:

$$(\bar{Y}_1 - \bar{Y}_4) \pm T\sqrt{\text{MSE}}$$
$$(25.9 - 19.6) \pm 3.765$$
$$6.3 \pm 3.765$$

or

$$(2.535, 10.065)$$

Because this confidence interval does not contain the value 0, we can conclude (based on an overall significance level of $\alpha = .05$) that $\mu_1 \neq \mu_4$.

Next, we look at 1 versus 3 as follows:

$$(\bar{Y}_1 - \bar{Y}_3) \pm T\sqrt{\text{MSE}}$$
$$(25.9 - 20.0) \pm 3.765$$

or

$$(2.135, 9.665)$$

Because this confidence interval also does not contain the value 0, we can conclude (with overall significance level $\alpha = .05$) that $\mu_1 \neq \mu_3$.

Next, we compare 1 and 2, obtaining

$$3.7 \pm 3.765$$

or

$$(-0.065, 7.465)$$

which contains the value 0. We thus conclude that there is no evidence to reject H_0: $\mu_1 = \mu_2$ and also that all remaining comparisons, which involve smaller (in absolute value) pairwise mean differences, are nonsignificant. This conclusion supports the hypothesis that $\mu_2 = \mu_3 = \mu_4$. In summary, we may schematically represent the results based on applying Tukey's method as shown in Figure 17-3.

(In general, it is possible that a pairwise difference between, say, the largest mean and the second largest is not significant, even though there is another pairwise difference (say, between the third- and fourth-largest means) that is significant. Thus, in general, one should not stop making *all* remaining comparisons when the first nonsignificant pairwise difference is encountered unless all remaining pairwise differences are smaller than that involved in the first nonsignificant pair. As an example, if it was found that $\bar{Y}_1 = 100$, $\bar{Y}_2 = 99$, $\bar{Y}_3 = 80$, and $\bar{Y}_4 = 20$, then $\bar{Y}_1 - \bar{Y}_2 = 1$ would be quite small (and possibly nonsignificant) while $\bar{Y}_3 - \bar{Y}_4 = 60$ would be large.)

FIGURE 17-3   Tukey comparison of sample means for potency data

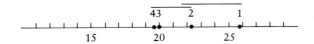

As with the LSD method, some ambiguity remains here, since substance 2 has again been associated with substance 1 and also with substances 3 and 4. Again, it should be noted that such an ambiguity is not uncommon. In this instance, the ambiguity suggests that the amount of data collected was not sufficient to clearly categorize substance 2.

17-7-3 Scheffé's Method

Scheffé's method is generally recommended when:

1. The sizes of the samples selected from the different populations are not all equal, or

2. comparisons other than simple pairwise comparisons between two means are of interest; these more general types of comparisons are referred to as *contrasts*.

To illustrate what we mean by a contrast, suppose the investigator who collected the potency data of Table 17-7 suspected that substances 1 and 3 had similar potencies, that substances 2 and 4 also had similar potencies, and that the potencies of 1 and 3, on the average, differed significantly from those of 2 and 4. Then it would be of interest to compare the average results obtained for 1 and 3 with the average results for 2 and 4 in order to assess whether $(\mu_1 + \mu_3)/2$ really differed from $(\mu_2 + \mu_4)/2$. In other words, one could consider the contrast

$$L_1 = \frac{\mu_1 + \mu_3}{2} - \frac{\mu_2 + \mu_4}{2}$$

which would be 0 if the null hypothesis $H_0: (\mu_1 + \mu_3)/2 = (\mu_2 + \mu_4)/2$ was true. Notice further that we can rewrite L_1 as follows:

$$L_1 = \frac{\mu_1 + \mu_3}{2} - \frac{\mu_2 + \mu_4}{2} = \frac{1}{2}\mu_1 - \frac{1}{2}\mu_2 + \frac{1}{2}\mu_3 - \frac{1}{2}\mu_4$$

or

$$L_1 = \sum_{i=1}^{4} c_{1i}\mu_i$$

so that L_1 is a *linear* function of the population means, with $c_{11} = \frac{1}{2}$, $c_{12} = -\frac{1}{2}$, $c_{13} = \frac{1}{2}$, and $c_{14} = -\frac{1}{2}$. Also, notice that

$$c_{11} + c_{12} + c_{13} + c_{14} = \tfrac{1}{2} - \tfrac{1}{2} + \tfrac{1}{2} - \tfrac{1}{2} = 0$$

In general, a *contrast* is defined to be any linear function of the population means, say

$$L = \sum_{i=1}^{k} c_i\mu_i$$

such that

$$\sum_{i=1}^{k} c_i = 0$$

The associated null hypothesis is

$$H_0: \sum_{i=1}^{k} c_i \mu_i = 0$$

and the two-tailed alternative hypothesis is

$$H_A: \sum_{i=1}^{k} c_i \mu_i \neq 0$$

As another example, the data in Table 17-7 suggest that the mean potency of substance 1 is definitely higher than the mean potencies of the other three substances. Such an observation invites a comparison of the mean potency of substance 1 with the average potency of substances 2, 3, and 4; in this case the appropriate contrast to consider would be

$$L_2 = \mu_1 - \frac{\mu_2 + \mu_3 + \mu_4}{3}$$

or

$$L_2 = \sum_{i=1}^{4} c_{2i} \mu_i$$

where $c_{21} = 1$, $c_{22} = c_{23} = c_{24} = -\frac{1}{3}$. (Notice again that $\sum_{i=1}^{4} c_{2i} = 0$.)

A third possible contrast of interest is a comparison of the average results for substances 1 and 2 with those for 3 and 4. The contrast here would then be

$$L_3 = \frac{\mu_1 + \mu_2}{2} - \frac{\mu_3 + \mu_4}{2} = \sum_{i=1}^{4} c_{3i} \mu_i$$

where $c_{31} = c_{32} = \frac{1}{2}$ and $c_{33} = c_{34} = -\frac{1}{2}$.

Finally, any pairwise comparison is also a contrast, since, for example, a comparison of 1 with 4 takes the form

$$L_4 = \mu_1 - \mu_4 = \sum_{i=1}^{4} c_{4i} \mu_i$$

where $c_{41} = 1$, $c_{42} = c_{43} = 0$, and $c_{44} = -1$.

Scheffé's method provides a family of confidence intervals for evaluating *all possible contrasts* that can be defined given a fixed number k of population means, such that the overall confidence coefficient associated with the entire family is $1 - \alpha$, where α is specified by the investigator. In other words, the probability is $1 - \alpha$ that these confidence intervals simultaneously contain the true values of all the contrasts being considered. We can further say that the overall significance level is α for testing hypotheses of the general form $H_0: L = \sum_{i=1}^{k} c_i \mu_i = 0$ concerning all possible contrasts; that is, the probability is α that at least one such null hypothesis will falsely be rejected.

The general form of a Scheffé-type confidence interval is as follows. Let $L = \sum_{i=1}^{k} c_i \mu_i$ be some contrast of interest. Then the appropriate confidence interval concerning L is given by

$$\sum_{i=1}^{k} c_i \bar{Y}_i \pm S \sqrt{\text{MSE} \left(\sum_{i=1}^{k} \frac{c_i^2}{n_i} \right)} \qquad (17.22)$$

where $\hat{L} = \sum_{i=1}^{k} c_i \bar{Y}_i$ is the unbiased point estimator of L and where $S^2 = (k-1) F_{k-1, n-k, 1-\alpha}$ with $n = \sum_{i=1}^{k} n_i$.

As a special case, when one is concerned only with pairwise comparisons, this formula simplifies to

$$(\bar{Y}_i - \bar{Y}_j) \pm S \sqrt{\text{MSE} \left(\frac{1}{n_i} + \frac{1}{n_j} \right)} \qquad (17.23)$$

when considering inferences regarding $\mu_i - \mu_j$.

It is important to point out that if the investigator is interested only in pairwise comparisons, then Scheffé's method using (17.23) is not recommended because Tukey's method will always provide narrower confidence intervals (i.e., will give more precise estimates of the true pairwise differences). However, if the sample sizes are unequal and/or if contrasts other than pairwise comparisons are of interest, then Scheffé's method is to be preferred. Furthermore, Scheffé's method has the desirable property that whenever the overall F test of the null hypothesis that all k population means are equal is rejected, then at least one contrast will be found that differs significantly from 0. Tukey's method, on the other hand, may not turn up any significant pairwise comparisons even when the overall F statistic is significant.

To illustrate the use of Scheffé's method, let us first consider all pairwise comparisons for the data of Table 17-7 and follow the same procedure as used with Tukey's method; that is, we shall consider in order 1 versus 4, 1 versus 3, 1 versus 2, 2 versus 4, 2 versus 3, and 3 versus 4. We begin by first computing the quantity

$$S \sqrt{\text{MSE} \left(\frac{1}{n_i} + \frac{1}{n_j} \right)}$$

which will have the same value for all pairwise comparisons since all the sample sizes are equal to 10. So

$$\sqrt{\text{MSE} \left(\frac{1}{n_i} + \frac{1}{n_j} \right)} = \sqrt{9.747 \left(\frac{1}{10} + \frac{1}{10} \right)} = 1.3962$$

And, with $k = 4$, $n = 40$, and $\alpha = .05$, we have

$$S = \sqrt{(k-1) F_{k-1, n-k, 1-\alpha}} = \sqrt{3 F_{3, 36, 0.95}} = \sqrt{3(2.886)} = 2.9424$$

Thus, $S \sqrt{\text{MSE}(1/n_i + 1/n_j)} = 2.9424(1.3962) = 4.1082$. Now, to compare substances 1 and 4, we obtain the following confidence interval:

$$(\bar{Y}_1 - \bar{Y}_4) \pm S \sqrt{\text{MSE} \left(\frac{1}{n_1} + \frac{1}{n_4} \right)}$$

$$(25.9 - 19.6) \pm 4.1082$$

or

$$(2.192, 10.408)$$

Since this interval does not contain the value 0 (which was also the case for the corresponding Tukey interval), we would support the contention that $\mu_1 \neq \mu_4$ and so proceed with the 1 versus 3 comparison:

$$(\bar{Y}_1 - \bar{Y}_3) \pm S \sqrt{\text{MSE}\left(\frac{1}{n_1} + \frac{1}{n_3}\right)}$$

$$(25.9 - 20.0) \pm 4.1082$$

or

$$(1.792, 10.008)$$

Like the corresponding interval obtained with Tukey's method, this interval does not contain the value 0, which supports the rejection of H_0: $\mu_1 = \mu_3$.

The next comparison, between substances 1 and 2, yields the interval

$$(-0.408, 7.808)$$

which contains the value 0. Thus, we cannot reject H_0: $\mu_1 = \mu_2$, and noting that the remaining pairwise comparisons would be nonsignificant, we would favor both the conclusion that $\mu_2 = \mu_3 = \mu_4$ and that $\mu_1 = \mu_2$. Thus, when considering only pairwise comparisons for this data set, Tukey's and Scheffé's methods yield the same general conclusions regarding the relative potencies of the four substances, including the ambiguity associated with substance 2. However, on closer inspection of the pairwise intervals derived (see Table 17-9), it can be seen that the Tukey intervals are narrower and so provide more precise inferences.

Let us now illustrate how to use Scheffé's method to make inferences regarding two contrasts of interest. In particular, we shall consider

$$L_2 = \mu_1 - \frac{\mu_2 + \mu_3 + \mu_4}{3} = \sum_{i=1}^{4} c_{2i}\mu_i$$

where $c_{21} = 1$ and $c_{22} = c_{23} = c_{24} = -\frac{1}{3}$, and

$$L_3 = \frac{\mu_1 + \mu_2}{2} - \frac{\mu_3 + \mu_4}{2} = \sum_{i=1}^{4} c_{3i}\mu_i$$

where $c_{31} = c_{32} = \frac{1}{2}$ and $c_{33} = c_{34} = -\frac{1}{2}$.

TABLE 17-9   Comparison of some Tukey and Scheffé confidence intervals for the potency data of Table 17-7

Pairwise Comparison	Tukey		Scheffé	
	Lower Limit	Upper Limit	Lower Limit	Upper Limit
$\mu_1 - \mu_4$	2.535	10.065	2.192	10.408
$\mu_1 - \mu_3$	2.135	9.665	1.792	10.008
$\mu_1 - \mu_2$	-0.065	7.465	-0.408	7.808
$\mu_2 - \mu_4$	-1.165	6.365	-1.508	6.708
$\mu_2 - \mu_3$	-1.565	5.965	-1.908	6.308
$\mu_3 - \mu_4$	-3.365	4.165	-3.708	4.508

Using our previously computed value $S = \sqrt{(k-1)F_{k-1,n-k,1-\alpha}} = 2.9424$, we calculate from (17.22) as follows:

$$\left(\bar{Y}_1 - \frac{\bar{Y}_2 + \bar{Y}_3 + \bar{Y}_4}{3}\right) \pm S \sqrt{\text{MSE}\left[\frac{(1)^2}{10} + \frac{(-1/3)^2}{10} + \frac{(-1/3)^2}{10} + \frac{(-1/3)^2}{10}\right]}$$

$$\left(25.9 - \frac{22.2 + 20.0 + 19.6}{3}\right) \pm 2.9424 \sqrt{9.747\left(\frac{12}{90}\right)}$$

$$(25.9 - 20.6) \pm 2.9424(1.1400)$$

$$5.3 \pm 3.354$$

or

$$1.946 \le L_2 \le 8.654$$

Since this interval does not contain the value 0, we have evidence that the potency of substance 1 is different from the average potency of substances 2, 3, and 4.

Next we calculate the following Scheffé interval regarding L_3:

$$\left(\frac{\bar{Y}_1 + \bar{Y}_2}{2} - \frac{\bar{Y}_3 + \bar{Y}_4}{2}\right) \pm S \sqrt{\text{MSE}\left[\frac{(1/2)^2}{10} + \frac{(1/2)^2}{10} + \frac{(-1/2)^2}{10} + \frac{(-1/2)^2}{10}\right]}$$

$$\left(\frac{25.9 + 22.2}{2} - \frac{20.0 + 19.6}{2}\right) \pm 2.9424 \sqrt{9.747\left(\frac{1}{10}\right)}$$

$$(24.05 - 19.80) \pm 2.9424(0.9873)$$

$$4.24 \pm 2.9049$$

or

$$(1.345, 7.155)$$

Because this interval also does not contain the value 0, we would be inclined to support the conclusion that the average potency of substances 1 and 2 is different from the average potency of substances 3 and 4.

How do the above results concerning the contrasts L_2 and L_3 help in clearing up the ambiguity created by considering all pairwise comparisons? First, it is obvious that uncertainty regarding substance 2 still remains. Nevertheless, we should consider the fact that the confidence interval for comparing 1 with the average of 2, 3, and 4 (i.e., 1.946 to 8.654) is farther away from 0 than is the confidence interval for comparing the average of 1 and 2 with the average of 3 and 4 (i.e., 1.345 to 7.155), suggesting that substance 2 is closer to 3 and 4 in potency than it is to substance 1. The pairwise comparisons also support this contention, since the confidence interval for $\mu_1 - \mu_2$ is farther from 0 than is the interval for $\mu_2 - \mu_3$, indicating again that 2 is closer to 3 than to 1. Thus, if a definite decision regarding substance 2 was required for this set of data, the most logical thing to do would be to consider the potency of 1 as distinct from the potencies of 2, 3, and 4, which, as a threesome, are too similar to separate. Schematically, this conclusion is represented in Figure 17-4.

FIGURE 17-4   Conclusion regarding sample means for potency data

17-8 Choosing a Multiple-Comparison Technique

Figure 17-5 summarizes a suggested strategy for choosing a multiple-comparison technique for ANOVA. The first choice is to determine whether pairwise or nonpairwise comparisons are desired. If only pairwise comparisons are being considered and all cell-specific sample sizes are equal, one should use the Tukey method. Otherwise, the less sensitive Scheffé method must be used.

If any nonpairwise comparisons are to be considered, then a choice must be made based on whether the comparisons are planned (a priori) or unplanned. Since the Bonferroni method should be used for planned comparisons, any unplanned comparisons should be evaluated using the Scheffé method.

Although we have discussed only the Bonferroni (LSD), Tukey, and Scheffé methods, a large number of other multiple-comparison techniques have been suggested in the statistical literature. Miller (1981) provides the most comprehensive review of these procedures. The methods differ as to the target set of contrasts (e.g., pairwise versus nonpairwise), the cell-specific sample sizes, and other properties. The developers of such methods have sought to minimize the Type II error rate (i.e., to maximize power; see Chapter 3) for a particular set of comparisons while controlling the Type I error rate.

The reasons for considering other multiple-comparison methods are to improve power and to achieve robustness. As stated earlier in this chapter, a multiple-comparison procedure is robust if a violation of assumptions, such as nonhomogeneity of variance, does not seriously affect the validity of an analysis. That is, if a robust procedure is used, then even if some assumptions are violated, Type I and Type II error rates from statistical testing are not likely to be compromised. Both the Bonferroni and Scheffé methods tend to be fairly robust,

FIGURE 17-5   Recommended procedure for choosing a multiple-comparison technique in ANOVA

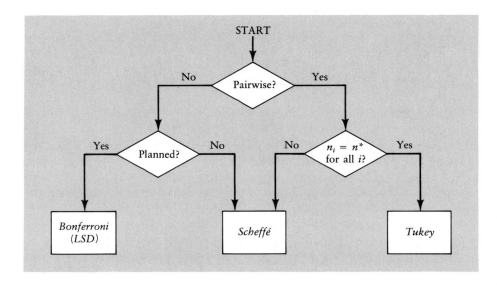

while the Tukey method is less so. Scheffé's method is always the least powerful, since all possible comparisons are considered. The LSD method has generally been presumed to be insensitive (i.e., to have low power). This presumption can be discounted if the sets of comparisons are well planned.

We mention these issues to make the reader aware that many acceptable multiple-comparison procedures are available, but that the range of application of most are limited in some way (e.g., Tukey's method requires equal cell-specific sample sizes). Both the Bonferroni and Scheffé procedures are completely general methods, the former for planned (a priori) and the latter for unplanned (a posteriori) multiple comparisons. To cover all multiple-comparison methods completely is beyond the scope of this book.

17-9 Orthogonal Contrasts and Partitioning an ANOVA Sum of Squares

From our previous discussions of multiple regression, we are familiar with the notion of a partitioned sum of squares in regression analysis, where the sum of squares due to regression (SSR) is broken down into various components reflecting the relative contributions of various terms in the fitted model. In an ANOVA framework, it is possible via the use of orthogonal contrasts to partition SST, the treatment sum of squares, into meaningful components associated with certain specific comparisons of particular interest. To illustrate how such a partitioning can be accomplished, we need to discuss two new concepts, *orthogonal contrasts* and the *sum of squares associated with a contrast*.

In the notation of Section 17-7, two estimated contrasts

$$\hat{L}_A = \sum_{i=1}^{k} c_{Ai} \bar{Y}_i \quad \text{and} \quad \hat{L}_B = \sum_{i=1}^{k} c_{Bi} \bar{Y}_i$$

are said to be *orthogonal* to one another (i.e., are orthogonal contrasts) if

$$\sum_{i=1}^{k} \frac{c_{Ai} c_{Bi}}{n_i} = 0 \tag{17.24}$$

In the special case where the n_i's are equal, then (17.24) reduces to the condition

$$\sum_{i=1}^{k} c_{Ai} c_{Bi} = 0 \tag{17.25}$$

As an illustration, consider the three contrasts discussed earlier with regard to the potency data of Table 17.7:

$$\hat{L}_1 = \frac{\bar{Y}_1 + \bar{Y}_3}{2} - \frac{\bar{Y}_2 + \bar{Y}_4}{2} = \frac{1}{2} \bar{Y}_1 - \frac{1}{2} \bar{Y}_2 + \frac{1}{2} \bar{Y}_3 - \frac{1}{2} \bar{Y}_4$$

where $c_{11} = c_{13} = \frac{1}{2}$ and $c_{12} = c_{14} = -\frac{1}{2}$,

$$\hat{L}_2 = \bar{Y}_1 - \frac{1}{3}(\bar{Y}_2 + \bar{Y}_3 + \bar{Y}_4) = \bar{Y}_1 - \frac{1}{3} \bar{Y}_2 - \frac{1}{3} \bar{Y}_3 - \frac{1}{3} \bar{Y}_4$$

where $c_{21} = 1$ and $c_{22} = c_{23} = c_{24} = -\frac{1}{3}$, and

$$\hat{L}_3 = \frac{\bar{Y}_1 + \bar{Y}_2}{2} - \frac{\bar{Y}_3 + \bar{Y}_4}{2} = \frac{1}{2}\bar{Y}_1 + \frac{1}{2}\bar{Y}_2 - \frac{1}{2}\bar{Y}_3 - \frac{1}{2}\bar{Y}_4$$

where $c_{31} = c_{32} = \frac{1}{2}$ and $c_{33} = c_{34} = -\frac{1}{2}$. Since $n_i = 10$ for every i, we need only verify that condition (17.25) holds to demonstrate orthogonality. In particular, we have, for $\hat{L}_1$ and $\hat{L}_2$,

$$\sum_{i=1}^{4} c_{1i}c_{2i} = \left(\frac{1}{2}\right)(1) + \left(-\frac{1}{2}\right)\left(-\frac{1}{3}\right) + \left(\frac{1}{2}\right)\left(-\frac{1}{3}\right) + \left(-\frac{1}{2}\right)\left(-\frac{1}{3}\right) = \frac{2}{3} \neq 0$$

For $\hat{L}_1$ and $\hat{L}_3$,

$$\sum_{i=1}^{4} c_{1i}c_{3i} = \left(\frac{1}{2}\right)\left(\frac{1}{2}\right) + \left(-\frac{1}{2}\right)\left(\frac{1}{2}\right) + \left(\frac{1}{2}\right)\left(-\frac{1}{2}\right) + \left(-\frac{1}{2}\right)\left(-\frac{1}{2}\right) = 0$$

And, for $\hat{L}_2$ and $\hat{L}_3$,

$$\sum_{i=1}^{4} c_{2i}c_{3i} = (1)\left(\frac{1}{2}\right) + \left(-\frac{1}{3}\right)\left(\frac{1}{2}\right) + \left(-\frac{1}{3}\right)\left(-\frac{1}{2}\right) + \left(-\frac{1}{3}\right)\left(-\frac{1}{2}\right) = \frac{2}{3} \neq 0$$

Thus, we conclude that $\hat{L}_1$ and $\hat{L}_3$ are orthogonal to one another but that neither is orthogonal to $\hat{L}_2$.

Orthogonality is a desirable property for the following reasons. Suppose that SST denotes the treatment sum of squares with $k - 1$ degrees of freedom for a fixed-effects one-way ANOVA, and suppose that $\hat{L}_1, \hat{L}_2, \ldots, \hat{L}_t$ are a set of t ($\leq k - 1$) mutually orthogonal contrasts of the k sample means (by mutually orthogonal, we mean that any two contrasts selected from the set of t contrasts are orthogonal to one another). Then it can be shown that SST can be partitioned into $t + 1$ statistically independent sums of squares, t of these sums of squares having 1 degree of freedom each and being associated with the t orthogonal contrasts, and the remaining sum of squares having $k - 1 - t$ degrees of freedom and being associated with what is left over after accounting for the t orthogonal contrast sums of squares. In other words, we can write

$$\text{SST} = \text{SS}(\hat{L}_1) + \text{SS}(\hat{L}_2) + \cdots + \text{SS}(\hat{L}_t) + \text{SS}(\text{remainder})$$

where $\text{SS}(\hat{L})$ is the notation for the sum of squares (with 1 degree of freedom) associated with the contrast $\hat{L}$. In particular, it can be shown that

$$\text{SS}(\hat{L}) = \frac{(\hat{L})^2}{\sum_{i=1}^{k}(c_i^2/n_i)} \qquad \text{when} \qquad \hat{L} = \sum_{i=1}^{k} c_i\bar{Y}_i \qquad (17.26)$$

and that

$$\frac{\text{SS}(\hat{L})}{\text{MSE}} \sim F_{1,n-k}$$

under $H_0: L = \sum_{i=1}^{k} c_i\mu_i = 0$

and, in general, that

$$\frac{[SS(\hat{L}_1) + SS(\hat{L}_2) + \cdots + SS(\hat{L}_t)]/t}{MSE} \sim F_{t,n-k}$$

under $H_0: L_1 = L_2 = \cdots = L_t = 0$ when $\hat{L}_1, \hat{L}_2, \ldots, \hat{L}_t$ are mutually orthogonal. Thus, by partitioning SST as above, we can test hypotheses concerning sets of orthogonal contrasts that are of more specific interest than the global hypothesis $H_0: \mu_1 = \mu_2 = \cdots = \mu_k$.

For example, to test $H_0: L_2 = \mu_1 - \frac{1}{3}(\mu_2 + \mu_3 + \mu_4) = 0$ for the potency data in Table 17-7, we first calculate, using (17.26),

$$SS(\hat{L}_2) = \frac{(5.30)^2}{[(1)^2 + (-1/3)^2 + (-1/3)^2 + (-1/3)^2]/10} = 210.675$$

and then form the ratio

$$\frac{SS(\hat{L}_2)}{MSE} = \frac{210.675}{9.747} = 21.614$$

which is highly significant ($P < .001$ based on the $F_{1,36}$ distribution). A modification of Table 17-8 to reflect this partitioning of SST (the sum of squares for "substances") gives

Source		df	SS	MS	F
Substances $\begin{cases} \\ \end{cases}$	1 vs. (2, 3, 4)	1	210.675	210.675	21.614 ($P < .001$)
	Remainder	2	39.200	Irrelevant	——
Error		36	350.900	9.747	
Total		39	600.775		

Similarly, the partitioned ANOVA table for testing

$$H_0: L_3 = \frac{\mu_1 + \mu_2}{2} - \frac{\mu_3 + \mu_4}{2} = 0$$

is as follows:

Source		df	SS	MS	F
Substances $\begin{cases} \\ \end{cases}$	(1, 2) vs. (3, 4)	1	179.776	179.776	18.444 ($P < .001$)
	Remainder	2	70.099	Irrelevant	——
Error		36	350.900	9.747	
Total		39	600.775		

This partition follows from the fact that

$$SS(\hat{L}_3) = \frac{(4.24)^2}{[(1/2)^2 + (1/2)^2 + (-1/2)^2 + (-1/2)^2]/10} = 179.776$$

Finally, since $\hat{L}_1$ and $\hat{L}_3$ are orthogonal, we can represent the independent contributions of these two contrasts to SST in one ANOVA table, as follows:

Source		df	SS	MS	F
Substances $\begin{cases} \\ \\ \\ \end{cases}$	(1, 3) vs. (2, 4)	1	42.025	42.025	4.312 (n.s.)
	(1, 2) vs. (3, 4)	1	179.776	179.776	18.444 ($P < .001$)
	Remainder	1	28.074	Irrelevant	——
Error		36	350.900	9.747	
Total		39	600.775		

Here

$$SS(\hat{L}_1) = \frac{\left(\dfrac{25.9 + 20.0}{2} - \dfrac{22.2 + 19.6}{2}\right)^2}{[(1/2)^2 + (-1/2)^2 + (1/2)^2 + (-1/2)^2]/10} = 42.025$$

It is important to point out that it would not be valid to present a partitioned ANOVA table simultaneously including partitions due to $\hat{L}_2$ and also to $\hat{L}_1$, $\hat{L}_3$, or both. This is because $\hat{L}_2$ is not orthogonal to these contrasts, and so its sum of squares does not represent a separate and independent contribution to SST; this can easily be seen by noting from the above tables that

$$SST \neq SS(\hat{L}_1) + SS(\hat{L}_2) + SS(\hat{L}_3)$$

A final example of using orthogonal contrasts involves the frequent need to assess whether the sample means exhibit a trend of some sort; this need often arises where the treatments (or populations) being studied represent, for example, different levels of the same factor (e.g., different concentrations of the same material or different temperature or pressure settings). In such situations it is of interest to be able to quantify how the sample means vary with changes in the level of the factor, that is, if the change in mean response takes place in a linear, quadratic, or other way as the level of the factor is increased or decreased.

A qualitative first step in assessing such a trend is to plot the observed treatment means as a function of the factor levels. This may yield some general idea of any pattern that is present. Although standard regression techniques can then be used to quantify any trends suggested by such plots, our goal here is not to fit a regression model but rather to evaluate a possible general trend in the means in a statistical manner rather than to form an opinion by simply examining a plot of the means.

(In a standard regression approach, the independent variable ("treatments") would be treated as an interval variable and the actual value of the variable at each treatment setting would be used. To test for a linear trend using regression, therefore, the appropriate model would be $Y = \beta_0 + \beta_1 X + E$, where X denotes the (interval) treatment variable. Using this model, the test for linear trend is the usual test for zero slope. This test is, in general, not equivalent to the test for a linear trend in *mean* response using the orthogonal polynomials described in this section except when the pure error mean square is used in place of the residual mean square in the denominator of the F statistic for testing for zero slope; this is because the regression pure error mean square and the one-way ANOVA error mean square are identical. It also follows that the usual test for lack of fit of the straight-line regression model is equivalent to the test for a nonlinear trend in mean response discussed in this section. (See Problem 10 at the end of the chapter for further considerations in this regard.))

A statistical trend analysis may be carried out by determining how much of the sum of squares due to treatments (SST) is associated with each of the terms (linear, quadratic, cubic, etc.) in a polynomial regression. If the various levels of the treatment or factor being studied are equally spaced, this determination is best carried out using the method of orthogonal polynomials (for a discussion, see Armitage, 1971, and also Chapter 13).

To illustrate the use of orthogonal polynomials, let us again turn to the potency data of Table 17-7. Further, let us suppose that the four substances actually represent four equally spaced concentrations of some toxic material, with substance 1 being the least concentrated solution and substance 4 the most concentrated. For example, substance 1 might represent a 10% solution of the toxic material, substance 2 a 20% solution, substance 3 a 30% solution, and substance 4 a 40% solution. In this case, a plot of the four sample means versus concentration would take the form shown in Figure 17-6. This plot suggests at least a linear and possibly a quadratic relationship between "concentration" and "response," and this general impression can be quantified via the use of orthogonal polynomials. In particular, since there are $k = 4$ sample means, it is possible to fit up to a third-order ($k - 1 = 3$) polynomial to these means; this cubic model (with four terms) would pass through all four points on the graph just given, thus explaining all the variation in the four sample means (or, equivalently, in SST). Also, because the concentrations are equally spaced, it is possible via the use of orthogonal polynomials to define three orthogonal contrasts of the four sample means, one (say, $\hat{L}_l$) measuring the strength of the linear component of the third-degree polynomial, one (say, $\hat{L}_q$) the quadratic component contribution, and one (say, $\hat{L}_c$) the cubic component effect. The sums of squares associated with these three orthogonal contrasts each have 1 degree of freedom, are statistically independent, and satisfy the relationship

$$\text{SST} = \text{SS}(\hat{L}_l) + \text{SS}(\hat{L}_q) + \text{SS}(\hat{L}_c)$$

Without further explanation here, it can be shown that the coefficients of these three orthogonal contrasts are as follows:

Contrast	\multicolumn{4}{c}{Coefficient of}	Calculated Value of Contrast
	$\bar{Y}_1 = 25.9$	$\bar{Y}_2 = 22.2$	$\bar{Y}_3 = 20.0$	$\bar{Y}_4 = 19.6$	
$\hat{L}_l$	-3	-1	1	3	-21.1
$\hat{L}_q$	1	-1	-1	1	3.3
$\hat{L}_c$	-1	3	-3	1	0.3

It should be clear that condition (17.25) holds for these three sets of coefficients, which is sufficient for mutual orthogonality here since the n_i's are all equal. These coefficients were taken from Table A-7 in Appendix A. The table may be used only for equally spaced treatment values and equal cell-specific sample sizes. Kirk (1969) summarized an algorithm for the general case not requiring these two restrictions. More conveniently, computer software can be used to obtain orthogonal contrast coefficients.

Now, from (17.26), the sums of squares for these three particular contrasts are

$$\text{SS}(\hat{L}_l) = \frac{(-21.1)^2}{[(-3)^2 + (-1)^2 + (1)^2 + (3)^2]/10} = 222.605$$

FIGURE 17-6 Plot of the sample means versus concentration

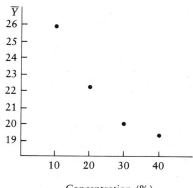

$$SS(\hat{L}_q) = \frac{(3.3)^2}{[(1)^2 + (-1)^2 + (-1)^2 + (1)^2]/10} = 27.225$$

$$SS(\hat{L}_c) = \frac{(0.3)^2}{[(-1)^2 + (3)^2 + (-3)^2 + (1)^2]/10} = 0.045$$

Note that

$$SST = 249.875 = 222.605 + 27.225 + 0.045$$

Finally, to assess the significance of these sums of squares, we form the following partitioned ANOVA table:

Source		df	SS	MS	F
Substances	$\hat{L}_l$	1	222.605	222.605	22.838 $(P < .001)$
	$\hat{L}_q$	1	27.225	27.225	2.793 $(.1 < P < .25)$
	$\hat{L}_c$	1	0.045	0.045	<1 (n.s.)
Error		36	350.900	9.747	
Total		39	600.775		

In summary, then, it is clear from the F tests above that the relationship between potency (as measured by the amount of injected material needed to cause death) and concentration is a strong linear one, with no real evidence of higher-order effects.

Problems

1. Five treatments for fever blisters, including a placebo, were randomly assigned to 30 patients. The following data give for each treatment the number of days from initial appearance of the blisters until healing is complete.

Treatment	No. of Days
Placebo (1)	5, 8, 7, 7, 10, 8
2	4, 6, 6, 3, 5, 6
3	6, 4, 4, 5, 4, 3
4	7, 4, 6, 6, 3, 5
5	9, 3, 5, 7, 7, 6

a. Compute the sample means and standard deviations for each treatment.
b. Determine the appropriate ANOVA table for the data given.
c. Are there significant differences among the effects of the five treatments with regard to the healing of fever blisters? In other words, test H_0: $\mu_1 = \mu_2 = \mu_3 = \mu_4 = \mu_5$ against H_A: "at least two treatments have different population means."
d. What are the estimates of the true effects $(\mu_i - \mu)$ of the treatments? Verify that the sum of these estimated effects is 0. (*Note*: μ_i is the population mean for the ith treatment and $\mu = \frac{1}{5}\sum_{i=1}^{5} \mu_i$ is the overall population mean.)
e. Using dummy variables, give an appropriate regression model that describes this experiment. Give two possible ways to define these dummy variables (one way using 0's and 1's and the other way using 1's and -1's), and describe for each of these coding schemes how the regression coefficients are related to the population means μ_1, μ_2, μ_3, μ_4, μ_5, and μ.
f. Carry out the Scheffé, Tukey, and LSD multiple-comparison procedures concerning pairwise differences between means for the data of this problem as described in Section 17.7. Also, compare the widths of the confidence intervals obtained by the three procedures.

2. The following data are replicate measurements of the sulfur dioxide concentration in each of three cities:

     City I: 2, 1, 3

     City II: 4, 6, 8

     City III: 2, 5, 2

a. Determine the appropriate ANOVA table for simultaneously comparing the mean sulfur dioxide concentrations in the three cities.
b. Test whether the three cities differ significantly in mean sulfur dioxide concentration levels.
c. What is the estimated effect associated with each city?
d. State precisely the appropriate ANOVA fixed-effects model for these data.
e. State precisely, using dummy variables, the regression model that corresponds to the fixed-effects model in part (d). What is the relationship between the coefficients in the regression model and the effects in the ANOVA model?
f. Using the t distribution, find a 90% confidence interval for the true difference between the effects of cities I and II (making sure to use the best estimate of σ^2 provided by the data).

3. Each of three chemical laboratories performed four replicate determinations of the concentration of suspended particulate matter in a certain area using the "Hi-Vol" method of analysis. The data are presented next.

Lab I	Lab II	Lab III
4	2	5
4	2	2
6	5	5
10	3	8

a. Determine the appropriate ANOVA table for these data.

b. Test the null hypothesis of no differences among the laboratories.

c. With large-scale interlaboratory studies, one usually makes inferences about a large population of laboratories of which only a random sample of labs (e.g., laboratories I, II, and III) is available for investigation. In such a case, describe the appropriate random-effects model for the data.

d. Two quantities of particular interest in a large-scale interlaboratory study are repeatability (i.e., a measure of the variability among replicate measurements within a single laboratory) and reproducibility (i.e., a measure of the variability between results from different laboratories). Using the random-effects model defined in part (c), define what you think are reasonable measures of repeatability and reproducibility, and obtain estimates of the quantities you have defined using the available data.

4. Ten randomly selected mental institutions were examined to determine the effects of three different antipsychotic drugs on patients with the same types of symptoms. Each institution used one and only one of the three drugs exclusively for a 1-year period. The proportion of treated patients in each institution who were discharged after 1 year of treatment is as follows for each drug used:

Drug 1: 0.10, 0.12, 0.08, 0.14 $(\bar{Y}_1 = 0.11, S_1 = 0.0192)$

Drug 2: 0.12, 0.14, 0.19 $(\bar{Y}_2 = 0.15, S_2 = 0.0361)$

Drug 3: 0.20, 0.25, 0.15 $(\bar{Y}_3 = 0.20, S_3 = 0.0500)$

a. Determine the appropriate ANOVA table for this data set.

b. Test to see if there are significant differences among drugs with regard to the average proportion of patients discharged.

c. What other factors should be considered when comparing the effects of the three drugs?

d. What basic ANOVA assumptions might be violated here?

5. Suppose that a random sample of five active members in each of four political parties in a certain western European country was given a questionnaire purported to measure (on a 100-point scale) the extent of "general authoritarian attitude toward interpersonal relationships." The means and standard deviations of the authoritarianism scores for each party are given in the table below.

	Party 1	Party 2	Party 3	Party 4
$\bar{Y}_i$	85	80	95	50
S_i	6	7	4	10
n_i	5	5	5	5

a. Determine the appropriate ANOVA table for this data set.
b. Test to see if there are significant differences among parties with respect to mean authoritarianism scores.
c. State, using dummy variables, an appropriate regression model for the above experimental situation.
d. Apply Tukey's method of multiple comparisons to determine those pairs in which the means are significantly different from one another. (Use $\alpha = .05$.)

6. A psychosociological questionnaire was administered to a random sample of 200 persons on an island in the South Pacific that has been increasingly westernized over the past 30 years. From the questionnaire data, each of the 200 persons was classified into one of three groups according to the discrepancy between the amount of prestige in that person's traditional culture and the amount of prestige in the modern (westernized) culture. The three groups were called HI-POS, NO-DIF, and HI-NEG. Also, using the questionnaire data, a measure of "anomie" (i.e., social disorientation), denoted as Y, was determined on a 100-point scale, with the summarized results given in the table below.

Group	n_i	$\bar{Y}_i$	S_i
HI-POS	50	65	9
NO-DIF	75	50	11
HI-NEG	75	55	10

a. Determine the appropriate ANOVA table.
b. Test whether the three different categories of prestige discrepancy have significantly different sample mean "anomie" scores.
c. How would you test whether there is a significant difference between the NO-DIF category and the other two categories combined?

7. To determine whether reading skills of the average high school graduate have declined over the past 10 years, the average verbal college aptitude scores (VSAT) were compared for a random sample of five big-city high schools for the years 1965, 1970, and 1975. The results are as follows:

Year	School 1	School 2	School 3	School 4	School 5
1965	550	560	535	545	555
1970	545	560	528	532	541
1975	536	552	526	527	530

a. Determine the sample means for each year.
b. Comment on the independence assumption required for using one-way ANOVA on the data given.
c. Determine the one-way ANOVA table for these data.
d. Test by means of one-way ANOVA whether there are any significant differences among the three VSAT average scores.
e. Use Scheffé's method to locate any significant differences between pairs of means. (Use $\alpha = .05$.)

8. Three persons (denoted A, B, and C) claiming to have unusual psychic ability underwent ESP tests at an eastern U.S. psychic research institute. On each of the five randomly selected days, each person was asked to specify for 26 pairs of cards whether both cards in a given pair were of the same color or not. The number of correct answers is given as follows:

Person	Day 1	Day 2	Day 3	Day 4	Day 5
A	20	22	20	21	18
B	24	21	18	22	20
C	16	18	14	13	16

a. Determine the mean score for each person and interpret the results.
b. Test whether the three persons have significantly different ESP ability.
c. Carry out Scheffé's multiple-comparison procedure to determine which pairs of persons, if any, are significantly different in ESP ability.
d. On the basis of the results in parts (b) and (c), can one conclude that any of these persons has statistically significant ESP ability? Explain.

9. The average generation times for four different strains of influenza virus were determined using six cultures for each strain. The data are summarized below.

Statistic	Strain A	Strain B	Strain C	Strain D
$\bar{Y}$	420.3	330.7	540.4	450.8
S	30.22	28.90	31.08	33.29

a. Test whether the true mean generation time is different among the four strains.
b. What is the appropriate ANOVA table for these data?
c. Use Tukey's multiple-comparison procedure to identify where there are any differences among the means. (Use $\alpha = .05$.)

10. Three replicate water samples were taken at each of four locations in a river to determine whether the quantity of dissolved oxygen, a measure of water pollution, varied from one location to another (the higher the level of pollution, the lower the dissolved oxygen reading). Location 1 was adjacent to the wastewater discharge point for a certain industrial plant, and locations 2, 3, and 4 were selected at points 10, 20, and 30 miles downstream from this discharge point. The data appear in the table. The quantity Y_{ij} denotes the value of the dissolved oxygen content for the jth replicate at location i ($j = 1, 2, 3; i = 1, 2, 3, 4$), and $\bar{Y}_i$ denotes the mean of the three replicates taken at location i.

Location	Dissolved Oxygen Content (Y_{ij})	Mean ($\bar{Y}_i$)
1	4, 5, 6	5
2	6, 6, 6	6
3	7, 8, 9	8
4	8, 9, 10	9

a. Do the data provide sufficient evidence to suggest a difference in mean dissolved oxygen content among the four locations? (Use $\alpha = .05$.) Make sure to write down the appropriate ANOVA table.

b. If μ_i is the true mean level of dissolved oxygen at location i ($i = 1, 2, 3, 4$), test the null hypothesis

$$H_0: -3\mu_1 - \mu_2 + \mu_3 + 3\mu_4 = 0$$

versus

$$H_A: -3\mu_1 - \mu_2 + \mu_3 + 3\mu_4 \neq 0$$

at the 2% level; the quantity $(-3\mu_1 - \mu_2 + \mu_3 + 3\mu_4)$ is a contrast based on orthogonal polynomials (see Section 17-9), which can be shown to be a measure of the linear relationship between "locations" (the four equally spaced distances of 0, 10, 20, and 30 miles downstream from the plant) and "dissolved oxygen content."

c. Another way to quantify the strength of this linear relationship is to fit by least squares the model

$$Y = \beta_0 + \beta_1 X + E$$

where

$$X = \begin{cases} 0 & \text{for location 1} \\ 10 & \text{for location 2} \\ 20 & \text{for location 3} \\ 30 & \text{for location 4} \end{cases}$$

Fitting such a regression model to the $n = 12$ data points yields the accompanying ANOVA table. Use this table to perform a test of $H_0: \beta_1 = 0$ at the 2% significance level.

Source		df	SS	
Regression		1	29.40	
Residual	Lack of fit	$\begin{cases} 2 \\ 10 \\ 8 \end{cases}$	6.60	$\begin{cases} 0.60 \\ 6.00 \end{cases}$
	Pure error			

d. Noting that the regression model in part (c) is actually saying that $\mu_i = \beta_0 + \beta_1 X_i$, show that the hypothesis tested in part (b) is equivalent to the hypothesis tested in part (c).

e. Can you supply a reason why the two test statistics calculated in parts (b) and (c) do *not* have the same numerical value? Can you suggest a reasonable modification of the test in part (c) that will yield the same F-value as that obtained in part (b)?

f. Using the results of part (b), a test for a nonlinear trend in mean response can be obtained by subtracting the sum of squares for the contrast

$$\hat{L} = -3\bar{Y}_1 - \bar{Y}_2 + \bar{Y}_3 + 3\bar{Y}_4$$

from the sum of squares for treatments and then dividing this difference by the appropriate degrees of freedom to yield an F statistic of the form

$$F(\text{nonlinear trend}) = \frac{[\text{SST} - \text{SS}(\hat{L})]/\text{df}}{\text{MSE}}$$

Carry out this test based on the results obtained in parts (a) and (b). (Use $\alpha = .05$.)

g. Carry out the usual regression lack-of-fit test for adequacy of the straight-line model fit in part (c) using $\alpha = .05$. Does the value of the F statistic equal the value obtained in part (f)?

11. Consider the data from Problem 15 of Chapter 5. The dependent variable of interest is LN_BRNTL. Use $\alpha = .05$.
    a. Conduct a one-way ANOVA, with a dosage level (PPM_TOLU) as a categorical predictor.
    b. Using the Bonferroni (LSD) technique, compute all pairwise comparisons.
    c. Repeat (b) using Tukey's method.
    d. Repeat (b) using Scheffé's method.
    e. The ANOVA overall test corresponds to fitting a polynomial model of order k. What is k?
    f. Examine the ANOVA assumptions by computing the estimates of the within-cell variances. Provide frequency histograms for each cell to aid in this.

12. Repeat Problem 11, skipping part (e), but use LN_BLDTL as the dependent variable.

13. The source table below concerns a study involving the effects of trimethyl-tin doses of 0, 3, 6, and 9 mg/kg on a group of 48 rats. Note that each rat received only one dose. The response variable was the log of the activity counts in 1 hour in a residential maze. Compute the unknown letters above, a, b, c, d, e, f, and g. Show your work, even if it seems obvious and trivial to you.

Source	SS	df	MS	F
Dosage	a	d	g	14.71
Within dosage	b	e	12.84	
Total	c	f		

14. For each of the following contrasts, indicate the null hypothesis being tested.

	$\bar{X}_1$	$\bar{X}_2$	$\bar{X}_3$	$\bar{X}_4$	$\bar{X}_5$
c_1	1	0	-1	0	0
c_2	0	$\frac{1}{2}$	0	$\frac{1}{2}$	-1
c_3	$\frac{1}{3}$	$\frac{1}{3}$	$\frac{1}{3}$	0	-1

References

Allen, D. M., and Grizzle, J. E. (1969). "Analysis of Growth and Dose Response Curves." *Biometrics, 25*: 357–382.

Armitage, P. (1971). *Statistical Methods in Medical Research.* Oxford: Blackwell Scientific Publications.

Daly, M. B. (1973). "The Effect of Neighborhood Racial Characteristics on the Attitudes, Social Behavior, and Health of Low Income Housing Residents." Ph.D. dissertation, Department of Epidemiology, University of North Carolina, Chapel Hill, N.C.

Guenther, W. C. (1964). *Analysis of Variance.* Englewood Cliffs, N.J.: Prentice-Hall, Inc.

Hollander, M., and Wolfe, D. A. (1973). *Nonparametric Statistical Methods.* New York: John Wiley & Sons.

Kirk, R. E. (1969). *Experimental Design: Procedures for the Behavioral Sciences.* Belmont, Calif.: Wadsworth Publishing Company, Inc.

Lehmann, E. L. (1975). *Non-parametrics: Statistical Methods Based on Ranks.* San Francisco, Calif.: Holden-Day, Inc.

Lindman, H. R. (1974). *Analysis of Variance in Complex Experimental Designs.* San Francisco: W. H. Freeman and Company.

Miller, R. G., Jr. (1966). *Simultaneous Statistical Inference.* New York: McGraw-Hill Book Company.

Miller, R. G., Jr. (1981). *Simultaneous Statistical Inference,* 2nd ed. New York: Springer-Verlag.

Neter, J., and Wasserman, W. (1974). *Applied Linear Statistical Models.* Homewood, Ill.: Richard D. Irwin, Inc.

Siegel, S. (1956). *Nonparametric Statistics for the Behavioral Sciences.* New York: McGraw-Hill Book Company.

Steele, R. G. D., and Torrie, J. H. (1960). *Principles and Procedures of Statistics, with Special Reference to Biological Sciences.* New York: McGraw-Hill Book Company.

Randomized Blocks: Special Case of Two-Way ANOVA

18-1 Preview

In Chapter 17 we considered the simplest kind of ANOVA problem, that involving a single factor (or independent variable). We now focus on the two-factor case, which is generally referred to as two-way ANOVA. This extension is by no means trivial. In fact, we will devote three chapters (Chapters 18, 19, and 20) to different aspects of the two-factor case. In this first chapter we shall describe how a two-factor situation may be classified according to the data pattern. We shall then restrict our attention to a specific type of pattern, which will lead us to consider the randomized-blocks design, the main topic of this chapter.

18-1-1 Two-Way Data Patterns

Several different types of data patterns for two-way ANOVA are illustrated in Figure 18-1. Each of these tables describes a two-factor study with four levels of factor 1 (the "row" factor) and three levels of factor 2 (the "column" factor). The Y's in each table correspond to individual observations on a single dependent variable Y. The number of Y's in a given cell is denoted by n_{ij} for the ith level of factor 1 and the jth level of factor 2. The marginal total for the ith row is denoted by $n_{i \cdot}$ and for the jth column by $n_{\cdot j}$. The total number of observations is denoted by $n_{\cdot \cdot}$.

The simplest two-factor pattern, which is illustrated in Figure 18-1a, arises when there is a single observation in each cell (i.e., $n_{ij} = 1$ for all i and j). This pattern incorporates the randomized-blocks design to be discussed in this chapter, although there are other ways in which such single-observation-per-cell data may arise.

A second type of pattern, illustrated in Figure 18-1b, occurs when there are equal numbers of observations in each cell. Here, $n_{ij} = 4$ for all i and j. Chapter 19 will focus on this equal-replications situation.

FIGURE 18-1   Some two-way data patterns for a 4 × 3 table

(a) Single observation
per cell ($n_{ij} = 1$)

(b) Equal replications per cell
($n_{ij} = 4$)

(c) Equal replications by column,
proportionate replications by
row ($n_{ij} = n_{.j}/4$)

(d) Proportionate row and column
replications ($n_{ij} = n_{i.}n_{.j}/n_{..}$)

(e) Nonsystematic replications

The patterns in the rest of Figure 18-1 present different problems in statistical analysis.[1] We shall discuss these patterns in Chapter 20. The common property of the latter three patterns is that all cells do not have the same number of observations. Unequal cell numbers often arise in observational studies in which the levels of certain factors are determined after, rather than before, the data are collected.

For the pattern in Figure 18-1c, cells in the same column have the same number of observations, whereas cells in the same row are in the ratio 4 : 2 : 3. For this table each of the

---

[1] In Chapters 18, 19, and 20 we shall assume that each cell in a table contains *at least* one observation. When there are cells with no observations, the analysis required is considerably more complicated. The reader is referred to texts by Ostle (1963), Peng (1967), and Armitage (1971) for further discussion of this situation.

four cell frequencies in the jth column is equal to the same fraction of the corresponding total column frequency (i.e., $n_{ij} = n_{\cdot j}/4$ in this case). Note, for example, that $n_{\cdot 1}/4 = 16/4 = 4$, which is the number of observations in any cell in column 1.

For Figure 18-1d the cells in a given column are in the ratio $1:2:3:2$, whereas the cells in a given row are in the ratio $4:2:3$. This pattern results because n_{ij} is determined as

$$n_{ij} = \frac{n_{i\cdot} n_{\cdot j}}{n_{\cdot\cdot}}$$

which means that any cell frequency can be obtained by multiplying the corresponding row and column marginal frequencies together and then dividing by the total number of observations. Thus, for cell $(1, 2)$ in Figure 18-1d, we have $n_{1\cdot} n_{\cdot 2}/n_{\cdot\cdot} = 9(16)/72 = 2$, which equals n_{12}. Similarly, for cell $(4, 3)$, $n_{4\cdot} n_{\cdot 3}/n_{\cdot\cdot} = 18(24)/72 = 6$, which equals n_{43}.

There is no mathematical rule for describing the pattern of cell frequencies in Figure 18-1e, and so we say that such a pattern is nonsystematic. As we will see in Chapter 20, the ANOVA procedures required for the patterns in Figure 18-1c and 18-1d differ from those required for the irregular pattern in Figure 18-1e. For the former two patterns, the same computational procedure may be used as when there are an equal number of observations per cell. For the nonsystematic case, a different procedure is required.

18-1-2 Single-Observation-per-Cell Case

A two-way table with a single observation in each cell can arise in a number of different experimental situations. For instance, consider the following three examples:

1. Six hypertensive individuals, matched pairwise by age and sex, are randomly assigned (within each pair) to either a treatment or a control group. For each individual, a measure of change in self-perception of health is determined after 1 year. The main question of interest is whether the true mean change in self-perception is different for the treatment group than for the control group.

2. Six growth-inducing treatment combinations are randomly assigned to six mice from the same litter. The treatment combinations are defined by the cross-classification of the levels of two factors: factor A (drug A1 or placebo A0) and factor B (drug B2 [high dose], drug B1 [low dose], or placebo B0). The dependent variable of interest is weight gain 1 week after treatment is initiated. The questions to be considered include (a) whether the effect of drug A1 is different from that of placebo A0, (b) whether there are differences among the effects of drugs B1 and B2 and placebo B0, and (c) whether the drug A1 effect differs from that of placebo A0 in the same way at each level of factor B.

3. Scores of satisfaction with medical care are recorded for six hypertensive patients assigned to one of six categories depending on whether the nurse practitioner assigned to the patient was measured to have high or low autonomy (factor A) and high, medium, or low knowledge of hypertension (factor B). The main questions of interest are (a) whether mean satisfaction scores differ between patients with high-autonomy nurses and those with low-autonomy nurses, and (b) whether these differences in mean satisfaction scores differ in the same way at each level of knowledge grouping.

FIGURE 18-2   Different experimental situations resulting in two-way tables with a single observation per cell

	Pair 1	Pair 2	Pair 3
Treatment	Y	Y	Y
Control	Y	Y	Y

Y = change in self-perception of health after 1 year

(a)

	Placebo B0	Drug B1	Drug B2
Placebo A0	Y	Y	Y
Drug A1	Y	Y	Y

Y = weight gain after 1 week

(b)

	Low knowledge	Medium knowledge	High knowledge
Low autonomy	Y	Y	Y
High autonomy	Y	Y	Y

Y = patient satisfaction with medical care

(c)

Each of these experiments may be represented by a 2 × 3 two-way table, as given in Figure 18-2.

In the figure the third example involves two factors whose levels (or categories) were determined *after* the data were gathered. Such a study is often referred to as an observational study rather than as an experiment, since the latter term is usually reserved for studies involving factors whose levels are decided beforehand. Epidemiologic, sociological, and psychological studies are usually of an observational rather than an experimental nature. For such studies, the levels of the various factors of interest are determined after the frequency distributions of these factors have been considered.[2] For this example, the autonomy groupings were determined after the frequency distribution of autonomy scores on nurses was considered. Similarly, the knowledge categories were determined using the observed knowledge scores. Thus, one patient was in each of the six groups because of a posteriori (rather than a priori) considerations. In actual practice it is often not possible to arrange things so nicely, especially when large samples are involved. That is why most such observational studies have an unequal number of observations per cell.

---

[2] In this chapter we shall focus on the case where the levels of each factor are considered fixed; nevertheless, even if one or both factors were considered as random, the tests of hypotheses of interest, as with one-way ANOVA, would be computed in exactly the same way as in the fixed-factor case. This is not so where there is more than one observation per cell, discussed in Chapter 19.

The second example (Figure 18-2b) involves two factors whose levels were determined before the data were collected, and the resulting six treatment combinations can be viewed, in one sense, as representing the different levels of a single factor (called "treatment combination") that have been randomly assigned to the six individuals. Although it may be necessary because of limited experimental material or because of prohibitive cost to apply or assign each treatment combination to only a single experimental unit, considerably more information is obtained if there are several replications at each treatment combination.

This brings us to Figure 18-2a, which is of the general type to be focused on in this chapter. As with Figure 18-2b, this example represents a designed rather than an observational study. However, it differs from the other tables in Figure 18-2 in the allocation of individuals to cells (i.e., treatment combinations). The allocation here was done by randomization within each pair rather than, say, by randomization among all six individuals, ignoring any pairing. Another feature unique to this table is that the effect of only one of the two factors involved (in this case "treatment") is of primary interest. The other factor, "pair" (with three levels corresponding to the three pairs), is used only to help make more precise the comparison between the effects of the treatment and control groups. That is, if pair matching (on age and sex) is used and significant differences are found between the change scores for treatment and control groups, such differences cannot solely be attributed, for example, to one group being older or having a different sex composition than the other. The pairing, therefore, serves to eliminate or block out the noise affecting the comparison of treatment and control groups due to the confounding effects of age and sex. Such pairs are often referred to as *blocks*, and the associated experimental design is called a *randomized-blocks design*.

The analysis required for data as in Figure 18-2a is described in most introductory statistics texts. Since two groups are involved (treatment and control) and since matching has been done, the generally recommended method of analysis involves the use of the paired-difference t test, which is based on using the differences (changes) in scores within pairs. That is, the key test statistic involved is of the form $T = \bar{d}\sqrt{n}/S_d$, where $\bar{d}$ is the difference between the treatment group mean and the control group mean, S_d is the standard deviation of the difference scores for all pairs, and n is the number of pairs (three in Figure 18-2a). This statistic has the t distribution with $n - 1$ degrees of freedom under H_0 (so that the critical region is of the form $|T| \geq t_{n-1,1-\alpha/2}$ for a two-sided test of the null hypothesis of no difference in true average change score for treatment and control groups).

Actually, it can be shown that the paired-difference t test can be looked upon as a special case of the general F test used in a randomized-blocks ANOVA, involving more than two treatments per block. In fact, this randomized-blocks F test represents a generalization of the paired-difference t test in the same way that the one-way ANOVA F test is a generalization of the two-sample t test.

18-2 Equivalent Analysis of a Matched-Pairs Experiment

Suppose that we make the matched-pairs example of the previous section more realistic by considering the data given in Table 18-1, which involves 15 pairs of individuals matched on age and sex. The main inferential question for these data concerns whether the treatment

TABLE 18-1   Matched-pairs design concerning change scores in self-perception of health (Y) among hypertensives

Group	\multicolumn{15}{c	}{Pair}	Total	Mean
	1	2	3	4	5	6	7	8	9	10	11	12	13	14	15		
Treatment	10	12	8	8	13	11	15	16	4	13	2	15	5	6	8	146	9.73
Control	6	5	7	9	10	12	9	8	3	14	6	10	1	2	1	103	6.87
Total	16	17	15	17	23	23	24	24	7	27	8	25	6	8	9	249	8.30
Difference	4	7	1	−1	3	−1	6	8	1	−1	−4	5	4	4	7	43	2.86

group has a mean change score that is significantly different from that of the control group. Stated in terms of population means, the null hypothesis is

$$H_0:\ \mu_T = \mu_C$$

where μ_T and μ_C denote the population means of the treatment and control groups, respectively. The alternative hypothesis would thus be

$$H_A:\ \mu_T \neq \mu_C$$

if it were not theorized in advance that one particular group would have a higher or lower population mean than the other. (If, however, it was thought a priori that the treatment group would have a significantly higher mean change score than the control group, the alternative would be one sided: $H_A:\ \mu_T > \mu_C$.)

18-2-1 Paired-Difference t Test

One method for testing the null hypothesis H_0 was described in the previous section, the paired-difference t test. To perform this test, we first determine from Table 18-1 that $\bar{d} = \bar{Y}_T - \bar{Y}_C = 2.86$ and

$$S_d^2 = \frac{1}{14}\left[\sum_{i=1}^{15} d_i^2 - \frac{\left(\sum_{i=1}^{15} d_i\right)^2}{15}\right] = 12.695$$

where d_i is the observed difference between the scores for the treatment and control groups for the ith pair. Then the test statistic is computed to be

$$T = \frac{\bar{d}\sqrt{n}}{S_d} = \frac{2.86\sqrt{15}}{\sqrt{12.695}} = 3.109$$

Since $t_{14,0.995} = 2.976$ and $t_{14,0.9995} = 4.140$, the P-value for this two-sided test is given by

$$.001 < P < .01$$

We therefore reject H_0 and conclude that the mean change score for the treatment group is significantly different from that for the control group.

TABLE 18-2 ANOVA table for matched-pairs data of
Table 18-1

Source	SS	df	MS	F
Treatment	61.63	1	61.63	9.68 $(.005 < P < .01)$
Pairs (blocks)	391.80	14	27.97	4.39 $(.001 < P < .005)$
Error	89.17	14	6.37	
Total	542.30	29		

18-2-2 Randomized-Blocks *F* Test

Another way to test the null hypothesis H_0: $\mu_T = \mu_C$ for the data of Table 18-1 is to use an *F* test, which is based on the ANOVA table given in Table 18-2.

This ANOVA table differs in form from that used for one-way ANOVA in that the total sum of squares is partitioned into three components instead of just two (between and within). The treatments component here is similar to the between (i.e., treatment) source in the one-way ANOVA case. The error component here is analogous to the within component in the one-way ANOVA case, because this source is used as an estimate of the population variance σ^2. Finally, we have a pairs (or blocks) component in Table 18-2, for which there is no corresponding component in the one-way ANOVA case.

Thus, whereas the total sum of squares in one-way ANOVA may be split up into two components, as indicated by the equation

$$SS(total) = SS(between) + SS(within)$$

the total sum of squares in a randomized-blocks ANOVA can be partitioned into three meaningful components,

$$SS(total) = SS(treatments) + SS(blocks) + SS(error) \tag{18.1}$$

The last two components on the right-hand side of (18.1) can be looked upon as representing a partition of the experimental error (or within) sum of squares associated with one-way ANOVA (i.e., with the blocking ignored). By separating the blocking effect, a more precise estimate of experimental error is obtained, one not contaminated by any noise due to the effects of the blocking variables (in our case, age and sex).

The computation of the sums of squares in expression (18.1) is a fairly straightforward exercise. To obtain SS(treatments), for example, we divide the sum of the squares of the treatment totals by the number of pairs (i.e., blocks) and then subtract the usual correction factor (which is the squared total of all observations divided by the total number of observations). In other words, using the row totals in Table 18-1, we obtain

$$SS(treatments) = \frac{1}{15}(146^2 + 103^2) - \frac{249^2}{30}$$
$$= 2{,}128.33 - 2.066.70$$
$$= 61.63$$

The SS(blocks) is obtained similarly by dividing the sum of the squares of the block totals by the number of treatments and then subtracting the correction factor:

$$\text{SS(blocks)} = \frac{1}{2}(16^2 + 17^2 + \cdots + 8^2 + 9^2) - \frac{249^2}{30}$$

$$= 2{,}458.50 - 2{,}066.70$$

$$= 391.80$$

The total sum of squares is obtained by subtracting the correction factor from the total of the squares of all the observations:

$$\text{SS(total)} = \sum_{\text{all observations}} [\text{observation}]^2 - [\text{correction factor}]$$

$$= (10^2 + 6^2 + 12^2 + 5^2 + \cdots + 8^2 + 1^2) - \frac{249^2}{30}$$

$$= 2{,}609.00 - 2{,}066.70$$

$$= 542.30$$

We then obtain SS(error) by difference:

$$\text{SS(error)} = \text{SS(total)} - \text{SS(treatments)} - \text{SS(blocks)}$$

$$= 542.30 - 61.63 - 391.80$$

$$= 89.17$$

With these sums of squares now computed, the next step is to determine the degrees of freedom associated with each. The degrees of freedom for "treatments" is always equal to 1 less than the number of treatments (in our example, $2 - 1 = 1$). The degrees of freedom for "blocks" is equal to 1 less than the number of blocks (in our example, $15 - 1 = 14$). The degrees of freedom for the total sum of squares is equal to 1 less than the number of observations (in our example, $30 - 1 = 29$). Finally, the degrees of freedom for "error" is obtained by subtraction (i.e., $29 - 14 - 1 = 14$). Alternatively, the error degrees of freedom in a randomized-blocks analysis can be obtained as the product of the treatment degrees of freedom and the block degrees of freedom:

$$\text{error df} = [\text{treatment df}][\text{block df}] = 1(14) = 14$$

The mean squares, as usual, are obtained by dividing the sums of squares by their corresponding degrees of freedom. Finally, the F statistic for the test of the null hypothesis of no differences among the treatments is given by the formula

$$F = \frac{\text{MS(treatments)}}{\text{MS(error)}} \qquad (18.2)$$

In particular, to test H_0: $\mu_T = \mu_C$ for the data of Table 18-1, we calculate

$$F = \frac{61.63/1}{89.17/14} = \frac{61.63}{6.37} = 9.68$$

which is the observed value of an F with 1 and 14 degrees of freedom under H_0. Thus, H_0 is rejected at significance level α if the observed value of F is greater than $F_{1,14,1-\alpha}$. Since $F_{1,14,0.99} = 8.86$ and $F_{1,14,0.995} = 11.06$, the P-value for this test satisfies the inequality $.005 < P < .01$. Since this is quite small, it is reasonable to reject H_0 and to conclude that the treatment group has a significantly different mean change score from that of the control group.

Thus, the conclusions reached via the paired-difference t test and the randomized-blocks F test are exactly the same: Reject H_0. This is not due to mere coincidence but because the two tests are completely equivalent. In fact, it can be shown mathematically that the square of a paired-difference T statistic with ν degrees of freedom is exactly equal to a randomized-blocks F statistic with 1 and ν degrees of freedom (i.e., $F_{1,\nu} = T_\nu^2$), and that $F \geq F_{1,\nu,1-\alpha}$ whenever $|T| \geq t_{\nu,1-\alpha/2}$, and vice versa.

In our example,

$$T_{14}^2 = (3.109)^2 = 9.67 = F_{1,14}$$

and

$$t_{14,0.995}^2 = (2.977)^2 = 8.86 = F_{1,14,0.99}$$

An F test of the null hypothesis H_0: "no significant differences among the blocks" may also be performed. This test is not of primary interest, since the very use of a randomized-blocks design is based on the a priori assumption that there is significant block-to-block variation. Nevertheless, an a posteriori F test may be used to check on the reasonableness of this assumption. The test statistic to be used in this case is

$$F = \frac{\text{MS(blocks)}}{\text{MS(error)}}$$

which, for the example of Table 18-1, has the F distribution under H_0 with 14 degrees of freedom in the numerator and 14 degrees of freedom in the denominator. From Table 18-2 this F statistic is computed to be 4.39, with a P-value satisfying $.001 < P < .005$. The conclusion for this test, as expected, is to reject the null hypothesis of no block differences.

18-3 Principle of Blocking

For the matched-pairs example, we indicated that the primary reason for "pairing up" the data was so that the confounding factors age and sex would not blur the comparison between the treatment and control groups. In other words, the use of the matched-pairs design represented an attempt to account for the fact that the experimental units (i.e., the subjects) were not homogeneous with regard to factors (other than experimental group membership status) that were likely to affect the response variable. The key point here is not simply that subjects of different age and sex are different, but more precisely that age and sex are likely to affect the response variable. In another experimental situation, age and sex might not be important covariables and so would not have to be controlled or adjusted for.

One motivation, then, for using a randomized-blocks design is that the experimental units under study are heterogeneous relative to certain concomitant variables that affect the response variable but are not of primary interest. In such a case, to use a randomized-blocks design requires the following two steps: (1) *The experimental units (e.g., people, animals) that are homogeneous are collected together to form a block, and* (2) *the various treatments are assigned at random to the experimental units within each block.*

These steps are illustrated in Figure 18-3, where six blocks are formed, each consisting of three homogeneous experimental units. Three treatments (labeled A, B, and C) are then assigned at random to the three units within each block.

Figure 18-4, on the other hand, provides an example of incorrect blocking. In this case, one type of experimental unit might get predominantly one kind of treatment and another type might not get that treatment at all. In Figure 18-4, for example, the experimental unit

FIGURE 18-3   Steps used in forming randomized blocks

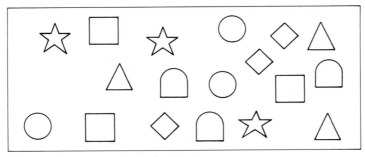

(a) Heterogeneous experimental units

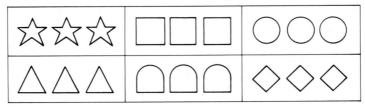

(b) Formation of blocks

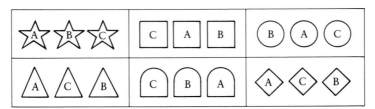

(c) Randomization of treatments A, B, and C within each block

type ☆ was assigned treatment B twice, whereas experimental unit types ○ and △ were not assigned treatment B at all. If the blocking had been done correctly, every distinct type of experimental unit would have been assigned each treatment exactly once.

     With regard to our matched-pairs design of Table 18-1, if we had not blocked on age, the treatment could have been assigned mostly to older subjects, and the control group might

FIGURE 18-4   Example of incorrect blocking

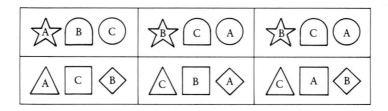

have consisted mostly of younger subjects. Then any differences that might have been found could very well have been due entirely to age differences in the two groups and not to the effect of the treatment itself.

18-4 Analysis of a Randomized-Blocks Experiment

In this section we shall describe the appropriate analysis for a randomized-blocks experiment. We shall focus on the case where both factors (i.e., treatments and blocks) are considered as fixed, although the tests of hypotheses of interest are computed in exactly the same way even if one or both factors are considered as random; that is, practically speaking, it does not matter how the factors are defined.[3]

18-4-1 Data Configuration

Table 18-3 gives the general layout of the data for a randomized-blocks design involving k treatments and b blocks. In this table Y_{ij} denotes the value of the observation on the

TABLE 18-3   Data layout for a randomized-block design

Treatment	Block 1	2	3	...	b	Total	Mean
1	Y_{11}	Y_{12}	Y_{13}	...	Y_{1b}	T_1	$\bar{Y}_{1\cdot}$
2	Y_{21}	Y_{22}	Y_{23}	...	Y_{2b}	T_2	$\bar{Y}_{2\cdot}$
3	Y_{31}	Y_{32}	Y_{33}	...	Y_{3b}	T_3	$\bar{Y}_{3\cdot}$
$\vdots$	$\vdots$	$\vdots$	$\vdots$		$\vdots$	$\vdots$	$\vdots$
k	Y_{k1}	Y_{k2}	Y_{k3}	...	Y_{kb}	T_k	$\bar{Y}_{k\cdot}$
Total	B_1	B_2	B_3	...	B_b	G	—
Mean	$\bar{Y}_{\cdot 1}$	$\bar{Y}_{\cdot 2}$	$\bar{Y}_{\cdot 3}$	...	$\bar{Y}_{\cdot b}$	—	$\bar{Y}_{\cdot\cdot}$

dependent variable Y corresponding to the ith treatment in the jth block (e.g., Y_{23} denotes the value of the observation associated with treatment 2 in block 3). The (row) total for treatment i is denoted by T_i; that is, $T_i = \sum_{j=1}^{b} Y_{ij}$. The (column) total for block j is denoted by B_j; that is, $B_j = \sum_{i=1}^{k} Y_{ij}$. The grand total of all bk observations is $G = \sum_{i=1}^{k} \sum_{j=1}^{b} Y_{ij}$. Finally, the treatment (row) means are denoted by $\bar{Y}_{i\cdot}$ for the ith treatment, the block (column) means are denoted by $\bar{Y}_{\cdot j}$ for the jth block, and the grand mean is denoted by $\bar{Y}_{\cdot\cdot}$.

A special case of this general format was given in Table 18-1, where $k = 2$ and $b = 15$. Another example (with $k = 4$ and $b = 8$) is given in Table 18-4. This example (although based on artificial data) illustrates the type of information being considered in several ongoing, large-scale U.S. intervention studies dealing with the risk factors associated with heart disease. Table 18-4 presents the data for a small experiment designed to assess the effects of four different cholesterol-reducing diets on persons having hypercholesterolemia. In such a

---

[3] In Chapter 19, which discusses the situation of equal replications (at least two) per cell, the F tests required differ depending on how the factors are defined.

TABLE 18-4 Randomized-blocks experiment for comparing the effects of four cholesterol-reducing diets on persons with hypercholesterolemia (Y = reduction in cholesterol level after 1 year)

| Treatment (Diet) | Block | | | | | | | | Total | Mean |
	1 (Male, Age > 50, QUET > 3.5)	2 (Male, Age > 50, QUET < 3.5)	3 (Male, Age < 50, QUET > 3.5)	4 (Male, Age < 50, QUET < 3.5)	5 (Female, Age > 50, QUET > 3.5)	6 (Female, Age > 50, QUET < 3.5)	7 (Female, Age < 50, QUET > 3.5)	8 (Female, Age < 50, QUET < 3.5)		
1	11.2	6.2	16.5	8.4	14.1	9.5	21.5	13.2	100.6	12.57
2	9.3	4.1	14.2	6.9	14.2	8.9	15.2	10.1	82.9	10.36
3	10.4	5.1	14.0	6.2	11.1	8.4	17.3	11.2	83.7	10.46
4	9.0	4.9	13.7	6.1	11.8	8.4	15.9	9.7	79.5	9.94
Total	39.9	20.3	58.4	27.6	51.2	35.2	69.9	44.2	346.7	—
Mean	9.97	5.07	14.60	6.90	12.80	8.80	17.47	11.05	—	10.83

study it would seem logical to try to take into account (or adjust for) the effects of age, sex, and body size on the dependent variable Y, which is "reduction in cholesterol level after 1 year." One way to do this is through the use of a randomized-blocks design, where the blocks are chosen to represent combinations of the various age–sex–body size categories of interest. Thus, in Table 18-4, block 1 consists of males over 50 years of age with a quetelet index (= $100[\text{weight}]/[\text{height}]^2$) above 3.5; block 3 consists of males under 50 with a quetelet index above 3.5; and so on. For each block of subjects, the four diets are randomly assigned to the sample of four persons in the block. Each subject is then followed for one year, after which the change in cholesterol level (Y) is recorded.

The primary research question of interest in this study concerns whether there are significant differences among the average reductions in cholesterol level achieved by the four diets. From inspection of Table 18-4, it can be seen that diet 1 appears to be the best, since it is associated with the largest average reduction (12.57 units). This statement, nevertheless, needs to be examined statistically by determining whether the observed differences among the mean reductions for the four diets can be attributed solely to chance. The randomized-blocks F test provides us with a method for making such a statistical evaluation.

Further inspection of Table 18-4 indicates that there are considerable differences among the block means. This is to be expected, since the very reason for blocking was based on the assumption that different blocks (or, equivalently, different categories of the covariates age, sex, and body size) would have different effects on the response. We can nevertheless perform a statistical test to satisfy ourselves that such block differences are statistically significant. In fact, if this test proves to be nonsignificant, it means that we probably should not have used these blocks in the first place, since they cost us degrees of freedom for estimating σ^2 and consequently gave us a less sensitive comparison of treatment effects.

18-4-2 Hypotheses Tested in a Randomized-Blocks Analysis

The primary null hypothesis of interest in a randomized-blocks analysis is the equality of treatment means:[4]

$$H_0: \mu_{1.} = \mu_{2.} = \cdots = \mu_{k.} \tag{18.3}$$

where $\mu_{i.}$ denotes the population mean response associated with the ith treatment. The alternative hypothesis may be stated

$$H_A: \text{"not all the } \mu_{i.}\text{'s are equal"}$$

If performing this test leads to rejection of H_0, it becomes of interest to determine where the important differences are among the treatment effects. One qualitative way to make such a determination is simply to look at the observed treatment means and to make visual comparisons. Thus, from Table 18-4, rejection of the null hypothesis of no differences among the diets leads to the conclusion that diet 1 is the best in reducing cholesterol level, that diets 2 and 3 are next in line and are about equally effective, and that diet 4 is the worst of the four. To verify such suspicions statistically, appropriate tests can be made using multiple-comparison techniques, as described in Chapter 17.

As mentioned earlier, another null hypothesis sometimes of interest is that there are no

---

[4] Again, we point out that we are considering the null hypothesis for the fixed-effects case. In one-way ANOVA, this hypothesis would be stated differently if the treatment effects were considered random, but the test of hypothesis would be computed exactly the same way.

significant differences among the block means. Since the use of a randomized-blocks design is based on the a priori conviction that the block means will be different, generally such a hypothesis is tested only to check that the blocking was justified.

18-4-3 *F* Test for the Equality of Treatment Means

The test of H_0: $\mu_{1\cdot} = \mu_{2\cdot} = \cdots = \mu_{k\cdot}$ is performed using the following F statistic (defined in terms of the notation of Table 18-3):[5]

$$F = \frac{\text{MST}}{\text{MSE}} \tag{18.4}$$

where

$$\text{MST} = \frac{1}{k-1} \left(\frac{1}{b} \sum_{i=1}^{k} T_i^2 - \frac{G^2}{bk} \right)$$

and

$$\text{MSE} = \frac{1}{(k-1)(b-1)} \left(\sum_{i=1}^{k} \sum_{j=1}^{b} Y_{ij}^2 - \frac{1}{k} \sum_{j=1}^{b} B_j^2 - \frac{1}{b} \sum_{i=1}^{k} T_i^2 + \frac{G^2}{bk} \right)$$

The assumptions required for this test are as follows:

1. The observations are statistically independent of one another.

2. Each observation is selected from a population that is normally distributed.

3. Each observation is selected from a population with variance σ^2 (i.e., variance homogeneity is assumed).

4. There is no block–treatment interaction effect; that is, the true extent to which treatments differ is the same regardless of the block considered. (We will discuss this in more detail in Section 18-5.)

If H_0 is true, the F statistic in (18.4) has the F distribution with $k - 1$ degrees of freedom in the numerator and $(k - 1)(b - 1)$ degrees of freedom in the denominator. Thus, for a given α, we would reject H_0 and conclude that not all treatments have the same effect on the response provided that

$$F \geq F_{k-1,(k-1)(b-1),1-\alpha}$$

For the data of Table 18-4, we find that

$$\text{MST} = \frac{1}{4-1} \left\{ \frac{1}{8} \left[(100.6)^2 + (82.9)^2 + (83.7)^2 + (79.5)^2 \right] - \frac{(346.7)^2}{32} \right\}$$

$$= 11.187$$

---

[5] Although (18.4) provides an explicit expression for MSE, in practice one would ordinarily calculate MSE as

$$\text{MSE} = \frac{1}{(k-1)(b-1)} (\text{TSS} - \text{SST} - \text{SSB})$$

where TSS, SST, and SSB are, respectively, the total, treatment, and block sums of squares.

$$MSE = \frac{1}{(4-1)(8-1)} \left\{ [(11.2)^2 + (9.3)^2 + (10.4)^2 + \cdots + (11.2)^2 + (9.7)^2] \right.$$

$$- \frac{1}{4} [(9.97)^2 + (5.07)^2 + \cdots + (11.05)^2]$$

$$\left. - \frac{1}{8} [(100.6)^2 + (82.9)^2 + (83.7)^2 + (79.5)^2] + \frac{(346.7)^2}{32} \right\}$$

$$= 0.939$$

so that

$$F = \frac{11.187}{0.939} = 11.914$$

Since $F_{3,21,0.999} = 7.94$, the P-value for this test satisfies $P < .001$. Thus, the null hypothesis H_0 should be rejected and the conclusion be made that there are significant differences among the four diets.

18-4-4 F Test for the Equality of Block Means

As previously mentioned, this test is rarely used except as an a posteriori check that blocking was effective. To perform this test, the following F statistic is calculated:

$$F = \frac{MSB}{MSE} \tag{18.5}$$

where

$$MSB = \frac{1}{b-1} \left(\frac{1}{k} \sum_{j=1}^{b} B_j^2 - \frac{G^2}{bk} \right)$$

and where MSE is calculated as before (see footnote 5).

Under the null hypothesis H_0: "there are no differences among the true block means," the F statistic (18.5) has the F distribution with $b - 1$ and $(k - 1)(b - 1)$ degrees of freedom. So H_0 is rejected at significance level α when F exceeds $F_{b-1,(k-1)(b-1),1-\alpha}$. For the data in Table 18-4, this test yields the following F:

$$F = \frac{66.123}{0.939} = 70.419$$

The P-value for this test satisfies $P \ll .001$, so H_0 is rejected, as expected.

18-5 ANOVA Table for a Randomized-Blocks Experiment

The procedures necessary to test the equality of the treatment and block means can be summarized as in Table 18-5. For the data given in Table 18-4, the corresponding ANOVA table is given by Table 18-6. From these ANOVA tables it can be seen that the total (corrected) sum of squares TSS is split up into the three components SST(treatments), SSB(blocks), and SSE(error) using the fundamental equation

$$TSS(\text{total}) = SST(\text{treatments}) + SSB(\text{blocks}) + SSE(\text{error}) \tag{18.6}$$

TABLE 18-5   ANOVA table for a randomized-blocks experiment (with k treatments and b blocks)

Source	df	SS	MS	F
Treatments	$k-1$	$\text{SST} = \dfrac{1}{b} \sum_{i=1}^{k} T_i^2 - \dfrac{G^2}{bk}$	$\text{MST} = \dfrac{\text{SST}}{k-1}$	$\dfrac{\text{MST}}{\text{MSE}}$
Blocks	$b-1$	$\text{SSB} = \dfrac{1}{k} \sum_{j=1}^{b} B_j^2 - \dfrac{G^2}{bk}$	$\text{MSB} = \dfrac{\text{SSB}}{b-1}$	$\dfrac{\text{MSB}}{\text{MSE}}$
Error	$(k-1)(b-1)$	$\text{SSE} = \text{TSS} - \text{SST} - \text{SSB}$	$\text{MSE} = \dfrac{\text{SSE}}{(k-1)(b-1)}$	
Total	$kb-1$	$\text{TSS} = \sum_{i=1}^{k} \sum_{j=1}^{b} Y_{ij}^2 - \dfrac{G^2}{bk}$		

TABLE 18-6   ANOVA table for data of Table 18-4

Source	df	SS	MS	F
Treatments	3	33.561	11.187	11.914 ($P < .001$)
Blocks	7	462.860	66.123	70.419 ($P \ll .001$)
Error	21	19.711	0.939	
Total	31	516.132		

Equivalent formulae for the total sum of squares are:

$$\text{TSS(total)} = \sum_{i=1}^{k} \sum_{j=1}^{b} (Y_{ij} - \bar{Y}_{..})^2 = \sum_{i=1}^{k} \sum_{j=1}^{b} Y_{ij}^2 - \frac{G^2}{bk} \qquad (18.7)$$

The second formula in (18.7) is preferable for computation. As is the case in any ANOVA situation, TSS measures the total unexplained variation in the data, corrected for the grand mean. The degrees of freedom associated with TSS is $kb - 1$. For the data of Table 18-4, TSS is computed to be

$$\text{TSS} = [(11.2)^2 + (9.3)^2 + (10.4)^2 + (9.0)^2 + (6.2)^2 + \cdots + (11.2)^2 + (9.7)^2]$$
$$- \frac{(346.7)^2}{32}$$
$$= 516.132$$

The treatment sum of squares, SST, is defined as

$$\text{SST} = b \sum_{i=1}^{k} (\bar{Y}_{i.} - \bar{Y}_{..})^2 = \frac{1}{b} \sum_{i=1}^{k} T_i^2 - \frac{G^2}{bk} \qquad (18.8)$$

The second formula of (18.8) is preferred for computation. The expression in the middle of (18.8) illustrates why SST reflects the variation among the treatment means, since its basic components are of the form $(\bar{Y}_{i.} - \bar{Y}_{..})$, which is the difference between the ith treatment mean and the grand mean. For the data of Table 18-4, SST is computed as

$$\text{SST} = \frac{1}{8}[(100.6)^2 + (82.9)^2 + (83.7)^2 + (79.5)^2] - \frac{(346.7)^2}{32}$$

$$= 33.561$$

TSS has $k - 1$ degrees of freedom.

The block sum of squares, SSB, is defined as

$$\text{SSB} = k \sum_{j=1}^{b} (\bar{Y}_{\cdot j} - \bar{Y}_{\cdot \cdot})^2 = \frac{1}{k} \sum_{j=1}^{b} B_j^2 - \frac{G^2}{bk} \tag{18.9}$$

with $b - 1$ degrees of freedom. Again, the second formula of (18.9) is the computational formula, whereas the middle expression explains why SSB measures variation among blocks. For the data of Table 18-4, we get

$$\text{SSB} = \frac{1}{4}[(39.9)^2 + (20.3)^2 + (58.4)^2 + (27.6)^2 + (51.2)^2 + (35.2)^2 + (69.9)^2$$

$$+ (44.2)^2] - \frac{(346.7)^2}{32}$$

$$= 462.860$$

Finally, the residual sum of squares, SSE, is given by

$$\text{SSE} = \sum_{i=1}^{k} \sum_{j=1}^{b} (Y_{ij} - \bar{Y}_{i\cdot} - \bar{Y}_{\cdot j} + \bar{Y}_{\cdot\cdot})^2 = \text{TSS} - \text{SST} - \text{SSB} \tag{18.10}$$

with degrees of freedom equal to $(k - 1)(b - 1)$.

Applying the computational formula on the far right-hand side of (18.10) to our example yields

$$\text{SSE} = 516.132 - 33.561 - 462.860$$

$$= 19.711$$

The complexity of the middle expression in (18.10) indicates that SSE is not easily recognizable as an estimate of σ^2. In fact, its basic component, $(Y_{ij} - \bar{Y}_{i\cdot} - \bar{Y}_{\cdot j} + \bar{Y}_{\cdot\cdot})$ can be written in the form

$$(Y_{ij} - \bar{Y}_{\cdot j}) - (\bar{Y}_{i\cdot} - \bar{Y}_{\cdot\cdot})$$

which is an estimate of the *difference between the effect of the ith treatment relative to the average effect of all treatments in the jth block* (i.e., $Y_{ij} - \bar{Y}_{\cdot j}$) *and the overall effect of the ith treatment relative to the overall mean* (i.e., $\bar{Y}_{i\cdot} - \bar{Y}_{\cdot\cdot}$). Actually, SSE here measures block–treatment interaction in the sense that SSE will be large when the treatment effects vary from block to block. Although we will discuss the concept of interaction more thoroughly in Chapter 19, it is imperative to note here that *the use of a randomized-blocks design requires the assumption that no such block–treatment interaction exists;* as a result, the residual variation reflected in SSE can be attributed solely to experimental error.[6] The expected effect

---

[6] A method for testing the validity of the assumption of no block–treatment interaction, developed by Tukey (1949), is described in Problem 3(f) at the end of the chapter.

of violating the no-interaction assumption is a reduction in the power of the tests of treatment and block effects. It is mandatory that this assumption not be violated, because otherwise there would be no way to obtain an unbiased estimate of σ^2 that could be used in the denominator of the F statistic. After all, for each block in a randomized-blocks design, no treatment is applied more than once, making it impossible to obtain a pure error estimate of σ^2 associated with a particular treatment in a given block. When we consider the two-way ANOVA case with more than one observation per cell, we will see that a pure error estimate of σ^2 can be developed by utilizing the available information on within-cell variability.

Each mean-square term in Table 18-5 is, as usual, obtained by dividing the corresponding sum-of-squares term by its degrees of freedom. Also, the F statistics are formed as the ratios of mean-square terms, with MSE in the denominator in each case (regardless of whether each factor is considered as fixed or random).

18-6 Regression Models for a Randomized-Blocks Experiment

The randomized-blocks experiment, like the one-way ANOVA situation, can be described by either a regression model or a classical ANOVA effects model. The regression formulation, which we consider in this section, is technically equivalent to the fixed-effects ANOVA approach in terms of estimating the unknown parameters in each model. Nevertheless, for testing purposes, mean-square terms in the ANOVA table obtained from the fit of a proper regression model can be used to compute appropriate F statistics regardless of whether the actual ANOVA model is fixed, random, or mixed.

An appropriate regression model for the randomized-blocks experiment should contain $k - 1$ dummy variables for the k treatments and $b - 1$ dummy variables for the b blocks. One such model formulation is given as follows:

$$Y = \mu + \sum_{i=1}^{k-1} \alpha_i X_i + \sum_{j=1}^{b-1} \beta_j Z_j + E$$

where

$$X_i = \begin{cases} -1 & \text{if treatment } k \\ 1 & \text{if treatment } i \\ 0 & \text{otherwise} \end{cases} \quad \text{and} \quad Z_j = \begin{cases} -1 & \text{if block } b \\ 1 & \text{if block } j \\ 0 & \text{otherwise} \end{cases} \qquad (18.11)$$

$(i = 1, 2, \ldots, k - 1; j = 1, 2, \ldots, b - 1)$.

(As in one-way ANOVA, other coding schemes for the independent variables are possible; for example, we may let

$$X_i = \begin{cases} 1 & \text{if treatment } i \\ 0 & \text{otherwise} \end{cases} \quad \text{and} \quad Z_j = \begin{cases} 1 & \text{if block } j \\ 0 & \text{otherwise} \end{cases}$$

$(i = 1, 2, \ldots, k - 1; j = 1, 2, \ldots, b - 1)$. For any such coding scheme, the F tests obtained from fitting the regression model are exactly equivalent to the randomized-blocks

ANOVA F tests described previously. The regression coefficients, however, represent different functions of the cell population means.)

The regression coefficients in this model may, as in the one-way ANOVA situation considered in Chapter 17, be expressed in terms of underlying cell (i.e., block–treatment combination) means. To see this we need to consider the matrix of population cell means associated with a general randomized-blocks layout, as presented in Table 18-7. In this table, the (cell) mean for the ith treatment in the jth block is denoted by μ_{ij}, the mean for the ith treatment (averaged over the b blocks) is denoted by $\mu_{i\cdot}$, the mean for the jth block (averaged over the k treatments) is denoted by $\mu_{\cdot j}$, and the overall mean is denoted by $\mu_{\cdot\cdot}$. That is, $\mu_{i\cdot}$, $\mu_{\cdot j}$, and $\mu_{\cdot\cdot}$ satisfy

$$\mu_{i\cdot} = \frac{1}{b} \sum_{j=1}^{b} \mu_{ij}, \qquad \mu_{\cdot j} = \frac{1}{k} \sum_{i=1}^{k} \mu_{ij}, \qquad \mu_{\cdot\cdot} = \frac{1}{bk} \sum_{i=1}^{k} \sum_{j=1}^{b} \mu_{ij}$$

For model (18.11), the coefficients α_i and β_j can then be expressed as follows:

$$
\begin{aligned}
\mu &= \mu_{\cdot\cdot} \\
\alpha_i &= \mu_{i\cdot} - \mu_{\cdot\cdot} \qquad i = 1, 2, \ldots, k-1 \\
\beta_j &= \mu_{\cdot j} - \mu_{\cdot\cdot} \qquad j = 1, 2, \ldots, b-1 \\
-\sum_{i=1}^{k-1} \alpha_i &= \mu_{k\cdot} - \mu_{\cdot\cdot} \\
-\sum_{j=1}^{b-1} \beta_j &= \mu_{\cdot b} - \mu_{\cdot\cdot}
\end{aligned}
\qquad (18.12)
$$

Thus, the coefficient μ is the overall mean, the coefficient α_i represents the difference between the ith treatment mean and the overall mean, and the coefficient β_j represents the difference between the jth block mean and the overall mean. Furthermore, the negative sum of the α_i (i.e., $-\sum_{i=1}^{k-1} \alpha_i$) gives the difference between the kth treatment mean and the overall mean, and the negative sum of the β_j (i.e., $-\sum_{j=1}^{b-1} \beta_j$) gives the difference between the bth block mean and the overall mean.

TABLE 18-7   Matrix of cell means for a
randomized-blocks layout

Treatment (i)	Block (j)					Mean
	1	2	3	$\cdots$	b	
1	μ_{11}	μ_{12}	μ_{13}	$\cdots$	μ_{1b}	$\mu_{1\cdot}$
2	μ_{21}	μ_{22}	μ_{23}	$\cdots$	μ_{2b}	$\mu_{2\cdot}$
3	μ_{31}	μ_{32}	μ_{33}	$\cdots$	μ_{3b}	$\mu_{3\cdot}$
$\vdots$	$\vdots$	$\vdots$	$\vdots$		$\vdots$	$\vdots$
k	μ_{k1}	μ_{k2}	μ_{k3}	$\cdots$	μ_{kb}	$\mu_{k\cdot}$
Mean	$\mu_{\cdot 1}$	$\mu_{\cdot 2}$	$\mu_{\cdot 3}$	$\cdots$	$\mu_{\cdot b}$	$\mu_{\cdot\cdot}$

These results can also be used to express the cell, treatment, and block means as a function of the regression parameters. The cell means are

$$\mu_{ij} = \mu + \alpha_i + \beta_j \qquad i = 1, 2, \ldots, k - 1; \quad j = 1, 2, \ldots, b - 1$$

$$\mu_{kj} = \mu - \sum_{i=1}^{k-1} \alpha_i + \beta_j \qquad j = 1, 2, \ldots, b - 1$$

$$\mu_{ib} = \mu + \alpha_i - \sum_{j=1}^{b-1} \beta_j \qquad i = 1, 2, \ldots, k - 1$$

$$\mu_{kb} = \mu - \sum_{i=1}^{k-1} \alpha_i - \sum_{j=1}^{b-1} \beta_j$$

In turn, the treatment means are

$$\mu_{i.} = \mu + \alpha_i \qquad i = 1, 2, \ldots, k - 1$$

$$\mu_{k.} = \mu - \sum_{i=1}^{k-1} \alpha_i$$

and the block means are

$$\mu_{.j} = \mu + \beta_j \qquad j = 1, 2, \ldots, b - 1$$

$$\mu_{.b} = \mu - \sum_{j=1}^{b-1} \beta_j$$

Similar results are obtained by using reference cell coding. The model statement remains the same, but the dummy variables change as follows:

$$X_i = \begin{cases} 1 & \text{if treatment } i \\ 0 & \text{otherwise} \end{cases} \qquad i = 1, 2, \ldots, k - 1$$

$$Z_j = \begin{cases} 1 & \text{if block } j \\ 0 & \text{otherwise} \end{cases} \qquad j = 1, 2, \ldots, b - 1$$

Note that the same parameter symbols are used, but their meanings change. In particular, the cell means (for reference cell coding) are

$$\mu_{ij} = \mu + \alpha_i + \beta_j \qquad i = 1, 2, \ldots, k - 1; \quad j = 1, 2, \ldots, b - 1$$

$$\mu_{kj} = \mu + \beta_j \qquad j = 1, 2, \ldots, b - 1$$

$$\mu_{ib} = \mu + \alpha_i \qquad i = 1, 2, \ldots, k - 1$$

$$\mu_{kb} = \mu$$

The treatment means are

$$\mu_{i.} = \mu + \alpha_i + \sum_{j=1}^{b-1} \frac{\beta_j}{b} \qquad i = 1, 2, \ldots, k - 1$$

$$\mu_{k.} = \mu + \sum_{j=1}^{b-1} \frac{\beta_j}{b}$$

and the block means are

$$\mu_{.j} = \mu + \beta_j + \sum_{i=1}^{k-1} \frac{\alpha_i}{k} \qquad j = 1, 2, \ldots, b-1$$

$$\mu_{.b} = \mu + \sum_{i=1}^{k-1} \frac{\alpha_i}{k}$$

These results demonstrate that estimates of means and contrasts between means can be expressed as linear combinations of estimated regression coefficients. One word of caution: When using a computer package to estimate effects in ANOVA, one must understand exactly what parameters the program is estimating.

As with one-way ANOVA, the F tests resulting from the fit of regression model (18.11) or from any other properly coded regression model (used for testing the hypotheses of equality of treatment means and of equality of block means) are exactly the same as obtained for the ANOVA procedures presented earlier. That is, the multiple-partial F test of H_0: $\alpha_1 = \alpha_2 = \cdots = \alpha_{k-1} = 0$ under model (18.11) provides exactly the same F-value as that given by

$$F = \frac{\text{MST}}{\text{MSE}}$$

of (18.4). Similarly, the multiple-partial F test of H_0: $\beta_1 = \beta_2 = \cdots = \beta_{b-1} = 0$ in model (18.11) yields exactly the same F-value as given by

$$F = \frac{\text{MSB}}{\text{MSE}}$$

of (18.5).

Thus, it really makes no difference which, if any, regression model (i.e., coding scheme) is used if we are interested only in performing the above global F tests. Furthermore, if we want to make certain specific comparisons of means, we can always calculate such comparisons directly without using a regression model.

18-7 Fixed-Effects ANOVA Model for a Randomized-Blocks Experiment

If both the block and treatment effects are considered to be fixed, a classical ANOVA model may be written in terms of such effects as was done in the one-way ANOVA case. The effects for a randomized-blocks experiment are defined in terms of differences between a given treatment mean and the overall mean (i.e., $\mu_{i.} - \mu_{..}$) and differences between a given block mean and the overall mean (i.e., $\mu_{.j} - \mu_{..}$). The fixed-effects model may be written in the form

$$Y_{ij} = \mu + \alpha_i + \beta_j + E_{ij} \qquad i = 1, 2, \ldots, k; \quad j = 1, 2, \ldots, b \qquad (18.13)$$

where

Y_{ij} = observed response associated with the ith treatment in the jth block

$\mu = \mu_{..}$ = grand (overall) mean

$\alpha_i = \mu_{i.} - \mu_{..}$ = effect of treatment i

$\beta_j = \mu_{.j} - \mu_{..}$ = effect of block j

$E_{ij} = Y_{ij} - \mu - \alpha_i - \beta_j$ = error component associated with the ith treatment in the jth block

The key characteristic of model (18.13) is that the effect of any given treatment is the same regardless of the block and that, similarly, the effect of any given block is the same regardless of the treatment. Another way of saying this is that, to determine the mean response for a given cell, it is necessary to know only the treatment effect and block effect associated with that cell and not the particular contribution of the cell itself. Still a third way of saying this is that there is no block–treatment interaction. A model incorporating such an interaction would be of the form

$$Y_{ij} = \mu + \alpha_i + \beta_j + \gamma_{ij} + E_{ij}$$

which contains the (interaction) term γ_{ij} specific to cell (i, j).

A few additional properties of the fixed-effects model (18.13) are worth noting. First, it can be shown that

$$\sum_{i=1}^{k} \alpha_i = 0 \quad \text{and} \quad \sum_{j=1}^{b} \beta_j = 0$$

since $\alpha_i = \mu_{i.} - \mu_{..}$ and $\beta_j = \mu_{.j} - \mu_{..}$.

Another property of interest is that the various treatment and block effects can be simply estimated from the data. These estimated effects are given by

$$\hat{\alpha}_i = \bar{Y}_{i.} - \bar{Y}_{..} \quad \text{and} \quad \hat{\beta}_j = \bar{Y}_{.j} - \bar{Y}_{..}$$

A final property of the fixed-effects model (18.13) is that it corresponds in structure to the regression model given by (18.11). That is, the coefficients α_1 through α_{k-1} of model (18.11) correspond to the first $k - 1$ treatment effects in model (18.13). Also, the coefficients β_1 through β_{b-1} correspond to the first $b - 1$ block effects in model (18.13). Finally, the negative sums $-\sum_{i=1}^{k-1} \alpha_i$ and $-\sum_{j=1}^{b-1} \beta_j$ represent the effects of the kth treatment and the bth block, respectively.

Problems

1. An experiment was conducted by a private research corporation to investigate the toxic effects of three chemicals (I, II, and III) used in the tire manufacturing industry. In this experiment 1-inch squares of skin on rats were treated with the chemicals and then scored from 0 to 10, depending on the degree of irritation. Three adjacent 1-inch squares were marked on the back of each of eight rats, and each of the three chemicals was applied to each rat. The data are as shown in the table that follows.

Chemical	Rat								Total
	1	2	3	4	5	6	7	8	
I	6	9	6	5	7	5	6	6	50
II	5	9	9	8	8	7	7	7	60
III	3	4	3	6	8	5	5	6	40
Total	14	22	18	19	23	17	18	19	150

a. What are the blocks and what are the treatments in this randomized-blocks design?

b. Determine the appropriate ANOVA table for the data set given.

c. Do the data provide sufficient evidence to indicate a significant difference in the toxic effects of the three chemicals?

d. Using a confidence interval of the form

$$(\bar{Y}_{1\cdot} - \bar{Y}_{2\cdot}) \pm t_{\nu,1-\alpha/2} \sqrt{MSE \left(\frac{1}{n_1} + \frac{1}{n_2} \right)}$$

where ν is the degrees of freedom, find a 98% confidence interval for the true difference in the toxic effects of chemicals I and II.

e. Using the ANOVA table obtained in part (b), provide a reasonable measure of the proportion of total variation explained by the particular statistical model used in analyzing this data set.

f. State the fixed-effects ANOVA model and the corresponding regression model for this analysis.

g. State the assumptions on which the validity of the analysis depends.

2. In a study of the psychosocial changes in individuals participating in a community-based intervention program, individuals who were clinically identified as hypertensives were randomly assigned to one of two treatment groups (assume that there are n individuals in each group). Group 1 was given the usual care provided for hypertensives by existing community facilities, whereas group 2 was given the special care provided by the intervention study team. Among the variables measured on each individual were: SP1, an index of the individual's self-perception of health immediately after being identified as hypertensive but before assignment to one of the two groups; SP2, an index of the individual's self-perception of health 1 year after assignment to one of the two groups; AGE; and SEX. Restricting attention to these variables only, one main research question of interest was whether the change in self-perception of health after 1 year would be greater for individuals in group 2 than for those in group 1. To examine this question, several different analytical approaches are possible, depending on the dependent variable chosen and on the treatment of the variables SP1, AGE, and SEX in the analysis. Some of these approaches are:

(1) matching pairwise on AGE and SEX (which we assume is possible) and then performing a paired-difference t test to determine whether the mean of group 1 change scores is significantly different from the mean of group 2 change scores (*Note*: This is equivalent to doing a randomized-blocks analysis, where the blocks are the pairs of individuals.)

(2) matching pairwise on AGE and SEX and then performing a regression analysis with the change score $Y = SP2 - SP1$ as the dependent variable and with SP1 as one of the independent variables

(3) matching pairwise on AGE and SEX and then performing a regression analysis with SP2 as the dependent variable and with SP1 as one of the independent variables

(4) controlling for AGE and SEX (without any prior matching) via analysis of covariance, with the change score $Y = SP2 - SP1$ as the dependent variable

(5) controlling for AGE and SEX via analysis of covariance, with the change score $Y = SP2 - SP1$ as the dependent variable and with SP1 as one of the independent variables

(6) controlling for AGE and SEX via analysis of covariance, with SP2 as the dependent variable and with SP1 as one of the independent variables

a. What is the appropriate regression model associated with each of the above six approaches? (Make sure to define your variables carefully.)

b. For each of the regression models above, state the appropriate null hypothesis (in terms of regression coefficients) for testing for group differences with respect to self-perception scores. Indicate for each regression model how to set up the appropriate ANOVA table to carry out the desired test.

c. Assuming that you have decided to match pairwise on AGE and SEX, which of the above regression models would you prefer to use and why? (*Note*: Actually, you have only two models to choose from, since the second and third models described above will produce exactly the same test statistic for comparing the two groups.)

3. An experiment was conducted at the University of North Carolina to see whether the BOD test for water pollution is biased by the presence of copper. In this test, the amount of dissolved oxygen in a sample of water is measured at the beginning and at the end of a 5-day period; the difference in dissolved oxygen content is ascribed to the action of bacteria on the impurities in the sample and is called the *biochemical oxygen demand* (BOD). The question is whether dissolved copper retards the bacterial action and results in an artificially low response for the test.

   The following data are partial results from this experiment. The three samples are from different sources. They are split into five subsamples, and the concentration of copper ion in each subsample is given. The BOD measurements are given for each subsample–copper ion concentration combination.

Sample	Copper Ion Concentration (ppm)					Mean
	0	0.1	0.3	0.5	0.75	
1	210	195	150	148	140	168.60
2	194	183	135	125	130	153.40
3	138	98	89	90	85	100.00
Mean	180.67	158.67	124.67	121.00	118.33	140.67

a. Using dummy variables and treating the copper ion concentration as a categorical (or nominal) variable, give an appropriate regression model for this experiment. Is this a randomized-blocks experiment?

b. If "copper ion concentration" is treated as an interval (continuous) variable, one appropriate regression model would be

$$Y = \beta_0 + \beta_1 Z_1 + \beta_2 Z_2 + \beta_3 X + \beta_4 Z_1 X + \beta_5 Z_2 X + E$$

in which

$$Z_1 = \begin{cases} 1 & \text{for sample 1} \\ 0 & \text{for sample 2} \\ -1 & \text{for sample 3} \end{cases} \qquad Z_2 = \begin{cases} 0 & \text{for sample 1} \\ 1 & \text{for sample 2} \\ -1 & \text{for sample 3} \end{cases}$$

and X = copper ion concentration. Comment on the advantages and disadvantages of using the models in parts (a) and (b). Which model would you prefer to use and why?

c. Compare (without doing any statistical tests) the average BOD responses at the various copper ion concentration levels.

d. Use the table, which is based on a randomized-blocks analysis, to test (at $\alpha = .05$) the null hypothesis that copper ion concentration has no effect on the BOD test.

Source	df	SS	MS	F
Samples	2	12,980.9333	6,490.4667	56.83
Concentrations	4	9,196.6667	2,299.1667	20.13
Error	8	913.7332	114.2166	

e. Based on the ANOVA table and on the observed block means, does blocking appear to be justified?

f. The randomized-blocks analysis assumes that the relative differences in BOD responses at different copper ion concentration levels are the same regardless of the sample used; in other words, there is no copper ion concentration–sample interaction. One method (see Tukey, 1949) for testing whether such an interaction effect actually exists is called *Tukey's test for additivity*, which addresses the null hypothesis

H_0: "no interaction exists" (i.e., the model is additive in the block and treatment effects)

versus the alternative hypothesis

H_A: "the model is not additive, and there exists a transformation $f(Y)$ that removes the nonadditivity in the model for Y"

Tukey's test statistic is given by

$$F = \frac{\text{SSN}}{(\text{SSE} - \text{SSN})/[(k-1)(b-1) - 1]}$$

(which is distributed as $F_{1,(k-1)(b-1)-1}$ under H_0), where

$$\text{SSN} = \frac{\left[\sum_{i=1}^{k} \sum_{j=1}^{b} Y_{ij}(\bar{Y}_{i\cdot} - \bar{Y}_{\cdot\cdot})(\bar{Y}_{\cdot j} - \bar{Y}_{\cdot\cdot}) \right]^2}{\sum_{i=1}^{k} (\bar{Y}_{i\cdot} - \bar{Y}_{\cdot\cdot})^2 \sum_{j=1}^{b} (\bar{Y}_{\cdot j} - \bar{Y}_{\cdot\cdot})^2}$$

using the notation in this chapter. (Since the numerator degrees of freedom for this test statistic is 1, the square root of F will have the t distribution with $(k - 1)(b - 1) - 1$ degrees of freedom.)

Given that the computed t statistic is $T = 2.090$ in Tukey's test for additivity, is there significant evidence of nonadditivity? (Use $\alpha = .05$.)

g. Use the following tables based on fitting the multiple regression model given in part (b) to test whether there is evidence of a significant effect due to copper ion concentration (i.e., test $H_0: \beta_3 = \beta_4 = \beta_5 = 0$). Use the result $R^2 = .888$.

Variable	Regression Coefficient	Beta Coefficient
(intercept)	167.200	
Z_1	32.513	0.67659
Z_2	16.972	0.35319
X	-80.389	-0.55585
Z_1X	-13.877	-0.12337
Z_2X	-12.844	-0.11419

Source	df	MS
Z_1	1	11,764.90
$Z_2 \vert Z_1$	1	1,216.03
$X \vert Z_1, Z_2$	1	7,134.57
$Z_1X \vert Z_1, Z_2, X$	1	303.26
$Z_2X \vert Z_1, Z_2, X, Z_1X$	1	91.07
Error	9	286.83

4. For the data in Problem 7 of Chapter 17, conduct a randomized-blocks analysis, treating the high schools as blocks, to test whether there are significant differences among the mean VSAT scores for the years 1965, 1970, and 1975. Do the results obtained from this randomized-blocks analysis differ from those previously obtained from the one-way ANOVA?

5. For the data in Problem 8 of Chapter 17, conduct a randomized-blocks analysis, treating the days as blocks, to test whether the three persons have significantly different ESP ability. Comment on whether blocking on days seems appropriate.

6. The promotional policies of four companies in a certain industry were compared to determine whether their rates for promoting blacks and whites differed. Data on the variable "rate discrepancy," defined as

$$d = \hat{p}_W - \hat{p}_B$$

where

$$\hat{p}_W = \frac{[\text{no. of whites promoted}] \, (100)}{[\text{no. of whites eligible for promotion}]}$$

and

$$\hat{p}_B = \frac{[\text{no. of blacks promoted}] \, (100)}{[\text{no. of blacks eligible for promotion}]}$$

were obtained for the four companies in each of three different 2-year periods. These data are presented below in tabular form.

Period	Company			
	1	2	3	4
1	3	5	5	4
2	4	4	3	5
3	8	12	10	9

a. Using dummy variables, write down an appropriate regression model for this data set. Is this a randomized-blocks design?

b. Use the accompanying table to test whether there are any significant differences among the average rate discrepancies for the four companies. Tukey's T for testing additivity equals 1.893 with 5 degrees of freedom.

Source	df	SS	MS	F
Periods	2	84.5000	42.2500	33.800
Companies	3	6.0000	2.0000	1.600
Error	6	7.5000	1.2500	

c. Does Tukey's test for these data indicate that a "removable" interaction effect is present?

d. If one finds no significant differences among the rate discrepancies for the four companies, would this support the contention that none of the companies has a discriminatory promotional policy?

e. Comment on the suitability of this analysis in view of the fact that the response variable is a difference in proportions.

7. Suppose that in a study to compare body sizes of three genotypes of fourth-instar silkworm, the mean length (in millimeters) for separately reared cocoons of heterozygous (HET), homozygous (HOM), and wild (WLD) silkworms was determined at five laboratory sites; the data are given in the table below.[7]

Variable	Site				
	1	2	3	4	5
HOM	29.87	28.16	32.08	30.84	29.44
HET	32.51	30.82	34.17	33.46	32.99
WLD	35.76	33.14	36.29	34.95	35.89

a. Assuming that this is a randomized-blocks type of experiment, what are the blocks and what are the treatments?

b. Comment on whether you think a randomized-blocks analysis is appropriate for this experiment.

---

[7] Adapted from a study by Sokal and Karlen (1964).

c. Carry out an appropriate analysis of the data for this experiment and state your conclusions. Make sure to present the ANOVA table and to state the null hypothesis for each test performed.

8. In Problem 3, the independent variable of interest is an interval-scale variable, copper ion concentration.

   a. Treating copper ion concentration as a five-level categorical predictor, as in the randomized-blocks analysis, corresponds to including a collection of polynomial terms for copper ion concentration (e.g., X, X^2, . . .) in the model. What is the highest order of such implicit terms?
   b. State a general principle in terms of k levels of a predictor, corresponding to the example in part (a).
   c. State the expansion of the regression model given in Problem 3(b) that encompasses the polynomial terms discussed in part (a) above.
   d. Use a computer program to fit the model stated in part (c). Center X by subtracting 0.330. (Why?) Provide estimated regression coefficients.
   e. Provide a multiple-partial F test comparing the model in Problem 3(b) and 3(g) to that fitted here; the latter is equivalent to the model fitted in 3(d).

9. Assume that the following table came from the analysis of a randomized-blocks design ANOVA:

Source	df	SS	MS	F
Treatments	4	b	e	5.00
Blocks	a	c	48.00	6.00
Error	20	d	f	

   a–f. Determine the values for a through f to complete the source table. Show your work.
   g. Provide tests of treatments and blocks with $\alpha = .05$. What are your conclusions?

10. An educator has class average scores on a standardized test for classes from each grade from first to seventh at each of four schools in a city. To assess differences between schools, the educator decides to conduct a randomized-blocks ANOVA, treating grade as a blocking factor.

   a. Conduct the appropriate analysis. Provide a source table.
   b. Provide F tests of school and grade differences. Use $\alpha = .05$. What do you conclude?
   c. Explain how using the children's original scores may be better than using means.

Observation	School	Grade	Score
1	1	1	70.4
2	1	2	74.5
3	1	3	49.6
4	1	4	72.9
5	1	5	60.3
6	1	6	59.5
7	1	7	77.0
8	2	1	73.8
9	2	2	71.0
10	2	3	65.0
11	2	4	73.2
12	2	5	79.7
13	2	6	82.5
14	2	7	60.9
15	3	1	71.3
16	3	2	62.7
17	3	3	71.9
18	3	4	99.4
19	3	5	74.4
20	3	6	82.8
21	3	7	74.4
22	4	1	68.5
23	4	2	79.5
24	4	3	72.1
25	4	4	77.0
26	4	5	86.5
27	4	6	101.8
28	4	7	100.7

References

Armitage, P. (1971). *Statistical Methods in Medical Research*. Oxford: Blackwell Scientific Publications.

Ostle, B. (1963). *Statistics in Research*, 2nd ed. Ames, Iowa: Iowa State University Press.

Peng, K. C. (1967). *Design and Analysis of Scientific Experiments*. Reading, Mass.: Addison-Wesley Publishing Company.

Sokal, R. R., and Karlen, I. (1964). "Competition Among Genotypes in *Tibolium castaneum* at Varying Densities and Gene Frequencies (the Black Locus)." *Genetics*, 49: 195–211.

Tukey, J. W. (1949). "One Degree of Freedom for Nonadditivity." *Biometrics*, 5: 232.

Two-Way ANOVA with Equal Cell Numbers

19-1 Preview

In this chapter we shall consider the analysis of the simplest two-way data layout involving more than one observation per cell, that is, the layout for which the number of observations in any given cell is at least two and is exactly the same as in any other cell (as previously illustrated in Figure 18-1b). We will see in this chapter that having equal cell numbers makes for a straightforward analysis involving only slightly more involved calculations than those for a randomized-blocks experiment. On the other hand, when there are unequal cell numbers, the analysis is much more complicated (see Chapter 20).

Two-way layouts with equal cell numbers are rarely seen in observational studies but are often generated by design in experimental studies where the investigator chooses the levels of the factors and the allocation of subjects to the various factor combinations. Such two-way layouts with equal cell numbers can be obtained by:

1. *blocking* so that several (but equal numbers of) observations on each treatment are made in each block,

2. *stratifying* according to the levels of the two factors of interest and then sampling within each stratum, and

3. *forming treatment combinations* (i.e., cells) and then allocating these combinations to individuals.

The design chosen depends, of course, on the study characteristics. If, for example, we wanted to eliminate the effects of a confounding factor, we could use blocking. If, on the other hand, we were interested in measuring the respiratory function of industrial workers in different plants (factor 1) subject to different levels of exposure to some substance (factor 2), a stratified sampling procedure would be appropriate. Or, if we were interested in the effects of combinations of different dosages of two different drugs, it would be appropriate to randomly assign the different drug combinations to different groups of subjects.

Regardless of the experimental design, the importance of having more than one observation at each combination of factor levels should be apparent. For example, in the respiratory function example mentioned above, if only one person subject to a certain level of exposure was examined from a given plant, there would be no direct way to determine how the responses of other persons in the same circumstances differed from the response of that individual. Similarly, for the drug example, if no more than one individual received a specific treatment combination of drugs, there would be no way to assess the variation in response among persons receiving that same treatment combination. Thus, a major reason for having more than one observation at each combination of factor levels (i.e., in each cell) is to be able to compute a pure estimate of experimental error (σ^2).

Pure in this context means a within-cell measure. The use of a randomized-blocks design (with a single observation per cell) precludes the possibility of obtaining a within-cell estimate of σ^2. However, if the blocking does *only* what it is assumed to do (i.e., eliminate the effects of confounding factors), it is still possible to obtain an estimate of σ^2 (although not pure) from what would ordinarily measure block–treatment interaction (which is assumed not to exist).

The detection of an interaction effect between two factors, although not of interest for the randomized-blocks design (in which the blocks are not considered as the levels of a factor or an independent variable), is an important reason for having repeated observations in each cell in two-way layouts. This notion of interaction will be elaborated on later in the chapter.

Example   In Table 19-1 the Cornell Medical Index (CMI) data of Table 17-3 from Daly's (1973) study has been categorized according to the levels of two factors:

> factor 1: percent black in the surrounding neighborhood (PSN);
> level 1 = low (≤50%), level 2 = high (>50%)

> factor 2: number of households in Turnkey neighborhood (NHT);
> level 1 = low (≤100), level 2 = high (>100)

Each of the four cells in Table 19-1 represents a combination of a level of factor 1 and a level of factor 2. There are 25 observations in each cell, which constitute random samples of 25 women who were heads of household selected from the four Turnkey neighborhoods as defined by the stratification of the two factors NHT and PSN.

> (It may be argued that the categorization scheme in Table 19-1 is inappropriate, since the neighborhood with 98 households (Cherryview) almost qualifies to be large ("high") in size. That is, perhaps the cut point for dichotomizing the variable NHT should be less than 100. Such a decision is often subjective and is by no means an easy one to make. Realizing this problem, we nevertheless have proceeded on the assumption that this categorization scheme is reasonable.)

We have already seen for these data that the *F* test for one-way ANOVA, which treats the four cells of Table 19-1 as the levels of a single variable "neighborhood," was nonsignificant. Thus, it was concluded that the mean CMI scores for the four neighborhoods, when compared simultaneously, are not significantly different from one another.

Because of this result, a researcher would justifiably tend not to expect much from further analysis, especially if no additional variables were taken into account. On the other hand, should the one-way ANOVA *F* test have led to the conclusion that the four neighbor-

TABLE 19-1   Cornell Medical Index scores for a sample of women from four
Turnkey neighborhoods

No. of Households (NHT)	% Black in Surrounding Neighborhoods (PSN)		Total
	Low ($\leq$50%)	High ($>$50%)	
Low ($\leq$100)	Cherryview: 49, 12, 28, 24, 16, 28, 21, 48, 30, 18, 10, 10, 15, 7, 6, 11, 13, 17, 43, 18, 6, 10, 9, 12, 12 ($n_{11} = 25$, $\bar{Y}_{11.} = 18.92$)	Easton: 13, 10, 20, 22, 14, 10, 8, 21, 35, 17, 23, 17, 23, 83, 21, 17, 41, 20, 25, 49, 41, 27, 37, 57 ($n_{12} = 25$, $\bar{Y}_{12.} = 26.84$)	$n_{1.} = 50$ $\bar{Y}_{1..} = 22.88$
High ($>$100)	Northhills: 20, 31, 19, 9, 7, 16, 11, 17, 9, 14, 10, 5, 15, 19, 29, 23, 70, 25, 6, 62, 2, 14, 26, 7, 55 ($n_{21} = 25$, $\bar{Y}_{21.} = 20.84$	Morningside: 5, 1, 44, 11, 4, 3, 14, 2, 13, 68, 34, 40, 36, 40, 22, 25, 14, 23, 26, 11, 20, 4, 16, 25, 17 ($n_{22} = 25$, $\bar{Y}_{22.} = 20.72$)	$n_{2.} = 50$ $\bar{Y}_{2..} = 20.78$
Total	$n_{.1} = 50$, $\bar{Y}_{.1.} = 19.88$	$n_{.2} = 50$, $\bar{Y}_{.2.} = 23.78$	$n_{..} = 100$ $\bar{Y}_{...} = 21.83$

hoods had significantly different mean CMI scores, it would have been of considerable interest to examine the nature of these differences. For example, do neighborhoods with a high percentage of blacks in the surrounding environs have significantly smaller mean CMI scores than those with a low percentage of blacks in surrounding environs? Or do neighborhoods with a large number of households have significantly smaller mean CMI scores than neighborhoods with a small number of households? Or does the amount and direction of the difference in CMI scores between neighborhoods of different size depend significantly on the racial makeup of the surrounding environs (i.e., is their an interaction effect)?

We shall demonstrate how these questions can be answered using two-way ANOVA. In fact, in spite of the nonsignificance of the one-way ANOVA F test, it will be of interest to perform a two-way ANOVA anyhow in order to quantify the separate effects of the factors PSN and NHT and, even more important, to examine the possibility of an interaction between these two factors.

19-2 Use of a Table of Cell Means

The important first step in examining a two-way layout should always be to construct a table of cell means. For our CMI data, such a table is presented in Table 19-2. From the table it can be observed that:

1.   The mean CMI score for low NHT is larger than that for high NHT:

$$\hat{\mu}_{1.} - \hat{\mu}_{2.} = \bar{Y}_{1..} - \bar{Y}_{2..} = 22.88 - 20.78 = 2.08$$

(This comparison measures what is called the _main effect_ of NHT.)

TABLE 19-2   Cell means for CMI data

NHT	PSN		Row Mean
	Low	High	
Low	$\hat{\mu}_{11} = \bar{Y}_{11\cdot} = 18.92$	$\hat{\mu}_{12} = \bar{Y}_{12\cdot} = 26.84$	$\hat{\mu}_{1\cdot} = \bar{Y}_{1\cdot\cdot} = 22.88$
High	$\hat{\mu}_{21} = \bar{Y}_{21\cdot} = 20.84$	$\hat{\mu}_{22} = \bar{Y}_{22\cdot} = 20.72$	$\hat{\mu}_{2\cdot} = \bar{Y}_{2\cdot\cdot} = 20.78$
Column Mean	$\hat{\mu}_{\cdot 1} = \bar{Y}_{\cdot 1\cdot} = 19.88$	$\hat{\mu}_{\cdot 2} = \bar{Y}_{\cdot 2\cdot} = 23.78$	$\hat{\mu}_{\cdot\cdot} = \bar{Y}_{\cdot\cdot\cdot} = 21.83$

2. The mean CMI score for low PSN is smaller than for high PSN:

$$\hat{\mu}_{\cdot 1} - \hat{\mu}_{\cdot 2} = \bar{Y}_{\cdot 1\cdot} - \bar{Y}_{\cdot 2\cdot} = 19.88 - 23.78 = -3.90$$

(This comparison measures what is called the *main effect* of PSN.)

3. There is little difference between high PSN and low PSN when NHT is high:

$$\hat{\mu}_{22} - \hat{\mu}_{21} = \bar{Y}_{22\cdot} - \bar{Y}_{21\cdot} = 20.72 - 20.84 = -0.12$$

whereas there is considerable difference between high PSN and low PSN when NHT is low:

$$\hat{\mu}_{12} - \hat{\mu}_{11} = \bar{Y}_{12\cdot} - \bar{Y}_{11\cdot} = 26.84 - 18.92 = 7.92$$

(These two comparisons measure what is called the *interaction between NHT and PSN*.)

Observation 1 suggests that persons from small Turnkey neighborhoods might not be as healthy as persons from large Turnkey neighborhoods (remember that the lower the CMI score, the healthier), which was as Daly theorized. Observation 2 suggests that persons from Turnkey neighborhoods with a high percentage of blacks in the surrounding neighborhood might not be as healthy as persons from Turnkey neighborhoods with a low percentage of blacks in the surrounding neighborhood. This observation is counter to Daly's theory. Finally, observation 3 suggests that there is little difference between neighborhoods with high and low black percentages in the surroundings when the Turnkey neighborhood size is large, whereas there is considerable difference between neighborhoods with high and low black percentages in the surroundings when the size of the Turnkey neighborhood is small.

Another way of describing the interaction effect pointed out in observation 3 is that when PSN is low, persons from neighborhoods with low NHT seem to be healthier than persons from neighborhoods with high NHT, but that when PSN is high, persons from neighborhoods with high NHT seem to be healthier than persons from neighborhoods with low NHT.

Of all the remarks above, clearly the most important is observation 3, which suggests some kind of interaction between NHT and PSN. That is, any difference in the health of persons from different PSN categories seems to depend on what NHT category is being considered. Or, equivalently, any difference between persons from different NHT categories appears to depend on the PSN category.

Nevertheless, we must remember that the differences found above were obtained from a sample and that, consequently, such differences could have occurred solely by chance. In other words, it is necessary to determine whether the above differences are statistically significant, and this can be done using two-way ANOVA.

19-2-1 Fixed, Random, or Mixed Model

To determine the appropriate significance tests for a two-way ANOVA, it is first necessary to specify whether each of the two factors is fixed or random. Although such a specification in the one-way ANOVA situation altered only the statement of the null and alternative hypotheses and not the form of the F test used, how the factors are classified in the two-way case affects the F test as well.

In fact, there are three different cases to be considered, depending on the classification of the two factors:

1. the fixed-effects case, where both factors are fixed,

2. the random-effects case, where both factors are random, and

3. the mixed-effects case, where one factor is fixed and the other is random.

In our example (i.e., Table 19-1), the classification of the factors NHT and PSN depends on one's point of view. The fixed-effects case would apply if the researcher was interested only in the particular Turnkey neighborhoods selected for study or (in terms of the two factors NHT and PSN) did not wish to make inferences to neighborhoods of different sizes or to different black percentages for surrounding neighborhoods. The random-effects case would apply if the Turnkey neighborhood sizes chosen were considered representative of a larger population of sizes of interest and the black percentages were representative of a larger population of black percentages of interest. The mixed-effects case would be applicable if one of the factors was considered fixed and the other random. Of these three cases, we believe that the random-effects case best represents the true situation. Nevertheless, the appropriate analysis for each case will be discussed.

19-2-2 Two-Way ANOVA Table for the Data of Table 19-1

Table 19-3 gives the two-way ANOVA table layout for the CMI data of Table 19-1. There are four sources of variation in this table, corresponding to the two main effects (for NHT and PSN, respectively), the interaction effect, and the error variation. Corresponding to these four sources, there are three null hypotheses that may be tested:

1. H_0: "no main effect of NHT"

2. H_0: "no main effect of PSN"

3. H_0: "no interaction effect between NHT and PSN"

Each of these hypotheses can be stated more precisely in terms of population cell means and/or variances, depending on whether the fixed-, random-, or mixed-effects case applies. For example, in the fixed-effects case, the null hypotheses may be given in terms of cell means (see Table 19-2) as follows:

1. H_0: $\mu_{1.} = \mu_{2.}$ (no main effect of NHT)

2. H_0: $\mu_{.1} = \mu_{.2}$ (no main effect of PSN)

3. H_0: $\mu_{22} - \mu_{21} - \mu_{12} + \mu_{11} = 0$ (no interaction effect between NHT and PSN)

TABLE 19-3   ANOVA table for CMI data of Table 19-1

Source	df	SS	MS	F Fixed	F Random	F NHT Fixed, PSN Random	F NHT Random, PSN Fixed
NHT	1	110.25	110.25	$0.43_{(1,96)}$	$0.27_{(1,1)}$	$0.27_{(1,1)}$	$0.43_{(1,96)}$
PSN	1	380.25	380.25	$1.50_{(1,96)}$	$0.94_{(1,1)}$	$1.50_{(1,96)}$	$0.94_{(1,1)}$
NHT × PSN	1	404.01	404.01	$1.60_{(1,96)}$	$1.60_{(1,96)}$	$1.60_{(1,96)}$	$1.60_{(1,96)}$
Error	96	24,302.00	253.14				
Total	99	25,196.51					

More will be said about the null hypotheses being tested in the fixed-, random-, and mixed-effects cases in Sections 19-4 and 19-7. Also, we shall describe in Section 19-3 how the sum-of-squares and degrees-of-freedom terms are determined for the general two-way ANOVA case. For now, we focus entirely on the F statistics given in Table 19-3, which differ according to how the factors are classified. The two numbers in parentheses next to each F statistic indicate the appropriate degrees of freedom to be used for that F test. None of the tests turns out to be significant, as might be expected from the previous results for one-way ANOVA.

Fixed-Effects Tests

Each F test for the fixed-effects case *always* involves dividing the mean square for the particular source being considered by the error mean square. The degrees of freedom correspond to the particular mean squares that are used. Thus,

$$F(\text{NHT}) = \frac{\text{MS(NHT)}}{\text{MS(error)}} = \frac{110.25}{253.14} = 0.43_{(1,96)}$$

$$F(\text{PSN}) = \frac{\text{MS(PSN)}}{\text{MS(error)}} = \frac{380.25}{253.14} = 1.50_{(1,96)}$$

$$F(\text{NHT} \times \text{PSN}) = \frac{\text{MS(NHT} \times \text{PSN)}}{\text{MS(error)}} = \frac{404.01}{253.14} = 1.60_{(1,96)}$$

Random-Effects Tests

In the random-effects case, the F test for each main effect involves dividing the mean square for the particular main effect being considered by the interaction mean square. Again, the degrees of freedom are based on the particular mean squares being used. Thus,

$$F(\text{NHT}) = \frac{\text{MS(NHT)}}{\text{MS(NHT} \times \text{PSN)}} = \frac{110.25}{404.01} = 0.27_{(1,1)}$$

$$F(\text{PSN}) = \frac{\text{MS(PSN)}}{\text{MS(NHT} \times \text{PSN)}} = \frac{380.25}{404.01} = 0.94_{(1,1)}$$

The F test for interaction is the same for the random-effects case as for the fixed-effects case.

Mixed-Effects Test: NHT Fixed, PSN Random

In the mixed-effects case, the F test for the main effect of the fixed factor involves dividing the mean square for that factor by the mean square for the interaction. The F test for the main effect of the random factor involves dividing the mean square for that factor by the mean square for the error. Thus, when NHT is fixed and PSN is random, we compute

$$F(\text{NHT}) = \frac{\text{MS(NHT)}}{\text{MS(NHT} \times \text{PSN)}} = \frac{110.25}{404.01} = 0.27_{(1,1)}$$

$$F(\text{PSN}) = \frac{\text{MS(PSN)}}{\text{MS(error)}} = \frac{380.25}{253.14} = 1.50_{(1,96)}$$

The F test for interaction is the same as for the fixed-effects and random-effects cases.

Mixed-Effects Test: NHT Random, PSN Fixed

In this situation the fixed and random factors have simply been reversed from the previous case. It thus follows that

$$F(\text{NHT}) = \frac{\text{MS(NHT)}}{\text{MS(error)}} = \frac{110.25}{253.14} = 0.43_{(1,96)}$$

$$F(\text{PSN}) = \frac{\text{MS(PSN)}}{\text{MS(NHT} \times \text{PSN)}} = \frac{380.25}{404.01} = 0.94_{(1,1)}$$

Again, the F test for interaction is the same as for the other cases.

19-3 General Methodology

In this section we shall describe the data configuration, computational formulae, and ANOVA table for the general balanced (i.e., having equal cell numbers) two-way situation, for which there are r levels of one factor (which we call the *row factor*), c levels of the other factor (which we call the column factor), and n observations in each of the rc cells.

19-3-1 General Layout of Data for Two-Way ANOVA

As with one-way ANOVA, the computations necessary for a two-way ANOVA are easily performed using an ordinary desk calculator when the data are suitably arranged. Table 19-4 gives a useful presentation of the data for the general two-way situation when there are an equal number of observations in each cell. Table 19-5 gives the corresponding table of (sample) cell means.

In Table 19-4 we have used three subscripts to differentiate the individual observations. The first two subscripts index the row and column (i.e., the cell), and the third subscript denotes the observation number within that cell. For example, Y_{122} denotes the second observation in cell $(1, 2)$ corresponding to row 1 and column 2. In general, Y_{ijk} denotes the kth observation in the (i, j)th cell of the table. Also, the cell total for the (i, j)th cell is denoted by T_{ij}, the ith row total is R_i, the jth column total is C_j, and the grand total is G. In other words,

$$R_i = \sum_{j=1}^{c} \sum_{k=1}^{n} Y_{ijk}, \qquad C_j = \sum_{i=1}^{r} \sum_{k=1}^{n} Y_{ijk}, \qquad G = \sum_{i=1}^{r} \sum_{j=1}^{c} \sum_{k=1}^{n} Y_{ijk}$$

In Table 19-5 we have denoted the mean of the n observations in cell (i, j) by $\bar{Y}_{ij\cdot}$. This sample mean estimates the population cell mean μ_{ij}. Also, the ith row mean is $\bar{Y}_{i\cdot\cdot}$, the jth

TABLE 19-4 General layout of data for two-way ANOVA with equal numbers of observations per cell

Row Factor	Column Factor				Row Total
	1	2	$\cdots$	c	
1	$(Y_{111}, Y_{112}, \ldots, Y_{11n})$ T_{11}	$(Y_{121}, Y_{122}, \ldots, Y_{12n})$ T_{12}	$\cdots$	$(Y_{1c1}, Y_{1c2}, \ldots, Y_{1cn})$ T_{1c}	R_1
2	$(Y_{211}, Y_{212}, \ldots, Y_{21n})$ T_{21}	$(Y_{221}, Y_{222}, \ldots, Y_{22n})$ T_{22}	$\cdots$	$(Y_{2c1}, Y_{2c2}, \ldots, Y_{2cn})$ T_{2c}	R_2
$\vdots$	$\vdots$	$\vdots$	$\vdots$	$\vdots$	$\vdots$
r	$(Y_{r11}, Y_{r12}, \ldots, Y_{r1n})$ T_{r1}	$(Y_{r21}, Y_{r22}, \ldots, Y_{r2n})$ T_{r2}	$\cdots$	$(Y_{rc1}, Y_{rc2}, \ldots, Y_{rcn})$ T_{rc}	R_r
Column Total	C_1	C_2	$\cdots$	C_c	G

TABLE 19-5 Sample cell means for two-way ANOVA

Row Factor	Column Factor				Row Mean
	1	2	$\cdots$	c	
1	$\bar{Y}_{11\cdot}$	$\bar{Y}_{12\cdot}$	$\cdots$	$\bar{Y}_{1c\cdot}$	$\bar{Y}_{1\cdot\cdot}$
2	$\bar{Y}_{21\cdot}$	$\bar{Y}_{22\cdot}$	$\cdots$	$\bar{Y}_{2c\cdot}$	$\bar{Y}_{2\cdot\cdot}$
$\vdots$	$\vdots$	$\vdots$		$\vdots$	$\vdots$
r	$\bar{Y}_{r1\cdot}$	$\bar{Y}_{r2\cdot}$	$\cdots$	$\bar{Y}_{rc\cdot}$	$\bar{Y}_{r\cdot\cdot}$
Column Mean	$\bar{Y}_{\cdot1\cdot}$	$\bar{Y}_{\cdot2\cdot}$	$\cdots$	$\bar{Y}_{\cdot c\cdot}$	$\bar{Y}_{\cdots}$

column mean is $\bar{Y}_{\cdot j\cdot}$, and the grand (overall) mean is $\bar{Y}_{\cdots}$. Thus, we have

$$\bar{Y}_{ij\cdot} = \frac{1}{n}\sum_{k=1}^{n} Y_{ijk}, \qquad \bar{Y}_{i\cdot\cdot} = \frac{R_i}{cn}, \qquad \bar{Y}_{\cdot j\cdot} = \frac{C_j}{rn}, \qquad \bar{Y}_{\cdots} = \frac{G}{rcn}$$

Example In our earlier example (Table 19-1), $r = c = 2$ and $n = 25$. Tables 19-6 and 19-7 give the data layout and table of cell means for an example in which $r = 3$, $c = 3$, and $n = 12$. This example (although artificial) considers one kind of data set used in occupational health studies for evaluating the health status of industrial workers. The dependent variable here is forced expiratory volume (FEV), which is a measure of respiratory function. Very low FEV indicates possible respiratory dysfunction, whereas high FEV indicates good respiratory function. In this example observations are taken on $n = 12$ individuals in each of three plants in a given industry where workers are exposed primarily to one of three toxicological substances. Thus, we have two factors, each with three levels. The categories of the row factor (PLANT) are labeled 1, 2, and 3 in Table 19-6. The categories of the column factor (TOXSUB) are labeled A, B, and C. Among the questions of interest here are:

1. Does the mean FEV level differ among plants (i.e., is there a main effect due to PLANT)?

TABLE 19-6   Forced expiratory volume classified by plant and toxicological exposure

| | Toxic Substance | | | |
Plant	A	B	C	Row Total
1	4.64, 5.92, 5.25, 6.17, 4.20, 5.90, 5.07, 4.13, 4.07, 5.30, 4.37, 3.76 ($T_{11} = 58.78$)	3.21, 3.17, 3.88, 3.50, 2.47, 4.12, 3.51, 3.85, 4.22, 3.07, 3.62, 2.95 ($T_{12} = 41.57$)	3.75, 2.50, 2.65, 2.84, 3.09, 2.90, 2.62, 2.75, 3.10, 1.99, 2.42, 2.37 ($T_{13} = 32.98$)	$R_1 = 133.33$
2	5.12, 6.10, 4.85, 4.72, 5.36, 5.41, 5.31, 4.78, 5.08, 4.97, 5.85, 5.26 ($T_{21} = 62.81$)	3.92, 3.75, 4.01, 4.64, 3.63, 3.46, 4.01, 3.39, 3.78, 3.51, 3.19, 4.04 ($T_{22} = 45.33$)	2.95, 3.21, 3.15, 3.25, 2.30, 2.76, 3.01, 2.31, 2.50, 2.02, 2.64, 2.27 ($T_{23} = 32.37$)	$R_2 = 140.51$
3	4.64, 4.32, 4.13, 5.17, 3.77, 3.85, 4.12, 5.07, 3.25, 3.49, 3.65, 4.10 ($T_{31} = 49.56$)	4.95, 5.22, 5.16, 5.35, 4.35, 4.89, 5.61, 4.98, 5.77, 5.23, 4.86, 5.15 ($T_{32} = 61.52$)	2.95, 2.80, 3.63, 3.85, 2.19, 3.32, 2.68, 3.35, 3.12, 4.11, 2.90, 2.75 ($T_{33} = 37.65$)	$R_3 = 148.73$
Column Total	$C_1 = 171.15$	$C_2 = 148.42$	$C_3 = 103.00$	$G = 422.57$

TABLE 19-7   Cell means for data of Table 19-6

| | Toxic Substance | | | Row Mean |
Plant	A	B	C	
1	4.90	3.46	2.75	3.70
2	5.23	3.78	2.70	3.90
3	4.13	5.13	3.13	4.13
Column Mean	4.75	4.12	2.86	3.91

2. Does the mean FEV level differ among exposure categories (i.e., is there a main effect due to TOXSUB)?

3. Do the differences in mean FEV levels among plants depend on the exposure category, and vice versa (i.e., is there an interaction effect between PLANT and TOXSUB)?

   A preliminary evaluation of these questions can be made by examining the cell means in Table 19-7. Of the three plants, plant 1 has the lowest mean FEV (3.70), followed by plant 2 (3.90), and then plant 3 (4.13). This suggests that the workers in plant 1 have poorer respiratory health than those in plant 2, and so on. Nevertheless, if the 3.70 value for plant 1 is considered clinically normal, then despite the differences observed, all plants will be given a "clean bill of health." Furthermore, these differences might have occurred solely by chance (i.e., might not be statistically significant).

In addition, it can be seen from Table 19-7 that exposure to substance C is associated with the poorest respiratory health (2.86), whereas exposures to substances A (4.75) and B (4.12) are associated with considerably better respiratory health. Again, it is necessary to decide whether the 2.86 value should be considered meaningfully low in a practical sense and to determine whether the differences among substances A, B, and C are statistically significant.

Finally, it can be observed from Table 19-7 that the differences among plants depend somewhat on the toxicological exposure being considered. For example, when considering toxic substance A, plant 3 has the lowest mean FEV (4.13). However, for toxic substance B, plant 1 has the lowest mean (3.46); for toxic substance C, plant 2 has the lowest (2.70). Furthermore, the magnitude of the differences among plants also varies with the toxic substance. For toxic substance B, the difference between the highest and lowest plant means is $5.13 - 3.46 = 1.67$, whereas for toxic substances A and C the maximum differences are smaller ($5.23 - 4.13 = 1.10$ and $3.13 - 2.70 = 0.43$, respectively). Such fluctuations suggest the possibility of a significant interaction effect, although this must be verified statistically.

19-3-2 ANOVA Table for Two-Way ANOVA

Table 19-8 gives the general form of the two-way ANOVA table when there are r levels of the row factor and c levels of the column factor. Table 19-9 gives the corresponding ANOVA table associated with the FEV data of Table 19-6. From these tables it can be seen that the total (corrected) sum of squares (TSS) has been split up into the four components, SSR (rows), SSC (columns), SSRC (R × C interaction), and SSE (error), based on the following fundamental equation:

$$TSS = SSR + SSC + SSRC + SSE$$

or

$$\sum_{i=1}^{r} \sum_{j=1}^{c} \sum_{k=1}^{n} (Y_{ijk} - \bar{Y}_{...})^2 = \sum_{i=1}^{r} \sum_{j=1}^{c} \sum_{k=1}^{n} (\bar{Y}_{i..} - \bar{Y}_{...})^2 + \sum_{i=1}^{r} \sum_{j=1}^{c} \sum_{k=1}^{n} (\bar{Y}_{.j.} - \bar{Y}_{...})^2$$

$$+ \sum_{i=1}^{r} \sum_{j=1}^{c} \sum_{k=1}^{n} (\bar{Y}_{ij.} - \bar{Y}_{i..} - \bar{Y}_{.j.} + \bar{Y}_{...})^2$$

$$+ \sum_{i=1}^{r} \sum_{j=1}^{c} \sum_{k=1}^{n} (Y_{ijk} - \bar{Y}_{ij.})^2 \qquad (19.1)$$

or, equivalently,

$$\sum_{i=1}^{r} \sum_{j=1}^{c} \sum_{k=1}^{n} (Y_{ijk} - \bar{Y}_{...})^2 = cn \sum_{i=1}^{r} (\bar{Y}_{i..} - \bar{Y}_{...})^2 + rn \sum_{j=1}^{n} (\bar{Y}_{.j.} - \bar{Y}_{...})^2$$

$$+ n \sum_{i=1}^{r} \sum_{j=1}^{c} (\bar{Y}_{ij.} - \bar{Y}_{i..} - \bar{Y}_{.j.} + \bar{Y}_{...})^2 + \sum_{i=1}^{r} \sum_{j=1}^{c} \sum_{k=1}^{n} (Y_{ijk} - \bar{Y}_{ij.})^2$$

Computational formulae for the sums of squares for each of these sources are given in the "SS" column of Table 19-8. These formulae involve the use of the grand total (G), the

TABLE 19-8 General (balanced) two-way ANOVA table

Source	df	SS	MS	F Fixed	F Random	F Row Fixed, Column Random	F Row Random, Column Fixed
Row (main effect)	$r-1$	$SSR = \dfrac{1}{cn}\sum_{i=1}^{r} R_i^2 - \dfrac{G^2}{rcn}$	$MSR = \dfrac{SSR}{r-1}$	$\dfrac{MSR}{MSE}$	$\dfrac{MSR}{MSRC}$	$\dfrac{MSR}{MSRC}$	$\dfrac{MSR}{MSE}$
Column (main effect)	$c-1$	$SSC = \dfrac{1}{rn}\sum_{j=1}^{c} C_j^2 - \dfrac{G^2}{rcn}$	$MSC = \dfrac{SSC}{c-1}$	$\dfrac{MSC}{MSE}$	$\dfrac{MSC}{MSRC}$	$\dfrac{MSC}{MSE}$	$\dfrac{MSC}{MSRC}$
Row × column (interaction)	$(r-1)(c-1)$	$SSRC = \dfrac{1}{n}\sum_{i=1}^{r}\sum_{j=1}^{c} T_{ij}^2 - SSR - SSC - \dfrac{G^2}{rcn}$	$MSRC = \dfrac{SSRC}{(r-1)(c-1)}$	$\dfrac{MSRC}{MSE}$	$\dfrac{MSRC}{MSE}$	$\dfrac{MSRC}{MSE}$	$\dfrac{MSRC}{MSE}$
Error	$rc(n-1)$	$SSE = TSS - SSR - SSC - SSRC$	$MSE = \dfrac{SSE}{rc(n-1)}$				
Total	$rcn-1$	$TSS = \sum_{i=1}^{r}\sum_{j=1}^{c}\sum_{k=1}^{n} Y_{ijk}^2 - \dfrac{G^2}{rcn}$					

TABLE 19-9 Two-way ANOVA for data of Table 19-6

Source	df	SS	MS	F			
				Fixed	Random	PLANT Fixed, TOXSUB Random	PLANT Random, TOXSUB Fixed
PLANT	2	3.299	1.649	$\dfrac{1.649}{0.2684} = 6.14^{**}$	$\dfrac{1.649}{6.128} = 0.27$	$\dfrac{1.649}{6.128} = 0.27$	$\dfrac{1.649}{0.2684} = 6.14^{**}$
TOXSUB	2	66.889	33.445	$\dfrac{33.445}{0.2684} = 124.60^{**}$	$\dfrac{33.445}{6.128} = 5.46$	$\dfrac{33.445}{0.2684} = 124.60^{**}$	$\dfrac{33.445}{6.128} = 5.46$
PLANT×TOXSUB	4	24.510	6.128	$\dfrac{6.128}{0.2684} = 22.83^{**}$	$\dfrac{6.128}{0.2684} = 22.83^{**}$	$\dfrac{6.128}{0.2684} = 22.83^{**}$	$\dfrac{6.128}{0.2684} = 22.83^{**}$
Error	99	26.576	0.2684				
Total	107	121.274					

row totals (R_i), the column totals (C_j), the cell totals (T_{ij}), and the individual observations (Y_{ijk}). An efficient stepwise procedure for making these calculations is as follows:

1. Calculate

$$\text{TSS} = \sum_{i=1}^{r} \sum_{j=1}^{c} \sum_{k=1}^{n} Y_{ijk}^2 - \frac{G^2}{rcn}$$

$$\text{SSR} = \frac{1}{cn} \sum_{i=1}^{r} R_i^2 - \frac{G^2}{rcn} \qquad (19.2)$$

$$\text{SSC} = \frac{1}{rn} \sum_{j=1}^{c} C_j^2 - \frac{G^2}{rcn}$$

2. Calculate the sum of squares for cells as

$$\text{SS(cells)} = \frac{1}{n} \sum_{i=1}^{r} \sum_{j=1}^{c} T_{ij}^2 - \frac{G^2}{rcn}$$

(This sum of squares, given by the general formula

$$\text{SS(cells)} = \sum_{i=1}^{r} \sum_{j=1}^{c} \sum_{k=1}^{n} (\bar{Y}_{ij\cdot} - \bar{Y}_{\cdots})^2$$

is a measure of the variation in the cell means about the overall mean.)

3. Calculate SSRC by subtraction:

$$\text{SSRC} = \text{SS(cells)} - \text{SSR} - \text{SSC} \qquad (19.3)$$

4. Calculate SSE by subtraction:

$$\text{SSE} = \text{TSS} - \text{SSR} - \text{SSC} - \text{SSRC} = \text{TSS} - \text{SS(cells)} \qquad (19.4)$$

For the FEV data of Table 19-6, the procedure is illustrated as follows:

1. $$\text{TSS} = \sum_{i=1}^{r} \sum_{j=1}^{c} \sum_{k=1}^{n} Y_{ijk}^2 - \frac{G^2}{rcn}$$

$$= \underbrace{[(4.64)^2 + (5.92)^2 + \cdots + (2.90)^2 + (2.75)^2]}_{\text{a sum of 108 terms}} - \frac{(422.57)^2}{108}$$

$$= 1{,}774.657 - 1{,}653.383$$

$$= 121.274$$

$$\text{SSR(PLANT)} = \frac{1}{cn} \sum_{i=1}^{r} R_i^2 - \frac{G^2}{rcn}$$

$$= \frac{1}{36} [(133.33)^2 + (140.51)^2 + (148.73)^2] - 1{,}653.383$$

$$= 3.299$$

$$SSC(TOXSUB) = \frac{1}{rn} \sum_{j=1}^{c} C_j^2 - \frac{G^2}{rcn}$$

$$= \frac{1}{36} [(171.15)^2 + (148.42)^2 + (103.00)^2] - 1,653.383$$

$$= 66.889$$

2.
$$SS(cells) = \frac{1}{n} \sum_{i=1}^{r} \sum_{j=1}^{c} T_{ij}^2 - \frac{G^2}{rcn}$$

$$= \frac{1}{12} [(58.78)^2 + (41.57)^2 + \cdots + (61.52)^2 + (37.65)^2] - 1,653.383$$

$$= 94.698$$

3.
$$SSRC(PLANT \times TOXSUB) = SS(cells) - SSR - SSC$$
$$= 94.698 - 3.299 - 66.889$$
$$= 24.510$$

4.
$$SSE = TSS - SS(cells)$$
$$= 121.274 - 94.698$$
$$= 26.576$$

The degrees of freedom associated with these sums of squares are as follows:

SSR has $r - 1$ df
SSC has $c - 1$ df
SSRC has $(r - 1)(c - 1)$ df   (19.5)
SSE has $rc(n - 1)$ df
TSS has $rcn - 1$ df

Each mean-square term is obtained (as usual) by dividing the corresponding sum of squares by its associated degrees of freedom. The appropriate *F* statistics to use, as discussed earlier, will depend on whether the row and column factors are classified as fixed or random. These *F* tests are described in the next section.

19-4 *F* Tests for Two-Way ANOVA

The null hypotheses of interest for two-way ANOVA, as well as the basic statistical assumptions required for validly testing them, can be stated quite generally so as to encompass the four possible schemes of factor classification.[1] These are given as follows:

The Null Hypotheses

1. $H_0(R)$: *There is no row factor (main) effect* (i.e., there are no differences among the effects of the levels of the row factor).

---

[1] However, the null hypotheses are quite different when they are more precisely stated in terms of population cell means and/or variances.

2. $H_0(C)$: *There is no column factor (main) effect* (i.e., there are no differences among the effects of the levels of the column factor).

3. $H_0(RC)$: *There is no interaction effect between rows and columns* (i.e., the row level effects within any one column are the same as within any other column, and the column level effects within any one row are the same as within any other row).

The Assumptions

1. All observations are *statistically independent* of one another.

2. Each observation comes from a *normally distributed population*.

3. Each observation has the same population variance (i.e., there is the usual assumption of *variance homogeneity*).

As previously stated, the appropriate F statistics depend on whether the row and column factors are classified as fixed or random. We will see later, when specifying the different statistical models for two-way ANOVA, that the mean-square term associated with a given source of variation will estimate different quantities, depending on whether the row and column factors are fixed or random; this is the reason for the different denominators used in the various F tests of two-way ANOVA.

Nevertheless, regardless of how the factors are classified, the F statistic used to test $H_0(RC)$ of no row–column interaction is always of the form

$$F(\mathrm{RC}) = \frac{\mathrm{MSRC}}{\mathrm{MSE}}$$

with $(r - 1)(c - 1)$ and $rc(n - 1)$ degrees of freedom.[2]

The tests for main effects, however, differ as follows with respect to the factor classification scheme:

1. *Rows and columns fixed.* Divide the mean squares for rows and columns by the mean square for error:

$$F(\mathrm{R}) = \frac{\mathrm{MSR}}{\mathrm{MSE}}$$

with $r - 1$ and $rc(n - 1)$ degrees of freedom;

$$F(\mathrm{C}) = \frac{\mathrm{MSC}}{\mathrm{MSE}}$$

with $c - 1$ and $rc(n - 1)$ degrees of freedom.

2. *Rows and columns random.* Divide the mean squares for rows and columns by the mean square for interaction:

$$F(\mathrm{R}) = \frac{\mathrm{MSR}}{\mathrm{MSRC}}$$

[2] $F(\mathrm{RC})$ denotes the F test of $H_0(\mathrm{RC})$. Similarly, $F(\mathrm{R})$ and $F(\mathrm{C})$ denote the F tests of $H_0(\mathrm{R})$ and $H_0(\mathrm{C})$, respectively.

with $r - 1$ and $(r - 1)(c - 1)$ degrees of freedom;

$$F(C) = \frac{MSC}{MSRC}$$

with $c - 1$ and $(r - 1)(c - 1)$ degrees of freedom.

3. *Rows fixed and columns random.* Divide the mean square for rows by the mean square for interaction:

$$F(R) = \frac{MSR}{MSRC}$$

with $r - 1$ and $(r - 1)(c - 1)$ degrees of freedom. Divide the mean square for columns by the mean square for error:

$$F(C) = \frac{MSC}{MSE}$$

with $c - 1$ and $rc(n - 1)$ degrees of freedom.

4. *Rows random and columns fixed.* Divide the mean square for rows by the mean square for error:

$$F(R) = \frac{MSR}{MSE}$$

with $r - 1$ and $rc(n - 1)$ degrees of freedom. Divide the mean square for columns by the mean square for interaction:

$$F(C) = \frac{MSC}{MSRC}$$

with $c - 1$ and $(r - 1)(c - 1)$ degrees of freedom.

For the FEV data, the classification of the factors would, as in the previous examples, depend on the point of view of the researcher. If, for example, the plants and toxicological substances were the only ones of interest, both factors would be considered fixed. However, if the plants were considered to be a sample from a large population of plants of interest and if the toxicological substances were representative of a population of toxic agents of interest, both factors would be considered random. Of course, the classification would be mixed if one of these factors was considered fixed and the other random.

We will not attempt here to defend any particular classification scheme for the factors in the FEV example, especially since the example is artificial to begin with. However, it is important to notice from Table 19-9 that the decisions regarding certain null hypotheses will be different depending on how the factors are classified. This can be seen from Table 19-9 as follows:

1. *Both factors fixed.* Both main effects are significant, since $F(\text{PLANT}) = 6.14^{**}$ (with 2 and 99 degrees of freedom) and $F(\text{TOXSUB}) = 124.60^{**}$ (with 2 and 99 degrees of freedom).

2. *Both factors random.* Neither main effect is significant, since $F(\text{PLANT}) = 0.27$ (with 2 and 4 degrees of freedom) and $F(\text{TOXSUB}) = 5.46$ (with 2 and 4 degrees of freedom).

3. *PLANT fixed, TOXSUB random.* The PLANT main effect is not significant and the TOXSUB main effect is significant, since $F(\text{PLANT}) = 0.27$ (with 2 and 4 degrees of freedom) and $F(\text{TOXSUB}) = 124.60^{**}$ (with 2 and 99 degrees of freedom).

4. *PLANT random, TOXSUB fixed.* The PLANT main effect is significant and the TOXSUB main effect is not significant, since $F(\text{PLANT}) = 6.14^{**}$ (with 2 and 99 degrees of freedom) and $F(\text{TOXSUB}) = 5.46$ (with 2 and 4 degrees of freedom).

Despite these differences among the main-effect test results, the most important finding from this analysis is that the interaction effect is significant:

$$F(\text{PLANT} \times \text{TOXSUB}) = 22.83^{**} \text{ (with 4 and 99 df)}$$

In Section 19-6 we shall discuss the interpretation of such interaction effects. For now it suffices to say that the presence of this interaction means that it does not make much sense to talk about the separate or independent effects (i.e., main effects) of PLANT and TOXSUB on FEV, since there is strong evidence that these factors do not affect FEV independently of one another.

Whether the factors are fixed or random, eight distinct patterns of significance may result, depending upon the significance (or nonsignificance) of the three tests involved (two main-effect tests and the interaction test). Labeling one factor A and the second factor B, the following table summarizes the possible outcomes (with significant results indicated by asterisks).

				Pattern				
Source	I	II	III	IV	V	VI	VII	VIII
A		*		*		*		*
B			*	*			*	*
A × B					*	*	*	*

In case I, the F tests for the A effect, the B effect, and the A × B interaction are all nonsignificant. In case VIII, all three are significant.

Each case deserves some comment. Case I leads us to conclude that there is no evidence of any relationship between the response variable and the predictors (either when considered separately or together). Case II implies that only factor A is related to the response variable. Similarly, Case III implies that only factor B is related to the response variable. In case IV, we conclude that both A and B affect the response. Furthermore, the nature of the relationship between factor A and the response is the same at all levels of factor B studied, and conversely. Cases V, VI, VII, and VIII all involve significant interaction. Most statisticians recommend discussing only this interaction in such cases. A significant interaction implies that both factors are important and that the level of the other factor must be known to characterize the effect of either factor on the response. See Section 19-6 for help in interpreting interactions.

19-5 Regression Model for Fixed-Effects Two-Way ANOVA

In this section we shall describe a particular regression model[3] and a related classical fixed-effects ANOVA model for two-way ANOVA when both factors are considered fixed. Random-effects and mixed-effects models are discussed in detail in Section 19-7. As in the one-way ANOVA and randomized-blocks ANOVA cases, a regression model for two-way ANOVA can be interpreted in terms of the cell, marginal, and overall means associated with the two-way layout (see Table 19-10). In the table,

$$\mu_{i\cdot} = \frac{1}{c} \sum_{j=1}^{c} \mu_{ij} \qquad i = 1, 2, \ldots, r$$

$$\mu_{\cdot j} = \frac{1}{r} \sum_{i=1}^{r} \mu_{ij} \qquad j = 1, 2, \ldots, c$$

$$\mu_{\cdot\cdot} = \frac{1}{rc} \sum_{i=1}^{r} \sum_{j=1}^{c} \mu_{ij}$$

TABLE 19-10 Table of population cell means for two-way layout

Rows	Column 1	Column 2	$\cdots$	Column c	Row Mean
1	μ_{11}	μ_{12}	$\cdots$	μ_{1c}	$\mu_{1\cdot}$
2	μ_{21}	μ_{22}	$\cdots$	μ_{2c}	$\mu_{2\cdot}$
$\vdots$	$\vdots$	$\vdots$		$\vdots$	$\vdots$
r	μ_{r1}	μ_{r2}	$\cdots$	μ_{rc}	$\mu_{r\cdot}$
Column Mean	$\mu_{\cdot 1}$	$\mu_{\cdot 2}$	$\cdots$	$\mu_{\cdot c}$	$\mu_{\cdot\cdot}$

19-5-1 Regression Model

When there are r rows and c columns, a regression model can be formulated involving $r - 1$ dummy variables for the row factor, $c - 1$ dummy variables for the column factor, and $(r - 1)(c - 1)$ interaction dummy variables constructed by forming products of each of the row dummy variables with each of the column dummy variables. Such a model can be expressed as follows:

$$Y = \mu + \sum_{i=1}^{r-1} \alpha_i X_i + \sum_{j=1}^{c-1} \beta_j Z_j + \sum_{i=1}^{r-1} \sum_{j=1}^{c-1} \gamma_{ij} X_i Z_j + E \qquad (19.6)$$

---

[3] Several other regression models can be defined, of course, depending on the coding scheme for the dummy variables. The regression model given here is the one most commonly used because of its natural connection with the classical fixed-effects two-way ANOVA model.

in which

$$X_i = \begin{cases} -1 & \text{for level } r \text{ of the row factor} \\ 1 & \text{for level } i \text{ of the row factor} \\ 0 & \text{otherwise} \end{cases}$$

and

$$Z_j = \begin{cases} -1 & \text{for level } c \text{ of the column factor} \\ 1 & \text{for level } j \text{ of the column factor} \\ 0 & \text{otherwise} \end{cases}$$

$(i = 1, 2, \ldots, r - 1; j = 1, 2, \ldots, c - 1)$.

The formulae relating the coefficients α_i, β_j, and γ_{ij} to the various means of Table 19-10 are given as follows:

$$
\begin{aligned}
\mu &= \mu_{..} \\
\alpha_i &= \mu_{i.} - \mu_{..} & i = 1, 2, \ldots, r - 1 \\
\beta_j &= \mu_{.j} - \mu_{..} & j = 1, 2, \ldots, c - 1 \\
\gamma_{ij} &= \mu_{ij} - \mu_{i.} - \mu_{.j} + \mu_{..} & i = 1, 2, \ldots, r - 1; \quad j = 1, 2, \ldots, c - 1 \\
-\sum_{i=1}^{r-1} \alpha_i &= \mu_{r.} - \mu_{..} \\
-\sum_{j=1}^{c-1} \beta_j &= \mu_{.c} - \mu_{..} \\
-\sum_{i=1}^{r-1} \gamma_{ij} &= \mu_{rj} - \mu_{r.} - \mu_{.j} + \mu_{..} & j = 1, 2, \ldots, c - 1 \\
-\sum_{j=1}^{c-1} \gamma_{ij} &= \mu_{ic} - \mu_{i.} - \mu_{.c} + \mu_{..} & i = 1, 2, \ldots, r - 1
\end{aligned}
$$

$$(19.7)$$

As with the other ANOVA regression analogies made in earlier chapters, the same F tests given in Table 19-8 when both factors are fixed can be obtained using the appropriate multiple-partial F tests concerning subsets of the coefficients in the regression model (19.6). Specifically, the multiple-partial F test of H_0: $\alpha_1 = \alpha_2 = \cdots = \alpha_{r-1} = 0$ for model (19.6) yields exactly the same F statistic as that used in standard (balanced) two-way fixed-effects ANOVA for testing the significance of the main effect of the row factor (i.e., $F = \text{MSR/MSE}$). Similarly, the multiple-partial F test of H_0: $\beta_1 = \beta_2 = \cdots = \beta_{c-1} = 0$ in model (19.6) yields exactly the same F statistic as that used in standard (balanced) two-way fixed-effects ANOVA for testing the significance of the main effect of the column factor (i.e., $F = \text{MSC/MSE}$). Finally, the multiple-partial F test of H_0: $\gamma_{ij} = 0$ ($i = 1, 2, \ldots, r - 1; j = 1, 2, \ldots, c - 1$) is identical to the (balanced) two-way fixed-effects ANOVA F test for interaction (i.e., $F = \text{MSRC/MSE}$).

The formulae given above may also be used to express the cell, row marginal, and column marginal means as functions of the regression coefficients, as follows:

$$\mu_{ij} = \mu + \alpha_i + \beta_j + \gamma_{ij} \qquad i = 1, 2, \ldots, r-1; \quad j = 1, 2, \ldots, c-1$$

$$\mu_{rj} = \mu - \sum_{i=1}^{r-1} \gamma_{ij} - \sum_{i=1}^{r-1} \alpha_i + \beta_j \qquad j = 1, 2, \ldots, c-1$$

$$\mu_{ic} = \mu - \sum_{j=1}^{c-1} \gamma_{ij} - \sum_{j=1}^{c-1} \beta_j + \alpha_i \qquad i = 1, 2, \ldots, r-1$$

$$\mu_{rc} = \mu - \sum_{i=1}^{r-1} \alpha_i - \sum_{j=1}^{c-1} \beta_j + \sum_{i=1}^{r-1}\sum_{j=1}^{c-1} \gamma_{ij}$$

The row marginal means are

$$\mu_{i\cdot} = \mu + \alpha_i \qquad i = 1, 2, \ldots, r-1$$

$$\mu_{r\cdot} = \mu - \sum_{i=1}^{r-1} \alpha_i$$

and the column marginal means are

$$\mu_{\cdot j} = \mu + \beta_j \qquad j = 1, 2, \ldots, c-1$$

$$\mu_{\cdot c} = \mu - \sum_{j=1}^{c-1} \beta_j$$

If a reference cell coding is used, the general form of model (19.6) stays the same, but the definitions of X_i and Z_j change. For example, we can define

$$X_i = \begin{cases} 1 & \text{for level } i \text{ of the row factor} \\ 0 & \text{otherwise} \end{cases} \qquad i = 1, 2, \ldots, r-1$$

and

$$Z_j = \begin{cases} 1 & \text{for level } j \text{ of the column factor} \\ 0 & \text{otherwise} \end{cases} \qquad j = 1, 2, \ldots, c-1$$

With these definitions, the parameters in model (19.6) have different interpretations than those given earlier. In particular, the cell means (for the above reference cell coding) can be expressed as the following functions of the parameters in model (19.6):

$$\mu_{ij} = \mu + \alpha_i + \beta_j + \gamma_{ij} \qquad i = 1, 2, \ldots, r-1; \quad j = 1, 2, \ldots, c-1$$
$$\mu_{rj} = \mu + \beta_j \qquad j = 1, 2, \ldots, c-1$$
$$\mu_{ic} = \mu + \alpha_i \qquad i = 1, 2, \ldots, r-1$$
$$\mu_{rc} = \mu$$

The row marginal means are

$$\mu_{i\cdot} = \mu + \alpha_i + \sum_{j=1}^{c-1} \frac{\beta_j + \gamma_{ij}}{c} \qquad i = 1, 2, \ldots, r-1$$

$$\mu_{r\cdot} = \mu + \sum_{j=1}^{c-1} \frac{\beta_j}{c}$$

Finally, the column marginal means are

$$\mu_{\cdot j} = \mu + \beta_j + \sum_{i=1}^{r-1} \frac{\alpha_i + \gamma_{ij}}{r} \qquad j = 1, 2, \ldots, c - 1$$

$$\mu_{\cdot c} = \mu + \sum_{i=1}^{r-1} \frac{\alpha_i}{r}$$

The reader should be cautioned that these expressions must be modified when cell-specific sample sizes are not all equal. Chapter 20 is devoted to two-way ANOVA with unequal cell-specific sample sizes.

19-5-2 Classical Two-Way Fixed-Effects ANOVA Model

When both factors are considered fixed, there are three types of effects to consider:

1. *Main effects of the row factor,* which are the differences between the various row means and the overall mean (i.e., $\mu_{i\cdot} - \mu_{\cdot\cdot}$, $i = 1, 2, \ldots, r$).

2. *Main effects of the column factor,* which are the differences between the various column means and the overall mean (i.e., $\mu_{\cdot j} - \mu_{\cdot\cdot}$, $j = 1, 2, \ldots, c$).

3. *Interaction effects,* which are differences between differences, of the form $(\mu_{ij} - \mu_{i\cdot}) - (\mu_{\cdot j} - \mu_{\cdot\cdot})$ or $(\mu_{ij} - \mu_{\cdot j}) - (\mu_{i\cdot} - \mu_{\cdot\cdot})$, $i = 1, 2, \ldots, r$ and $j = 1, 2, \ldots, c$.

The classical two-way fixed-effects ANOVA model involving such effects is of the following form:

$$Y_{ijk} = \mu + \alpha_i + \beta_j + \gamma_{ij} + E_{ijk} \tag{19.8}$$

where

$$\mu = \mu_{\cdot\cdot} = \text{overall mean}$$
$$\alpha_i = \mu_{i\cdot} - \mu_{\cdot\cdot} = \text{effect of row } i$$
$$\beta_j = \mu_{\cdot j} - \mu_{\cdot\cdot} = \text{effect of column } j$$
$$\gamma_{ij} = \mu_{ij} - \mu_{i\cdot} - \mu_{\cdot j} + \mu_{\cdot\cdot} = \text{interaction effect associated with cell } (i, j)$$
$$E_{ijk} = Y_{ijk} - \mu - \alpha_i - \beta_j - \gamma_{ij} = \text{error (or residual) associated with the } k\text{th}$$
$$\text{observation in cell } (i, j)$$

$(i = 1, 2, \ldots, r; j = 1, 2, \ldots, c; k = 1, 2, \ldots, n)$.

The following relationships are clearly satisfied by the effects in the model above:

$$\sum_{i=1}^{r} \alpha_i = 0, \qquad \sum_{j=1}^{c} \beta_j = 0, \qquad \sum_{i=1}^{r} \gamma_{ij} = 0, \qquad \sum_{j=1}^{c} \gamma_{ij} = 0 \tag{19.9}$$

It is also clear when comparing the regression coefficients in model (19.6) with the ANOVA effects in model (19.8) that the models are completely equivalent.

Finally, it should be pointed out that each of the effects in model (19.8) can be simply estimated using sample means, as follows:

$$\hat{\mu} = \bar{Y}_{...}$$
$$\hat{\alpha}_i = \bar{Y}_{i..} - \bar{Y}_{...} \qquad\qquad i = 1, 2, \ldots, r$$
$$\hat{\beta}_j = \bar{Y}_{.j.} - \bar{Y}_{...} \qquad\qquad j = 1, 2, \ldots, c \qquad\qquad (19.10)$$
$$\hat{\gamma}_{ij} = \bar{Y}_{ij.} - \bar{Y}_{i..} - \bar{Y}_{.j.} + \bar{Y}_{...} \qquad i = 1, 2, \ldots, r; \quad j = 1, 2, \ldots, c$$

19-6 Interactions in Two-Way ANOVA

In this section we shall describe several ways to evaluate interaction in two-way ANOVA. We focus, for convenience, on the fixed-effects case. Nevertheless, even though the parameters involved and the test statistics used for making inferences in the fixed-effects case will be different from those used in the random- and mixed-effects cases, the interpretations of interactions will generally be the same.

19-6-1 Concept of Interaction

Generally speaking, an interaction exists between two factors if the relationship among the effects associated with the levels of one factor differs according to the levels of the second factor. Another way of saying this is that an interaction represents an effect due to the joint influence of two factors, over and above the effects of each factor considered separately.

More specifically, there are three equivalent ways of describing or representing an interaction in statistical terms if we consider the two-way table of cell means and the various ways of writing the statistical model in the fixed-effects case.

Method 1: Interaction as a Difference in Differences of Means   In two-way ANOVA, an interaction exists between the row and column factors if any of the following equivalent statements are true:

1. For some pair of columns, the difference between the means in these columns for a given row is not equal to the difference between these means for some other row. For example, for rows 1 and 2 and columns 1 and 2, $\mu_{11} - \mu_{12} \neq \mu_{21} - \mu_{22}$.

2. For some pair of rows, the difference between the means in these rows for a given column is not equal to the difference between these means for some other column. For example, for rows 1 and 2 and columns 1 and 2, $\mu_{11} - \mu_{21} \neq \mu_{12} - \mu_{22}$.

3. For some cell in the table, the difference between that cell mean and its associated marginal row mean is not equal to the difference between its associated marginal column mean and the overall mean. For example, for the (i, j)th cell, $\mu_{ij} - \mu_{i\cdot} \neq \mu_{\cdot j} - \mu_{\cdot\cdot}$, or $\mu_{ij} - \mu_{i\cdot} - \mu_{\cdot j} + \mu_{\cdot\cdot} \neq 0$.

4. For some cell in the table, the difference between that cell mean and its associated marginal column mean is not equal to the difference between its associated marginal row mean and the overall mean. For example, $\mu_{ij} - \mu_{\cdot j} \neq \mu_{i\cdot} - \mu_{\cdot\cdot}$, or $\mu_{ij} - \mu_{i\cdot} - \mu_{\cdot j} + \mu_{\cdot\cdot} \neq 0$.

Another way of saying all this is that when there is no interaction, the relationship among the column effects (β_j's) is the same regardless of the row being considered, and vice versa. Also, since from statements 3 and 4

$$\mu_{ij} - \mu_{i\cdot} - \mu_{\cdot j} + \mu_{\cdot\cdot} = 0$$

when there is no interaction, it should be apparent that

$$\text{MSRC} = \frac{n}{(r-1)(c-1)} \sum_{i=1}^{r} \sum_{j=1}^{c} (Y_{ij} - \bar{Y}_{i\cdot} - \bar{Y}_{\cdot j} + \bar{Y}_{\cdot\cdot})^2,$$

which estimates

$$\frac{n}{(r-1)(c-1)} \sum_{i=1}^{r} \sum_{j=1}^{c} (\mu_{ij} - \mu_{i\cdot} - \mu_{\cdot j} + \mu_{\cdot\cdot})^2,$$

will be small when there is no interaction and large when there is interaction.

Method 2: *Interaction as an Effect in the Fixed-Effects Model*   An interaction exists if the appropriate fixed-effects ANOVA model is of the form

$$Y_{ijk} = \mu + \alpha_i + \beta_j + \gamma_{ij} + E_{ijk}$$

where $\gamma_{ij} \neq 0$ for at least one (i, j) pair.

(Representations 1 and 2 are completely equivalent. For example, if there is no interaction, then $\mu_{ij} = \mu + \alpha_i + \beta_j$, so that $\mu_{1j} - \mu_{2j} = (\mu + \alpha_1 + \beta_j) - (\mu + \alpha_2 + \beta_j) = \alpha_1 - \alpha_2$, which is independent of j. Thus, $\mu_{11} - \mu_{21} = \mu_{12} - \mu_{22} = \cdots = \mu_{1c} - \mu_{2c}$.)

Method 3: *Interaction as a Term in a Regression Model*   An interaction exists if the appropriate regression model (using dummy variables) contains a term that involves the product (or, in general, any function) of variables from different factors, for example, if the appropriate model is of the form

$$Y = \mu + \sum_{i=1}^{r-1} \alpha_i X_i + \sum_{j=1}^{c-1} \beta_j Z_j + \sum_{i=1}^{r-1} \sum_{j=1}^{c-1} \gamma_{ij} X_i Z_j + E$$

where at least one of the γ_{ij} is not 0.

(When $r = c = 2$, the model simplifies to

$$Y = \mu + \alpha_1 X_1 + \beta_1 Z_1 + \gamma_{11} X_1 Z_1 + E$$

where

$$X_1 = \begin{cases} -1 & \text{if level 2 of the row factor} \\ 1 & \text{if level 1 of the row factor} \end{cases}$$

$$Z_1 = \begin{cases} -1 & \text{if level 2 of the column factor} \\ 1 & \text{if level 1 of the column factor} \end{cases}$$

Then, $\mu = \mu_{\cdot\cdot}$, $\alpha_1 = \mu_{1\cdot} - \mu_{\cdot\cdot}$, $\beta_1 = \mu_{\cdot 1} - \mu_{\cdot\cdot}$, and $\gamma_{11} = \mu_{11} - \mu_{1\cdot} - \mu_{\cdot 1} + \mu_{\cdot\cdot}$.
    Representation 3 is equivalent to the other two representations provided that both independent variables (i.e., factors) are considered nominal and so are represented by dummy variables. If, however, both independent variables are continuous, a regression model with any kind of product term will exhibit an interaction effect of a somewhat different type not necessarily characterized by a nonzero difference of mean differences.)

19-6-2 Some Hypothetical Examples

We shall now consider some hypothetical two-way tables of population cell means that illustrate different patterns of interaction. These tables pertain to the example in Section 19-2, for which the factors were NHT and PSN and the dependent variable was CMI score. Subsequently, we shall examine the table of sample cell means actually obtained (Table 19-2), keeping in mind that the statistical test for interaction may negate whatever tentative trends are suggested by the sample means. We shall also examine the example of Section 19-3 in this light.

Row and Column Main Effects But No Interaction Effect

Table 19-11 presents three alternative layouts, each representing the general situation in which there is *both* a row main effect and a column main effect but no interaction effect. Keep in mind that each of these tables gives *population* (and not sample) mean values, so there is no sampling variation to consider.

The main effects are reflected in the differences between the marginal row means and between the marginal column means in each table. The lack of an interaction effect can be established by comparing the differences among the cell means, as discussed in Section 19-6-1. From Table 19-11a, for example, we have $\mu_{11} - \mu_{12} = \mu_{21} - \mu_{22}$, since $26 - 23 = 3 = 20 - 17$. Also, for this table, $\mu_{11} - \mu_{1.} - \mu_{.1} + \mu_{..} = 26 - 24.5 - 23 + 21.5 = 0$, and similar terms associated with the other three cells in the table are also 0. Furthermore, for this table, the model (19.8) can be shown to have the specific structure:

$$\mu_{ij} = 21.5 + \alpha_i + \beta_j$$

where

$$\alpha_i = \begin{cases} 3 & \text{if } i = 1 \\ -3 & \text{if } i = 2 \end{cases} \quad \text{and} \quad \beta_j = \begin{cases} 1.5 & \text{if } j = 1 \\ -1.5 & \text{if } j = 2 \end{cases}$$

Note that this model does not involve any γ_{ij} term (i.e., there is no interaction term in the model). Thus, we have

$$\mu_{11} = 21.5 + 3 + 1.5 = 26$$
$$\mu_{12} = 21.5 + 3 - 1.5 = 23$$
$$\mu_{21} = 21.5 - 3 + 1.5 = 20$$
$$\mu_{22} = 21.5 - 3 - 1.5 = 17$$

TABLE 19-11   Main effects but no interaction

(a)

NHT	PSN Low	PSN High	Row Mean
Low	26	23	24.5
High	20	17	18.5
Column Mean	23	20	21.5

(b)

NHT	PSN Low	PSN High	Row Mean
Low	18	26	22
High	20	28	24
Column Mean	19	27	23

(c)

NHT	PSN Low	PSN High	Row Mean
Low	18	26	22
High	16	24	20
Column Mean	17	25	21

The models for tables (b) and (c) are also no-interaction models; they have the particular forms, respectively, of

$$\mu_{ij} = 23 + \alpha_i + \beta_j$$

where

$$\alpha_i = \begin{cases} -1 & \text{if } i = 1 \\ 1 & \text{if } i = 2 \end{cases} \quad \text{and} \quad \beta_j = \begin{cases} -4 & \text{if } j = 1 \\ 4 & \text{if } j = 2 \end{cases}$$

and

$$\mu_{ij} = 21 + \alpha_i + \beta_j$$

where

$$\alpha_i = \begin{cases} 1 & \text{if } i = 1 \\ -1 & \text{if } i = 2 \end{cases} \quad \text{and} \quad \beta_j = \begin{cases} -4 & \text{if } j = 1 \\ 4 & \text{if } j = 2 \end{cases}$$

Exactly One Main Effect and No Interaction Effect

This situation is depicted in Table 19-12. Table 19-12a contains a main effect due to PSN but no main effect due to NHT. Table 19-12b contains a main effect due to NHT but no main effect due to PSN. One can also verify that there is no PSN × NHT interaction.

Same-Direction Interaction

Three examples of same-direction interaction are given in Table 19-13. In Table 19-13a we can see that $\mu_{11} - \mu_{12} = 26 - 23 = 3$, whereas $\mu_{21} - \mu_{22} = 20 - 13 = 7$. Also, $\mu_{11} - \mu_{1.} = 26 - 24.5 = 1.5$ and $\mu_{.1} - \mu_{..} = 23 - 20.5 = 2.5$, so that $\mu_{11} - \mu_{1.} - \mu_{.1} + \mu_{..} = -1.0$. The other interactions of the general form $(\mu_{ij} - \mu_{i.} - \mu_{.j} + \mu_{..})$ are similarly determined to be either 1.0 (when $i = 1$, $j = 2$ or when $i = 2$, $j = 1$) or -1.0 (when $i = 2$, $j = 2$).

These hypothetical results indicate that in *both* low and high NHT neighborhoods, persons in friendly surroundings (i.e., high PSN) are healthier (i.e., have a lower CMI score) than persons in unfriendly surroundings (i.e., low PSN), but that the extent of this difference is greater when there is a large number of households (high NHT) than when there is a small number (low NHT). In other words, then, at each level of NHT, the difference between the PSN level effects is in the same direction (i.e., high PSN is associated with a lower mean CMI score than is low PSN), but the magnitude of the difference depends on the NHT level. This is what is meant by same-direction interaction.

TABLE 19-12   One main effect and no interaction

(a)

NHT	PSN Low	PSN High	Row Mean
Low	18	26	22
High	18	26	22
Column Mean	18	26	22

(b)

NHT	PSN Low	PSN High	Row Mean
Low	19	19	19
High	24	24	24
Column Mean	21.5	21.5	21.5

TABLE 19-13 Same-direction interaction

(a)

NHT	PSN Low	PSN High	Row Mean
Low	26	23	24.5
High	20	13	16.5
Column Mean	23	18	20.5

(b)

NHT	PSN Low	PSN High	Row Mean
Low	18	26	22
High	20	36	28
Column Mean	19	31	25

(c)

NHT	PSN Low	PSN High	Row Mean
Low	18	26	22
High	12	24	18
Column Mean	15	25	20

The model for Table 19-13a may be given as follows:

$$\mu_{ij} = 20.5 + \alpha_i + \beta_j + \gamma_{ij}$$

where

$$\alpha_i = \begin{cases} 4.0 & \text{if } i = 1 \\ -4.0 & \text{if } i = 2 \end{cases} \qquad \beta_j = \begin{cases} 2.5 & \text{if } j = 1 \\ -2.5 & \text{if } j = 2 \end{cases}$$

$$\gamma_{ij} = \begin{cases} -1.0 & \text{if } i = 1, j = 1 \\ 1.0 & \text{if } i = 1, j = 2 \\ 1.0 & \text{if } i = 2, j = 1 \\ -1.0 & \text{if } i = 2, j = 2 \end{cases}$$

Thus,

$$\mu_{11} = 20.5 + 4.0 + 2.5 - 1.0 = 26$$
$$\mu_{12} = 20.5 + 4.0 - 2.5 + 1.0 = 23$$
$$\mu_{21} = 20.5 - 4.0 + 2.5 + 1.0 = 20$$
$$\mu_{22} = 20.5 - 4.0 - 2.5 - 1.0 = 13$$

The reader may verify that the same type of model holds for tables (b) and (c).

Reverse Interaction

Two examples of reverse interaction are given in Table 19-14. By *reverse* we mean that the direction of the difference between two cell means for one row (column) is opposite to, or reversed from, the direction of the difference between the corresponding cell means for some other row (column).

In Table 19-14a, we can see that $\mu_{11} - \mu_{12} = 18 - 26 = -8$, whereas $\mu_{21} - \mu_{22} = 22 - 20 = 2$. Also, $\mu_{21} - \mu_{2.} = 22 - 21 = 1$ and $\mu_{.1} - \mu_{..} = 20 - 21.5 = -1.5$, so that $\mu_{21} - \mu_{2.} - \mu_{.1} + \mu_{..} = 2.5$.

These hypothetical results indicate (for this table) that for neighborhoods with a small number of households (low NHT), persons in unfriendly surroundings (low PSN) are healthier than persons in friendly surroundings, but that for neighborhoods with a large number of households, this situation is reversed. In other words, the difference between the effects of the high and low PSN levels is positive for low NHT but negative for high NHT (i.e., there is a reversal in sign). This, then, is what we mean by reverse interaction.

TABLE 19-14   Reverse interaction

(a)

NHT	PSN Low	PSN High	Row Mean
Low	18	26	22
High	22	20	21
Column Mean	20	23	21.5

(b)

NHT	PSN Low	PSN High	Row Mean
Low	26	22	24
High	18	24	21
Column Mean	22	23	22.5

The model in this case is given as follows:

$$\mu_{ij} = 21.5 + \alpha_i + \beta_j + \gamma_{ij}$$

where

$$\alpha_i = \begin{cases} 0.5 & \text{if } i = 1 \\ -0.5 & \text{if } i = 2 \end{cases} \qquad \beta_j = \begin{cases} -1.5 & \text{if } j = 1 \\ 1.5 & \text{if } j = 2 \end{cases}$$

$$\gamma_{ij} = \begin{cases} -2.5 & \text{if } i = 1, j = 1 \\ 2.5 & \text{if } i = 1, j = 2 \\ 2.5 & \text{if } i = 2, j = 1 \\ -2.5 & \text{if } i = 2, j = 2 \end{cases}$$

Thus,

$$\mu_{11} = 21.5 + 0.5 - 1.5 - 2.5 = 18$$
$$\mu_{12} = 21.5 + 0.5 + 1.5 + 2.5 = 26$$
$$\mu_{21} = 21.5 - 0.5 - 1.5 + 2.5 = 22$$
$$\mu_{22} = 21.5 - 0.5 + 1.5 - 2.5 = 20$$

It can be shown that a reverse interaction is indicated in Table 19-14b as well.

19-6-3 Interaction Effects for Data of Table 19-2

The table of sample cell means actually obtained for the CMI example of Section 19-2 is given in Table 19-15. From this table the following differences of means can be determined:

$$\bar{Y}_{11.} - \bar{Y}_{12.} = 18.92 - 26.84 = -7.92 \quad \text{whereas} \quad \bar{Y}_{21.} - \bar{Y}_{22.} = 20.84 - 20.72 = 0.12$$
$$\bar{Y}_{11.} - \bar{Y}_{21.} = 18.92 - 20.84 = -1.92 \quad \text{whereas} \quad \bar{Y}_{12.} - \bar{Y}_{22.} = 26.84 - 20.72 = 6.12$$
$$\bar{Y}_{11.} - \bar{Y}_{1..} - \bar{Y}_{.1.} + \bar{Y}_{...} = 18.92 - 22.88 - 19.88 + 21.83 = -2.01$$
$$\bar{Y}_{12.} - \bar{Y}_{1..} - \bar{Y}_{.2.} + \bar{Y}_{...} = 26.84 - 22.88 - 23.78 + 21.83 = 2.01$$
$$\bar{Y}_{21.} - \bar{Y}_{2..} - \bar{Y}_{.1.} + \bar{Y}_{...} = 20.84 - 20.78 - 19.88 + 21.83 = 2.01$$
$$\bar{Y}_{22.} - \bar{Y}_{2..} - \bar{Y}_{.2.} + \bar{Y}_{...} = 20.72 - 20.78 - 23.78 + 21.83 = -2.01$$

These comparisons suggest a possible reverse interaction. More specifically, it is indicated that (1) for small Turnkey neighborhoods, persons from friendly surroundings (high PSN) appear to have worse health (higher mean CMI scores) than persons from unfriendly

TABLE 19-15 Cell means for data of Table 19-1

NHT	PSN Low	PSN High	Row Mean
Low	18.92	26.84	22.88
High	20.84	20.72	20.78
Column Mean	19.88	23.78	21.83

surroundings, but that (2) there is little difference in mean CMI scores for large Turnkey neighborhoods. As mentioned in Section 19-2, this pattern is counter to that expected by Daly (1973). However, these observed differences are subject to sampling variation (i.e., they are sample values and not population values), and, as we know, the test for interaction for these data is not significant.

19-6-4 Interaction Effects for Data of Table 19-6

The table of sample cell means for the FEV example (Table 19-6) is as shown in Table 19-16. This table of means is slightly more difficult to interpret than the one for the CMI data, because now there are three rows and columns instead of two. Nevertheless, we can see immediately by observation that the relative magnitudes of the means vary from column to column. For example, for TOXSUB A, the order of PLANTS by increasing mean FEV is 3, 1, 2. For TOXSUB B, on the other hand, the order is 1, 2, 3, and for TOXSUB C the order is 2, 1, 3. These differences in ordering indicate some interaction, the significance of which was established earlier. The following comparisons of cell means should help in interpreting the nature of this significant interaction effect:

$$\bar{Y}_{11.} - \bar{Y}_{12.} = 4.90 - 3.46 = 1.44 \quad \text{whereas} \quad \bar{Y}_{31.} - \bar{Y}_{32.} = 4.13 - 5.13 = -1.00$$

$$\bar{Y}_{21.} - \bar{Y}_{31.} = 5.23 - 4.13 = 1.10 \quad \text{whereas} \quad \bar{Y}_{22.} - \bar{Y}_{32.} = 3.78 - 5.13 = -1.35$$

$$\bar{Y}_{21.} - \bar{Y}_{.1.} = 5.23 - 4.75 = 0.48 \quad \text{whereas} \quad \bar{Y}_{2..} - \bar{Y}_{...} = 3.90 - 3.91 = -0.01$$

The set of interaction effects of the form $\hat{\gamma}_{ij} = \bar{Y}_{ij.} - \bar{Y}_{i..} - \bar{Y}_{.j.} + \bar{Y}_{...}$ is given in Table 19-17. These patterns demonstrate that some plants are associated with better respiratory health than others for one kind of toxic exposure but are worse for other kinds of exposure. It would not be possible, therefore, to conclude that one plant was better overall than

TABLE 19-16 Cell means for data of Table 19-6

PLANT	TOXSUB A	TOXSUB B	TOXSUB C	Row Mean
1	4.90	3.46	2.75	3.70
2	5.23	3.78	2.70	3.90
3	4.13	5.13	3.13	4.13
Column Mean	4.75	4.12	2.86	3.91

TABLE 19-17 Interaction effects ($\hat{\gamma}_{ij}$'s) for data
of Table 19-6

PLANT	TOXSUB			Row Total
	A	B	C	
1	0.35	−0.45	0.10	0.00
2	0.49	−0.34	−0.15	0.00
3	−0.84	0.78	0.06	0.00
Column Total	0.00	0.00	0.00	0.00

another. Rather, the differences in respiratory health among plants depend on which toxic substance is being considered.

19-7 Random- and Mixed-Effects Two-Way ANOVA Models

In this section we present the classical two-way ANOVA statistical models appropriate when both factors are random and when one factor is fixed and the other is random. We will specify the appropriate null hypotheses of interest and the expected mean squares associated with each model. The expected mean square for a particular source is the true average (population) value of the mean-square term in the ANOVA table.

19-7-1 Random-Effects Model

When both factors are random, the two-way ANOVA model is given as follows:

$$Y_{ijk} = \mu + A_i + B_j + C_{ij} + E_{ijk} \tag{19.11}$$

where A_i, B_j, C_{ij}, and E_{ijk} are mutually independent random variables satisfying

$$\begin{aligned} A_i &\sim N(0, \sigma_R^2) \\ B_j &\sim N(0, \sigma_C^2) \\ C_{ij} &\sim N(0, \sigma_{RC}^2) \\ E_{ijk} &\sim N(0, \sigma^2) \end{aligned} \qquad i = 1, 2, \ldots, r; \quad j = 1, 2, \ldots, c; \quad k = 1, 2, \ldots, n$$

19-7-2 Mixed-Effects Model with Fixed Row Factor and Random Column Factor

One particular model[4] is

$$Y_{ijk} = \mu + \alpha_i + B_j + C_{ij} + E_{ijk} \tag{19.12}$$

---

[4] There are several ways to define a two-way mixed model; for an excellent discussion of these various models, see Hocking (1973).

where each α_i is a constant such that $\sum_{i=1}^{r} \alpha_i = 0$, and where B_j, C_{ij}, and E_{ijk} are random variables satisfying $\sum_{i=1}^{r} C_{ij} = 0$ for each j and

$$B_j \sim N(0, \sigma_C^2)$$
$$C_{ij} \sim N\left[0, \frac{(r-1)\sigma_{RC}^2}{r}\right] \quad i = 1, 2, \ldots, r; \quad j = 1, 2, \ldots, c; \quad k = 1, 2, \ldots, n$$
$$E_{ijk} \sim N(0, \sigma^2)$$

Also, $\mathrm{Cov}(C_{ij}, C_{i'j}) = -\sigma_{RC}^2/r$ for $i \neq i'$, and all other covariances are 0.

19-7-3 Mixed-Effects Model with Random Row Factor and Fixed Column Factor

One particular model (see footnote 4) is

$$Y_{ijk} = \mu + A_i + \beta_j + C_{ij} + E_{ijk} \qquad (19.13)$$

where each β_j is a constant such that $\sum_{j=1}^{c} \beta_j = 0$, and where A_j, C_{ij}, and E_{ijk} are random variables satisfying $\sum_{j=1}^{c} C_{ij} = 0$ for each i and

$$A_i \sim N(0, \sigma_R^2)$$
$$C_{ij} \sim N\left[0, \frac{(c-1)\sigma_{RC}^2}{c}\right] \quad i = 1, 2, \ldots, r; \quad j = 1, 2, \ldots, c; \quad k = 1, 2, \ldots, n$$
$$E_{ijk} \sim N(0, \sigma^2)$$

Also, $\mathrm{Cov}(C_{ij}, C_{ij'}) = -\sigma_{RC}^2/c$ for $j \neq j'$, and all other covariances are 0.

19-7-4 Null Hypotheses and Expected Mean Squares for Two-Way ANOVA Models

Table 19-18 gives (for fixed-, random-, and mixed-effects models) the specific null hypotheses being tested regarding row main effects, column main effects, and interaction effects. Table 19-19 gives the expected mean square for each factor in each of the models. These two tables emphasize why different F statistics are required for testing the various hypotheses of interest. In this regard, the primary consideration is the choice of the appropriate *denominator* mean squares to use in the various F statistics. The numerator mean square *always* corresponds to the factor being considered; for example, if the factor is "rows," the numerator mean square is MSR, regardless of the type of model. Similarly, if the factor is "columns" or "interaction," the numerator mean square is MSC or MSRC, respectively. *The denominator mean square, however, is chosen to correspond to that expected mean square to which the numerator expected mean square reduces under the null hypothesis of interest.* For example, when testing for significant row effects in a random-effects model, the numera-

TABLE 19-18   Null hypotheses for two-way ANOVA

	Model Type			
			Mixed Effects	
Source	Fixed Effects	Random Effects	Rows Fixed, Columns Random	Rows Random, Columns Fixed
Rows	$\alpha_1 = \alpha_2 = \cdots = \alpha_r = 0$	$\sigma_R^2 = 0$	$\alpha_1 = \alpha_2 = \cdots = \alpha_r = 0$	$\sigma_R^2 = 0$
Columns	$\beta_1 = \beta_2 = \cdots = \beta_c = 0$	$\sigma_C^2 = 0$	$\sigma_C^2 = 0$	$\beta_1 = \beta_2 = \cdots = \beta_c = 0$
Interaction	$\gamma_{ij} = 0$ for all i, j	$\sigma_{RC}^2 = 0$	$\sigma_{RC}^2 = 0$	$\sigma_{RC}^2 = 0$

TABLE 19-19   Expected mean squares for two-way ANOVA (r rows, c columns, n replications per cell)

	Model Type			
			Mixed Effects	
Source	Fixed Effects	Random Effects	Rows Fixed, Columns Random	Rows Random, Columns Fixed
Rows	$\sigma^2 + cn \sum_{i=1}^{r} \dfrac{\alpha_i^2}{r-1}$	$\sigma^2 + n\sigma_{RC}^2 + cn\sigma_R^2$	$\sigma^2 + n\sigma_{RC}^2 + cn \sum_{i=1}^{r} \dfrac{\alpha_i^2}{r-1}$	$\sigma^2 + cn\sigma_R^2$
Columns	$\sigma^2 + m \sum_{j=1}^{c} \dfrac{\beta_j^2}{c-1}$	$\sigma^2 + n\sigma_{RC}^2 + m\sigma_C^2$	$\sigma^2 + m\sigma_C^2$	$\sigma^2 + n\sigma_{RC}^2 + m \sum_{j=1}^{c} \dfrac{\beta_j^2}{c-1}$
Interaction	$\sigma^2 + n \sum_{i=1}^{r}\sum_{j=1}^{c} \dfrac{\gamma_{ij}^2}{(r-1)(c-1)}$	$\sigma^2 + n\sigma_{RC}^2$	$\sigma^2 + n\sigma_{RC}^2$	$\sigma^2 + n\sigma_{RC}^2$
Error	σ^2	σ^2	σ^2	σ^2

tor expected mean square for this test, $\sigma^2 + n\sigma_{RC}^2 + cn\sigma_R^2$ from Table 19-19, reduces to $\sigma^2 + n\sigma_{RC}^2$ under H_0: $\sigma_R^2 = 0$. This requires that the denominator mean square be MSRC, since the expected mean square of MSRC under the random-effects model is exactly $\sigma^2 + n\sigma_{RC}^2$.

In this way, the ratio of expected mean squares

$$\frac{\text{EMS(R)}}{\text{EMS(RC)}} = \frac{\sigma^2 + n\sigma_{RC}^2 + cn\sigma_R^2}{\sigma^2 + n\sigma_{RC}^2}$$

reduces to $(\sigma^2 + n\sigma_{RC}^2)/(\sigma^2 + n\sigma_{RC}^2) = 1$ under H_0: $\sigma_R^2 = 0$, so the F statistic MSR/MSRC is the ratio of two estimates of the same variance under H_0.

As another example, let us consider the F test for significant row effects based on the mixed-effects model with the row factor fixed and column factor random. The test statistic in this case, $F = \text{MSR/MSRC}$, concerns the following ratio of expected mean squares (see Table 19-19):

$$\frac{\text{EMS(R)}}{\text{EMS(RC)}} = \frac{\sigma^2 + n\sigma_{RC}^2 + cn \sum\limits_{i=1}^{r} \alpha_i^2/(r-1)}{\sigma^2 + n\sigma_{RC}^2}$$

Under H_0: $\alpha_1 = \alpha_2 = \cdots = \alpha_r = 0$, this ratio simplifies to $(\sigma^2 + n\sigma_{RC}^2)/(\sigma^2 + n\sigma_{RC}^2) = 1$. Thus, the F statistic is the ratio of two estimates of the same variance under H_0.

As a final example, we consider the F test for significant row effects based on the mixed-effects model with the row factor random and column factor fixed. The test statistic is $F = \text{MSR/MSE}$, which concerns

$$\frac{\text{EMS(R)}}{\text{EMS(E)}} = \frac{\sigma^2 + cn\sigma_R^2}{\sigma^2}$$

Under H_0: $\sigma_R^2 = 0$, this ratio simplifies to $\sigma^2/\sigma^2 = 1$, as desired.

Problems

1. The following data come from an animal experiment designed to investigate whether levorphanol reduces stress as reflected in the cortical sterone level. There were five animals in each of the four treatment groups, and the data are given in the table below.

Control	Levorphanol Only	Epinephrine Only	Levorphanol and Epinephrine
1.90	0.82	5.33	3.08
1.80	3.36	4.84	1.42
1.54	1.64	5.26	4.54
4.10	1.74	4.92	1.25
1.89	1.21	6.07	2.57

   a. These data may be analyzed by means of two-way ANOVA. What are the two factors?

   b. Classify each factor as either fixed or random.

c. Rearrange the data into a two-way table appropriate for two-way ANOVA.
d. Form the table of sample means and comment.
e. Determine the appropriate ANOVA table for this data set.
f. Analyze the data to determine whether there are significant main effects due to levorphanol and epinephrine and whether there is a significant interaction effect between epinephrine and levorphanol.

2. The table gives the performance competency scores for a random sample of family nurse practitioners (FNPs) with different specialties from hospitals in three cities.

Specialty	City 1	City 2	City 3
Pediatrics	91.7, 74.9,	86.3, 88.1,	82.3, 78.7,
	88.2, 79.5	92.0, 69.5	89.8, 84.5
Obstetrics and	80.1, 76.2,	71.3, 73.4,	90.1, 65.6,
gynecology	70.3, 89.5	76.9, 87.2	74.6, 79.1
Diabetes and	71.5, 49.8,	80.2, 76.1,	48.7, 54.4,
hypertension	55.1, 75.4	44.2, 50.5	60.1, 70.8

a. Classify each factor as either fixed or random and justify your classification.
b. Form the table of sample means (you may use the computer printout given) and then comment.
c. Using the computer results given, compute the appropriate F statistic for each of the four possible factor classification schemes (i.e., both factors fixed, both random, and one factor of each type).
d. Analyze the data based on each possible factor classification scheme. How do the results compare?
e. Using Scheffé's method as described in Chapter 17, find a 95% confidence interval for the true difference in mean scores between pediatric FNPs and obgyn FNPs.
f. Assuming each factor to be fixed, state a regression model appropriate for the two-way ANOVA table and provide estimates of the regression coefficients associated with the factor main effects using the sample means obtained in part (b).

COMPUTER RESULTS FOR PROBLEM 2

Specialty	City	N	Mean	Standard Deviation
PEDIATRI	1	4	8.3575E + 01	7.73019E + 00
PEDIATRI	2	4	8.3975E + 01	9.93894E + 00
PEDIATRI	3	4	8.3825E + 01	4.64569E + 00
OBSGYN	1	4	7.9025E + 01	8.06200E + 00
OBSGYN	2	4	7.7200E + 01	7.05550E + 00
OBSGYN	3	4	7.7350E + 01	1.01858E + 01
DIABHYP	1	4	6.2950E + 01	1.24184E + 01
DIABHYP	2	4	6.2750E + 01	1.80452E + 01
DIABHYP	3	4	5.8500E + 01	9.42868E + 00

Overall mean
(mean of means):   7.43500E + 01

ANOVA TABLE FOR PROBLEM 2

Source	df	SS	MS
SPEC	2	3.2299E + 03	1.6149E + 03
CITY	2	2.4542E + 01	1.2271E + 01
SPEC × CITY	4	3.4537E + 01	8.6342E + 00
Error	27	2.9022E + 03	1.0749E + 02

3. The table gives the average patient waiting time in minutes for a sample of 16 physicians classified by type of practice and type of physician.

	Type of Practice	
Physician Type	Group	Solo
General practitioner	15, 20, 25, 20	20, 25, 30, 25
Specialist	30, 25, 30, 35	25, 20, 30, 30

a. Classify each factor as either fixed or random and justify your classification scheme.
b. Using the computer results given, compute the F statistic corresponding to each of the four possible factor classification schemes.
c. Discuss the analysis of the data when both factors are considered to be fixed.
d. What is the estimate of the (fixed) effect due to "general practitioner," the (fixed) effect due to "group practice," and the interaction effect $\mu_{11} - \mu_{12} - \mu_{21} + \mu_{22}$, in which μ_{ij} denotes the cell mean in the ith row and jth column of the table of cell means?
e. Interpret the interaction effect observed.
f. What is an appropriate regression model for this two-way ANOVA?
g. How might one modify the model in part (f) to reflect the conclusions made in part (c)?

COMPUTER RESULTS FOR PROBLEM 3

PHYSTP	TYPRAC	N	Mean	Standard Deviation
GP	GROUP	4	2.0000E + 01	4.08248E + 00
GP	SOLO	4	2.5000E + 01	4.08248E + 00
SPEC	GROUP	4	3.0000E + 01	4.08248E + 00
SPEC	SOLO	4	2.6250E + 01	4.78714E + 00

Overall mean
(mean of means):   2.53125E + 01

ANOVA TABLE FOR PROBLEM 3

Source	df	SS	MS
PHYSTP	1	1.2656E + 02	1.2656E + 02
TYPRAC	1	1.5625E + 00	1.5625E + 00
PHYSTP × TYPRAC	1	7.6563E + 01	7.6563E + 01
Error	12	2.1875E + 02	1.8229E + 01

4. A study was undertaken to measure and compare the sexist attitudes of students at various types of colleges. Random samples of 10 undergraduate seniors of each sex were selected from each of three types of colleges. A questionnaire was then administered to each student, from which a "degree of sexism" score was determined (the higher the score, the more sexist the attitude). Here, "degree of sexism" reflected the extent to which a student considered males and females to have different life roles. The resulting data are given in the table.

College Type	Male	Female
Coed with 75% or more males	50, 35, 37, 32, 46, 38, 36, 40, 38, 41	38, 27, 34, 30, 22, 32, 26, 24, 31, 33
Coed with less than 75% males	30, 29, 31, 27, 22, 20, 31, 22, 25, 30	28, 31, 28, 26, 20, 24, 31, 24, 31, 26
Not coed	45, 40, 32, 31, 26, 28, 39, 27, 37, 35	40, 35, 32, 29, 24, 26, 36, 25, 35, 35

a. Form the table of cell means and interpret the results obtained (see the computer printout).
b. Using the computer results given, compute the F statistics corresponding to a model with both factors considered fixed.
c. Discuss the analysis of the data for this fixed-effects model case.

COMPUTER RESULTS FOR PROBLEM 4

TYPCOL	Sex	N	Mean	Standard Deviation
75PLCOED	MALE	10	3.9300E + 01	5.31350E + 00
75PLCOED	FEMALE	10	2.9700E + 01	4.92274E + 00
LS75COED	MALE	10	2.6700E + 01	4.16467E + 00
LS75COED	FEMALE	10	2.6900E + 01	3.63471E + 00
NOCOED	MALE	10	3.4000E + 01	6.27163E + 00
NOCOED	FEMALE	10	3.1700E + 01	5.41705E + 00

Overall mean
(mean of means):   3.13833E + 01

ANOVA TABLE FOR PROBLEM 4

Source	df	SS	MS
TYPCOL	2	6.5743E + 02	3.2872E + 02
SEX	1	2.2815E + 02	2.2815E + 02
TYPCOL × SEX	2	2.5930E + 02	1.2965E + 02
Error	54	1.3653E + 03	2.5283E + 01

5. Random samples of 100 persons awaiting trial on felony charges were selected from rural, urban, and suburban court locations in each of two states, one (state 1) in the Northeast and the other (state 2) in the South. The table summarizes the data on the time (in months) between arrest and the beginning of trial for these random samples.

State	Court Location		
	Rural	Suburban	Urban
1	$\bar{Y} = 3.4, S = 1.3$	$\bar{Y} = 5.8, S = 1.2$	$\bar{Y} = 6.8, S = 1.5$
2	$\bar{Y} = 2.4, S = 1.5$	$\bar{Y} = 3.5, S = 1.7$	$\bar{Y} = 4.7, S = 1.7$

a. Do the sample means in the table suggest that the average waiting times for state 1 vary by court location differently than they do for state 2; in other words, is there an interaction effect?

b. Analyze these data using the ANOVA table below, assuming that both factors are fixed.

Source	df	SS	MS
States	1	486.00	486.00
Court locations	2	826.33	413.17
Interaction	2	49.00	24.50
Error	594	1,327.591	2.235

c. Define an appropriate regression model for this two-way ANOVA.

d. How might one revise the model in part (c) and the associated ANOVA table in order to investigate whether there is a linear trend in waiting time with the degree of urbanization (defined by treating the categories rural, suburban, and urban on an ordinal scale)? What difficulty does one encounter when considering such a model?

6. An experiment was conducted at a large state university to determine whether two different instructional methods for teaching a beginning statistics course would result in different levels of cognitive achievement. One instructional method involved the use of a self-instructional format, including a sequence of slide-tape presentations; the other method utilized the standard lecture format. The 100 students who registered for the course were randomly assigned to one of four sections, 25 per section, corresponding to the combinations of one of the two methods with one of two instructors. The results obtained from identical final exams given to each section are summarized in the table below.

Instructor	Method	
	Lecture	Self-Instruction
A	$\bar{Y} = 71.2, S = 13.8$	$\bar{Y} = 80.2, S = 12.1$
B	$\bar{Y} = 73.8, S = 11.7$	$\bar{Y} = 77.5, S = 14.1$

a. What do the results suggest about the comparative effects of the two instructional methods?

b. Classify each factor as either fixed or random, and explain your classification.

c. Using the ANOVA table, perform the appropriate F tests for each of the four types of factor classification schemes possible. Compare the conclusions reached under each scheme.

ANOVA TABLE FOR PROBLEM 6

Source	df	SS	MS
INSTRUC	1	6.2500E − 02	6.2500E − 02
METHOD	1	1.0081E + 03	1.0081E + 03
INSTRUC × METHOD	1	1.7556E + 02	1.7556E + 02
Error	96	1.6141E + 04	1.6814E + 02

d. What are some factors that should be controlled for in this experiment?

e. Given a continuous variable C to be controlled for, write down an appropriate regression model for this data set that takes C into account. What general method of analysis is characterized by such a model?

7. The table presents data on the uric acid level found in the bloodstreams of mongoloids and in the bloodstreams of normal control subjects or nonmongoloid mentally retarded subjects. All subjects were between the ages of 21 and 25. Analyze these data using the ANOVA table to determine whether there is evidence of a higher uric acid level in the mongoloid group, making sure to characterize any sex relationships that exist.

Group	Sex Male	Sex Female
Mongoloid	5.84, 6.30, 6.95, 5.92, 7.94	4.90, 6.95, 6.73, 5.32, 4.81
Others	5.50, 6.08, 5.12, 7.58, 6.78	4.94, 7.20, 5.22, 4.60, 3.88

ANOVA TABLE FOR PROBLEM 7

Source	df	SS	MS
Groups	1	1.1329	1.1329
Sex	1	4.4746	4.4746
Interaction	1	0.4802	0.4802
Error	16	17.3150	1.0822

8. An experiment was conducted to investigate the survival of diplococcus pneumonia bacteria in chick embryos under relative humidities (RH) of 0%, 25%, 50%, and 100% and under temperatures (TEMP) of 10°C, 20°C, 30°C, and 40°C using 10 chicks for each RH–TEMP combination.[5] The partially completed ANOVA table is as given.

Source	df	MS
RH		2.010
TEMP		7.816
Interaction		1.642
Error		0.775
Total		

[5] Adapted from a study by Price (1954).

a. Should the two factors RH and TEMP be considered as fixed or random? Explain.
b. Carry out the analysis of variance for both the fixed-effects case and the random-effects case. Are your conclusions different in the two cases?
c. Write down both the fixed-effects and the random-effects models that could describe this experiment.
d. Provide a regression model using dummy variables that can be used to obtain the results in the ANOVA table.
e. What regression model would be appropriate for describing the relationship of RH and TEMP to survival time (Y) if the independent variables are to be treated intervally rather than nominally?

9. The diameters (Y) of three species of pine trees were compared at each of four locations using samples of five trees per species at each location. The data are given in the table.

Species	Location			
	1	2	3	4
A	23	25	21	14
	15	20	17	17
	26	21	16	19
	13	16	24	20
	21	18	27	24
B	28	30	19	17
	22	26	24	21
	25	26	19	18
	19	20	25	26
	26	28	29	23
C	18	15	23	18
	10	21	25	12
	12	22	19	23
	22	14	13	22
	13	12	22	19

a. Comment on whether each of the two factors should be considered as fixed or random.
b. Use the partially completed ANOVA table below to carry out your analysis, first considering both factors fixed and then considering a mixed model with "locations" as random. Compare your conclusions.

Source	df	SS
Species		314.10
Locations		55.80
Interaction		103.10
Error		945.60

10. Consider an ANOVA table of the following form.

Source	df	SS	MS	F	P
A					P_1
B					P_2
A × B					P_3
Error					

Use $\alpha = .05$ for all parts of this problem. In each case, decide what effects, if any, are significant, and what conclusions to draw, based on a two-way ANOVA table with the following P-values:

a. $P_1 = .03$, $P_2 = .51$, $P_3 = .31$
b. $P_1 = .001$, $P_2 = .63$, $P_3 = .007$
c. $P_1 = .093$, $P_2 = .79$, $P_3 = .02$
d. $P_1 = .56$, $P_2 = .38$, $P_3 = .24$

11. Assume that a total of 75 subjects were tested in a balanced two-way fixed-effects factorial. A plot of the means from the study is below. The dependent variable is Y, and the factors are A and B.

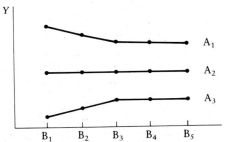

a. Give the left two columns (the source and degrees-of-freedom columns) of the source table for the data plotted.
b. Assume that the above plot shows the population means. Indicate which significance tests should be significant in the ANOVA table.

12. Assume the following source table came from a balanced two-way fixed-effects ANOVA. Show the formula used, the numbers filled in, and the value for each letter in the table.

Source	df	SS	MS	F
A	a	5.12	g	6.40*
B	b	e	0.76	3.80*
A × B	c	4.32	0.36	i
Error	20	4.00	h	
Total	d	f		

j. Indicate which, if any, family or families of means should be evaluated with multiple comparisons.

13. The following data are from a hypothetical study of human body temperature as affected by air temperature and a diet supplement that is hoped to increase heat tolerance. Body temperatures (in degrees Celsius) were measured for 36 athletes immediately following a standard exercise routine in a room controlled to a fixed air temperature (in degrees Celsius). Each subject had been receiving a steady, fixed dose (in milligrams per kilogram body weight) of the diet supplement.

   a. Provide an appropriate two-way ANOVA source table.
   b. Provide F tests of the two main effects and the interaction. Use $\alpha = .05$. What do you conclude?
   c. Define dummy variables, and specify an appropriate multiple regression model corresponding to the analysis done in part (a). (*Hint*: Use dummy variables that have the value -1 for AIRTEMP $= 21$ and for DOSE $= 0$. Why is this a scientifically sensible choice?)
   d. Since both factors are interval-scale variables, specify a corresponding natural-polynomial multiple regression model.

Observation	AIRTEMP	DOSE	BODYTEMP
1	21	0.00	37.2
2	21	0.00	37.2
3	21	0.00	36.8
4	21	0.05	37.1
5	21	0.05	36.9
6	21	0.05	36.8
7	21	0.10	37.1
8	21	0.10	37.1
9	21	0.10	37.1
10	25	0.00	36.9
11	25	0.00	37.0
12	25	0.00	37.1
13	25	0.05	37.1
14	25	0.05	36.7
15	25	0.05	37.0
16	25	0.10	36.9
17	25	0.10	37.0
18	25	0.10	37.3
19	29	0.00	36.9
20	29	0.00	37.0
21	29	0.00	36.8
22	29	0.05	36.9
23	29	0.05	37.0
24	29	0.05	36.9
25	29	0.10	36.9
26	29	0.10	37.0
27	29	0.10	37.2
28	33	0.00	37.1
29	33	0.00	37.3
30	33	0.00	36.7
31	33	0.05	36.9

Observation	AIRTEMP	DOSE	BODYTEMP
32	33	0.05	37.0
33	33	0.05	37.0
34	33	0.10	36.9
35	33	0.10	36.8
36	33	0.10	37.2

References

Daly, M. B. (1973). "The Effect of Neighborhood Racial Characteristics on the Attitudes, Social Behavior, and Health of Low Income Housing Residents." Ph.D. dissertation, Department of Epidemiology, University of North Carolina, Chapel Hill, N.C.

Hocking, R. R. (1973). "A Discussion of the Two-Way Mixed Model." *Amer. Statistician*, 27(4): 148–152.

Price, R. D. (1954). "The Survival of Bacterium Tularense in Lice and Louse Feces." *Amer. J. Trop. Med. Hyg.*, 3: 179–186.

Two-Way ANOVA with Unequal Cell Numbers

20-1 Preview

When we first began our discussion of two-way ANOVA in Chapter 18, we indicated in Figure 18-1 several ways to classify a two-factor problem according to the observed data pattern. We have already described methods for handling the single-observation-per-cell case (Chapter 18) and the equal-cell-number case (Chapter 19). In this chapter we shall present procedures for analyzing two-factor patterns having unequal cell numbers. This latter case exhibits special problems in statistical analysis. Problems in computation and, more important, interpretation arise when dealing with unequal cell-specific sample sizes.

In treating these problems, we shall find it necessary to make several distinctions among patterns of cell frequency. The first distinction is between *balanced* and *unbalanced* patterns. A balanced design has an equal number of observations in each and every cell; an unbalanced design does not. A *complete* design has at least one observation per cell; an *incomplete* design has zero observations in one or more cells. All incomplete designs are unbalanced. Finally, some unbalanced designs exhibit *proportional cell frequencies* (detailed below).

The general data configuration for the unequal-cell-number case in two-way ANOVA is presented in Table 20-1. A numerical example is given in Table 20-2, which we shall refer to throughout this chapter. In the table we have

Y_{ijk} = kth observation in the cell associated with the ith row and jth column

n_{ij} = number of observations in the cell associated with the ith row and jth column

Also,

$$\bar{Y}_{ij\cdot} = \frac{1}{n_{ij}} \sum_{k=1}^{n_{ij}} Y_{ijk}$$

$$\bar{Y}_{i\cdot\cdot} = \frac{1}{n_{i\cdot}} \sum_{j=1}^{c} \sum_{k=1}^{n_{ij}} Y_{ijk} \quad \text{where} \quad n_{i\cdot} = \sum_{j=1}^{c} n_{ij}$$

457

$$\bar{Y}_{\cdot j \cdot} = \frac{1}{n_{\cdot j}} \sum_{i=1}^{r} \sum_{k=1}^{n_{ij}} Y_{ijk} \quad \text{where} \quad n_{\cdot j} = \sum_{i=1}^{r} n_{ij}$$

$$\bar{Y}_{\cdots} = \frac{1}{n_{\cdot\cdot}} \sum_{i=1}^{r} \sum_{j=1}^{c} \sum_{k=1}^{n_{ij}} Y_{ijk} \quad \text{where} \quad n_{\cdot\cdot} = \sum_{i=1}^{r} \sum_{j=1}^{c} n_{ij}$$

TABLE 20-1   Data layout for the unequal-cell-number case (two-way ANOVA)

Row Factor	Column Factor				Row Marginals
	1	2	...	c	
1	$Y_{111}, Y_{112}, \ldots, Y_{11n_{11}}$ (sample size $= n_{11}$) (cell mean $= \bar{Y}_{11\cdot}$)	$Y_{121}, Y_{122}, \ldots, Y_{12n_{12}}$ (sample size $= n_{12}$) (cell mean $= \bar{Y}_{12\cdot}$)	...	$Y_{1c1}, Y_{1c2}, \ldots, Y_{1cn_{1c}}$ (sample size $= n_{1c}$) (cell mean $= \bar{Y}_{1c\cdot}$)	$n_{1\cdot}, \bar{Y}_{1\cdot\cdot}$
2	$Y_{211}, Y_{212}, \ldots, Y_{21n_{21}}$ (sample size $= n_{21}$) (cell mean $= \bar{Y}_{21\cdot}$)	$Y_{221}, Y_{222}, \ldots, Y_{22n_{22}}$ (sample size $= n_{22}$) (cell mean $= \bar{Y}_{22\cdot}$)	...	$Y_{2c1}, Y_{2c2}, \ldots, Y_{2cn_{2c}}$ (sample size $= n_{2c}$) (cell mean $= \bar{Y}_{2c\cdot}$)	$n_{2\cdot}, \bar{Y}_{2\cdot\cdot}$
⋮	⋮	⋮	...	⋮	⋮
r	$Y_{r11}, Y_{r12}, \ldots, Y_{r1n_{r1}}$ (sample size $= n_{r1}$) (cell mean $= \bar{Y}_{r1\cdot}$)	$Y_{r21}, Y_{r22}, \ldots, Y_{r2n_{r2}}$ (sample size $= n_{r2}$) (cell mean $= \bar{Y}_{r2\cdot}$)	...	$Y_{rc1}, Y_{rc2}, \ldots, Y_{rcn_{rc}}$ (sample size $= n_{rc}$) (cell mean $= \bar{Y}_{rc\cdot}$)	$n_{r\cdot}, \bar{Y}_{r\cdot\cdot}$
Column Marginals	$n_{\cdot 1}, \bar{Y}_{\cdot 1\cdot}$	$n_{\cdot 2}, \bar{Y}_{\cdot 2\cdot}$	...	$n_{\cdot c}, \bar{Y}_{\cdot c\cdot}$	$n_{\cdot\cdot}, \bar{Y}_{\cdots}$

TABLE 20-2   Satisfaction with medical care (Y) by patient worry and affective communication between patient and physician

Affective Communication	Worry		Row Marginals
	Negative	Positive	
High	2, 5, 8, 6, 2, 4, 3, 10 ($n_{11} = 8$) ($\bar{Y}_{11\cdot} = 5$)	7, 5, 8, 6, 3, 5, 6, 4, 5, 6, 8, 9 ($n_{12} = 12$) ($\bar{Y}_{12\cdot} = 6$)	$n_{1\cdot} = 20$ $\bar{Y}_{1\cdot\cdot} = 5.6$
Medium	4, 6, 3, 3 ($n_{21} = 4$) ($\bar{Y}_{21\cdot} = 4$)	7, 7, 8, 6, 4, 9, 8, 7 ($n_{22} = 8$) ($\bar{Y}_{22\cdot} = 7$)	$n_{2\cdot} = 12$ $\bar{Y}_{2\cdot\cdot} = 6$
Low	8, 7, 5, 9, 9, 10, 8, 6, 8, 10 ($n_{31} = 10$) ($\bar{Y}_{31\cdot} = 8$)	5, 8, 6, 6, 9, 7, 7, 8 ($n_{32} = 8$) ($\bar{Y}_{32\cdot} = 7$)	$n_{3\cdot} = 18$ $\bar{Y}_{3\cdot\cdot} = 7.56$
Column Marginals	$n_{\cdot 1} = 22$ $\bar{Y}_{\cdot 1\cdot} = 6.18$	$n_{\cdot 2} = 28$ $\bar{Y}_{\cdot 2\cdot} = 6.57$	$n_{\cdot\cdot} = 50$ $\bar{Y}_{\cdots} = 6.40$

The unequal-cell-number case arises quite frequently in observational studies. In such studies, one or more of the following are typically true:

1.   All the variables of interest are not categorized before the data are collected.

2. New variables are often considered after the data are collected.

3. When all the variables are separately categorized, it is often not practical or even possible to control in advance how the various categories will combine to form combinations of interest.

The unequal-cell-number case can also arise in experimental studies when a posteriori consideration is given to variables other than those of primary interest, even if the design based on the primary variables calls for equal cell numbers. Furthermore, unequal cell numbers will generally result whenever there are missing data points, which, for example, can occur because of study dropouts or incomplete records.

The example presented in Table 20-2 is derived from Thompson's (1972) study concerning the relationship of patient perception of pregnancy and physician–patient communication to patient satisfaction with medical care. Two main variables of interest were the patient's WORRY and a measure of affective communication (AFFCOM). These variables were developed from scales based on questionnaires administered to patients and their physicians. Based on the distribution of scores, the WORRY variable was grouped into the categories "positive" and "negative," and the AFFCOM variable was grouped into the categories "high," "medium," and "low." Table 20-2 presents artificial data of this type, showing satisfaction-with-medical-care scores ($Y = $ TOTSAT) classified according to these six combinations of levels of the factors WORRY and AFFCOM.

As can be seen from the table, the categorization scheme used leads to a two-way table with unequal cell numbers. For WORRY, there are 22 negatives and 28 positives; for AFFCOM, there are 20 high, 12 medium, and 18 low scores. When the separate categories for the two variables are considered together, the resulting six categories have different cell sample sizes, ranging from 4 (medium AFFCOM, negative WORRY) to 12 (high AFFCOM, positive WORRY).

20-2 Problem with Unequal Cell Numbers: Nonorthogonality

The key statistical concept associated with the special analytical problems encountered in the unequal-cell-number case pertains to the *nonorthogonality* of the sums of squares usually used to describe the sources of variation in a two-way ANOVA table. To explain what we mean by *orthogonality*, we first note that the general formulae for these sums of squares (given in Section 19-3 for the equal-cell-number case) in terms of unequal cell numbers are as follows:

$$\text{SSR} = \sum_{i=1}^{r} \sum_{j=1}^{c} \sum_{k=1}^{n_{ij}} (\bar{Y}_{i\cdot\cdot} - \bar{Y}_{\cdots})^2, \qquad \text{SSC} = \sum_{i=1}^{r} \sum_{j=1}^{c} \sum_{k=1}^{n_{ij}} (\bar{Y}_{\cdot j\cdot} - \bar{Y}_{\cdots})^2,$$

$$\text{SSRC} = \sum_{i=1}^{r} \sum_{j=1}^{c} \sum_{k=1}^{n_{ij}} (\bar{Y}_{ij\cdot} - \bar{Y}_{i\cdot\cdot} - \bar{Y}_{\cdot j\cdot} + \bar{Y}_{\cdots})^2, \qquad (20.1)$$

$$\text{SSE} = \sum_{i=1}^{r} \sum_{j=1}^{c} \sum_{k=1}^{n_{ij}} (Y_{ijk} - \bar{Y}_{ij\cdot})^2, \qquad \text{TSS} = \sum_{i=1}^{r} \sum_{j=1}^{c} \sum_{k=1}^{n_{ij}} (Y_{ijk} - \bar{Y}_{\cdots})^2$$

These formulae for SSR, SSC, and SSRC are often referred to as the *unconditional* sums of squares for rows, columns, and interaction, respectively; by *unconditional* we mean that each of these sums of squares may be separately defined from basic principles to describe the variability associated with the estimated effects $(\bar{Y}_{i\cdot\cdot} - \bar{Y}_{\cdots})$, $(\bar{Y}_{\cdot j\cdot} - \bar{Y}_{\cdots})$, and $(\bar{Y}_{ij\cdot} - \bar{Y}_{i\cdot\cdot} - \bar{Y}_{\cdot j\cdot} + \bar{Y}_{\cdots})$ for rows, columns, and interaction, respectively. (There is an equivalent way to illustrate the meaning of the term *unconditional* using regression analysis methodology, as we shall soon see.)

If the collection of sums of squares in (20.1) are orthogonal, the following fundamental equation holds:

$$SSR + SSC + SSRC + SSE = TSS$$

That is, the terms on the left-hand side must partition the total sum of squares into nonoverlapping sources of variation.

We have already seen that this fundamental equation holds true for the equal-cell-number case (Chapter 19). Unfortunately, with unequal cell numbers, the unconditional sums of squares will no longer represent completely separate (i.e., orthogonal) sources of variation; thus,

$$\begin{matrix} \text{unequal cell} \\ \text{numbers} \end{matrix} \Rightarrow SSR + SSC + SSRC + SSE \neq TSS$$

To see why this is so, it helps to consider the general regression formulation for two-way ANOVA, which incorporates the unequal-cell-number case as well as the equal-cell-number case; the general regression equation is

$$Y = \mu + \sum_{i=1}^{r-1} \alpha_i X_i + \sum_{j=1}^{c-1} \beta_j Z_j + \sum_{i=1}^{r-1} \sum_{j=1}^{c-1} \gamma_{ij} X_i Z_j + E \tag{20.2}$$

where μ, α_i, β_j, and γ_{ij} are regression coefficients and X_i and Z_j are appropriately defined dummy variables. Recall that the general form of the fundamental regression equation for this model may be written

$$TSS = SSReg + SSE$$

where

$$TSS = \Sigma\, (Y_l - \bar{Y})^2$$
$$SSReg = \Sigma\, (\hat{Y}_l - \bar{Y})^2 = \text{regression } SS(X_1, X_2, \ldots, X_{r-1}; Z_1, Z_2, \ldots, Z_{c-1};$$
$$X_1 Z_1, X_1 Z_2, \ldots, X_{r-1} Z_{c-1})$$
$$SSE = \Sigma\, (Y_l - \hat{Y}_l)^2$$

and where the summation is over all $n_{\cdot\cdot}$ observations.

Now, using the *extra-sum-of-squares principle* (see Chapter 9), we can partition the regression sum of squares in various ways to emphasize the contribution due to adding sets of variables to a regression model already containing other sets of variables. In particular, we can further partition the fundamental regression equation above as follows with regard to model (20.2):

$$TSS = \text{regression } SS(\overbrace{X_1, X_2, \ldots, X_{r-1}}^{R})$$

$$+ \text{ regression } SS(\overbrace{Z_1, Z_2, \ldots, Z_{c-1}}^{C} \bigg| \overbrace{X_1, X_2, \ldots, X_{r-1}}^{R})$$

$$+ \text{ regression } SS(\overbrace{X_1 Z_1, X_1 Z_2, \ldots, X_{r-1} Z_{c-1}}^{RC} \bigg|$$

$$\overbrace{X_1, X_2, \ldots, X_{r-1}, Z_1, Z_2, \ldots, Z_{c-1}}^{R, C}) + SSE \qquad (20.3)$$

On the other hand, if we wish to enter the column effects into the model first, the equation would be as follows:

$$TSS = \text{regression } SS(\overbrace{Z_1, Z_2, \ldots, Z_{c-1}}^{C})$$

$$+ \text{ regression } SS(\overbrace{X_1, X_2, \ldots, X_{r-1}}^{R} \bigg| \overbrace{Z_1, Z_2, \ldots, Z_{c-1}}^{C})$$

$$+ \text{ regression } SS(\overbrace{X_1 Z_1, X_1 Z_2, \ldots, X_{r-1} Z_{c-1}}^{RC} \bigg|$$

$$\overbrace{X_1, X_2, \ldots, X_{r-1}, Z_1, Z_2, \ldots, Z_{c-1}}^{R, C}) + SSE \qquad (20.4)$$

As suggested by (20.3) and (20.4), it can be shown that

$$\text{regression } SS(X_1, X_2, \ldots, X_{r-1}) \equiv SSR$$
$$\text{regression } SS(Z_1, Z_2, \ldots, Z_{c-1}) \equiv SSC \qquad (20.5)$$
$$\text{regression } SS(X_1 Z_1, X_1 Z_2, \ldots, X_{r-1} Z_{c-1}) \equiv SSRC$$

where SSR, SSC, and SSRC are the unconditional sums of squares given by (20.1). For example, we can express (20.3) and (20.4) as

$$SSR + SS(C|R) + SS(RC|R, C) + SSE = TSS$$

and

$$SSC + SS(R|C) + SS(RC|R, C) + SSE = TSS$$

respectively. Note that both of these equations involve conditional sums of squares. However, when all the cell sample sizes are equal, it is also true that

$$\begin{matrix} \text{equal cell} \\ \text{numbers} \end{matrix} \Rightarrow \begin{cases} SSR = SS(R|C) \\ SSC = SS(C|R) \\ SSRC = SS(RC|R, C) \end{cases}$$

Consequently, when all the cell sample sizes are equal, the extra sums of squares are not actually affected by variables in the model, and the following holds:

$$\begin{matrix} \text{equal cell} \\ \text{numbers} \end{matrix} \Rightarrow SSR + SSC + SSRC + SSE = TSS \qquad (20.6)$$

FIGURE 20-1   Flow diagram for two-way ANOVA

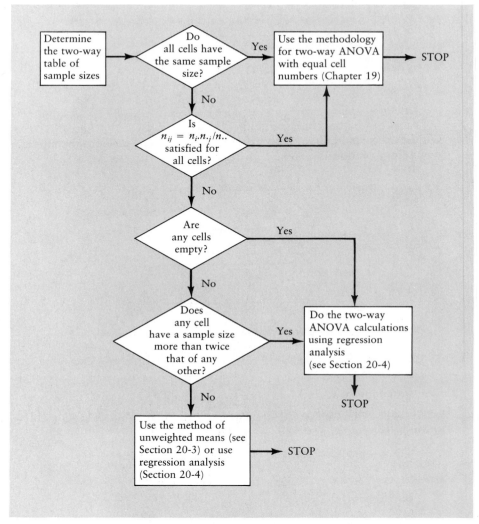

However, in the unequal-cell-number case, we have the following:

$$\begin{array}{c}\text{unequal cell}\\\text{numbers}\end{array} \Rightarrow \left\{\begin{array}{l}\text{SSR} \neq \text{SS(R}|\text{C)}\\\text{SSC} \neq \text{SS(C}|\text{R)}\\\text{SSRC} \neq \text{SS(RC}|\text{R, C)}\end{array}\right.$$

Therefore, in the unequal-cell-number case, (20.6) does not hold, and it is necessary to consider such expressions as (20.3) and (20.4), which reflect the importance of the order in which the effects are entered into the model. As we shall see in Section 20-4, the unequal-cell-number case is best handled by using regression analysis to carry out the two-way ANOVA calculations.

There is, however, one exception to this, which occurs when there are proportional cell frequencies satisfying

$$n_{ij} = \frac{n_{i.} n_{.j}}{n_{..}} \tag{20.7}$$

When (20.7) holds, it turns out that the following statement can be made:

$$n_{ij} = \frac{n_{i.} n_{.j}}{n_{..}} \Rightarrow \begin{cases} SSR = SS(R|C) \\ SSC = SS(C|R) \\ SSRC \ne SS(RC|R, C) \end{cases}$$

Thus, although (20.6) still does not hold in this case, (20.3) and (20.4) simplify to the single equation

$$SSR + SSC + SS(RC|R, C) + SSE = TSS \tag{20.8}$$

Thus, (20.8) contains only one term, $SS(RC|R, C)$, which is different from the terms in (20.6). This sum of squares, however, can be easily obtained by subtraction, as is clear from (20.8). Thus, *when the proportional cell frequency allocation of (20.7) is used, the standard equal-cell-number ANOVA calculations can be performed*, without the need to resort to regression analysis or alternative methods (e.g., the method of unweighted means, to be described in the next section).

To summarize, we have the flow diagram for two-way ANOVA shown in Figure 20-1.

20-3 Method of Unweighted Means

In one method for analyzing a two-way data array with unequal cell numbers, a new two-way table is constructed involving *only* the cell means for the original data set; then this new table is analyzed as a two-way data layout with one observation (the cell mean) in each cell.[1] This method is only an *approximate* procedure, however, because it provides ANOVA test statistics that are only *approximately* F statistics under the null hypotheses of interest.[2] The approximation results from the fact that one of the basic assumptions of ANOVA, homoscedasticity, does not hold when this method is used. The effect of violating this assumption is almost negligible (i.e., the approximation is quite good), however, provided two conditions are satisfied:

1. No cell is empty.

2. The cell sample sizes are not too far from being equal. A rough rule of thumb is to use this method only if the sample size in any one cell is not more than twice the sample size in any other cell.

---

[1] The discussion of the method of unweighted means, as well as the regression approach of the next section, treats both factors as fixed. Nevertheless, mixed and random models can also be considered and are discussed briefly in Section 20-5.

[2] An alternative analysis of means is that known as the "weighted squares of means" (see Searle, 1971); it is computationally complex but provides *exact* F tests.

Nevertheless, we recommend using the unweighted-means procedure only when a regression computer program is not available. Access to standard statistical packages avoids the need to use the method of unweighted means.

A look at the sample sizes in Table 20-2 indicates that condition 2 is not satisfied, since $n_{21} = 4$ and $n_{12} = 12$. Thus, in this example, it would be best to perform a regression analysis. We will nevertheless analyze these data using the method of unweighted means as well as regression analysis.

Four steps are necessary for the method of unweighted means. These are:

Step 1. Compute the usual (pooled) within-cell estimate of experimental error. This estimate is calculated using the formula

$$\text{MSE} = \frac{1}{n_{..} - rc} \sum_{i=1}^{r} \sum_{j=1}^{c} \sum_{k=1}^{n_{ij}} (Y_{ijk} - \bar{Y}_{ij.})^2$$

$$= \frac{1}{n_{..} - rc} \sum_{i=1}^{r} \sum_{j=1}^{c} \left[\sum_{k=1}^{n_{ij}} Y_{ijk}^2 - \frac{\left(\sum_{k=1}^{n_{ij}} Y_{ijk} \right)^2}{n_{ij}} \right]$$

where

$n_{..}$ = total number of observations

r = number of rows

c = number of columns

In our example (Table 20-2), $n_{..} = 50$, $r = 3$, and $c = 2$, and MSE is calculated to be

$$\text{MSE} = 3.4091$$

with $n_{..} - rc = 44$ degrees of freedom.

Step 2. Compute the estimated average variance of the cell means, $\hat{\sigma}_{av}^2$. This estimate is calculated using the formula

$$\hat{\sigma}_{av}^2 = \frac{1}{rc} (\text{MSE}) \sum_{i=1}^{r} \sum_{j=1}^{c} \frac{1}{n_{ij}} \tag{20.9}$$

(The variance $\hat{\sigma}_{av}^2$ can be equivalently written as

$$\hat{\sigma}_{av}^2 = \frac{\text{MSE}}{\text{harmonic mean of the } n_{ij}\text{'s}}$$

where

$$\frac{1}{(1/rc) \sum_{i=1}^{r} \sum_{j=1}^{c} 1/n_{ij}}$$

is the harmonic mean.)

To provide some motivation for (20.9), note that any cell mean $\bar{Y}_{ij.}$ has a variance equal to σ^2/n_{ij}, and these variances will not all be equal unless all the n_{ij} are equal. Then the average variance of these rc means is defined as

$$\sigma_{\bar{a}v}^2 = \frac{1}{rc} \sum_{i=1}^{r} \sum_{j=1}^{c} \frac{\sigma^2}{n_{ij}} = \frac{1}{rc} \sigma^2 \sum_{i=1}^{r} \sum_{j=1}^{c} \frac{1}{n_{ij}}$$

and the estimate (20.9) is obtained by substituting MSE for σ^2. If conditions 1 and 2 stated earlier are satisfied, the assumption that each mean has the same variance (estimated by $\hat{\sigma}_{\bar{a}v}^2$), although not actually true, holds to a reasonable degree of approximation.

In our example (Table 20-2), the estimated average variance is computed as follows:

$$\hat{\sigma}_{\bar{a}v}^2 = \frac{1}{3(2)} (\text{MSE}) \left(\frac{1}{8} + \frac{1}{12} + \frac{1}{4} + \frac{1}{8} + \frac{1}{10} + \frac{1}{8} \right)$$

$$= \frac{97}{720} (\text{MSE})$$

$$= \frac{\text{MSE}}{7.42} = \frac{3.4091}{7.42}$$

$$= 0.4594$$

Step 3. Form the table of cell means. In our example this table is as follows:

AFFCOM	WORRY Negative	WORRY Positive	Total
High	$\bar{Y}_{11.} = 5$	$\bar{Y}_{12.} = 6$	$T_{1.} = 11$
Medium	$\bar{Y}_{21.} = 4$	$\bar{Y}_{22.} = 7$	$T_{2.} = 11$
Low	$\bar{Y}_{31.} = 8$	$\bar{Y}_{32.} = 7$	$T_{3.} = 15$
Total	$T_{.1} = 17$	$T_{.2} = 20$	$T_{..} = 37$

Notice that in this table the row and column marginal totals (the $T_{i.}$'s and $T_{.j}$'s, respectively) are obtained simply by adding up the cell means in each row and column; the grand total $T_{..} = 37$ is the sum of the six cell means. These marginals are clearly not the same as those based on the original table (Table 20-2). This follows since we are treating these cell means as if they were the only observations available, whereas in Table 20-2 there are multiple observations in each cell.

Step 4. Analyze the table of cell means using two-way, one-observation-per-cell ANOVA, with $\hat{\sigma}_{\bar{a}v}^2$ as the denominator in all F tests. The appropriate ANOVA table associated with this step is given in general by Table 20-3 and, for our example, by Table 20-4.

The sums of squares given in Table 20-4 are computed from the table of cell means as follows:

$$\text{SS(AFFCOM)} = \frac{1}{2} (11^2 + 11^2 + 15^2) - \frac{37^2}{6} = 5.33$$

$$\text{SS(WORRY)} = \frac{1}{3} (17^2 + 20^2) - \frac{37^2}{6} = 1.50$$

$$\text{SS(AFFCOM} \times \text{WORRY)} = (5^2 + 4^2 + 8^2 + 6^2 + 7^2 + 7^2)$$
$$- \frac{37^2}{6} - 5.33 - 1.50 = 4.00$$

TABLE 20-3   ANOVA table based on the method of unweighted means

Source	df	SS	MS	F
Rows	$r - 1$	$\text{SSR} = \dfrac{1}{c} \sum\limits_{i=1}^{r} T_{i\cdot}^2 - \dfrac{T_{\cdot\cdot}^2}{rc}$	$\text{MSR} = \dfrac{\text{SSR}}{r - 1}$	$\dfrac{\text{MSR}}{\hat{\sigma}_{\text{av}}^2}$
Columns	$c - 1$	$\text{SSC} = \dfrac{1}{r} \sum\limits_{j=1}^{c} T_{\cdot j}^2 - \dfrac{T_{\cdot\cdot}^2}{rc}$	$\text{MSC} = \dfrac{\text{SSC}}{c - 1}$	$\dfrac{\text{MSC}}{\hat{\sigma}_{\text{av}}^2}$
Interaction	$(r - 1)(c - 1)$	$\text{SSRC} = \sum\limits_{i=1}^{r} \sum\limits_{j=1}^{c} \bar{Y}_{ij}^2 - \dfrac{T_{\cdot\cdot}^2}{rc} - \text{SSR} - \text{SSC}$	$\text{MSRC} = \dfrac{\text{SSRC}}{(r - 1)(c - 1)}$	$\dfrac{\text{MSRC}}{\hat{\sigma}_{\text{av}}^2}$
Error	$n_{\cdot\cdot} - rc$	$\cdots$	$\hat{\sigma}_{\text{av}}^2$	

Note: $T_{i\cdot}$ = sum of cell means in the ith row; $T_{\cdot j}$ = sum of cell means in the jth column; $T_{\cdot\cdot}$ = sum of all cell means.

TABLE 20-4   Application of method of unweighted means to data of Table 20-2

Source	df	SS	MS	F
AFFCOM	2	5.33	2.6650	5.81**
WORRY	1	1.50	1.5000	3.27
AFFCOM × WORRY	2	4.00	2.0000	4.35*
Error	44	$\cdots$	0.4594	

The conclusions based on Table 20-4 are that AFFCOM has a highly significant main effect (since an F of 5.81 with degrees of freedom of 2 and 44 has $P < .01$) and that the AFFCOM × WORRY interaction is mildly significant (since an F of 4.35 with degrees of freedom of 2 and 44 has $P < .025$). Thus, if the researcher prefers a conservative significance level, the primary conclusion from this ANOVA is that patient satisfaction with medical care differs only according to the level of affective patient–physician communication. If, on the other hand, the researcher works at the .05 significance level, the conclusion is that patient satisfaction depends on the combined effect of AFFCOM and WORRY. From Table 20-2, we see that when the AFFCOM level is high or medium, patients with a positive score on WORRY are more satisfied with care than patients with a negative score; when the AFFCOM level is low, patients with negative WORRY are more satisfied than patients with positive WORRY.

Rather than using the ANOVA table given in Table 20-3, some researchers prefer another form, for which *all* the sum-of-squares and mean-square terms are multiplied by the harmonic mean

$$\frac{1}{(1/rc) \sum\limits_{i=1}^{r} \sum\limits_{j=1}^{c} 1/n_{ij}}$$

The primary reason for using this alternative ANOVA table is that the error term turns out to be MSE, as can be seen from (20.9).

TABLE 20-5 Alternative ANOVA table for the method of unweighted means

Source	df	SS	MS	F
AFFCOM	2	39.588	19.794	5.81**
WORRY	1	11.134	11.134	3.27
AFFCOM × WORRY	2	29.691	14.845	4.35*
Error	44	150.000	3.4091	

Thus, the alternative table (with MSE as the error term) is more like the usual two-way ANOVA table for equal sample sizes in each cell. Nevertheless, it is important to note that using this alternative table does not affect the F statistics. This is because *both* mean-square terms in any F statistic of Table 20-3 are multiplied by the same harmonic mean; for example, the F statistic for AFFCOM in the alternative ANOVA table (Table 20-5) is given by

$$F_{(AFFCOM)} = \frac{19.794}{3.4091} = \frac{(2.6650)(7.42)}{(0.4594)(7.42)} = \frac{2.6650}{0.4594} = 5.81$$

which was previously obtained in Table 20-4.

In any case, this alternative form of ANOVA table is most often given by computer programs performing the method of unweighted means.

20-4 Regression Approach for Unequal Cell Sample Sizes

As described in Section 20-2, the general regression model that is applicable to both the equal- and unequal-cell-number cases is

$$Y = \mu + \sum_{i=1}^{r-1} \alpha_i X_i + \sum_{j=1}^{c-1} \beta_j Z_j + \sum_{i=1}^{r-1} \sum_{j=1}^{c-1} \gamma_{ij} X_i Z_j + E$$

in which the $\{X_i\}$ and $\{Z_j\}$ are sets of dummy variables representing the r levels of the row factor and the c levels of the column factor, respectively. In general, *any* two-way ANOVA problem can be analyzed by a regression approach utilizing such a model. However, when there are unequal cell numbers, the order in which effects (row, column, or interaction) are tested is an important decision that, if not made carefully, can lead to inappropriate conclusions. The procedure we recommend is a backward-type algorithm in which interaction is considered before main effects (see Applebaum and Cramer, 1974). The steps involved in this algorithm are as follows:

Step 1. After fitting the full model, perform a chunk test for interaction (i.e., test H_0: $\gamma_{ij} = 0$ for all i and j in the above model). The F statistic is given by

$$F(X_1 Z_1, X_1 Z_2, \ldots, X_{r-1} Z_{c-1} | X_1, X_2, \ldots, X_{r-1}, Z_1, Z_2, \ldots, Z_{c-1})$$

and its numerator and denominator degrees of freedom are $(r - 1)(c - 1)$ and $n_{..} - rc$, respectively.

Step 2.

a. If the step 1 test is significant, two primary options are available:
    i. Do no further testing and use the above full model as the final model.
    or
    ii. Do individual testing to eliminate any nonsignificant interaction terms. The final model will then contain all main effects and those product terms found to be significant.[3]

b. If the step 1 is not significant, reduce the model by eliminating all interaction terms; this reduced model is of the form

$$Y = \mu + \sum_{i=1}^{r-1} \alpha_i X_i + \sum_{j=1}^{c-1} \beta_j Z_j + E \qquad (20.10)$$

Then conduct two main-effect chunk tests of H_0: $\alpha_i = 0$ for all i and H_0: $\beta_j = 0$ for all j using the F statistics $F(X_1, X_2, \ldots, X_{r-1} | Z_1, Z_2, \ldots, Z_{c-1})$ and $F(Z_1, Z_2, \ldots, Z_{c-1} | X_1, X_2, \ldots, X_{r-1})$, respectively.[4] These tests consider, respectively, the significance of the row effects given the column effects, and the column effects given the row effects.

Step 3.

a. If step 2b yields nonsignificant results for both tests, reduce the model further by eliminating the chunk of variables (the set of row or column main effects) corresponding to the less significant chunk test (i.e., that test having the larger P-value). Thus, if the test of H_0: $\alpha_i = 0$ for all i (i.e., the test for "rows" given "columns") has the larger P-value, then the new reduced model is

$$Y = \mu + \sum_{j=1}^{c-1} \beta_j Z_j + E$$

Alternatively, if H_0: $\beta_j = 0$ for all j (i.e., the test for "columns" given "rows") is less significant, the new reduced model is

$$Y = \mu + \sum_{i=1}^{r-1} \alpha_i X_i + E$$

After reducing the model, conduct a chunk test for the main effects in this new reduced model, using either $F(X_1, X_2, \ldots, X_{r-1})$ or $F(Z_1, Z_2, \ldots, Z_{c-1})$, depending on which set of main effects remains.[5] If this final test is nonsignificant,

---

[3] Another possible component of option ii is to allow for the possible removal of main-effect terms that are not components of significant interaction terms.

[4] As mentioned elsewhere (e.g., Chapter 9), some statisticians prefer to use the mean-square residual for the full (interaction) model in these main-effect F tests rather than the mean-square residual for the reduced model (20.10).

[5] As in step 2, some statisticians prefer to use for the denominator of these tests the mean-square residual for the full interaction model, rather than the mean-square residual for the reduced model.

then the overall conclusion is that the row, column, and interaction effects are all unimportant. If this test is significant, the final model contains only these significant (row or column) main effects, and the conclusion is that only these effects are important.

b. If step 2b yields significant results for both tests, then the reduced model (20.10) is the final model, and the conclusion is that both row and column effects are important but that there are no important interaction terms.

c. If step 2b produces exactly one significant test, there is no need to reduce the model further; the conclusion is that one of the two sets of main effects is important and that there is no significant interaction.[6]

Figure 20-2 provides a flow diagram for the above strategy.

20-4-1 Example of Regression Approach to Analyzing Unbalanced Two-Way ANOVA Data

For the satisfaction-with-medical-care data of Table 20-2, two regression-model-based ANOVA tables (Tables 20-6 and 20-7) can be produced depending on whether rows precede columns or columns precede rows into the model. Following the strategy for regression analysis that we have outlined, the first step is to conduct a chunk test for interaction. The regression model we are using for this test is

$$Y = \mu + \alpha_1 X_1 + \alpha_2 X_2 + \beta_1 Z_1 + \gamma_{11} X_1 Z_1 + \gamma_{21} X_2 Z_1 + E$$

TABLE 20-6   ANOVA table when rows precede columns for regression analysis of data in Table 20-2

Source	df	SS	MS	F	P
X_1, X_2	2	38.756	19.378	$4.97_{2,47}$	$.01 < P < .025$
$Z_1 \vert X_1, X_2$	1	5.861	5.861	$1.52_{1,46}$	$.10 < P < .25$
$X_1 Z_1, X_2 Z_1 \vert X_1, X_2, Z_1$	2	27.384	13.692	$4.02_{2,44}$	$.01 < P < .025$
Residual	44	149.996	3.409		

TABLE 20-7   ANOVA table when columns precede rows for regression analysis of data in Table 20-2

Source	df	SS	MS	F	P
Z_1	1	1.870	1.870	$0.41_{1,48}$	$P > .25$
$X_1, X_2 \vert Z_1$	2	42.746	21.373	$5.54_{2,46}$	$.005 < P < .01$
$X_1 Z_1, X_2 Z_1 \vert X_1, X_2, Z_1$	2	27.384	13.692	$4.02_{2,44}$	$.01 < P < .025$
Residual	44	149.996	3.409		

---

[6] An alternative here is to reduce the model further by eliminating the nonsignificant set of main-effect variables. However, the unconditional test for the remaining set of main-effect variables may then be nonsignificant, even though the corresponding conditional test under model (20.10) is significant. In this situation, we believe that the conclusions based on model (20.10) are more appropriate.

FIGURE 20-2   Flow diagram for regression analysis of unbalanced two-way ANOVA data

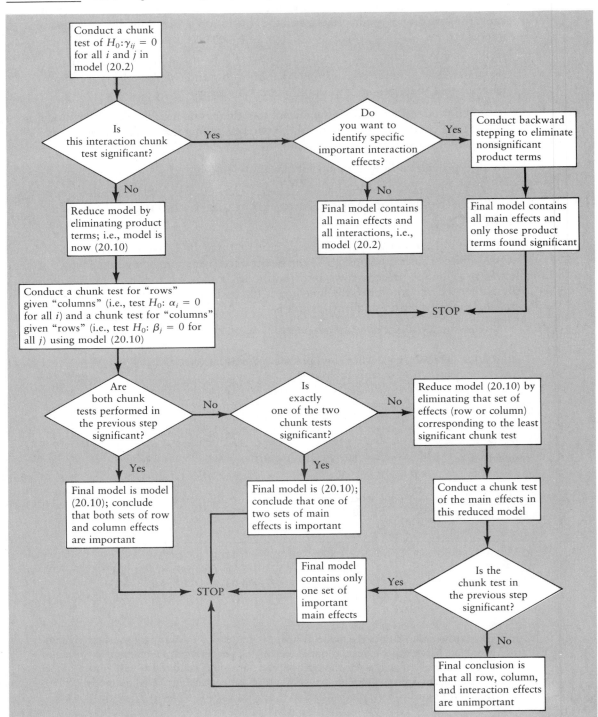

where X_1 and X_2 are dummy variables for the row effects corresponding to the variable AFFCOM and Z_1 is a dummy variable for the column effects corresponding to the variable WORRY. The null hypothesis of no interaction is H_0: $\gamma_{11} = \gamma_{21} = 0$. As expected, both Tables 20-6 and 20-7 provide the same numerical results for this conditional test. The P-value for this interaction test indicates significance at the .05 level but not at the .01 level. If the investigator is using a .05 significance level, then, based on our strategy, the analysis would stop at this point; one would conclude that there is significant interaction and that main-effect interpretations are not relevant. If the significance level is .01, however, then one would conclude that there are no significant interaction effects and then consider the reduced model

$$Y = \mu + \alpha_1 X_1 + \alpha_2 X_2 + \beta_1 Z_1 + E$$

The conditional test for "rows" given "columns" (i.e., H_0: $\alpha_1 = \alpha_2 = 0$) based on fitting the above model leads to an F ratio of 5.54, as given in the "$X_1, X_2 | Z_1$" row of Table 20-7. The corresponding P-value indicates significance at the .01 level. The conditional test for "columns" given "rows" (i.e., H_0: $\beta_1 = 0$) in the above reduced model results in an F-value of 1.52 (see the "$Z_1 | X_1, X_2$" row of Table 20-6). The corresponding P-value ($>.10$) clearly indicates nonsignificance. Thus, based on our strategy, we would conclude (using $\alpha = .01$) that there is a significant AFFCOM main effect and a nonsignificant WORRY main effect.

If the model is further reduced to contain only the AFFCOM main effects (X_1 and X_2), the corresponding unconditional main effect F test, as found in the "X_1, X_2" row of Table 20-6, is no longer significant at the .01 level. Nevertheless, since the model containing both sets of main effects indicates a significant AFFCOM main effect, we would argue that AFFCOM is an important variable because this latter model is taking the WORRY variable into account (i.e., the AFFCOM main-effect F test is a conditional one).

20-4-2 Using Computer Programs

Nearly all complex data analysis is now conducted using computer programs. Statistical packages usually contain ANOVA programs or ANOVA options within regression programs. With unbalanced data, and especially with incomplete data, the user of such packages must make the extra effort to understand the dummy variable coding schemes being used and the partial F tests being computed (e.g., which linear combinations of parameters are being tested in which model). Without this understanding, the user may inadvertently conduct tests that are quite different from those desired.

As mentioned earlier, ANOVA may be thought of as an analysis of means. As discussed in the three previous chapters, we have found it helpful to consider such means as linear combinations of regression parameters. This approach can be particularly helpful with unbalanced data and is a lifesaver with incomplete data.

20-5 Random- and Mixed-Effects Models

The method of unweighted means (Section 20-3) can easily be adapted to deal with random- and mixed-effects models. All that is necessary is to use the same ANOVA format as given in Table 19-8 but to use the mean-square terms of Table 20-3. For example, to test for a significant row effect in the random-effects model, we simply use

$$F = \frac{MSR}{MSRC}$$

where MSR and MSRC are obtained from the unweighted-means approach. Other, more detailed approaches for dealing with random- and mixed-effects models for two-way data arrays with unequal cell numbers are discussed in Searle (1971).

20-6 Higher-Way ANOVA

It should not be surprising to learn that ANOVA, as a special case of regression analysis, can be generalized to any number of factors (i.e., independent variables). We contend, nevertheless, that too much emphasis on complex ANOVA models and the associated testing procedures is likely to be unwarranted, especially for researchers in the health, medical, social, and behavioral sciences. This is because of the following:

1. As more independent variables are considered, wanting to treat them all as nominal variables becomes more unlikely.

2. Even if all independent variables are treated as nominal, either sufficient numbers of observations are not available in all cells (e.g., there are some empty cells) or equal sample sizes cannot be placed in each cell.

Methods are available, however, for designing and analyzing experimental studies in which only a *fraction* of the total number of possible cells need be used; these methods still permit the researcher to estimate the effects of primary interest. The reader is referred to texts by

TABLE 20-8   General three-way ANOVA table for equal cell sample sizes

			F			
Source	df	MS	All Fixed	A, B Fixed; C Random	A, B Random; C Fixed	All Random
A (a levels)	$a - 1$	MSA	MSA/MSE	MSA/MSAC	MSA/MSAB	No exact test
B (b levels)	$b - 1$	MSB	MSB/MSE	MSB/MSBC	MSB/MSAB	No exact test
C (c levels)	$c - 1$	MSC	MSC/MSE	MSC/MSE	No exact test	No exact test
AB	$(a - 1)(b - 1)$	MSAB	MSAB/MSE	MSAB/MSABC	MSAB/MSE	MSAB/MSABC
AC	$(a - 1)(c - 1)$	MSAC	MSAC/MSE	MSAC/MSE	MSAC/MSABC	MSAC/MSABC
BC	$(b - 1)(c - 1)$	MSBC	MSBC/MSE	MSBC/MSE	MSBC/MSABC	MSBC/MSABC
ABC	$(a - 1)(b - 1)(c - 1)$	MSABC	MSABC/MSE	MSABC/MSE	MSABC/MSE	MSABC/MSE
Error	$n - abc$	MSE				
Total	$n - 1$					

Note: See Ostle (1963) for more information.
    Those cases for which there is no exact F test result from the lack of equivalent numerator and denominator expected mean squares under the null hypothesis. Four other factor categorizations have been omitted from this table: (1) A fixed, B and C random; (2) B fixed, A and C random; (3) A and C fixed, B random; and (4) B and C fixed, A random. The F statistics for each of these cases can nevertheless be derived from one of the cases in the table by an appropriate permutation of the letters A, B, and C. For example, when considering A fixed, B and C random, switch A with C in the "Source" and third F columns.

Ostle (1963), Snedecor and Cochran (1967), and Peng (1967) for applications of such methods.

In general, however, regression analysis should be the predominant analysis for higher-way ANOVA situations, especially since so much research in the health, medical, social, and behavioral sciences is observational. We thus will not extend our discussion to three-way or higher ANOVA situations, but rather suggest the use of regression analysis for such situations. Nevertheless, since the three-way ANOVA case is more prevalent than other higher-way cases, we present for reference purposes Table 20-8, the general three-way ANOVA table for the equal-cell-number situation. We have also provided an exercise at the end of the chapter dealing with the three-way case.

Problems

1. Consider hypothetical data based on a study concerning the effects of rapid cultural change on blood pressure levels for native citizens of an island in Micronesia. Blood pressures were taken on a random sample of 30 males over age 40 from a certain province who commuted to work in the nearby westernized capital city. These persons were also given a sociological questionnaire from which their social rankings in both their traditional and modern (i.e., westernized) cultures were determined. The results are summarized in the table below.

Modern Rank (Factor A)	Traditional Rank (Factor B)		
	HI	MED	LO
HI	130, 140, 135	150, 145	175, 160, 170, 165, 155
MED	145, 140, 150	150, 160, 155	165, 155, 165, 170, 160
LO	180, 160, 145	155, 140, 135	125, 130, 110

   a. Discuss the table of sample means for this data set.
   b. Analyze this data set by the method of unweighted means, making sure to give the appropriate ANOVA table.
   c. Give an appropriate regression model for this data set, treating the two factors as nominal variables.
   d. Using the regression ANOVA tables that follow (where X pertains to factor A and Z to factor B), carry out two different main-effect tests for each factor, and also test for interaction. Compare the results of the two main-effect tests for each factor. Also, compare the regression results with the results obtained using the method of unweighted means.
   e. How might one modify the regression model given in part (c) so that any trends in blood pressure levels could be quantified in terms of increasing social rankings for the two factors? (Note that this requires assigning numerical values to the categories of each factor.) What difficulty does one encounter in defining such a model?

REGRESSION RESULTS FOR PROBLEM 1

Source	df	MS
X_1	1	469.17985
$X_2 \mid X_1$	1	508.52217
$Z_1 \mid X_1, X_2$	1	187.97673
$Z_2 \mid X_1, X_2, Z_1$	1	7.54570
$X_1 Z_1 \mid X_1, X_2, Z_1, Z_2$	1	3,925.29395
$X_1 Z_2 \mid X_1, X_2, Z_1, Z_2, X_1 Z_1$	1	9.70621
$X_2 Z_1 \mid X_1, X_2, Z_1, Z_2, X_1 Z_1, X_1 Z_2$	1	633.17613
$X_2 Z_2 \mid X_1, X_2, Z_1, Z_2, X_1 Z_1, X_1 Z_2, X_2 Z_1$	1	2.67593
Residual	21	75.83333

Source	df	MS
Z_1	1	278.59213
$Z_2 \mid Z_1$	1	22.71129
$X_1 \mid Z_1, Z_2$	1	391.77041
$X_2 \mid Z_1, Z_2, X_1$	1	480.15062
$X_1 Z_1 \mid Z_1, Z_2, X_1, X_2$	1	3,925.29395
$X_1 Z_2 \mid Z_1, Z_2, X_1, X_2, X_1 Z_1$	1	9.70621
$X_2 Z_1 \mid Z_1, Z_2, X_1, X_2, X_1 Z_1, X_1 Z_2$	1	633.17613
$X_2 Z_2 \mid Z_1, Z_2, X_1, X_2, X_1 Z_1, X_1 Z_2, X_2 Z_1$	1	2.67593
Residual	21	75.83333

2. A study was conducted to assess the combined effects of patient attitude and patient–physician communication on patient satisfaction with medical care during pregnancy. A random sample of 110 pregnant women under the care of private physicians was followed from the first visit with the physician until delivery. Using specially devised questionnaires, the following variables were measured for each patient: Y = satisfaction score, X_1 = attitude score, and X_2 = communication score. Each score was developed as an interval variable, but there was some question as to whether the analysis should treat the attitude and/or communication scores as nominal variables.

   a. What would be an appropriate regression model for describing the joint effect of X_1 and X_2 on Y if an interaction between communication and attitude is possible and if all variables are treated as interval variables?

   b. What would be an appropriate regression model (using dummy variables) if one still wished to allow for an interaction effect but desired only to compare high values versus low values (i.e., to make group comparisons) for both the communication and attitude variables? What kind of ANOVA model would this regression model correspond to?

   c. When would one prefer the model in part (a) to that in (b), and vice versa?

   d. If both independent variables are treated nominally as in part (b), would one expect the associated 2×2 table to have equal numbers in each of the four cells?

3. The data listed at the end of this problem are from an unpublished study by Harbin and others (1985). Subjects were young (ages 18 through 29) or old (ages 60

through 86) men. Each subject was exposed to 0 or 100 parts per million (PPM) of carbon monoxide (CO) for a period before and during testing. Median reaction times for 30 trials are reported for two different tasks: (1) simple reaction time (no choice), REACTIM1, and (2) two-choice reaction time, REACTIM2. Pilot data analysis from an earlier study established that this dependent variable followed an appropriate Gaussian distribution. For this problem, consider only REACTIM1 as a dependent variable. Use $\alpha = .01$.

a. Tabulate the number of subjects in each AGEGROUP × CO combination. Explain why this might be called a nearly orthogonal two-way design. Any missing data were due to technical problems unrelated to the response variable or treatments.

b. Using dummy variables coded -1 (young, PPM_CO = 0) and 1, define dummy variables and a corresponding multiple regression model for a two-way ANOVA.

c. Use an appropriate computer program to fit the regression model in (b). Report appropriate tests of the AGEGROUP × CO interaction and tests of their main effects in an appropriate summary source table.

d. Use an appropriate computer program that automatically codes dummy variables (and properly treats unequal cell sizes) to conduct a two-way ANOVA. Compare the results to those in (c). Explain any differences.

4. a–d. Repeat Problem 3 using REACTIM2 as the dependent variable.

DATA FOR PROBLEM 3

Observation	AGEGROUP	PPM_CO	REACTIM1	REACTIM2
1	Young	0	291.5	632.0
2	Young	0	471.0	607.5
3	Young	0	692.0	859.0
4	Young	0	376.0	484.0
5	Young	0	372.5	501.0
6	Young	0	307.0	381.0
7	Young	0	501.0	559.0
8	Young	0	466.0	632.0
9	Young	0	375.0	434.0
10	Young	0	425.0	454.0
11	Young	0	343.0	542.0
12	Young	0	348.0	471.0
13	Young	0	503.0	521.0
14	Young	0	382.5	519.0
15	Young	100	472.5	515.0
16	Young	100	354.0	521.0
17	Young	100	350.0	456.0
18	Young	100	486.0	522.0
19	Young	100	402.0	472.0
20	Young	100	347.0	414.0
21	Young	100	320.0	363.0
22	Young	100	446.0	591.0
23	Young	100	410.0	539.5

Observation	AGEGROUP	PPM_CO	REACTIM1	REACTIM2
24	Young	100	302.0	472.5
25	Young	100	692.5	656.0
26	Young	100	447.5	548.0
27	Young	100	525.5	527.0
28	Young	100	322.5	574.0
29	Young	100	468.5	559.5
30	Young	100	378.0	499.5
31	Young	100	497.5	529.5
32	Old	0	542.0	595.0
33	Old	0	599.0	606.0
34	Old	0	562.0	598.0
35	Old	0	586.0	744.0
36	Old	0	674.0	724.0
37	Old	0	762.0	836.5
38	Old	0	697.0	834.0
39	Old	0	583.0	698.5
40	Old	0	533.5	668.0
41	Old	0	524.5	670.0
42	Old	0	500.0	587.0
43	Old	0	680.0	912.5
44	Old	0	563.5	619.0
45	Old	100	523.5	646.5
46	Old	100	770.0	862.5
47	Old	100	712.0	829.0
48	Old	100	653.0	697.0
49	Old	100	699.5	818.0
50	Old	100	561.0	819.5
51	Old	100	751.0	872.0
52	Old	100	520.5	889.0
53	Old	100	523.0	601.0

5. A crime victimization study was undertaken in a medium-sized southern city. The main purpose was to determine the effects of being a crime victim on confidence in law enforcement authority and in the legal system itself. A questionnaire was administered to a stratified random sample of 40 city residents, from which was obtained data on the number of times victimized, a measure of social class status (SCLS), and a measure of the respondent's confidence in law enforcement and in the legal system. The data are as given in the table.

No. of Times Victimized	Social Class Status		
	LO	MED	HI
0	4, 14, 15, 19, 17, 17, 16	7, 10, 12, 15, 16	8, 19, 10, 17
1	2, 7, 18	6, 19, 12, 12	7, 6, 5, 3, 16
2+	7, 8, 2, 11, 12	1, 2, 4	4, 2, 8, 9

a. Determine the table of sample means and comment on any patterns noted.
b. Analyze this data set using the ANOVA computer results given.
c. How would you analyze this data set using the regression computer output?
d. What ANOVA assumption(s) might not hold for these data?

ANOVA TABLE FOR PROBLEM 5

Source	df	SS	MS
VICTIM	2	$4.0000E + 02$	$2.0000E + 02$
SCLS	2	$2.2739E + 01$	$1.1370E + 01$
VICTIM $\times$ SCLS	4	$1.0993E + 02$	$2.7483E + 01$
Error	31	$7.0408E + 02$	$2.2712E + 01$

REGRESSION RESULTS FOR PROBLEM 5

Source	df	MS
Z_1	1	44.03235
$Z_2|Z_1$	1	1.03496
$X_1|Z_1, Z_2$	1	395.75734
$X_2|Z_1, Z_2, X_1$	1	0.06778
$X_1Z_1|Z_1, Z_2, X_1, X_2$	1	1.68985
$X_1Z_2|Z_1, Z_2, X_1, X_2, X_1Z_1$	1	3.31635
$X_2Z_1|Z_1, Z_2, X_1, X_2, X_1Z_1, X_1Z_2$	1	0.40190
$X_2Z_2|Z_1, Z_2, X_1, X_2, X_1Z_1, X_1Z_2, X_2Z_1$	1	94.59353
Residual	31	22.71229

Source	df	MS
X_1	1	407.86993
$X_2|X_1$	1	0.52174
$Z_1|X_1, X_2$	1	27.98766
$Z_2|X_1, X_2, Z_1$	1	4.51309
$X_1Z_1|X_1, X_2, Z_1, Z_2$	1	1.68985
$X_1Z_2|X_1, X_2, Z_1, Z_2, X_1Z_1$	1	3.31635
$X_2Z_1|X_1, X_2, Z_1, Z_2, X_1Z_1, X_1Z_2$	1	0.40190
$X_2Z_2|X_1, X_2, Z_1, Z_2, X_1Z_1, X_1Z_2, X_2Z_1$	1	94.59353
Residual	31	22.71229

Note: X pertains to number of times victimized; Z pertains to social class status.

6. The effect of a new antidepressant drug on reducing the severity of depression was studied in manic–depressive patients at two state mental hospitals. In each hospital all such patients were randomly assigned to either a treatment (new drug) or control (old drug) group. The results of this experiment are summarized in the table, where a high mean score indicates more of a lowering in depression level than does a low mean score.

Hospital	Group	
	Treatment	Control
A	$n = 25, \bar{Y} = 8.5, S = 1.3$	$n = 31, \bar{Y} = 4.6, S = 1.8$
B	$n = 25, \bar{Y} = 2.3, S = 0.9$	$n = 31, \bar{Y} = -1.7, S = 1.1$

a. Without performing any statistical tests, interpret the means in the table.

b. Analyze the data using the method of unweighted means. Is the new drug effective? (Make sure to give the appropriate ANOVA table.)

c. What regression model is appropriate for analyzing the data? For this model, describe how to test whether there is a significant effect due to the new drug.

7. A study was conducted by a television network to evaluate the viewing characteristics of adult females in a certain state. Each individual in a stratified random sample of 480 women was sent a questionnaire; the strata were formed on the basis of the following three factors: season (winter or summer), region (eastern, central, or western), and residence (rural or urban). The averages of the total time reported watching TV (hours per day) are summarized in the accompanying table of sample means and standard deviations.

Residence and Region	Summer			Winter			Marginals	
	n	$\bar{Y}$	S	n	$\bar{Y}$	S	n	$\bar{Y}$
Rural								
East	40	2.75	1.340	40	4.80	0.851	80	3.78
Central	40	2.75	1.380	40	4.85	0.935	80	3.80
West	40	2.65	1.180	40	4.78	0.843	80	3.71
Marginals	120	2.72		120	4.81		240	3.76
Urban								
East	40	3.38	0.958	40	3.65	0.947	80	3.52
Central	40	3.15	1.130	40	4.50	0.743	80	3.83
West	40	3.65	0.779	40	4.05	0.781	80	3.85
Marginals	120	3.39		120	4.07		240	3.73
Marginals	240	3.06		240	4.44		480	3.75

a. Suppose that the questionnaire contained items concerning additional factors such as occupation (categorized as housewife, blue-collar worker, white-collar worker, or professional), age (categorized as 20 to 34, 35 to 50, and over 50), and number of children (categorized as 0, 1–2, and 3+). Comment on the likelihood of having equal cell numbers when carrying out an ANOVA considering these additional variables.

b. Examine the table of sample means for main effects and interactions. (You may want to form two-factor summary tables for looking at two-factor interactions.)

c. Using the ANOVA computer results given, carry out appropriate F tests and discuss your results.

d. State a regression model appropriate for obtaining information equivalent to the ANOVA results presented next.

ANOVA TABLE FOR PROBLEM 7

Source	df	SS	MS
RESID	1	1.3333E − 01	1.3333E − 01
REGION	2	2.5527E + 00	1.2763E + 00
SEASON	1	2.2963E + 02	2.2963E + 02
RESID × REGION	2	3.3247E + 00	1.6623E + 00
RESID × SEASON	1	6.0492E + 01	6.0492E + 01
REGION × SEASON	2	7.2247E + 00	3.6123E + 00
RESID × REGION × SEASON	2	6.7460E + 00	3.3730E + 00
Error	468	4.7821E + 02	1.0218E + 00

8. Suppose that the following data were obtained by an investigator studying the influence of estrogen injections on change in pulse rate of adolescent chimpanzees.

Male $\begin{cases} \text{Control:} & 5.1, -2.3, 4.2, 3.8, 3.2, -1.5, 6.1, -2.5, 1.9, -3.0, -2.8, 1.7 \\ \text{Estrogen:} & 15.0, 6.2, 4.1, 2.3, 7.6, 14.8, 12.3, 13.1, 3.4, 8.5, 11.2, 6.9 \end{cases}$

Female $\begin{cases} \text{Control:} & -2.3, -5.8, -1.5, 3.8, 5.5, 1.6, -2.4, 1.9 \\ \text{Estrogen:} & 7.3, 2.4, 6.5, 8.1, 10.3, 2.2, 12.7, 6.3 \end{cases}$

a. What are the factors in this experiment? Should they be designated as fixed or random?
b. Demonstrate that for this problem the cell frequency for two-way ANOVA is proportional; that is, $n_{ij} = n_{i.}n_{.j}/n_{..}$ for each of the four cells.
c. Using the following table of sample means

	Control	Estrogen
Male	1.158333	8.783333
Female	0.100000	6.975000

and the general formulae

$$\text{SS(rows)} = \sum_{i=1}^{r} \sum_{j=1}^{c} \sum_{k=1}^{n_{ij}} (\bar{Y}_{i..} - \bar{Y}_{...})^2$$

$$\text{SS(columns)} = \sum_{i=1}^{r} \sum_{j=1}^{c} \sum_{k=1}^{n_{ij}} (\bar{Y}_{.j.} - \bar{Y}_{...})^2$$

$$\text{SS(cells)} = \sum_{i=1}^{r} \sum_{j=1}^{c} \sum_{k=1}^{n_{ij}} (\bar{Y}_{ij.} - \bar{Y}_{...})^2$$

analyze the data for this problem using the usual methodology for equal-cell-number two-way ANOVA and the fact that SSE = 530.24078 (i.e., compute the sums of squares for rows, columns, and cells directly, and then obtain the sum of squares for interaction by subtraction).
d. Analyze the data for this problem using the method of unweighted means and compare your results with those obtained in part (c).
e. What regression model is appropriate for analyzing this data set?
f. Using the regression analysis results given, check whether SS(rows) = regression SS(SEX) = regression SS(SEX|TREATMENT) and SS(columns) = regres-

sion SS(TREATMENT) = regression SS(TREATMENT|SEX), where SS(rows) and SS(columns) are as obtained in part (c). What has been demonstrated here?

ANOVA FOR PROBLEM 8

Source	df	SS
SEX	1	19.72667
TREATMENT|SEX	1	536.55619
SEX × TREATMENT|SEX, TREATMENT	1	1.35000
Residual	36	530.24078

Source	df	SS
TREATMENT	1	536.55619
SEX|TREATMENT	1	19.72667
SEX × TREATMENT|SEX, TREATMENT	1	1.35000
Residual	36	530.24078

Note: $\text{SEX} = \begin{cases} -1 & \text{if male} \\ 1 & \text{if female} \end{cases}$ and $\text{TREATMENT} = \begin{cases} -1 & \text{if control} \\ 1 & \text{if estrogen} \end{cases}$

9. The data listed at the end of this problem concern a study by Reiter and others (1981) of the effects of injecting triethyl-tin (TET) into rats once at age 5 days. The animals were injected with either 0, 3, or 6 mg per kilogram body weight. The response was the log of the activity count for 1 hour counting at 21 days of age. The rat was left to move about freely in a figure 8 maze. Analysis of other studies with this type of activity count assures us that log counts should yield Gaussian errors if the model is correct.
   a. Tabulate the DOSAGE × SEX cell sample sizes. Explain why this might be called a nearly orthogonal design.
   b. Conduct a two-way ANOVA with SEX and DOSAGE as factors. Provide an appropriate source table.
   c. Using $\alpha = .05$, report your conclusions based on the ANOVA.
   d. Which, if any, families of means should be followed up with multiple-comparison tests? What type of comparisons would you recommend?

DATA FOR PROBLEM 9

Observation	LOGACT21	DOSAGE	SEX	CAGE
1	2.636	0	Male	5
2	2.736	0	Male	6
3	2.775	0	Male	7
4	2.672	0	Male	9
5	2.653	0	Male	11
6	2.569	0	Male	12
7	2.737	0	Male	15
8	2.588	0	Male	16
9	2.735	0	Male	17
10	2.444	3	Male	3

Observation	LOGACT21	DOSAGE	SEX	CAGE
11	2.744	3	Male	5
12	2.207	3	Male	6
13	2.851	3	Male	7
14	2.533	3	Male	9
15	2.630	3	Male	11
16	2.688	3	Male	12
17	2.665	3	Male	15
18	2.517	3	Male	16
19	2.769	3	Male	17
20	2.694	6	Male	3
21	2.845	6	Male	5
22	2.865	6	Male	6
23	3.001	6	Male	7
24	3.043	6	Male	9
25	3.066	6	Male	11
26	2.747	6	Male	12
27	2.894	6	Male	15
28	1.851	6	Male	16
29	2.489	6	Male	17
30	2.494	0	Female	3
31	2.723	0	Female	5
32	2.841	0	Female	6
33	2.620	0	Female	7
34	2.682	0	Female	9
35	2.644	0	Female	11
36	2.684	0	Female	12
37	2.607	0	Female	15
38	2.591	0	Female	16
39	2.737	0	Female	17
40	2.220	3	Female	3
41	2.371	3	Female	5
42	2.679	3	Female	6
43	2.591	3	Female	7
44	2.942	3	Female	9
45	2.473	3	Female	11
46	2.814	3	Female	12
47	2.622	3	Female	15
48	2.730	3	Female	16
49	2.955	3	Female	17
50	2.540	6	Female	3
51	3.113	6	Female	5
52	2.468	6	Female	6
53	2.606	6	Female	7
54	2.764	6	Female	9
55	2.859	6	Female	11
56	2.763	6	Female	12
57	3.000	6	Female	15
58	3.111	6	Female	16
59	2.858	6	Female	17

10. The experimenters described in Problem 9 of this chapter hoped that home cage would not affect activity level in any systematic fashion. Explore this question by repeating Problem 9 but replacing SEX with CAGE in your analysis.

References

Appelbaum, M. I., and Cramer, E. M. (1974). "Some Problems in the Non-Orthogonal Analysis of Variance." *Psych. Bull.*, *81*(6): 335–343.

Harbin, T. J., Benignus, V. A., Muller, K. E., and Barton, C. N. (1985). "Effects of Low-Level Carbon Monoxide Exposure upon Evoked Cortical Potentials in Young and Elderly Men." Manuscript in review, U.S. Environmental Protection Agency, Washington, D.C.

Ostle, B. (1963). *Statistics in Research*, 2nd ed. Ames, Iowa: Iowa State University Press.

Peng, K. C. (1967). *Design and Analysis of Scientific Experiments*. Reading, Mass.: Addison-Wesley Publishing Company, Inc.

Reiter, L. W., Heavner, G. G., Dean, K. F., and Ruppert, P. H. (1981). "Developmental and Behavioral Effects of Early Postnatal Exposure to Triethyltin in Rats." *Neurobehav. Toxicol. Teratol.*, *3*: 285–293.

Searle, S. R. (1971). *Linear Models*. New York: John Wiley & Sons, Inc.

Snedecor, G. W., and Cochran, W. G. (1967). *Statistical Methods*, 6th ed. Ames, Iowa: Iowa State University Press.

Thompson, S. J. (1972). "The Doctor–Patient Relationship and Outcomes of Pregnancy." Ph.D. dissertation, Department of Epidemiology, University of North Carolina, Chapel Hill, N.C.

Maximum Likelihood Methods: Theory and Applications

21-1 Preview

The purpose of this chapter is to describe the methodology of maximum likelihood (ML) estimation and its associated inference-making procedures. The term *maximum likelihood* refers to a very general algorithm for obtaining estimators of population parameters, such estimators having excellent (large-sample) statistical properties. One major advantage of the ML method of estimating parameters is that it is applicable to a wide variety of situations, and certain situations of considerable practical importance will be described in this chapter.

This chapter begins with a general discussion of the basic theoretical principles underlying the method of maximum likelihood. Next, inference-making procedures based on the use of ML estimators will be described. Finally, several important applications of ML methodology will be presented. Certain connections between ML estimation and other well-known algorithms for estimating parameters (e.g., unweighted- and weighted-least-squares procedures) will be made.

21-2 The Principle of Maximum Likelihood

To introduce the underlying principle of maximum likelihood estimation, let us focus on a relatively simple problem of statistical estimation. Suppose that a large population contains a certain unknown proportion θ $(0 \le \theta \le 1)$ of individuals with a particular genetic disorder. In epidemiology, the parameter θ is referred to as the "prevalence of the disorder of interest in the population under study." Further, suppose that a random sample of m individuals is selected from this population, and that Y represents the random variable denoting the number of individuals in the random sample of size m who have this genetic disorder. The possible values of Y are the $m + 1$ integer values 0, 1, 2, . . . , m. If y denotes the observed

483

value of Y for the particular random sample selected (i.e., y is the realization of Y in our data), then the statistical estimation problem concerns how to use m and y to obtain an estimate of θ with good statistical properties. By "good," we mean that the chosen estimator of θ (which is itself a random variable since it is a function of m and the random variable Y) has little or no bias (i.e., its expected value is equal to θ) and has a small variance.

Given the above scenario, it is reasonable to assume that the underlying probability distribution of the discrete random variable Y is binomial; in particular, we have

$$p_Y(y;\,\theta) = \text{pr}(Y = y;\,\theta) = {}_mC_y\theta^y(1-\theta)^{m-y} \qquad y = 0, 1, 2, \ldots, m \tag{21.1}$$

where ${}_mC_y = m!/y!(m-y)!$ denotes the number of combinations of m things taken y at a time. To simplify our discussion for now, let us assume (unrealistically, of course) that only two values of θ are possible; specifically, let us assume that θ is equal either to .2 or to .6. Then, the true underlying probability distribution of Y is, from (21.1), either

$$p_Y(y;\,.2) = {}_mC_y(.2)^y(.8)^{m-y} \qquad y = 0, 1, \ldots, m \tag{21.2}$$

or

$$p_Y(y;\,.6) = {}_mC_y(.6)^y(.4)^{m-y} \qquad y = 0, 1, \ldots, m \tag{21.3}$$

Given this situation, the method of maximum likelihood will choose that point estimate of θ (namely, either .2 or .6 since these are the only two permissible values) which, loosely speaking, agrees more with the observed data (i.e., with the observed value y or the observed sample proportion y/m).

To illustrate the above concept numerically, assume that we have a random sample of $m = 5$ individuals. Then the possible observed values y of Y and their associated probabilities based on (21.2) and (21.3) are given in the following table:

	y					
θ	0	1	2	3	4	5
.2	.328	.409	.205	.051	.007	.000
.6	.010	.077	.230	.346	.259	.078

As an illustration of a typical computation, if $\theta = .2$, then the probability that Y takes the value $y = 4$ is, from (21.2),

$$p_Y(4;\,.2) = {}_5C_4(.2)^4(.8)^{5-4} = 5(.2)^4(.8)^1 \approx .007$$

In contrast, if $\theta = .6$, then the probability that Y takes the value $y = 4$ is, from (21.3),

$$p_Y(4;\,.6) = {}_5C_4(.6)^4(.4)^{5-4} \approx .259$$

Based on these computations, if the observed value of Y is actually $y = 4$, the method of maximum likelihood will choose $\hat{\theta} = .6$ as the estimate of θ, since the probability of observing $y = 4$ is higher when $\theta = .6$ than when $\theta = .2$ (i.e., the observed data agree, or are consistent, more with the value .6 than with the value .2). More generally, then, it follows by inspection of the entries in the preceding table that we will estimate θ by .2 when $y = 0$ or 1 and by .6 when $y = 2, 3, 4, 5$; in compact notation, the ML estimator $\hat{\theta}$ of θ in this simple example is defined as

$$\hat{\theta} = \begin{cases} .2 & y = 0, 1 \\ .6 & y = 2, 3, 4, 5 \end{cases} \tag{21.4}$$

From (21.4), it is clear that the principle of ML estimation dictates selecting for each value of y that value of θ, denoted as $\hat{\theta}$, which satisfies the inequality

$$p_Y(y; \hat{\theta}) > p_Y(y; \theta')$$

where θ' is the alternative value of θ.

In general, in the realistic situation when all possible values of θ satisfying $0 \leq \theta \leq 1$ are possible, it makes sense to do the following. If the observed value of Y based on (21.1) is equal to y, then the ML estimator of θ will be that value of θ, denoted $\hat{\theta}$, for which the expression

$$p_Y(y; \theta) = {}_mC_y\theta^y(1 - \theta)^{m-y} \tag{21.5}$$

attains its **maximum value** as a function of θ. Using calculus, the specific value $\hat{\theta}$ of θ that maximizes the function (21.5) can be found by setting the derivative of (21.5) with respect to θ equal to 0 and then solving the resulting equation for $\hat{\theta}$. In particular,

$$\frac{d}{d\theta}[p_Y(y; \theta)] = {}_mC_y[y\theta^{y-1}(1 - \theta)^{m-y} - (m - y)\theta^y(1 - \theta)^{m-y-1}]$$
$$= {}_mC_y\theta^{y-1}(1 - \theta)^{m-y-1}[y(1 - \theta) - (m - y)\theta]$$
$$= {}_mC_y\theta^{y-1}(1 - \theta)^{m-y-1}(y - m\theta) \tag{21.6}$$

Equating (21.6) to 0 gives the three solutions 0, 1, and y/m. The first two solutions minimize (21.5). However, the solution y/m maximizes (21.5), as can be verified by checking that the second derivative of (21.5) with respect to θ is negative when θ is replaced by the value y/m. Thus, the ML estimator of θ is $\hat{\theta} = Y/m$ when Y has the binomial distribution (21.1); the reader will recognize $\hat{\theta}$ as the *sample proportion* of subjects with the genetic disorder in question. Figure 21-1 below illustrates graphically that the ML estimator $\hat{\theta} = y/m$ maximizes the function $p_Y(y; \theta)$ given by (21.1).

The ML estimator $\hat{\theta} = Y/m$ determined above has the property that

$$p_Y(y; \hat{\theta}) > p_Y(y; \theta^*) \tag{21.7}$$

FIGURE 21-1   Illustration of the principle of maximum likelihood using the function (21.1)

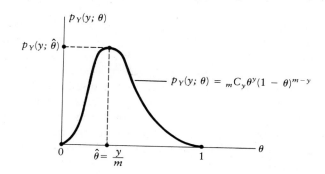

where θ^* is any other value of θ satisfying $0 \leq \theta^* \leq 1$. In other words, since $p_Y(y; \theta)$ gives the likelihood (or probability) that Y takes the value y, the estimator $\hat{\theta} = Y/m$, having been chosen to maximize this likelihood function, is naturally called the ML estimator of θ.

In the particular example that we have been considering, the data set consists simply of the observed value y of Y, and so $p_Y(y; \theta)$ is the likelihood function for (or, equivalently, the probability distribution of) the observed data. Thus, as mentioned earlier, the ML estimator $\hat{\theta}$ of θ is that estimator which agrees most closely with the observed data in the sense of (21.7). With this concept in mind, we are now in a position to make a more general definition of the principle of maximum likelihood.

Let $L(y; \theta)$ denote the *likelihood function* for a data set $y = (y_1, y_2, \ldots, y_n)$ of n observations from some population characterized by the parameter set $\theta = (\theta_1, \theta_2, \ldots, \theta_p)$.

(Note that, for the example considered above, our data set consists of only the single observation $y = y_1 = y$, the number of individuals with a certain genetic disorder in a sample of size m; the parameter set θ consists of the single parameter $\theta_1 = \theta$. Now we are generalizing the situation so that the data set can consist of several observations and we can consider more than one parameter of interest.)

The likelihood function $L(y; \theta)$ can, loosely speaking, be thought of as the probability distribution of the multivariate variable $Y = (Y_1, Y_2, \ldots, Y_n)$ involving the n random variables $Y_1, Y_2, \ldots, Y_n$. By a straightforward extension of the general principles of maximum likelihood discussed earlier, the ML estimator of θ is that set $\hat{\theta} = (\hat{\theta}_1, \hat{\theta}_2, \ldots, \hat{\theta}_p)$ of estimators for which

$$L(y; \hat{\theta}) > L(y; \theta^*) \qquad (21.8)$$

where θ^* denotes any other set of estimators of the elements of θ; the similarity between expressions (21.7) and (21.8) is clear.

In practice, to find the set $\hat{\theta}$ of numerical functions of the observed data for which $L(y; \hat{\theta})$ is a maximum requires solving a system of p equations in p unknowns. In particular, since maximizing $L(y; \theta)$ is equivalent to the computationally easier problem of maximizing the natural logarithm of $L(y; \theta)$, the elements of $\hat{\theta}$ are typically found as the solutions of the p equations obtained by setting the partial derivative of $\ln L(y; \theta)$ with respect to each θ_j ($j = 1, 2, \ldots, p$) equal to 0; in particular, this set of ML equations can be written in the form

$$\frac{\partial}{\partial \theta_j} [\ln L(y; \theta)] = 0 \qquad j = 1, 2, \ldots, p \qquad (21.9)$$

Except for some special cases, the system of ML equations (21.9) does not generally lead to closed-form expressions for the ML estimators; thus these equations must usually be solved via the use of sophisticated computer algorithms. This complexity results because the set of equations (21.9) typically involves nonlinear (as opposed to linear) functions of the elements of θ, thus requiring the use of so-called *iterative* computational procedures. However, such calculations are not a major problem since sophisticated computer algorithms are available that are designed specifically to perform such numerical manipulations.

At several points later in this chapter, the reader will be introduced to some particular forms of $L(y; \theta)$ that have important practical applications. For now, it is sufficient for the reader to appreciate the fact that ML estimation requires the specification of a likelihood function $L(\mathbf{y}; \boldsymbol{\theta})$, which is then used to produce the estimator set $\hat{\boldsymbol{\theta}}$ of ML estimators of the elements of $\boldsymbol{\theta}$ via the system of equations (21.9).

21-3 Statistical Inference via Maximum Likelihood

Once the ML estimates have been obtained, the next step is to use the elements of $\hat{\boldsymbol{\theta}}$ to make statistical inferences about the corresponding elements of $\boldsymbol{\theta}$. The computations involved in making such inferences are based on the use of two quantities that are generally included as a part of the output provided by standard ML computer programs. The first of these two quantities is the *maximized likelihood value.* This statistic is simply the numerical value of the likelihood function $L(\mathbf{y}; \boldsymbol{\theta})$ when the numerical values of the ML estimates (i.e., the elements of $\hat{\boldsymbol{\theta}}$) are substituted for the corresponding elements of $\boldsymbol{\theta}$ in the expression for $L(\mathbf{y}; \boldsymbol{\theta})$; in particular, this ML value is simply $L(\mathbf{y}; \hat{\boldsymbol{\theta}})$, the quantity appearing in expression (21.8). The second quantity of interest is the *estimated covariance matrix* $\hat{\mathbf{V}}(\hat{\boldsymbol{\theta}})$ of the ML estimators; this $p \times p$ symmetric matrix has as its elements the p estimated variances of the ML estimators (appearing on the diagonal) and the $p(p - 1)/2$ estimated covariances between pairs of estimators (appearing off the diagonal).

To illustrate how these two quantities are to be used in making inferences about the elements of $\boldsymbol{\theta}$, let us consider the following simple regression analysis setting. Suppose that we are interested in the following three models:

model 0: $E(Y) = \beta_0$

model 1: $E(Y) = \beta_0 + \beta_1 x$

model 2: $E(Y) = \beta_0 + \beta_1 x + \beta_2 x^2$

Here, $E(Y)$ denotes the expectation (i.e., the true mean) of the random variable Y, assumed to be at most a second-degree polynomial function of a single regressor variable x. Given the n pairs of data points (x_i, y_i), $i = 1, 2, \ldots, n$, the objective of the analysis is to decide which of the above three models best fits the data by using statistical estimation and inference-making methodology based on the principle of maximum likelihood.

To fit any of the above models by the method of maximum likelihood, first a likelihood function must be specified. As an important illustration, suppose that we make the typical regression analysis assumptions regarding the above models. In particular, we will assume that Y_i ($i = 1, 2, \ldots, n$) is normally distributed with mean $\mu_i = E(Y_i)$ possibly depending on x_i and with variance Var $Y_i = \sigma^2$ not varying with i. Also, we will assume that x_i is measured without error (i.e., x_i is nonstochastic) and that the n random variables $Y_1, Y_2, \ldots, Y_n$ are mutually independent. In more compact notation, the above set of assumptions implies, for model 1, say, that

$$Y_i \sim N(\beta_0 + \beta_1 x_i, \sigma^2) \qquad i = 1, 2, \ldots, n \tag{21.10}$$

and that the $\{Y_i\}_{i=1}^{n}$ are mutually independent. (As used elsewhere in this book, the symbol $\sim$ means "is distributed as," and the notation "$N(\beta_0 + \beta_1 x_i, \sigma^2)$" refers to a normal distribution with mean $\beta_0 + \beta_1 x_i$ and variance σ^2.

Assuming that the reader is familiar with the expression for the distribution (density function) of a normally distributed random variable, it follows from (21.10) that the distribution of Y_i is

$$f_{Y_i}(y_i; \beta_0, \beta_1, \sigma^2) = \frac{1}{\sqrt{2\pi\sigma^2}} \exp\left\{-\frac{1}{2\sigma^2}[y_i - (\beta_0 + \beta_1 x_i)]^2\right\} \tag{21.11}$$

where $-\infty < y_i < +\infty$. Note that (21.11) is a function of three parameters, namely, β_0, β_1, and σ^2. Under the assumption that the $\{Y_i\}_{i=1}^{n}$ are mutually independent,[1] it can be shown (from certain principles of statistical theory) that the joint distribution of $Y_1, Y_2, \ldots, Y_n$ (i.e., the likelihood function) is, from (21.11),

$$L(\mathbf{y}; \beta_0, \beta_1, \sigma^2) = \prod_{i=1}^{n} f_{Y_i}(y_i; \beta_0, \beta_1, \sigma^2)$$

$$= \frac{1}{(2\pi\sigma^2)^{n/2}} \exp\left\{-\frac{1}{2\sigma^2}\sum_{i=1}^{n}[y_i - (\beta_0 + \beta_1 x_i)]^2\right\} \tag{21.12}$$

where $-\infty < y_i < +\infty$, $i = 1, 2, \ldots, n$.

By solving simultaneously the three ML equations

$$\frac{\partial}{\partial \beta_0}[\ln L(\mathbf{y}; \beta_0, \beta_1, \sigma^2)] = 0, \quad \frac{\partial}{\partial \beta_1}[\ln L(\mathbf{y}; \beta_0, \beta_1, \sigma^2)] = 0,$$

and

$$\frac{\partial}{\partial(\sigma^2)}[\ln L(\mathbf{y}; \beta_0, \beta_1, \sigma^2)] = 0$$

one can show that the ML estimators of β_0, β_1, and σ^2 under model 1 are, respectively,

$$\hat{\beta}_0 = \bar{y} - \hat{\beta}_1 \bar{x}, \quad \hat{\beta}_1 = \frac{\sum_{i=1}^{n}(x_i - \bar{x})(y_i - \bar{y})}{\sum_{i=1}^{n}(x_i - \bar{x})^2} \tag{21.13}$$

and

$$\hat{\sigma}_1^2 = \frac{1}{n}\sum_{i=1}^{n}[y_i - (\hat{\beta}_0 + \hat{\beta}_1 x_i)]^2 = \frac{SSE_1}{n} \quad \text{biased} \tag{21.14}$$

where SSE_1 is the sum of squares of residuals about the fitted straight line.

The ML estimators $\hat{\beta}_0$ and $\hat{\beta}_1$ given by (21.13) are exactly the same estimators of the intercept and slope of a straight-line model that would be obtained by the method of least squares.

---

[1] The assumption of mutual independence within a set of random variables is the strongest assumption that one can make about the joint behavior of this set. Such an assumption allows one to describe precisely the joint distribution of the variables (i.e., the likelihood function) from a knowledge *solely* of the *separate* behavior (i.e., the so-called marginal distribution) of each variable in the set. In particular, this joint distribution under mutual independence is simply the product of these marginal distributions.

(This equivalence between maximum likelihood and least-squares estimation carries over to the more general situation of multiple *linear* regression analysis. In particular, assume (for $i = 1, 2, \ldots, n$) that Y_i is normal with mean $E(Y_i) = \beta_0 + \sum_{j=1}^{k} \beta_j x_{ij}$; further, assume that Var $Y_i = \sigma^2$ and that the $\{Y_i\}_{i=1}^{n}$ are mutually independent. Then it can be shown that the set $\hat{\boldsymbol{\beta}} = (\hat{\beta}_0, \hat{\beta}_1, \ldots, \hat{\beta}_k)$ of ML estimators of the elements of $\boldsymbol{\beta} = (\beta_0, \beta_1, \ldots, \beta_k)$ is identical to the set of unweighted-least-squares estimators of $\boldsymbol{\beta}$, which are chosen to minimize the quantity $\sum_{i=1}^{n} (y_i - \beta_0 - \sum_{j=1}^{k} \beta_j x_{ij})^2$ with respect to $\beta_0, \beta_1, \ldots, \beta_k$. More generally, under the same assumptions as above except that Var $Y_i = \sigma_i^2$, it can be shown that ML estimation and least-squares estimation using the function $\sum_{i=1}^{n} (y_i - \beta_0 - \sum_{j=1}^{k} \beta_j x_{ij})^2 / \sigma_i^2$ both lead to the same set of weighted-least-squares estimators.)

The ML estimator $\hat{\sigma}_1^2$ of σ^2, given by (21.14), is a biased estimator of σ^2; the unbiased estimator is

$$\left(\frac{n}{n-2} \right) \hat{\sigma}_1^2 = \frac{\mathrm{SSE}_1}{n-2}$$

under model 1. What this means, in general, is that the method of maximum likelihood does not always produce unbiased estimators of the parameters of interest,[2] although the extent of such bias generally decreases as the sample size increases.

Once the ML estimators (21.13) and (21.14) of the parameters in the likelihood function (21.12) have been obtained, the next step is to calculate the value of the maximized likelihood function $L(\mathbf{y}; \hat{\beta}_0, \hat{\beta}_1, \hat{\sigma}_1^2)$ and to determine the 3×3 estimated covariance matrix $\hat{\mathbf{V}}(\hat{\beta}_0, \hat{\beta}_1, \hat{\sigma}_1^2)$ of the three estimators $\hat{\beta}_0, \hat{\beta}_1,$ and $\hat{\sigma}_1^2$. The maximized likelihood $L(\mathbf{y}; \hat{\beta}_0, \hat{\beta}_1, \hat{\sigma}_1^2)$ and the estimated covariance matrix $\hat{\mathbf{V}}(\hat{\beta}_0, \hat{\beta}_1, \hat{\sigma}_1^2)$ are needed to make statistical inferences.

In general, the numerical value of the maximized likelihood function and the numerical values of the entries in the estimated covariance matrix must be obtained via the use of an ML computer program. This is necessitated by the fact that the form of likelihood function typically under consideration will not lead (as we will later illustrate) to explicit algebraic expressions either for the ML estimators or for their variances and covariances. However, for our particular example, the likelihood function (21.12) is sufficiently well-behaved to allow us to obtain the explicit algebraic expressions (21.13) and (21.14) for the ML estimators, from which exact algebraic expressions for $L(\mathbf{y}; \hat{\beta}_0, \hat{\beta}_1, \hat{\sigma}_1^2)$ and $\hat{\mathbf{V}}(\hat{\beta}_0, \hat{\beta}_1, \hat{\sigma}_1^2)$ can be obtained.

A specific algebraic expression for the maximized likelihood $L(\mathbf{y}; \hat{\beta}_0, \hat{\beta}_1, \hat{\sigma}_1^2)$ can be specified by substituting (21.13) and (21.14) into (21.12) and then simplifying; the resulting maximized likelihood function can be written in the form

$$L(\mathbf{y}; \hat{\beta}_0, \hat{\beta}_1, \hat{\sigma}_1^2) = (2\pi \hat{\sigma}_1^2 e)^{-n/2} \tag{21.15}$$

[2] Without giving formal mathematical arguments, it is important to emphasize that the ML method produces estimators whose properties are optimal for *large* samples (under certain conditions of mathematical regularity) when the assumed likelihood function is correct. ML estimators are said to be *asymptotically optimal* in the sense that desirable properties such as unbiasedness, minimum variance, and normality hold exactly in the limit only as the amount of data becomes infinitely large. In practice, this means that it is reasonable to assume *for large data sets* that an ML estimator will be essentially unbiased, have a small variance, and be approximately normally distributed when the appropriate likelihood function is being used.

As a numerical example, let us refer to the data in Table 13-2 in Chapter 13; for convenience, these data are reproduced below:

Dosage level (x)	1	2	3	4	5	6	7	8
Weight gain (y) (dag)	1.0	1.2	1.8	2.5	3.6	4.7	6.6	9.1

For this particular set of $n = 8$ data points (which is admittedly too small to justify the use of asymptotic theory, but is appropriate for illustrative purposes), the use of formulae (21.13) and (21.14) gives $\hat{\beta}_0 = -1.20$, $\hat{\beta}_1 = 1.11$, $SSE_1 = 5.03$, and $\hat{\sigma}_1^2 = SSE_1/n = 5.03/8 = 0.6288$. (Note from Table 13-2 that $MSE = [n/(n-2)]\hat{\sigma}_1^2 = SSE_1/6 = 5.03/6 = 0.84$.) Hence, from (21.15),

SSE
MSE
$\hat{\sigma}_1^2$

$$L(\mathbf{y}; \hat{\beta}_0, \hat{\beta}_1, \hat{\sigma}_1^2) = L(\mathbf{y}; -1.20, 1.11, 0.6288)$$
$$= [2(3.1416)(0.6288)(2.7183)]^{-8/2} = (10.7397)^{-4}$$

This is the number that would be included in the computer output from an ML estimation program using the likelihood function (21.12) and the data set under consideration.

The estimated covariance matrix for the ML estimators $\hat{\beta}_0$, $\hat{\beta}_1$, and $\hat{\sigma}_1^2$ will be of the following general form:

$$\hat{\mathbf{V}}(\hat{\beta}_0, \hat{\beta}_1, \hat{\sigma}_1^2) = \begin{bmatrix} \widehat{Var}\, \hat{\beta}_0 & \widehat{Cov}(\hat{\beta}_0, \hat{\beta}_1) & \widehat{Cov}(\hat{\beta}_0, \hat{\sigma}_1^2) \\ \widehat{Cov}(\hat{\beta}_0, \hat{\beta}_1) & \widehat{Var}\, \hat{\beta}_1 & \widehat{Cov}(\hat{\beta}_1, \hat{\sigma}_1^2) \\ \widehat{Cov}(\hat{\beta}_0, \hat{\sigma}_1^2) & \widehat{Cov}(\hat{\beta}_1, \hat{\sigma}_1^2) & \widehat{Var}\, \hat{\sigma}_1^2 \end{bmatrix}$$

Note that the estimated variances of the ML estimators appear on the diagonal of this matrix, and that the estimated covariances are given by the off-diagonal elements of this symmetric matrix.

Without going into the mathematical development, it can be shown that the elements of $\hat{\mathbf{V}}(\hat{\beta}_0, \hat{\beta}_1, \hat{\sigma}_1^2)$ satisfy the following relationships:

$$\widehat{Var}\, \hat{\beta}_0 = \hat{\sigma}_1^2 \left[\frac{1}{n} + \frac{\bar{x}^2}{\sum_{i=1}^{n} (x_i - \bar{x})^2} \right]$$

$$\widehat{Var}\, \hat{\beta}_1 = \frac{\hat{\sigma}_1^2}{\sum_{i=1}^{n} (x_i - \bar{x})^2}$$

$$\widehat{Var}\, \hat{\sigma}_1^2 = \frac{2\hat{\sigma}_1^4}{n}$$

$$\widehat{Cov}(\hat{\beta}_0, \hat{\beta}_1) = \frac{-\bar{x}\hat{\sigma}_1^2}{\sum_{i=1}^{n} (x_i - \bar{x})^2}$$

$$\widehat{Cov}(\hat{\beta}_0, \hat{\sigma}_1^2) = \widehat{Cov}(\hat{\beta}_1, \hat{\sigma}_1^2) = 0$$

For the data under consideration, $\bar{x} = 4.50$ and $\sum_{i=1}^{n} (x_i - \bar{x})^2 = 42$. Thus, the reader can verify that

$$\hat{V}(\hat{\beta}_0, \hat{\beta}_1, \hat{\sigma}_1^2) = \begin{bmatrix} 0.3818 & -0.0674 & 0 \\ -0.0674 & 0.0150 & 0 \\ 0 & 0 & 0.0988 \end{bmatrix} \qquad (21.16)$$

This is the matrix that would be printed out based on the use of the likelihood function (21.12) for the data set under consideration.

The maximized likelihood value of $(10.7397)^{-4}$ will be used to carry out so-called likelihood ratio tests; these tests, to be discussed shortly, involve comparing the ratios of maximized likelihoods for different models. For now, let us focus on the estimated covariance matrix (21.16) and use it to perform certain exercises in statistical inference making.

21-3-1 Hypothesis Testing and Interval Estimation

Hypothesis Testing

Based on the large-sample properties of ML estimators, it can be shown that (under model 1) the quantity

$$\frac{\hat{\beta}_1 - \beta_1}{\sqrt{\widehat{\mathrm{Var}}\,\hat{\beta}_1}}$$

will behave approximately as a standard normal random variable (i.e., a Z variate) when the sample size is large. Hence, a test of $H_0: \beta_1 = 0$ versus $H_1: \beta_1 \neq 0$ can be based on the statistic $\hat{\beta}_1 / \sqrt{\widehat{\mathrm{Var}}\,\hat{\beta}_1}$, which will (for large n) have an approximate standard normal distribution under $H_0: \beta_1 = 0$.

For the data under analysis, the test statistic $\hat{\beta}_1 / \sqrt{\widehat{\mathrm{Var}}\,\hat{\beta}_1}$ has the numerical value $1.11/\sqrt{0.0150} = 9.063$, which is highly significant. Note that the value 0.0150 is the second diagonal element of the matrix (21.16), and it represents the ML-based estimate of $\mathrm{Var}\,\hat{\beta}_1$ under the likelihood function (21.12). An equivalent test can be based on the chi-square distribution; in particular, since $Z^2 \sim \chi_1^2$ when $Z \sim N(0, 1)$, it follows that $\hat{\beta}_1^2/\widehat{\mathrm{Var}}\,\hat{\beta}_1$ will have an approximate χ_1^2 distribution under $H_0: \beta_1 = 0$ when the sample size is large. In our numerical example, then, this χ_1^2 statistic has the highly significant value[3] of $(9.063)^2 = 82.14$.

More generally, ML-based test statistics are usually assumed to have *large-sample* chi-square distributions (with appropriately specified degrees of freedom) under the null hypotheses of interest. The large-sample requirement is crucial to the validity of such *asymptotic* procedures. In such large-sample situations, the P-values associated with the use of ML-

---

[3] The reader, after inspecting Table 13-3 is probably wondering about the difference between the ML-based $\chi_1^2 (= Z^2)$ statistic just discussed and the $F_{1,6} (= t_6^2)$ statistic in Table 13-3 (with the value 61.95), both of which have been used to test $H_0: \beta_1 = 0$ versus $H_1: \beta_1 \neq 0$ under model 1. For this particular example, the difference essentially lies in the fact that the estimated variance of $\hat{\beta}_1$ for the $t_6^2 = F_{1,6}$ statistic is $n/(n-2)$ times that for the $Z^2 = \chi_1^2$ statistic, a point that has been made earlier. Thus, $[n/(n-2)](61.95) = [8/(8-2)](61.95) = 82.60$, which agrees closely with the χ_1^2-value of 82.14.

based chi-square statistics will be comparable to those based on other methods of parameter estimation and statistical inference. In small-sample situations (e.g., as in the present example with $n = 8$), discrepancies among different analysis procedures can be large.[4]

Interval Estimation

To obtain confidence intervals for parameters using the method of maximum likelihood, we can again appeal to the approximate normality of ML estimators when the sample size is large. In our example, since

$$\frac{\hat{\beta}_1 - \beta_1}{\sqrt{\widehat{\text{Var}}\,\hat{\beta}_1}}$$

is approximately $N(0, 1)$ for large samples under model 1, it follows that an approximate $100(1 - \alpha)\%$ large-sample ML-based confidence interval for β_1 is of the general form

$$\hat{\beta}_1 \pm Z_{1-(\alpha/2)} \sqrt{\widehat{\text{Var}}\,\hat{\beta}_1}$$

where $\text{pr}[Z > Z_{1-(\alpha/2)}] = \alpha/2$ when $Z \sim N(0, 1)$. For instance, a 95% ML-based confidence interval for β_1 using our data set of size $n = 8$ is

$$1.11 \pm 1.96\sqrt{0.0150} = 1.11 \pm 0.24$$

giving the interval $(0.87, 1.35)$.

As another example, let us obtain a $100(1 - \alpha)\%$ large-sample confidence interval for the true mean of Y when x is set at some value x_0 within the range of x-values for the data under consideration (i.e., interpolation, but not extrapolation, is permitted). Denoting this parameter as $E(Y|x = x_0)$, the ML estimator of $E(Y|x = x_0)$ based on fitting model 1 is

$$\hat{E}(Y|x = x_0) = \hat{Y}_0 = \hat{\beta}_0 + \hat{\beta}_1 x_0$$

Since $\hat{Y}_0$ is a linear combination of random variables (namely, $\hat{\beta}_0$ and $\hat{\beta}_1$), it follows that $\hat{Y}_0 = \hat{\beta}_0 + \hat{\beta}_1 x_0$ has estimated variance[5]

$$\widehat{\text{Var}}\,\hat{Y}_0 = \widehat{\text{Var}}\,\hat{\beta}_0 + x_0^2\,\widehat{\text{Var}}\,\hat{\beta}_1 + 2x_0\,\widehat{\text{Cov}}(\hat{\beta}_0, \hat{\beta}_1)$$

Under the assumption that the quantity

$$\frac{\hat{Y}_0 - E(Y|x = x_0)}{\sqrt{\widehat{\text{Var}}\,\hat{Y}_0}}$$

---

[4] So far in our discussion, we have been dealing with so-called *unconditional* ML methods. *Conditional* ML procedures, which will be discussed at the very end of this chapter, offer a viable alternative when one is faced with a small-sample problem.

[5] In general, if $\hat{L} = \sum_{j=1}^{k} a_j X_j$ is a linear function of the k random variables $X_1, X_2, \ldots, X_k$, and if $a_1, a_2, \ldots, a_k$ are known constants, then

$$\text{Var}\,\hat{L} = \sum_{j=1}^{k} a_j^2 \,\text{Var}\, X_j + 2 \sum_{\text{all}} \sum_{j<j'} a_j a_{j'} \,\text{Cov}(X_j, X_{j'})$$

For example, when $k = 2$, we have

$$\text{Var}\,\hat{L} = a_1^2 \,\text{Var}\, X_1 + a_2^2 \,\text{Var}\, X_2 + 2a_1 a_2 \,\text{Cov}(X_1, X_2)$$

is approximately $N(0, 1)$ in large samples, it follows that a $100(1 - \alpha)\%$ confidence interval for $E(Y|x = x_0)$ is

$$\hat{Y}_0 \pm Z_{1-(\alpha/2)} \sqrt{\widehat{\text{Var }} \hat{Y}_0}$$

For our particular numerical example, suppose that $x_0 = \bar{x} = 4.50$. Then, using (21.16), it follows that

$$\widehat{\text{Var }} \hat{Y}_0 \approx 0.3818 + (4.50)^2(0.0150) + 2(4.50)(-0.0674)$$
$$\approx 0.3818 + 0.3038 - 0.6066 = 0.0790$$

Since $\hat{Y}_0 = -1.20 + 1.11(4.50) = -1.20 + 5.00 = 3.80$, it follows that the 95% ML-based large-sample confidence interval for $E(Y|x = 4.50)$ is

$$3.80 \pm 1.96\sqrt{0.0790} = 3.80 \pm 0.55$$

giving the interval $(3.25, 4.35)$.

21-3-2 Hypothesis Testing Using Likelihood Ratio Tests

As the name itself suggests, a *likelihood ratio test* involves a *ratio* comparison of (maximized) likelihood values. To illustrate how such a comparison is used to make statistical inferences, let us again focus on models 0, 1, and 2 introduced at the beginning of Section 21-3. For models 0 and 2, we can obtain the value of the maximized likelihood function in the same way that we did for model 1. Recall, for model 1, that maximization of the likelihood function (21.12) led to the estimators (21.13) and (21.14) and then to the maximized likelihood expression (21.15); for the data under consideration, (21.15) had the specific numerical value of $(10.7397)^{-4}$.

Without going through all the mathematical details,[6] it can be shown that $L(\mathbf{y}; \hat{\beta}_0, \hat{\sigma}_0^2)$ and $L(\mathbf{y}; \hat{\beta}_0, \hat{\beta}_1, \hat{\beta}_2, \hat{\sigma}_2^2)$, the maximized likelihood values for models 0 and 2 under the same general assumptions used for the ML-fitting of model 1, are equal to $(121.8426)^{-4}$ and $(0.4270)^{-4}$, respectively.

---

[6] Under model 0, for example,

$$L(\mathbf{y}; \beta_0, \sigma^2) = \frac{1}{(2\pi\sigma^2)^{n/2}} \exp\left[-\frac{1}{2\sigma^2} \sum_{i=1}^{n} (y_i - \beta_0)^2\right]$$

The ML estimators of β_0 and σ^2 using $L(\mathbf{y}; \beta_0, \sigma^2)$ are $\hat{\beta}_0 = \bar{y}$ and $\hat{\sigma}_0^2 = (1/n)\sum_{i=1}^{n}(y_i - \bar{y})^2$, and so

$$L(\mathbf{y}; \hat{\beta}_0, \hat{\sigma}_0^2) = (2\pi\hat{\sigma}_0^2 e)^{-n/2}$$

Similarly, if $\hat{\beta}_0, \hat{\beta}_1$, and $\hat{\beta}_2$ are the ML estimators of β_0, β_1, and β_2 under model 2, then

$$L(\mathbf{y}; \hat{\beta}_0, \hat{\beta}_1, \hat{\beta}_2, \hat{\sigma}_2^2) = (2\pi\hat{\sigma}_2^2 e)^{-n/2}$$

where

$$\hat{\sigma}_2^2 = \frac{1}{n} \sum_{i=1}^{n} [y_i - (\hat{\beta}_0 + \hat{\beta}_1 x_i + \hat{\beta}_2 x_i^2)]^2$$

Note that

$$L(\mathbf{y}; \hat{\beta}_0, \hat{\sigma}_0^2) < L(\mathbf{y}; \hat{\beta}_0, \hat{\beta}_1, \hat{\sigma}_1^2) < L(\mathbf{y}; \hat{\beta}_0, \hat{\beta}_1, \hat{\beta}_2, \hat{\sigma}_2^2) \tag{21.17}$$

for the particular data set under consideration. This result reflects the principle of multiple regression analysis that the squared multiple correlation coefficient R^2 will always increase somewhat (for any set of data) whenever another regression parameter is included in the model under consideration. In a completely analogous way, since model 0 is a special case of model 1 when $\beta_1 = 0$, with model 1 in turn a special case of model 2 when $\beta_2 = 0$, it is necessarily true that (21.17) will hold for any set of data.

The fact that models 0, 1, and 2 constitute a so-called hierarchical class of models is an important prerequisite for the use of likelihood ratio tests.[7] This is because the magnitude of the ratio of two maximized likelihood values essentially reflects how much the maximized likelihood value for one specific model has changed based on the addition (or deletion) of one or more parameters to (or from) that given model. A decision to reject some null hypothesis based on a likelihood ratio test will depend upon whether some appropriate function of the ratio of maximized likelihoods for the two models being compared (i.e., the test statistic) is large enough to claim that there is a statistically significant disparity between the two maximized likelihood values. This likelihood-ratio-testing philosophy of assessing the significance of a change in maximized likelihood values (via a test statistic that is a function of their ratio) is completely analogous to the philosophy in standard multiple regression analysis of assessing the significance of change in R^2 based on the addition of one or more parameters to a given model.

In earlier chapters, we described how to assess whether an increase in R^2 is statistically significant by using a partial F test (or, equivalently, a t test) when a parameter is added to a given regression model or by using a multiple-partial F test when more than one new parameter is introduced into that model.

A likelihood ratio test is performed analogously. However, instead of having an F distribution, the likelihood ratio statistic to be used will have, in large samples, an approximate chi-square distribution under the null hypothesis. The degrees of freedom for this distribution will equal the number of parameters in the more complex model that must be set equal to 0 to obtain the less complex model as a special case.[8]

The specific test statistic to be used in performing a likelihood ratio test is of the general form $-2 \ln(\hat{L}_1/\hat{L}_2)$, where $\hat{L}_1$ is the maximized likelihood value for the less complex model and $\hat{L}_2$ is the maximized likelihood value for the more complex model. Thus, this statistic is a function of the ratio $\hat{L}_1/\hat{L}_2$ of maximized likelihoods. Since $\hat{L}_2$ corresponds to the more

---

[7] For our purposes, a hierarchical class of models will be defined as a group of models in which each model in the class, except for the most complex (i.e., that one containing the largest number of regression coefficients), can be obtained as a special case of another more complex model in the class by setting one or more regression coefficients equal to 0 in the more complex model. As an example, the three models $E_1(Y) = \beta_0 + \beta_1 x_1$, $E_2(Y) = \beta_0 + \beta_1 x_1 + \beta_2 x_2$, and $E_3(Y) = \beta_0 + \beta_1 x_1 + \beta_2 x_2 + \beta_3 x_3 + \beta_{12} x_1 x_2$ constitute a hierarchical class since $E_2(Y)$ becomes $E_1(Y)$ when $\beta_2 = 0$, and $E_3(Y)$ reduces to $E_2(Y)$ when $\beta_3 = \beta_{12} = 0$.

[8] The reader should keep in mind that a given likelihood ratio test will always entail a comparison between two models that are members of a hierarchical class, so that the terms *more complex* and *less complex* are well defined.

complex model, we have $\hat{L}_2 > \hat{L}_1 > 0$, so that $0 < \hat{L}_1/\hat{L}_2 < 1$, $-\infty < \ln(\hat{L}_1/\hat{L}_2) < 0$, and hence that $0 < -2\ln(\hat{L}_1/\hat{L}_2) < +\infty$. Clearly, the larger $\hat{L}_2$ is relative to $\hat{L}_1$ (so that the more complex model is in much better agreement with the data than is the less complex model), the larger will be the value of the test statistic $-2\ln(\hat{L}_1/\hat{L}_2)$, and so the likelier it is that this value will be large enough to fall in the rejection region (i.e., the specified area in the *upper* tail of the appropriate chi-square distribution).

We will now provide some numerical examples of likelihood ratio tests. Again, we will consider models 0, 1, and 2 and the data set used earlier. To start, let us compare models 0 and 1 via a likelihood ratio test. Since model 0 is a special case of model 1 when $\beta_1 = 0$, it should be apparent that this likelihood ratio test is addressing $H_0: \beta_1 = 0$ versus $H_A: \beta_1 \neq 0$. The appropriate chi-square distribution for the likelihood ratio statistic under this null hypothesis has 1 degree of freedom (since one parameter is being restricted to be equal to 0 under H_0). Using the previous notation (see expression (21.15) and footnote 6), we are claiming, for large n, that

$$-2\ln\left[\frac{L(\mathbf{y}; \hat{\beta}_0, \hat{\sigma}_0^2)}{L(\mathbf{y}; \hat{\beta}_0, \hat{\beta}_1, \hat{\sigma}_1^2)}\right] = n\ln\left(\frac{\hat{\sigma}_0^2}{\hat{\sigma}_1^2}\right), \tag{21.18}$$

has an approximate χ_1^2 distribution under $H_0: \beta_1 = 0$.

Based on our previous numerical work, we can compute the numerical value of (21.18) for our particular data set as

$$-2\ln\left[\frac{(121.8426)^{-4}}{(10.7397)^{-4}}\right] = 8\ln\left(\frac{121.8426}{10.7397}\right) = 8\ln 11.3451 = 19.43$$

which corresponds to a P-value of less than .0005 (based on upper-tail χ_1^2-values) and thus provides strong evidence in favor of $H_A: \beta_1 \neq 0$.[9]

As another illustration, a test of $H_0: \beta_2 = 0$ versus $H_A: \beta_2 \neq 0$ involves a comparison of the maximized likelihoods for models 1 and 2. In particular, we will assume that

$$-2\ln\left[\frac{L(\mathbf{y}; \hat{\beta}_0, \hat{\beta}_1, \hat{\sigma}_1^2)}{L(\mathbf{y}; \hat{\beta}_0, \hat{\beta}_1, \hat{\beta}_2, \hat{\sigma}_2^2)}\right] = n\ln\left(\frac{\hat{\sigma}_1^2}{\hat{\sigma}_2^2}\right)$$

has an approximate χ_1^2 distribution for large samples under $H_0: \beta_2 = 0$. For our data, the numerical value of the above test statistic is

$$-2\ln\left[\frac{(10.7397)^{-4}}{(0.4270)^{-4}}\right] = 25.80$$

which is strongly in favor of $H_A: \beta_2 \neq 0$.

As a final numerical illustration, a likelihood ratio test involving the maximized likelihood values for models 0 and 2 provides a test of $H_0: \beta_1 = \beta_2 = 0$ versus $H_A:$ "at least one of

---

[9] The discrepancy between this value of 19.43 and the value of 82.14 obtained earlier via the statistic $\hat{\beta}_1^2/\widehat{\text{Var}}\,\hat{\beta}_1$ is alarming. However, such a discrepancy is entirely plausible because of the small sample size ($n = 8$). Because the two test statistics are only *asymptotically* equivalent, their numerical values will be reasonably close only when n is large. Thus, although we have chosen this data set for pedagogical purposes, it is actually inappropriate for the application of large-sample statistical procedures!

the parameters β_1 and β_2 is different from 0." The appropriate likelihood ratio test statistic is

$$-2 \ln \left[\frac{L(\mathbf{y}; \hat{\beta}_0, \hat{\sigma}_0^2)}{L(\mathbf{y}; \hat{\beta}_0, \hat{\beta}_1, \hat{\beta}_2, \hat{\sigma}_2^2)} \right] = n \ln \left(\frac{\hat{\sigma}_0^2}{\hat{\sigma}_2^2} \right)$$

which, for large samples and under H_0: $\beta_1 = \beta_2 = 0$, will have an approximate chi-square distribution with *2 degrees of freedom*. (Note that the null hypothesis restricts *two* independent model parameters to be 0, thus leading to the 2 degrees of freedom for the chi-square statistic.) The computed likelihood ratio test statistic is

$$-2 \ln \left[\frac{(121.8426)^{-4}}{(0.4270)^{-4}} \right] = 45.23$$

which is highly significant. Note that the test just performed is completely analogous to an overall F test in standard regression analysis.

Note that even though a sample size of $n = 8$ is too small to justify the use of statistical inference-making procedures whose desirable statistical properties are asymptotic ones (i.e., they hold only for very large samples), the decisions made about the importance of the linear (β_1) and quadratic (β_2) effects in the data agree with the conclusions drawn based on the standard regression analysis given in Section 13-6.

One way to assess the goodness of fit of this second-degree model in x is to employ a likelihood ratio statistic to examine whether the addition of a cubic term in x (i.e., the term $\beta_3 x^3$) to the second-degree model significantly improves the prediction of Y (i.e., a test of H_0: $\beta_3 = 0$ versus H_A: $\beta_3 \neq 0$). This particular statistic has the form $n \ln(\hat{\sigma}_2^2 / \hat{\sigma}_3^2)$, where $\hat{\sigma}_3^2 = (1/n) \sum_{i=1}^{n} [y_i - (\hat{\beta}_0 + \hat{\beta}_1 x_i + \hat{\beta}_2 x_i^2 + \hat{\beta}_3 x_i^3)]^2$. For our data, it can be seen from Table 13-5 in Chapter 13 that $\hat{\sigma}_3^2 = 0.056/8 = 0.007$; hence, $n \ln(\hat{\sigma}_2^2 / \hat{\sigma}_3^2) = 8 \ln(0.025/0.007) = 10.18$, which corresponds to a P-value of between .0005 and .005 based on the chi-square distribution with 1 degree of freedom. Although this result argues for the addition of the term $\beta_3 x^3$ to the quadratic model, the following information suggests otherwise: The change in R^2 in going from a quadratic to a cubic model is negligible ($\Delta R^2 = .999 - .997 = .002$), a plot of the data clearly suggests no more than a second-degree function in x, and the above likelihood ratio test is based only on a data set with $n = 8$ observations and so is not completely reliable. Given all these considerations, the correct conclusion is that a second-degree model in x is sufficient to describe the $x-Y$ relationship with high precision.

In summary, this section has illustrated how statistical inference based on ML estimation of parameters makes use of comparisons of maximized likelihood values and of the elements of the estimated covariance matrix of parameter estimators. We have provided the mechanics for carrying out likelihood ratio tests and for constructing confidence intervals, after appropriately specified likelihood functions have been analyzed via readily available programs for estimating maximum likelihoods.

The final two sections of this chapter will deal with the application of the principles of maximum likelihood to some important problems in general data analysis. We will focus on two ML-based procedures that are both popular and widely applicable, namely, Poisson regression analysis and logistic regression analysis. The general principles and inference-making procedures described previously carry over directly to the more complex likelihood functions characteristic of these analysis procedures.

21-4 Poisson Regression Analysis

The reader will recall from our earlier discussions that, in the method of maximum likelihood, a likelihood function $L(\mathbf{y};\ \boldsymbol{\theta})$ must be specified that depends both on the observed set of responses $\mathbf{y} = (y_1, y_2, \ldots, y_n)$ and on a set of unknown parameters $\boldsymbol{\theta} = (\theta_1, \theta_2, \ldots, \theta_p)$. In this section, we shall discuss specific likelihood functions and associated inference-making procedures that are appropriate for the application of Poisson regression analysis methods.

21-4-1 The Poisson Distribution

The methodology of Poisson regression analysis assumes that the underlying distribution of the response variable Y under consideration is Poisson. The Poisson probability distribution with parameter μ is given by the formula

$$p_Y(y; \mu) = \text{pr}(Y = y; \mu) = \frac{\mu^y e^{-\mu}}{y!} \qquad y = 0, 1, \ldots, \infty \qquad (21.19)$$

Note that a Poisson random variable can theoretically take any nonnegative integer value. From (21.19), for example, the probability that Y takes the value $y = 10$ is

$$\text{pr}(Y = 10; \mu) = \frac{\mu^{10} e^{-\mu}}{10!} = \frac{\mu^{10} e^{-\mu}}{3{,}628{,}800}$$

Note that this probability will change as a function of the value of μ.

The Poisson distribution is often used to model the occurrence of rare events, such as the number of new cases of lung cancer developing in some population over a certain period of time, the number of automobile accidents occurring at a certain location per year, and so on. The Poisson distribution possesses an interesting statistical attribute; in particular, it can be shown theoretically that $E(Y) = \text{Var}(Y) = \mu$ when Y has the Poisson distribution (21.19).

21-4-2 An Example of Poisson Regression

To illustrate the utility of Poisson regression analysis, we shall consider a data analysis situation where Poisson regression has been used quite successfully. Table 21-1 gives non-melanoma skin cancer data for women stratified by age in two metropolitan areas: Dallas–Ft. Worth and Minneapolis–St. Paul (Scotto, Kopf, and Urbach, 1974). In this example, the dependent variable Y is a count, the number of cases of skin cancer. Since there are eight age strata and two metropolitan areas, we let Y_{ij} denote the count for the ith age stratum in the jth area, where i ranges from 1 to 8 for the eight age groups and $j = 0$ (Minneapolis–St. Paul) or $j = 1$ (Dallas–Ft. Worth). We also let ℓ_{ij} denote the population size for the ith age stratum in the jth area. For these data, one analysis goal is to determine whether the risk for skin cancer adjusted for age is higher in one metropolitan area than in the other. The term *risk* in this context essentially means the probability associated with an event of interest, e.g., the probability of developing skin cancer. We will let λ_{ij} denote the true (i.e., population) risk in the (i, j)th group. The ratio

$$\text{RR}_i = \frac{\lambda_{i1}}{\lambda_{i0}}$$

TABLE 21-1 Comparison of incidence of nonmelanoma skin cancer among women in Minneapolis–St. Paul and Dallas–Ft. Worth

| Age Group (yr) | Minneapolis–St. Paul | | | Dallas–Ft. Worth | | | Estimated Risk Ratio* |
	No. of Cases	Population Size	PoP F.R.	No. of Cases	Population Size	PoP F.R.	
15–24	1	172,675	.265	4	181,343	.246	3.81
25–34	16	123,065	.189	38	146,207	.199	2.00
35–44	30	96,216	.148	119	121,374	.165	3.14
45–54	71	92,051	.141	221	111,353	.151	2.57
55–64	102	72,159	.111	259	83,004	.113	2.21
65–74	130	54,722	.084	310	55,932	.076	2.33
75–84	133	32,185	.049	226	29,007	.039	1.89
85+	40	8,328	.013	65	7,538	.010	1.80
Total	523	651401		1242	735758		

Source: Adapted from Scotto, Kopf, and Urbach (1974).

* With Minneapolis–St. Paul as the reference group.

is commonly referred to as the *relative risk* or *risk ratio*, which here is the population risk for Dallas–Ft. Worth in the ith age group *divided by* the risk for Minneapolis–St. Paul in the ith age group. If $RR_i = 1$, then the population risks are the same in the ith age group; however, if $RR_i > 1$, then the risk for Dallas–Ft. Worth is higher than the risk for Minneapolis–St. Paul in this age group.

On inspection of the last column of Table 21-1, it can be seen that the estimated risk ratios in all age groups are greater than 1, which clearly suggests that the Dallas–Ft. Worth area seems to have a higher overall incidence of skin cancer than Minneapolis–St. Paul. Our objective here is to use Poisson regression analysis to determine whether such a data pattern is statistically significant and to obtain an estimate of the overall risk ratio that adjusts for the effect of age.

Where does the Poisson distribution enter into this problem? To answer this question, we first point out that the count Y_{ij} is, in theory, a binomial random variable with mean $\mu_{ij} = \ell_{ij}\lambda_{ij}$. It also follows from statistical theory that the binomial distribution can be approximated by a Poisson distribution with the same mean provided the population size is large and the binomial probability parameter is small, so that the expected binomial count (i.e., the mean μ) is small relative to the population size. In other words, the Poisson distribution provides a good approximation to the binomial distribution for rare events. The data in Table 21-1 reflect the above requirement reasonably well since all stratum-specific counts are quite small relative to the corresponding population sizes.

To develop a Poisson regression model for the above situation, we need to define a model for the expected number of skin cancer cases, $E(Y_{ij})$, in terms of the variables of interest. Here, there are two underlying variables of interest: "age" and "area." Since "age" has been categorized into eight groups, we will use eight dummy variables to index them;[10]

[10] The model can alternatively be defined by using seven dummy variables for "age" and one dummy variable for "area"; such a coding scheme requires a constant term (e.g., α) in the model. When using eight dummy variables for "age," the use of a constant term is redundant. We have chosen the eight-dummy-variable alternative in order to be consistent with the published analysis of this data set.

"area," which contains two categories, will require only one dummy variable. Thus, one possible model for the expected number of skin cancer cases in the (i, j)th group can be written in the form

$$E(Y_{ij}) = \mu_{ij} = \ell_{ij}\lambda_{ij} \quad i = 1, 2, \ldots, 8; \quad j = 0, 1$$

where

$$\ln \lambda_{ij} = \sum_{i=1}^{8} \alpha_i U_i + \beta E$$

with

$$U_i = \begin{cases} 1 & \text{if age group } i \\ 0 & \text{otherwise} \end{cases} \quad i = 1, 2, \ldots, 8$$

$$E = \begin{cases} 1 & \text{if } j = 1 \quad \text{(Dallas–Ft. Worth)} \\ 0 & \text{if } j = 0 \quad \text{(Minneapolis–St. Paul)} \end{cases}$$

Using this model, we can write the risks, λ_{ij}, in terms of the parameters α_i and β to obtain

$$\ln \lambda_{i0} = \alpha_i$$

and

$$\ln \lambda_{i1} = \alpha_i + \beta \quad i = 1, 2, \ldots, 8$$

Hence,

$$\ln \lambda_{i1} - \ln \lambda_{i0} = (\alpha_i + \beta) - \alpha_i = \beta$$

or, in other words,

$$RR_i = \frac{\lambda_{i1}}{\lambda_{i0}} = \frac{\exp(\alpha_i + \beta)}{\exp(\alpha_i)} = \exp(\beta) = e^\beta$$

Thus, using the above model, the risk ratio for any age group can be estimated by fitting the model, estimating the coefficient of the E-variable, and then exponentiating this estimate. Since the estimated risk ratio, $e^{\hat\beta}$, is independent of i, we can interpret $e^{\hat\beta}$ as an estimate of an overall risk ratio adjusted for age.

The example we have just described provides an illustration of the type of model used when performing a Poisson regression analysis. In general, instead of having two variables (like age and area) to consider, we may have several (say k) predictor variables $x_1, x_2, \ldots, x_k$ to examine. Nevertheless, the general method of fitting a Poisson regression model is still to use the Poisson model formulation to derive a likelihood function that can then be maximized so that parameter estimates, estimated standard errors, maximized likelihood statistics, and other information can be produced. With the availability of packaged programs to carry out such analyses, the user need only specify the model to be fit; the program will then determine the likelihood function, maximize it, and then compute relevant statistics. We shall return later to the above example to illustrate numerically methods of Poisson regression analysis.

Before describing the general Poisson regression model, we point out that, whereas the above example (strictly speaking) involved a model for estimating the *risk* of developing a disease, a more general and popular application of Poisson regression concerns the modeling of *failure rates* for different subgroups of interest. The estimated failure rate, or more simply the estimated rate, is generally defined as

$$\hat{\lambda} = \frac{y}{\ell}$$

where y is the observed count of health failures (e.g., the number of cases of skin cancer, the number of new cases of heart disease, etc., for a subgroup of interest), and ℓ denotes the accumulated length of (disease-free) follow-up time for all persons in the subgroup. Thus, $\hat{\lambda}$ measures the number of failures relative to the total amount of follow-up time for all persons in a given subgroup. If, for example, the data in Table 21-1 were based on one year of follow-up of the Minneapolis–St. Paul and Dallas–Ft. Worth populations, then the numbers in the table giving population size might be considered as *person-years* of follow-up time for the age–area subgroups. The ratio of two rates, e.g., $\lambda_{i1}/\lambda_{i0}$, is commonly referred to as a *rate ratio*. Other terms used are *incidence density ratio* (abbreviated *IDR*) and *hazard ratio*.

21-4-3 Poisson Regression: General Considerations

We are now ready to describe the general Poisson regression analysis framework. The dependent variable Y is, as already mentioned, typically a count of health failures obtained for each of a number of subgroups that are described by a set of predictor variables $x_1, x_2, \ldots, x_k$. For subgroup i, $i = 1, 2, \ldots, n$, let y_i denote the observed number of failures and let ℓ_i denote the total length of follow-up time for all persons in that subgroup. Let $\mathbf{x}_i = (x_{i1}, x_{i2}, \ldots, x_{ik})$ denote the set of values of $x_1, x_2, \ldots, x_k$ specific to subgroup i, let $\boldsymbol{\beta} = (\beta_0, \beta_1, \ldots, \beta_k)$ be a set of unknown parameters, and let $\lambda(\mathbf{x}_i, \boldsymbol{\beta})$ denote some specific function of $\mathbf{x}_i$ and $\boldsymbol{\beta}$ (e.g., $\exp(\beta_0 + \sum_{j=1}^{k} \beta_j x_{ij})$) that represents the failure rate for subgroup i (i.e., $\lambda(\mathbf{x}_i, \boldsymbol{\beta})$ measures the rate at which failures occur per unit of follow-up time). Then the expected number of failures in the ith subgroup is

$$E(Y_i) = \mu_i = \ell_i \lambda(\mathbf{x}_i, \boldsymbol{\beta}) \qquad i = 1, 2, \ldots, n \tag{21.20}$$

where Y_i is the Poisson random variable whose realization in the data is y_i. Note that it is required that $\lambda(\mathbf{x}_i, \boldsymbol{\beta}) > 0$.[11]

Under the assumption that Y_i is Poisson with mean μ_i,[12] so that

$$p_{Y_i}(y_i; \mu_i) = \frac{\mu_i^{y_i} e^{-\mu_i}}{y_i!} \qquad i = 1, 2, \ldots, n \tag{21.21}$$

it follows from (21.20) and (21.21) that

$$p_{Y_i}(y_i; \boldsymbol{\beta}) = \text{pr}(Y_i = y_i; \boldsymbol{\beta}) = \frac{[\ell_i \lambda(\mathbf{x}_i, \boldsymbol{\beta})]^{y_i} e^{-\ell_i \lambda(\mathbf{x}_i, \boldsymbol{\beta})}}{y_i!} \tag{21.22}$$

where $y_i = 0, 1, \ldots, \infty$ and $i = 1, 2, \ldots, n$.

---

[11] Formula (21.20) is analogous (but not equivalent) to the formula for the mean $\mu = np$ of a binomial random variable; here, ℓ_i is similar to n, and $\lambda(\mathbf{x}_i, \boldsymbol{\beta})$ is similar to p.

[12] That the Poisson distribution is useful for modeling certain types of health count data can be loosely argued on the basis of the well-known Poisson approximation to the binomial distribution (e.g., see Remington and Schork, 1985, chap. 5). If $Y \sim \text{Bin}(n, p)$, and if n is large and p is very small, then $Y \sim \text{Poi}(\mu = np)$. For many health outcomes (e.g., the development of a rare disease), the length ℓ_i of follow-up time (analogous to n) is large and the rate $\lambda(\mathbf{x}_i, \boldsymbol{\beta})$ of occurrence of the health outcome in question (analogous to p) is small, thus suggesting the Poisson model.

By comparing expressions (21.20) and (21.22) above and expressions (21.10) and (21.11) given earlier, we can see that the only real conceptual difference between Poisson regression and standard multiple regression is that the former involves the Poisson distribution and the latter the normal distribution. In each instance, the analysis goal is the same, namely, to fit to the data a regression equation that will accurately model $E(Y)$ as a function of a set of predictor variables $x_1, x_2, \ldots, x_k$.[13] This is exactly what expression (21.20) is saying!

In the most general sense, then, regression analysis pertains to modeling the mean of the dependent variable under consideration as a function of certain predictor variables. The form of likelihood function that is used to estimate the regression coefficient set $\boldsymbol{\beta}$ is determined by the assumptions made about the distribution of that dependent variable.

As we did earlier to obtain the likelihood function (21.12), let us assume that $Y_1, Y_2, \ldots, Y_n$ constitute a mutually independent set of Poisson random variables, with Y_i having the probability distribution (21.22). Then, the *likelihood function for Poisson regression analysis* is of the general form

$$
\begin{aligned}
L(\mathbf{y}; \boldsymbol{\beta}) &= \prod_{i=1}^{n} p_{Y_i}(y_i; \boldsymbol{\beta}) \\
&= \prod_{i=1}^{n} \left\{ \frac{[\ell_i \lambda(\mathbf{x}_i, \boldsymbol{\beta})]^{y_i} e^{-\ell_i \lambda(\mathbf{x}_i, \boldsymbol{\beta})}}{y_i!} \right\} \\
&= \frac{\left\{ \prod_{i=1}^{n} [\ell_i \lambda(\mathbf{x}_i, \boldsymbol{\beta})]^{y_i} \right\} \exp\left[-\sum_{i=1}^{n} \ell_i \lambda(\mathbf{x}_i, \boldsymbol{\beta}) \right]}{\prod_{i=1}^{n} y_i!}
\end{aligned}
\tag{21.23}
$$

where $E(Y_i) = \mu_i = \ell_i \lambda(\mathbf{x}_i, \boldsymbol{\beta})$, $i = 1, 2, \ldots, n$.

To utilize the likelihood function (21.23) in practice, the investigator must specify a particular form for the rate function $\lambda(\mathbf{x}_i, \boldsymbol{\beta})$. Such a specification should be based on the process under study and on previous knowledge and experience concerning the relationships among the variables under consideration. Examples of possible choices for $\lambda(\mathbf{x}_i, \boldsymbol{\beta})$ are $e^{\lambda_i^*}$ when $\lambda_i^* = \beta_0 + \sum_{j=1}^{k} \beta_j x_{ij}$, λ_i^* when $\lambda_i^* > 0$, and $\ln \lambda_i^*$ when $\lambda_i^* > 1$.

Recall using (21.9) that the maximum likelihood estimators $\hat{\beta}_0, \hat{\beta}_1, \ldots, \hat{\beta}_k$ of $\beta_0, \beta_1, \ldots, \beta_k$ are obtained from (21.23) as the solutions of the $k + 1$ equations

$$
\frac{\partial}{\partial \beta_j}[\ln L(\mathbf{y}; \boldsymbol{\beta})] = 0 \qquad j = 0, 1, \ldots, k
\tag{21.24}
$$

The solution to the set of ML equations given by (21.24) must generally be obtained by some computer-based iteration procedure. Frome (1983) discusses the use of algorithms for solving the system of equations (21.24). In particular, he argues for the use of a computational

---

[13] If, in the likelihood (21.12) we replace $E(Y_i) = \beta_0 + \beta_1 x_i$ by, say, $E(Y_i) = \beta_0 + e^{\beta_1 x_i}$, then we change from a *linear* (in the coefficients) model to a *nonlinear* model, and hence from a *linear* regression analysis to a *nonlinear* one. The major effect of this change is that we will have to solve a set of nonlinear (as opposed to linear) likelihood equations in the β's. This solution will generally require some sort of computer-assisted iteration procedure.

algorithm referred to as *iteratively reweighted least squares* (IRLS).[14] Certain statistical packages such as GLIM (Baker and Nelder, 1978) and CATMAX (Stokes and Koch, 1983) can be utilized to find the ML estimator $\hat{\boldsymbol{\beta}}$ of $\boldsymbol{\beta}$ based on the likelihood (21.23). In addition, the estimated covariance matrix $\hat{\mathbf{V}}(\hat{\boldsymbol{\beta}})$ of $\hat{\boldsymbol{\beta}}$, measures of goodness of fit of the model under consideration, and certain regression diagnostic statistics (i.e., indices useful for detecting influential observations and multicollinearity) can be obtained as part of the computer output.

As an application of the above procedures, we return to the data in Table 21-1 describing nonmelanoma skin cancer data for women stratified by age in Minneapolis–St. Paul and Dallas–Ft. Worth (adapted from Scotto, Kopf, and Urbach, 1974, and reanalyzed by Frome and Checkoway, 1985). We previously considered the following Poisson regression model for the expected number of skin cancer cases in subgroup (i, j), $i = 1, 2, \ldots, 8$ and $j = 0, 1$:

$$E(Y_{ij}) = \mu_{ij} = \ell_{ij}\lambda_{ij}$$

where

$$\ln \lambda_{ij} = \sum_{i=1}^{8} \alpha_i U_i + \beta E$$

Here, the U_i's were 0–1 dummy variables indexing the age strata, and E was a 0–1 variable delineating metropolitan area (1 = Dallas–Ft. Worth, 0 = Minneapolis–St. Paul). For this model, the risk (or, more generally, rate) ratio

$$RR_i = \frac{\lambda_{i1}}{\lambda_{i0}}$$

reduced to the expression

$$RR_i = e^{\beta}$$

where e^{β} is independent of i and represents an overall rate ratio estimate adjusted for age.

The likelihood function L for the above model, based on the assumption that the count Y_{ij} follows the Poisson distribution with mean $\mu_{ij} = \ell_{ij}\lambda_{ij}$, is given by the expression:

$$L = \prod_{i=1}^{8} \left\{ \left[\frac{(\ell_{i0}\lambda_{i0})^{y_{i0}} e^{-\ell_{i0}\lambda_{i0}}}{y_{i0}!} \right] \left[\frac{(\ell_{i1}\lambda_{i1})^{y_{i1}} e^{-\ell_{i1}\lambda_{i1}}}{y_{i1}!} \right] \right\}$$

where $\lambda_{i0} = \exp \alpha_i$ and $\lambda_{i1} = \exp(\alpha_i + \beta)$. The use of a Poisson regression computer package would then maximize the above likelihood function to produce the nine parameter estimates

$$\{\hat{\alpha}_1, \hat{\alpha}_2, \ldots, \hat{\alpha}_8, \hat{\beta}\}$$

along with a 9×9 estimated covariance matrix. For these data, the following results are obtained:

$$\hat{\beta} = 0.804, \quad \widehat{s.e.}(\hat{\beta}) = 0.0522$$

---

[14] The fact that $E(Y_i) = \text{Var}(Y_i) = \mu_i = \ell_i\lambda(\mathbf{x}_i, \boldsymbol{\beta})$ means that the variance of the response variable is *not* constant (i.e., it varies as a function of ℓ_i and $\mathbf{x}_i$, thus requiring a weighted-least-squares regression analysis); also, because this variance is a mathematical function of $\boldsymbol{\beta}$, the weights in such a weighted regression analysis will necessarily change as a function of the change in the estimate $\hat{\boldsymbol{\beta}}$ at each step of the iteration process (i.e., a reweighting is required at each step). This is the reason for the terminology "iteratively reweighted least squares," or IRLS for short.

Thus, the point estimate of the adjusted rate ratio is given by

$$e^{\hat{\beta}} = e^{0.804} = 2.2345$$

An approximate large-sample 95% confidence interval for e^{β} is calculated as follows:

$$\exp[\hat{\beta} \pm 1.96\,\widehat{s.e.}(\hat{\beta})] = \exp[0.804 \pm 1.96(0.0522)]$$
$$= \exp(0.804 \pm 0.1023)$$

which gives the 95% confidence limits

$$(e^{0.7017}, e^{0.9063}) = (2.0172, 2.4751)$$

Also, a large-sample test of $H_0: \beta = 0$ versus $H_A: \beta \neq 0$ can be based on the statistic

$$Z = \frac{\hat{\beta} - 0}{\widehat{s.e.}(\hat{\beta})}$$

which is approximately $N(0, 1)$ under $H_0: \beta = 0$.
For our example,

$$Z = \frac{0.804 - 0}{0.0522} \approx 15.40 \qquad (P \approx 0)$$

Thus, the above Poisson regression analysis indicates that there is a statistically significant effect due to area and that the overall (adjusted for age) rate for nonmelanoma skin cancer in women in Dallas–Ft. Worth is approximately 2.2 times the corresponding adjusted rate in women in Minneapolis–St. Paul; a 95% confidence interval for the (adjusted) rate ratio is (2.0172, 2.4751). We will return to this example to illustrate how to evaluate confounding and goodness of fit.

21-4-4 Measures of Goodness of Fit

Measures of goodness of fit of Poisson regression models are obtained from comparisons of maximized likelihood values. In particular, suppose that Y_i has the Poisson distribution (21.21) and that $Y_1, Y_2, \ldots, Y_n$ are mutually independent; then, expressed as a general function of $\mu_1, \mu_2, \ldots, \mu_n$ (i.e., the predictors $x_1, x_2, \ldots, x_k$ are ignored completely), the likelihood function takes the form

$$L(\mathbf{y}; \boldsymbol{\mu}) = \prod_{i=1}^{n} \frac{\mu_i^{y_i} e^{-\mu_i}}{y_i!}$$

$$= \frac{\left(\prod_{i=1}^{n} \mu_i^{y_i}\right) \exp\left(-\sum_{i=1}^{n} \mu_i\right)}{\prod_{i=1}^{n} y_i!} \qquad (21.25)$$

where $\boldsymbol{\mu} = (\mu_1, \mu_2, \ldots, \mu_n)$. The system of ML equations

$$\frac{\partial}{\partial \mu_i}[\ln L(\mathbf{y}; \boldsymbol{\mu})] = 0 \qquad i = 1, 2, \ldots, n$$

leads to the solution $\hat{\mu}_i = y_i$, $i = 1, 2, \ldots, n$. Thus, the maximized likelihood value for the likelihood function (21.25) is

$$L(\mathbf{y}; \hat{\boldsymbol{\mu}}) = \frac{\left(\prod\limits_{i=1}^{n} y_i^{y_i}\right) \exp\left(-\sum\limits_{i=1}^{n} y_i\right)}{\prod\limits_{i=1}^{n} y_i!} \tag{21.26}$$

where $\hat{\boldsymbol{\mu}} = (\hat{\mu}_1, \hat{\mu}_2, \ldots, \hat{\mu}_n) = (y_1, y_2, \ldots, y_n)$.

The maximized likelihood $L(\mathbf{y}; \hat{\boldsymbol{\mu}})$ based on (21.25) will be larger in value (for any set of data) than that achieved by maximizing a likelihood like (21.23) when $(k + 1) < n$. This is because (21.25) has imposed no restrictions on the structure of μ_i, whereas (21.23) has imposed the restriction $\mu_i = \ell_i \lambda(\mathbf{x}_i, \boldsymbol{\beta})$. In other words, (21.23) can be thought of as the likelihood function under H_0: $\mu_i = \ell_i \lambda(\mathbf{x}_i, \boldsymbol{\beta})$, $i = 1, 2, \ldots, n$, whereas (21.25) is the likelihood under H_A: "μ_i unrestricted in structure, $i = 1, 2, \ldots, n$."

Thus, if $L(\mathbf{y}; \hat{\boldsymbol{\beta}})$ is the maximized likelihood value under (21.23), where $\hat{\boldsymbol{\beta}}$ is the ML estimator of $\boldsymbol{\beta}$, then

$$-2 \ln \left[\frac{L(\mathbf{y}; \hat{\boldsymbol{\beta}})}{L(\mathbf{y}; \hat{\boldsymbol{\mu}})}\right] \tag{21.27}$$

is a likelihood-ratio-type statistic reflecting the goodness of fit of the model $\mu_i = \ell_i \lambda(\mathbf{x}_i, \boldsymbol{\beta})$ relative to that model where no structure has been imposed on μ_i. Since the objective of any regression analysis is to obtain a parsimonious description of the data, it is hoped that the model $\mu_i = \lambda(\mathbf{x}_i, \boldsymbol{\beta})$ involving $k + 1$ parameters will provide a maximized likelihood value almost as large as can be obtained by that baseline (and uninformative) model involving as many parameters (namely, n) as there are data points. By "almost as large," we mean that $L(\mathbf{y}; \hat{\boldsymbol{\beta}})$ will not be significantly smaller than $L(\mathbf{y}; \hat{\boldsymbol{\mu}})$ based on a likelihood ratio test using (21.27).

The quantity

$$D(\hat{\boldsymbol{\beta}}) = -2 \ln \left[\frac{L(\mathbf{y}; \hat{\boldsymbol{\beta}})}{L(\mathbf{y}; \hat{\boldsymbol{\mu}})}\right] \tag{21.28}$$

is known as the goodness-of-fit statistic employed to assess whether $L(\mathbf{y}; \hat{\boldsymbol{\beta}})$ is significantly less than $L(\mathbf{y}; \hat{\boldsymbol{\mu}})$ and thus to suggest meaningful lack of fit to the data of the assumed regression model $\mu_i = \ell_i \lambda(\mathbf{x}_i, \boldsymbol{\beta})$. The quantity $D(\hat{\boldsymbol{\beta}})$ is also called the *deviance* for the Poisson regression model $\mu_i = \ell_i \lambda(\mathbf{x}_i, \boldsymbol{\beta})$, and (as we shall see presently) it can be thought of as a measure of residual variation about (or deviation from) the fitted model. Under H_0: $\mu_i = \ell_i \lambda(\mathbf{x}_i, \boldsymbol{\beta})$, $D(\hat{\boldsymbol{\beta}})$ is typically (although not strictly legitimately) assumed to have (for large samples) an approximate chi-square distribution with $n - k - 1$ degrees of freedom, where n is the number of parameters (i.e., number of subgroups, cells, or categories) specified in the likelihood (21.25), and $k + 1$ is the number of parameters (i.e., β_j's) in the likelihood (21.23). Thus, a very approximate test for goodness of fit of the model $\mu_i = \ell_i \lambda(\mathbf{x}_i, \boldsymbol{\beta})$ to a given data set can be performed by comparing the calculated value of $D(\hat{\boldsymbol{\beta}})$ to an appropriate upper-tail value of the chi-square distribution with $n - k - 1$ degrees of freedom.

With $\hat{y}_i = \ell_i \lambda(\mathbf{x}_i, \hat{\boldsymbol{\beta}})$ denoting the predicted response in cell i under model (21.20), it can be shown (we omit the details) that (21.28) can be written in the form

$$D(\hat{\boldsymbol{\beta}}) = 2 \sum_{i=1}^{n} \left[y_i \ln \left(\frac{y_i}{\hat{y}_i}\right) - (y_i - \hat{y}_i)\right] \tag{21.29}$$

Hence, $D(\hat{\boldsymbol{\beta}})$ behaves like SSE $= \sum_{i=1}^{n} (y_i - \hat{y}_i)^2$ in standard multiple linear regression analysis. When the fitted model exactly predicts the observed data (i.e., $y_i = \hat{y}_i$, $i = 1, 2, \ldots, n$), then $D(\hat{\boldsymbol{\beta}}) = 0$, and the larger the discrepancy between observed and predicted responses, the larger will be the value of $D(\hat{\boldsymbol{\beta}})$.

When the predicted values are all of reasonable size (i.e., $\hat{y}_i > 3$, $i = 1, 2, \ldots, n$), then (21.29) can be reasonably approximated by the more familiar Pearson-type observed-versus-predicted chi-square statistic of the form

$$\chi^2 = \sum_{i=1}^{n} \frac{(y_i - \hat{y}_i)^2}{\hat{y}_i} \tag{21.30}$$

As a word of caution, the statistic (21.30) can be quite misleadingly large when certain $\hat{y}_i$-values are very small.

The deviances for various models in a hierarchical class can be used to produce likelihood ratio tests. In particular, consider again the likelihood (21.23) involving the parameter set $\boldsymbol{\beta} = (\beta_0, \beta_1, \ldots, \beta_k)$, with deviance $D(\hat{\boldsymbol{\beta}})$ given by (21.28). Now, for $0 < r < k$, suppose we wish to test whether the last $k - r$ parameters in $\boldsymbol{\beta}$ are equal to 0; i.e., our null hypothesis is H_0: $\beta_{r+1} = \beta_{r+2} = \cdots = \beta_k = 0$. Under H_0, the (null hypothesis) likelihood can be obtained by replacing $\boldsymbol{\beta}$ in (21.23) by $\boldsymbol{\beta}_r$, where

$$\boldsymbol{\beta}_r = (\beta_0, \beta_1, \ldots, \beta_r; 0, 0, \ldots, 0)$$

If we denote this likelihood function by $L(\mathbf{y}; \boldsymbol{\beta}_r)$, then the likelihood ratio test of H_0 is performed using the test statistic

$$-2 \ln \left[\frac{L(\mathbf{y}; \hat{\boldsymbol{\beta}}_r)}{L(\mathbf{y}; \hat{\boldsymbol{\beta}})} \right] \tag{21.31}$$

which has an approximate chi-square distribution with $k - r$ degrees of freedom for large samples when H_0 is true.

Furthermore, expression (21.31) is exactly equal to the deviance difference

$$D(\hat{\boldsymbol{\beta}}_r) - D(\hat{\boldsymbol{\beta}}) \tag{21.32}$$

To see this, recall the general definition of $D(\hat{\boldsymbol{\beta}})$ given by expression (21.28). Using (21.28) and (21.32), we have

$$D(\hat{\boldsymbol{\beta}}_r) - D(\hat{\boldsymbol{\beta}}) = -2 \ln \left[\frac{L(\mathbf{y}; \hat{\boldsymbol{\beta}}_r)}{L(\mathbf{y}; \hat{\boldsymbol{\mu}})} \right] + 2 \ln \left[\frac{L(\mathbf{y}; \hat{\boldsymbol{\beta}})}{L(\mathbf{y}; \hat{\boldsymbol{\mu}})} \right]$$

$$= -2 \ln \left[\frac{L(\mathbf{y}; \hat{\boldsymbol{\beta}}_r)}{L(\mathbf{y}; \hat{\boldsymbol{\beta}})} \right]$$

which is exactly the likelihood ratio test statistic (21.31). Under H_0: $\beta_{r+1} = \beta_{r+2} = \cdots = \beta_k = 0$, the difference $D(\hat{\boldsymbol{\beta}}_r) - D(\hat{\boldsymbol{\beta}})$ will have an approximate chi-square distribution with $k - r$ degrees of freedom in large samples.

Thus, when using Poisson regression to analyze a set of data, members of a set of candidate models within a hierarchical class can be compared by considering differences between pairs of deviances for these models.

21-4-5 Continuation of Skin Cancer Data Example

We again consider the data in Table 21-1 giving skin cancer counts for women stratified by age in Minneapolis–St. Paul and Dallas–Ft. Worth. For these data, we used ML estimation to fit the following Poisson regression model for the expected number of skin cancer cases:

$$E(Y_{ij}) = \mu_{ij} = \ell_{ij}\lambda_{ij}$$

where

$$\ln \lambda_{ij} = \sum_{i=1}^{8} \alpha_i U_i + \beta E$$

The set $\{U_i\}$ are 0–1 dummy variables indexing the age strata and E contrasts metropolitan areas (1 = Dallas–Ft. Worth, 0 = Minneapolis–St. Paul). We will refer to this model as model 1. For this model, the estimated rate ratio adjusted for age was $e^{\hat{\beta}} = 2.2345$, and a 95% confidence interval for the true adjusted rate ratio was (2.0172, 2.4751). Also, a large-sample test for $H_0: \beta = 0$ versus $H_A: \beta \neq 0$ yielded a Z statistic of 15.40 ($P \approx 0$), which is highly significant.

Two additional questions of interest for these data are:

1. Is "age" an effect modifier; that is, is the "area" effect (as measured by a rate ratio parameter) different for different age strata?

2. If the answer to question 1 is no, then is "age" a confounder; that is, does "age" need to be in the model in some form in order to obtain a valid estimate of the "area" effect?

(At this point, the reader may wish to review the definitions and properties of effect modifiers and confounders discussed in Chapter 11.)

To answer question 1 directly, one could modify model 1 to include interaction terms as follows:

$$\text{model 2:} \quad \ln \lambda_{ij} = \sum_{i=1}^{8} \alpha_i U_i + \beta E + \sum_{i=1}^{7} \delta_i(EU_i)$$

(In order to avoid a singularity (i.e., perfect collinearity), we have added product terms involving only seven of the eight U_i's.)

For the revised interaction model above, we can test for effect modification by testing $H_0: \delta_1 = \delta_2 = \cdots = \delta_7 = 0$ using a likelihood ratio χ^2 statistic with 7 degrees of freedom. This test involves comparing model 1 (without interaction terms) to model 2 (with the seven EU_i terms).

An equivalent way to carry out this particular interaction test is to use the deviance described in the previous section to evaluate how well model 1, involving nine parameters, fits the data. In particular, the deviance for this model, namely

$$2 \sum_{i=1}^{8} \sum_{j=0}^{1} \left[y_{ij} \ln \left(\frac{y_{ij}}{\hat{y}_{ij}} \right) - (y_{ij} - \hat{y}_{ij}) \right]$$

which is approximately equal to

$$\sum_{i=1}^{8} \sum_{j=0}^{1} \frac{(y_{ij} - \hat{y}_{ij})^2}{\hat{y}_{ij}}$$

has the value 8.17, where

$$\hat{y}_{i0} = \ell_{i0} \, e^{\hat{\alpha}_i} \quad \text{and} \quad \hat{y}_{i1} = \ell_{i1} \, e^{\hat{\alpha}_i + \hat{\beta}}$$

When compared with upper-tail χ^2-values with

df = [number of y_{ij}'s] − [number of parameters in model 1]

= 16 − 9 = 7

there is clearly no lack of fit of model 1 (i.e., there are no large deviations of observed y_{ij}-values from predicted $\hat{y}_{ij}$-values). This indicates that adding further terms (e.g., of the form EU_i) to model 1 will not significantly improve the fit of that model.

To answer question 2 about confounding, we need to see whether $\hat{\beta}$, or $e^{\hat{\beta}}$, changes meaningfully if we ignore (i.e., don't control for) "age." In particular, we need to drop the age terms (i.e., the "$\Sigma \alpha_i U_i$" component) from model 1 to see if the estimated coefficient of E changes meaningfully from its value of 0.804 (or if the estimate of the rate ratio changes from its value of $e^{\hat{\beta}} = 2.2345$). If we fit

model 3: $\ln \lambda_{ij} = \alpha + \beta_0 E$

to these data, we obtain $\hat{\beta}_0 = 0.743$ and a crude rate ratio estimate of

$$\widehat{RR}_c = e^{\hat{\beta}_0} = e^{0.743} = 2.1022$$

There is enough change here to suggest that "age" (the group) is a confounder and so should be controlled at the analysis stage.

Up to this point, we have treated age as a *categorical* variable, using eight terms of the form $\alpha_i U_i$ in model 1 to reflect the eight age strata. A graph of $\ln \hat{\lambda}_{ij}$ versus $\ln([\text{age group } i \text{ midpoint}] - 15)$, $i = 1, 2, \ldots, 8$, plots as a straight line for each j, with the two lines essentially parallel. This suggests the use of a parsimonious model involving only a single *linear* effect for "age," with no interaction terms involving "age" and E.

In particular, consider the following model:

model 4: $\ln \lambda_{ij} = \alpha + \theta \ln T_i + \beta E$

where

$$T_i = \frac{[\text{midpoint of } i\text{th age interval}] - 15}{35}$$

for $i = 1, 2, \ldots, 8$, and E is defined as before. Note that $T_1 = (20 - 15)/35 = \frac{1}{7}$, $T_2 = \frac{3}{7}$, $T_3 = \frac{5}{7}$, $T_4 = 1$, $T_5 = \frac{9}{7}$, $T_6 = \frac{11}{7}$, $T_7 = \frac{13}{7}$, and $T_8 = \frac{15}{7}$.

Model 4 says that

$$\lambda_{i0} = e^{\alpha} T_i^{\theta} \quad \text{and} \quad \lambda_{i1} = e^{\alpha} T_i^{\theta} e^{\beta}$$

so that

$$RR_i = \frac{\lambda_{i1}}{\lambda_{i0}} = e^{\beta}$$

Note that when $i = 4$ (so that $T_4 = 1$) and $j = 0$, then $\lambda_{40} = e^{\alpha} T_4^{\theta} = e^{\alpha}(1)^{\theta} = e^{\alpha}$, so that $\ln \lambda_{40} = \alpha$. Hence, α, the intercept term in model 4, is the natural logarithm of the rate for the 45-to-54-year age group ($i = 4$) in Minneapolis–St. Paul ($j = 0$). Fitting model 4 using

Poisson regression methods gives the following point estimates and estimated standard errors:

$$\hat{\alpha} = -7.076, \quad \widehat{s.e.}(\hat{\alpha}) = 0.048$$
$$\hat{\theta} = 2.290, \quad \widehat{s.e.}(\hat{\theta}) = 0.068$$
$$\hat{\beta} = 0.803, \quad \widehat{s.e.}(\hat{\beta}) = 0.052$$

The deviance value for model 4 is 14.36, with $16 - 3 = 13$ degrees of freedom, indicating a good fit to these data. The estimated adjusted rate ratio for model 4 is

$$e^{\hat{\beta}} = e^{0.803} = 2.2322$$

The 95% confidence interval is given by

$$\exp[0.803 \pm 1.96(0.052)] = \exp(0.803 \pm 0.1019)$$
$$= (e^{0.7011}, e^{0.9049})$$
$$= (2.0160, 2.4717)$$

Thus, the use of model 4 leads to the same answers obtained from model 1, with a very small gain in precision (since the confidence interval for model 4 is slightly narrower than that for model 1). Since model 4 contains fewer parameters than model 1 without compromising on validity and precision, and since model 4 describes the linear effect of "age" in a clear-cut manner, it is the model of choice!

A tabular summary of the results of fitting various models to the data set under consideration can be presented in a Poisson ANOVA table, as illustrated in Table 21-2. From this table, model 1 and model 4 clearly have the best deviance values, but model 4 becomes our final choice. Note that model 1 has a deviance of 8.2, whereas model 4 has a deviance of 14.4. Model 1 should have a smaller deviance (i.e., should fit the data better) than model 4, since model 1 contains nine parameters, whereas model 4 contains only three. The issue is really whether model 1 fits the data *significantly* better than model 4. To address this issue, one can look at the difference in the two deviance values, namely $14.4 - 8.2 = 6.2$, just as one looks at the difference in SSE values in standard multiple regression analysis.

Specifically, under $H_0: \alpha_i = \alpha + \theta T_i$, where $i = 1, 2, \ldots, 8$,

deviance for model 4 − deviance for model 1

is approximately a chi-square random variable for large n with degrees of freedom of $13 - 7 = 6$. The alternative hypothesis is H_A: "$\{\alpha_i\}$ have unspecified structure." Since

TABLE 21-2 Poisson ANOVA table for skin cancer data
of Table 21-1

Model for $\ln \lambda_{ij}$	No. of Parameters	$D(\beta)$	df
α	1	2,790.3	15
Model 3: $\alpha + \beta E$	2	2,569.7	14
$\alpha + \theta \ln T_i$	2	272.7	14
Model 4: $\alpha + \theta \ln T_i + \beta E$	3	14.4	13
$\sum_{i=1}^{8} \alpha_i U_i$	8	266.9	8
Model 1: $\sum_{i=1}^{8} \alpha_i U_i + \beta E$	9	8.2	7
α_{ij}	16	0.0	0

$\chi^2_{6,0.60} = 6.211$, the P-value for this test is about .40. Hence, there is absolutely no evidence that model 1 provides a better fit to the data than model 4.

Finally, a pseudo-R^2 for model 4 can be computed as

$$\text{pseudo-}R^2 = \frac{2{,}790.3 - 14.4}{2.790.3} = .9948,$$

which indicates a superb fit to the data.

21-4-6 A Second Illustration of Poisson Regression Analysis

We now consider an example given by Frome (1983). The data to be used appear in Table 21-3. The basic response variable Y is "number of lung cancer deaths," which is assumed to have a Poisson distribution. More specifically, y_{ij} denotes the observed number of lung cancer deaths in row i and column j, $i = 1, 2, \ldots, 9$ and $j = 1, 2, \ldots, 7$; thus there are $n = 9 \times 7 = 63$ subgroups. The rows represent "years of smoking" (defined as age minus 20 years) in five-year categories from 15 through 19 to 55 through 59; the columns represent "number of cigarettes smoked per day," starting at 0 for nonsmokers and going up to 35 or more for the heaviest smokers. The variable ℓ_{ij} denotes the number of man-years at risk for cell (i, j). The variable t_i, defined as the midpoint of the ith "years of smoking" category divided by 42.5, will be employed when we fit some dose–response models to the data in Table 21-3; the variable d_j will denote the dosage level variable for the jth "number of cigarettes smoked per day" category.

One particular form of failure rate model $\lambda(\mathbf{x}_{ij}, \boldsymbol{\beta})$ to be fit to these data is the standard two-way cross-classification model *in exponentiated form*. (Note that a rate is nonnegative, and the use of an exponential function assures that this is so.) In particular, consider modeling the rate $\lambda(\mathbf{x}_{ij}, \boldsymbol{\beta}) \equiv \lambda_{ij}$ for cell (i, j) as

$$\lambda_{ij} = e^{\mu + \alpha_i + \delta_j} \tag{21.33}$$

TABLE 21-3   Man-years at risk (ℓ_{ij}) and observed number (y_{ij}) of lung cancer deaths (in parentheses)

Years of Smoking*	$42.5 t_i$	No. of Cigarettes Smoked per Day						
		0 ($d_1 = 0$)	1–9 ($d_2 = 5.2$)	10–14 ($d_3 = 11.2$)	15–19 ($d_4 = 15.9$)	20–24 ($d_5 = 20.4$)	25–34 ($d_6 = 27.4$)	35+ ($d_7 = 40.8$)
15–19	17.5	10,366 (1)	3,121 (0)	3,577 (0)	4,317 (0)	5,683 (0)	3,042 (0)	670 (0)
20–24	22.5	8,162 (0)	2,937 (0)	3,286 (1)	4,214 (0)	6,385 (1)	4,050 (1)	1,166 (0)
25–29	27.5	5,969 (0)	2,288 (0)	2,546 (1)	3,185 (0)	5,483 (1)	4,290 (4)	1,482 (0)
30–34	32.5	4,496 (0)	2,015 (0)	2,219 (2)	2,560 (4)	4,687 (6)	4,268 (9)	1,580 (4)
35–39	37.5	3,512 (0)	1,648 (1)	1,826 (0)	1,893 (0)	3,646 (5)	3,529 (9)	1,336 (6)
40–44	42.5	2,201 (0)	1,310 (2)	1,386 (1)	1,334 (2)	2,411 (12)	2,424 (11)	924 (10)
45–49	47.5	1,421 (0)	927 (0)	988 (2)	849 (2)	1,567 (9)	1,409 (10)	556 (7)
50–54	52.5	1,121 (0)	710 (3)	684 (4)	470 (2)	857 (7)	663 (5)	255 (4)
55–59	57.5	826 (2)	606 (0)	449 (3)	280 (5)	416 (7)	284 (3)	104 (1)

Note: If ℓ_{ij} was actually the number of people in cell (i, j) from which the observed number y_{ij} of lung cancer cases developed, then Y_{ij} could be considered to be a binomial random variable with sample size ℓ_{ij} and unknown probability (or risk) of lung cancer death Π_{ij}, in which case the categorical data analysis methods of Chapter 22 could possibly be utilized (although having several cells with no deaths can be quite problematic).

The quantity t_i is the midpoint of the ith "years of smoking" category divided by 42.5; d_j is the dosage level variable for the jth "cigarettes per day" category.

* Age minus 20 years.

where μ is the overall mean, α_i is the effect of the ith row, and δ_j is the effect of the jth column. Here, as in standard two-way ANOVA, we impose the constraints $\sum_{i=1}^{9} \alpha_i = \sum_{j=1}^{7} \delta_j = 0$, so that there will be a total of $1 + (9 - 1) + (7 - 1) = 15$ parameters to be estimated using model (21.33).

(As with standard regression model representations of ANOVA-type data (see elsewhere in this book), the "x" variables underlying model (21.33) are the usual dummy variables used to index the various rows and columns; their appearance has been suppressed for notational convenience. An equivalent way to write (21.33) in dummy variable regression notation is $\lambda_{ij} = \exp(\mu + \sum_{i=1}^{8} \alpha_i R_i + \sum_{j=1}^{6} \delta_j C_j)$, where the R_i's and C_j's denote dummy variables indexing the rows and columns, respectively.)

The Poisson-model-based likelihood function for the data in Table 21-1 and under model (21.33) is

$$\prod_{i=1}^{9}\prod_{j=1}^{7}\left[\frac{(\lambda_{ij})^{y_{ij}}e^{-\lambda_{ij}}}{(y_{ij})!}\right]$$

where $\lambda_{ij} = \exp(\mu + \alpha_i + \delta_j)$, $\alpha_9 = -\sum_{i=1}^{8}\alpha_i$, and $\delta_7 = -\sum_{j=1}^{6}\delta_j$. For the data in Table 21-1, the use of IRLS methods leads to the following estimates (in exponential form):

$e^{\hat{\mu}} = 7.69 \times 10^{-5}$

$e^{\hat{\alpha}_1} = 0.039, \quad e^{\hat{\alpha}_2} = 0.117, \quad e^{\hat{\alpha}_3} = 0.247, \quad e^{\hat{\alpha}_4} = 1.105, \quad e^{\hat{\alpha}_5} = 1.144,$

$e^{\hat{\alpha}_6} = 3.017, \quad e^{\hat{\alpha}_7} = 3.823, \quad e^{\hat{\alpha}_8} = 6.047, \quad e^{\hat{\alpha}_9} = 10.052$

$e^{\hat{\delta}_1} = 1.00, \quad e^{\hat{\delta}_2} = 3.39, \quad e^{\hat{\delta}_3} = 8.16, \quad e^{\hat{\delta}_4} = 10.10,$

$e^{\hat{\delta}_5} = 18.20, \quad e^{\hat{\delta}_6} = 22.60, \quad e^{\hat{\delta}_7} = 36.80$

Given these estimates, the predicted number $\hat{y}_{ij}$ of cancer deaths in cell (i, j) is $\hat{y}_{ij} = \ell_{ij}\, e^{\hat{\mu}+\hat{\alpha}_i+\hat{\delta}_j}$. For example, when $i = 4$ and $j = 5$, then

$\hat{y}_{45} = \ell_{45}e^{\hat{\mu}}e^{\hat{\alpha}_4}e^{\hat{\delta}_5}$

$\qquad = (4{,}687)(7.69 \times 10^{-5})(1.105)(18.20)$

$\qquad = 7.25$

which should be compared with the actual observed value $y_{45} = 6$. It can be shown that the deviance for this fitted model, as calculated using (21.29), has the numerical value of 51.47. (Formula (21.30) is not appropriate for these data, since several $\hat{y}_{ij}$'s are close to 0 in value.) When compared with critical values of the chi-square distribution with $63 - 15 = 48$ degrees of freedom, the value 51.47 does not suggest the presence of any significant lack of fit of the cross-classification model (21.33).

However, one criticism of model (21.33) is that its use involves estimating *15* parameters (whereas n is only 63); hence it does not really provide a parsimonious description of the data. In addition, it is of considerable interest with these data to attempt to fit a model for which the parameters have some realistic interpretation with regard to the mathematical theory of carcinogenesis.

In what follows, we will consider the *four-parameter nonlinear* model described by Frome (1983), namely,

$$\lambda_{ij} \equiv \lambda(t_i, d_j; \gamma, \alpha, \theta, \delta) = (\gamma + \alpha d_j^\theta) t_i^\delta \tag{21.34}$$

where t_i and d_j are as defined earlier.[15]

By using the mathematical identity $e^{\ln a} = a$ for $a > 0$, we can write (21.34) in an equivalent form considered by Frome (1983), namely,

$$\lambda_{ij} = [e^{\ln \gamma} + e^{(\ln \alpha + \theta \ln d_j)}]e^{\delta \ln t_i} = [e^{\beta_3} + e^{(\beta_1 + \beta_2 x_{2j})}]e^{\beta_0 x_{1i}} \equiv \lambda(\mathbf{x}_{ij}, \boldsymbol{\beta}) \tag{21.35}$$

where $\mathbf{x}_{ij} = (x_{1i}, x_{2j}) = (\ln t_i, \ln d_j)$ and $\boldsymbol{\beta} = (\beta_0, \beta_1, \beta_2, \beta_3) = (\delta, \ln \alpha, \theta, \ln \gamma)$. Expressing (21.34) in the exponential form (21.35) insures that predicted rates will always be positive.

The fitting of model (21.35) by IRLS gives the estimates $\hat{\beta}_0 = 4.46$, $\hat{\beta}_1 = 1.82$, $\hat{\beta}_2 = 1.29$, and $\hat{\beta}_3 = 2.94$. The estimated standard errors for these four estimators (obtained as the square roots of the appropriate diagonal elements of the estimated covariance matrix) are, respectively, 0.33, 0.66, 0.20, and 0.58. For example, an approximate large-sample 95% confidence interval for α $(= e^{\beta_1})$ would be

$$\exp(\hat{\beta}_1 \pm 1.96 \sqrt{\widehat{\mathrm{Var}}\ \hat{\beta}_1})$$

giving $\exp[1.82 \pm 1.96(0.66)] = \exp(1.82 \pm 1.29)$, and thus the interval $(e^{0.53}, e^{3.11}) = (1.70, 22.40)$.

Finally, a summary of the results of fitting various subset models of (21.34) is provided in Table 21-4.

As discussed earlier, calculation of the various differences between deviances in the ANOVA-type table of Table 21-4 will provide likelihood ratio tests about the importance of the parameters γ, δ, α, and θ. First of all, the difference $445.10 - 180.82 = 264.28$, when compared with upper-tail values of the χ_1^2 distribution, clearly argues strongly for rejecting $H_0: \delta = 0$ in favor of $H_A: \delta \neq 0$. This means, first, that the (multiplicative) effect of time since first exposure is important. Second, the difference $180.82 - 61.84 = 118.98$ rejects $H_0: \alpha = 0$ in favor of $H_A: \alpha \neq 0$, suggesting that the amount smoked is an important variable. Finally, the difference $61.84 - 59.58 = 2.26$ does not lead to rejection of $H_0: \theta = 1$, so that the first

TABLE 21-4   Summary of analyses of
                data in Table 21-3

Model for λ_{ij}	No. of Parameters	$D(\beta)$	df*
γ	1	445.10	62
γt_i^δ	2	180.82	61
$(\gamma + \alpha d_j)t_i^\delta$	3	61.84	60
$(\gamma + \alpha d_j^\theta)t_i^\delta$	4	59.58	59

* df = 63 − (number of parameters in fitted model)

[15] In model (21.34), γ represents the background (i.e., $d_j = 0$ for a nonsmoker) rate at age 62.5 (i.e., $t_i = [\text{age} - 20]/42.5 = 1$ at age 62.5), αd_j^θ describes the effect of dosage (i.e., amount smoked) on lung cancer death rates, and t_i^δ is the multiplicative effect (on $\gamma + \alpha d_j^\theta$) of the time elapsed since the smoking habit was started. Frome (1983) provides further discussion concerning scientific evidence for the use of a model like (21.34) to describe the incidence of lung cancer.

power of dosage seems most appropriate. Since the deviance value of 61.84 with 60 degrees of freedom for the model $\lambda_{ij} = (\gamma + \alpha d_j)t_i^\delta$ does not suggest any lack of fit, this model seems preferable.[16]

21-5 Logistic Regression Analysis

As in the previous section where the Poisson distribution provided the underlying structure for the statistical methodology described, the procedures of logistic regression analysis require a key assumption about the distribution of the basic response variable Y of interest. Similarly, the regression analysis goal is to describe the mean (or expected value) of Y as a (logistic) function of a set of predictor variables $x_1, x_2, \ldots, x_k$.

21-5-1 Theoretical Considerations

For logistic regression, the basic random variable Y of interest is a dichotomous variable taking the value 1 with probability θ and the value 0 with probability $1 - \theta$. Such a random variable is called a Bernoulli (or point-binomial) variable, and it has the simple discrete probability distribution

$$p_Y(y; \theta) = \text{pr}(Y = y; \theta) = \theta^y(1 - \theta)^{1-y} \qquad y = 0, 1 \tag{21.36}$$

The name *point-binomial* arises because (21.36) is that special case of the binomial distribution $_nC_y\theta^y(1 - \theta)^{n-y}$ when $n = 1$.

Although (21.36) is a very simply structured probability distribution, it can serve as a useful model in many important practical situations where the response variable takes only one of two possible values. As a key example, a study of the development of a particular disease in some human population could employ (21.36) to describe in probabilistic terms whether a given individual in the study group will ($Y = 1$) or will not ($Y = 0$) develop the disease in question over the follow-up period of interest.

In general, suppose (for $i = 1, 2, \ldots, n$) that Y_i has the Bernoulli distribution

$$p_{Y_i}(y_i; \theta_i) = \text{pr}(Y_i = y_i; \theta_i) = \theta_i^{y_i}(1 - \theta_i)^{1-y_i} \qquad y_i = 0, 1 \tag{21.37}$$

For example, then, θ_i could represent the probability that individual i in a random sample of n individuals from some population will develop some particular disease during the follow-up period in question.

Given that $Y_1, Y_2, \ldots, Y_n$ are mutually independent, the likelihood function based on (21.37) is obtained as the product of the marginal distributions for the Y_i's, namely,

$$L(\mathbf{y}; \boldsymbol{\theta}) = \prod_{i=1}^n p_{Y_i}(y_i; \theta_i) = \prod_{i=1}^n [\theta_i^{y_i}(1 - \theta_i)^{1-y_i}] \tag{21.38}$$

where $\boldsymbol{\theta} = (\theta_1, \theta_2, \ldots, \theta_n)$. Now suppose, without loss in generality, that the first n_1 out of the n individuals in our random sample actually develop the disease in question (so that $y_1 = y_2 = \cdots = y_{n_1} = 1$), and that the remaining $n - n_1$ individuals do not (so that $y_{n_1+1} = y_{n_1+2} =$

---

[16] It is important to note that the reduction in the deviance due to adding one parameter (or a group of parameters) is *order dependent*; consequently, a different order of parameter additions can lead to a different final model.

$\cdots = y_n = 0$). Given this set of observed outcomes, the likelihood expression (21.38) takes the specific form

$$L(\mathbf{y}; \boldsymbol{\theta}) = \left(\prod_{i=1}^{n_1} \theta_i\right)\left[\prod_{i=n_1+1}^{n} (1 - \theta_i)\right] \tag{21.39}$$

In what follows, we will work with $L(\mathbf{y}; \boldsymbol{\theta})$ as expressed in (21.39).

Now, it follows from standard statistical theory that the expected value (i.e., population mean) of Y_i, where Y_i has the distribution (21.37), is simply $E(Y_i) = (1)\theta_i + (0)(1 - \theta_i) = \theta_i$, $i = 1, 2, \ldots, n$.

With notation completely similar to that used in Section 21-4, if $\mathbf{x}_i = (x_{i1}, x_{i2}, \ldots, x_{ik})$ is the set of values of the k predictors $x_1, x_2, \ldots, x_k$ specific to individual i, and if $\boldsymbol{\beta} = (\beta_0, \beta_1, \ldots, \beta_k)$ is the set of unknown parameters, then the goal of regression analysis is to postulate (based on knowledge and experience concerning the process under study) a mathematical model describing θ_i, the mean of Y_i, as a function of the x_{ij}'s and β_j's. Then the model is fitted to the data (e.g., by maximum likelihood), and finally appropriate statistical inferences are made (after, of course, the adequacy of the fit of said model is verified, including consideration of the relevant regression diagnostic indices).

The methodology of logistic regression analysis assumes that the relationship between θ_i and the x_{ij}'s and β_j's is as described by the logistic function; in particular, this relationship is of the specific form

$$\theta_i = \frac{1}{1 + \exp\left[-\left(\beta_0 + \sum_{j=1}^{k} \beta_j x_{ij}\right)\right]} \qquad i = 1, 2, \ldots, n \tag{21.40}$$

This form is well suited as a model for a probability parameter like θ_i since (21.40) varies from 0 to 1 as $\beta_0 + \sum_{j=1}^{k} \beta_j x_{ij}$ varies from $-\infty$ to $+\infty$. The graph in Figure 21-2 depicts the general *sigmoid* shape of the logistic function (21.40).

The logistic model has been used both extensively and successfully to describe the probability (or risk) of developing some disease over a specified time period as a function of certain risk factors $x_1, x_2, \ldots, x_k$. Kleinbaum, Kupper, and Morgenstern (1982) devote the last five chapters of their book on epidemiologic research methods to a detailed discussion of both the theoretical and applied aspects of logistic regression as it relates to the analysis of observational study data on human populations. The reader is encouraged to consult that

FIGURE 21-2   The logistic function

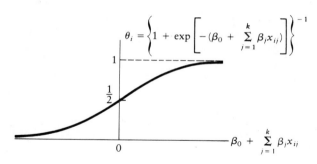

reference for detailed numerical examples of the application of logistic regression methods to epidemiologic data sets.

If we replace θ_i in (21.39) by the logistic function expression (21.40), we obtain the so-called unconditional likelihood function characterizing standard logistic regression analysis,[17] namely,

$$L(\mathbf{y}; \boldsymbol{\beta}) = \left(\prod_{i=1}^{n_1} \left\{ 1 + \exp\left[-\left(\beta_0 + \sum_{j=1}^{k} \beta_j x_{ij} \right) \right] \right\}^{-1} \right) \times$$

$$\left(\prod_{i=n_1+1}^{n} \exp\left[-\left(\beta_0 + \sum_{j=1}^{k} \beta_j x_{ij} \right) \right] \left\{ 1 + \exp\left[-\left(\beta_0 + \sum_{j=1}^{k} \beta_j x_{ij} \right) \right] \right\}^{-1} \right)$$

With a little algebraic manipulation, the reader can verify that an equivalent expression for the above likelihood function is[18]

$$L(\mathbf{y}; \boldsymbol{\beta}) = \frac{\displaystyle\prod_{i=1}^{n_1} \exp\left(\beta_0 + \sum_{j=1}^{k} \beta_j x_{ij} \right)}{\displaystyle\prod_{i=1}^{n} \left[1 + \exp\left(\beta_0 + \sum_{j=1}^{k} \beta_j x_{ij} \right) \right]} \tag{21.41}$$

Because (21.41) is a complex nonlinear function of the elements of $\boldsymbol{\beta}$, the maximization of (21.41) to find the ML estimator $\hat{\boldsymbol{\beta}}$ of $\boldsymbol{\beta}$ must involve the use of appropriate computer algorithms. Such programs will also produce the maximized likelihood value $L(\mathbf{y}; \hat{\boldsymbol{\beta}})$ and the estimated (asymptotic) covariance matrix $\hat{\mathbf{V}}(\hat{\boldsymbol{\beta}})$ for a given model, which can then be used to make appropriate statistical inferences.

A general analysis strategy for selecting the variables to retain in a logistic model is provided in Kleinbaum, Kupper, and Morgenstern (1982, chap. 21). This strategy has been formulated with the goal of obtaining a *valid* estimate (based on the analysis of epidemiologic research data) of the relationship between a specified exposure variable and a particular disease variable, while controlling or adjusting for other covariates which, if not correctly taken into account, can lead to an incorrect assessment of the strength of the exposure–disease relationship of interest. These covariates can act as confounders and/or effect modifiers, terms that are discussed in detail in Chapter 11.

[17] The term *unconditional likelihood* refers to the unconditional probability of obtaining the particular set of data under consideration. More specifically, the unconditional likelihood function is the joint probability distribution for discrete data or the joint density function for continuous data. A *conditional likelihood* gives the conditional probability of obtaining the data configuration *actually observed* given all possible configurations (i.e., permutations) of the data values (see Section 21-5-2).

[18] The likelihood (21.41) is based on the responses of *individual* subjects, with (21.37) pertaining to the ith subject. In contrast, a categorical data analysis (Chapter 22) involving the binomial distribution is based on the responses of *groups* of subjects, with, say, y_i subjects developing the disease in question out of n_i in the ith group. In this situation, the underlying distribution of Y_i is binomial, namely,

$$p_{Y_i}(y_i; \theta_i) = {}_{n_i}C_{y_i} \theta_i^{y_i} (1 - \theta_i)^{n_i - y_i}$$

and the validity of such a categorical data analysis requires that n_i be fairly large in each group. The effects of small samples on the validity of ML analyses based on (21.41) will be briefly discussed in Section 21-5-2.

21-5-2 Conditional ML Methods

An alternative to the use of the so-called unconditional likelihood function (21.41) for estimating the elements of $\boldsymbol{\beta}$ is to employ a conditional likelihood function. The primary reason for using conditional likelihood methods is that unconditional methods can lead to seriously biased estimates of the elements of $\boldsymbol{\beta}$ when the amount of data available for analysis is not large. Although "not large" is admittedly vague, we are referring to the potential problem of using large-sample-based statistical procedures like maximum likelihood when the number of parameters to be estimated constitutes a fair proportion of the total sample size.

Kleinbaum, Kupper, and Morgenstern (1982, chap. 24) provide examples of epidemiologic data analysis in which the methods of unconditional and conditional logistic regression analysis are compared. In situations where the data are sparse for the intended model-fitting exercise (e.g., as when the data are cast into strata whose effects must be estimated, and stratum-specific sample sizes are small), these examples clearly demonstrate that estimates of parameters based on unconditional logistic model likelihoods will generally be too high.

The conditional likelihood function corresponding to (21.41) takes the form

$$L_c(\mathbf{y}; \boldsymbol{\beta}) = \frac{\displaystyle\prod_{i=1}^{n_1} \exp\left(\beta_0 + \sum_{j=1}^{k} \beta_j x_{ij}\right)}{\displaystyle\sum_{\mathbf{u}} \left(\prod_{\ell=1}^{n_1} \exp\left(\beta_0 + \sum_{j=1}^{k} \beta_j x_{u_{\ell j}}\right)\right)} \tag{21.42}$$

where the sum in the denominator is over all partitions of the set $\{1, 2, \ldots, n\}$ into two subsets, the first of which contains n_1 elements; there are $_nC_{n_1} = n!/(n_1)!(n - n_1)!$ of such partitions and hence that many terms in the summation.

In words, the conditional likelihood function (21.42) can be considered analogous to a conditional probability. More specifically, $L_c(\mathbf{y}; \boldsymbol{\beta})$ is the conditional probability (based on the underlying logistic model assumption) that the first n_1 of the set $\mathbf{x}_1, \mathbf{x}_2, \ldots, \mathbf{x}_n$ actually go with the n_1 subjects who developed the disease in question (as we know they do), *given* (or conditional on) the observed set $\mathbf{x}_1, \mathbf{x}_2, \ldots, \mathbf{x}_n$ and *given* the fact that exactly n_1 of the n subjects under study actually developed the disease in question. To express it yet another way, what (21.42) is doing is comparing the likelihood of what we actually observed (the numerator) relative to the likelihood for all possible arrangements of the given data set (the denominator). The reason that (21.42) is referred to as a conditional likelihood is that it is completely analogous to a conditional probability and is conditional on (arrangements of) the data actually observed.[19]

With regard to data analysis, the conditional likelihood (21.42) is employed just like any other likelihood function discussed in this chapter. The computational aspects are, however, somewhat more involved because of the permutations of the data required to

---

[19] The conditional likelihood (21.42) has a structure somewhat similar to that of Cox's (1975) partial likelihood, a likelihood function used to analyze data where the response variable, "time to development of disease" (i.e., "time to failure" or "survival time"), is continuous rather than discrete. Although we will not discuss the statistical methodology known as *survival analysis*, the reader can appreciate the fact that it is simply another maximum-likelihood-based procedure to which the general principles discussed in this chapter apply and for which appropriate computer programs are available.

evaluate the denominator in (21.42). For example, if $n = 20$ and $n_1 = 3$, then $_{20}C_3 = 1,140$, and so the denominator in (21.42) will involve the sum of over a thousand terms. A small number of computer algorithms have been developed to carry out conditional logistic regression analyses (e.g., see Smith et al., 1981).

21-6 Summary

This chapter has discussed the maximum likelihood method of estimating parameters and associated inference-making procedures. In maximum likelihood estimation, a likelihood function must be specified that is to be maximized. Procedures for testing hypotheses and constructing confidence intervals utilize maximized likelihood values and estimated covariance matrices for the various models under study.

The intimate connection between standard weighted- and unweighted-least-squares methods and ML estimation under the usual assumption of normality was discussed. Methods of analysis employing Poisson regression and logistic regression were described in detail, and several numerical examples were used to illustrate the principles involved.

Problems

For each of three age groups (25–34, 35–44, and 45–54), suppose that we have recorded yearly sex-specific lung cancer mortality rates for the five-year period 1960 through 1964. These data are to be analyzed by Poisson regression methods to see whether the change (if any) in log rate over time varies by age–sex group and, if so, to quantify that variation. In what follows, we will index the six age–sex groups as follows:

group 1 ($i = 1$):   25–34-year-old females
group 2 ($i = 2$):   35–44-year-old females
group 3 ($i = 3$):   45–54-year-old females
group 4 ($i = 4$):   25–34-year-old males
group 5 ($i = 5$):   35–44-year-old males
group 6 ($i = 6$):   45–54-year-old males

Consider the following model for the expected cell count $E(Y_{ik})$ for age–sex group i in year $(1960 + k)$, where $i = 1, 2, 3, 4, 5, 6$ and $k = 0, 1, 2, 3, 4$:

$$E(Y_{ik}) = \ell_{ik}\lambda_{ik}$$

where

$$\ln \lambda_{ik} = \sum_{i=1}^{6} \alpha_i A_i + \beta k \tag{1}$$

Here, ℓ_{ik} and λ_{ik} are, respectively, the person-years at risk and the (unknown) population lung cancer mortality rate in cell (i, k). The independent variables in model (1) are defined as follows:

$$A_i = \begin{cases} 1 & \text{if age–sex group } i \\ 0 & \text{otherwise} \end{cases}$$

$$k = \text{year} - 1960$$

1. What is the total number n of data points, or pairs (ℓ_{ik}, y_{ik}), for this data set?

2. Based on model (1), what is the expected cell count for a 40-year-old male in 1962?

3. For the ith age–sex group, how does model (1) describe the *change* in log rate over time?

4. What does model (1) assume about the effect of age–sex group on the *change* in log rate over time?

5. Find a general expression for

$$\text{IDR}_{ik} = \frac{\lambda_{ik}}{\lambda_{10}}$$

the incidence density ratio (IDR) comparing the mortality rate for age group i and year $1960 + k$ to the mortality rate for the reference category "25–34-year-old females in 1960."

6. Suppose that it is of interest to assess whether the *change* in log rate over time actually varies by age–sex group (i.e., there is a group-by-time interaction). By adding appropriate cross-product terms to model (1), construct an appropriate model that will permit such an assessment and then discuss how you would interpret such a model.

7. Consider the following Poisson regression ANOVA table based on fitting various models to the data under study:

Model for $\ln \lambda_{ik}$	No. of Parameters	$D(\hat{\boldsymbol{\beta}})$	df
(1) α	1	300	29
(2) $\alpha + \beta k$	2	200	28
(3) $\sum_{i=1}^{6} \alpha_i A_i$	6	175	24
(4) $\sum_{i=1}^{6} \alpha_i A_i + \beta k$	7	60	23
(5) $\sum_{i=1}^{6} \alpha_i A_i + \beta_1 k + \beta_2 k^2$	8	59	22
(6) $\sum_{i=1}^{6} \alpha_i A_i + \beta k + \gamma_1(A_1 k)$	8	25	22
(7) $\sum_{i=1}^{6} \alpha_i A_i + \beta k + \sum_{i=1}^{5} \gamma_i(A_i k)$	12	20	18
(8) α_{ik}	30	0	0

Using the above table, answer the following questions:

a. Ignoring (for now) the variable "time," is there evidence that average mortality rates differ among the six age–sex groups?

  b. Is the addition of the linear time term (βk) to model (3) worthwhile?

  c. Assuming that the change in log rate over time is the same for all age–sex groups, do the data argue for the addition of a quadratic time term to a model already containing a linear component of time?

  d. Is there evidence in the data that the change in log rate over time is different for different age–sex groups?

  e. Carry out a test of H_0: "all six age–sex groups have the same slope" versus H_A: "all age–sex groups except group 1 have the same slope."

  f. Of those models that fit the data well (use $\alpha = .1$ for any test of lack of fit that you perform), which of these models would you choose as your final model and why?

  g. For your final model chosen in part (f), calculate a pseudo-R^2-value.

  h. Now, assume that model (6) has been fit to the data, resulting in the following point estimates and standard errors:

Parameter	Point Estimate	Standard Error
$\hat{\alpha}_1$	0.50	0.25
$\hat{\alpha}_2$	1.00	0.40
$\hat{\alpha}_3$	1.50	0.30
$\hat{\alpha}_4$	1.25	0.30
$\hat{\alpha}_5$	1.50	0.50
$\hat{\alpha}_6$	1.75	0.40
$\hat{\beta}$	0.50	0.20
$\hat{\gamma}_1$	-3.00	0.50

  i. Find and interpret approximate 95% confidence intervals for the following parameters: (1) the common slope for age–sex groups 2 through 6 and (2) the slope for age–sex group 1. (*Note*: The estimated covariance between $\hat{\beta}$ and $\hat{\gamma}_1$ is $\widehat{\text{Cov}}(\hat{\beta}, \hat{\gamma}_1) = -.10$.)

References

Baker, R. J., and Nelder, J. A. (1978). *Generalized Linear Interactive Modeling (GLIM), Release 3*. Oxford: Numerical Algorithms Group.

Cox, D. R. (1975). "Partial Likelihood." *Biometrika*, 62: 269–276.

Frome, E. L. (1983). "The Analysis of Rates Using Poisson Regression Models." *Biometrics*, 39: 665–674.

Frome, E. L., and Checkoway, H. (1985). "Use of Poisson Regression Models in Estimating Incidence Rates and Ratios." *Amer. J. Epidem.*, 121(2): 309–323.

Kleinbaum, D. G., Kupper, L. L., and Morgenstern, H. (1982). *Epidemiologic Research: Principles and Quantitative Methods*. Belmont, Calif.: Lifetime Learning Publications.

Remington, R. D., and Schork, M. A. (1985). *Statistics with Applications to the Biological and Health Sciences*. Englewood Cliffs, N.J.: Prentice-Hall, Inc.

Scotto, J., Kopf, A. W., and Urbach, F. (1974). "Non-Melanoma Skin Cancer among Caucasians in Four Areas of the United States." *Cancer*, 34: 1333–1338.

Smith, P. G., Pike, M. C., Hill, A. P., Breslow, N. E., and Day, N. E. (1981). "Multivariate Conditional Logistic Analysis of Stratum-Matched Case-Control Studies." *J. Royal Stat. Soc., Series C,* *30*(2): 190–197.

Stokes, M. E., and Koch, G. G. (1983). "A Macro for Maximum Likelihood Fitting of Log-Linear Models to Poisson and Multinomial Counts with Contrast Matrix Capability for Hypothesis Testing." *Proceedings of the Eighth Annual SAS Users Group International Conference,* pp. 795–800.

Analysis of Categorical Data

22-1 Preview

In this chapter we describe a method for analyzing multivariable data that are in the form of counts or frequencies corresponding to categories or combinations of categories of nominally treated variables. A simple form of such a categorical data layout is the standard (2×2) two-way contingency table of the form shown in Table 22-1. Here a, b, c, and d are the frequencies of observations (i.e., the observed cell counts) in each of the four cells. As we shall discuss in Section 22-2, the two basic kinds of questions generally asked about such two-way layouts concern independence or homogeneity, and the usual method of analysis in either case involves the use of the chi-square statistic of the general form

$$\chi^2 = \Sigma \frac{(O - E)^2}{E}$$

where the O's refer to observed cell counts and the E's refer to expected cell counts specified under a given null hypothesis of interest.

We shall describe how the analysis of such a simple 2×2 contingency table can be considered a special case of a more general method of analysis of categorical data that utilizes a linear-models approach similar to that of regression analysis. We mentioned this methodology briefly in Section 2-3 as a method for describing the relationship between a nominal dependent variable and several nominally or ordinally treated independent variables. Our intention at that early point in the text was to distinguish the classification of the variables under study used in this method from the classification used in the more common methods of regression analysis and discriminant analysis. In this regard, categorical data analysis involves nominally treated dependent variables, whereas regression analysis generally considers continuous dependent variables. Also, categorical data analysis treats the independent

520

TABLE 22-1   A 2 × 2
               contingency
               table

| | Factor B | |
Factor A	Level 1	Level 2
Level 1	a	b
Level 2	c	d

variables as nominal or ordinal, whereas discriminant analysis allows the independent variables to be continuous. In this chapter we shall go much further in our discussion of this method for analyzing categorical data to describe its general applicability (even to situations involving several dependent variables), the data configurations considered, the assumptions required, and the general methodology used.

The need for multivariable analysis of categorical data techniques has increased steadily in recent years, since present-day studies in the health, social science, and other fields now typically generate data on several qualitative (i.e., nominal or categorical) variables. Simple chi-square statistics based on looking at such variables two at a time do not do justice to the multivariable nature of such data. Thus, general and powerful methods must be considered for analyzing multidimensional contingency tables (i.e., tables arising from cross-classifications of several qualitative variables), methods that provide the flexibility to fit multivariable models and to test appropriate hypotheses. Also, the qualitative nature of the response variable in a categorical data analysis precludes the use of standard multivariable techniques (e.g., analysis of variance and dummy variable regression procedures) that rely on the assumptions of normality and homogeneity of variance for testing purposes.

As an example of a categorical data set that benefits from the type of sophisticated analysis we are advocating, we will describe a set of data obtained from a large-scale study (Hogue et al., 1974) on postabortion fertility control (PAFC) in several countries conducted by the International Fertility Research Program (IFRP). The dependent variable in this study is dichotomous: An individual either *accepts* or *does not accept* PAFC at the time of follow-up. The independent variables of interest are age (A), number of living children (C), years of education completed (E), and country or location (L); each of these demographic variables is categorized into two groups. Cross-classification according to these four factors then results in $2^4 = 16$ cells and data on 1,052 cases distributed as in Table 22-2.

The IFRP personnel were concerned with the relationship between the four independent variables A, C, E, and L and the binomial response PAFC, and they were particularly interested in the main effects of and the interactions among the independent variables. Their analysis objective would be best met by finding a simple predictive model with as few parameters as possible that would efficiently represent the multivariable relationship between the four demographic factors and the dichotomous response. In this regard, Grizzle, Starmer, and Koch (1969) have described how to use linear regression techniques and weighted-least-squares analysis to build such models when the data are categorical; we shall adopt their approach in this chapter. Their methodology involves the use of test statistics that have chi-square distributions in large samples when the corresponding null hypotheses of interest are true. The theoretical justification for their approach can be found in papers by Wald (1943) and Neyman (1949).

TABLE 22-2   Proportion of cases accepting postabortion fertility
control by age (A), number of living children (C), years
of education completed (E), and location (L)

Variable				Fertility Control		
A	C	E	L	No. Accepting	No. Not Accepting	Proportion Accepting
<25	0–1	0–6	India	4	8	0.333
25+	0–1	0–6	India	2	1	0.667
<25	0–1	7+	India	14	7	0.667
25+	0–1	7+	India	7	3	0.700
<25	2+	0–6	India	17	1	0.944
25+	2+	0–6	India	175	9	0.951
<25	2+	7+	India	12	2	0.857
25+	2+	7+	India	90	5	0.947
<25	0–1	0–6	Yugoslavia	12	21	0.364
25+	0–1	0–6	Yugoslavia	13	9	0.591
<25	0–1	7+	Yugoslavia	112	155	0.419
25+	0–1	7+	Yugoslavia	40	39	0.506
<25	2+	0–6	Yugoslavia	8	3	0.727
25+	2+	0–6	Yugoslavia	63	36	0.636
<25	2+	7+	Yugoslavia	13	9	0.591
25+	2+	7+	Yugoslavia	96	66	0.593
				678	374	0.588

(Two other approaches to the analysis of categorical data are that based on maximum likelihood as formulated by Bishop (1969) and Goodman (1970) and that based on minimum discrimination information as formulated by Ku, Varner, and Kullback (1971). In large-sample situations, all three of these methods are essentially equivalent. In most applications, which method to use is a matter of practical and computational convenience, although we prefer the Grizzle–Starmer–Koch approach because it is operationally so similar to that of multiple regression.)

In Section 22-2, we shall provide a general discussion of categorical data (at which time we will establish notation for representing the data), and we shall examine a particular data set. In Sections 22-3 and 22-4 we shall describe the Grizzle–Starmer–Koch (GSK) linear-models approach to the analysis of categorical data. The main purpose of working through several diverse examples will be to illustrate how to specify an underlying model, how to formulate (in terms of this model) hypotheses to be tested, and how to interpret (with respect to the model) the results of these statistical tests.

The reader is not expected to become an expert in analyzing complex categorical data sets after working through this chapter. Indeed, the purpose of the chapter is simply to introduce the reader to a general and powerful approach for analyzing such data sets, and this will be accomplished via a general (nonmathematical) discussion of the methodology and a consideration of some applications.

22-2 Representation of Categorical Data

In any experimental situation, data of two general types are collected on each subject or unit under study:

1. Data on the group or subpopulation to which the subject belongs or on the set of experimental conditions imposed on the subject.

2. Data on what happens to the subject during the course of the experiment.

In the case of categorical data, we shall use the term *factor* when referring to type 1 classification; in other words, a *factor* in categorical data analysis is analogous to an *independent variable* in regression analysis. When referring to type 2 classification, we shall use the term *response*, which is akin to the term *dependent variable* in regression analysis. Hence, the data in any multidimensional table are simply the observed frequencies with which subjects belonging to the same combination of factor categories (i.e., the same subpopulation) give the same combination of responses (i.e., fall in the same response category).

To firm up these concepts, consider the hypothetical data set shown in Table 22-3. Here n_{ij} is the observed number (or frequency) of individuals in the ith subpopulation giving the jth category of response, $i = 1, 2, \ldots, s$ and $j = 1, 2, \ldots, r$. Here $n_{i.} = \sum_{j=1}^{r} n_{ij}$ is the total number of individuals in the ith subpopulation. At this time we note that all marginal frequencies that depend on factor combinations only (i.e., the $n_{i.}$'s) are considered *fixed* numbers known before the study was performed, while all other frequencies are random variables.

The "factor–independent variable/response–dependent variable" analogy that we alluded to above leads to a method for classifying the types of tables one may encounter. Tables can be classified according to how the data have been gathered and/or according to the general types of questions one should be asking of the data. There are four principal types of multidimensional tables of usual interest: multiresponse, no-factor tables; multiresponse, one-factor tables; one-response, multifactor tables; and multiresponse, multifactor tables.

TABLE 22-3   Hypothetical set of categorical data

Subpopulation (Factor Combination)	Category of Response (Response Combination)				Total
	1	2	...	r	
1	n_{11}	n_{12}	...	n_{1r}	$n_{1.}$
2	n_{21}	n_{22}	...	n_{2r}	$n_{2.}$
.	.	.		.	.
.	.	.		.	.
.	.	.		.	.
s	n_{s1}	n_{s2}	...	n_{sr}	$n_{s.}$
					$n_{..} = \sum_{i=1}^{s} n_{i.}$

In multiresponse, no-factor tables (for which $s = 1$ in Table 22-3), only one subpopulation is sampled, and two or more responses (i.e., dependent variables) are measured on each sampling unit, each response being categorized to yield r combinations of response categories in total. For such a table, the researcher is interested in the relationships among the different responses, and the analysis problems here are analogous to those concerning questions of independence and correlation among several continuous dependent variables, $Y_1, Y_2, \ldots, Y_k$, say.

If the table is of the multiresponse, one-factor type, samples are considered to have been taken from each of s (≥ 2) subpopulations corresponding to the categories of a single factor. The interest of the investigator in this case is to study the effects of the factor on the various interrelationships among the responses, as, for example, in the multivariate regression of several continuous dependent variables ($Y_1, Y_2, \ldots, Y_k$, say) on a single independent variable X (an issue we have not dealt with in this text).

With one-response, multifactor tables, samples are considered to have come from each of s (≥ 2) subpopulations corresponding to combinations of categories of two or more factors, and one response with r (≥ 2) categories is measured in each subpopulation. Here one would be interested in how the factors jointly influence the response, and the analysis problems here are analogous to those arising in regression analysis involving one continuous dependent variable Y and several independent variables, $X_1, X_2, \ldots, X_p$, say.

Finally, for multiresponse, multifactor tables, samples are considered to have come from each of s (≥ 2) subpopulations corresponding to combinations of categories of *several* factors, and *several* responses are measured in each subpopulation. For such tables, both the relationships among the responses and how the factors combine to affect these relationships are of interest, as they are when performing a multivariate regression of several continuous dependent variables ($Y_1, Y_2, \ldots, Y_k$) on several independent variables ($X_1, X_2, \ldots, X_p$).

As an example, Table 22-2 is best looked upon (for analysis purposes) as a one-response, multifactor table. There are four factors (A, C, E, and L), each at two levels, which results in $s = 2^4 = 16$ subpopulations (or factor combinations); the one response variable is binomial, giving $r = 2$ response categories.

(It is important to note that, based on the sampling scheme actually used to generate these data, a multiresponse, no-factor classification appears more technically correct. Thus, as this example illustrates, the sampling scheme actually used may not correspond to the sampling method appropriate for answering the primary research questions of interest. For example, for the IFRP data (Table 22-2), it is desirable to assume that the marginal totals for each row (i.e., subpopulation) were known (or fixed) before the data were collected, even though, in actuality, only the total sample size of 1,052 was fixed. In such a case it is necessary to consider that the subpopulation (conditional) distributions based on the condition of *assuming* fixed marginal totals a priori adequately mimic the unconditional distributions based on actually sampling with fixed marginals a priori in order to justify the desired analysis. Thus, for the IFRP data, we must appeal to such a conditional argument in order to justify a one-response, multifactor analysis. In most applications, such an argument is usually quite tenable.)

In particular, if Π_{i1} is the true (but unknown) probability of an individual in the ith subpopulation accepting postabortion fertility control (so that Π_{i2} [$= 1 - \Pi_{i1}$] is the true but

TABLE 22-4   Cell probabilities for hypothetical data
set in Table 22-3

Subpopulation (Factor Combination)	Category of Response (Response Combination)				Total
	1	2	. . .	r	
1	Π_{11}	Π_{12}	. . .	Π_{1r}	1
2	Π_{21}	Π_{22}	. . .	Π_{2r}	1
.	.	.		.	.
.	.	.		.	.
.	.	.		.	.
s	Π_{s1}	Π_{s2}	. . .	Π_{sr}	1

unknown probability of that individual *not* accepting PAFC), then the parameter of interest for the ith subpopulation is Π_{i1}.[1]

In the general case when $r > 2$, there are r probability parameters $\Pi_{i1}, \Pi_{i2}, \ldots, \Pi_{ir}$ associated with the ith subpopulation, and the (joint) distribution of the observed cell frequencies $n_{i1}, n_{i2}, \ldots, n_{ir}$ is *multinomial*,[2] where $\sum_{j=1}^{r} n_{ij} = n_{i\cdot}$ and $\sum_{j=1}^{r} \Pi_{ij} = 1$.

In Table 22-4 the data layout of Table 22-3 is given in terms of these unknown multinomial cell probability parameters. In the analysis of data of the general form given in Table 22-3, usually functions of the unknown cell probabilities given in Table 22-4 are estimated and hypotheses concerning these functions are tested. In fact, as will be illustrated in the next section, most analysis questions can be handled by considering either linear functions or log-linear functions of the Π_{ij}'s. The GSK approach to be discussed in Section 22-3 allows for fitting such functions to a linear model, for evaluating the goodness of fit of the model, and for testing hypotheses about the parameters of the fitted model.

Finally, before proceeding to an in-depth discussion of the GSK methodology, we will present another example. This example will illustrate some of the concepts presented in this section and also emphasize the fact that the conditions under which an experiment is conducted (e.g., the sampling scheme used to collect the data) determine whether particular dimensions in a contingency table should be considered as factors or responses (which would affect the subsequent data analysis and interpretations thereof).

Example   Consider Table 22-5, which summarizes hypothetical data from a political science study concerning opinion in a particular city of a new governmental policy by party

---

[1] The (joint) probability distribution of the responses for the binomial distribution in the ith subpopulation is given by

$$P(n_{i1}, n_{i2}) = \frac{n_{i\cdot}!}{n_{i1}! n_{i2}!} \Pi_{i1}^{n_{i1}} \Pi_{i2}^{n_{i2}}$$

[2] The (joint) probability distribution of the responses for the multinomial distribution in the ith subpopulation is given by

$$P(n_{i1}, n_{i2}, \ldots, n_{ir}) = \frac{n_{i\cdot}!}{n_{i1}! n_{i2}! \cdots n_{ir}!} \Pi_{i1}^{n_{i1}} \Pi_{i2}^{n_{i2}} \cdots \Pi_{ir}^{n_{ir}}$$

TABLE 22-5   Hypothetical data from a political
science study

Party Affiliation	Favor Policy	Do Not Favor Policy	No Opinion	Total
Democrats	200	200	100	500
Republicans	250	175	75	500
Total	450	375	175	1,000

affiliation. We shall consider two distinct sampling procedures that could have produced the data array in Table 22-5:

1.   A random sample of 1,000 individuals in the city was selected and each individual asked his or her party affiliation (for simplicity, we assume that "Democrat" and "Republican" were the only two responses) and his or her opinion concerning the new policy ("favor," "do not favor," or "no opinion").

2.   A random sample of 500 Democrats was selected from a list of registered Democrats in the city and each Democrat was asked his or her opinion concerning the new policy ("favor," "do not favor," or "no opinion"), and a completely analogous procedure was used on 500 Republicans.

Sampling scheme 1 would have elicited *two responses* from each individual (political party and policy opinion), and no marginal frequencies would have been fixed other than the total sample size of 1,000. Under this sampling scheme, then, we would consider a two-response, no-factor table with $s = 1$ subpopulation and $r = 2 \times 3 = 6$ response categories. Using the format of Tables 22-3 and 22-4, the data of Table 22-5 would look like that in Table 22-6. Here we note that $\sum_{j=1}^{6} \Pi_{1j} = 1$, since we would be dealing with only *one* multinomial population.

Sampling scheme 2 would result in *one response* (policy opinion) from each individual, and both row marginal totals would be fixed a priori. Under this sampling scheme we would consider a one-response, one-factor table with $s = 2$ subpopulations and $r = 3$ response categories. Using the format of Tables 22-3 and 22-4, we would have the representation given in Table 22-7. Here we note that $\sum_{j=1}^{3} \Pi_{1j} = 1$ and $\sum_{j=1}^{3} \Pi_{2j} = 1$, since we would be dealing with *two* multinomial populations.

To analyze the data in Table 22-5, a person with a knowledge of basic statistics would naturally think of using the standard chi-square test involving a comparison of observed (O)

TABLE 22-6   Representation of data in Table 22-5 based on a one-multinomial-population model

Subpopulation	Democrat and Favor	Democrat and Do Not Favor	Democrat and No Opinion	Republican and Favor	Republican and Do Not Favor	Republican and No Opinion	Total
No. in subpopulation	$n_{11} = 200$	$n_{12} = 200$	$n_{13} = 100$	$n_{14} = 250$	$n_{15} = 175$	$n_{16} = 75$	$n_{1.} = 1,000$
Probability	Π_{11}	Π_{12}	Π_{13}	Π_{14}	Π_{15}	Π_{16}	1

TABLE 22-7   Representation of data in Table 22-5 based on a
two-multinomial-population model

| | Three Response Categories | | | |
| | Favor Policy | Do Not Favor Policy | No Opinion | Total |
Two Subpopulations				
Democrats (subpopulation 1)				
No. in subpopulation	$n_{11} = 200$	$n_{12} = 200$	$n_{13} = 100$	$n_{1.} = 500$
Probability	Π_{11}	Π_{12}	Π_{13}	1
Republicans (subpopulation 2)				
No. in subpopulation	$n_{21} = 250$	$n_{22} = 175$	$n_{23} = 75$	$n_{2.} = 500$
Probability	Π_{21}	Π_{22}	Π_{23}	1

versus expected (E) cell frequencies. In particular, for Table 22-5, suppose that we let O_{ij} be the observed cell frequency in the ith row ($i = 1$ for Democrats, $i = 2$ for Republicans) and jth column ($j = 1$ for "favor policy," $j = 2$ for "do not favor policy," $j = 3$ for "no opinion"). Further, suppose that we let $O_{i.} = \Sigma_{j=1}^{3} O_{ij}$ be the ith row marginal total and $O_{.j} = \Sigma_{i=1}^{2} O_{ij}$ be the jth column marginal total; then it can be seen from Table 22-5 that $O_{11} = 200$, $O_{12} = 200$, $O_{13} = 100$, $O_{21} = 250$, $O_{22} = 175$, $O_{23} = 75$, $O_{1.} = O_{2.} = 500$, $O_{.1} = 450$, $O_{.2} = 375$, and $O_{.3} = 175$.

Then the expected cell frequency in the ith row and jth column under the null hypothesis of "no association between the two methods of classification" is computed as

$$E_{ij} = \frac{O_{i.}O_{.j}}{O_{..}}$$

where $O_{..} = \Sigma_{j=1}^{3} O_{.j} = \Sigma_{i=1}^{2} O_{i.} = n_{..}$ ($= 1,000$ in this example). For the data in Table 22-5 one can check that

$$E_{11} = E_{21} = \frac{500(450)}{1,000} = 225$$

$$E_{12} = E_{22} = \frac{500(375)}{1,000} = 187.5$$

$$E_{13} = E_{23} = \frac{500(175)}{1,000} = 87.5$$

Finally, the test statistic to be used here is of the form

$$\chi^2 = \sum_{i=1}^{2} \sum_{j=1}^{3} \frac{(O_{ij} - E_{ij})^2}{E_{ij}}$$

which, for our example, has the value

$$\chi^2 = \frac{(200 - 225)^2}{225} + \frac{(200 - 187.5)^2}{187.5} + \frac{(100 - 87.5)^2}{87.5} + \frac{(250 - 225)^2}{225}$$

$$+ \frac{(175 - 187.5)^2}{187.5} + \frac{(75 - 87.5)^2}{87.5}$$

$$= 2.78 + 0.83 + 1.79 + 2.78 + 0.83 + 1.79 = 10.80$$

This computed value of χ^2 is significant at the 1% level since $\chi^2_{2,0.01} = 9.21$, the degrees of freedom for χ^2 being determined as the [number of rows $-$ 1] $\times$ [number of columns $-$ 1] $=$ $(2 - 1) \times (3 - 1) = 1 \times 2 = 2$.

Thus, the null hypothesis of "no association between the two methods of classification" is rejected, but what does this statement really mean? What do we mean by the phrase "no association"? In other words, what does the vague statement "no association" mean in terms of the unknown parameters in the problem, the cell probabilities (the Π_{ij}'s)? The answers to these questions depend on the sampling scheme used to generate the data. In particular, even though the $(O - E)^2/E$ chi-square statistic given above happens to be computationally the same for both sampling schemes described earlier, the particular null hypotheses being tested via this statistic are markedly different in the two instances.

For sampling scheme 1, the null hypothesis of no association is really one of independence (i.e., we wish to investigate the possibility of a contingency between the two methods of classification and hence the term *contingency table*). More specifically, as when two events A and B are said to be independent in a probabilistic sense if $P(A \text{ and } B) = P(A)P(B)$, so here the two responses (or methods of classification) are said to be independent when the probability of an individual being in a particular response category (i.e., a cell in the original contingency table, Table 22-5) is the product of the corresponding marginal probabilities associated with that response category. This is why $E_{ij} = O_{i\cdot}O_{\cdot j}/O_{\cdot\cdot}$, since E_{ij} can be expressed as

$$E_{ij} = O_{\cdot\cdot} \left(\frac{O_{i\cdot}}{O_{\cdot\cdot}} \right)\left(\frac{O_{\cdot j}}{O_{\cdot\cdot}} \right)$$

where $O_{i\cdot}/O_{\cdot\cdot}$ and $O_{\cdot j}/O_{\cdot\cdot}$ can be looked upon as the estimated marginal probabilities of being in the ith row and jth column, respectively.

Let us pursue this point a little further. From Table 22-6, it follows that

$\Pi_D \ (= \Pi_{11} + \Pi_{12} + \Pi_{13})$ is the true (marginal) probability of being a Democrat

$\Pi_R \ (= \Pi_{14} + \Pi_{15} + \Pi_{16})$ is the true (marginal) probability of being a Republican

$\Pi_F \ (= \Pi_{11} + \Pi_{14})$ is the true (marginal) probability of favoring the policy

$\Pi_{NF} \ (= \Pi_{12} + \Pi_{15})$ is the true (marginal) probability of not favoring the policy

$\Pi_{NO} \ (= \Pi_{13} + \Pi_{16})$ is the true (marginal) probability of having no opinion on
 the policy

Then the null hypothesis H_I of independence specifies that the unknown true cell probabilities have a particular structure, namely that they are formed as products of the above marginal probabilities as follows:

H_I: $\Pi_{11} = \Pi_D \cdot \Pi_F$

  $\Pi_{12} = \Pi_D \cdot \Pi_{NF}$

  $\Pi_{13} = \Pi_D \cdot \Pi_{NO}$

  $\Pi_{14} = \Pi_R \cdot \Pi_F$

  $\Pi_{15} = \Pi_R \cdot \Pi_{NF}$

  $\Pi_{16} = \Pi_R \cdot \Pi_{NO}$

For sampling scheme 2, the null hypothesis of no association is really one of homogeneity (i.e., the two multinomial populations involved are identical in structure and hence the term *homogeneous*). More specifically, the null hypothesis H_H of homogeneity (or, equivalently, the null hypothesis of identicalness of the two multinomial distributions) means that the corresponding cell probabilities in Table 22-7 are equal:

$$H_H: \quad \Pi_{11} = \Pi_{21}$$
$$\Pi_{12} = \Pi_{22}$$
$$\Pi_{13} = \Pi_{23}$$

This is why $E_{ij} = O_{i.}O_{.j}/O_{..}$ can be expressed as

$$E_{ij} = O_{i.} \left(\frac{O_{1j} + O_{2j}}{O_{1.} + O_{2.}} \right)$$

where $(O_{1j} + O_{2j})/(O_{1.} + O_{2.})$ is the *pooled* estimate of the probability of giving the jth response under the null hypothesis that $\Pi_{1j} = \Pi_{2j}$, $j = 1, 2, 3$.

Of course, since $\Pi_{11} + \Pi_{12} + \Pi_{13} = 1$ and $\Pi_{21} + \Pi_{22} + \Pi_{23} = 1$, it follows that the conditions $\Pi_{11} = \Pi_{21}$ and $\Pi_{12} = \Pi_{22}$ together imply that $\Pi_{13} = \Pi_{23}$. Consequently, the hypothesis H_H above can be more compactly stated as

$$H'_H: \quad \Pi_{11} = \Pi_{21}$$
$$\Pi_{12} = \Pi_{22}$$

In this regard it is important in what follows to express any null hypothesis of interest with a minimum number of parameters. An equivalent but more rigorous way to say this is that the constraints on the parameters imposed by the null hypothesis must be mathematically independent of one another (e.g., the constraints given by H_H are linearly dependent, while those given by H'_H are not). Such mathematical independence is required under the GSK formulation.

Thus, it is clear that the null hypotheses of independence and homogeneity are quite different, even though the usual chi-square test is computed exactly the same way for these two hypotheses.

There is one other important point to be made about this particular example. As mentioned earlier, the GSK approach to the analysis of categorical data involves estimating and testing appropriate hypotheses about linear and/or log-linear functions of the unknown cell probabilities, since many questions in categorical data analysis involve such functions. Hence, it is of interest here to see just how the null hypotheses of independence and homogeneity discussed earlier can be represented in terms of such functions of the cell probabilities.

Now, the null hypothesis H_I of independence given earlier can be shown (e.g., see Bhapkar and Koch, 1968) to be equivalent to the null hypothesis

$$H'_I: \quad \Delta_{11} = 1$$
$$\Delta_{12} = 1$$

where

$$\Delta_{11} = \frac{\Pi_{11}\Pi_{16}}{\Pi_{14}\Pi_{13}} \quad \text{and} \quad \Delta_{12} = \frac{\Pi_{12}\Pi_{16}}{\Pi_{15}\Pi_{13}}$$

Note that under the null hypothesis H_I,

$$\Delta_{11} = \frac{(\Pi_D \cdot \Pi_F)(\Pi_R \cdot \Pi_{NO})}{(\Pi_R \cdot \Pi_F)(\Pi_D \cdot \Pi_{NO})} = 1$$

and

$$\Delta_{12} = \frac{(\Pi_D \cdot \Pi_{NF})(\Pi_R \cdot \Pi_{NO})}{(\Pi_R \cdot \Pi_{NF})(\Pi_D \cdot \Pi_{NO})} = 1$$

It is then immediately apparent that H_I' is equivalent to

H_I'': $\ln \Delta_{11} = 0$

$\ln \Delta_{12} = 0$

where

$$\ln \Delta_{11} = \ln \Pi_{11} + \ln \Pi_{16} - \ln \Pi_{14} - \ln \Pi_{13}$$

and

$$\ln \Delta_{12} = \ln \Pi_{12} + \ln \Pi_{16} - \ln \Pi_{15} - \ln \Pi_{13}$$

Thus, H_I can be converted to an equivalent null hypothesis H_I'' involving log-linear functions of the unknown cell probabilities. By a log-linear function of the Π_{ij}'s, we mean that the function is of the form $\sum_{i=1}^{s} \sum_{j=1}^{r} k_{ij} \ln \Pi_{ij}$ for suitable choices of the constants k_{ij} (for the previous example, each k_{ij} is either 1, -1, or 0, depending on the values of i and j). As we will see, the GSK approach involves estimating the log functions in H_I'' and then testing H_I'' using these estimates (which of course requires variance estimates as well).

(There is a more general form of log-linear function considered in the GSK methodology; this is a function that is linear in the logs of linear functions of the Π_{ij}, that is, a function of the general form

$$\sum_{\gamma=1}^{u} k_\gamma \ln \left(\sum_{i=1}^{s} \sum_{j=1}^{r} a_{\gamma, ij} \Pi_{ij} \right)$$

In the next section we shall illustrate the utility of such a function with some examples.)

As for the hypothesis H_H of homogeneity, it is easy to see that

H_H': $\Pi_{11} = \Pi_{21}$

$\Pi_{12} = \Pi_{22}$

can be equivalently written in the form

H_H'': $\Pi_{11} - \Pi_{21} = 0$

$\Pi_{12} - \Pi_{22} = 0$

which involves linear functions of the unknown cell probabilities.

To summarize, then, we have indicated how the hypotheses H_I and H_H of independence and homogeneity can be equivalently expressed as hypotheses H_I'' and H_H'' stating that certain linear and/or log-linear functions of the unknown cell probabilities are equal to 0. The GSK approach can be used to test these and even more complex hypotheses (e.g., those concerning regression coefficients produced by means of a weighted-least-squares fitting of a linear model to the cell probabilities or functions thereof).

22-3 GSK Linear-Models Approach

As we mentioned previously, the Grizzle–Starmer–Koch (GSK) linear-models approach to the analysis of categorical data is based on the application of general weighted-least-squares regression techniques to estimates of appropriate (linear and/or log-linear) functions of the cell proportions in complex layouts of categorical data. From these results, statistical tests of particular hypotheses of interest are directly produced. For instance, the relationship between the dichotomous dependent variable PAFC and the independent variables A, C, E, and L in the example of Section 22-1 can be investigated with this methodology, and questions regarding variable selection, model appropriateness, and interaction can be pursued in the same spirit as used in analysis of variance and regression procedures for quantitative data.

The basic requirement here for any test of hypothesis is that the sample size be sufficiently large. By this we mean that the observed frequencies (i.e., the n_{ij}'s) in Table 22-3 must each be at least 10 in value (and preferably 25 or more).[3] The only other requirement is that there be either a sampling or observational basis for arguing that the inherent variability in the set of observed frequencies for any subpopulation can be characterized in terms of a multinomial distribution (as discussed previously).[4]

We shall now describe the steps in using the GSK methodology, inserting examples where necessary for clarification.

Step 1. Formulate the appropriate multidimensional contingency table (in the form of Table 22-3) *for your problem*, specifying s (the number of subpopulations) and r (the number of response categories). The n_{ij} of this table are the data to be used in the analysis.

Step 2. Letting the Π_{ij} denote the unknown cell probabilities (see Table 22-4), *specify the functions of the* Π_{ij} *that you wish to study.* These functions will be defined as either of the following:

1. u functions $a_1(\Pi)$, $a_2(\Pi)$, . . . , $a_u(\Pi)$, each of which is a *linear function of the* Π_{ij}, that is, a function of the form

$$a_\gamma(\Pi) = \sum_{i=1}^{s} \sum_{j=1}^{r} a_{\gamma, ij} \Pi_{ij} \qquad \gamma = 1, 2, \ldots, u$$

for suitable choices of the $a_{\gamma, ij}$, or

---

[3] The GSK weighted-least-squares approach to estimation and testing suffers when several of the n_{ij} are equal to 0. The effects of replacing such zero values for observed cell counts by some small positive value, as has been advocated by Grizzle, Starmer, and Koch, has not been definitively investigated.

[4] Under the multinomial structure, the research questions of interest are based entirely on the study of selected functions of estimates of the Π_{ij}, where the estimate of Π_{ij} is given by the relative frequency $p_{ij} = n_{ij}/n_{i\cdot}$, and the variances and covariances of these functions of the p_{ij} are themselves functions of the variances and covariances of the p_{ij}.

2. t functions $f_1(\Pi), f_2(\Pi), \ldots, f_t(\Pi)$, each of which is a *log-linear function of the Π_{ij}*, that is, a function of the form

$$f_\alpha(\Pi) = \sum_{\gamma=1}^{u} k_{\alpha\gamma} \ln a_\gamma(\Pi) \qquad \alpha = 1, 2, \ldots, t$$

for suitable choices of the $k_{\alpha\gamma}$, where the $a_\gamma(\Pi)$ are the linear functions defined in definition 1.

Examples

1. For the IFRP study data of Table 22-2, $s = 16$, $r = 2$, and the $u = 16$ linear functions of interest are $a_\gamma(\Pi) = \Pi_{\gamma 1}$, $\gamma = 1, 2, \ldots, 16$, which are the true probabilities of accepting fertility control for the 16 subpopulations. The $a_{\gamma, ij}$ for each linear function is given by

$$a_{\gamma, ij} = \begin{cases} 1 & \text{if } i = \gamma, j = 1 \\ 0 & \text{otherwise} \end{cases}$$

2. For the political science data of Table 22-7, $s = 2$, $r = 3$, and the homogeneity null hypothesis H''_H involves the $u = 2$ *linear functions* $a_1(\Pi) = \Pi_{11} - \Pi_{21}$ and $a_2(\Pi) = \Pi_{12} - \Pi_{22}$. Each function is of the form, $a_\gamma(\Pi) = \sum_{i=1}^{2} \sum_{j=1}^{3} a_{\gamma, ij} \Pi_{ij}$, where the $a_{\gamma, ij}$ are given by $a_{1,11} = 1$, $a_{1,21} = -1$, $a_{1,12} = a_{1,13} = a_{1,22} = a_{1,23} = 0$ for $a_1(\Pi)$, and $a_{2,12} = 1$, $a_{2,22} = -1$, $a_{2,11} = a_{2,13} = a_{2,21} = a_{2,23} = 0$ for $a_2(\Pi)$.

3. For the political science data of Table 22-6, $s = 1$, $r = 6$, and the independence null hypothesis H''_I involves the $t = 2$ *log-linear functions*

$$f_1(\Pi) = \ln \Pi_{11} + \ln \Pi_{16} - \ln \Pi_{14} - \ln \Pi_{13}$$

and

$$f_2(\Pi) = \ln \Pi_{12} + \ln \Pi_{16} - \ln \Pi_{15} - \ln \Pi_{13}$$

Each of these functions is of the general form

$$f_\alpha(\Pi) = \sum_{\gamma=1}^{u} k_{\alpha\gamma} \ln a_\gamma(\Pi)$$

where $u = 6$, $a_\gamma(\Pi) = \sum_{j=1}^{6} a_{\gamma, 1j} \Pi_{1j} = \Pi_{1\gamma}$, so

$$a_{\gamma, 1j} = \begin{cases} 1 & \text{if } j = \gamma \\ 0 & \text{otherwise} \end{cases}$$

and the $k_{\alpha\gamma}$ are

$$\begin{cases} \text{for } f_1(\Pi): & k_{11} = 1, k_{12} = 0, k_{13} = -1, k_{14} = -1, k_{15} = 0, k_{16} = 1 \\ \text{for } f_2(\Pi): & k_{21} = 0, k_{22} = 1, k_{23} = -1, k_{24} = 0, k_{25} = -1, k_{26} = 1 \end{cases}$$

Step 3. Determine the point estimates of the functions $a_\gamma(\Pi)$ or $f_\alpha(\Pi)$ defined in step 2 by substituting $p_{ij} = n_{ij}/n_i$. for Π_{ij} everywhere it appears in each function. (These point estimates are printed out by the GSK computer program GENCAT, which we shall discuss in the next section.)

Step 4 (optional).[5] When considering a single or multifactor problem, *postulate a linear model for the parameters of interest* (i.e., the functions of the Π_{ij} defined in step 2) *in terms of factor effects,* as follows:

Let $g_1(\Pi), g_2(\Pi), \ldots, g_m(\Pi)$ denote the set of functions specified in step 2 (i.e., $m = u$ or t, depending on whether the functions are linear or log-linear, respectively). Define your model by specifying v (the number of parameters in your model) and the $x_{\xi l}$ ($\xi = 1, 2, \ldots, m; l = 1, 2, \ldots, v$), so that the model is of the form

$$g_\xi(\Pi) = x_{\xi 1}\beta_1 + x_{\xi 2}\beta_2 + \cdots + x_{\xi v}\beta_v \qquad \xi = 1, 2, \ldots, m$$

Weighted-least-squares procedures are used in the GSK approach to estimate the unknown β's in the above model. Even the multiresponse, no-factor situation can be couched in this linear-model framework by considering the trivial case where $v = 0$ or all $x_{\xi l}$ are 0. The $x_{\xi l}$'s are constants whose specification depends on the nature of the data set, on the structure of the g_ξ's, and on the particular research questions of interest. The weighted-least-squares analysis produces estimates $b_1, b_2, \ldots, b_v$ of $\beta_1, \beta_2, \ldots, \beta_v$ by fitting the model above to the observed responses $g_\xi(p)$, which are obtained by replacing Π_{ij} by p_{ij} in each $g_\xi(\Pi)$.

> (We have briefly discussed weighted least squares in its most elementary form (straight-line regression) in Chapter 12. For the more mathematically sophisticated, we point out that the general form of the weighted-least-squares solution referred to above can be expressed in matrix notation as
>
> $$\mathbf{b} = (\mathbf{X}'\hat{\boldsymbol{\Sigma}}^{-1}\mathbf{X})^{-1}\mathbf{X}'\hat{\boldsymbol{\Sigma}}^{-1}\mathbf{g}(p)$$
>
> where $\mathbf{b} = (b_1, b_2, \ldots, b_v)'$ is the vector of estimated coefficients, $\mathbf{X}$ is the $m \times v$ matrix of the $x_{\xi l}$'s, $\hat{\boldsymbol{\Sigma}}$ is the matrix of estimated variances and covariances of the estimated functions $g_\xi(p)$, and $\mathbf{g}(p) = [g_1(p), g_2(p), \ldots, g_m(p)]'$. For a brief introduction to matrix mathematics, the reader is referred to Appendix B.)

The *predicted value* $\hat{g}_\xi(p)$ using the fitted model is then given by

$$\hat{g}_\xi(p) = x_{\xi 1}b_1 + x_{\xi 2}b_2 + \cdots + x_{\xi v}b_v$$

Example   For the IFRP data (see Table 22-2), one postulated model for the Π_{i1} contains an *overall mean effect* (μ), *the four main effects* ($\alpha_A, \alpha_C, \alpha_E, \alpha_L$), and the six *two-factor interaction effects* ($\gamma_{AC}, \gamma_{AE}, \gamma_{AL}, \gamma_{CE}, \gamma_{CL}, \gamma_{EL}$). In equation form, this model is written

$$\Pi_{i1} = x_0\mu + x_A\alpha_A + x_C\alpha_C + x_E\alpha_E + x_L\alpha_L + x_Ax_C\gamma_{AC} + x_Ax_E\gamma_{AE}$$
$$+ x_Ax_L\gamma_{AL} + x_Cx_E\gamma_{CE} + x_Cx_L\gamma_{CL} + x_Ex_L\gamma_{EL} \qquad i = 1, 2, \ldots, 16$$

where

$$x_0 \equiv 1 \qquad x_A = \begin{cases} 1 & \text{if } A = 25+ \\ 0 & \text{if } A = {<}25 \end{cases} \qquad x_C = \begin{cases} 1 & \text{if } C = 2+ \\ 0 & \text{if } C = 0\text{–}1 \end{cases}$$

$$x_E = \begin{cases} 1 & \text{if } E = 7+ \\ 0 & \text{if } E = 0\text{–}6 \end{cases} \qquad x_L = \begin{cases} 1 & \text{if } L = \text{Yugoslavia} \\ 0 & \text{if } L = \text{India} \end{cases}$$

---

[5] This step may be skipped, for example, in a multiresponse, no-factor situation (e.g., a test for independence), since it is only of interest in such a case to test whether the functions of the Π_{ij} defined in step 2 are equal to 0.

Note that the x's are dummy variables, defined as in regression analysis. Analogously, then, for a factor having k categories, $k - 1$ dummy variables would be needed.

Step 5. State the null hypotheses that you wish to test in terms of the m functions $g_1(\Pi)$, $g_2(\Pi)$, ..., $g_m(\Pi)$ or in terms of the β's if you have postulated a linear model in step 4, and then perform the appropriate tests using the GSK methodology.

If you hypothesize $H_0: g_1(\Pi) = g_2(\Pi) = \cdots = g_m(\Pi) = 0$, the test of this hypothesis uses a large-sample chi-square statistic with m degrees of freedom (which is printed out by the GSK computer program).

If you have postulated a linear model in step 4:

1. A test of the goodness of fit (GOF) of this model can be carried out using a large-sample chi-square statistic (printed out by the program) with $m - v$ degrees of freedom, where v is the number of parameters in the model. The null hypothesis for this test is H_0: "the model provides a good fit to the data"; if the model does actually describe the data well, the differences between the observed $g_\xi(p)$'s and the predicted $\hat{g}_\xi(p)$'s based on the fitted model will be small, and the goodness-of-fit chi-square statistic will be nonsignificant. If the goodness-of-fit statistic is significant, however, the discrepancies between observed and predicted values (i.e., the residuals) are sufficiently large to warrant fitting an alternative model.

2. A test of any linear null hypothesis about the β's in the postulated model may be carried out using a chi-square statistic with d degrees of freedom when the null hypothesis is specified in terms of d (linearly independent) functions of the β's of the following form:

$$H_0: \quad c_{11}\beta_1 + c_{12}\beta_2 + \cdots + c_{1v}\beta_v = 0$$
$$c_{21}\beta_1 + c_{22}\beta_2 + \cdots + c_{2v}\beta_v = 0$$
$$\vdots$$
$$c_{d1}\beta_1 + c_{d2}\beta_2 + \cdots + c_{dv}\beta_v = 0$$

where the c's are constants whose values can be specified to form particular null hypotheses of interest. For example, the null hypothesis $H_0: \beta_1 = 0$ can be specified

TABLE 22-8   Typical ANOVA table for categorical data analysis

Source	df	χ^2
Main effect A (l_A levels)	$l_A - 1$	$\chi^2(A)$
Main effect B (l_B levels)	$l_B - 1$	$\chi^2(B)$
$\vdots$	$\vdots$	$\vdots$
Interaction effect AB	$(l_A - 1)(l_B - 1)$	$\chi^2(AB)$
$\vdots$	$\vdots$	$\vdots$
Residual (goodness of fit)	$m - v$	χ^2

by letting $d = 1$, $c_{11} = 1$, and all the other c's equal 0; the null hypothesis $H_0: \beta_1 = \beta_2 = \beta_3$ (or, equivalently, $H_0: \beta_1 - \beta_2 = 0$, $\beta_1 - \beta_3 = 0$) can be specified by letting $d = 2$, $c_{11} = 1$, $c_{12} = -1$, $c_{21} = 1$, $c_{23} = -1$, and all the other c's equal 0.

Thus, if the linear model is of the usual ANOVA regression model form, separate chi-square tests for main effects and interactions can be made by specifying the appropriate subsets of the β's that represent each effect and then assigning appropriate values for the c's. Furthermore, the results of each test may be summarized in the form of an ANOVA table (Table 22-8), as with regression analysis (although it should be recognized that the separate χ^2-values in such a table are not additive).

Examples

1. For the political science data of Table 22-5, a test of the null hypothesis of homogeneity $H_H'': \Pi_{11} - \Pi_{21} = 0$ and $\Pi_{12} - \Pi_{22} = 0$ using the GSK methodology yields a chi-square statistic (with $u = 2$ degrees of freedom) having the value 10.91. A GSK test of the null hypothesis of independence

    $$H_I'': \quad \ln \Pi_{11} + \ln \Pi_{16} - \ln \Pi_{14} - \ln \Pi_{13} = 0$$
    $$\ln \Pi_{12} + \ln \Pi_{16} - \ln \Pi_{15} - \ln \Pi_{13} = 0$$

    yields a chi-square statistic (with $t = 2$ degrees of freedom) having the value 10.74. Note that the $(O - E)^2/E$ chi-square statistic described in Section 22-2 had a value 10.80 based on 2 degrees of freedom.

    (The reader is no doubt wondering at this time why the two GSK statistics and the $(O - E)^2/E$ statistic all have slightly different values in this example. Without being too mathematical, we will try to explain. First, the GSK statistics are actually based on the use of an $(O - E)^2/O$ criterion, which was suggested by Neyman (1949) as a modification of the usual $(O - E)^2/E$ statistic. This modified statistic has a chi-square distribution for large samples if the postulated model is true. Bhapkar (1966) has pointed out the relationship between the modified chi-square statistic and a Wald statistic (1943) calculated using a weighted-least-squares regression algorithm (as employed in the GSK approach). When the constraints on the cell probabilities imposed under a particular null hypothesis are linear (e.g., as they are under H_H''), the Neyman modified chi-square statistic is algebraically the same as the Wald statistic. On the other hand, when the constraints on the cell probabilities imposed under a particular null hypothesis are nonlinear (e.g., as with the log-linear functions given under H_I''), the Wald statistic is algebraically equivalent to the Neyman statistic based on estimates of these nonlinear functions using a linear approximation to these constraints.)

2. For the IFRP data (Table 22-2) and the model postulated in the example given after step 4, separate tests for main effects and interactions can be carried out using the GSK approach to yield the summary table given as Table 22-9. From this table it may be concluded that only the main effect of variable C is significant ($\chi^2 = 21.86$

TABLE 22-9   ANOVA table for the data in Table 22-2

Source		df	χ^2	P
Main effects	A $(H_0: \alpha_A = 0)$	1	2.59	.11
	C $(H_0: \alpha_C = 0)$	1	21.86**	.00
	E $(H_0: \alpha_E = 0)$	1	1.76	.19
	L $(H_0: \alpha_L = 0)$	1	0.94	.33
Two-factor interactions	AC $(H_0: \gamma_{AC} = 0)$	1	2.32	.13
	AE $(H_0: \gamma_{AE} = 0)$	1	0.00	1.00
	AL $(H_0: \gamma_{AL} = 0)$	1	0.28	.60
	CE $(H_0: \gamma_{CE} = 0)$	1	1.58	.21
	CL $(H_0: \gamma_{CL} = 0)$	1	2.71	.10
	EL $(H_0: \gamma_{EL} = 0)$	1	1.49	.22
All two-factor interactions		6	10.58	.10
Residual (goodness of fit)		5	3.92	.56

with 1 degree of freedom), and that the goodness-of-fit test is not significant ($\chi^2 = 3.92$ with 5 degrees of freedom).[6]

Step 6. Estimate parameters of interest by computing appropriate confidence intervals.
If you have tested $H_0: g_1(\Pi) = g_2(\Pi) = \cdots = g_m(\Pi) = 0$ in step 5 and also wish to estimate one or more of the m parameters $g_1(\Pi), g_2(\Pi), \ldots, g_m(\Pi)$, the following general confidence interval formula may be used:

$$g_\xi(p) \pm z_{1-\alpha/2} S_{g_\xi(p)} \qquad \xi = 1, 2, \ldots, m$$

where $g_\xi(p)$ is the estimate of $g_\xi(\Pi)$ obtained by replacing Π_{ij} by $p_{ij} = n_{ij}/n_{i\cdot}$, and where $S_{g_\xi(p)}$ is the estimated standard error of $g_\xi(p)$ (the value of which is printed out by the GSK program GENCAT).

If you have postulated a linear model in step 4, then GENCAT will print out weighted-least-squares estimates $b_1, b_2, \ldots, b_v$ of $\beta_1, \beta_2, \ldots, \beta_v$. Confidence intervals for the β's may be obtained from the b's using the formula

$$b_\ell \pm z_{1-\alpha/2} S_{b\ell}$$

where $S_{b\ell}$ denotes the estimated standard error of b_ℓ.

Also, a comparison of the observed values $g_\xi(p)$ with the predicted values $\hat{g}_\xi(p)$ can be made using the residuals

$$g_\xi(p) - \hat{g}_\xi(p) \qquad \xi = 1, 2, \ldots, m$$

---

[6] As with the unequal-cell-number ANOVA situation (see Chapter 20), care must be taken in interpreting main-effect test results when interaction effects are included in the fitted model. In particular, it is recommended (as before) that in model building the highest-order effects in the model be examined at each stage and then the model reduced to simpler form only when such effects are nonsignificant.

22-4 GSK Computer Program

The GSK linear-models approach to the analysis of categorical data, as described above in conceptual terms, can be implemented in practice by employing a computer program called GENCAT, written and used extensively in the biostatistics department at the University of North Carolina.[7] The user of this program must first enter the categorical data set under study into the system according to the layout in Table 22-3. Then, to generate estimates and tests of hypotheses concerning appropriate functions of interest, the user has to specify values for the various sets of constants described in Section 22-3. In particular, the examination of linear functions (the $a_\gamma(\Pi)$'s) requires specification of the $\{a_{\gamma,ij}^-\}$, while consideration of log-linear functions (the $f_\alpha(\Pi)$'s) involves giving values to the $\{k_{\alpha\gamma}\}$ as well. Fitting a linear model to either the $a_\gamma(p)$'s or the $f_\alpha(p)$'s means assigning values to the $\{x_{\xi\ell}\}$. The subsequent testing of hypotheses about the β's necessitates choosing appropriate values for the c's.

Once the user has appropriately specified these sets of constants, the computer will perform the necessary computations and then print out the desired estimates and chi-square test statistics directly. Of course, to specify these sets of constants properly requires some sophistication, which can only be cultivated through understanding and experience. Indeed, to perform an in-depth analysis of a complex set of categorical data using the GSK methodology requires considerable expertise, and we hope that by studying examples of some of these analysis techniques, the reader will appreciate the care needed in interpreting the test results.

Example   The purpose here is to briefly summarize the analysis of the political science data in Table 22-5, which has already been discussed in considerable detail. Specifically, the null hypothesis of independence,

$$H_I'': \quad f_1(\Pi) = \ln \Delta_{11} = 0, \quad f_2(\Pi) = \ln \Delta_{12} = 0$$

and the null hypothesis of homogeneity,

$$H_H'': \quad a_1(\Pi) = (\Pi_{11} - \Pi_{21}) = 0, \quad a_2(\Pi) = (\Pi_{12} - \Pi_{22}) = 0$$

involve log-linear and linear functions of the Π_{ij}'s, respectively, and the sets of constants $\{a_{\gamma,ij}\}$ and $\{k_{\alpha\gamma}\}$ needed to form these particular functions have been specified earlier. The GSK methodology produces chi-square statistics (each with 2 degrees of freedom) having the values 10.74 and 10.91 for testing H_I'' and H_H'', respectively; the commonly used $(O - E)^2/E$ statistic has the value 10.80. We have already discussed why these three statistics differ slightly in value.

In epidemiology, a *case–control study* is an inquiry involving groups of individuals selected on the basis of whether they do (the cases) or do not (the controls) have a particular disease of interest; the two groups of cases and controls are then compared with respect to an

---

[7] This computer program can be obtained from the Program Librarian, Department of Biostatistics, School of Public Health, University of North Carolina, Chapel Hill, N.C. 27514. For a program description, see Landis et al. (1976).

TABLE 22-10 Data layout for a case–control study

Two Subpopulations	Two Response Categories		Total
	Exposed	Not Exposed	
Cases			
No. in subpopulation	n_{11}	n_{12}	$n_{1\cdot}$
Probability	Π_{11}	Π_{12}	1
Controls			
No. in subpopulation	n_{21}	n_{22}	$n_{2\cdot}$
Probability	Π_{21}	Π_{22}	1

existing or a past exposure to a particular characteristic judged to be of possible relevance to the etiology of the disease. Commonly, in a case–control study, a specific hypothesis is tested—for example, that a connection exists between lung cancer and cigarette smoking or between congenital malformation and maternal rubella during pregnancy.

The data from a case–control study can best be represented via a two-way layout, as indicated in Table 22-10. This is a one-response, one-factor table, where the dichotomous response ($r = 2$) is either "exposed" or "not exposed," and the factor has two levels corresponding to the ($s = 2$) subpopulations of cases and controls. Here we have random samples of $n_{1\cdot}$ cases and $n_{2\cdot}$ controls; n_{11} of the $n_{1\cdot}$ cases have the exposure characteristic, while n_{21} of the $n_{2\cdot}$ controls have been exposed. Note that $\Pi_{11} + \Pi_{12} = 1$ and that $\Pi_{21} + \Pi_{22} = 1$, in accord with our two-population model.

Now, a parameter of more than just casual interest to epidemiologists involved in a case–control study is the *odds ratio* (see Cornfield, 1956), which is defined as

$$\text{OR} = \frac{\Pi_{11}/\Pi_{12}}{\Pi_{21}/\Pi_{22}} = \frac{\Pi_{11}\Pi_{22}}{\Pi_{12}\Pi_{21}}$$

As its name implies, OR is the ratio of the odds for a case being exposed as opposed to not being exposed (Π_{11}/Π_{12}) to the corresponding odds for a control (Π_{21}/Π_{22}).

The specific null hypothesis to be tested here is

$$H_0: \quad \text{OR} = \frac{\Pi_{11}\Pi_{22}}{\Pi_{12}\Pi_{21}} = 1$$

versus the alternative

$$H_A: \quad \text{OR} > 1$$

By taking natural logarithms, we can write H_0 as

$$H_0': \quad \ln \text{OR} = \ln \Pi_{11} + \ln \Pi_{22} - \ln \Pi_{12} - \ln \Pi_{21} = 0$$

which is in a form amenable to investigation by the GSK approach.

In particular, H_0' involves exactly one log-linear function of the Π_{ij}'s, which (for $s = 2$, $r = 2$, and $u = 4$) is of the general form

$$f_1(\Pi) = \sum_{\gamma=1}^{4} k_{1\gamma} \ln a_\gamma(\Pi)$$

where

$$a_\gamma(\Pi) = \sum_{i=1}^{2} \sum_{j=1}^{2} a_{\gamma,ij} \Pi_{ij} = a_{\gamma,11}\Pi_{11} + a_{\gamma,12}\Pi_{12} + a_{\gamma,21}\Pi_{21} + a_{\gamma,22}\Pi_{22}$$

for $\gamma = 1, 2, 3, 4$. Now $a_1(\Pi) = \Pi_{11}$ if $a_{1,11} = 1$ and $a_{1,12} = a_{1,21} = a_{1,22} = 0$, and we can similarly arrange things so that $a_2(\Pi) = \Pi_{12}$, $a_3(\Pi) = \Pi_{21}$, and $a_4(\Pi) = \Pi_{22}$. So $f_1(\Pi) =$ ln OR if we then take $k_{11} = 1$, $k_{12} = -1$, $k_{13} = -1$, and $k_{14} = 1$. Based on these specifications, the GSK methodology produces a chi-square statistic based on 1 degree of freedom for testing H_0'. Also, because the point estimate $\widehat{\ln OR}$ of ln OR is given, along with its estimated standard error, a confidence interval for OR can be developed. As we know, this point estimate is obtained by using $p_{ij} = n_{ij}/n_{i\cdot}$ for Π_{ij} ($i = 1, 2$ and $j = 1, 2$) in the expression for ln OR:

$$f_1(p) = \widehat{\ln OR} = \ln p_{11} + \ln p_{22} - \ln p_{12} - \ln p_{21}$$

It can also be shown that an *explicit* expression for the estimated standard error of $\widehat{\ln OR}$ as computed by the GSK approach is

$$\text{s.e.}(\widehat{\ln OR}) = \sqrt{\frac{1}{n_{11}} + \frac{1}{n_{12}} + \frac{1}{n_{21}} + \frac{1}{n_{22}}}$$

which is a large-sample approximation to the actual standard error. Then, based on the reasonable assumption that $\widehat{\ln OR}$ is approximately normally distributed for large samples, it follows that an approximate $100(1 - \alpha)\%$ confidence interval for ln OR is given by

$$P(L \leq \ln OR \leq U) = 1 - \alpha$$

where

$$L = \widehat{\ln OR} - z_{1-\alpha/2} \sqrt{\frac{1}{n_{11}} + \frac{1}{n_{12}} + \frac{1}{n_{21}} + \frac{1}{n_{22}}}$$

and

$$U = \widehat{\ln OR} + z_{1-\alpha/2} \sqrt{\frac{1}{n_{11}} + \frac{1}{n_{12}} + \frac{1}{n_{21}} + \frac{1}{n_{22}}}$$

so that

$$P(e^L \leq OR \leq e^U) = 1 - \alpha$$

gives the corresponding large sample $100(1 - \alpha)\%$ confidence interval for OR. We shall now consider a specific numerical example.

Example   Table 22-11 records the frequency of past tonsillectomy (the exposure variable) among cases of Hodgkin's disease and their siblings (the controls).[8] Interest is in

---

[8] These data were taken from Johnson and Johnson (1972). Since the controls were actually siblings of the cases, a matched analysis (see Fleiss, 1973) would likely be more appropriate than an analysis in which all observations in the table are assumed to be independent. Nevertheless, we have proceeded under this assumption for illustrative purposes.

TABLE 22-11   Frequency of past tonsillectomy among cases of Hodgkin's disease and their siblings

Two Subpopulations	History of Tonsillectomy		Total
	Yes	No	
Cases	$n_{11} = 90$	$n_{12} = 84$	$n_{1.} = 174$
Controls	$n_{21} = 165$	$n_{22} = 307$	$n_{2.} = 472$

determining whether there is evidence that having had a tonsillectomy increases the risk of developing Hodgkin's disease.

The GSK chi-square statistic for testing H_0' has the value 14.72 with 1 degree of freedom, which is highly significant and thus provides strong evidence that the true value of the OR is greater than 1. Because the point estimate of ln OR as provided by the GSK methodology is

$$\ln OR = \ln \frac{90}{174} + \ln \frac{307}{472} - \ln \frac{84}{174} - \ln \frac{165}{472} = 0.688$$

the point estimate of OR is taken to be

$$e^{0.688} = \frac{90(307)}{84(165)} = 1.99$$

and the estimated standard error of ln OR is

$$\text{s.e.}(\ln OR) = \sqrt{\frac{1}{90} + \frac{1}{84} + \frac{1}{165} + \frac{1}{307}} = 0.180$$

Then, for an approximate 95% confidence interval for OR, we have

$$e^L = 1.99e^{-1.96(0.180)} = 1.99e^{-0.353} = 1.40$$

and

$$e^U = 1.99e^{1.96(0.180)} = 1.99e^{0.353} = 2.83$$

so that (approximately)

$$1.40 \le OR \le 2.83$$

Before leaving this example, one final concept deserves some attention. Our reason for working with H_0', in which a log-linear function is considered, is that we obtain directly as output the value of ln OR along with an estimate of its standard error, thus enabling us to construct a confidence interval for OR as indicated above. However, if we are interested only in testing the null hypothesis H_0 and not concerned specifically with making inferences regarding the odds ratio parameter OR, we can test H_0 in a way that involves a linear function of the cell probabilities. This approach is to be preferred because no approximations are required when dealing with linear functions as they are with log-linear functions. In particular, since OR = 1 is equivalent to $\Pi_{11}\Pi_{22} = \Pi_{12}\Pi_{21}$, or $\Pi_{11}(1 - \Pi_{21}) = (1 - \Pi_{11})\Pi_{21}$, it follows that testing H_0: OR $= 1$ is equivalent to testing the null hypothesis H_0'': $\Pi_{11} = \Pi_{21}$. Thus, H_0'' is easily seen to be a null hypothesis of homogeneity of the binomial populations for cases and controls, and the GSK chi-square statistic for testing H_0'' has the value 14.67 with 1 degree of freedom. Again, because of the reasons given previously, this value is slightly different from both the GSK chi-square statistic for testing H_0' (which has the value 14.72) and from the $(O - E)^2/E$ chi-square statistic (which has the value 14.96).

Example   This example is taken from the 1969 paper of Grizzle, Starmer, and Koch. The data in Table 22-12 are a tabulation of the severity of the "dumping syndrome," an undesirable consequence of surgery for duodenal ulcer. The four surgical procedures considered are: drainage and vagotomy (variable A), 25% resection (antrectomy) and vagotomy (variable B), 50% resection (hemigastrectomy) and vagotomy (variable C), and 75% resection (variable D). Table 22-12 describes a one-response, two-factor situation, the factors being "hospitals" (designated 1, 2, 3, 4) and "surgical procedures" (designated A, B, C, D), and the response being the "clinical evaluation of severity" of the dumping syndrome (with the $r = 3$ levels "none," "slight," and "moderate"). Thus, we have a total of $s = 4 \times 4 = 16$ trinomial populations, corresponding to the four hospitals performing each of the four surgical procedures.

The purpose of the analysis is to assess the relationship between the two factors and the response, as one would strive to do via analysis of variance or regression procedures applied to quantitative data. To proceed further we need to convert the three responses for each hospital–surgical procedure combination to a single score. If we assign the response categories "none," "slight," and "moderate" the weights 1, 2, and 3, respectively, a reasonable score for the γth hospital–surgical procedure combination ($\gamma = 1, 2, \ldots, 16$) is the linear function

$$a_\gamma(p) = 1p_{\gamma1} + 2p_{\gamma2} + 3p_{\gamma3}$$

where $p_{\gamma1}$ is the estimated probability of the response category "none," $p_{\gamma2}$ is the estimated probability of the response category "slight," and $p_{\gamma3}$ is the estimated probability of the

TABLE 22-12   Data from study of severity of dumping syndrome

	Hospital															
	1				2				3				4			
Clinical Evaluation of Severity	Surgical Procedure				Surgical Procedure				Surgical Procedure				Surgical Procedure			
	A	B	C	D	A	B	C	D	A	B	C	D	A	B	C	D
None	23	23	20	24	18	18	13	9	8	12	11	7	12	15	14	13
Slight	7	10	13	10	6	6	13	15	6	4	6	7	9	3	8	6
Moderate	2	5	5	6	1	2	2	2	3	4	2	4	1	2	3	4
Total	32	38	38	40	25	26	28	26	17	20	19	18	22	20	25	23
Score*	1.3	1.5	1.6	1.6	1.3	1.4	1.6	1.7	1.7	1.6	1.5	1.8	1.5	1.4	1.6	1.6

* Equals $1p_{\gamma1} + 2p_{\gamma2} + 3p_{\gamma3}$.

response category "moderate."[9] Here $a_\gamma(p)$ is estimating the parametric function $a_\gamma(\Pi) = 1\Pi_{\gamma 1} + 2\Pi_{\gamma 2} + 3\Pi_{\gamma 3}$.

The particular form for $a_\gamma(p)$ is obtained from the general expression

$$a_\gamma(p) = \sum_{i=1}^{16} \sum_{j=1}^{3} a_{\gamma, ij} p_{ij}$$

given earlier by taking

$$a_{\gamma, ij} = \begin{cases} 1 & \text{when } \gamma = i, j = 1 \\ 2 & \text{when } \gamma = i, j = 2 \\ 3 & \text{when } \gamma = i, j = 3 \end{cases}$$

and setting $a_{\gamma, ij} = 0$ if $\gamma \neq i$. For example, when $\gamma = 1$, then $a_{1,11} = 1$, $a_{1,12} = 2$, $a_{1,13} = 3$, and $a_{1,21} = a_{1,22} = a_{1,23} = \cdots = a_{1,16,1} = a_{1,16,2} = a_{1,16,3} = 0$, so

$$a_1(p) = 1p_{11} + 2p_{12} + 3p_{13}$$

which has the specific value

$$a_1(p) = 1\left(\frac{23}{32}\right) + 2\left(\frac{7}{32}\right) + 3\left(\frac{2}{32}\right) = 1.3$$

The numerical values for the $u = 16$ linear functions $a_1(p), a_2(p), \ldots, a_{16}(p)$ are given in the last line of Table 22-12. We shall attempt next to describe the variation in these 16 scores by means of a regression model involving the "hospitals" and "surgical procedures" factors.

The regression model that we propose to fit by weighted least squares to the $a_\gamma(p)$'s is of the general form

$$a_\gamma(\Pi) = x_{\gamma 0}\mu + x_{\gamma 1}\alpha_1 + x_{\gamma 2}\alpha_2 + x_{\gamma 3}\alpha_3 + x_{\gamma A}\tau_A + x_{\gamma B}\tau_B + x_{\gamma C}\tau_C$$

where μ = overall mean effect; α_1, α_2, and α_3 are the differential effects of hospitals 1, 2, and 3, respectively, and τ_A, τ_B, and τ_C are the differential effects of surgical procedures A, B, and C, respectively.

An explanation of how the x's are defined is best given after seeing Table 22-13, which lists the values of the x's associated with each of the 16 scores. From this tabulation one can see that the regression model we are considering is completely equivalent to a two-way ANOVA model with no interaction, where the two ways of classification are by "hospitals" and by "surgical procedures." For example, the true score for hospital 1 and surgical procedure A, $a_1(\Pi) = 1\Pi_{11} + 2\Pi_{12} + 3\Pi_{13}$, is modeled as

$$a_1(\Pi) = (1)\mu + (1)\alpha_1 + (0)\alpha_2 + (0)\alpha_3 + (1)\tau_A + (0)\tau_B + (0)\tau_C = \mu + \alpha_1 + \tau_A$$

and, similarly, $a_{11}(\Pi) = \mu + \alpha_3 + \tau_C$.

---

[9] We recognize that using equally spaced numbers to describe the response categories may be too crude a scoring scheme to accurately compare these categories on an ordinal scale. Other scoring procedures might be considered, one of which is called *ridit analysis* (see Bross, 1958). Also, the "moderate" and "slight" categories could be pooled together to create a simple binomial response.

TABLE 22-13   Values of dependent and independent variables for data in Table 22-12

Hospital–Surgical Procedure Combination	Score $(a_\gamma(p))$	Independent Variable						
		$x_{\gamma 0}$	$x_{\gamma 1}$	$x_{\gamma 2}$	$x_{\gamma 3}$	$x_{\gamma A}$	$x_{\gamma B}$	$x_{\gamma C}$
1,A	$a_1(p) = 1.3$	1	1	0	0	1	0	0
1,B	$a_2(p) = 1.5$	1	1	0	0	0	1	0
1,C	$a_3(p) = 1.6$	1	1	0	0	0	0	1
1,D	$a_4(p) = 1.6$	1	1	0	0	-1	-1	-1
2,A	$a_5(p) = 1.3$	1	0	1	0	1	0	0
2,B	$a_6(p) = 1.4$	1	0	1	0	0	1	0
2,C	$a_7(p) = 1.6$	1	0	1	0	0	0	1
2,D	$a_8(p) = 1.7$	1	0	1	0	-1	-1	-1
3,A	$a_9(p) = 1.7$	1	0	0	1	1	0	0
3,B	$a_{10}(p) = 1.6$	1	0	0	1	0	1	0
3,C	$a_{11}(p) = 1.5$	1	0	0	1	0	0	1
3,D	$a_{12}(p) = 1.8$	1	0	0	1	-1	-1	-1
4,A	$a_{13}(p) = 1.5$	1	-1	-1	-1	1	0	0
4,B	$a_{14}(p) = 1.4$	1	-1	-1	-1	0	1	0
4,C	$a_{15}(p) = 1.6$	1	-1	-1	-1	0	0	1
4,D	$a_{16}(p) = 1.6$	1	-1	-1	-1	-1	-1	-1

What about the -1 values appearing in the table? These values arise because of the restrictions

$$\alpha_1 + \alpha_2 + \alpha_3 + \alpha_4 = 0 \quad \text{and} \quad \tau_A + \tau_B + \tau_C + \tau_D = 0$$

or, equivalently,

$$\alpha_4 = -\alpha_1 - \alpha_2 - \alpha_3 \quad \text{and} \quad \tau_D = -\tau_A - \tau_B - \tau_C$$

The reader will remember from our earlier discussions of analysis of variance that these standard linear restrictions are imposed because there are only 3 degrees of freedom for "hospitals" and 3 degrees of freedom for "surgical procedures." Consequently, only three α effects (in our case, α_1, α_2, and α_3) and only three τ effects (in our case, τ_A, τ_B, and τ_C) can appear *explicitly* in the model; the remaining effects (in our case, α_4 and τ_D) are linear functions of the effects in the model. For example, from Table 22-13, the true score $a_{16}(\Pi)$ for hospital 4 and surgical procedure D is given by

$$a_{16}(\Pi) = (1)\mu + (-1)\alpha_1 + (-1)\alpha_2 + (-1)\alpha_3 + (-1)\tau_A + (-1)\tau_B + (-1)\tau_C$$
$$= \mu + (-\alpha_1 - \alpha_2 - \alpha_3) + (-\tau_A - \tau_B - \tau_C) = \mu + \alpha_4 + \tau_D$$

A fitting of this model by the GSK weighted-least-squares methodology produces the following estimates:

$$\hat{\mu} = 1.54$$

$$\hat{\alpha}_1 = -0.04, \quad \hat{\alpha}_2 = -0.04, \quad \hat{\alpha}_3 = 0.11, \quad \hat{\alpha}_4 = -\hat{\alpha}_1 - \hat{\alpha}_2 - \hat{\alpha}_3 = -0.03$$

$$\hat{\tau}_A = -0.11, \quad \hat{\tau}_B = -0.07, \quad \hat{\tau}_C = 0.05, \quad \hat{\tau}_D = -\hat{\tau}_A - \hat{\tau}_B - \hat{\tau}_C = 0.13$$

Since (in the notation of Section 22-3) there are $m = u = 16$ responses (the $a_\gamma(p)$'s) and $v = 7$ parameters in the model, the chi-square goodness-of-fit test for assessing how well the linear model describes the variation in the $a_\gamma(p)$'s has $m - v = 16 - 7 = 9$ degrees of freedom. The value of this error (or residual) sum of squares for these data is 6.32, which does not approach statistical significance; this can be interpreted to mean that an additive model having only mean, hospital, and surgical procedure effects fits the data adequately.

(It is important to remember that a test of significance concerning one or more effects in a model is adjusted for the presence of other effects in the model; thus an ANOVA table presented under these conditions should *not* be read as describing *orthogonal* effects but rather as providing a summary of several tests on different *nonorthogonal* effects. In particular, such an ANOVA table for the data we are presently considering would contain the sum of squares for hospitals *adjusted for* surgical procedure effects and the sum of squares for surgical procedures *adjusted for* hospital effects, as well as the goodness-of-fit sum of squares.)

Next, to test hypotheses concerning the hospital and surgical procedure effects in the model, the c's must be specified in the general linear hypothesis

$$H_0: \quad c_{10}\mu + c_{11}\alpha_1 + c_{12}\alpha_2 + c_{13}\alpha_3 + c_{1A}\tau_A + c_{1B}\tau_B + c_{1C}\tau_C = 0$$
$$c_{20}\mu + c_{21}\alpha_1 + c_{22}\alpha_2 + c_{23}\alpha_3 + c_{2A}\tau_A + c_{2B}\tau_B + c_{2C}\tau_C = 0$$
$$\vdots$$
$$c_{d0}\mu + c_{d1}\alpha_1 + c_{d2}\alpha_2 + c_{d3}\alpha_3 + c_{dA}\tau_A + c_{dB}\tau_B + c_{dC}\tau_C = 0$$

Thus, to test

$$H_0: \quad \alpha_1 = 0$$
$$\alpha_2 = 0$$
$$\alpha_3 = 0$$

we simply take $c_{11} = c_{22} = c_{33} = 1$ and set the rest of the c's equal to 0; similarly, to test

$$H_0: \quad \tau_A = 0$$
$$\tau_B = 0$$
$$\tau_C = 0$$

we set $c_{1A} = c_{2B} = c_{3C} = 1$ and the rest of the c's equal to 0. The GSK chi-square statistics for testing these two hypotheses each have $d = 3$ degrees of freedom. There are no significant hospital effects, since the computed χ^2-value is 2.33. However, the surgical procedure effects are significant at less than the .05 level, the χ^2-value being 8.90. These results can be summarized in an ANOVA table (Table 22-14).

TABLE 22-14   ANOVA table for data in Table 22-12

Source	df	SS (or χ^2)
Hospitals (adjusted for "surgical procedures")	3	2.33
Surgical procedures (adjusted for "hospitals")	3	8.90
Residual (goodness of fit)	9	6.32

Finally, it is of interest to relate the severity of the dumping syndrome to the amount of stomach removed; since the fractions of stomach removed are approximately 0, $\frac{1}{4}$, $\frac{1}{2}$, and $\frac{3}{4}$ for surgical procedures A, B, C, and D, respectively, we may ask if the severity is *linearly* related to the fraction of stomach removed. This question can be answered by testing

$$H_0: \quad -3\tau_A - \tau_B + \tau_C + 3\tau_D = 0$$

due to the equal spacings in the fractions of stomach removed.[10] Expressed in terms of τ_A, τ_B, and τ_C, this hypothesis is of the form $-3\tau_A - \tau_B + \tau_C + 3(-\tau_A - \tau_B - \tau_C) = -6\tau_A - 4\tau_B - 2\tau_C = 0$, which is generated by taking $c_{1A} = -6$, $c_{1B} = -4$, and $c_{1C} = -2$ and setting all other c's equal to 0. The resulting chi-square statistic is 8.74 based on $d = 1$ degree of freedom, which is highly significant. Therefore, we reject the null hypothesis of no linear trend and conclude that the severity of the dumping syndrome tends to increase approximately linearly with the fraction of stomach removed.

Example   This example deals with the IFRP study introduced in Section 22-1, which focuses on the problem of describing the determinants of postabortion fertility control (PAFC) using the demographic variables age (A), number of living children (C), education (E), and location (L). The basic data has been given in Table 22-2, columns 5 and 6, which list the $r = 2$ observed cell counts for each of the $s = 16$ binomial populations (i.e., n_{i1} and n_{i2}, $i = 1, 2, \ldots, 16$) corresponding to the 2^4 combinations of the four (dichotomous) factors. The last column gives the estimated probability $p_{i1} = n_{i1}/(n_{i1} + n_{i2})$ of accepting PAFC for each of the 16 populations. As mentioned earlier, we are concerned here with a one-response, four-factor table, and the specific analysis objective is to find a simple predictive model with as few parameters as possible that best describes the relationship between the four demographic variables (A, C, E, and L) and the dependent variable p_{i1}, the observed proportion accepting PAFC. Techniques for selecting variables and simplifying the model will be given special attention in this analysis.

The first model to be fitted to the data is a complete (or *saturated*) model containing all 16 effects associated with a complete 2^4 factorial design. That is, the model contains an overall mean effect (μ), all four main effects (α_A, α_C, α_E, α_L), all six two-factor interactions (γ_{AC}, γ_{AE}, γ_{AL}, γ_{CE}, γ_{CL}, γ_{EL}), all four three-factor interactions (τ_{ACE}, τ_{ACL}, τ_{AEL}, τ_{CEL}), and the one four-factor interaction (δ_{AECL}). In equation form, this model is as follows:

$$\begin{aligned}
\Pi_{i1} = x_0\mu + x_A\alpha_A &= x_C\alpha_C + x_E\alpha_E + x_L\alpha_L + x_Ax_C\gamma_{AC} + x_Ax_E\gamma_{AE} + x_Ax_L\gamma_{AL} \\
&+ x_Cx_E\gamma_{CE} + x_Cx_L\gamma_{CL} + x_Ex_L\gamma_{EL} + x_Ax_Cx_E\tau_{ACE} + x_Ax_Cx_L\tau_{ACL} \\
&+ x_Ax_Ex_L\tau_{AEL} + x_Cx_Ex_L\tau_{CEL} + x_Ax_Cx_Ex_L\delta_{ACEL}
\end{aligned} \tag{22.1}$$

where

$$x_0 \equiv 1$$

$$x_A = \begin{cases} 1 & \text{if population } i \text{ has factor } A = 25+ \\ 0 & \text{if population } i \text{ has factor } A = <25 \end{cases}$$

$$x_C = \begin{cases} 1 & \text{if population } i \text{ has factor } C = 2+ \\ 0 & \text{if population } i \text{ has factor } C = 0-1 \end{cases}$$

---

[10] As mentioned in Section 17-9, the coefficients -3, -1, 1, and 3 in the contrast ($-3\tau_A - \tau_B + \tau_C + 3\tau_D$) are orthogonal polynomial coefficients (see Armitage, 1971, for a further discussion.)

$$x_E = \begin{cases} 1 & \text{if population } i \text{ has factor } E = 7+ \\ 0 & \text{if population } i \text{ has factor } E = 0\text{--}6 \end{cases}$$

$$x_L = \begin{cases} 1 & \text{if population } i \text{ has factor } L = \text{Yugoslavia} \\ 0 & \text{if population } i \text{ has factor } L = \text{India} \end{cases}$$

For example, for row (population) $i = 10$ of Table 22-2, $A = 25+$, $C = 0\text{--}1$, $E = 0\text{--}6$, and $L =$ Yugoslavia, so

$$\begin{aligned}
\Pi_{10,1} &= (1)\mu + (1)\alpha_A + (0)\alpha_C + (0)\alpha_E + (1)\alpha_L + (1)(0)\gamma_{AC} + (1)(0)\gamma_{AE} + (1)(1)\gamma_{AL} \\
&\quad + (0)(0)\gamma_{CE} + (0)(1)\gamma_{CL} + (0)(1)\gamma_{EL} + (1)(0)(0)\tau_{ACE} + (1)(0)(1)\tau_{ACL} \\
&\quad + (1)(0)(1)\tau_{AEL} + (0)(0)(1)\tau_{CEL} + (1)(0)(0)(1)\delta_{ACEL} \\
&= \mu + \alpha_A + \alpha_L + \gamma_{AL}
\end{aligned}$$

The value of the x's for each of the 16 populations as listed in Table 22-2 are given in Table 22-15.

A listing of the results of the significance tests concerning the various individual effects in model (22.1) (except for μ) is given in Table 22-16, together with the results of global tests concerning the joint effects of the second-order and third-order interactions. (It is assumed that the reader, after having worked through the previous examples, will be able to specify the values of the c's required to generate the appropriate hypotheses considered in Table 22-16.)

From Table 22-16 it is clear that the main effect of factor C (adjusted for the other effects in the model) is highly significant ($P < .001$) and that the CE interaction effect is also significant ($P < .05$). However, none of the three-factor interactions or the four-factor interactions are significant. This suggests that a simpler model than the saturated one just considered can be found that will describe the relationship between the four factors and the response more efficiently.

The most obvious choice for a simpler model will *not* contain three- and four-factor interaction terms; that is, it will contain only the main effects and the two-factor interactions corresponding to the first 11 columns of Table 22-15. A summary of the results of the significance tests associated with fitting this reduced model (which we presented in Section 22-3) is given in Table 22-17.

A goodness-of-fit chi-square test can be performed on this 11-term model; this residual sum of squares has the value 3.92 with $16 - 11 = 5$ degrees of freedom, which is not significant. The conclusions here are somewhat different from those implied by Table 22-16. As with the complete model, the main effect of factor C is highly significant, but now no other effects are significant at the .05 level. The two effects closest to significance are the A and CL effects, whereas the CE interaction, which was significant for the complete model, is now clearly nonsignificant.

From the estimated parameters (i.e., $\hat{\mu}$, the $\hat{\alpha}$'s, and the $\hat{\gamma}$'s) obtained from fitting this reduced model, predicted (or adjusted) values for the PAFC acceptance rates can be calculated.

(It is important to discuss the reason for preferring predicted or adjusted rates to the crude rates obtained for each population. If all 16 effects in the complete model had been significant, the crude rates would have been appropriate to use. However, since we have established that one or more of the 16 effects are not significant and thus should not be included in the model, the original crude rates must contain a certain amount of noise due

TABLE 22-15 Values of independent variables for model (22.1) based on data of Table 22-2

Independent Variable

Population	x_0	x_A	x_C	x_E	x_L	$x_A x_C$	$x_A x_E$	$x_A x_L$	$x_C x_E$	$x_C x_L$	$x_E x_L$	$x_A x_C x_E$	$x_A x_C x_L$	$x_A x_E x_L$	$x_C x_E x_L$	$x_A x_C x_E x_L$
1	1	0	0	0	0	0	0	0	0	0	0	0	0	0	0	0
2	1	1	0	0	0	0	0	0	0	0	0	0	0	0	0	0
3	1	0	0	1	0	0	0	0	0	0	0	0	0	0	0	0
4	1	1	0	1	0	0	1	0	0	0	0	0	0	0	0	0
5	1	0	1	0	0	0	0	0	0	0	0	0	0	0	0	0
6	1	1	1	0	0	1	0	0	0	0	0	0	0	0	0	0
7	1	0	1	1	0	0	0	0	1	0	0	0	0	0	0	0
8	1	1	1	1	0	1	1	0	1	0	0	1	0	0	0	0
9	1	0	0	0	1	0	0	0	0	0	0	0	0	0	0	0
10	1	1	0	0	1	0	0	1	0	0	0	0	0	0	0	0
11	1	0	0	1	1	0	0	0	0	0	1	0	0	0	0	0
12	1	1	0	1	1	0	1	1	0	0	1	0	0	1	0	0
13	1	0	1	0	1	0	0	0	0	1	0	0	0	0	0	0
14	1	1	1	0	1	1	0	1	0	1	0	0	1	0	0	0
15	1	0	1	1	1	0	0	0	1	1	1	0	0	0	1	0
16	1	1	1	1	1	1	1	1	1	1	1	1	1	1	1	1

TABLE 22-16 Summary of tests of hypotheses concerning parameters in model (22.1)

Source	df	χ^2	P
A	1	1.20	.27
C	1	17.42**	.00
E	1	3.82	.05
L	1	0.04	.85
AC	1	1.11	.29
AE	1	0.72	.40
AL	1	0.10	.75
CE	1	4.34*	.04
CL	1	1.32	.25
EL	1	2.08	.15
ACE	1	1.08	.30
ACL	1	0.00	1.00
AEL	1	0.17	.68
CEL	1	0.67	.41
ACEL	1	0.12	.73
All two-factor interactions	6	6.65	.35
All three-factor interactions	4	2.90	.58

TABLE 22-17 Summary of tests of hypotheses for a model reduced from model (22.1) corresponding to the first 11 columns of Table 22-15

Source	df	χ^2	P
A	1	2.59	.11
C	1	21.86**	.00
E	1	1.76	.19
L	1	0.94	.33
AC	1	2.32	.13
AE	1	0.00	1.00
AL	1	0.28	.60
CE	1	1.58	.21
CL	1	2.71	.10
EL	1	1.49	.22
All two-factor interactions	6	10.58	.10
Residual (goodness of fit)	5	3.92	.56

to the presence of these nonsignificant effects, while the predicted (or adjusted) rates are based on a more efficient model not containing these noise producers and so are "smoothed" with respect to such random variation.)

The $\hat{p}_{i1}$'s are calculated using the general expression

$$\hat{p}_{i1} = \hat{\mu} + x_A\hat{\alpha}_A + x_C\hat{\alpha}_C + x_E\hat{\alpha}_E + x_L\hat{\alpha}_L + x_Ax_C\hat{\gamma}_{AC} + x_Ax_E\hat{\gamma}_{AE} + x_Ax_L\hat{\gamma}_{AL} + x_Cx_E\hat{\gamma}_{CE} + x_Cx_L\hat{\gamma}_{CL} + x_Ex_L\hat{\gamma}_{EL} \tag{22.2}$$

Table 22-18 gives the observed p_{i1}'s, the predicted $\hat{p}_{i1}$'s, and the residual $(p_{i1} - \hat{p}_{i1})$ for each of the 16 populations based on model (22.2).

It should be noted in Table 22-18 that fairly large residuals with absolute values exceeding .05 arise for populations 1, 3, 7, and 10. Also, Johnson and Koch (1971) recommend that, regardless of the number of degrees of freedom associated with the goodness-of-fit test, the critical value of χ^2 for the test should be 3.84 (the $\alpha = .05$ value of the chi-square distribution with 1 degree of freedom) to ensure that the residual sum of squares does not contain any hidden but individually significant component, and we see that this conservative critical value of 3.84 is exceeded (although just barely) for the 11-term model by the goodness-of-fit χ^2-value of 3.92. Thus, even though this model fits the data fairly well, it is good strategy to consider other possible candidates, and this is what we shall do.

At this point it becomes important to realize that the search for a best model in categorical data analysis is philosophically similar in many respects to that conducted in the usual regression analysis, although unfortunately a well-documented algorithm for categorical data does not exist that operates like stepwise regression to incorporate or delete terms one step at a time according to a specified criterion (e.g., a partial F test). Nevertheless, through a comparison of several reasonable candidate models using the goodness-of-fit test, a careful examination of residual values, and chi-square tests of significance concerning the effects in the model, the best one or two models can usually be identified.

With this in mind, 14 candidate models were fitted to the data in Table 22-2; Table 22-19 presents a summary of the tests of hypotheses concerning the parameters associated with each of these fitted models. The entries in Table 22-19 are P-values; values greater than

TABLE 22-18 Values of p_{i1}, $\hat{p}_{i1}$, and $p_{i1} - \hat{p}_{i1}$ calculated using model (22.2)

	Population							
Probability	1	2	3	4	5	6	7	8
p_{i1}	.333	.667	.667	.700	.944	.951	.857	.947
$\hat{p}_{i1}$	.465	.628	.578	.737	.925	.951	.926	.947
$p_{i1} - \hat{p}_{i1}$	−.132	.039	.089	−.037	.019	.000	−.069	.000

	Population							
Probability	9	10	11	12	13	14	15	16
p_{i1}	.364	.591	.419	.506	.727	.636	.591	.593
$\hat{p}_{i1}$	.379	.495	.419	.530	.679	.658	.607	.581
$p_{i1} - \hat{p}_{i1}$	−.015	.096	.000	−.024	.048	−.022	−.016	.012

TABLE 22-19   P-values for tests of hypotheses concerning effects in several candidate models fitted to the data in Table 22-2

Source	Model													
	1	2	3	4	5	6	7	8	9	10	11	12	13	14
A	.27	.11	.11	.03	.11	—	.00	—	—	—	—	.11	.03	.33
C	.00	.00	.00	.00	.00	.00	—	.00	—	.00	.00	.00	.00	.00
E	.05	.19	.63	.01	—	.64	.23	—	—	—	—	—	—	—
L	.85	.33	.00	—	.00	.00	.00	.00	.00	—	.04	.05	.05	.13
AC	.29	.13	—	—	—	—	—	—	—	—	—	—	.13	.45
AE	.40	1.00	—	—	—	—	—	—	—	—	—	—	—	—
AL	.75	.60	—	—	—	—	—	—	—	—	—	—	—	.83
CE	.04	.21	—	—	—	—	—	—	—	—	—	—	—	—
CL	.25	.10	—	—	—	—	—	—	—	—	.02	.02	.02	.25
EL	.15	.22	—	—	—	—	—	—	—	—	—	—	—	—
ACE	.30	—	—	—	—	—	—	—	—	—	—	—	—	—
ACL	1.00	—	—	—	—	—	—	—	—	—	—	—	—	.90
AEL	.68	—	—	—	—	—	—	—	—	—	—	—	—	—
CEL	.41	—	—	—	—	—	—	—	—	—	—	—	—	—
ACEL	.73	—	—	—	—	—	—	—	—	—	—	—	—	—
Residual (goodness of fit)	1.00	.56	.21	.00	.26	.14	.00	.19	.00	.00	.46	.60	.73	.59

.05 indicate that the effects are nonsignificant or, in the case of the goodness-of-fit test, that the model provides a reasonable fit. Dashes in the table indicate effects that were *not* included in the particular model fitted.

From Table 22-19, models 4, 7, 9, and 10 can be eliminated from consideration because they do not provide a reasonable fit to the data. Note that each of these four models omits either the *C*-factor main effect or the *L*-factor main effect, suggesting that both of these two effects should be part of the predictive model ultimately selected.

Finally, by examining the goodness-of-fit statistics and residual patterns of the remaining 10 models and by adopting the policy that each effect in a given model should be significant at the 5% level, we select models 11, 12, and 13 as the three most qualified candidates. A choice among these three models would have to be based on interpretability and relevance.

Problems

1. The accompanying table displays information gathered from a demographic study concerning regional opinion about family planning.

Region	Approve	Disapprove	No Opinion	Total
East	500	350	150	1,000
South	400	300	300	1,000
Total	900	650	450	2,000

a. Describe the two kinds of sampling schemes that could have produced the table and identify which scheme is associated with a two-response, no-factor representation of the data and which scheme is associated with a one-response, one-factor representation. Which of the two sampling schemes do you think was actually used?

b. Which of the two null hypotheses of independence and homogeneity is appropriate for each of the sampling schemes identified?

c. Construct a table corresponding to the two-response, no-factor representation of the data. What are r (the number of response categories) and s (the number of subpopulations)?

d. For the two-response, no-factor case, specify the structure of each of the unknown true cell probabilities under the appropriate null hypothesis (i.e., describe the form of Π_{ij} under H_0 for each (i, j) pair).

e. For the two-response, no-factor case, express the appropriate null hypothesis in terms of linear and/or log-linear functions of the unknown cell probabilities. Also, specify the constants $\{a_{\gamma, ij}\}$ and $\{k_{a\gamma}\}$ needed to form these functions. (*Hint*: Use the fact that H_0 can be written $(\Pi_{11}\Pi_{16}/\Pi_{14}\Pi_{13} = 1$ and $\Pi_{12}\Pi_{16}/\Pi_{15}\Pi_{13} = 1$ to define your two functions.)

f. Construct a table corresponding to the one-response, one-factor representation of the data.

g. For the one-response, one-factor case, specify the structure of the unknown true cell probabilities under the appropriate null hypothesis.

h. For the one-response, one-factor case, express the appropriate null hypothesis in terms of linear and/or log-linear functions of the unknown cell probabilities. Also, specify the constants needed to form these functions.

i. Carry out the usual chi-square test for the data of this problem. (Use $\alpha = .05$.)

j. Compare the value of the usual χ^2 to the GSK-modified χ^2-values for independence and homogeneity ($\chi^2_{\text{INDEP}} = 62.89$ and $\chi^2_{\text{HOM}} = 67.14$).

2. The data below represent a cross-classification of 6,099 males, aged 20 through 30, according to the grip strength of each hand.

Right Hand	Left Hand				
	Highest Grade	Second Grade	Third Grade	Lowest Grade	Total
Highest grade	1,275	159	103	42	1,579
Second grade	206	1,233	361	49	1,849
Third grade	98	312	1,521	139	2,070
Lowest grade	23	60	120	398	601
Total	1,602	1,764	2,105	628	6,099

a. Describe the sampling scheme that produces a two-response, no-factor representation for this data set and construct the corresponding table of unknown cell probabilities. (For notational simplicity, let Π_{ij} denote the cell probability associated with the ith row and jth column in the table, where $i = 1, 2, 3, 4$ and $j = 1, 2, 3, 4$.)

b. Suppose that it is of interest to test whether the distribution of persons according to right-hand grip strength grade is the same as the corresponding distribution for left-hand grades. How can this null hypothesis be expressed in terms of homogeneity of marginal distributions? That is, specify H_0 in terms of the marginal probabilities $\Pi_{i \cdot}$ and $\Pi_{\cdot j}$ for all i and j, where $\Pi_{i \cdot} = \Sigma_{j=1}^{4} \Pi_{ij}$ and $\Pi_{\cdot j} = \Sigma_{i=1}^{4} \Pi_{ij}$.

c. Express the null hypothesis given in part (b) in terms of $u = 3$ linear functions $a_1(\Pi)$, $a_2(\Pi)$, and $a_3(\Pi)$ of the unknown cell probabilities. Also, specify the constants $\{a_{\gamma, ij}\}$ needed to form these linear functions.

d. Carry out the test of the null hypothesis in part (c) using the GSK methodology (computed $\chi^2 = 7.155$).

e. Determine the point estimates of the functions $a_\gamma(\Pi)$, $\gamma = 1, 2, 3$.

3. The data given in the table below come from a study of burglary and larceny arrests in Mecklenburg County, North Carolina (Clarke and Koch, 1975). The investigators wished to assess the relationship of prior arrest history and type of offense to the probability that an arrest results in a prison sentence.

		Outcome	
Prior Arrest History	Type of Offense	No Prison Sentence	Prison Sentence
None	Nonresidential burglary	17	38
None	Other	21	244
Some	Nonresidential burglary	42	67
Some	Other	67	302

a. Construct the table of unknown cell probabilities (Π_{ij}) associated with a one-response, two-factor representation of the data set. What kind of sampling scheme would produce such a table? Do the data suggest that such a sampling scheme was actually used? If not, how can the one-response, two-factor representation be justified?

b. Consider the following null hypothesis:

$$H_0: \Pi_{12} = \Pi_{22} = \Pi_{32} = \Pi_{42}$$

What does this null hypothesis mean? Express this hypothesis in terms of three linear functions of the unknown cell probabilities appropriate for the GSK methodology.

c. What does the null hypothesis

$$H_0: \quad \Pi_{12} = \Pi_{22}, \Pi_{32} = \Pi_{42}$$

mean? Express this null hypothesis in terms of two linear functions of the unknown cell probabilities.

d. Answer the same questions as in part (c) for the null hypothesis

$$H_0: \quad \Pi_{12} = \Pi_{32}, \Pi_{22} = \Pi_{42}$$

e. The GSK chi-square statistics obtained for the null hypotheses specified in parts (b) through (d) are given by

$$\chi^2(b) = 52.53, \quad \chi^2(c) = 28.83, \quad \chi^2(d) = 16.40$$

Using these results, carry out the corresponding statistical tests, making sure to specify in each case the appropriate degrees of freedom.

f. Letting $a_\gamma(\Pi) = \Pi_{\gamma 2}$, $\gamma = 1, 2, 3, 4$, postulate a linear regression model for these linear functions in terms of factor effects. Make sure to specify v (the number of parameters), m (the number of linear functions), and the $x_{\xi l}$ ($\xi = 1, 2, \ldots, m$; $l = 1, 2, \ldots, v$), using the general model form:

$$g_\xi(\Pi) = x_{\xi 1}\beta_1 + x_{\xi 2}\beta_2 + \cdots + x_{\xi v}\beta_v \qquad \xi = 1, 2, \ldots, m$$

(*Note*: Postulate a model that contains an overall effect, an effect due to prior arrest history, an affect due to type of offense, and an interaction effect.)

g. Fill in the blanks in the summary table of chi-square tests concerning the effects in the model defined in part (f).

Source	df	χ^2	P
Type of offense		12.705	
Prior arrest history		15.442	
Interaction		0.101	

h. For each chi-square statistic given in the table in part (g), specify the appropriate null hypothesis in terms of d linearly independent linear functions of the regression coefficients.

i. Based on your test results in part (g), how might you simplify the regression model to describe the data more efficiently?

j. Why is there no goodness-of-fit test presented in the table given in part (g)?

k. Compute a 95% confidence interval for each of the effects in the model postulated in part (f) by using the point estimates and standard errors printed out by the GSK computer program, as listed in the table.

	Effect			
	Overall	Type of Offense	Prior Arrest History	Interaction
Coefficient	0.0792	0.2299	0.1023	−0.0261
Standard error	0.0166	0.0645	0.0260	0.0821

4. A questionnaire developed for the family nurse practitioner study at the University of North Carolina (Wagner et al., 1976) was mailed to three types of physicians for the purpose of developing criteria for appropriate medical care. The information in the following table allows for an assessment of the extent to which certain factors are related to whether a physician responded to the questionnaire.

Year of Graduation	MD Type					
	Infectious-Disease Pediatricians		General Pediatricians		Family Physicians	
	Respondents	Nonrespondents	Respondents	Nonrespondents	Respondents	Nonrespondents
Before 1940	5	5	38	33	56	86
1940–1950	9	10	85	58	42	51
1951–1960	23	14	102	60	81	61
After 1960	8	2	49	21	27	35
Total	45	31	274	172	206	233

a. Construct the table of unknown cell probabilities corresponding to a one-response, two-factor representation of this data set.

b. Given that it is of interest to describe the response rate for physicians receiving the questionnaire in terms of the factors "MD type" and "year of graduation," specify the appropriate linear functions of the Π_{ij}'s that define the functions of interest for this study.

c. Determine the point estimates of the linear functions specified in part (b).

d. Postulate a linear model that describes the response rate in terms of the main effects and two-factor interaction effects for the factors "year of graduation" and "MD type"; that is, specify v, m, and the $x_{\xi l}$'s for a model of the general form

$$g_\xi(\Pi) = x_{\xi 1}\beta_1 + x_{\xi 2}\beta_2 + \cdots + x_{\xi v}\beta_v \qquad \xi = 1, 2, \ldots, m$$

e. Use the ANOVA table to carry out the appropriate significance tests regarding main effects and interactions. What do you conclude from these tests?

Source	df	χ^2	P
MD type		12.630	
Year of graduation		9.669	
Interaction		6.732	

f. For each chi-square statistic given in the table in part (e), specify the appropriate null hypothesis in terms of d independent linear functions of the regression coefficients.

g. Based on the test results in part (e), how would you modify the model postulated in part (d) to better predict the response rate?

5. The data in the next table were obtained from an occupational health study (Higgins and Koch, 1977) concerning factors related to the development of byssinosis in workers employed in a certain textile industry.

Workplace	Length of Employment (yr)	Smoker	Response (byssinotic or not)	
			Yes	No
	<10	No	7	119
		Yes	30	203
Dusty	10–20	No	3	17
		Yes	16	51
	>20	No	8	64
		Yes	41	110
	<10	No	12	1,004
		Yes	14	1,340
Not dusty	10–20	No	2	209
		Yes	5	409
	>20	No	8	777
		Yes	19	951

a. Identify the representation of the response factor that is most appropriate for analyzing this data set, and construct the corresponding table of unknown cell probabilities.

b. Specify 12 linear functions $a_\gamma(\Pi)$ of the Π_{ij} that identify the functions of interest for this data set.

c. What are the point estimates of the linear functions specified in part (b)?

d. Postulate a linear model appropriate for this data set that corresponds to the summary table below.

Source	df	χ^2
Workplace (W)	1	7.356
Length of employment (E)	2	0.156
Smoking (S)	1	2.705
$W \times E$	?	2.832
$W \times S$	?	8.402
$E \times S$	?	2.285
$W \times E \times S$	?	1.611

e. Carry out the appropriate test associated with each chi-square statistic presented in the table in part (d). What are your conclusions about the extent to which the selected factors are related to the development of byssinosis?

f. Specify an appropriate linear model corresponding to the following modified summary table of chi-square tests (which omits certain interaction effects found to be nonsignificant in part (e)):

Source	df	χ^2
W	1	20.619
E	2	1.017
S	1	0.625
$W \times E$	2	12.758
$W \times S$	1	13.324
Goodness of fit	?	4.414

g. Carry out the appropriate test associated with each chi-square statistic given in the table in part (f) and draw appropriate conclusions.

h. The predicted response, $\hat{g}_\xi(p)$, obtained for the modified model in part (f), is given by the expression:

$$\hat{g}_\xi(p) = 0.0128 + 0.1319x_1 - 0.0035x_2 - 0.0038x_3 + 0.0025x_4$$
$$- 0.0956x_1x_2 + 0.0045x_1x_3 + 0.0921x_1x_4,$$

where the variables in this model are defined as follows:

$$x_1 = \begin{cases} 1 & \text{if dusty} \\ 0 & \text{if not dusty} \end{cases} \qquad x_2 = \begin{cases} 1 & \text{if } <10 \text{ years of} \\ & \text{employment} \\ 0 & \text{otherwise} \end{cases}$$

$$x_3 = \begin{cases} 1 & \text{if } 10\text{--}20 \text{ years of} \\ & \text{employment} \\ 0 & \text{otherwise} \end{cases} \qquad x_4 = \begin{cases} 1 & \text{if smoker} \\ 0 & \text{if nonsmoker} \end{cases}$$

Use this result to determine the value of the predicted response $\hat{g}_\xi(p)$ for each subpopulation and compare the observed values $g_\xi(p)$ with these predicted values. Complete a table like the one shown below and evaluate the results.

	Subpopulation											
	1	2	3	4	5	6	7	8	9	10	11	12
$g_\xi(p) = p_{i1}$												
$\hat{g}_\xi(p) = \hat{p}_{i1}$												
$p_{i1} - \hat{p}_{i1}$												

i. Based on your results in parts (g) and/or (h), how would you modify the linear model, if at all, to possibly improve prediction?

6. In an experiment reported by Kastenbaum and Lamphiear (1959), litters of mice were treated in either one of two ways, and the distribution of the number of depletions per litter before weaning was observed to be as given in the table that follows.

Litter Size	Treatment	No. of Depletions			Total No. of Litters
		0	1	2+	
7	A	58	11	5	74
	B	75	19	7	101
8	A	49	14	10	73
	B	58	17	8	83
9	A	33	18	15	66
	B	45	22	10	77
10	A	15	13	15	43
	B	39	22	18	79
11	A	4	12	17	33
	B	5	15	8	28

a. Identify the representation of the response factor that characterizes this data set and construct the corresponding table of unknown cell probabilities. What are the values of r and s?

b. One function of the unknown cell probabilities for each subpopulation of interest has the form

$$a_\gamma(\Pi) = \Pi_{\gamma 1} + \Pi_{\gamma 2} \qquad \gamma = 1, 2, \ldots, 10$$

which is the probability of one or more depletions in the γth subpopulation. Express this function as a linear function of the Π_{ij}'s by determining appropriate values for $a_{\gamma, ij}$ in the general formula

$$a_\gamma(\Pi) = \sum_{i=1}^{s} \sum_{j=1}^{r} a_{\gamma, ij} \Pi_{ij}$$

c. Using the data given, fill in the blanks in the accompanying table to provide point estimates of the functions $a_\gamma(\Pi)$.

Litter Size	Treatment	$\hat{a}_\gamma(\Pi)$
7	A	0.216
	B	0.257
8	A	0.329
	B	0.301
9	A	0.500
	B	0.416
10	A	
	B	
11	A	
	B	

 d. Based on the point estimates obtained in part (c), does there appear to be a treatment effect? A litter effect? Explain.

 e. Write down a linear model that describes $a_\gamma(\Pi)$ as a function of the main effects of treatment and litter size (with no interaction terms).

 f. Use the ANOVA table below to carry out the appropriate significance tests regarding the effects in the model given in part (e). Make sure to state the appropriate null hypothesis for each test.

Source	df	χ^2	P
Treatments		1.2166	
Litters		141.7631	
Residual (goodness of fit)		3.2246	

 g. Based on the results in part (f), what conclusions do you draw?

References

 Armitage, P. (1971). *Statistical Methods in Medical Research*. Oxford: Blackwell Scientific Publications.

 Bhapkar, V. P. (1966). "A Note on the Equivalence of Two Test Criteria for Hypotheses in Categorical Data." *J. Amer. Statist. Assoc.*, 61: 228–235.

 Bhapkar, V. P., and Koch, G. G. (1968). "Hypotheses of No Interaction in Multi-Dimensional Contingency Tables." *Technometrics*, 10: 107–123.

 Bishop, Y. M. M. (1969). "Full Contingency Tables, Logits, and Split Contingency Tables." *Biometrics*, 25: 383–399.

 Bross, I. D. (1958). "How to Use Ridit Analysis." *Biometrics*, 14: 18–38.

 Clarke, S. H., and Koch, G. G. (1975). "Who Goes to Prison? The Likelihood of Receiving an Active Sentence." *Popular Government*, 41(2): 25–37.

 Cornfield, J. (1956). "A Statistical Problem Arising from Retrospective Studies." *Proc. Third Berkeley Symp.*, 4: 135–148.

 Fleiss, J. L. (1973). *Statistical Methods for Rates and Proportions*. New York: John Wiley & Sons, Inc.

 Goodman, J. A. (1970). "The Multivariate Analysis of Qualitative Data: Interactions Among Multiple Classifications." *J. Amer. Statist. Assoc.*, 65: 226–256.

 Grizzle, J. E., Starmer, C. F., and Koch, G. G. (1969). "Analysis of Categorical Data by Linear Models." *Biometrics*, 25: 489–504.

 Higgins, J. E., and Koch, G. G. (1977). "Variable Selection and Generalized Chi-Square Analysis of Categorical Data Applied to a Large Cross-Sectional Occupational Health Survey." *Internat. Statist. Rev.*, 45: 51–62.

 Hogue, C. J., Kleinbaum, D. G., Omran, A. R., Gruber, F. J., and Freeman, D. H., Jr. (1974). "The Impact of Personal Characteristics on Post-Abortion Contraceptive Acceptance." Paper presented to 102nd Annual Meeting of the American Public Health Association, New Orleans.

 Johnson, S. K., and Johnson, R. E. (1972). "Tonsillectomy History in Hodgkin's Disease." *New England J. Med.*, 287: 1122–1125.

Johnson, W. D., and Koch, G. G. (1971). "A Note on the Weighted Least Squares Analysis of the Ries–Smith Contingency Table Data." *Technometrics, 13*: 438–447.

Kastenbaum, M. A., and Lamphiear, D. E. (1959). "Calculation of Chi-Square to Test the No-Three-Factor-Interaction Hypothesis." *Biometrics, 15*: 107–115.

Ku, H. H., Varner, R., and Kullback, S. (1971). "Analysis of Multidimensional Contingency Tables." *J. Amer. Statist. Assoc., 66*: 55–64.

Landis, J. R., Stanish, W. M., Freeman, J. L., and Koch, G. G. (1976). "A Computer Program for the Generalized Chi-Square Analysis of Categorical Data Using Weighted Least Squares (GENCAT)." *Computer Programs in Biomedicine, 6*(4): 196–231.

Neyman, J. (1949). "Contributions to the Theory of the χ^2 Test." *Proc. Berkeley Symp. Math. Statist. Prob.*, pp. 239–273.

Wagner, E. H., Greenberg, R. A., Imrey, P. B., Williams, C. A., Wolf, S. H., and Ibrahim, M. A. (1976). "Influence of Training and Experience on Selecting Criteria to Evaluate Medical Care." *New England J. Med., 294*: 871–876.

Wald, A. (1943). "Tests of Statistical Hypotheses Concerning Several Parameters When the Number of Observations Is Large." *Trans. Amer. Math. Soc., 54*: 426–482.

Two-Group Discriminant Analysis

23-1 Preview

In this chapter we shall be concerned with the problem of distinguishing (discriminating) between two populations on the basis of observations on several variables. We will define the two populations before gathering the data and have available a sample of individuals from each population. The statistical problem is to develop a rule, or discriminant function, based on the measurements obtained on each of these individuals, that will help us to assign some new individual to the correct population when it is not known from which of the two populations the individual comes.

The data obtained on each individual will invariably consist of observed values of a set of mutually correlated random variables, and the presence of these intercorrelations will necessitate consideration of the variables together rather than one at a time. The general approach, as with regression analysis, is to construct in some *optimal* way a linear combination of these variables that is then used for classification. This transforms the basic problem from a complex multivariable one to an easier-to-handle univariable one, and the assignment of an individual to one of the two populations is then based simply on the value of the linear combination for that particular individual. The statistical manipulations associated with constructing a good (i.e., optimal) linear combination and with developing related methodology differ from those of regression analysis and are classed under the general heading *discriminant analysis*. This particular statistical procedure was first introduced by R. A. Fisher (1936) as a statistical technique useful in taxonomic problems, and much statistical research in the area has gone on since that time (see Lachenbruch, 1975, for a comprehensive review).

We previously (Section 2-3) pointed out that discriminant analysis involves a nominal dependent variable, whereas the classical regression analysis considers a continuous dependent variable. This distinction is somewhat oversimplified, since these two methods are actually based on different conceptual frameworks and require different statistical assumptions (e.g., classically, regression analysis requires normality of the dependent variable, whereas discriminant analysis requires multivariate normality of the independent variables). Nevertheless, the goals of these two methods are quite similar: Both attempt to describe by using a linear model the relationship between a dependent and several independent variables, one for the primary purpose of discrimination, the other for the primary purpose of prediction. Furthermore, as we shall see later in this chapter, the numerical results of a discriminant analysis can be computationally obtained using an appropriately defined regression formulation.

In this chapter we will be confined to the two-population (or two-group) situation, in which case just one linear combination—or *discriminant function*—is needed. The extension to discrimination problems involving more than two groups (in which case more than one discriminant function is needed) can be found in Anderson (1958), Morrison (1967), and Overall and Klett (1972), the last offering a more applied discussion than the former two.

Before developing the methodology of discriminant analysis, let us emphasize one important point. To perform a discriminant analysis, the groups to be delineated must be specified before the data are collected and analyzed without regard to the variables being studied. This philosophy is in contrast to that of classification-of-variables procedures (e.g., cluster analysis), which begin without the designation of the groups and attempt to form groups (e.g., clusters) as distinct as possible based on the data at hand (see Overall and Klett, 1972). This chapter thus discusses discrimination procedures (groups specified a priori) and not classification-of-variables procedures (groups determined a posteriori).

23-2 Real-Life Examples

To formalize somewhat the discriminant analysis problem of interest to us, let us suppose that there are two populations, designated population 1 and population 2, and that we have sets of n_1 and n_2 individuals, respectively, selected from each population. Let us suppose further that, for each individual, we have measured or observed values on p correlated random variables $X_1, X_2, \ldots, X_p$. The basic strategy in discriminant analysis is to form a linear combination of these variables, say

$$L = \beta_1 X_1 + \beta_2 X_2 + \cdots + \beta_p X_p$$

and then to assign a new individual to either group 1 or group 2 on the basis of the value of L obtained for that new individual.

Some real-life research investigations whose statistical analysis problems fit quite naturally into the above framework are as follows:

1. In a study of primary health care, it is of interest to be able to discriminate between those people with symptoms of illness who seek medical care (population 1, say) and those with symptoms who don't (population 2, say). The types of variables (the X's) used in describing differences between the two populations are measures concerning the duration of symptoms, the perceived seriousness of the symptoms, the amount of worry or anxiety concerning the symptoms, the person's feelings concerning the doctor's ability to relieve the complaint, the number of bed-loss days (days spent in bed and hence resulting in loss of workdays), and various socioeconomic and demographic variables. Discriminant analysis techniques can be useful in deciding which linear combination of these variables is most helpful in

predicting whether a person with symptoms will seek medical aid or not (see Hulka, Kupper, and Cassel, 1972).

2. An epidemiologic study concerning the incidence of coronary heart disease (CHD) is designed to follow for a number of years a group of people initially free of CHD. At the start of the follow-up period, variables (the X's) concerning cholesterol level, blood pressure, occupation, height, weight, and certain socioeconomic and demographic variables are recorded for each individual. At the end of the follow-up period, it is determined whether a subject has developed CHD (population 1) or not (population 2). Using techniques of discriminant analysis, the linear combination of variables that may help distinguish between the CHD and non-CHD populations can be identified (see Kleinbaum et al., 1971).

3. In a study of occupational health in a certain industry, the research goal is to determine whether the work history (e.g., the pattern of jobs held) for those employees (past or present) who now have leukemia (population 1) is different in any significant way from that of employees who do not have leukemia (population 2), while also considering other possible variables such as age, education, and smoking history. Discriminant analysis techniques can be useful in assessing the importance of an employee's work history in combination with several concomitant factors as a possible leukemia-associated agent (see McMichael et al., 1975).

The above examples emphasize most dramatically how many seemingly diverse research questions of a statistical nature can be viewed in the context of a problem in discriminant analysis. Clearly, the common thread running through these three examples is the need for a statistical rule to decide which combination of a number of important variables provides the best discriminator between two defined populations. The calculation of that best linear combination is the subject of the next section.

As a word of caution, discriminant analysis should be used only when the X-variables under consideration are continuous, with distributions that are not highly skewed. In the discussions to follow, we will assume that this is the case.

23-3 Calculation of the Discriminant Function

Let us first establish some notation. Recall that we are given the existence of two populations and that we have observed values on p random variables $X_1, X_2, \ldots, X_p$ for each of the n_1 individuals from population 1 and the n_2 individuals from population 2. In particular, for the ith population ($i = 1, 2$), suppose that we let x_{ijk} denote the observed value of variable X_j ($j = 1, 2, \ldots, p$) for the kth sampled individual ($k = 1, 2, \ldots, n_i$). Thus, the set of variable values

$$\{x_{i1k}, x_{i2k}, \ldots, x_{ipk}\}$$

represents the group of measurements obtained for the kth individual selected from population i.

Now, recall from Section 23-2 that the idea was to develop a linear combination L of the variables, say

$$L = \beta_1 X_1 + \beta_2 X_2 + \cdots + \beta_p X_p$$

with values for $\beta_1, \beta_2, \ldots, \beta_p$ chosen so as to provide maximum discrimination between the two populations. What do we mean by "maximum discrimination"? Well, if the function L is going to discriminate between the two groups, the variation in the values of L *between* the two groups should be much greater than the variation in the values of L *within* the two groups. This is not unlike the reasoning employed in analysis-of-variance procedures for detecting differences among population means.

More specifically, for any individual in the group of $n_1 + n_2$ individuals available, we can calculate (if we know the β's) the value of L for that individual. In particular, for the kth individual selected from population i, the associated value of L is

$$L_{ik} = \beta_1 x_{i1k} + \beta_2 x_{i2k} + \cdots + \beta_p x_{ipk}$$

Thus, for each individual, we can convert or transform from a set of p variable values to a single univariate score. Once this transformation is made, the problem is to distinguish between the two populations on the basis of values $\{L_{11}, L_{12}, \ldots, L_{1n_1}\}$ from population 1 and values $\{L_{21}, L_{22}, \ldots, L_{2n_2}\}$ from population 2. In an analysis-of-variance framework, then, the total amount of variation in these scores is measured by

$$\sum_{i=1}^{2} \sum_{k=1}^{n_i} (L_{ik} - \bar{L})^2$$

where

$$\bar{L} = \frac{1}{n_1 + n_2} \sum_{i=1}^{2} \sum_{k=1}^{n_i} L_{ik} = \frac{1}{n_1 + n_2} (n_1 \bar{L}_1 + n_2 \bar{L}_2)$$

As we know from our previous experience with analysis-of-variance procedures, the total sum of squares can be broken down into two interpretable components, a between-groups sum of squares

$$B = \sum_{i=1}^{2} n_i (\bar{L}_i - \bar{L})^2$$

and a within-group sum of squares

$$W = \sum_{i=1}^{2} \sum_{k=1}^{n_i} (L_{ik} - \bar{L}_i)^2$$

It is not too difficult to show that B can be written in the equivalent form

$$B = \frac{n_1 n_2}{n_1 + n_2} (\bar{L}_1 - \bar{L}_2)^2$$

so that B is large if the average value of L in group 1 is quite different from the average value of L in group 2. The statistic W is calculated by first obtaining the sum of squares of the L's about their mean within each group separately and then adding (or pooling) these within-group sums of squares together.

The ratio B/W, then, can be thought of as a measure of the discriminatory power of L, in the sense that the larger the value of B relative to W, the more L reflects between-populations variation (which is what we are interested in) as opposed to within-population variation.

Now, the function B/W depends on L and thus on the parameters $\beta_1, \beta_2, \ldots, \beta_p$. It is reasonable from our discussion up to this point to ask whether it is possible to choose particular values of $\beta_1, \beta_2, \ldots, \beta_p$, say $b_1, b_2, \ldots, b_p$ (which will depend, of course, on the observed x's), in order to maximize the function B/W. To maximize B/W with respect to $\beta_1, \beta_2, \ldots, \beta_p$ is certainly an intuitively appealing goal in choosing an optimal discriminator, but the question arises as to whether such a maximization procedure is mathematically tractable. The answer is, in fact, yes, and the techniques of calculus permit a fairly straightforward solution to the problem. Without going into all the mathematical details, we will simply present the solution to the maximization problem.

Let $\bar{x}_{ij} = \sum_{k=1}^{n_i} x_{ijk}/n_i$ be the observed mean value of variable j for the sample of n_i individuals from population i, and let

$$d_j = \bar{x}_{1j} - \bar{x}_{2j} \tag{23.1}$$

be the observed difference between the mean values of variable j in the two groups. Further, let

$$s_{jj'} = \frac{1}{n_1 + n_2 - 2} \sum_{i=1}^{2} \sum_{k=1}^{n_i} (x_{ijk} - \bar{x}_{ij})(x_{ij'k} - \bar{x}_{ij'}) \tag{23.2}$$

for $j, j' = 1, 2, \ldots, p$, so that (when $j = j'$) s_{jj} gives the usual *pooled sample variance* for the jth variable, and $s_{jj'}$ gives the *pooled sample covariance* between the variables j and j'. Then form the matrix of such pooled estimates of population variances and covariances of the x's,

$$S = \begin{bmatrix} s_{11} & s_{12} & \cdots & s_{1p} \\ s_{21} & s_{22} & \cdots & s_{2p} \\ \vdots & \vdots & \ddots & \vdots \\ s_{p1} & s_{p2} & \cdots & s_{pp} \end{bmatrix}$$

At this point it is informative to express B/W explicitly as a function of the β's, the d's, and the elements of the matrix S. With some algebraic manipulation, one can show that B/W can equivalently be written in the form

$$\frac{B}{W} = \frac{[n_1 n_2/(n_1 + n_2)] \sum_{j=1}^{p} \sum_{j'=1}^{p} \beta_j \beta_{j'} d_j d_{j'}}{(n_1 + n_2 - 2) \sum_{j=1}^{p} \sum_{j'=1}^{p} \beta_j \beta_{j'} s_{jj'}}$$

For example, when $p = 2$, the expression above becomes

$$\frac{B}{W} = \frac{[n_1 n_2/(n_1 + n_2)](\beta_1^2 d_1^2 + \beta_2^2 d_2^2 + 2\beta_1\beta_2 d_1 d_2)}{(n_1 + n_2 - 2)(\beta_1^2 s_{11} + \beta_2^2 s_{22} + 2\beta_1\beta_2 s_{12})}$$

Expressed in this way, it is clear that the value of B/W remains unchanged if, for some nonzero constant c, β_1 is replaced by $c\beta_1$, β_2 by $c\beta_2$, and so on. In practical terms, this means that if the set of values $\{b_1, b_2, \ldots, b_p\}$ maximizes B/W, so does the set $\{cb_1, cb_2, \ldots, cb_p\}$. As we shall see later, this "scale invariance" property relates to the fact that only the relative sizes (i.e., ratios) of the coefficients really matter when using a discriminant function for classifying individuals into groups.

With these remarks in mind, the particular set $\{b_1, b_2, \ldots, b_p\}$ of solutions that we choose to present explicitly is that set which is commonly printed out by means of standard discriminant analysis computer programs (e.g., those in the BMD, SPSS, and SAS series). In particular, if

$$S^{-1} = \begin{bmatrix} s^{11} & s^{12} & \cdots & s^{1p} \\ s^{21} & s^{22} & \cdots & s^{2p} \\ \cdot & \cdot & \cdot & \cdot \\ \cdot & \cdot & \cdot & \cdot \\ \cdot & \cdot & \cdot & \cdot \\ s^{p1} & s^{p2} & \cdots & s^{pp} \end{bmatrix} \tag{23.3}$$

is the inverse matrix associated with the matrix S given earlier, the values $b_1, b_2, \ldots, b_p$ that maximize B/W are given as follows:[1]

$$\begin{aligned} b_1 &= s^{11}d_1 + s^{12}d_2 + \cdots + s^{1p}d_p \\ b_2 &= s^{21}d_1 + s^{22}d_2 + \cdots + s^{2p}d_p \\ &\;\;\vdots \\ b_p &= s^{p1}d_1 + s^{p2}d_2 + \cdots + s^{pp}d_p \end{aligned} \tag{23.4}$$

The linear combination based on the b's we will call

$$\ell = b_1 X_1 + b_2 X_2 + \cdots + b_p X_p$$

and this linear combination maximizes the quantity B/W based on the sample at hand. Since we actually have only a sample from each population, the b's should be looked upon as estimates of the β's, and ℓ can be considered to be an estimate of the optimal linear combination L, which could be determined if all population parameters (i.e., the true values of the means, variances, and covariances for the X's) were known.

It is worthwhile to examine the form of the expressions for the b's. First, notice that the b's are constructed as *linear combinations of the differences* (23.1) between the variable means in the two groups. Second, the coefficients of these mean differences are functions of pooled sums of squares and cross-products (23.2), and the validity of such pooling is contingent on the assumption that the variances of and covariances among the p variables are the same in each of the two groups. Such an assumption will almost certainly never hold exactly

---

[1] The values s^{ij} ($i = 1, 2, \ldots, p$ and $j = 1, 2, \ldots, p$) of the inverse matrix are defined so as to satisfy the following general mathematical relationship:

$$s^{i1}s_{1j} + s^{i2}s_{2j} + \cdots + s^{ip}s_{pj} = \begin{cases} 1 & \text{if } i = j \\ 0 & \text{if } i \neq j \end{cases}$$

Further discussion of matrix mathematics, including the notion of an inverse of a matrix, can be found in Appendix B.

in actual practice, but moderate departures from homogeneity do not seem to affect the behavior of the discriminant function seriously (e.g., see Lachenbruch, 1975).

23-4 Calculation of ℓ Using Dummy-Dependent-Variable Regression

There is another approach to computing the discriminant function coefficients. This procedure is based on the use of multiple regression techniques to fit a model for which the dependent variable is dichotomous.

To be more specific, suppose that we define a dummy dependent variable Y that takes the value $n_2/(n_1 + n_2)$ for an individual in our sample from group 1 and the value $-n_1/(n_1 + n_2)$ for an individual from group 2. We could, in fact, use any two distinct values for Y to designate the two groups, but the particular values we have specified make things rather nice computationally.[2] Now, consider fitting the regression model

$$Y = \beta_0 + \beta_1 X_1 + \beta_2 X_2 + \cdots + \beta_p X_p + E$$

where these X's are exactly the same as those appearing in the expression for ℓ given earlier.

Since we have arranged by choice of coding that the overall mean of the dependent variable ($\bar{Y}$, say) is 0, it follows that the fitted model can be written as

$$\hat{Y} = \hat{\beta}_0 + \hat{\beta}_1 X_1 + \hat{\beta}_2 X_2 + \cdots + \hat{\beta}_p X_p$$

where $\hat{\beta}_0 = -\sum_{j=1}^{p} \hat{\beta}_j \bar{x}_j$; $\hat{\beta}_1, \hat{\beta}_2, \ldots, \hat{\beta}_p$ are the usual least-squares estimates of $\beta_0, \beta_1, \ldots, \beta_p$; and $\bar{x}_j = \sum_{i=1}^{2} \sum_{k=1}^{n_i} x_{ijk}/(n_1 + n_2)$ is the *overall mean* of the values on variable X_j for both groups.

The question of interest here is what is the relationship between the $\hat{\beta}$'s obtained by means of the multiple regression approach above and the b's whose computational formulae were presented earlier. Without going into the mathematical details, it can be shown that $\hat{\beta}_j = cb_j$ for every j, where c is some positive constant; in other words, the regression procedure above can be used to produce a linear combination that differs from ℓ only by a constant multiplier. For the particular coding of Y that we have advocated, this constant multiplier c can be shown to have the specific value

$$c^* = \frac{n_1 n_2/(n_1 + n_2)}{(n_1 + n_2 - 2) + [n_1 n_2/(n_1 + n_2)]D^2} \tag{23.5}$$

where

$$D^2 = \sum_{j=1}^{p} \sum_{j'=1}^{p} d_j d_{j'} s^{jj'} \tag{23.6}$$

---

[2] In general, Y can be defined as follows:

$$Y = \begin{cases} k_1 & \text{for an individual from population 1 sample} \\ k_2 & \text{for an individual from population 2 sample} \end{cases}$$

The values of k_1 and k_2 most frequently used other than the values already mentioned are $k_1 = 1$ and $k_2 = 0$.

In general, other choices of codings for Y (see footnote 2) will change the value of the constant multiplier above, and the estimate of β_0 will also depend (linearly) on the new value of $\bar{Y}$.

The quantity D^2 of (23.6) has special significance. It is called the "Mahalanobis D^2" (named after a famous Indian statistician), and it represents a generalized measure of the "distance" between the two populations (e.g., $D^2 = (\bar{x}_{11} - \bar{x}_{21})^2/s^{11}$ when $p = 1$). Now, the usual assumptions made in discriminant analysis are that (1) the X's have a multivariate normal distribution (which implies, among other things, that each X has a normal distribution and that any linear combination of the X's is normal) and (2) the variances of and covariances among the X's are the same in the two groups. If these assumptions hold, it can be shown that

$$F = \frac{n_1 n_2 (n_1 + n_2 - p - 1)}{(n_1 + n_2)(n_1 + n_2 - 2)p} D^2 \sim F_{p, n_1 + n_2 - p - 1} \tag{23.7}$$

which provides a test as to *whether there are significant differences between the group means for all variables considered together.*

The preceding test can be related to the test for the significance of the fitted regression model. In particular, the ANOVA table based on the regression is given in Table 23-1.

It can be shown that the usual ratio of mean squares

$$\frac{\text{SSR}/p}{\text{SSE}/(n_1 + n_2 - p - 1)}$$

is identical to

$$\frac{n_1 n_2 (n_1 + n_2 - p - 1)}{(n_1 + n_2)(n_1 + n_2 - 2)p} D^2$$

in value, and (given the stated discriminant analysis assumptions, which are an inversion of the usual regression assumptions of Y normal and the X's nonstochastic) provides an F test of the regression null hypothesis

$$H_0: \quad \beta_1 = \beta_2 = \cdots = \beta_p = 0$$

If H_0 is rejected, one can conclude that there is evidence of between-groups differences, although the utility of the discriminant function for assigning individuals to groups is a separate issue (to be discussed in Section 23-6).

TABLE 23-1   ANOVA table for the dummy-dependent-variable regression

Source	df	SS
Due to regression	p	$\text{SSR} = \dfrac{n_1 n_2}{n_1 + n_2} \displaystyle\sum_{j=1}^{p} \hat{\beta}_j d_j$
Deviations from regression	$n_1 + n_2 - p - 1$	$\text{SSE} = \dfrac{n_1 n_2}{n_1 + n_2} \left(1 - \displaystyle\sum_{j=1}^{p} \hat{\beta}_j d_j \right)$
Total	$n_1 + n_2 - 1$	$\dfrac{n_1 n_2}{n_1 + n_2}$

There is also an interesting relationship between R^2, the squared multiple correlation coefficient obtained by means of the regression analysis, and the quantity D^2. In particular, it can be shown that

$$D^2 = \frac{(n_1 + n_2)(n_1 + n_2 - 2)}{n_1 n_2} \left(\frac{R^2}{1 - R^2} \right) \tag{23.8}$$

or equivalently that

$$R^2 = c^* D^2$$

where c^* is given by (23.5).

23-5 Numerical Example

The methodology presented in the preceding sections will be demonstrated on data collected in a longitudinal epidemiologic study (see Kleinbaum et al., 1971) designed to investigate the joint effects of three factors on the risk of developing coronary heart disease (CHD). At the start of the study period, the age (X_1), the diastolic blood pressure (X_2), and the cholesterol level (X_3) were recorded for each of 832 white males free of CHD. By the end of the study period, 71 of these males had developed CHD. The statistical question of import here is whether it is possible to discriminate between the 71 white males who developed CHD and the remaining 761 who did not on the basis of available data on the three independent variables. In the discriminant analysis to follow, we will consider the CHD group of $n_1 = 71$ white males to be a sample from the (conceptual) aggregate (population 1, say) of all white males who develop CHD under circumstances like those in this study. Similarly, population 2 will denote the collection of all those who remain free of CHD under such circumstances, and the NCHD group of $n_2 = 761$ individuals will be looked upon as a sample from that conceptual population.

The computational aspects of discriminant analysis described earlier are best illustrated by relating them to the output of a typical computer program designed to perform a discriminant analysis (e.g., like the BMD or SPSS program). We shall first consider discriminant analysis procedures associated with the function $\ell = b_1 X_1 + b_2 X_2 + b_3 X_3$, which involves all three variables. We shall then illustrate the use of a regression analysis program to arrive at the same results. Finally, we shall illustrate the use of stepwise discriminant analysis as a tool in choosing a discriminant function, although this procedure for selecting variables will not necessarily lead us to a discriminant function involving all three variables.

23-5-1 Output from a Typical Discriminant Analysis Program

Initial computer output typically consists of various summary statistics. These appear in Table 23-2 for our particular data set.

Next, the discriminant function coefficients are calculated. First the inverse matrix S^{-1}, given by (23.3), is computed, which here has the form

$$S^{-1} = \begin{bmatrix} 0.00560 & -0.00153 & -0.00045 \\ -0.00153 & 0.00553 & -0.00037 \\ -0.00045 & -0.00037 & 0.00063 \end{bmatrix}$$

TABLE 23-2 Summary statistics for CHD study data

MEANS $\{\bar{x}_{ij}\}$

Variable	Group		$d_j = \bar{x}_{1j} - \bar{x}_{2j}$
	NCHD	CHD	
X_1	44.81	56.86	12.05
X_2	86.99	95.62	8.63
X_3	201.27	221.51	20.24

STANDARD DEVIATIONS

Variable	Group	
	NCHD	CHD
X_1	14.98	10.28
X_2	14.50	15.37
X_3	43.01	38.83

WITHIN-GROUP COVARIANCE MATRIX (S)

$$\begin{array}{c c c c}
 & X_1 & X_2 & X_3 \\
\begin{matrix} X_1 \\ X_2 \\ X_3 \end{matrix} & \begin{bmatrix} 214.26 & 72.37 & 195.61 \\ & 212.44 & 175.53 \\ & & 1{,}820.61 \end{bmatrix}
\end{array}$$

WITHIN-GROUP CORRELATION MATRIX $(r_{jj'} = s_{jj'}/\sqrt{s_{jj}s_{j'j'}})$

$$\begin{array}{c c c c}
 & X_1 & X_2 & X_3 \\
\begin{matrix} X_1 \\ X_2 \\ X_3 \end{matrix} & \begin{bmatrix} 1.00 & 0.34 & 0.31 \\ & 1.00 & 0.28 \\ & & 1.00 \end{bmatrix}
\end{array}$$

Then the coefficient b_j associated with X_j is calculated from (23.4) as

$$b_j = \sum_{j'=1}^{p} s^{jj'} d_j = \sum_{j'=1}^{p} s^{jj'} \bar{x}_{1j'} - \sum_{j'=1}^{p} s^{jj'} \bar{x}_{2j'}$$

We have expressed b_j as a *difference* between two quantities for a particular reason. Some computer programs (e.g., those specifically designed for two-group discriminant analysis only) provide the b_j's directly; others (e.g., those with the capacity to work with more than two groups) print out the quantity $\sum_{j'=1}^{p} s^{jj'} \bar{x}_{ij'}$ as the score for the jth variable in the ith group, so that b_j must be obtained by the subtraction process indicated above. For the data we are considering, the computer produced the table of scores given as Table 23-3. We have generated the column of discriminant function coefficients based on these scores. The constant b_0, which we have not previously defined, is important when using the calculated discriminant function for classification purposes, and we shall discuss this in the next section. To illustrate the calculations resulting in these scores and coefficients, we obtain b_1 from the

TABLE 23-3   Discriminant function coefficients
             for CHD study data

| Variable | Group | | b_j |
	NCHD	CHD	
X_1	0.027	0.072	$b_1 =$ 0.045
X_2	0.338	0.360	$b_2 =$ 0.022
X_3	0.075	0.079	$b_3 =$ 0.004
(intercept)	-23.561	-28.726	$b_0 = -5.165$

table of means and from the elements of the first row of the matrix S^{-1} as

$$b_1 = [0.00560(56.86) - 0.00153(95.62) - 0.00045(221.51)]$$
$$- [0.00560(44.81) - 0.00153(86.99) - 0.00045(201.27)]$$
$$= 0.072 - 0.027 = 0.045$$

Thus, the calculated discriminant function ℓ has the specific form

$$\ell = 0.045X_1 + 0.022X_2 + 0.004X_3$$

This is the basic function that we shall be using for assignment purposes.

The typical discriminant analysis program next prints out the F statistic given by (23.7), which in our case has $p = 3$ and $n_1 + n_2 - p - 1 = 828$ degrees of freedom and tests *whether there are significant differences between the groups with respect to X_1, X_2, and X_3.* For the data we are considering, F has the value 17.605, which is highly significant (probably due to the large value for the denominator degrees of freedom) but says little about the fit of the discriminant model or about the utility of the model for classification purposes. Partial F statistics (based on 1 and $n_1 + n_2 - p - 1$ degrees of freedom) reflecting the *relative* contribution of each variable in the final model are also printed out in addition to the overall F statistic. As we will see, these F's are identical in value to those obtained by means of a multiple regression approach to the problem and hence can be interpreted in this light. For our data, these F-values (with 1 and 828 degrees of freedom) are 22.657 for X_1, 5.282 for X_2, and 1.675 for X_3, which suggests that X_3 might be a superfluous variable. We shall explore this possibility later by means of stepwise-discriminant-analysis procedures.

Some programs do not print out the value of D^2. However, this is no problem since, from (21.7),

$$D^2 = \frac{(n_1 + n_2)(n_1 + n_2 - 2)p}{n_1 n_2(n_1 + n_2 - p - 1)} F$$

In our case, then, we find that

$$D^2 = \frac{832(830)(3)}{71(761)(828)}(17.605) = 0.815$$

We shall use this value of D^2 to relate the output above to that obtained by means of a multiple regression approach to the problem and also to estimate misclassification rates.

23-5-2 Output from a Typical Regression Analysis Program

The output of a typical multiple regression program provides much (but not all) of the information obtained by means of a discriminant analysis. In particular, no classification tables (to be discussed in Section 23-6) are output by a multiple regression program, but the discriminant function coefficients can be obtained from the regression coefficients. In addition, the F statistics generated by each program are completely comparable. To see all this, let us use a typical multiple regression program to fit the model

$$Y = \beta_0 + \beta_1 X_1 + \beta_2 X_2 + \beta_3 X_3 + E$$

where $Y = n_2/(n_1 + n_2) = 761/832$ for an individual in the CHD group and $Y = -n_1/(n_1 + n_2) = -71/832$ for an individual in the NCHD group, and where X_1, X_2, and X_3 are as defined earlier. Typical output consists first of the usual type of ANOVA table for a regression analysis, which for our data is as follows:

Source	df	SS	MS	F
Regression	3	3.894	1.298	17.605*
Residual	828	61.047	0.074	

* $R^2 = .06$.

As pointed out in Section 23-4, the F statistic in this ANOVA table, which tests the regression null hypothesis $H_0: \beta_1 = \beta_2 = \beta_3 = 0$, is identical in value to the overall discriminant analysis F statistic that tests for overall differences between the group means. Finally, the table of estimated regression coefficients, their standard errors, and the associated partial F statistics appear in Table 23-4. As mentioned earlier, these partial F's are always identical in value to those given by the discriminant analysis.

To transform to the b's from the $\hat{\beta}$'s, we first obtain the value of D^2 using (23.8); then we determine the value of c^* using (23.5) as

$$c^* = \frac{n_1 n_2/(n_1 + n_2)}{(n_1 + n_2 - 2) + [n_1 n_2/(n_1 + n_2)]D^2}$$

$$= \frac{71(761)/832}{830 + [71(761)/832](0.815)} = 0.0736$$

Then we obtain b_j as $\hat{\beta}_j/c^*$ for $j = 1, 2, 3$. For example,

$$b_1 = \frac{\hat{\beta}_1}{c^*} = \frac{0.00331}{0.0736} = 0.045$$

TABLE 23-4   Regression coefficients for CHD study data

Variable	Coefficient	Standard Error	F
X_1	$\hat{\beta}_1 = $ 0.00331	0.00070	22.657
X_2	$\hat{\beta}_2 = $ 0.00161	0.00070	5.282
X_3	$\hat{\beta}_3 = $ 0.00031	0.00024	1.675
(intercept)	$\hat{\beta}_0 = $ −0.355		

We also point out that the low value of R^2 obtained from the regression analysis (namely, .06) indicates that the fitted regression model is not a good predictor of the dichotomous response under consideration. This emphasizes the often-overlooked points that significant F values do not necessarily imply a high value of R^2, and that the range of values taken by the response variable (in this case, $n_2/(n_1 + n_2)$ and $-n_1/(n_1 + n_2)$ only) can have an effect on R^2.

23-6 Classification Using the Discriminant Function

In this section we shall discuss how to use the calculated discriminant function for classification purposes and how to evaluate its performance in this regard. Let us suppose that we have calculated the linear discriminant function

$$\ell = b_1 X_1 + b_2 X_2 + \cdots + b_p X_p$$

by means of the methodology discussed in Section 23-3 and that we are now interested in using ℓ to assign individuals to one of the two groups. Since such an assignment is a process of statistical decision making, it is imperative to have a measure of the goodness of the decision rule. An excellent statistic in this regard is the *error rate* or *misclassification rate* (i.e., the probability of assigning an individual to the wrong population); there are, of course, two such rates, one for each population. We shall now discuss assignment rules and their associated error rates.

23-6-1 Cutoff Points

As before, it is helpful to think of associating with each individual a score that is the value of ℓ based on that particular individual's set of observed variable values. Then, in order to use ℓ for allocating individuals to one of the two groups, we need to specify a critical score, or *cutoff point,* such that an individual is assigned to one group if his score exceeds the cutoff point and to the other if it does not. If we begin with

$$\bar{\ell}_1 = b_1 \bar{x}_{11} + b_2 \bar{x}_{12} + \cdots + b_p \bar{x}_{1p}$$

and

$$\bar{\ell}_2 = b_1 \bar{x}_{21} + b_2 \bar{x}_{22} + \cdots + b_p \bar{x}_{2p}$$

which are the observed mean scores for our samples from groups 1 and 2, respectively, we can construct a completely symmetric classification rule by using as the cutoff point the mean of $\bar{\ell}_1$ and $\bar{\ell}_2$,

$$\frac{1}{2}(\bar{\ell}_1 + \bar{\ell}_2) = b_1 \frac{\bar{x}_{11} + \bar{x}_{21}}{2} + b_2 \frac{\bar{x}_{12} + \bar{x}_{22}}{2} + \cdots + b_p \frac{\bar{x}_{1p} + \bar{x}_{2p}}{2}$$

Then, if $\bar{\ell}_1 > \bar{\ell}_2$, the allocation rule is: Assign an individual with observed variable values $x_1, x_2, \ldots, x_p$ (and score $\ell = b_1 x_1 + b_2 x_2 + \cdots + b_p x_p$) to group 1 if $\ell > \frac{1}{2}(\bar{\ell}_1 + \bar{\ell}_2)$ and to group 2 if $\ell < \frac{1}{2}(\bar{\ell}_1 + \bar{\ell}_2)$. In discriminant analysis terminology, the quantity $-\frac{1}{2}(\bar{\ell}_1 + \bar{\ell}_2)$ is designated b_0; equivalently, then, we allocate to group 1 or 2 depending on whether the function

$$\ell^* = b_0 + b_1 x_1 + b_2 x_2 + \cdots + b_p x_p \tag{23.9}$$

is greater than or less than 0. Since the sign of ℓ^* becomes the sole criterion for classification, multiplying ℓ^* by any positive constant does not alter the allocation rule. In fact, if $n_1 = n_2$ so that $\bar{x}_j = \frac{1}{2}(\bar{x}_{1j} + \bar{x}_{2j})$, the classification rule using ℓ^* above is identical to one based on the sign of the fitted regression model discussed in Section 23-4. Using just the sign of ℓ^* would not be correct, of course, if the cutoff point for ℓ^* was some value other than 0.

In certain instances, important considerations may warrant changing the cutoff point to something other than the symmetric one discussed above. For example, suppose it is known a priori that an individual selected at random has probability p_1 of being from population 1 and $p_2 = 1 - p_1$ of being from population 2. Then, one might consider using some cutoff point that reflects the relative sizes of p_1 and p_2 so that an individual is likelier to be assigned to the population with the larger a priori probability. For example, using (23.9), we might choose $\ln(p_2/p_1)$ instead of 0 as the cutoff point. Then, if $p_1 > p_2$ (so that a randomly selected individual is likelier to come from population 1 than population 2), $\ln(p_2/p_1)$ will be strictly negative and we will assign an individual to population 1 as long as ℓ^* exceeds this negative value (thus allowing for more persons to be assigned to population 1 than if zero were the cutoff point). Analogously, the condition $p_1 < p_2$ implies that ℓ^* must exceed some positive quantity to merit assignment to group 1. In general, p_1 will be unknown; however, if the samples are selected randomly in such a way that $n_1/(n_1 + n_2)$ provides a good estimate of p_1, then $\ln(p_2/p_1)$ can be approximated by $\ln(n_2/n_1)$.

A modification of the cutoff point is also warranted when the cost or seriousness of an incorrect assignment is group-dependent. For example, it would be a much more serious error to classify an individual with cancer as not having cancer (often called a "false negative" finding) than it would be to treat a cancer-free individual unnecessarily (a "false positive"). More quantitatively, if c_{12} is the cost of misclassifying a member of group 1 as a member of group 2 and if c_{21} is the cost of incorrectly assigning an individual from group 2 to group 1, then $\ln(c_{21}/c_{12})$ has properties completely analogous to those of $\ln(p_2/p_1)$. The choice of the cost ratio c_{21}/c_{12} must very often be based on a subjective evaluation of the situation by the experimenter.

A generalized cutoff point that takes both the ratio of a priori probabilities and the cost ratio into account is

$$\ln \frac{p_2}{p_1} + \ln \frac{c_{21}}{c_{12}} = \ln \frac{p_2}{p_1}\left(\frac{c_{21}}{c_{12}}\right)$$

and a purely mathematical (and not just heuristic) justification for its use can be found in Anderson (1958). Note that this generalized cutoff point has the value 0 when $p_2/p_1 = c_{12}/c_{21}$, which includes the special case when $p_1 = p_2$ and $c_{12} = c_{21}$.

23-6-2 Error Rates

For any specified cutoff point, it is possible to count how many individuals in the sample have been incorrectly classified using ℓ^*. In particular, the following table can be constructed:

Actual Group	Assigned Group		Row Totals
	Group 1	Group 2	
1	n_{11}	n_{12}	n_1
2	n_{21}	n_{22}	n_2

The ratios n_{12}/n_1 and n_{21}/n_2 are often used to estimate the misclassification (or error) rates for the two groups, but these *apparent error rates* generally underestimate the true misclassification rates. This is because the function ℓ^* has been determined using these two particular samples and so should perform better with them than with a new group of individuals. Better misclassification rate estimates are available, and we will present them shortly.

Incidentally, when groups 1 and 2 designate sets of diseased and nondiseased individuals, respectively, it is common practice to refer to n_{11}/n_1 as the *sensitivity* and n_{22}/n_2 as the *specificity*. Naturally, the sensitivity and specificity (and also the two apparent error rates) vary inversely with one another as the cutoff point is altered; that is, if the sensitivity goes up with a change in the cutoff point, the specificity must go down, and vice versa.

Now, calculating the apparent error rates does not require any assumptions concerning distribution. However, under the usual assumptions of discriminant analysis mentioned earlier, it is possible to obtain more realistic estimates of the misclassification rates than those provided by the apparent error rates. In particular, if the decision rule is to assign an individual to group 1 if

$$\ell^* > \ln \frac{p_2}{p_1}\left(\frac{c_{21}}{c_{12}}\right) \tag{23.10}$$

and to group 2 otherwise, then an estimate (say, $\hat{P}_1$) of the probability of incorrectly assigning to population 2 an individual from population 1 can be shown to be

$$\hat{P}_1 = \Phi\left[\frac{\ln(p_2/c_{21})(p_1/c_{12}) - D^2/2}{\sqrt{D^2}}\right] \tag{23.11}$$

where D^2 is the Mahalanobis distance defined earlier and where Φ represents the standard normal cumulative distribution function (i.e., $\Phi(z_0) = \mathrm{pr}(Z < z_0)$ when $Z \sim N(0, 1)$). Similarly, $\hat{P}_2$, an estimate of the probability of incorrectly assigning an individual from population 2 to population 1, is given by the expression

$$\hat{P}_2 = \Phi\left[-\frac{\ln(p_2/c_{21})(p_1/c_{12}) + D^2/2}{\sqrt{D^2}}\right] \tag{23.12}$$

Note that when $\ln[(p_2/p_1)(c_{21}/c_{12})] = 0$, yielding the symmetric cutoff point introduced earlier, we have

$$\hat{P}_1 = \hat{P}_2 = \Phi\left(\frac{-\sqrt{D^2}}{2}\right)$$

In his discussion on error rates, Lachenbruch (1975) pointed out that these estimates themselves may be misleading for small sample sizes. On the basis of sampling experiments, he has suggested using

$$\left(\frac{n_1 + n_2 - p - 3}{n_1 + n_2 - 2}\right) D^2$$

in place of D^2 in the formulae for $\hat{P}_1$ and $\hat{P}_2$ given by (23.11) and (23.12), respectively.

23-6-3 Continuation of Our Numerical Example

As an assessment of the utility of the previously calculated discriminant function

$$\ell^* = -5.165 + 0.045X_1 + 0.022X_2 + 0.004X_3$$

in classifying individuals correctly, the usual discriminant analysis program provides a table giving for each sample the frequency distribution of the number of individuals assigned to each group. This enables one to calculate the apparent error rates discussed in Section 23-6-2. Based on the rule that an individual is assigned to the CHD group or the NCHD group according as ℓ^* is positive or negative, the following classification table is obtained:

Sample	No. Classified into CHD Group ($\ell^* > 0$)	No. Classified into NCHD Group ($\ell^* < 0$)	Total
CHD	51	20	71
NCHD	272	489	761

Thus, $20/71 = .282$ is the apparent error rate for the CHD group, and $272/761 = .357$ is the apparent error rate associated with the NCHD group. The alternative error rate estimates $\hat{P}_1$ and $\hat{P}_2$ designed to overcome the drawbacks of the apparent error rates have the values

$$\hat{P}_1 = \hat{P}_2 = \Phi\left(-\frac{\sqrt{D^2}}{2}\right) = \Phi\left(-\frac{\sqrt{0.815}}{2}\right)$$
$$= \Phi(-0.451) = .326$$

Incidentally, Lachenbruch's correction to D^2 has the value $826/830 = 0.995$, which is so close to 1 that it can be ignored.

Some discriminant analysis programs allow the user to specify values for the a priori probabilities p_1 and $p_2 = 1 - p_1$; thus classification can be based on whether ℓ^* is greater than or less than $\ln(p_2/p_1)$. From the way in which these data were gathered, it is reasonable to estimate $\ln(p_2/p_1)$ as

$$\ln\frac{n_2}{n_1} = \ln\frac{761}{71} = 2.372$$

and the classification table based on this new cutoff point is as follows:

Sample	No. Classified into CHD Group ($\ell^* > 2.372$)	No. Classified into NCHD Group ($\ell^* < 2.372$)	Total
CHD	0	71	71
NCHD	0	761	761

The values of $\hat{P}_1$ and $\hat{P}_2$ based on this modified cutoff point are

$$\hat{P}_1 = \Phi\left(\frac{2.372 - 0.815/2}{\sqrt{0.815}}\right) = .985$$

and

$$\hat{P}_2 = \Phi\left(-\frac{2.372 + 0.815/2}{\sqrt{0.815}}\right) = .001$$

These error rates are so one-sided because the information in our sample suggests that an individual is much likelier to be a member of the NCHD group (761/832) than of the CHD group (71/832). In addition, the one-sidedness reflects the fact that the cutoff point, $\ln(p_2/p_1)$, assumes (unrealistically) equal costs of misclassification, whereas a cutoff point of 0 can be interpreted as having arisen from the more realistic condition that $p_2/p_1 = c_{12}/c_{21}$. In any case, the example above dramatically illustrates the effect of a change in the value of the cutoff point on the misclassification rates, and it suggests that an appropriate cutoff point should be carefully chosen.

23-7 Stepwise Discriminant Analysis

We shall now analyze these data using *stepwise discriminant analysis*. This procedure operates in principle like that of stepwise multiple regression in the sense that one variable is included in the discriminant function at each step. This variable is the one that results in the most significant F-value after adjusting for variables already included in the model. This step-by-step procedure continues until no further significant gain in discrimination (as reflected by these partial F's) is achieved by adding more variables to the discriminant function. At every step of stepwise discriminant analysis, the importance of the variables that have been included and those that are candidates for inclusion in the discriminant function can be examined. This is quite important since a variable that may have appeared to be fairly important at an early stage may become superfluous at a later stage because of the relationship between it and other variables already in the model. In general, such redundancy among variables often goes unnoticed when these variables are forced into the discriminant function (as they were in the analysis just completed) and can even lead to a loss in discriminatory power.

At each step in a stepwise discriminant analysis, the computer output consists of the overall F-value, the partial F for each variable already in the model (often called the "F-to-remove" since a variable with an insignificant value is dropped from the discriminant function), the partial F for each variable still a candidate for inclusion (which is called the "F-to-enter"), the discriminant function coefficients, and the usual classification table. At the final step when there are exactly p variables in the model, the overall F has p and $n_1 + n_2 - p - 1$ degrees of freedom, an F-to-remove has 1 and $n_1 + n_2 - p - 1$ degrees of freedom, and an F-to-enter for a particular variable has 1 and $n_1 + n_2 - p - 2$ degrees of freedom, since it is the partial F obtained when that variable is the next variable added to the model.

For our data from Table 23-2, the initial F's-to-enter (with 1 and 830 degrees of freedom) are 43.969 for X_1, 22.768 for X_2, and 14.607 for X_3. Thus, the variable X_1 (age) is the first to be included in the discriminant function and the associated output for the first step is as follows:

Step number: 1

Variable entered: X_1

Variables included and F-to-remove: X_1: 43.969 (1, 830 df)

Variables not included and F-to-enter: X_2: 6.820, X_3: 3.202 (1, 829 df)

Overall F: $F_{1,830} = 43.969$

DISCRIMINANT FUNCTION SCORES

Variable	Group NCHD	Group CHD	b_j
X_1	0.209	0.265	$b_1 =$ 0.056
(intercept)	−5.380	−8.238	$b_0 =$ −2.858

CLASSIFICATION TABLE (BASED ON $\ell^* = -2.858 + 0.056X_1$)

Sample	No. Classified into CHD Group ($\ell^* > 0$)	No. Classified into NCHD Group ($\ell^* < 0$)	Total
CHD	48	23	71
NCHD	279	482	761

Examination of the F's-to-enter indicates that variable X_2 (diastolic blood pressure) is to be included in the model at the second step:

Step number: 2

Variable entered: X_2

Variables included and F-to-remove: X_1: 27.599, X_2: 6.819 (1, 829 df)

Variables not included and F-to-enter: X_3: 1.675 (1, 828 df)

Overall F: $F_{2,829} = 25.548$

DISCRIMINANT FUNCTION SCORES

Variable	Group NCHD	Group CHD	b_j
X_1	0.080	0.128	0.048
X_2	0.382	0.406	0.024
(intercept)	−19.111	−23.768	−4.657

CLASSIFICATION TABLE (BASED ON $\ell^* = -4.657 + 0.048X_1 + 0.024X_2$)

Sample	No. Classified into CHD Group ($\ell^* > 0$)	No. Classified into NCHD Group ($\ell^* < 0$)	Total
CHD	53	18	71
NCHD	267	494	761

It is not necessary, of course, to present step 3, since the output would be identical to that obtained earlier when the full model was considered.

Examination of the classification tables suggests that the model containing X_1 and X_2 does a better job of classifying individuals than does the model with X_1 alone or even the model containing all three variables. The partial F statistics reflect this situation in the sense that X_1 and X_2 are always associated with significant F-to-enter and F-to-remove values, whereas X_3's F-to-enter value becomes nonsignificant once X_1 and X_2 are in the model. This example illustrates how forcing an essentially superfluous variable into the discriminant function can lead to a loss in discriminatory power. Finally, in a manner completely analogous to that given earlier, a stepwise-multiple-regression program can be used to produce the same information at each step provided by the stepwise-discriminant-analysis program. Incidentally, the R^2-values associated with steps 1, 2, and 3 are .050, .058, and .060, respectively, indicating minimal predictive power.

23-8 Miscellaneous Remarks

In this final section some special points of interest will be briefly discussed. First, the discriminant function need not be linear in the X's. In other words, as long as the discriminant function is linear in its coefficients, all the procedures that we have described can be used no matter how the X's are defined. As an example, for the data of the previous section, we could have considered a model containing an interaction term of the form $X_1 X_2$ or such terms as $\ln X_3$ or $X_3^{1/2}$. The analogy to multiple regression should be apparent.

Second, in typical situations where the purpose is to discriminate between diseased and nondiseased groups of individuals, it is often desirable to use a function of ℓ^* as a measure of the probability (or risk) of developing the disease. In particular, the *multiple logistic function*

$$P(\ell^*) = \frac{1}{1 + (p_2/p_1)(c_{21}/c_{12})e^{-\ell^*}}$$

is often used for this purpose. Note that as ℓ^* ranges from $-\infty$ to $+\infty$ the value of $P(\ell^*)$ increases monotonically from 0 to 1. The above form of the multiple logistic function is based on a cutoff point of $\ln[(p_2/p_1)(c_{21}/c_{12})]$, so a value of ℓ^* exactly equal to $\ln[(p_2/p_1)(c_{21}/c_{12})]$ implies a 50–50 chance of developing the disease. For a cutoff point of 0, then, $P(\ell^*)$ equals $1/(1 + e^{-\ell^*})$, and so $P(0) = \frac{1}{2}$. Based on the assumptions of multivariate normality and of variance and covariance homogeneity in the two populations, it is possible to argue on theoretical grounds alone in favor of the function $P(\ell^*)$. Although we will not present the argument here, it is given, for example, by Cornfield (1962), and applications based on the use of the multiple logistic function can be found in Cornfield (1962), Truett, Cornfield, and Kannel (1967), and Kleinbaum and others (1971).

In practice, however, it is generally the case that one or more X-variables will not be continuous, much less normally distributed; for example, categorical variables like "sex" (male or female), "race" (black or white), "socioeconomic status" (high, medium, or low), "exposure level" (none, moderate, or heavy) are typical in many studies. When such variables are to be used in a logistic regression model, discriminant analysis should *not* be employed to estimate the model parameters, since the resulting estimates can be very positively biased. Instead, either unconditional or conditional maximum likelihood estimation methods should be used. For further discussion, see Section 21-5 of Chapter 21 and also Kleinbaum, Kupper, and Morgenstern (1982, chaps. 20–24).

Finally, we have not explicitly discussed how to estimate the variance of one of the discriminant function coefficients b_j, $j = 1, 2, \ldots, p$. A large-sample expression for this estimated variance that has been suggested in the literature is

$$s^{jj}\left(\frac{1}{n_1} + \frac{1}{n_2}\right)$$

For the example we have been considering, the estimated variance of b_1 based on the formula above is

$$0.00560 \left(\frac{1}{71} \times \frac{1}{761}\right) = 8.62 \times 10^{-5}$$

Another possible method for obtaining an estimate of the variance of b_j can be based on the fact that $b_j = \hat{\beta}_j / c^*$, so that

$$\text{Var } b_j = \frac{\text{Var } \hat{\beta}_j}{(c^*)^2}$$

Using this expression, we estimate the variance of b_1 as

$$\text{Var } b_1 = \frac{(0.00070)^2}{(0.0736)^2} = 9.05 \times 10^{-5}$$

which is close to the large-sample value above.

Problems

1. A retrospective study of risk factors for coronary heart disease (CHD) involved a sample of $n_1 = 15$ persons free of CHD and a second sample of $n_2 = 10$ recent CHD cases selected from the employment files for workers with at least 10 years of continuous service. For each person sampled, values of systolic blood pressure (SBP), cholesterol (CHOL), and age (AGE) were obtained from the company's medical records of 10 years ago. The goal of the study was to determine the extent to which the variables SBP, CHOL, and AGE can help to discriminate future cases from noncases. The data are as follows for noncases:

Variable	Control														
	1	2	3	4	5	6	7	8	9	10	11	12	13	14	15
SBP	135	122	130	148	146	129	162	160	144	166	138	152	138	140	134
CHOL	227	228	219	245	223	215	245	262	230	255	222	250	264	271	220
AGE	45	41	49	52	54	47	60	48	44	64	59	51	54	56	50

For cases the data are the following:

Variable	Case									
	1	2	3	4	5	6	7	8	9	10
SBP	145	142	135	149	180	150	161	170	152	164
CHOL	238	232	225	230	255	240	253	280	271	260
AGE	60	64	54	48	43	43	63	63	62	65

a. Consider the accompanying summary table of sample means and standard deviations for each variable in each group. What standard statistical procedure might you use, given the information in the table, to evaluate whether each variable individually discriminates cases from noncases? Carry out this procedure for SBP, CHOL, and AGE separately, and comment on your findings.

Variable	Noncases		Cases	
	$\bar{X}$	S	$\bar{X}$	S
SBP	142.93	12.83	154.80	13.77
CHOL	238.40	18.61	248.40	18.39
AGE	51.60	6.37	56.50	8.81

b. In assessing the combined ability of SBP, CHOL, and AGE to discriminate cases from noncases, how would you define the dummy dependent variable (for sample sizes $n_1 = 15, n_2 = 10$) if you were using multiple regression to carry out the discriminant analysis computations?

c. Using (23.5), determine for the regression model containing SBP, CHOL, and AGE that constant c^* which, when divided into the regression coefficients, yields the corresponding linear discriminant function coefficients. (*Note:* The quantity D^2 can be computed directly from the regression printout using the F statistic (with p and $n_1 + n_2 - p - 1$ degrees of freedom) for testing the significant overall regression by means of the following formula:

$$D^2 = \frac{(n_1 + n_2)(n_1 + n_2 - 2)p}{n_1 n_2 (n_1 + n_2 - p - 1)} F$$

The F statistic for this data set has the value 1.871.)

d. Given the multiple regression solution

$$\hat{Y} = 2.23602 - 0.01406(SBP) + 0.00242(CHOL) - 0.01393(AGE)$$

compute the linear discriminant function using your answer to part (c), and then use this function to define two distinct classification rules according to the following specifications of a priori probabilities and costs of misclassification:

(1) $p_1 = .6, p_2 = .4, c_{12} = c_{21}$
(2) $p_2/p_1 = c_{12}/c_{21}$

(*Note:* Each rule should be of the following general form: If $\ell^* > C$, assign to group 1 (NCHD); if $\ell^* < C$, assign to group 2 (CHD).)

e. For classification rule (2) in part (d), determine the appropriate group assignment for noncase 1 and case 10 based on this rule.

f. Consider the following classification tables based on the two classification rules specified in part (d):

(1)

Actual Group	Assigned Group	
	NCHD	CHD
NCHD	13	2
CHD	5	5

(2)

Actual Group	Assigned Group	
	NCHD	CHD
NCHD	11	4
CHD	3	7

Which table is preferable and why?

g. For classification table (2) in part (f), determine the apparent error rates and the more precise estimated probabilities of misclassification discussed in the chapter.

h. Using the regression ANOVA table, test whether there are significant differences between the two groups when all variables are considered together.

Source	df	SS
SBP	1	1.0441
AGE|SBP	1	0.1956
CHOL|SBP, AGE	1	0.0259
Residual	21	4.7345

i. The order of variables presented in the ANOVA table is precisely that order obtained from forward-stepwise-regression analysis. Based on this information, how would you modify the linear discriminant function to eliminate variables not making a significant contribution to discrimination?

j. Given the fitted regression model

$$\hat{Y} = 2.1656 - 0.01466(\text{SBP})$$

determine an appropriate decision rule for discriminating cases from noncases based on this model. How might one evaluate whether this reduced discriminant function does a better job of discriminating than the function containing all three variables?

k. What is the multiple logistic function corresponding to the discriminant function found in part (j)? Using this function, what is the estimated risk of developing CHD for noncase 1 and for case 10?

2. An anthropologist was interested in developing a procedure for classifying skulls from one of two Egyptian historical periods (I and II). She applied discriminant analysis to the measurement data given here, which were obtained on a sample of five skulls from each period.

Period	Maximum Breadth	Basialveolar Length	Nasal Height	Basibregmatic Height
I	134.825	98.703	50.538	133.110
I	134.203	97.667	52.380	134.201
I	133.714	98.976	52.100	134.628
I	135.901	99.203	50.875	133.765
I	133.976	99.521	49.615	132.861
II	135.603	95.400	52.903	131.646
II	135.422	97.832	52.200	130.876
II	136.721	95.212	51.875	133.888
II	134.750	95.861	53.769	130.773
II	137.100	96.500	54.010	129.998

Period	Group Means			
I	134.5238	98.8140	51.1016	133.7130
II	135.9192	96.1610	52.9514	131.4362

a. Determine the linear discriminant function ℓ involving all four variables utilizing the following computer results of scores for each variable in each group:

Variable	Period I	Period II
Maximum breadth	178.8834	180.6715
Basialveolar length	417.4353	410.6387
Nasal height	224.4495	223.8206
Basibregmatic height	284.1536	279.4690
(intercept)	$-57{,}389.2539$	$-56{,}314.6914$

b. Determine the value of b_0, and then define a classification rule assuming that the ratio of prior probabilities is inversely related to the ratio of costs of misclassifications, that is, $p_2/p_1 = c_{12}/c_{21}$.

c. Given that the F statistic of (23.7) has the value 11.730 for the model involving all four variables, test whether there are significant differences between the two groups based on this model.

d. Use the classification rule developed in part (b) to assign the first skull in each group to one of the two periods.

e. The classification table based on the rule in part (b) is as shown. Comment on the discrimination achieved.

Actual Group	Assigned Group	
	I	II
I	5	0
II	0	5

f. The use of a stepwise-discriminant-function program on these data turned up only two variables, basialveolar length and basibregmatic height, as making a significant contribution to discrimination based on partial F tests. The scores based on this two-variable model are given in the table below. Form the discriminant function, the classification rule (based on the cutoff point b_0), and the classification table for this model. Comment on the difference, if any, in discrimination achieved using only these two variables as opposed to using all four.

Variable	Period I	Period II
Basialveolar length	273.3677	267.0010
Basibregmatic height	211.6374	207.3149
Constant	−27,656.2852	−26,462.5547

3. A methodological study was undertaken in a large metropolitan area to describe the organizational climate of public high schools in the area. In this study, several variables describing various components of a school's organizational personality were derived by factor analysis and then used in a two-group discriminant analysis to classify schools into one of two categories of organizational climate: satisfactory (A) and unsatisfactory (B). A discriminant function involving four variables was developed from the data: disengagement (i.e., anomie) of teachers (DIS), hindrance from principal as perceived by teachers (HIN), morale (MOR), and aloofness (formality) exhibited by principal (ALO). The data are given below.

Climate Type	DIS	HIN	MOR	ALO
A	2.58	1.86	1.77	1.86
A	3.12	2.97	1.28	1.71
A	1.87	1.99	0.72	1.89
A	1.65	1.92	1.86	0.98
A	1.42	2.31	1.25	0.85
A	2.41	3.12	0.82	1.28
A	1.25	1.78	1.06	2.01
A	−2.78	2.21	0.25	1.72
A	1.25	0.51	0.56	−0.67
A	−1.28	3.02	0.65	−0.97
B	0.85	1.21	0.48	1.65
B	0.96	1.35	0.38	1.57
B	−2.16	0.18	−0.17	−1.28
B	−1.85	−2.61	0.76	−2.01
B	−1.66	−1.89	0.85	−3.10

a. To use multiple regression to carry out the discriminant analysis computations, how would you define the dummy dependent variable?

b. The multiple regression solution involving all four variables is as follows:

$$\hat{Y} = -0.59047 - 0.03022(\text{DIS}) + 0.23192(\text{HIN})$$
$$+ 0.38134(\text{MOR}) - 0.04191(\text{ALO})$$

The discriminant function scores for each group obtained from a standard stepwise-discriminant-analysis program are given in the table below. Form the linear discriminant function using the stepwise-discriminant-analysis computer output, and verify that the discriminant function coefficients are constant multiples of the corresponding regression coefficients.

Variable	Climate A	Climate B
DIS	−1.22922	−0.87242
HIN	3.28807	0.54974
MOR	8.29556	3.79302
ALO	−0.58024	−0.08542
(intercept)	−7.48262	−1.83262

c. The F statistic of (23.7) has the value 5.0687 for this data set. Carry out the appropriate test of hypothesis using this statistic, and then state your conclusions clearly.

d. Assuming that $\ln[(p_2/p_1)(c_{21}/c_{12})] = 0$, define an appropriate decision rule for classifying schools as either climate type A or climate type B.

e. Use the rule developed in part (d) to classify the first school listed as climate type A in the data set and the first school listed as climate type B.

f. Use the classification table given, which is based on the rule in part (d), to estimate the probabilities of misclassification.

Actual Group	Assigned Group	
	A	B
A	9	1
B	0	5

4. The data given below were used in a discriminant analysis to describe the relationship between pregnancy characteristics of the mother and the prematurity status of her baby. Use the SPSS regression computer printout on the next page to carry out two discriminant analyses of this data set, one using a model involving only prepregnant maternal weight and one using a model involving prepregnant maternal weight and maternal age at delivery.

Variable	Prematurity Status*									
	NP	NP	NP	NP	NP	NP	NP	NP	NP	NP
Prepregnant maternal weight	105	110	115	120	125	130	135	140	125	115
Maternal age at delivery	22	25	27	30	33	35	37	30	25	25

* P = premature delivery; NP = nonpremature delivery.

Variable	Prematurity Status									
	NP	NP	NP	NP	NP	P	P	P	P	P
Prepregnant maternal weight	135	110	120	105	115	145	125	130	140	135
Maternal age at delivery	22	26	32	28	24	35	30	32	40	37

Computer Printout (SPSS) for Problem 4

```
DEPENDENT VARIABLE.. PREM        (Y = .25 if NP,  Y = -.75 if P)

VARIABLE(S) ENTERED ON STEP NUMBER 1..  MW  (Pre-pregnant Maternal Weight)

MULTIPLE R          0.61551       ANALYSIS OF VARIANCE   D.F.   SUM OF SQUARES   MEAN SQUARE        F
R SQUARE            0.37885       REGRESSION             1.     1.42068          1.42068     10.97842
ADJUSTED R SQUARE   0.37885       RESIDUAL               18.    2.32932          0.12941
STANDARD ERROR      0.35973

                            ———— VARIABLES IN THE EQUATION ————
VARIABLE            B            BETA          STD ERROR B           F
MW                -0.00029      -0.61551        0.00009          10.978
(CONSTANT)         1.07384

VARIABLE(S) ENTERED ON STEP NUMBER 2..  MAGE  (Maternal Age at Delivery)

MULTIPLE R          0.62360       ANALYSIS OF VARIANCE   D.F.   SUM OF SQUARES   MEAN SQUARE        F
R SQUARE            0.38888       REGRESSION             2.     1.45831          0.72915      5.40892
ADJUSTED R SQUARE   0.35493       RESIDUAL               17.    2.29169          0.13481
STANDARD ERROR      0.36716

                            ———— VARIABLES IN THE EQUATION ————
VARIABLE            B            BETA          STD ERROR B           F
MW                -0.00045      -0.97191        0.00033           1.924
MAGE               0.03109       0.37021        0.05885           0.279
(CONSTANT)         0.77074
```

From *Statistical Package for the Social Sciences* by Nie et al. Copyright © 1975 by McGraw-Hill, Inc. Used with permission of McGraw-Hill Book Company and Dr. Norman Nie, President, SPSS, Inc.

Computer Printout (BMD excerpt) for Problem 5

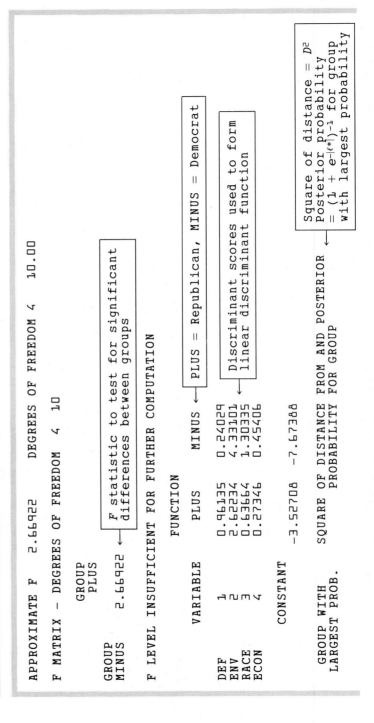

APPROXIMATE F   2.66922   DEGREES OF FREEDOM 4   10.00

F MATRIX - DEGREES OF FREEDOM 4   10

	GROUP PLUS	

F statistic to test for significant differences between groups

GROUP MINUS	2.66922	→

F LEVEL INSUFFICIENT FOR FURTHER COMPUTATION

FUNCTION

		PLUS	MINUS ←	PLUS = Republican, MINUS = Democrat
VARIABLE				

Discriminant scores used to form linear discriminant function

VARIABLE		PLUS	MINUS
DEF	1	0.96135	0.24029
ENV	2	2.62234	4.33101
RACE	3	0.63664	1.30335
ECON	4	0.27346	0.45406
CONSTANT		-3.52708	-7.67388

SQUARE OF DISTANCE FROM AND POSTERIOR PROBABILITY FOR GROUP

GROUP WITH LARGEST PROB.

Square of distance = D^2
Posterior probability
= $(1 + e^{-(\ell^*)})^{-1}$ for group with largest probability

GROUP PLUS CASE		PLUS		MINUS	
1	PLUS	6.450	0.974	13.719	0.026
2	PLUS	1.197	0.901	5.605	0.099
3	PLUS	2.395	0.986	10.918	0.014
4	MINUS	3.038	0.312	1.455	0.688
5	PLUS	4.496	0.751	6.701	0.249

GROUP MINUS CASE		PLUS		MINUS	
1	PLUS	5.663	0.559	6.140	0.441
2	MINUS	7.066	0.358	5.901	0.642
3	MINUS	9.451	0.054	3.707	0.946
4	MINUS	5.779	0.118	1.764	0.882
5	MINUS	5.022	0.193	2.157	0.807
6	MINUS	8.818	0.040	2.480	0.960
7	MINUS	13.026	0.007	3.150	0.993
8	MINUS	14.026	0.006	3.927	0.994
9	PLUS	2.285	0.772	4.721	0.228
10	MINUS	4.927	0.097	0.476	0.903

NUMBER OF CASES CLASSIFIED INTO GROUP

GROUP	PLUS	MINUS
PLUS	4	1
MINUS	2	8

Classication table for which $\ln\left[\frac{p_2}{p_1}\left(\frac{c_{21}}{c_{12}}\right)\right] = 0$

Source: Data from study supported by the Health Sciences Computing Facility, University of California, Los Angeles, California 90024, under NIH Grant RR-00003.

Make sure to complete the following steps:
- **a.** Determine the linear discriminant function from the regression output.
- **b.** Determine the cutoff point b_0, and define your classification rule assuming that $p_2/p_1 = c_{12}/c_{21}$.
- **c.** Test for significant differences between the groups based on the fitted models.
- **d.** Assign each individual to one of the two groups based on the rule in part (b).
- **e.** Estimate the probabilities of misclassification.
- **f.** Draw some overall conclusions about the quality of discrimination achieved.

5. A discriminant analysis was carried out on the data given in the accompanying table to determine the extent to which a person's views on political issues could be used to predict that person's political party affiliation. The independent variables represent indices derived from questionnaire responses regarding certain political issues. Use the computer printout given to perform a discriminant analysis of this data set using a discriminant function model containing all four variables. Complete the same steps as outlined in Problem 4.

DATA FOR PROBLEM 5

Political Party	Defense Spending	Environment	Race	Economy
R	2.42	1.82	−0.65	−2.32
R	2.25	1.76	0.42	1.54
R	2.84	1.32	−0.93	1.65
D	1.79	2.25	1.72	1.15
D	1.01	2.86	0.10	2.79
D	−1.85	2.75	1.78	−1.11
D	−1.72	2.21	1.65	0.21
R	−1.01	1.86	1.41	0.51
D	−0.84	1.92	2.32	0.81
D	−0.77	3.17	1.39	2.31
D	−0.71	3.85	2.55	1.63
D	−0.80	3.65	3.10	1.75
D	−0.45	1.01	1.02	1.10
R	−0.17	0.92	1.25	2.86
D	−1.21	2.35	1.82	1.50

6. For the data in the accompanying table, use the computer printout to carry out a discriminant analysis to describe how the four symptom factors can be used jointly to classify psychiatric patients as either depressive or schizophrenic. Complete the same steps as outlined in Problem 4.

DATA FOR PROBLEM 6

Psychopathological Group*	Thinking Disturbance (TD)	Withdrawal-Retardation (WR)	Hostile-Suspiciousness (HS)	Anxious-Depression (AD)
D	5.0	6.4	6.6	10.2
D	2.9	5.3	2.5	9.2
D	2.7	5.0	2.5	10.3
D	2.5	4.7	3.5	8.6
D	1.9	4.3	4.5	11.1
S	7.8	6.9	3.8	6.0
S	7.6	5.6	4.2	7.3
S	7.4	5.1	5.4	4.9
S	7.2	6.0	6.0	3.6
S	3.5	5.6	3.0	11.8

* D = depressive patient, S = schizophrenic patient.

7. A botanist was interested in classifying hybrid roses as either of the floribunda or the multiflora type by using numerical characteristics of the leaves from a sample of bushes of each parental type to determine a classification index.[3] She selected 15 leaves from each of 10 bushes of each parental type and measured each leaf for two characteristics: the number of stomata on the outer face (X_1) and the number of stomata on the inner face (X_2). She then determined the means of the measurements for each bush and obtained the results given in the table.

Bush	Floribunda X_1	Floribunda X_2	Multiflora X_1	Multiflora X_2
1	128.5	133.8	95.2	88.4
2	133.7	133.6	91.0	95.7
3	126.5	131.1	100.0	103.9
4	131.3	130.5	110.3	112.6
5	119.4	120.6	88.7	93.8
6	123.8	133.6	110.6	108.7
7	132.1	122.3	115.5	116.0
8	134.3	134.6	102.6	101.8
9	129.8	133.5	94.8	98.6
10	130.1	128.7	114.8	115.9
Mean	128.95	130.23	102.35	103.54

[3] Adapted from a study by Mergen (1958).

Computer Printout (BMD excerpt) for Problem 6

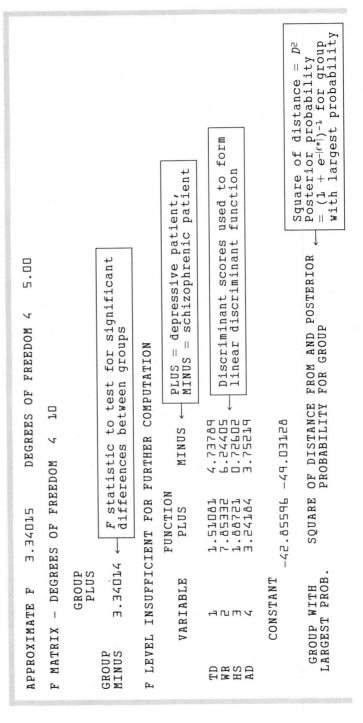

APPROXIMATE F   3.34015   DEGREES OF FREEDOM 4   5.00

F MATRIX – DEGREES OF FREEDOM 4   10

     GROUP
     PLUS

GROUP
MINUS   3.34014 ←   *F* statistic to test for significant
              differences between groups

F LEVEL INSUFFICIENT FOR FURTHER COMPUTATION

              FUNCTION   PLUS = depressive patient,
    VARIABLE   PLUS   MINUS ← MINUS = schizophrenic patient

TD   1   1.51081   4.73789
WR   2   7.85332   6.24405   Discriminant scores used to form
HS   3   1.88721   0.72602   linear discriminant function
AD   4   3.24184   3.75219

    CONSTANT   -42.85596 -49.03128

GROUP WITH   SQUARE OF DISTANCE FROM AND POSTERIOR
LARGEST PROB.   PROBABILITY FOR GROUP

Square of distance = D^2
Posterior probability
$= (1 + e^{-|(*)|})^{-1}$ for group
with largest probability

```
GROUP
PLUS
CASE            PLUS                MINUS
  1         5.932  0.943        11.527  0.057
  2         1.712  0.972         8.819  0.028
  3         1.186  0.959         7.496  0.041
  4         1.248  0.995        11.941  0.005
  5         2.374  0.999        15.422  0.001

GROUP
MINUS
CASE            PLUS                MINUS
  1        16.208  0.001         3.125  0.999
  2        19.921  0.000         3.547  1.000
  3        14.039  0.003         2.583  0.997
  4         8.890  0.093         4.342  0.907
  5         3.243  0.795         5.951  0.205
```

NUMBER OF CASES CLASSIFIED INTO GROUP

GROUP	PLUS	MINUS
PLUS	5	0
MINUS	1	4

> Classication table for
> which $\ln\left[\dfrac{p_2}{p_1}\left(\dfrac{c_{21}}{c_{12}}\right)\right] = 0$

Source: Data from study supported by the Health Sciences Computing Facility, University of California, Los Angeles, California 90024, under NIH Grant RR-00003.

a. Determine the linear discriminant function ℓ involving X_1 and X_2 by utilizing the table below of computer results of scores for each variable in each group.

Variable	Floribunda	Multiflora
X_1	1.04420	0.82099
X_2	1.34694	1.07863
(intercept)	-155.72366	-98.54755

b. Determine the value of b_0 and then define a rule to classify hybrid roses as either of the floribunda type or multiflora type, assuming equal costs of misclassification and equal prior probabilities.

c. If you were using multiple regression to carry out your discriminant analysis computations, how would you define the dummy dependent variable?

d. Using (23.5) and the note given in part (c) of Problem 1, determine the constant c^* which, when divided into the regression coefficients, yields the corresponding linear discriminant function coefficients. (The F statistic for testing for significant overall regression for these data has the value 30.9277.)

e. Given the multiple regression solution

$$\hat{Y} = -3.4240 + 0.01337X_1 + 0.01607X_2$$

compute the linear discriminant function ℓ using your answer to part (d) and then check that the coefficients are the same as those obtained in part (a).

f. Use your classification rule defined in part (b) to classify the hybrids in the table below as either of the floribunda or the multiflora type.

Hybrid	X_1	X_2
1	119.8	117.6
2	121.9	122.4
3	117.2	114.9
4	112.8	119.6

8. To discriminate between two species of clover that are quite difficult to distinguish, measurements were taken on three characteristics for samples of each type, with the results shown in the table that follows.[4]

---

[4] Adapted from a study by Whitehead (1954).

Sample	Species 1			Sample	Species 2		
	X_1	X_2	X_3		X_1	X_2	X_3
1	0.547	0.696	0.217	1	0.062	0.540	0.090
2	0.695	0.624	0.158	2	0.155	0.485	0.120
3	0.321	0.620	0.149	3	0.100	0.550	0.012
4	0.307	0.542	0.126	4	0.141	0.420	0.046
5	0.429	0.562	0.130	5	0.126	0.550	0.006
6	0.210	0.877	0.228	6	0.051	0.511	0.062
7	0.592	0.571	0.148	7	0.131	0.604	0.026
8	0.565	0.612	0.103	8	−0.016	0.510	0.002
9	0.627	0.703	0.115	9	0.061	0.612	0.008
10	0.275	0.614	0.099	10	0.188	0.549	0.064
11	0.702	0.608	0.087	11	0.128	0.544	0.011
12	0.389	0.815	0.190	12	0.112	0.570	−0.008
Mean	0.472	0.654	0.146	Mean	0.103	0.537	0.037

a. In using a multiple regression program to determine a discriminant function involving all three variables, how would you define the dummy dependent variable?

b. Using the stepwise regression results given below, assess the order of importance and significance of each variable. Based on these results, which variables would you use in a discriminant function for discriminating between species?

Source	df	SS	Fitted Equation
X_1	1	4.1771	$\hat{Y} = -0.5432 + 1.89007X_1$
$X_3\|X_1$	1	0.7955	$\hat{Y} = -0.6668 + 1.25118X_1 + 3.36758X_3$
$X_2\|X_1, X_3$	1	0.0828	$\hat{Y} = -1.0726 + 1.25614X_1 + 0.79186X_2 + 2.63219X_3$
Residual	20	0.9446	

c. Using the estimated multiple regression coefficients $\hat{\beta}_1$, $\hat{\beta}_2$, and $\hat{\beta}_3$ for the model containing all three independent variables, determine the cutoff point,

$$\frac{1}{2}(\bar{\ell}_1 + \bar{\ell}_2)c^* = \hat{\beta}_1\left(\frac{\bar{X}_{11} + \bar{X}_{21}}{2}\right) + \hat{\beta}_2\left(\frac{\bar{X}_{12} + \bar{X}_{22}}{2}\right) + \hat{\beta}_3\left(\frac{\bar{X}_{13} + \bar{X}_{23}}{2}\right)$$

that can be used to define a classification rule based on the assumption of equal a priori probabilities and equal costs of misclassification.

d. Give the appropriate classification rule based on the cutoff point obtained in part (c). Why is it not necessary to actually compute c^* in defining this rule? When would it be necessary to compute c^*?

e. Determine an appropriate cutoff point and classification rule when only those variables determined to be significant discriminators in part (b) are considered, again assuming equal a priori probabilities and equal costs of misclassification.

f. For the model in part (e), use the appropriate F test to determine whether there are significant differences between groups.

g. The classification table obtained for both the classification rule involving all three variables and the rule involving only X_1 and X_3 is the same, as is shown below. What does this imply about which model is more appropriate?

Actual Species	Assigned Species	
	1	2
1	12	0
2	0	12

References

Anderson, T. W. (1958). *An Introduction to Multivariate Statistical Analysis*. New York: John Wiley & Sons, Inc.

Cornfield, J. (1962). "Joint Dependence of Risk on Coronary Heart Disease Function Analysis." *Fed. Proc.*, 21: 58–61.

Fisher, R. A. (1936). "The Use of Multiple Measurement in Taxonomic Problems." *Ann. Eugenics*, 7: 179–188.

Hulka, B. S., Kupper, L. L., and Cassel, J. C. (1972). "Determinants of Physician Utilization: Approach to a Service-Oriented Classification of Symptoms." *Medical Care*, 10(4): 300–309.

Kleinbaum, D. G., Kupper, L. L., Cassel, J. C., and Tyroler, H. A. (1971). "Multivariate Analysis of Risk of Coronary Heart Disease in Evans County, Georgia." *Arch. Inter. Med.*, 128: 943–948.

Kleinbaum, D. G., Kupper, L. L., and Morgenstern, H. (1982). *Epidemiologic Research*. Belmont, Calif.: Lifetime Learning Publications.

Lachenbruch, P. A. (1975). *Discriminant Analysis*. New York: Hafner Press.

McMichael, A. J., Spirtas, R., Kupper, L. L., and Gamble, J. F. (1975). "Solvent Exposure and Leukemia Among Rubber Workers: An Epidemiologic Study." *J. Occup. Med.*, 17(4): 234–239.

Mergen, F. (1958). "Genetic Variation in Needle Characteristics of Slash Pine and in Some of Its Hybrids." *Silvae Genetica*, 7: 1–9.

Morrison, D. F. (1967). *Multivariate Statistical Methods*. New York: McGraw-Hill Book Company.

Overall, J. E., and Klett, C. J. (1972). *Applied Multivariate Analysis*. New York: McGraw-Hill Book Company.

Truett, J., Cornfield, J., and Kannel, W. (1967). "A Multivariate Analysis of the Risk of Coronary Heart Disease in Framingham." *J. Chron. Dis.*, 20: 511–524.

Whitehead, F. H. (1954). "An Example of Taxonomic Discrimination by Biometric Methods." *New Phytologist*, 53: 496–510.

Variable Reduction
and Factor Analysis

24-1 Preview

Researchers often face data involving a large number of correlated variables. These variables may constitute a set of potential predictors, a set of potential responses, or simply a set of variables needing to be described or interpreted together. In any case, two types of questions usually arise:

1. Can a smaller set of variables be used to replace the entire original set?

2. What are the underlying dimensions (or characteristics) being measured by the entire set of data?

The first question concerns *variable reduction*. The goal of reduction may be to eliminate collinearity, to simplify data analysis, or to obtain a parsimonious and conceptually meaningful summary of the data. The reduced set of variables may be either a subset of the original set or a newly defined set of variables derived from the original set.

The second question concerns *factor analysis*. The goal of factor analysis is to understand conceptually, and as parsimoniously as possible, what the data are measuring. Ideally, we hope to find that a large collection of variables measures essentially only a small number of "dimensions." Consequently, factor analysis is usually interrelated with variable reduction. That is, using factor analysis, the researcher may be able to understand what is being measured and, at the same time, reduce the number of original variables to more basic factors for analysis. Note, however, that there are methods for variable reduction other than factor analysis.

In this chapter, we first focus on methods for variable reduction that identify an important subset of the original set of variables. We illustrate such methods using an example involving pulmonary function.

We then discuss an application of factor analysis in which underlying factors are conceptually identified and the number of variables is reduced. This application uses data

595

from an epidemiologic study of the effect of socioecologic stress on mortality. The particular type of factor analysis illustrated through this example is intimately related to a method called principal-components analysis.

24-2 Variable Reduction: The Problem

The variable reduction problem is defined as follows: We begin with an original set of k variables, $X_1, X_2, \ldots, X_k$. We seek to find a subset of these variables $X_1, X_2, \ldots, X_p$, with $p < k$, that best represents the information contained in the original set. This problem may be stated equivalently as finding a partition of $X_1, X_2, \ldots, X_k$ consisting of two disjoint sets, the p-variable "in-set" and the $(k - p)$-variable "out-set." Each X-variable is to belong to one and only one of these two sets. The out-set variables are considered to contain information that is redundant to that contained in the in-set.

As a simple example, suppose $k = 10$, so that our original variable set is $\{X_1, X_2, \ldots, X_{10}\}$. After variable reduction, we might find that $p = 3$ and that our in-set consists of X_2, X_5, and X_7; the remaining seven variables, $X_1, X_3, X_4, X_6, X_8, X_9$, and X_{10}, fall into the out-set.

Note that this example presents a somewhat restricted situation. That is, the reduced set of variables can only be a proper subset of the original variables. A more general formulation of the problem would permit the reduced set to consist of *new* variables, say $F_1, F_2, \ldots, F_p$, that are more complicated functions of the original variables. For example, each F_i could be a linear function of the original X-variables, i.e.,

$$F_i = C_{i1}X_1 + C_{i2}X_2 + \cdots + C_{ik}X_k \qquad i = 1, 2, \ldots, p$$

This is a special case of the formulation used in factor analysis, a procedure that we will discuss in later sections.

(Note that if the C_{ij} terms are defined so that, for each i, C_{ij} equals 1 for only one of the X's and is 0 for the remaining X's, then each F_i will simply be one of the X's; this corresponds to the in-set/out-set formulation of the problem. However, in general, the C_{ij} will not have this restricted definition. In fact, if factor analysis is used, the resulting factors will be linear combinations like F_i above, in which some or all of the C_{ij} are nonzero. Nevertheless, using factor analysis, each *new* variable usually gives most weight (high C_{ij}'s) to only a few of the X's. Thus, only those X's with high "loadings" (i.e., nonnegligible C_{ij}-values) can be used to replace the full set of X's. This means that, rather than the methods for variable reduction described earlier in this section, it may be possible to use factor analysis to determine an in-set.)

The methods for variable reduction discussed here assume that we have interval-scale variables. We also assume that the population pattern of correlations among variables is the same for all subjects in the sample. Methods for non-interval-scale variables fall under the general topic of "multidimensional scaling." (See Muller, Hosking, and Helms, 1979, for a review of some methods for handling the problem of a nonconstant correlation pattern. More general procedures are given by Jöreskog, 1978.)

The exploratory nature of the methods discussed here is their most important limitation. Mulaik (1972) discussed the distinctions between using factor analysis for exploratory analysis and using it for confirmatory analysis. In exploratory analysis, the data are examined to make decisions about subsequent reanalysis of the data. Like methods of variable selection in regression, methods of variable reduction suffer from instability. Hence, the same approaches are recommended for demonstrating the reliability of conclusions, namely, split-sample and cross-validation methods. (See Chapter 16 for more details.)

24-3 Methods for Variable Reduction

What is the best in-set? (Equivalently, what is the best out-set?) We choose to address this question in small steps. Recall that we are restricting our attention to interval-scale variables. This leads us to consider the squared multiple correlation coefficient. In general, given k X's, a "best" in-set of size p, $p = 1, 2, \ldots, k - 1$, can be determined; the associated "best" out-set is then of size $k - p$. For each p,

$$_kC_p = \frac{k!}{p!(k - p)!}$$

possible in-set choices can be made. Altogether, $2^k - 1$ choices of in-set/out-set partitions are possible.

Consider, for the moment, in-sets of size $p = k - 1$, so that the corresponding out-set is of size 1. Only k such in-sets are possible, each with one of the original variables deleted. With the squared multiple correlation for variable j, given all others, denoted $R_j^2 = R^2(X_j | X_1, X_2, \ldots, X_{j-1}, X_{j+1}, \ldots, X_k)$, it seems natural to consider deleting that variable with the largest R_j^2. Indeed, we encountered R_j^2 in Chapter 12, where we also used it as a measure of redundancy. Equivalently, we can consider the tolerance value, $\text{Tol}_j = 1 - R_j^2$. The suggested rule for defining an out-set of size 1, then, is to choose as the out-set variable that one with the maximum squared multiple correlation (minimum tolerance) with the in-set. By doing so, we insure that, at least for the sample being analyzed, we have deleted the variable with the most redundant information.

The above rule may be generalized in many plausible ways. Beale, Kendall, and Mann (1967) suggested maximizing the *minimum* multiple correlation between the out-set (as responses, taken one at a time) and the in-set (as predictors). McCabe (1984) suggested four more general multivariate criteria. One of them is equivalent to maximizing the *average* multiple correlation between the out-set (as responses, taken one at a time) and the in-set (as predictors). This criterion is also equivalent to maximizing the "total redundancy" statistic, defined in another context by Stewart and Love (1968; also see Muller, 1981).

24-4 Choosing Fewer Variables: An Example

An unpublished study by McDonnell and Davis (1982) will be used to highlight the basic issues involved in choosing a subset of variables from a larger basic set. Plethysmography and spirometry maneuver variables were used as indicators of pulmonary function in young, white human males. A total of 18 variables were measured. The goal of the analysis was to find a small number of variables that would provide a profile of pulmonary function without sacrificing significant information contained in the original 18 variables. A potential advantage of such variable reduction was to simplify subsequent analysis and interpretation of studies in which plethysmography and spirometry were used as responses. Removing superfluous variables could also substantially help control both Type I and II error rates in future studies (see Muller, Barton, and Benignus, 1984).

Table 24-1 provides a summary of the variables studied. The demographic variables are not of primary interest, since they can be assumed to be always measurable as control variables. Thus, they will be treated separately in subsequent analysis. The remaining variables have been grouped into "volume," "flow," "airway resistance," and "miscellaneous" measurement groups.

Since substantial exploratory analysis was anticipated, a split-sample, cross-validation approach was used. In Chapter 16, we discussed the split-sample approach in detail. Here, the sample was split into an exploratory sample ($n = 77$) and a confirmatory sample ($n = 78$).

TABLE 24-1 Selected study variables in the pulmonary function example ($n = 155$)

Variable	Description	Mean	S.D.
Demographic measures			
HEIGHT	Height (cm)	180	7.1
WEIGHT	Weight (kg)	72.9	8.6
AGE	Age (yr)	27.4	3.7
Volume measures			
FRC	Functional residual capacity (ml)	3,589	653
TGV	Thoracic gas volume (ml)	3,937	668
FVC	Forced vital capacity (ml)	5,422	798
Flow measures			
FEV05	Forced expiratory volume in 0.5 sec (ml)	3,243	446
FEV1	Forced expiratory volume in 1 sec (ml)	4,448	588
FEV2	Forced expiratory volume in 2 sec (ml)	5,152	718
FEV3	Forced expiratory volume in 3 sec (ml)	5,329	765
FEF1	Midmaximal expiratory flow (l/min)	274	66.5
FEF2	Average flow from 200 to 1,200 (ml/sec)	555	88.9
PEF	Peak expiratory flow (ml/sec)	10,601	1,519
FEF25	Expiratory flow at 25% of volume (ml/sec)	2,340	693
FEF50	Expiratory flow at 50% of volume (ml/sec)	5,155	1,270
Airway resistance measures			
RAW	Resistance in airways (cm H_2O/sec/l)	1.1	0.3
ADJRAW	RAW adjusted for TGV (cm H_2O/sec)	4.3	1.5
FEV1FVC	Ratio of FEV1 to FVC	82.5	6.9
Miscellaneous measures			
TMFVC	Time to FVC (msec)	413	117
TMPEF	Time to PEF (sec)	8.1	3.1
VOLRDC	Volume from regression decay curve (ml)	753	184

Table 24-2 presents the matrix of intercorrelations among all variables for the exploratory sample. A matrix of this size can be understood only by considering meaningful submatrices. For example, the volume measures are all highly intercorrelated, as are the flow measures and the airway resistance measures. In contrast, neither the miscellaneous measures nor the demographic measures are highly intercorrelated. Also note that some correlations are in the .95 to .99 range (e.g., r(FVC, FEV2) = .96, r(FEV2, FEV3) = .99). When selecting variables in multiple regression, such high correlations argue for deleting some redundant variables to avoid collinearity. That is exactly what we want to accomplish here.

The analysis of the exploratory sample follows a backward deletion strategy, similar to that used in Chapter 16 when selecting a best regression model. However, rather than maximizing a single multiple correlation (or other criterion such as C_p) based on a single response variable (as in Chapter 16), the goal here is to pick the "best" out-set. Here such an out-set is that for which the minimum multiple squared correlation based on predicting each out-set member using the in-set variables is maximized. This process typically consists of the following steps:

1. Delete the single variable with the smallest tolerance ($\text{Tol}_j = 1 - R_j^2$), thus defining the out-set of size 1.

2. Within the in-set, delete that single variable with the smallest tolerance (this calculation ignores the variable deleted in the previous step).

3. Compute the set of multiple squared correlations based on predicting each out-set variable from the set of in-set variables.

4. If the minimum multiple squared correlation from step 3 is too small, return the last deleted variable to the in-set and stop. Otherwise, deleting exactly one variable each time, repeat steps 2 through 4 until the minimum multiple squared correlation is too small.

For the example under consideration, the deletion decisions at each step involved the subjective preferences of the pulmonary function specialist. For example, if two variables were close in tolerance, then the least interesting one was deleted, even if it had a slightly larger tolerance. Furthermore, since a confirmatory analysis was planned, many additional exploratory analyses were conducted, including a principal-components analysis.

Throughout the exploratory analysis, HEIGHT, WEIGHT, and AGE were assumed to be mandatory in-set members. Based on the above four-step process (modified somewhat by subjective preference), an in-set consisting of six variables was chosen to define the pulmonary function profile (i.e., the set of in-set variables): FRC, FVC, FEF1, PEF, ADJRAW, and TMPEF. The first two are volume measures, the second two are flow measures, ADJRAW is an airway resistance measure, and TMPEF is a miscellaneous measure. A conscious effort was made to include measures from each of the four groups.

Table 24-3 summarizes the ability of the profile variables to predict each of the out-set variables. The second column provides squared correlations for each out-set variable with the pulmonary function profile variables and the demographic covariates. The third column provides squared correlations for each out-set variable and the pulmonary variables, but with the covariates partialed out of the relationship. The analysts required that approximately 60% of the variance of each out-set variable be predicted by in-set variables. Such a requirement was clearly met.

TABLE 24-2 Correlation matrix for exploratory sample ($n = 77$) in pulmonary function example

Variable	FVC	FRC	TGV	FEF1	FEV1	PEF	FEF50	FEF25	FEF2	FEV05	FEV2	FEV3	ADJRAW	FEV1FVC	RAW	TMPEF	TMFVC	VOLRDC	HEIGHT	WEIGHT	AGE
Volume measures																					
FVC	1.00																				
FRC	.63	1.00																			
TGV	.67	.83	1.00																		
Flow measures																					
FEF1	.16	.09	.13	1.00																	
FEV1	.80	.51	.56	.71	1.00																
PEF	.46	.12	.16	.44	.59	1.00															
FEF50	.16	.02	.08	.97	.69	.45	1.00														
FEF25	.19	.20	.24	.93	.69	.28	.84	1.00													
FEF2	.47	.15	.18	.50	.64	.94	.52	.33	1.00												
FEV05	.60	.29	.33	.83	.92	.74	.84	.71	.79	1.00											
FEV2	.96	.66	.69	.39	.92	.49	.37	.42	.53	.74	1.00										
FEV3	.99	.67	.69	.26	.85	.47	.25	.29	.50	.66	.99	1.00									
Airway resistance measures																					
ADJRAW	.27	.30	.40	-.47	-.11	-.15	-.50	-.37	-.24	-.29	.15	.22	1.00								
FEV1FVC	-.34	-.21	-.20	.85	.29	.18	.82	.78	.25	.48	-.08	-.23	-.62	1.00							
RAW	-.03	.02	-.03	-.55	-.37	-.23	-.56	-.49	-.35	-.46	-.17	-.09	.90	-.55	1.00						
Miscellaneous measures																					
TMPEF	-.01	.05	.20	.24	.11	-.33	.21	.28	-.29	.01	.06	.02	-.01	.18	-.10	1.00					
TMFVC	.18	-.25	-.13	-.53	-.23	.02	-.46	-.62	-.04	-.24	-.07	.05	.24	-.65	.32	-.20	1.00				
VOLRDC	.64	.69	.64	-.09	.42	.09	-.14	.07	.10	.17	.64	.66	.47	-.34	.19	.00	-.26	1.00			
Demographic measures																					
HEIGHT	.59	.55	.55	.35	.63	.35	.35	.35	.34	.51	.64	.62	.15	.04	-.10	.05	-.13	.43	1.00		
WEIGHT	.55	.23	.30	.18	.47	.37	.19	.16	.34	.40	.51	.53	.16	-.13	.02	.08	.18	.19	.58	1.00	
AGE	.02	-.03	-.07	-.26	-.15	.05	-.24	-.29	-.05	-.16	-.09	-.04	.01	-.26	.04	-.12	.34	-.03	-.11	.09	1.00

TABLE 24-3	Squared multiple correlations when predicting each out-set variable using in-set variables for exploratory sample ($n = 77$) in pulmonary function example	

	R^2 with In-Set	
Out-Set Variable	Including Covariates	Partialing Covariates
FEV2	.981	.966
FEV3	.992	.987
FEF50	.946	.935
FEF25	.899	.876
FEF2	.916	.902
FEV05	.974	.962
FEV1FVC	.955	.950
RAW	.938	.937
TGV	.763	.659
FEV1	.983	.970
TMFVC	.620	.529
VOLRDC	.629	.543

TABLE 24-4	Squared multiple correlations when predicting each out-set variable using in-set variables for validation sample ($n = 78$) in pulmonary function example	

	R^2 with In-Set	
Out-Set Variable	Including Covariates	Partialing Covariates
FEV2	.984	.965
FEV3	.993	.985
FEF50	.931	.925
FEF25	.820	.809
FEF2	.913	.885
FEV05	.978	.964
FEV1FVC	.953	.948
RAW	.929	.929
TGV	.905	.821
FEV1	.979	.957
TMFVC	.594	.507
VOLRDC	.700	.569

Turning next to the validation component of the analysis, the question was whether the exploratory sample results would be replicated in the validation sample. Table 24-4 answers this question. Shrinkage values (Chapter 16) for the 12 out-set variables (comparing Tables 24-3 and 24-4) are never more than 0.15, indicating good reliability. One can thus be reasonably confident that comparable results would occur for similar samples of subjects. Since the scientist intended to conduct many more studies with such variables, such high reliability was very comforting. In future studies, apparently only 6 response variables (the in-set) need to be considered, rather than the original 18.

24-5 Overview of Factor Analysis

Factor analysis is a multivariable method intended to explain relationships among several difficult-to-interpret, correlated variables in terms of a few conceptually meaningful, relatively independent factors. Pictorially, this purpose of factor analysis is represented by Figure 24-1, in which the mass of several overlapping circles of various shades is reconstituted into two relatively nonoverlapping circles with different shading patterns.

For a specific set of data, the goal of factor analysis is usually interrelationship modeling, score replacement, or both. By "interrelationship modeling," we mean the assessment of

FIGURE 24-1   General purpose of factor analysis

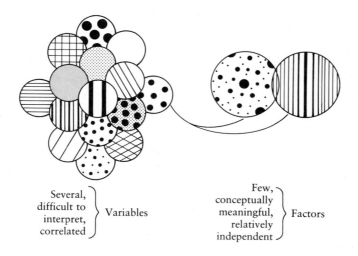

Several, difficult to interpret, correlated } Variables

Few, conceptually meaningful, relatively independent } Factors

the underlying relationships or dimensions in the data; that is, the focus is on understanding what the data are measuring. By "score replacement," we mean the replacement of the original variables with new variables or scores that summarize the data parsimoniously; that is, the focus is on variable reduction.

In Chapter 1 we introduced the reader to factor analysis with the example of the Ponape study, where an application of this method led to the construction of two measures of cultural incongruity. We also gave a brief general description of factor analysis in Chapter 2, in which we pointed out that the method involves the construction of new factors that may be used as independent or dependent variables in later analyses. Thus, factor analysis is frequently used as a data analysis tool in conjunction with other methods such as regression or discriminant analysis.

In the rest of this chapter, we shall describe in some detail the essential terminology and mechanics of factor analysis. We shall motivate our discussion of this subject with an example based on a study by James and Kleinbaum (1976) concerning the relationship between socioecologic stress and hypertension-related mortality in North Carolina.

Prior research (e.g., Stamler, 1967, and Harburg et al., 1973) has indicated that black Americans tend to have higher average blood pressure levels than white Americans and that the death rate due to hypertension and related disorders is considerably higher among blacks than among whites. The reasons for these differences are not fully understood, although most researchers attribute them to a combination of genetic and socioenvironmental factors, including access to good medical care. A study by James and Kleinbaum (1976) considered the socioenvironmental aspect by focusing on the issue of whether blacks residing in high-stress areas tend to have higher hypertension-related mortality rates than either blacks in low-stress areas or whites in high- or low-stress areas. Using 86 counties in North Carolina as their sampling units, James and Kleinbaum computed a three-year hypertension-related death rate for the period 1959 through 1961 for each county and then related these county rates to an index of socioecologic stress for each county that was derived by factor analysis.

The primary question of interest was whether high-stress counties had significantly higher mortality rates than low-stress counties.

The socioecologic stress index was constructed using 15 variables (whose 1960 values were obtained separately by race for each county) that appeared to reflect the economic and social well-being of the counties during the study period. These 15 variables and their code names are given in Table 24-5. Also there are two factors that resulted from the factor analysis and were used in combination to define the index of socioecologic stress. These two factors are called the *socioeconomic status* (SES) factor and the *socioinstability* (SIS) factor. As can be seen in Table 24-5, certain variables (e.g., per capita income and median years of education) were expected to be more related to SES than SIS; other variables (e.g., juvenile delinquency and correction school measures) were expected to be SIS-type variables. Also, some variables (e.g., unemployment) were expected to be related to both the SES and SIS factors. It was expected, therefore, that each of the factors emerging from the factor analysis would correlate highly with those variables thought a priori to be related to that factor and would be essentially uncorrelated with those variables hypothesized to be unrelated to that factor. Later in this chapter we shall examine whether this actually occurred.

(Although in this study the two factors of primary interest had been identified before the factor analysis, such an a priori identification is not a prerequisite for applying this method. In fact, sometimes the researcher does not have the resulting factors firmly in mind and wishes to utilize factor analysis to help characterize the meaningful factors describing the data.)

Figure 24-2 illustrates the purpose of using factor analysis in the James–Kleinbaum study. The circles on the left denote the 15 intercorrelated variables considered in the study, which were used to construct the two relatively independent factors (SES and SIS) on the right.

TABLE 24-5   Variables used in the James–Kleinbaum (1976)
study of 1960 North Carolina counties

1	Per capita income	PCI	⊙
2	Median years of education	MED	⊙
3	Percent unemployment	UEM	⊙ ⊠
4	Percent families earning over $8,000/year	P8K+	⊙
5	Percent male white-collar jobs	WCM	⊙
6	Percent male blue-collar jobs	BC	⊙
7	Percent families earning under $3,000/year	P3K−	⊙ ⊠
8	Percent females separated or divorced	PSDF	⊠
9	Juvenile-delinquency-index males	JDM	⊠
10	Juvenile-delinquency-index females	JDF	⊠
11	Percent of males in correction school	CSM	⊠
12	Percent of females in correction school	CSF	⊠
13	Percent of males in prisons	PM	⊠
14	Homicide rate	HR	⊠
15	Percent of children under 18 not with parent	PCBH	⊠

Note: ⊙ indicates expected SES-type variables; ⊠ indicates expected SIS-type variables. See discussion in text.

FIGURE 24-2   Purpose of factor analysis in the James–Kleinbaum (1976) study

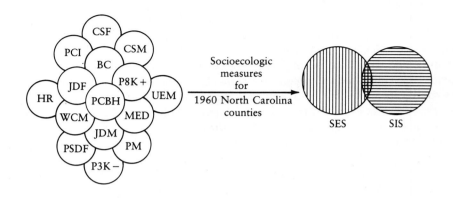

24-6 The Model for Factor Analysis

The classical model used for factor analysis is called the *common-factor* (or *shared-factor*) *model*. Consider a set of mean zero, interval-scale variables, $X_1, X_2, \ldots, X_p$, each observed on n subjects. The common-factor model states that, for variable $j, j = 1, 2, \ldots, p$,

$$X_j = \lambda_{j1}F_1 + \lambda_{j2}F_2 + \cdots + \lambda_{jc}F_c + U_j \tag{24.1}$$

Here, the $\{F_m\}$, $m = 1, 2, \ldots, c$, with $c \le p$, denote a collection of variables called the *common factors*. The λ's are unknown regression-type coefficients called *factor loadings*. Finally, U_j denotes a component of X_j that is called the *unique factor*. Note that the F-variables are common to (shared by) each X_j-variable, whereas U_j is associated only with X_j and not with any of the other X-variables.

Two assumptions about unique factors are always necessary: (1) Unique factors must be mutually uncorrelated, and (2) unique factors must be uncorrelated with the common factors. Although not necessary, it is usually assumed initially that the common factors are also mutually uncorrelated; this last assumption is like a requirement for having *orthogonal* factors. Nonorthogonal factors are said to be *oblique*, which means that scores on such factors are correlated.

A special case of the common-factor model is the *principal-components model*. This model has the form

$$X_j = \lambda_{j1}F_1 + \lambda_{j2}F_2 + \cdots + \lambda_{jc}F_c \tag{24.2}$$

where $j = 1, 2, \ldots, p$. This model differs from the common-factor model (24.1) in that there are no unique factors U_j. An advantage of the principal-components model over the common-factor model is that the former leads to *unique* expressions for the factor variables $\{F_m\}$ as linear functions of the original X-variables. This is why the principal-components model is preferred if the focus of the analysis is score replacement and variable reduction.

The main disadvantage of the principal-components model is that, by assuming no unique factors, the underlying factor structure may not be correctly determined (Mulaik,

1972). This is why the common-factor model is preferred if the focus is on modeling inter-relationships.

The James–Kleinbaum study will provide an example for using principal-components analysis. Since score replacement was the goal there (see Figure 24-2), a principal-components analysis is preferred. Note that if any common-factor model were used for analysis, unique factor scores could not be determined.

In this chapter, we have chosen to examine one application in detail to illustrate one analysis path. The reader must recognize that exploratory factor analysis is a data-driven process. By this, we mean that particular outcomes later in the analysis will depend directly on earlier results. Consequently, this particular example will not (indeed cannot) serve as a template for all analyses.

24-7 Basic Terminology of Factor Analysis

Although a long list of terms concerning factor analysis could be presented, we think that one can understand this method more clearly by focusing on only five basic terms: (1) factor loadings, (2) factor cosines, (3) factor weights (or factor score coefficients), (4) factor scores, and (5) factors. Examples of these terms using data from the James–Kleinbaum study are given in Table 24-6.

24-7-1 Factor Loadings

Factor loadings describe the correlations between the factors emerging from a factor analysis and the original variables used in the construction of the factors. As shown in Table 24-6, the loadings associated with a given factor-analytic solution can be represented by a matrix display, in which the numbers in each column are the correlations of a specific factor with the original variables. The primary use of such a matrix is to pinpoint those variables that are highly correlated (i.e., "load high") with a given factor, so that the factor can be conceptually interpreted. For example, PCI loads high (.88) with SES, as originally hypothesized, whereas PSDF (.75) loads high with SIS. We shall provide a more thorough discussion of these results later. For now, the important thing to remember is that *a factor loading is a correlation of a variable with a factor*.

24-7-2 Factor Cosines

A *factor cosine* is also a correlation, but, in contrast to a factor loading, relates one factor to another factor. Such correlations are important because they quantify the degree to which different factors are related. Ideally, as indicated earlier, one hopes that the factors will be relatively uncorrelated, that is, that the factor cosines will be relatively close to zero. If this is so, each factor can be thought of as representing a distinctly different underlying component of information contained in the original set of variables. On the other hand, if two factors are very *highly* correlated, each would be describing essentially the same component of information; therefore only one of them would have to be considered. This desire for independence, however, does not necessarily call for complete independence among factors. If two factors are considered on empirical or theoretical grounds to be necessarily related to some extent, there should be allowance for some correlation between the two factors. For

TABLE 24-6   Examples of basic terminology of factor analysis (all data for nonwhites in North Carolina counties, 1960)

FACTOR LOADINGS[a]

Variable	Factor	
	SES	SIS
PCI	.88	.22
MED	.88	−.07
UEM	.03	−.03
P8K+	.65	.03
WCM	.56	.23
BC	.81	.12
P3K−	−.88	−.03
PSDF	.18	.75
JDM	.34	.73
JDF	.44	.62
CSM	.38	.60
CSF	.50	.35
PM	−.06	.71
HR	−.41	.26
PCBH	−.12	.75

FACTOR WEIGHTS (SCORE COEFFICIENTS)[b]

Variable	Factor	
	SES	SIS
PCI	.19	−.00
MED	.22	−.10
UEM	.01	−.01
P8K+	.15	−.05
WCM	.11	.03
BC	.18	−.03
P3K−	−.21	.07
PSDF	−.02	.24
JDM	.01	.22
JDF	.05	.17
CSM	.04	.17
CSF	.09	.07
PM	−.08	.25
HR	−.12	.13
PCBH	−.10	.27

FACTOR COSINES[c]

Factor	Factor	
	SES	SIS
SES	1.0	0.17
SIS	0.17	1.0

STANDARDIZED FACTOR SCORES[d]

County	Factor	
	SES	SIS
1	0.87	−0.57
2	1.04	−0.52
3	−0.72	−0.08
⋮	⋮	⋮
85	−1.04	0.94
86	0.25	−0.29

[a] Correlations between factors and variables.

[b] Weights assigned to variables to determine factor scores.

[c] Correlations between factors.

[d] Standardized values of new variables (the factors) defined as weighted sums of the original standardized variables.

example, empirical evidence from other studies about socioeconomic status (SES) and socio-instability (SIS) suggested that these factors would be negatively correlated, with low SES scores corresponding to high SIS scores and high SES scores corresponding to low SIS scores. As Table 24-6 shows, the direction of the factor cosine actually obtained (.17) was in the opposite direction to that predicted, although the amount of correlation was nonsignificant.

24-7-3 Factor Weights

A *factor weight* is not a correlation, but rather a number assigned to a variable (usually standardized into Z score form) for use in determining the scores for a given factor. If one compares the matrix of factor weights in Table 24-6 with that for factor loadings, one should notice that the corresponding entries are not equal. Nevertheless, for a given column (i.e., factor), high factor loadings tend to correspond to high weights, and this will generally be the case for any factor analysis solution. Thus, factor weights and factor loadings give similar information except that they are measured on different scales and are used for different purposes: the weights to compute factor scores and the loadings to describe correlations.

24-7-4 Factor Scores

A *factor score* is a specific value of a factor calculated for a particular sampling unit and is formed as a weighted sum of the values of the (usually standardized) variables for that sampling unit. Since the sampling units for the James–Kleinbaum study are 86 counties in North Carolina, there will be 86 factor scores for each factor, each score corresponding to a specific county. In Table 24-6, the scores for any one factor (e.g., SES) are scaled to Z score form, so that a county (e.g., county 2) with a positive score (e.g., 1.04 on SES) is above average with regard to that factor, whereas a county with a negative score (e.g., -1.04 on SES for county 85) is below average with regard to that factor.

24-7-5 Factors

A factor F is a weighted linear combination of the original variables:

$$F = \sum_{j=1}^{p} w_j X_j = w_1 X_1 + w_2 X_2 + \cdots + w_p X_p$$

in which the w's are the factor weights (to be estimated from the data) and the X's are the original variables (generally expressed in standardized form). When there are k factors under consideration (say, $F_1, F_2, \ldots, F_k$), the ith of these factors is denoted as

$$F_i = \sum_{j=1}^{p} w_{ij} X_j = w_{i1} X_1 + w_{i2} X_2 + \cdots + w_{ip} X_p \qquad i = 1, 2, \ldots, k$$

As an illustration, from Table 24-6 (with $p = 15$), the (estimated) SES factor is

0.19(PCI) + 0.22(MED) + 0.01(UEM) + 0.15(P8K+) + 0.11(WCM)
  + 0.18(BC) − 0.21(P3K−) − 0.02(PSDF) + 0.01(JDM) + 0.05(JDF)
  + 0.04(CSM) + 0.09(CSF) − 0.08(PM) − 0.12(HR) − 0.10(PCBH)

and the (estimated) SIS factor is

−0.00(PCI) − 0.10(MED) − 0.01(UEM) − 0.05(P8K+) + 0.03(WCM)
  − 0.03(BC) + 0.07(P3K−) + 0.24(PSDF) + 0.22(JDM) + 0.17(JDF)
  + 0.17(CSM) + 0.07(CSF) + 0.25(PM) + 0.13(HR) + 0.27(PCBH)

Note that for the SES factor the weights are higher (in absolute value) for most of the SES-type variables (the first seven variables) than for the SIS-type variables; the reverse is true for the SIS factor.

Now, let us distinguish between a factor and a factor score. The term *factor* refers to the general expression for F given earlier, which takes a specific value for each sampling unit selected. The term *factor score* refers to a specific value of the factor obtained by plugging into the general expression for F the values of the original variables for the particular sampling unit selected. When the X's are standardized (e.g., so that they are dimensionless), it makes sense to say that the higher the factor weight for a given variable, the more that variable contributes to the overall factor score and the higher the corresponding factor loading.

It should be mentioned that the factor weights calculated for whites (which are not presented in Table 24-6) were, as expected, quite different from the corresponding weights calculated for nonwhites, since different values of the original variables were obtained for each group. The problem of whether to use a separate set of weights for each racial group rather than a single set for both groups was a methodological issue faced by James and Kleinbaum, and the choice of using separate sets of weights reflected their goal of obtaining the most conceptually meaningful factors for each group. A single analysis of the two groups' data gives factor loadings that are not conceptually meaningful. This demonstrates the need to control for nuisance variables, in this case "race."

24-8 Analysis Strategy for the James–Kleinbaum Study

The primary methodological issue in the James–Kleinbaum study concerned how to use the factor scores to test the primary research question of interest: Do high-stress counties have higher mortality rates than low-stress counties? The steps in the analysis strategy finally adopted were as follows:

1. The factor scores for the counties were first rank-ordered separately for each race. This yielded four orderings of 86 scores, one ordering for each combination of factor (SES and SIS) and race (white and nonwhite).

2. For each ordering, the county factor scores were divided at the median score into two groups, one group representing counties with high scores on the factor and the other representing counties with low scores on the factor. This yielded four groupings:

SES–W	SIS–W	SES–NW	SIS–NW
HI	HI	HI	HI
LO	LO	LO	LO

3. For each race, the two HI–LO groups were then cast into a two-way layout (see Figure 24-3) consisting of the following four cells: HI–SES, HI–SIS; HI–SES,

FIGURE 24-3   Representation of the analysis strategy for
the James–Kleinbaum (1976) study

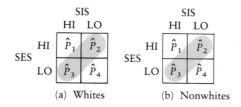

(a) Whites   (b) Nonwhites

LO–SIS; LO–SES, HI–SIS; and LO–SES, LO–SIS. Of these four cells, two were of primary interest: HI–SES, LO–SIS (which was designated the *low-stress group*) and "LO–SES, HI–SIS" (which was designated the *high-stress group*). The rationale for these designations was that a county considered to be high in socioeconomic status (SES) and to have low socioinstability (SIS) was a healthy county in terms of social stress and so was hypothesized to have low mortality. On the other hand, a county low in socioeconomic status and high in socioinstability was considered an unhealthy county and so was hypothesized to have high mortality. The other two cells (HI–SES, HI–SIS) and (LO–SES, LO–SIS), because of their intermediate characteristics, were not considered of principal interest in the analysis.

4. For each of the cells in these two-way tables (one for whites and the other for nonwhites), crude hypertension-related three-year mortality rates were computed for certain age–sex specific groups (we shall report on only those rates computed for males between 45 and 54 years of age).

The primary comparison of interest was between $\hat{P}_2$ and $\hat{P}_3$, the "off-diagonal" rates in each of the two-way tables, as illustrated in Figure 24-3.

24-9 Methodological Steps in Factor Analysis

Having briefly described some of the basic terminology of factor analysis and some of the important methodological issues, we are now ready to discuss in detail the factor-analytic method using this example. The method generally proceeds in four steps.[1]

The first step is to set up the data for input. To this point we have described the data as the set of values of the original variables $X_1, X_2, \ldots, X_p$ for each of the units of study (in the James–Kleinbaum example, $p = 15$, and the units of study are 86 counties in North Carolina). Actually, although considering the data in this form is sufficient, the factor-analytic

[1] There are, however, factor-analytic approaches (e.g., the orthogonal power vector method) that cannot be described as four-step procedures. We will not discuss such approaches here, but the reader is referred to the text by Overall and Klett (1972).

method requires only the correlation matrix among the variables.[2] Many computer programs for factor analysis therefore give the user the option to input the correlation matrix (rather than the original data) if it has already been computed elsewhere. This option is particularly useful if the researcher is concerned about costs, because using the correlation matrix directly involves less storage space and less computer time.

There is one option concerning the correlation matrix $\mathbf{R}$ that the researcher must consider at this first step: whether to replace the 1's on the diagonal of $\mathbf{R}$ with some other numbers, which are called *communalities*. The need to consider this option is why the first step is often called "preparing the correlation matrix." In the next section we shall discuss the implications of this decision in more detail.

The second step involves using the correlation matrix $\mathbf{R}$ to determine a set of initial factors. This is often accomplished by the method of principal components. The important point here is that the determination of these initial factors accomplishes the first two of the three goals of factor analysis: (1) parsimony, (2) approximate independence, and (3) conceptual meaningfulness.

Parsimony is achieved by representing the information contained in several original variables in terms of a much smaller number of factors. *Approximate independence* is achieved by constructing the factors so that they are essentially statistically independent.

> (Actually, the initial factors, if constructed as principal components, are completely independent rather than only approximately independent. The term *approximate* becomes meaningful after the third step, depending on the procedure used at that step.)

Unfortunately, although the initial factors provide a useful first impression, they are usually difficult to interpret directly without some further manipulation. Consequently, a third step, which entails rotation, is usually required to achieve *conceptual meaningfulness*. The concept of rotation will be discussed in detail in Section 24-12.

After rotation, the researcher will have the set of factor weights to be used to determine the factor scores. This is what is done in the fourth step. The researcher will obtain a set of factor scores on the units of study for each derived factor, which may then be used as (newly constructed) variables in further analyses. We will examine the factor scores for the James–Kleinbaum study and describe how these scores were used to test the primary research hypothesis.

24-10 Step 1: Preparation of the Correlation Matrix

As described in the previous section, factor analysis works on the correlation matrix $\mathbf{R}$ of the original variables to obtain the desired factors. This should be an intuitively appealing

---

[2] Recall from Chapter 11 that the correlation matrix $\mathbf{R}$ has the form

$$\mathbf{R} = \begin{array}{c} \\ X_1 \\ X_2 \\ \vdots \\ \vdots \\ X_p \end{array} \begin{array}{c} \begin{array}{cccc} X_1 & X_2 & \cdots & X_p \end{array} \\ \left[\begin{array}{cccc} 1 & r_{12} & \cdots & r_{1p} \\ & 1 & \cdots & r_{2p} \\ & & \ddots & \vdots \\ & & & \vdots \\ & & & 1 \end{array} \right] \end{array}$$

approach, since the relationships among the variables, which we seek to succinctly describe, are represented by the information in the correlation matrix.

> (A simple matrix equation describes how the **R** matrix is worked on; in particular, this equation is of the form
>
> $$\mathbf{RW} = \mathbf{L}$$
>
> where **W** stands for a matrix of factor weights and **L** stands for a matrix of factor loadings. Although this matrix equation can be interpreted in a purely mathematical way, we hope that the reader can view the expression here in a conceptual way, without being concerned with the matrix mathematics (see Appendix B). What this equation says quite succinctly is that the factor-analytic method determines a weight matrix **W** that can be applied to the correlation matrix **R** to get a desirable factor-loading matrix **L** (by "desirable," we mean an **L** matrix that reflects parsimony, approximate independence, and conceptual meaningfulness). This particular **L** can be thought of as resulting from a two-stage procedure. The first stage (or, equivalently, the second step in our four-step procedure) is to determine (e.g., by means of principal-components analysis) an initial weight matrix $\mathbf{W}_I$ and an initial factor-loading matrix $\mathbf{L}_I$ such that $\mathbf{RW}_I = \mathbf{L}_I$. The second stage (or third step) involves the use of a rotation matrix **T** to generate the desired factor-loading matrix $\mathbf{L}$ ($= \mathbf{L}_I \mathbf{T}$) and the corresponding weight matrix $\mathbf{W}$ ($= \mathbf{W}_I \mathbf{T}$).)

In fact, it is only through the off-diagonal elements that the relationships among the variables are reflected. The diagonal elements (which are 1's in the **R** matrix) are, in fact, superfluous in factor analysis. This suggests an alternative approach: Replace these 1's with some other numbers (called *communalities*) and then factor-analyze the resulting adjusted correlation matrix in order to find a more parsimonious and conceptually meaningful factor-analytic solution than can be obtained using the unadjusted **R**.

Of course, some questions are associated with the use of such an approach. What values for the communalities should be used? Does such an alteration of **R** affect the basic structure and/or meaning of the original variables? If so, is the goal of the analysis subverted?

Researchers have provided some theoretical answers to these questions. We shall briefly discuss these answers below; we wish to point out, however, that many applied researchers using factor analysis have not been entirely satisfied with these theoretical explanations and consequently have worked primarily with the original correlation matrix. James and Kleinbaum considered three different correlation matrices, the original and two adjusted correlation matrices. They eventually settled on the factor solution based on the original **R** both because the use of **R** provided the only conceptually meaningful solution and because they could not really justify theoretically the use of an adjusted correlation matrix.

24-10-1 Theoretical Discussion of Communalities

Communalities can be incorporated into the theoretical framework of factor analysis by consideration of the underlying structure of the original variables $X_1, X_2, \ldots, X_p$. Recall that the common-factor model (24-1) was given by

$$X_j = \lambda_{j1}F_1 + \lambda_{j2}F_2 + \cdots + \lambda_{jc}F_c + U_j = C_j + U_j$$

where $F_1, F_2, \ldots, F_c$ are the c factors (which are present in or common to the expressions for all the X's), the λ's are the factor loadings, and U_j is a random component unique to X_j and statistically independent of the F's. In words, any variable can be represented as the sum of a linear combination of the F's (namely, C_j) and a random quantity (namely, U_j) unique to that variable.

Furthermore, the variance of X_j (which is equal to 1 since X_j is in standardized form) can be written as the sum of two variances:

$$\text{Var } X_j = 1 = \sigma^2_{C_j} + \sigma^2_{U_j}$$

where $\sigma^2_{C_j}$ is the variance of the linear combination C_j of the common factors and where $\sigma^2_{U_j}$ is the variance of the unique component U_j.

It is $\sigma^2_{C_j}$ that we call the *communality* of the variable X_j since this quantity measures the information (in terms of variance) that the variable has in common (through the common factors) with all the other variables. Thus, replacing the 1's on the diagonal of the correlation matrix **R** with some other numbers can be shown to be theoretically equivalent to using estimates of the $\sigma^2_{C_j}$'s. Thus, using communalities is equivalent to restricting attention to the common parts (i.e., the C_j's) of the variables (i.e., ignoring the unique components). Keeping 1's on the diagonal, on the other hand, can be viewed as focusing attention on the original variables in their entirety and/or assuming that the variables have no unique components at all.

Thus, the choice of whether to use communalities depends on whether the researcher wishes to consider only the common parts of the variables or to work directly with the original variables. Because such considerations are difficult to quantify directly, many researchers tend to prefer using the original correlation matrix **R**, as did James and Kleinbaum in their study.

24-10-2 Two Methods for Estimating Communalities

The two methods most often used to estimate communalities are as follows:[3]

Method 1   For each variable, use the squared multiple correlation coefficient (R^2) relating that variable to all other variables in the set.

Method 2   For each variable, use the largest (in absolute value) off-diagonal element in **R** associated with that variable (i.e., use the largest correlation involving that variable).

The reader is referred to other texts on factor analysis (e.g., Overall and Klett, 1972) for a discussion of the rationale behind these methods.

24-10-3 Examination of the Correlation Matrices in the James–Kleinbaum Study

Table 24-7 gives the two correlation matrices (one for whites and one for nonwhites) that were used in the James–Kleinbaum study. There are several markings on these matrices (letters, circles, and boxes) that need some interpretation. These markings are intended to indicate how a correlation matrix can be examined to give the researcher a preliminary idea

---

[3] Both methods will yield communalities having values less than 1. In general, the true communality for a given variable will be a number between R^2 and 1, where R^2 is defined as in method 1.

TABLE 24-7 Correlation matrices for the James–Kleinbaum study

WHITES

Variable	1	2	3	4	5	6	7	8	9	10	11	12	13	14	15
1 (PCI)	1.000 (SES)														
2 (MED)	.646	1.000													
3 (UEM)	-.168	-.227	1.000												
4 (P8K+)	.931	.673	-.201	1.000											
5 (WCM)	.734	.865	-.040	.765	1.000										
6 (BC)	.286	-.281	.278	.096	-.170	1.000									
7 (P3K−)	-.905	-.492	.129	-.799	-.567	-.481	1.000								
8 (PSDF)	.649	.225	.207	.592	.484	.447	-.650	1.000							
9 (JDM)	.629	.396	.025	.574	.534	.303	-.596	.566 (SIS)	1.000						
10 (JDF)	.442	.172	.084	.396	.274	.439	-.477	.483	.625	1.000					
11 (CSM)	.237	.034	.175	.199	.245	.348	-.270	.480	.369	.397	1.000				
12 (CSF)	.250	.076	.241	.219	.269	.395	-.253	.451	.401	.567	.694	1.000			
13 (PM)	.158	.038	.138	.189	.225	.005	-.206	.310	.308	.206	.406	.192	1.000		
14 (HR)	-.137	-.176	.209	-.135	-.138	.137	.100	.127	-.009	.052	.071	.050	-.005	1.000	
15 (PCBH)	-.158	-.175	.265	-.103	.003	.111	.116	.402	-.007	.071	.363	.351	.235	.306	1.000

NONWHITES

Variable	1	2	3	4	5	6	7	8	9	10	11	12	13	14	15
1	1.000 (SES)														
2	.720	1.000													
3	.010	.053	1.000												
4	.633	.446	.118	1.000											
5	.519	.587	.311	.433	1.000										
6	.802	.702	.043	.436	.377	1.000									
7	-.930	-.693	.088	-.568	-.420	-.732	1.000								
8	.425	.289	.111	.197	.473	.295	-.269	1.000 (SIS)							
9	.542	.337	-.021	.251	.437	.464	-.387	.579	1.000						
10	.590	.376	-.044	.402	.441	.496	-.463	.519	.881	1.000					
11	.537	.415	-.166	.247	.301	.458	-.404	.575	.581	.499	1.000				
12	.533	.437	-.225	.257	.235	.496	-.456	.370	.427	.492	.570	1.000			
13	.276	-.011	-.189	.117	.132	.083	-.213	.432	.463	.397	.445	.226	1.000		
14	-.144	-.292	.022	-.070	-.126	-.170	.142	.008	-.073	-.118	-.142	-.220	.200	1.000	
15	.178	-.024	.224	.104	.206	.209	.017	.559	.433	.366	.359	.245	.283	.182	1.000

FIGURE 24-4   Ideal correlation matrix structure
for James–Kleinbaum (1976)
study

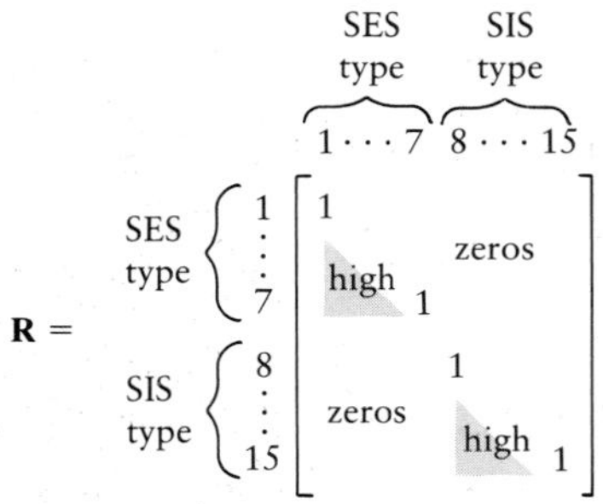

$$
\mathbf{R} = \begin{array}{c}
 \\
 \\
\text{SES}\left\{\begin{matrix}1\\ \vdots\\ 7\end{matrix}\right. \\
\text{type} \\
\text{SIS}\left\{\begin{matrix}8\\ \vdots\\ 15\end{matrix}\right. \\
\text{type}
\end{array}
\begin{array}{c}
\overbrace{1\cdots 7}^{\text{SES type}}\ \overbrace{8\cdots 15}^{\text{SIS type}} \\
\begin{bmatrix}
1 & & & \text{zeros} \\
\text{high} & 1 & & \\
 & & 1 & \\
\text{zeros} & & \text{high} & 1
\end{bmatrix}
\end{array}
$$

of what to expect from a factor analysis. For the James–Kleinbaum study, where the goal of the factor analysis was to construct an SES index and an SIS index from several correlated variables, the ideal correlation matrix would be of the form shown in Figure 24-4.

Thus, it was hoped that the SES-type variables would be highly intercorrelated, the SIS-type variables would be highly intercorrelated, and the correlations among the SES-type and SIS-type variables would be close to zero. If these conditions were satisfied, it could be reasoned that the entire set of original variables were separated into two essentially independent factors, one describing SES and the other describing SIS.

The differences between this ideal pattern and the one actually obtained are highlighted by the markings in Table 24-7. The triangular-shaped boxes marked "SES" enclose the set of intercorrelations of SES-type variables, which were expected to be high. Similarly, the triangular-shaped boxes marked "SIS" identify the set of expected high SIS intercorrelations. A look inside these boxes indicates that most of these correlations are reasonably high (for both whites and nonwhites), although some correlations are quite close to zero.

The correlations or sets of correlations that are circled are between SES-type and SIS-type variables that did not conform to the ideal; that is, they are the correlations that were hoped to be negligible but actually turned out to be considerably different from zero. The fact that there are several circles within both matrices indicates that the final factor solutions for both whites and nonwhites are not likely to be entirely satisfactory. This will be borne out later when we examine the factor loadings obtained from the analysis.

(A factor that does not measure what it was expected to measure is often described as "lacking construct validity." In the James–Kleinbaum example, this occurs if the computed factors are describing phenomena different from those usually measured by indices like SES and SIS or if the variables themselves are not strongly associated with SES and/or SIS measures.)

It should be pointed out that rarely, if ever, does factor analysis produce an ideal result. On the other hand, the fact that most of the intercorrelations among SES variables are fairly high, most of the intercorrelations among SIS variables are also reasonably high, and several of the correlations between SES variables and SIS variables are somewhat close to zero suggests that the factor solutions are not likely to be bad enough to preclude their use in testing the study hypothesis of primary interest.

24-11 Step 2: Determining Initial Factors by Principal-Components Analysis

Initial factors for the example were determined by the method of principal-components analysis. If some communalities are taken not to be 1's, the method emanates from the (general) common-factor model and is usually called *principal-axis analysis*. The term *principal components* is the name given to the set of components resulting from the application of this method.

24-11-1 Total Variation in $X_1, X_2, \ldots, X_p$

To describe the method of principal components, we first need to introduce the concept of the *total variation* in the data with regard to the variables $X_1, X_2, \ldots, X_p$. The total variation is mathematically defined to be the sum of the sample variances of the k variables:[4]

$$\text{total variation} = S_1^2 + S_2^2 + \cdots + S_p^2$$

where S_j^2 is the sample variance of X_j, $j = 1, 2, \ldots, p$.

Conceptually, the total variation is a measure of the amount of uncertainty associated with the observations on all p variables. By "uncertainty," we refer to how much the observations on the units of study differ from one another. For example, if all the observations on a given variable are exactly the same, there is no uncertainty about what value to expect for that variable, whereas if the observations are quite different from one another, there is considerable uncertainty about what value to expect.

24-11-2 Definition of Principal Components

The purpose of principal-components analysis is to explain as much of the total variation in the data as possible with as few factors (i.e., principal components) as possible.

The *first principal component*, PC(1), is the weighted linear combination of the variables that accounts for the largest amount of the total variation in the data. That is, PC(1) is that linear combination of the X's, say

$$\text{PC}(1) = w_{(1)1}X_1 + w_{(1)2}X_2 + \cdots + w_{(1)p}X_p$$

where the weights $w_{(1)1}, w_{(1)2}, \ldots, w_{(1)p}$ have been chosen to maximize the quantity

$$\frac{\text{variance of PC}(1)}{\text{total variation}}$$

In other words, no other linear combination of the X's will have as large a variance as PC(1).

(When the X's are in standardized form (so that the analysis is based on the correlation matrix), the proportion of the total variation in the data accounted for by PC(1) is

$$\frac{\text{variance of PC}(1)}{p}$$

---

[4] If, as is usually the case, the variables are in standardized form (so that $S_j^2 = 1$ for every j), the total variation is simply equal to p, the number of variables.

Also, the weights are chosen subject to the restriction $\sum_{j=1}^{p} w_{(1)j}^2 = 1$ so that the variance of PC(1) will not exceed the total variation.)

The *second principal component*, PC(2), is the weighted linear combination of the variables that is uncorrelated with PC(1) and that accounts for the maximum amount of the remaining total variation not already accounted for by PC(1). In other words,

$$PC(2) = w_{(2)1}X_1 + w_{(2)2}X_2 + \cdots + w_{(2)p}X_p$$

is the linear combination of the X's that has the largest variance of all linear combinations which are uncorrelated with PC(1).

In general, the ith *principal component* PC(i) is the linear combination

$$PC(i) = w_{(i)1}X_1 + w_{(i)2}X_2 + \cdots + w_{(i)p}X_p$$

that has the largest variance of all linear combinations which are uncorrelated with all of the previously determined $i - 1$ principal components. Actually, it is possible to determine as many principal components as there are original variables. However, in most practical applications, most of the total variation in the data is usually accounted for by the first few components. Furthermore, these components are chosen to be mutually uncorrelated. Thus, the analytic goals of parsimony and independence are quite often achieved via this method.

24-11-3 Interpretation of Principal Components

Components are often difficult to interpret directly; as a result, further manipulation (e.g., via rotation) may be desired. Some analysts object to rotating principal components because after such an operation these components are no longer mathematically optimal.

In any case, it is often found that the first principal component, PC(1), represents an overall measure of the information contained in all the variables.[5] Such a general index usually has *large factor loadings* (in absolute value) on almost all the variables. Consequently, it is usually difficult to determine whether such a factor is more related to one particular interpretable subset of the variables than to another. Nevertheless, if the primary aim of factor analysis is data reduction by index construction, with no major emphasis on interpretability, the final factor solution is likely to consist of principal components themselves.

On the other hand, if it is desired to find meaningful underlying factors that describe the variation in a set of variables, then another step, involving rotation, is generally required.

24-11-4 Principal-Components Solutions for the James–Kleinbaum Study

The principal-components solutions for both whites and nonwhites in the James–Kleinbaum study are given in Table 24-8. As can be seen from the table, the first two principal components for whites explain 57% of the total variation, and those for nonwhites explain 55% of the total variation. The proportions .38, .19, .41, and .14 are easily obtained from the variances above them (5.70, 2.85, 6.18, and 2.13, respectively) by dividing each of the latter numbers by $p = 15$. This, of course, follows because the variables have been standardized prior to analysis.

---

[5] The second principal component can often be interpreted as a contrast between a particular subset of variables and the remaining subset.

TABLE 24-8   Principal-components solutions in the James–Kleinbaum (1976) study

Variable	Principal Components			
	Whites		Nonwhites	
	SES	SIS	SES	SIS
PCI	.90	−.31	.90	−.26
MED	.60	−.61	.72	−.50
UEM	.00	.58	.01	−.04
P8K+	.85	−.37	.58	−.31
WCM	.76	−.38	.63	−.10
BC	.37	.55	.78	−.31
P3K−	−.86	.16	−.78	.42
PSDF	.79	.32	.64	.49
JDM	.79	.05	.76	.39
JDF	.65	.30	.79	.26
CSM	.51	.55	.72	.27
CSF	.54	.55	.66	.02
PM	.35	.25	.41	.58
HR	−.06	.42	−.18	.41
PCBH	.09	.63	.38	.63
Variance of principal component (eigenvalue)	5.70	2.85	6.18	2.13
Proportion of total variation explained by principal component	.38	.19	.41	.14

Notice also in Table 24-8 that the first component for both whites and nonwhites involves mostly high loadings. We will see in the next section how rotation techniques can be used to provide more interpretable factors by reducing the number of high loadings associated with each factor.

24-12 Step 3: Rotation of Initial Factors

Rotation is a method of altering the initial factors in order to achieve more interpretability. For example, in rotating the first two principal components obtained in the James–Kleinbaum study, it was hoped that the resulting factors would be more interpretable by creating some meaningful subset of the original variables (e.g., the SES-type variables would ideally be highly correlated with one rotated factor but not with the other, and similarly for the SIS-type variables).

The primary objective of obtaining conceptually meaningful factors by rotation may be translated into more quantitative terms via the concept of *simple structure*. A factor structure is considered to be simple if each of the original variables relates highly to only one factor and each factor can be identified as representing what is common to a relatively small number of variables. Thus, simple structure is said to be achieved when, for each factor, the factor loadings for most variables are near 0 and the remaining factor loadings are relatively large. If so, the factor can be conceived as describing the variation shared in common by the subset of variables highly related to it and not describing the variation in the other variables.

It is important, nevertheless, to realize that obtaining simple structure by rotation does not guarantee that the variables that "hang together" on a given factor will describe a conceptually meaningful factor. It is always possible that a relatively good simple structure will still yield factors that are difficult to interpret. Nevertheless, without such simple structure, the interpretation of factors is virtually impossible.

The two best ways to describe how rotation attempts to achieve simple structure are *geometrically*, by rotating the coordinate axes, and *numerically*, by improving the structure of the factor loadings.

24-12-1 Geometric Illustration of Rotation

Figure 24-5 illustrates geometrically how rotation may help to achieve simple structure.[6] The figure portrays an essentially ideal result due to rotation that is not based on the James–Kleinbaum data. In the figure there are as many dots as there are variables, and each dot corresponds to a particular variable. The coordinates associated with each dot (or variable) are the two factor loadings on that variable for the two factors chosen to be rotated. The values of these coordinates, of course, are based on the two axes used to define the scales of measurement, which represent the two factors considered for rotation. If these axes are rotated, we then have defined two new rotated factors.

Since the initial factors determined are the principal components, the prerotation axes represent two principal components. These are labeled I(PC) and II(PC) in Figure 24-5 for the first and second components, respectively, being considered. In the figure, for example, the coordinates of the two dots that are circled are (0.6, 0.6) and (0.8, −0.8), based on axes I(PC) and II(PC). The first 0.6 is the factor loading on a particular variable for the first component, and the second 0.6 is the factor loading on that same variable for the second component. Similarly, the 0.8 and −0.8 are the factor loadings on another variable for the first and second components, respectively.

The second set of coordinates given for each of these two circled dots are relative to a new pair of axes, labeled I'(R) and II'(R). These two axes were determined by rotating the original axes clockwise (through an angle ϕ). Notice what has been achieved by this rotation. Each one of these circled dots is now close to only one of the two new (rotated) axes. In fact, the coordinates of the two circled dots have now been changed as follows:

$$(0.6, 0.6)_{PC} \rightarrow (0.0, 0.61)_R$$

and

$$(0.8, -0.8)_{PC} \rightarrow (0.82, 0.0)_R$$

---

[6] The rotation of two factors is all that is illustrated here. When more than two factors are rotated, rotation is performed pairwise until satisfactory simple structure is achieved for all factors under consideration.

<u>FIGURE 24-5</u>   The purpose of rotation (The goal here is to rotate the axes so that
(1) each dot is close to only one of the two rotated axes (simple structure) and (2) dots close to the same rotated axis define a meaningful
factor.)

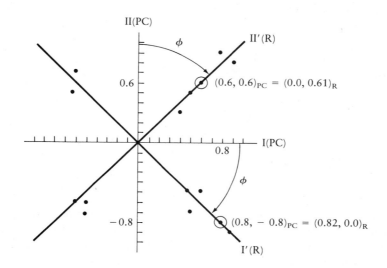

Similarly, the coordinates of the other dots will tend to be high for one coordinate and close to zero for the other. Thus, simple structure has been achieved!

The importance of this accomplishment in conceptual terms is that the dots (i.e., the variables) can now be seen to be clustered into two subgroups, one subgroup lying close to one rotated axis and the other subgroup lying close to the other. Since these new axes represent new (rotated) factors, we now can interpret each new factor in terms of the particular subgroup of variables lying close to that factor.

24-12-2   Methods of Rotation

There are two ways in which the axes can be rotated. First, the axes may be kept in the same orientation to one another during rotation so that they are still perpendicular after rotation (i.e., there is a 90° angle between the two new axes); this is called *orthogonal rotation*. Second, each axis may be rotated independently, so that they are *not* necessarily perpendicular after rotation; this is called *oblique rotation*.

The difference between these two types of rotation can be described by the angle of rotation. In general, there are two such angles: the angle between the original first axis (I) and its corresponding rotated axis (I'), and the angle between the original second axis (II) and its rotated axis (II'). Under orthogonal rotation, only one angle has to be specified—this is the angle ϕ in Figure 24-6a. Under oblique rotation, two angles must be specified—these are the angles ϕ_1 (between axes I and I') and ϕ_2 (between II and II') in Figure 24-6b.

An important statistical difference between these two methods of rotation is that the factors resulting from orthogonal rotation remain statistically uncorrelated (i.e., the factor cosines are all zero), whereas factors resulting from oblique rotation are usually correlated to some extent (i.e., some or all of the factor cosines will be nonzero). To generate statistically

FIGURE 24-6   Orthogonal and oblique rotation

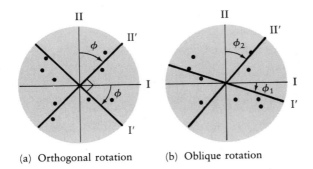

(a) Orthogonal rotation   (b) Oblique rotation

uncorrelated factors is a desirable goal, primarily because of the advantages associated with representing a complex set of interrelationships among several correlated variables in terms of a few uncorrelated indices. Another desirable property of orthogonal rotation is that the amount of the total variation accounted for by the factors under consideration is unaffected by the rotation. For example, under orthogonal rotation of the first two principal components (for whites) obtained in the James–Kleinbaum study, the two rotated factors account for 57% of the total variation, which is precisely the percentage accounted for by the first two principal components.

Unfortunately, however, orthogonal rotation may not result in finding the best set of rotated factors. Often the researcher can reason empirically what characteristics the factors should be measuring, and orthogonal rotation is often not sufficient to determine factors with the desired attributes. Since the primary goals of rotation are simple structure and meaningful factors, these goals are likelier to be achieved if oblique as well as orthogonal rotations are considered. In fact, by permitting the factor axes to become oblique, it is frequently possible to arrive at much more interpretable factors.

Although the most simplistic approach to rotation may be to examine the data graphically and then decide upon the proper rotation visually, this involves considerable subjectivity. Because of this, rotation is generally performed using computerized algorithms based on well-defined quantitative criteria. Nevertheless, we recommend that the researcher consider looking at the data graphically, since a geometric picture often provides additional insight.

24-12-3 Computer Algorithms for Orthogonal Rotation

Three algorithms for orthogonal rotation that are available as options in most factor analysis computer programs are the *varimax*, *quartimax*, and *equimax* methods. Varimax is the most often used. The essential differences among these methods are as follows: Varimax attempts to achieve simple structure with respect to the columns of the factor-loading matrix, quartimax with respect to the rows, and equimax with respect to both the rows and columns. For more detailed descriptions of these methods, we refer the reader to other texts on factor analysis (e.g., see Overall and Klett, 1972, and Rummel, 1970).

24-12-4 Computer Algorithms for Oblique Rotation

A large number of computer algorithms have been developed to perform oblique rotation. The most common are the *oblimin*, *quartimin*, *biquartimin*, and *covarimin* algorithms. All represent algorithms designed to satisfy various types of simple-structure criteria. Unfortunately, because no one algorithm always produces a superior solution, several different algorithms may need to be tried on the same data set. (See Rummel, 1970, and Harman, 1960, for detailed descriptions of these algorithms.)

The user of computer algorithms for factor analysis should also be aware of some additional complexities of oblique rotation. In particular, when considering the results of an oblique rotation performed by using some computer algorithm, it is necessary to know that there are two alternative representations for the factor loadings (i.e., the coordinates of the points [or variables] with respect to the rotated axes) depending on how each point is projected onto the rotated axes. One alternative, yielding what are called *pattern loadings* (Figure 24-7), is based on projecting each point onto each rotated axis by lines *parallel to these two axes* (so that the factor loadings are then defined as the two projected coordinates). The other alternative, yielding what are called *structure loadings* (Figure 24-8), is based on projecting each point onto each rotated axis by lines *perpendicular to these two axes*. As the diagrams suggest, corresponding pattern and structure loadings will generally be different under oblique rotation. Consequently, some computer programs print two factor-loading matrices (called the *pattern matrix* and the *structure matrix*) for the same data set. Other programs let the user specify which of these matrices is to be printed.

An important difference between these loadings is that pattern loadings are not really correlation coefficients between variables and factors, whereas structure loadings do represent such correlations. Nevertheless, the pattern matrix is often much more useful than the structure matrix in interpreting the rotated factors.

An additional characteristic of oblique rotation here is the option to consider an adjusted (usually called *reference*) solution obtained from the original (or *primary*) oblique solution. This reference solution is obtained by determining two new adjusted axes that are perpendicular to the primary axes (see Figure 24-9). As with the primary axes, there are pattern and structure matrices associated with the reference axes. In contrast with the primary loadings, however, the reference pattern loadings are correlation coefficients (as were the primary structure loadings), whereas the reference structure loadings (like the primary pattern loadings) are not correlations but nevertheless are often more useful for interpreting the rotated factors. Some programs, in fact, print only the primary pattern or the reference structure solution and ignore the primary structure and the reference pattern solutions, because the former two solutions are generally more useful for interpretation.

FIGURE 24-7   Pattern loadings

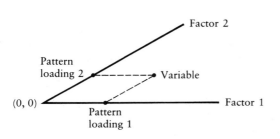

FIGURE 24-8   Structure loadings

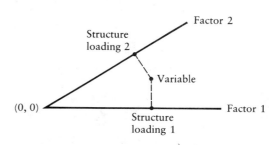

FIGURE 24-9   Relationship between reference axes
and primary axes

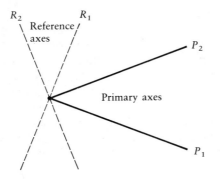

24-12-5 Rotation in the James–Kleinbaum Study

Figures 24-10 and 24-11 illustrate what was achieved by oblique rotation of the first two principal components in the James–Kleinbaum study. Those variables that are not strongly associated with either (rotated) factor are circled; these variables will be identified when we examine the factor-loading matrices.

Although Figures 24-10 and 24-11 illustrate oblique rotation, examination of these graphs indicates that the rotated axes are not far from being perpendicular. This suggests that orthogonal rotation yields similar results, which indeed was the case.

FIGURE 24-10   Oblique rotation of first two principal components for
whites in James–Kleinbaum (1976) study

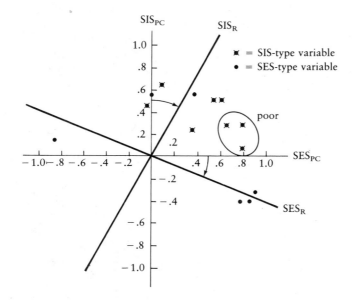

FIGURE 24-11   Oblique rotation of first two principal components for nonwhites
in James–Kleinbaum (1976) study

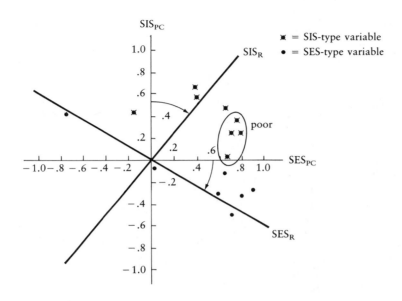

For both whites and nonwhites, notice that the results of rotation are far from ideal, since the SES-type variables do not closely hug one rotated axis and the SIS-type variables do not closely hug the other rotated axis in either figure. Nevertheless, it can be seen that the SES-type variables are at least approximately clustered along one axis, and the SIS-type variables along the other.

24-12-6   Examination of Factor-Loading Matrices

The results of rotation are usually quantified and evaluated by using factor-loading matrices. The goal of simple structure is achieved by using a factor-loading matrix that contains high loadings on only one (or very few) factor(s) for each variable and/or on only a few variables for each factor, with close-to-zero loadings otherwise.

Table 24-9 illustrates the idealized simple structure for the James–Kleinbaum factor analysis. Note that the SES factor was expected to load high on the SES-type variables and to load zero on the SIS-type variables; the SIS factor, on the other hand, was expected to load high on the SIS-type variables and to load zero on the SES-type variables. Some variables, such as BC and P3K−, did not have clearly predictable factor loadings and were so designated with question marks.

Comparing Factor Loadings of Initial Factors with Rotated Factors

Tables 24-10 and 24-11 illustrate how rotation helps to simplify the structure of the factor-loading matrix. In these tables, the factor loadings for the principal-components solution and for the oblique factor solution are presented for the data on whites and nonwhites from the James–Kleinbaum study. For each racial group, the principal components

TABLE 24-9   Idealized factor
              loadings for
              James–Kleinbaum
              (1976) study

	Factor Loading	
Variable	SES	SIS
PCI	High (+)	0
MED	High (+)	0
UEM	High (−)	High (+)
P8K+	High (+)	0
WCM	High (+)	0
BC	High (+)	?
P3K−	High (−)	?
PSDF	0	High (+)
JDM	0	High (+)
JDF	0	High (+)
CSM	0	High (+)
CSF	0	High (+)
PM	0	High (+)
HR	0	High (+)
PCBH	0	High (+)

TABLE 24-10   Principal components and
              oblique-rotated factor loadings for
              whites in James–Kleinbaum (1976)
              study

	Principal Component		Oblique Factor	
Variable	SES	SIS	SES	SIS
PCI	.90	−.31	.95	.04
MED	.60	−.61	.82	−.34
UEM	.00	.58	−.27	.54
P8K+	.85	−.37	.93	−.03
WCM	.76	−.38	.86	−.06
BC	.37	.55	.06	.65
P3K−	−.86	.16	−.84	−.17
PSDF	.79	.32	.55	.59
JDM	.79	.05	.67	.34
JDF	.65	.30	.44	.53
CSM	.51	.55	.19	.71
CSF	.54	.55	.21	.71
PM	.35	.25	.19	.37
HR	−.06	.42	−.25	.37
PCBH	.09	.63	−.22	.62

TABLE 24-11   Principal components and oblique-rotated
factor loadings for nonwhites in
James–Kleinbaum (1976) study

Variable	Principal Component		Oblique Factor	
	SES	SIS	SES	SIS
PCI	.90	−.26	.88	.22
MED	.72	−.50	.88	−.07
UEM	.01	−.04	.03	−.03
P8K+	.58	−.31	.65	.03
WCM	.63	−.10	.56	.23
BC	.78	−.31	.81	.12
P3K−	−.78	.42	−.88	−.03
PSDF	.64	.49	.18	.75
JDM	.76	.39	.34	.73
JDF	.79	.26	.44	.62
CSM	.72	.27	.38	.60
CSF	.66	.02	.50	.35
PM	.41	.58	−.06	.71
HR	−.18	.41	−.41	.26
PCBH	.38	.63	−.12	.75

are compared with the rotated factors with regard to simple structure; loadings that would be considered poor in comparison with the idealized loadings have been circled. The main feature to notice here is that the number of circled factor loadings has been considerably reduced by rotation. Thus, although the rotated factors do not have perfect simple structure, they are nevertheless much more interpretable as SES and SIS factors after rotation than before.

Comparing Various Rotational Methods

Tables 24-12 and 24-13 describe the results of three different methods of rotation: reference oblique, orthogonal varimax, and orthogonal quartimax. James and Kleinbaum felt it appropriate to consider different methods of rotation before deciding on the best simple-structure solution.

Examination of Tables 24-12 and 24-13 shows that, as expected, there is little difference among the results obtained from each of these rotations. For whites, the same number of poor loadings is circled for each rotational method; for nonwhites, the oblique solution has one less circled loading than either of the other two solutions. Also, the factor solution for whites appears somewhat better in terms of simple structure than the factor solution for nonwhites.

Because an oblique rotation was considered preferable on empirical grounds and also simply because little difference in factor loadings was obtained from other methods of rotation, James and Kleinbaum decided to consider as the final factor loadings those obtained by means of the (reference) oblique rotation method; the values of these factor loadings for whites and nonwhites are given in Tables 24-12 and 24-13.

TABLE 24-12   Factor loadings for whites obtained from three rotational methods in James–Kleinbaum (1976) study

Variable	Reference Oblique		Orthogonal Varimax		Orthogonal Quartimax	
	SES	SIS	SES	SIS	SES	SIS
PCI	.95	.04	.95	.10	.95	.10
MED	.82	−.34	.80	−.30	.80	−.29
UEM	−.27	.54	−.24	.52	−.24	.52
P8K+	.93	−.03	.93	.02	.93	.03
WCM	.86	−.06	.85	−.01	.85	−.02
BC	(.06)	.65	(.10)	.66	(.10)	.66
P3K−	−.84	−.17	−.85	−.22	−.85	−.22
PSDF	(.55)	.59	(.59)	.62	(.56)	.62
JDM	(.67)	.34	(.69)	.38	(.69)	.38
JDF	(.44)	.53	(.47)	.55	(.46)	.55
CSM	.19	.71	.23	.72	.22	.72
CSF	.21	.71	.25	.73	.25	.73
PM	.19	.37	.21	.38	.20	.38
HR	−.25	.37	−.23	.36	−.23	.36
PCBH	−.22	.62	−.19	.60	−.19	.60

TABLE 24-13   Factor loadings for nonwhites obtained from three rotational methods in James–Kleinbaum (1976) study

Variable	Reference Oblique		Orthogonal Varimax		Orthogonal Quartimax	
	SES	SIS	SES	SIS	SES	SIS
PCI	.88	.22	.88	(.34)	.92	.19
MED	.88	−.07	.87	.05	.87	−.09
UEM	(.03)	(−.03)	(.03)	(−.03)	(.03)	(−.03)
P8K+	.65	.03	.65	.12	.66	.01
WCM	.56	.23	.56	.31	.61	.21
BC	.81	.12	.80	.23	.83	.10
P3K−	−.88	−.03	−.87	−.15	−.88	−.01
PSDF	.18	.75	.21	.78	(.33)	.73
JDM	(.34)	.73	(.36)	.78	(.48)	.71
JDF	(.44)	.62	(.46)	.69	(.57)	.60
CSM	(.38)	.60	(.40)	.68	(.50)	.58
CSF	(.50)	.35	(.51)	.42	(.51)	.33
PM	−.06	.71	−.04	.70	.08	.70
HR	−.41	.26	−.40	.21	−.36	.27
PCBH	−.12	.75	−.09	.73	.03	.74

24-13 Step 4: Determination of the Component Scores

Table 24-14 gives the weights associated with the component loadings obtained via oblique rotation in the James–Kleinbaum study. As previously described, a component score is a numerical value F obtained by substituting specific values for the standardized X's into the expression

$$F = w_1 X_1 + w_2 X_2 + \cdots + w_p X_p$$

In computing such scores in the James–Kleinbaum study, variable values and weights for whites were used to compute the SES and SIS factor scores for whites, and similarly for nonwhites.

TABLE 24-14   Factor weights obtained from oblique rotation in the James–Kleinbaum (1976) study

Variable	Whites SES	Whites SIS	Nonwhites SES	Nonwhites SIS
PCI	.19	−.02	.19	.00
MED	.18	−.14	.22	−.10
UEM	−.07	.18	.01	−.01
P8K+	.19	−.04	.15	−.05
WCM	.17	−.05	.11	.03
BC	−.01	.20	.18	−.03
P3K−	−.16	−.02	−.21	.07
PSDF	.09	.16	−.02	.24
JDM	.12	.08	.01	.22
JDF	.07	.15	.05	.17
CSM	.01	.21	.04	.17
CSF	.02	.21	.09	.07
PM	.02	.11	−.08	.25
HR	−.06	.13	−.12	.13
PCBH	−.07	.20	−.10	.27

With these scores the analysis strategy described in Section 24-8 can be conducted to test the major study question: Do high-stress counties have higher mortality rates than low-stress counties?

Recall that the steps in analyzing these factor scores are as follows:

1. Rank-order the factor scores (by factor) separately for each race.

2. Divide each ordered set of scores into a high-score group and a low-score group.

3. For each race separately, form a two-way table based on the HI and LO groups defined for each factor:

```
            SIS
          HI   LO
      ┌─────┬─────┐
   HI │ P̂₁  │ P̂₂  │
SES   ├─────┼─────┤
   LO │ P̂₃  │ P̂₄  │
      └─────┴─────┘
```

4. Compute crude hypertension-related mortality rates for each of the cells in the table and then compare $\hat{P}_2$ (low-stress rate) with $\hat{P}_3$ (high-stress rate).

Tables 24-15 and 24-16 summarize the numerical results of this procedure up through the third step for males 45 through 54 years of age. These tables give the standardized factor scores for each of the four "stress" groups defined in the third step. Mean factor scores are also given for each group.

TABLE 24-15   Standardized factor scores in the four "stress" groups for white males, 45–54 years of age, in James–Kleinbaum (1976) study

N = 25		N = 18		N = 18		N = 25	
HI SES	HI SIS	HI SES	LO SIS	LO SES	HI SIS	LO SES	LO SIS
1.51	0.03	0.25	−0.61	−0.13	0.71	−1.00	−0.37
1.09	1.37	0.63	−0.79	−0.83	0.53	−0.64	−0.38
0.79	1.54	0.45	−0.25	−1.08	0.40	−0.41	−1.70
0.15	2.75	2.58	−0.25	−0.35	0.99	−0.59	−0.35
0.23	0.91	0.59	−1.84	−0.14	−0.04	−0.57	−1.52
0.84	0.73	0.59	−0.15	−0.84	0.06	−0.24	−0.76
1.07	0.11	3.09	−0.42	−0.35	−0.03	−0.90	−0.10
0.32	0.76	0.25	−1.54	−0.24	0.12	−1.05	−1.12
2.73	0.35	1.75	−1.31	−0.84	0.13	−0.75	−0.43
0.50	1.39	0.76	−0.79	−1.10	1.92	−0.63	−1.65
2.29	0.72	0.39	−1.29	−0.36	0.07	−1.13	−1.84
0.33	0.55	−0.11	−0.30	−0.54	1.41	−1.08	−1.53
0.40	0.48	0.01	−0.73	−0.45	0.91	−1.31	−0.52
0.58	0.76	2.77	−0.27	−0.50	0.56	−1.64	−0.87
0.34	0.61	−0.07	−1.32	−0.16	2.24	−0.38	−0.74
0.45	0.30	−0.08	−0.68	−1.44	2.70	−0.57	−0.71
0.30	0.23	0.38	−0.38	−1.40	0.30	−0.92	−0.37
0.47	0.22	0.14	−0.48	−1.19	0.97	−1.18	−0.23
2.07	1.37					−0.65	−1.66
0.23	0.00					−0.33	−0.84
0.08	1.20					−1.11	−1.10
−0.06	1.54					−0.25	−0.32
0.03	0.69					−1.27	−0.59
0.74	0.58					−0.61	−0.24
0.12	0.59					−0.86	−0.39
Total 17.66	19.78	13.95	−13.40	−11.94	13.73	−20.07	−20.33
Mean 0.71	0.79	0.78	−0.74	−0.66	0.76	−0.80	−0.81

TABLE 24-16   Standardized factor scores in the four "stress" groups for nonwhite males, 45–54 years of age, in James–Kleinbaum (1976) study

N = 23		N = 20		N = 20		N = 23	
HI SES	HI SIS	HI SES	LO SIS	LO SES	HI SIS	LO SES	LO SIS
1.40	1.83	0.87	−0.57	−0.72	−0.08	−1.02	−0.74
0.51	1.52	1.04	−0.52	−1.30	0.33	−0.48	−0.61
1.31	0.12	1.55	−1.27	−0.60	0.51	−0.24	−1.02
1.40	0.51	1.11	−0.18	−1.05	0.25	−0.20	−1.28
−0.09	0.42	0.49	−1.34	−1.03	0.02	−0.80	−1.29
1.24	0.22	0.92	−1.06	−0.91	0.39	−0.30	−0.25
1.21	0.46	1.70	−1.28	−1.62	0.04	−0.64	−0.58
1.53	2.19	−0.04	−0.30	−0.99	0.28	−0.29	−1.44
1.93	2.68	1.28	−1.24	−1.46	0.11	−0.61	−0.55
0.81	0.38	1.23	−1.30	−1.10	0.82	−1.67	−0.40
2.47	1.06	0.96	−0.20	−0.37	0.72	−1.35	−0.26
0.16	2.61	0.57	−0.78	−1.05	0.56	−0.21	−0.69
0.15	0.08	0.40	−0.52	−0.45	−0.13	−1.67	−0.28
−0.10	1.46	1.03	−0.45	−1.33	0.37	−0.33	−0.83
1.61	2.38	0.82	−0.74	−1.29	0.35	−1.21	−0.60
−0.06	0.99	0.05	−0.87	−0.93	−0.13	−0.57	−0.82
0.94	3.29	0.23	−0.44	−0.79	0.36	−0.62	−0.84
−0.02	0.22	1.50	−1.75	−0.32	0.28	−0.87	−1.07
1.05	−0.14	0.17	−0.83	−0.86	−0.03	−0.55	−0.25
0.53	0.90	0.25	−0.29	−1.04	0.94	−0.79	−0.67
1.29	0.05					−1.04	−0.36
0.75	1.72					−1.24	−0.66
−0.02	1.34					−0.21	−0.65
Total 20.00	26.09	16.13	−15.93	−19.21	5.96	−16.91	−16.14
Mean 0.87	1.13	0.81	−0.80	−0.96	0.30	−0.74	−0.70

24-14 Study Results

James and Kleinbaum conjectured that the crude hypertension-related death rate for nonwhite males in high-stress counties would be significantly higher than in low-stress counties. Level of socioecologic stress was not expected, however, to mediate these rates for white males.

Tables 24-17 and 24-18 present the data necessary to test the appropriate hypotheses of interest. The rates of interest are

$$\text{whites} \begin{cases} \hat{P}_2 = 1,002.5 \text{ deaths per million} \\ \hat{P}_3 = 698.5 \text{ deaths per million} \end{cases} \quad \text{nonwhites} \begin{cases} \hat{P}_2 = 7,852.1 \text{ deaths per million} \\ \hat{P}_3 = 3,964.7 \text{ deaths per million} \end{cases}$$

Based on these rates, the conjecture concerning nonwhite males was confirmed using a simple

TABLE 24-17   Crude hypertension-related mortality rates in the four "stress" groups for white males, 45–54 years of age, North Carolina counties, 1960

SES	SIS		Marginal
	High	Low	
High	No. counties = 25 No. deaths = 75 PAR = 79,852 Mortality rate = 926.7×10^{-6}	No. counties = 18 No. deaths = 30 PAR = 42,946 Mortality rate = $\boxed{698.5 \times 10^{-6}}$	846.91×10^{-6}
Low	No. counties = 18 No. deaths = 24 PAR = 23,940 Mortality rate = $\boxed{1,002.5 \times 10^{-6}}$	No. counties = 25 No. deaths = 32 PAR = 24,298 Mortality rate = $1,316.9 \times 10^{-6}$	$1,160.91 \times 10^{-6}$
Marginal	944.19×10^{-6}	922.01×10^{-6}	

Note: PAR = population at risk.

TABLE 24-18   Crude hypertension-related mortality rates in the four "stress" groups for nonwhite males, 45–54 years of age, North Carolina counties, 1960

SES	SIS		Marginal
	High	Low	
High	No. counties = 23 No. deaths = 156 PAR = 22,190 Mortality rate = $7,030.1 \times 10^{-6}$	No. counties = 20 No. deaths = 18 PAR = 4,540 Mortality rate = $\boxed{3,964.7 \times 10^{-6}}$	$6,509.53 \times 10^{-6}$
Low	No. counties = 20 No. deaths = 113 PAR = 14,391 Mortality rate = $\boxed{7,852.1 \times 10^{-6}}$	No. counties = 23 No. deaths = 50 PAR = 9,336 Mortality rate = $5,355.6 \times 10^{-6}$	$6,869.81 \times 10^{-6}$
Marginal	$7,253.54 \times 10^{-6}$	$4,900.54 \times 10^{-6}$	

Note: PAR = population at risk.

one-tailed Z test ($P < .005$). Furthermore, the high-stress rate was nearly twice as large as that for the low-stress group! For white males the difference between the two observed rates was not significant ($P > .10$), as expected.

Thus, the results obtained by James and Kleinbaum provide some supporting evidence for the theory that socioecologic stress is a mediating factor in the determination of cerebrovascular disease for populations of the type considered in their study. Nevertheless, similar conclusions would have to be made independently by other researchers to solidify these findings. This characterizes any data-driven, exploratory analysis.

Also, other methods of analysis of these data might also be considered, such as the use of regression analysis with county death rate as the dependent variable and with the stress scores and other concomitant variables as independent variables. In any case the use of factor analysis has been shown here to be an important tool in data analysis.

24-15 Pitfalls

For the methods described in this chapter, three common pitfalls can trap the unwary. These problems arise in varying degrees in variable reduction, factor analysis, and component analysis. These pitfalls are (1) unreliability of conclusions due to extensive exploratory analysis without needed confirmatory analysis, (2) computational difficulties that invalidate statistical analyses, and (3) failure to control for nuisance variables.

The major pitfall is the first above. As with any method involving extensive exploratory analysis, the methods discussed in this chapter must yield conclusions that can be replicated (i.e., reliability must be shown). Without such reliability, any conclusions must be considered and reported as tentative. Split-sample techniques help to deal with this problem. See Mulaik (1972) for a discussion of methods for confirming factor analysis.

The second pitfall arises when computational difficulties invalidate statistical analyses. As described at the beginning of this chapter, interest centers on sets of variables that overlap, at least partially. As a result, collinearity or near collinearity may be present. This can lead to a variety of problems: Communalities may not be estimable, parameter estimates may be totally unreliable (e.g., negative variance estimates), or results may be very unstable (varying greatly with small perturbations in the data). Knowledge of the behavior of the model and analysis algorithm used is the only safe path around this pitfall.

With linear models, nuisance variables can be controlled by using the variables of interest as responses in a multivariate linear model with the nuisance variables as predictors (see Muller, Hosking, and Helms, 1979). This amounts to using residuals from the linear model as variables for exploratory and confirmatory analyses. The resulting correlation matrix will then involve partial correlations. (Less generally, if there is only one nuisance variable and it has only two values [such as "race" in the example], then analyzing the two groups' data separately gives a valid analysis.)

The need to analyze residuals can be appreciated by referring to the pulmonary function example. If a factor analysis or component analysis were conducted using the pulmonary function variables, then the demographic variables should be either included or preferably partialed from the pulmonary variables. If this were not done, then the factor structure could be seriously distorted by what Joiner (1981) called "lurking variables." For example, consider analyzing the same pulmonary function variables measured on a sample of males and females. If sex mean differences are not removed, any such differences may distort the factor structure, even if the structure is the same for males and females. One should suspect that the structure might not be the same (even after adjusting for mean differences). If so, then methods for comparing structures, such as Procrustes rotation or structural modeling, should be considered (Mulaik, 1972).

Problems

1. Determine whether each of the following statements about the method of factor analysis is generally true or false:
   a. Factor analysis may be used as a method for data reduction.

b. Factor analysis may be used to help determine the underlying dimensions that are measured by an instrument like a questionnaire.

c. Factor analysis differs from regression analysis in that the former considers nominal dependent variables, whereas the latter considers continuous dependent variables.

d. Factor analysis is inappropriate if the original (input) variables are measured on different scales (e.g., responses to item i in a questionnaire can range between 1 and 3, whereas responses to item j can range between 1 and 10).

e. Mathematically speaking, a factor is a variable defined as the sum of all the original variables input into the factor analysis.

f. Communalities are the 1's on the diagonal of the correlation matrix.

g. The use of communalities assumes that each variable can be described entirely in terms of factors common to all variables.

h. Inspection of the correlation matrix suggests the possibility of a good factor solution if the set of original variables can be partitioned into mutually exclusive groups for which there are low correlations between variables within the same group and high correlations between variables from different groups.

i. Factor loadings are correlations between factors.

j. It is generally true that the higher (in absolute value) a factor weight is, the higher (in absolute value) is the factor loading.

k. If a factor loading is highly negative, the variable involved is an important component of the associated factor.

l. If the factor loadings for a given variable on all factors obtained are near zero, the removal of this variable followed by a second factor analysis on the remaining variables will always yield a better factor solution.

m. The method of principal components usually results in factors that are conceptually meaningful.

n. If the main goal of factor analysis is data reduction, the method of principal components can often be used without subsequent rotation of the principal-components solutions.

o. If the proportion of total variation explained by the first principal component is small (e.g., below .20), the use of the first principal component as an overall general factor is always preferable to the use of the unweighted average of the (standardized) original variables.

p. There are as many principal components as there are original variables.

q. The second principal component is uncorrelated with the first principal component and always explains a lesser amount of the total variation.

r. Initial factors are rotated to achieve parsimony and independence.

s. Oblique rotation of two factors should be preferred to orthogonal rotation if the two factors are considered to be relatively independent on either theoretical or empirical grounds.

t. A simple-structure factor solution, even if obtained, does not guarantee that the resulting factors are conceptually meaningful.

u. In an oblique rotation of two factors, the starting (i.e., principal component) axes must be rotated at two unequal angles of rotation.

v. Geometrically speaking, the goal of any rotation is to rotate the original axes so that each point on the graph lies very close to only one of the two rotated axes.

w. A factor solution has good construct validity if, for each factor, most variables have nearly zero loadings, yet a few variables have very high loadings.

x. A factor obtained from a factor analysis has good construct validity if the size and direction of the factor loadings correspond favorably to the size and direction of such loadings as perceived theoretically or empirically by the investigator.

y. If a factor solution has poor construct validity, the only possible explanation is that the factors obtained measure different dimensions than those perceived by the investigator.

2. A questionnaire containing 19 statements was developed by Arkin (1976) to measure sex role orientation in women. Each item was scaled so that a response high on the scale reflected a modern, or self-actualized, sex role orientation (i.e., role behavior not determined by sex) and a response low on the scale reflected a more traditional attitude (i.e., strong differentiation between sex roles). To determine what were the underlying dimensions of sex role orientation as measured by the instrument, if any, factor analysis was applied to data collected using this instrument on 34 women reported to have an abnormal Pap smear test for cervical cancer. (The basic aim of the study was to determine whether sex role orientation was related to delay by the patient in pursuing further treatment for cancer.)

One of the factor analysis computer runs made on these data involved a reduced set of 13 items (after eliminating items with low loadings on all factors) with 1's on the diagonal of the correlation matrix. The loadings for the unrotated (i.e., principal components) solution and for the varimax rotated solution involving the first two factors are as follows:

Item	Principal Components		Varimax Rotated Factors	
	Factor 1	Factor 2	Factor 1	Factor 2
1	.770	−.064	.511	.580
2	.689	.484	.832	.129
3	.750	.031	.562	.497
4	.634	.262	.639	.251
5	.686	−.440	.190	.793
6	.661	.398	.753	.172
7	.682	−.424	.198	.778
8	.591	−.433	.126	.721
9	.581	−.553	.035	.801
10	.581	.410	.703	.107
11	.553	.397	.674	.097
12	.334	.266	.426	.040
13	.395	−.294	.081	.486

a. Based on the information provided, what would you say is the most serious drawback to the utility of the factor solution obtained in this study?
b. How would you contrast the principal-components solution with the varimax rotated solution in regard to simple structure?
c. Which items, if any, would you consider for elimination from the factor-analytic model?
d. Using the rotated solution, determine the cluster of variables that best describes the first factor and the cluster that best describes the second factor. Can you conclude from the information provided whether any of the factors are conceptually meaningful?
e. Plot each of the original 13 items as points on graph paper by using as the coordinates for each item the pair of factor loadings for the first two principal components associated with that item.
f. Using the graph constructed in part (e), rotate your axis orthogonally until you have obtained what you think is the best fit to the plotted points and then draw your rotated axes on the same graph. Using these rotated axes as your new coordinate frame of reference, roughly determine the coordinates of at least three of the points on the graph. Compare these rotated loadings with their corresponding varimax loadings.
g. Using the graph in part (e), rotate your axes obliquely until you are satisfied with the fit to the points and then draw your oblique axes on the graph. Using these oblique axes as your new coordinate frame of reference, determine the coordinates of the same points you selected in part (f) and compare these rotated loadings with the loadings obtained in (f).

	No. of Principal Components												
	1	2	3	4	5	6	7	8	9	10	11	12	13
Eigenvalue	5.00	1.82	1.17	1.12	0.83	0.62	0.59	0.55	0.46	0.34	0.23	0.16	0.10
P_v	0.385	0.140	0.090	0.086	0.064	0.048	0.046	0.042	0.036	0.026	0.018	0.013	0.007

h. A table containing eigenvalues and proportions of variance (P_v) is presented for the principal-components solution given earlier (note that 13 variables always yield 13 principal-component factors). What would you recommend to the investigator who finds only the first two factors to be conceptually meaningful in the example above?
i. For orthogonal rotation, the proportion of the total variation explained by each factor may always be computed from the factor loadings on that factor by summing the squares of these loadings and dividing by the number of original (standardized) variables considered in the analysis (i.e., $P_v = \sum_{i=1}^{p} l_i^2 / p$, where l_i denotes the loading on the ith variable, $i = 1, 2, \ldots, p$). Using this rule, determine the proportion of total variation explained by each of the rotated factors in the example given, and show that, even though these values are not the same as those for the unrotated factors (i.e., the principal components), their sum is the same as the sum for the unrotated factors (0.525).

3. Kleinbaum and Kleinbaum (1976) developed an instrument containing 28 items to measure three basic dimensions of attitudes toward statistics of students enrolled

in an introductory statistics course. The three attitudes were defined as (1) the *confidence* that the student has in his or her ability to learn, understand, or use statistics; (2) the *interest* that the student has in learning or using statistics; and (3) the *value* that the student places on the importance or use of statistics. The instrument was constructed so that each of the 28 items was supposed to tap some aspect of at least one of the three attitude dimensions. Initial attitude dimension scores were then derived as the sums of the scores on those items associated with the same attitude dimension. In terms of item numbers, the three attitude dimension clusters of items (identified before conducting any data analysis) are given in the accompanying figure.

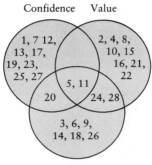

To evaluate whether the three attitude measures thus constructed actually reflect the underlying dimensions measured by the instrument and to determine whether any items should be eliminated from the instrument, a factor analysis was undertaken using data collected on 47 students at two different times during the introductory course. Oblique factor loadings for three factors based on data obtained at the beginning of the course (time 1) and 2 weeks later (time 2) are presented in the table on the next page.

a. Why does an oblique solution appear to be more appropriate than an orthogonal solution for this data set?

b. For each time point, label, *if possible*, each of the factors as C (for confidence), I (for interest), or V (for value) by considering the factor loadings in relation to the clusters of items suspected a priori to be associated with each of the three attitude dimensions. Discuss your choices of labels.

c. Which items, if any, would you consider removing from the instrument entirely? (Explain.) How might you substantiate by further analysis that certain items should be eliminated?

d. Based on the factor loadings given, how might you redefine or relabel the three clusters of items in view of the three factors obtained from the factor analysis?

e. An alternative approach for assessing the construct validity of a newly developed instrument is called *item analysis*. This approach considers the correlations of a set of individual items with that particular factor defined as the sum of those items; items that correlate highly with a factor involving those items are considered to be appropriate components of that factor, whereas those items with small (or nonsignificant) correlations with the factor are considered to be poor items that should be removed.

OBLIQUE SOLUTION TO KLEINBAUM QUESTIONNAIRE

Item	Time 1			Time 2		
	Factor 1	Factor 2	Factor 3	Factor 1	Factor 2	Factor 3
1	.165	−.839	−.077	.038	−.897	−.139
2	−.583	.123	.209	−.717	.180	−.068
3	−.112	.018	.721	−.291	−.035	.528
4	.052	.225	.429	−.258	−.196	−.574
5	−.082	−.051	.471	−.580	.135	.186
6	−.532	.060	.060	−.698	−.315	−.048
7	.183	−.821	.225	.091	−.850	.252
8	−.683	.199	.052	−.870	−.053	−.094
9	−.205	−.186	.116	−.585	−.089	−.056
10	−.811	.050	−.136	−.686	−.074	.085
11	−.580	−.069	.238	−.595	−.007	.180
12	.232	−.903	−.039	.082	−.910	.087
13	.201	−.874	.015	.066	−.931	−.058
14	−.076	−.510	.519	−.173	−.393	.476
15	−.093	.057	.400	−.320	−.119	.163
16	−.779	.129	−.035	−.825	.101	−.231
17	.208	−.100	.818	.056	−.076	.608
18	−.084	.042	.508	−.168	.030	.604
19	−.275	−.331	−.075	−.247	−.217	.242
20	−.368	−.070	.202	−.546	.134	.403
21	−.228	−.046	.457	−.611	.006	−.059
22	−.872	.065	.041	−.887	.009	−.048
23	−.516	−.529	−.161	−.430	−.554	−.024
24	−.785	−.133	−.002	−.784	.212	.211
25	−.330	−.555	−.073	.043	−.469	.481
26	−.475	.038	.113	−.423	.361	.396
27	−.129	−.301	.433	.001	−.166	.564
28	−.442	−.278	.347	−.364	−.134	.599

For the data collected at time 1, the item analysis results listed in the table on the next page were obtained. What conclusions can you draw from the results above about the construct validity of the three factors in general and about the utility of specific items in particular? (In answering this question, consider an r below .500 to be small.)

f. What overall recommendations or conditions do you have concerning the attitude instrument discussed in this problem?

g. Based on the information provided, what would you say is the most serious drawback to the utility of the factor analysis?

ANALYSIS OF ITEMS FOR
KLEINBAUM QUESTIONNAIRE

Dimension	Item	r
C	1	.607
C	5	.419
C	7	.693
C	11	.399
C	12	.662
C	13	.678
C	17	.494
C	19	.437
C	20	.397
C	23	.635
C	25	.572
C	27	.598
I	3	.608
I	5	.564
I	6	.470
I	9	.485
I	11	.733
I	14	.590
I	18	.433
I	20	.531
I	24	.684
I	26	.535
I	28	.714
V	2	.725
V	4	.297
V	5	.545
V	8	.682
V	10	.615
V	11	.695
V	15	.389
V	16	.686
V	21	.582
V	22	.800
V	24	.731
V	28	.635

4. For residents on an island in the South Seas experiencing rapid social change (e.g., the Ponape study, Patrick et al., 1974), an index of cultural incongruity (called PIML) was developed by means of a factor analysis of data obtained from a sociological questionnaire administered to a random sample of residents. This index was constructed by using the method of principal components to determine two factors, one measuring a person's preparedness for modern life (PML) and the other measuring his involvement in modern life (IML). PIML was defined as the difference between these two factors, PML − IML. The factors PML and IML are

the first principal components based on data from two disjoint subsets of questionnaire items suspected (a priori) to be related to these factors. The factor loadings associated with each factor (based on working with the correlation matrix for a sample of 496 males) are given in the table below.

PML

Item	Factor Loading
1. Where grew up	.397
2. Educational level	.761
3. Years in town	.656
4. Travel outside homeland	.599
5. Understand/read English	.792
Proportion of total variation	.430

IML

Item	Factor Loading
6. Source of financial support	.579
7. How often listen to news	.327
8. Zone living in currently	.604
9. Segment of economy	.829
10. Occupational rank	.739
Proportion of total variation	.408

a. For each of the factors, would you recommend the removal of any items? (Explain the rationale behind your answer.)

b. Based on the information given, comment on whether it would be more appropriate to use a simple average of the (standardized) scores for each item to represent the PML and IML indices or to use the principal components obtained.

c. When the principal components based on the correlation matrix are used (without further rotation) as the factors, the factor weights for a given factor may be determined directly from the factor loadings by dividing each loading by the square root of the variance associated with that factor. (*Note*: If P_1 denotes the proportion of total variation explained by the first principal component, p denotes the total number of input variables, and V_1 denotes the variance associated with the first principal component, then $P_1 = V_1/p$ when the correlation matrix is used.)

   For each set of factor loadings presented above, determine the corresponding factor weights; that is, fill in the following table:

PML		IML	
Item	Factor Weight	Item	Factor Weight
1		6	
2		7	
3		8	
4		9	
5		10	

   d. Based on these weights, compute and interpret the PIML incongruity scores based on the three sets of standardized item responses given.

	Item									
Person	1	2	3	4	5	6	7	8	9	10
1	3.10	2.70	2.50	2.20	3.80	1.60	1.20	0.76	−0.45	−0.36
2	1.60	0.81	0.90	1.10	0.25	1.50	0.93	0.36	0.85	0.95
3	1.60	−0.85	−0.75	0.58	−1.72	3.70	2.60	2.40	2.30	3.10

   e. Given the sample size of nearly 500, briefly sketch a split-sample strategy to the analysis. Should stratification be used?

5. The Brief Psychiatric Rating Scale (BPRS) is an instrument used by clinical psychologists for measuring and classifying persons with manifest psychopathology (see Overall and Klett, 1972). In the table is presented a fictitious (rotated) factor-loading matrix for three factors based on a subset of the items contained in this scale. Based on these loadings, identify three clusters of variables that describe the three factors and interpret each factor in conceptual terms (i.e., what type of psychopathology does each factor measure?). Also, what names would you use to describe these factors?

Item	Factor 1	Factor 2	Factor 3
Somatic concern	.15	.40	−.07
Anxiety	−.06	.70	−.15
Conceptual disorganization	−.12	−.05	.60
Guilt feelings	.09	.60	.03
Tension	.30	.35	.06
Depressive mood	.01	.75	.18
Hostility	.70	.15	−.08
Suspiciousness	.65	.10	.35
Hallucinatory behavior	.09	.03	.65
Uncooperativeness	.50	−.08	.06
Unusual thought content	.17	−.04	.80

6. Seashore (1966) reported factor analysis in a study of arm–hand precision. First a group of $n = 39$ was studied, then a follow-up group of $n = 100$. The correlation matrix for eight measures of arm–hand precision was as follows:

Measure	1	2	3	4	5	6	7
1 (Ataxiameter, horizontal)							
2 (Ataxiameter, vertical)	.47						
3 (Target register)	.44	.44					
4 (Straight trace)	.39	.20	.40				
5 (Curved trace)	.30	.04	.26	.61			
6 (Line trace)	.31	−.06	.26	.57	.72		
7 (3-dim trace)	.22	−.07	.34	.37	.72	.83	
8 (Thrust)	.31	.35	.43	.66	.38	.41	.49

a. Using the maximum-off-diagonal approach, provide communality estimates.

b. The unrotated factor loadings for three factors were as follows:

Measure	Factor 1	Factor 2	Factor 3
1 (Ataxiameter, horizontal)	.56	.31	.22
2 (Ataxiameter, vertical)	.37	.58	.24
3 (Target register)	.60	.32	.25
4 (Straight trace)	.78	.09	−.50
5 (Curved trace)	.72	−.40	−.09
6 (Line trace)	.74	.53	−.03
7 (3-dim trace)	.72	.56	.17
8 (Thrust)	.69	.17	−.17

Provide the three pairwise factor plots.

References

Arkin, N. C. (1976). "Diagnosis for Suspected Cervical Cancer." Master's thesis, Department of Epidemiology, University of North Carolina, Chapel Hill, N.C.

Beale, E. M. L., Kendall, M. G., and Mann, D. W. (1967). "The Discarding of Variables in Multivariate Analysis." *Biometrika*, 54: 357–366.

Harburg, E., Erfurt, J. C., Chapel, C., et al. (1973). "Socioecological Stressor Areas and Black–White Blood Pressure." *Detroit J. Chron. Dis.*, 26: 596–611.

Harman, H. H. (1960). *Modern Factor Analysis*. Chicago: University of Chicago Press.

James, S. A., and Kleinbaum, D. G. (1976). "Socioecologic Stress and Hypertension-Related Mortality Rates in North Carolina." *Amer. J. Public Health*, 66(4): 354–358.

Joiner, B. L. (1981). "Lurking Variables: Some Examples." *Amer. Statistician*, 35(4): 227–233.

Jöreskog, K. G. (1978). "Structural Analysis of Covariance and Correlation Matrices." *Psychometrika*, 43: 443–477.

Kleinbaum, D. G., and Kleinbaum, A. (1976). "A Team Approach for Systematic Design and Evaluation of Visually Oriented Modules." In *Modular Instruction in Statistics—Report of ASA Study*, ed. J. R. O'Fallon and J. Service. Washington, D.C.: American Statistical Association, pp. 115–121.

McCabe, G. P. (1984). "Principal Variables." *Technometrics*, 26: 137–144.

McDonnell, W. E., and Davis, G. W. (1982). "Construction of a Pulmonary Function Profile Using a Variable Reduction Analysis." Unpublished manuscript, University of North Carolina, Chapel Hill, N.C.

Mulaik, S. A. (1972). *The Foundations of Factor Analysis*. New York: McGraw-Hill Book Company.

Muller, K. E. (1981). "Relationships Between Redundancy Analysis, Canonical Correlation, and Multivariate Regression." *Psychometrika*, 46: 139–142.

Muller, K. E., Barton, C. N., and Benignus, V. A. (1984). "Recommendations for Appropriate Statistical Practice in Toxicologic Experiments." *Neurotoxicology*, 5(2): 113–126.

Muller, K. E., Hosking, J. D., and Helms, R. W. (1979). "Using LINMOD to Adjust for Treatment Effects When Analyzing the Covariance Matrix." *Proceedings of the Statistical Computing Section*, American Statistical Association, pp. 136–140.

Overall, J. E., and Klett, C. J. (1972). *Applied Multivariate Analysis*. New York: McGraw-Hill Book Company.

Patrick, R., Cassel, J. C., Tyroler, H. A., Stanley, L., and Wild, J. (1974). "The Ponape Study of the Health Effects of Cultural Change." Paper presented to annual meeting, Society for Epidemiologic Research, Berkeley, Calif.

Rummel, R. J. (1970). *Applied Factor Analysis*. Evanston, Ill.: Northwestern University Press.

Seashore R. H. (1966). "Work and Motor Performance." In *Handbook of Experimental Psychology*, ed. S. S. Stevens. New York: John Wiley & Sons, Inc.

Stamler, J. (1967). *Lectures in Preventive Cardiology*. New York: Grune & Stratton.

Stewart, D., and Love, W. (1968). "A General Canonical Correlation Index." *Psychological Bulletin*, 70: 160–163.

A

Appendix - Tables

TABLE A-1 Standard normal cumulative probabilities

z	0.00	0.01	0.02	0.03	0.04	0.05	0.06	0.07	0.08	0.09
-3.8	0.0001	0.0001	0.0001	0.0001	0.0001	0.0001	0.0001	0.0001	0.0001	0.0001
-3.7	0.0001	0.0001	0.0001	0.0001	0.0001	0.0001	0.0001	0.0001	0.0001	0.0001
-3.6	0.0002	0.0002	0.0001	0.0001	0.0001	0.0001	0.0001	0.0001	0.0001	0.0001
-3.5	0.0002	0.0002	0.0002	0.0002	0.0002	0.0002	0.0002	0.0002	0.0002	0.0002
-3.4	0.0003	0.0003	0.0003	0.0003	0.0003	0.0003	0.0003	0.0003	0.0003	0.0002
-3.3	0.0005	0.0005	0.0005	0.0004	0.0004	0.0004	0.0004	0.0004	0.0004	0.0003
-3.2	0.0007	0.0007	0.0006	0.0006	0.0006	0.0006	0.0006	0.0005	0.0005	0.0005
-3.1	0.0010	0.0009	0.0009	0.0009	0.0008	0.0008	0.0008	0.0008	0.0007	0.0007
-3.0	0.0014	0.0013	0.0013	0.0012	0.0012	0.0011	0.0011	0.0011	0.0010	0.0010
-2.9	0.0019	0.0018	0.0018	0.0017	0.0016	0.0016	0.0015	0.0015	0.0014	0.0014
-2.8	0.0026	0.0025	0.0024	0.0023	0.0023	0.0022	0.0021	0.0021	0.0020	0.0019
-2.7	0.0035	0.0034	0.0033	0.0032	0.0031	0.0030	0.0029	0.0028	0.0027	0.0026
-2.6	0.0047	0.0045	0.0044	0.0043	0.0041	0.0040	0.0039	0.0038	0.0037	0.0036
-2.5	0.0062	0.0060	0.0059	0.0057	0.0055	0.0054	0.0052	0.0051	0.0049	0.0048
-2.4	0.0082	0.0080	0.0078	0.0076	0.0073	0.0071	0.0069	0.0068	0.0066	0.0064
-2.3	0.0107	0.0104	0.0102	0.0099	0.0096	0.0094	0.0091	0.0089	0.0087	0.0084
-2.2	0.0139	0.0136	0.0132	0.0129	0.0125	0.0122	0.0119	0.0116	0.0113	0.0110
-2.1	0.0179	0.0174	0.0170	0.0166	0.0162	0.0158	0.0154	0.0150	0.0146	0.0143
-2.0	0.0228	0.0222	0.0217	0.0212	0.0207	0.0202	0.0197	0.0192	0.0188	0.0183
-1.9	0.0287	0.0281	0.0274	0.0268	0.0262	0.0256	0.0250	0.0244	0.0239	0.0233
-1.8	0.0359	0.0351	0.0344	0.0336	0.0329	0.0322	0.0314	0.0307	0.0301	0.0294
-1.7	0.0446	0.0436	0.0427	0.0418	0.0409	0.0401	0.0392	0.0384	0.0375	0.0367
-1.6	0.0548	0.0537	0.0526	0.0516	0.0505	0.0495	0.0485	0.0475	0.0465	0.0455
-1.5	0.0668	0.0655	0.0643	0.0630	0.0618	0.0606	0.0594	0.0582	0.0571	0.0559
-1.4	0.0808	0.0793	0.0778	0.0764	0.0749	0.0735	0.0721	0.0708	0.0694	0.0681
-1.3	0.0968	0.0951	0.0934	0.0918	0.0901	0.0885	0.0869	0.0853	0.0838	0.0823
-1.2	0.1151	0.1131	0.1112	0.1093	0.1075	0.1057	0.1038	0.1020	0.1003	0.0985
-1.1	0.1357	0.1335	0.1314	0.1292	0.1271	0.1251	0.1230	0.1210	0.1190	0.1170
-1.0	0.1587	0.1562	0.1539	0.1515	0.1492	0.1469	0.1446	0.1423	0.1401	0.1379
-0.9	0.1841	0.1814	0.1788	0.1762	0.1736	0.1711	0.1685	0.1660	0.1635	0.1611
-0.8	0.2119	0.2090	0.2061	0.2033	0.2005	0.1977	0.1949	0.1922	0.1894	0.1867
-0.7	0.2420	0.2389	0.2358	0.2327	0.2297	0.2266	0.2236	0.2206	0.2177	0.2148
-0.6	0.2743	0.2709	0.2676	0.2643	0.2611	0.2578	0.2546	0.2514	0.2483	0.2451
-0.5	0.3085	0.3050	0.3015	0.2981	0.2946	0.2912	0.2877	0.2843	0.2810	0.2776
-0.4	0.3446	0.3409	0.3372	0.3336	0.3300	0.3264	0.3228	0.3192	0.3156	0.3121
-0.3	0.3821	0.3783	0.3745	0.3707	0.3669	0.3632	0.3594	0.3557	0.3520	0.3483
-0.2	0.4207	0.4168	0.4129	0.4090	0.4052	0.4013	0.3974	0.3936	0.3897	0.3859
-0.1	0.4602	0.4562	0.4522	0.4483	0.4443	0.4404	0.4364	0.4325	0.4286	0.4247
-0.0	0.5000	0.4960	0.4920	0.4880	0.4840	0.4801	0.4761	0.4721	0.4681	0.4641

Note: Table entry is the area under the standard normal curve to the left of the indicated z-value, thus giving $P(Z < z)$.

Standard normal cumulative probabilities (*continued*)

z	0.00	0.01	0.02	0.03	0.04	0.05	0.06	0.07	0.08	0.09
0.0	0.5000	0.5040	0.5080	0.5120	0.5160	0.5199	0.5239	0.5279	0.5319	0.5359
0.1	0.5398	0.5438	0.5478	0.5517	0.5557	0.5596	0.5636	0.5675	0.5714	0.5753
0.2	0.5793	0.5832	0.5871	0.5910	0.5948	0.5987	0.6026	0.6064	0.6103	0.6141
0.3	0.6179	0.6217	0.6255	0.6293	0.6331	0.6368	0.6406	0.6443	0.6480	0.6517
0.4	0.6554	0.6591	0.6628	0.6664	0.6700	0.6736	0.6772	0.6808	0.6844	0.6879
0.5	0.6915	0.6950	0.6985	0.7019	0.7054	0.7088	0.7123	0.7157	0.7190	0.7224
0.6	0.7257	0.7291	0.7324	0.7357	0.7389	0.7422	0.7454	0.7486	0.7517	0.7549
0.7	0.7580	0.7611	0.7642	0.7673	0.7703	0.7734	0.7764	0.7794	0.7823	0.7852
0.8	0.7881	0.7910	0.7939	0.7967	0.7995	0.8023	0.8051	0.8078	0.8106	0.8133
0.9	0.8159	0.8186	0.8212	0.8238	0.8264	0.8289	0.8315	0.8340	0.8365	0.8389
1.0	0.8413	0.8438	0.8461	0.8485	0.8508	0.8531	0.8554	0.8577	0.8599	0.8621
1.1	0.8643	0.8665	0.8686	0.8708	0.8729	0.8749	0.8770	0.8790	0.8810	0.8830
1.2	0.8849	0.8869	0.8888	0.8907	0.8925	0.8943	0.8962	0.8980	0.8997	0.9015
1.3	0.9032	0.9049	0.9066	0.9082	0.9099	0.9115	0.9131	0.9147	0.9162	0.9177
1.4	0.9192	0.9207	0.9222	0.9236	0.9251	0.9265	0.9279	0.9292	0.9306	0.9319
1.5	0.9332	0.9345	0.9357	0.9370	0.9382	0.9394	0.9406	0.9418	0.9429	0.9441
1.6	0.9452	0.9463	0.9474	0.9484	0.9495	0.9505	0.9515	0.9525	0.9535	0.9545
1.7	0.9554	0.9564	0.9573	0.9582	0.9591	0.9599	0.9608	0.9616	0.9625	0.9633
1.8	0.9641	0.9649	0.9656	0.9664	0.9671	0.9678	0.9686	0.9693	0.9699	0.9706
1.9	0.9713	0.9719	0.9726	0.9732	0.9738	0.9744	0.9750	0.9756	0.9761	0.9767
2.0	0.9772	0.9778	0.9783	0.9788	0.9793	0.9798	0.9803	0.9808	0.9812	0.9817
2.1	0.9821	0.9826	0.9830	0.9834	0.9838	0.9842	0.9846	0.9850	0.9854	0.9857
2.2	0.9861	0.9864	0.9868	0.9871	0.9875	0.9878	0.9881	0.9884	0.9887	0.9890
2.3	0.9893	0.9896	0.9898	0.9901	0.9904	0.9906	0.9909	0.9911	0.9913	0.9916
2.4	0.9918	0.9920	0.9922	0.9924	0.9927	0.9929	0.9931	0.9932	0.9934	0.9936
2.5	0.9938	0.9940	0.9941	0.9943	0.9945	0.9946	0.9948	0.9949	0.9951	0.9952
2.6	0.9953	0.9955	0.9956	0.9957	0.9959	0.9960	0.9961	0.9962	0.9963	0.9964
2.7	0.9965	0.9966	0.9967	0.9968	0.9969	0.9970	0.9971	0.9972	0.9973	0.9974
2.8	0.9974	0.9975	0.9976	0.9977	0.9977	0.9978	0.9979	0.9979	0.9980	0.9981
2.9	0.9981	0.9982	0.9982	0.9983	0.9984	0.9984	0.9985	0.9985	0.9986	0.9986
3.0	0.9986	0.9987	0.9987	0.9988	0.9988	0.9989	0.9989	0.9989	0.9990	0.9990
3.1	0.9990	0.9991	0.9991	0.9991	0.9992	0.9992	0.9992	0.9992	0.9993	0.9993
3.2	0.9993	0.9993	0.9994	0.9994	0.9994	0.9994	0.9994	0.9995	0.9995	0.9995
3.3	0.9995	0.9995	0.9995	0.9996	0.9996	0.9996	0.9996	0.9996	0.9996	0.9997
3.4	0.9997	0.9997	0.9997	0.9997	0.9997	0.9997	0.9997	0.9997	0.9997	0.9998
3.5	0.9998	0.9998	0.9998	0.9998	0.9998	0.9998	0.9998	0.9998	0.9998	0.9998
3.6	0.9998	0.9998	0.9999	0.9999	0.9999	0.9999	0.9999	0.9999	0.9999	0.9999
3.7	0.9999	0.9999	0.9999	0.9999	0.9999	0.9999	0.9999	0.9999	0.9999	0.9999
3.8	0.9999	0.9999	0.9999	0.9999	0.9999	0.9999	0.9999	0.9999	0.9999	0.9999
3.9	1.0000									

TABLE A-1 Standard normal cumulative probabilities (*continued*)

z	$P(Z < z)$		z	$P(Z < z)$
-4.265	0.00001		0	0.50
-3.891	0.00005		0.126	0.55
-3.719	0.0001		0.253	0.60
-3.291	0.0005			
-3.090	0.001		0.385	0.65
-2.576	0.005		0.524	0.70
-2.326	0.01		0.674	0.75
			0.842	0.80
-2.054	0.02		1.036	0.85
-1.960	0.025			
-1.881	0.03		1.282	0.90
-1.751	0.04		1.341	0.91
-1.645	0.05		1.405	0.92
			1.476	0.93
-1.555	0.06		1.555	0.94
-1.476	0.07			
-1.405	0.08		1.645	0.95
-1.341	0.09		1.751	0.96
-1.282	0.10		1.881	0.97
			1.960	0.975
-1.036	0.15		2.054	0.98
-0.842	0.20			
-0.674	0.25		2.326	0.99
-0.524	0.30		2.576	0.995
-0.385	0.35		3.090	0.999
			3.291	0.9995
-0.253	0.40		3.719	0.9999
-0.126	0.45		3.891	0.99995
0	0.50		4.265	0.99999

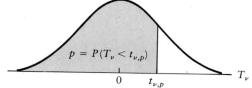

$$p = P(T_\nu < t_{\nu,p})$$

Student's *t* distribution

% df	55	65	75	85	90	95	97.5	99	99.5	99.95
1	0.158	0.510	1.000	1.963	3.078	6.314	12.706	31.821	63.657	636.619
2	0.142	0.445	0.816	1.386	1.886	2.920	4.303	6.965	9.925	31.599
3	0.137	0.424	0.765	1.250	1.638	2.353	3.182	4.541	5.841	12.924
4	0.134	0.414	0.741	1.190	1.533	2.132	2.776	3.747	4.604	8.610
5	0.132	0.408	0.727	1.156	1.476	2.015	2.571	3.365	4.032	6.869
6	0.131	0.404	0.718	1.134	1.440	1.943	2.447	3.143	3.707	5.959
7	0.130	0.402	0.711	1.119	1.415	1.895	2.365	2.998	3.499	5.408
8	0.130	0.399	0.706	1.108	1.397	1.860	2.306	2.896	3.355	5.041
9	0.129	0.398	0.703	1.100	1.383	1.833	2.262	2.821	3.250	4.781
10	0.129	0.397	0.700	1.093	1.372	1.812	2.228	2.764	3.169	4.587
11	0.129	0.396	0.697	1.088	1.363	1.796	2.201	2.718	3.106	4.437
12	0.128	0.395	0.695	1.083	1.356	1.782	2.179	2.681	3.055	4.318
13	0.128	0.394	0.694	1.079	1.350	1.771	2.160	2.650	3.012	4.221
14	0.128	0.393	0.692	1.076	1.345	1.761	2.145	2.624	2.977	4.140
15	0.128	0.393	0.691	1.074	1.341	1.753	2.131	2.602	2.947	4.073
16	0.128	0.392	0.690	1.071	1.337	1.746	2.120	2.583	2.921	4.015
17	0.128	0.392	0.689	1.069	1.333	1.740	2.110	2.567	2.898	3.965
18	0.127	0.392	0.688	1.067	1.330	1.734	2.101	2.552	2.878	3.922
19	0.127	0.391	0.688	1.066	1.328	1.729	2.093	2.539	2.861	3.883
20	0.127	0.391	0.687	1.064	1.325	1.725	2.086	2.528	2.845	3.850
21	0.127	0.391	0.686	1.063	1.323	1.721	2.080	2.518	2.831	3.819
22	0.127	0.390	0.686	1.061	1.321	1.717	2.074	2.508	2.819	3.792
23	0.127	0.390	0.685	1.060	1.319	1.714	2.069	2.500	2.807	3.768
24	0.127	0.390	0.685	1.059	1.318	1.711	2.064	2.492	2.797	3.745
25	0.127	0.390	0.684	1.058	1.316	1.708	2.060	2.485	2.787	3.725
26	0.127	0.390	0.684	1.058	1.315	1.706	2.056	2.479	2.779	3.707
27	0.127	0.389	0.684	1.057	1.314	1.703	2.052	2.473	2.771	3.690
28	0.127	0.389	0.683	1.056	1.313	1.701	2.048	2.467	2.763	3.674
29	0.127	0.389	0.683	1.055	1.311	1.699	2.045	2.462	2.756	3.659
30	0.127	0.389	0.683	1.055	1.310	1.697	2.042	2.457	2.750	3.646
35	0.127	0.388	0.682	1.052	1.306	1.690	2.030	2.438	2.724	3.591
40	0.126	0.388	0.681	1.050	1.303	1.684	2.021	2.423	2.704	3.551
45	0.126	0.388	0.680	1.049	1.301	1.679	2.014	2.412	2.690	3.520
50	0.126	0.388	0.679	1.047	1.299	1.676	2.009	2.403	2.678	3.496
60	0.126	0.387	0.679	1.045	1.296	1.671	2.000	2.390	2.660	3.460
70	0.126	0.387	0.678	1.044	1.294	1.667	1.994	2.381	2.648	3.435
80	0.126	0.387	0.678	1.043	1.292	1.664	1.990	2.374	2.639	3.416
90	0.126	0.387	0.677	1.042	1.291	1.662	1.987	2.368	2.632	3.402
100	0.126	0.386	0.677	1.042	1.290	1.660	1.984	2.364	2.626	3.390
120	0.126	0.386	0.677	1.041	1.289	1.658	1.980	2.358	2.617	3.373
140	0.126	0.386	0.676	1.040	1.288	1.656	1.977	2.353	2.611	3.361
160	0.126	0.386	0.676	1.040	1.287	1.654	1.975	2.350	2.607	3.352
180	0.126	0.386	0.676	1.039	1.286	1.653	1.973	2.547	2.603	3.345
200	0.126	0.386	0.676	1.039	1.286	1.653	1.972	2.345	2.601	3.340
∞	0.126	0.385	0.674	1.036	1.282	1.645	1.960	2.326	2.576	3.291

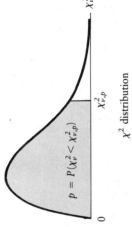

$$p = P(\chi_v^2 < \chi_{v,p}^2)$$

χ^2 distribution

TABLE A-3   Percentiles of the chi-square distribution

% df	0.5	1	2.5	5	10	20	30	40	50	60	70	80	90	95	97.5	99	99.5	99.95
1	0.0001	0.0002	0.001	0.004	0.016	0.064	0.148	0.275	0.455	0.708	1.074	1.642	2.706	3.841	5.024	6.635	7.879	12.116
2	0.010	0.020	0.051	0.103	0.211	0.446	0.713	1.022	1.386	1.833	2.408	3.219	4.605	5.991	7.378	9.210	10.597	15.202
3	0.072	0.115	0.216	0.352	0.584	1.005	1.424	1.869	2.366	2.946	3.665	4.642	6.251	7.815	9.348	11.345	12.838	17.730
4	0.207	0.297	0.484	0.711	1.064	1.649	2.195	2.753	3.357	4.045	4.878	5.989	7.779	9.488	11.143	13.277	14.860	19.997
5	0.412	0.554	0.831	1.145	1.610	2.343	3.000	3.655	4.351	5.132	6.064	7.289	9.236	11.070	12.833	15.086	16.750	22.105
6	0.676	0.872	1.237	1.635	2.204	3.070	3.828	4.570	5.348	6.211	7.231	8.558	10.645	12.592	14.449	16.812	18.548	24.103
7	0.989	1.239	1.690	2.167	2.833	3.822	4.671	5.493	6.346	7.283	8.383	9.803	12.017	14.067	16.013	18.475	20.278	26.018
8	1.344	1.646	2.180	2.733	3.490	4.594	5.527	6.423	7.344	8.351	9.524	11.030	13.362	15.507	17.535	20.090	21.955	27.868
9	1.735	2.088	2.700	3.325	4.168	5.380	6.393	7.357	8.343	9.414	10.656	12.242	14.684	16.919	19.023	21.666	23.589	29.666
10	2.156	2.558	3.247	3.940	4.865	6.179	7.267	8.295	9.342	10.473	11.781	13.442	15.987	18.307	20.483	23.209	25.188	31.420
11	2.603	3.053	3.816	4.575	5.578	6.989	8.148	9.237	10.341	11.530	12.899	14.631	17.275	19.675	21.920	24.725	26.757	33.137
12	3.074	3.571	4.404	5.226	6.304	7.807	9.034	10.182	11.340	12.584	14.011	15.812	18.549	21.026	23.337	26.217	28.300	34.821
13	3.565	4.107	5.009	5.892	7.042	8.634	9.926	11.129	12.340	13.636	15.119	16.985	19.812	22.362	24.736	27.688	29.819	36.478
14	4.075	4.660	5.629	6.571	7.790	9.467	10.821	12.078	13.339	14.685	16.222	18.151	21.064	23.685	26.119	29.141	31.319	38.109
15	4.601	5.229	6.262	7.261	8.547	10.307	11.721	13.030	14.339	15.733	17.322	19.311	22.307	24.996	27.488	30.578	32.801	39.719
16	5.142	5.812	6.908	7.962	9.312	11.152	12.624	13.983	15.338	16.780	18.418	20.465	23.542	26.296	28.845	32.000	34.267	41.308
17	5.697	6.408	7.564	8.672	10.085	12.002	13.531	14.937	16.338	17.824	19.511	21.615	24.769	27.587	30.191	33.409	35.718	42.879
18	6.265	7.015	8.231	9.390	10.865	12.857	14.440	15.893	17.338	18.868	20.601	22.760	25.989	28.869	31.526	34.805	37.156	44.434
19	6.844	7.633	8.907	10.117	11.651	13.716	15.352	16.850	18.338	19.910	21.689	23.900	27.204	30.144	32.852	36.191	38.582	45.973
20	7.434	8.260	9.591	10.851	12.443	14.578	16.266	17.809	19.337	20.951	22.775	25.038	28.412	31.410	34.170	37.566	39.997	47.498
21	8.034	8.897	10.283	11.591	13.240	15.445	17.182	18.768	20.337	21.991	23.858	26.171	29.615	32.671	35.479	38.932	41.401	49.011
22	8.643	9.542	10.982	12.338	14.041	16.314	18.101	19.729	21.337	23.031	24.939	27.301	30.813	33.924	36.781	40.289	42.796	50.511
23	9.260	10.196	11.689	13.091	14.848	17.187	19.021	20.690	22.337	24.069	26.018	28.429	32.007	35.172	38.076	41.638	44.181	52.000
24	9.886	10.856	12.401	13.848	15.659	18.062	19.943	21.752	23.337	25.106	27.096	29.553	33.196	36.415	39.364	42.980	45.559	53.479
25	10.520	11.524	13.120	14.611	16.473	18.940	20.867	22.616	24.337	26.143	28.172	30.675	34.382	37.652	40.646	44.314	46.928	54.947
26	11.160	12.198	13.844	15.379	17.292	19.820	21.792	23.579	25.336	27.179	29.246	31.795	35.563	38.885	41.923	45.642	48.290	56.407
27	11.808	12.879	14.573	16.151	18.114	20.703	22.719	24.544	26.336	28.214	30.319	32.912	36.741	40.113	43.195	46.963	49.645	57.858
28	12.461	13.565	15.308	16.928	18.939	21.588	23.647	25.509	27.336	29.249	31.391	34.027	37.916	41.337	44.461	48.278	50.993	59.300
29	13.121	14.256	16.047	17.708	19.768	22.475	24.577	26.475	28.336	30.283	32.461	35.139	39.087	42.557	45.722	49.588	52.336	60.735
30	13.787	14.953	16.791	18.493	20.599	23.364	25.508	27.442	29.336	31.316	33.530	36.250	40.256	43.773	46.979	50.892	53.672	62.162
35	17.192	18.509	20.569	22.465	24.797	27.836	30.178	32.282	34.336	36.475	38.859	41.778	46.059	49.802	53.203	57.342	60.275	69.199
40	20.707	22.164	24.433	26.509	29.051	32.345	34.872	37.134	39.335	41.622	44.165	47.269	51.805	55.758	59.342	63.691	66.766	76.095
45	24.311	25.901	28.366	30.612	33.350	36.884	39.585	41.995	44.335	46.761	49.452	52.729	57.505	61.656	65.410	69.957	73.166	82.876
50	27.991	29.707	32.357	34.764	37.689	41.449	44.313	46.864	49.335	51.892	54.723	58.164	63.167	67.505	71.420	76.154	79.490	89.561
60	35.534	37.485	40.482	43.188	46.459	50.641	53.809	56.620	59.335	62.135	65.227	68.972	74.397	79.082	83.298	88.379	91.952	102.695
70	43.275	45.442	48.758	51.739	55.329	59.898	63.346	66.396	69.334	72.358	75.689	79.715	85.527	90.531	95.023	100.425	104.215	115.578
80	51.172	53.540	57.153	60.391	64.278	69.207	72.915	76.188	79.334	82.566	86.120	90.405	96.578	101.879	106.629	112.329	116.321	128.261
90	59.196	61.754	65.647	69.126	73.291	78.558	82.511	85.993	89.334	92.761	96.524	101.054	107.565	113.145	118.136	124.116	128.299	140.782
100	67.328	70.065	74.222	77.929	82.358	87.945	92.129	95.808	99.334	102.946	106.906	111.667	118.498	124.342	129.561	135.807	140.169	153.167
120	83.852	86.923	91.573	95.705	100.624	106.806	111.419	115.465	119.334	123.289	127.616	132.806	140.233	146.567	152.211	158.950	163.648	177.603
140	100.655	104.034	109.137	113.659	119.029	125.758	130.766	135.149	139.334	143.604	148.269	153.854	161.827	168.613	174.648	181.840	186.847	201.683
160	117.679	121.346	126.870	131.756	137.546	144.783	150.158	154.856	159.334	163.898	168.876	174.828	183.311	190.516	196.915	204.530	209.824	225.481
180	134.884	138.820	144.741	149.969	156.153	163.868	169.588	174.580	179.334	184.173	189.446	195.743	204.704	212.304	219.044	227.056	232.620	249.048
200	152.241	156.432	162.728	168.279	174.835	183.003	189.049	194.319	199.334	204.434	209.985	216.609	226.021	233.994	241.058	249.445	255.264	272.423

TABLE A-4 Percentiles of the *F* distribution

Upper 25% point of the *F* distribution

$$p = P(F_{v_1, v_2} < F_{v_1, v_2; p})$$

F distribution

DEGREES OF FREEDOM FOR NUMERATOR

v_2 \ v_1	1	2	3	4	5	6	7	8	9	10	11	12	13	14	15	16	17	18	19	20	25	30	40	50	100	150	200
1	5.83	7.50	8.20	8.58	8.82	8.98	9.10	9.19	9.26	9.32	9.37	9.41	9.44	9.47	9.49	9.52	9.53	9.55	9.57	9.58	9.63	9.67	9.71	9.74	9.80	9.81	9.82
2	2.57	3.00	3.15	3.23	3.28	3.31	3.34	3.35	3.37	3.38	3.39	3.39	3.40	3.41	3.41	3.41	3.42	3.42	3.43	3.44	3.44	3.45	3.46	3.47	3.47	3.47	3.47
3	2.02	2.28	2.36	2.39	2.41	2.42	2.43	2.44	2.44	2.44	2.45	2.45	2.45	2.45	2.46	2.46	2.46	2.46	2.46	2.46	2.46	2.47	2.47	2.47	2.47	2.47	2.47
4	1.81	2.00	2.05	2.06	2.07	2.08	2.08	2.08	2.08	2.08	2.08	2.08	2.08	2.08	2.08	2.08	2.08	2.08	2.08	2.08	2.08	2.08	2.08	2.08	2.08	2.08	2.08
5	1.69	1.85	1.88	1.89	1.89	1.89	1.89	1.89	1.89	1.89	1.89	1.89	1.89	1.89	1.89	1.88	1.88	1.88	1.88	1.88	1.88	1.88	1.88	1.88	1.87	1.87	1.87
6	1.62	1.76	1.78	1.79	1.79	1.78	1.78	1.78	1.77	1.77	1.77	1.77	1.77	1.76	1.76	1.76	1.76	1.76	1.76	1.76	1.75	1.75	1.75	1.75	1.74	1.74	1.74
7	1.57	1.70	1.72	1.72	1.71	1.71	1.70	1.70	1.69	1.69	1.69	1.68	1.68	1.68	1.68	1.67	1.67	1.67	1.67	1.67	1.67	1.66	1.66	1.66	1.65	1.65	1.65
8	1.54	1.66	1.67	1.66	1.66	1.65	1.64	1.64	1.63	1.63	1.63	1.62	1.62	1.62	1.62	1.62	1.61	1.61	1.61	1.61	1.60	1.60	1.59	1.59	1.58	1.58	1.58
9	1.51	1.62	1.63	1.63	1.62	1.61	1.60	1.60	1.59	1.59	1.58	1.58	1.58	1.57	1.57	1.57	1.57	1.56	1.56	1.56	1.55	1.55	1.54	1.54	1.53	1.53	1.53
10	1.49	1.60	1.60	1.59	1.59	1.58	1.57	1.56	1.56	1.55	1.55	1.54	1.54	1.54	1.53	1.53	1.53	1.53	1.53	1.52	1.52	1.55	1.51	1.51	1.49	1.49	1.49
11	1.47	1.58	1.58	1.57	1.56	1.55	1.54	1.53	1.53	1.52	1.52	1.51	1.51	1.51	1.50	1.50	1.50	1.50	1.49	1.49	1.49	1.48	1.47	1.47	1.46	1.46	1.46
12	1.46	1.56	1.56	1.55	1.54	1.53	1.52	1.51	1.51	1.50	1.50	1.49	1.49	1.49	1.48	1.48	1.47	1.47	1.47	1.47	1.46	1.45	1.45	1.44	1.43	1.43	1.43
13	1.45	1.55	1.55	1.53	1.52	1.51	1.50	1.49	1.49	1.49	1.47	1.47	1.47	1.46	1.46	1.46	1.45	1.45	1.45	1.45	1.44	1.43	1.42	1.42	1.41	1.41	1.40
14	1.44	1.53	1.53	1.52	1.51	1.50	1.49	1.48	1.47	1.46	1.46	1.45	1.45	1.44	1.44	1.44	1.44	1.43	1.43	1.43	1.42	1.41	1.41	1.40	1.39	1.39	1.38
15	1.43	1.52	1.52	1.51	1.49	1.48	1.47	1.46	1.46	1.45	1.44	1.44	1.43	1.43	1.43	1.42	1.42	1.41	1.41	1.41	1.40	1.40	1.39	1.38	1.37	1.37	1.37
16	1.42	1.51	1.51	1.50	1.48	1.47	1.46	1.45	1.44	1.44	1.43	1.43	1.42	1.42	1.41	1.41	1.41	1.40	1.40	1.40	1.39	1.38	1.37	1.37	1.36	1.35	1.35
17	1.42	1.51	1.50	1.49	1.47	1.46	1.45	1.44	1.43	1.43	1.42	1.41	1.41	1.41	1.40	1.40	1.39	1.39	1.39	1.39	1.38	1.37	1.36	1.36	1.34	1.34	1.34
18	1.41	1.50	1.49	1.48	1.46	1.45	1.44	1.43	1.42	1.42	1.41	1.40	1.40	1.40	1.39	1.39	1.38	1.38	1.38	1.38	1.37	1.36	1.35	1.35	1.33	1.33	1.32
19	1.41	1.49	1.49	1.47	1.46	1.44	1.43	1.42	1.41	1.41	1.40	1.40	1.39	1.39	1.38	1.38	1.37	1.37	1.37	1.37	1.36	1.35	1.34	1.33	1.32	1.31	1.31
20	1.40	1.49	1.48	1.47	1.45	1.44	1.43	1.42	1.41	1.40	1.39	1.39	1.38	1.38	1.37	1.37	1.36	1.36	1.36	1.36	1.35	1.34	1.33	1.32	1.31	1.30	1.30
21	1.40	1.48	1.48	1.46	1.44	1.43	1.42	1.41	1.40	1.39	1.39	1.38	1.37	1.37	1.37	1.36	1.36	1.35	1.35	1.35	1.34	1.33	1.32	1.31	1.30	1.29	1.29
22	1.40	1.48	1.47	1.45	1.44	1.42	1.41	1.40	1.39	1.39	1.38	1.37	1.37	1.36	1.36	1.36	1.35	1.35	1.34	1.34	1.33	1.32	1.31	1.31	1.29	1.29	1.28
23	1.39	1.47	1.47	1.45	1.43	1.42	1.41	1.40	1.39	1.38	1.37	1.37	1.36	1.36	1.35	1.35	1.35	1.34	1.34	1.34	1.33	1.32	1.31	1.30	1.28	1.28	1.28
24	1.39	1.47	1.46	1.44	1.43	1.41	1.40	1.39	1.38	1.38	1.37	1.36	1.36	1.35	1.35	1.34	1.34	1.34	1.33	1.33	1.32	1.31	1.30	1.29	1.28	1.27	1.27
25	1.39	1.47	1.46	1.44	1.42	1.41	1.40	1.39	1.38	1.37	1.36	1.36	1.35	1.35	1.34	1.34	1.33	1.33	1.33	1.33	1.31	1.31	1.29	1.29	1.27	1.27	1.26
26	1.38	1.46	1.45	1.44	1.42	1.41	1.39	1.38	1.37	1.37	1.36	1.35	1.35	1.34	1.34	1.33	1.33	1.33	1.32	1.32	1.31	1.30	1.29	1.28	1.27	1.26	1.26
27	1.38	1.46	1.45	1.43	1.42	1.40	1.39	1.38	1.37	1.36	1.35	1.35	1.34	1.34	1.33	1.33	1.33	1.32	1.32	1.32	1.30	1.30	1.28	1.28	1.26	1.26	1.25
28	1.38	1.46	1.45	1.43	1.41	1.40	1.39	1.38	1.37	1.36	1.35	1.34	1.34	1.33	1.33	1.32	1.32	1.32	1.31	1.31	1.30	1.29	1.28	1.27	1.25	1.25	1.25
29	1.38	1.45	1.45	1.43	1.41	1.40	1.38	1.37	1.36	1.35	1.35	1.34	1.33	1.33	1.32	1.32	1.32	1.31	1.31	1.31	1.29	1.29	1.27	1.27	1.25	1.24	1.24
30	1.38	1.45	1.44	1.42	1.41	1.39	1.38	1.37	1.36	1.35	1.35	1.34	1.33	1.33	1.32	1.32	1.31	1.31	1.31	1.30	1.29	1.28	1.27	1.26	1.25	1.24	1.24
32	1.37	1.44	1.44	1.42	1.40	1.39	1.37	1.36	1.35	1.34	1.34	1.33	1.32	1.32	1.31	1.31	1.30	1.30	1.30	1.30	1.28	1.28	1.26	1.25	1.23	1.23	1.23
34	1.37	1.44	1.43	1.41	1.40	1.38	1.37	1.36	1.35	1.34	1.33	1.32	1.32	1.31	1.31	1.30	1.30	1.29	1.29	1.29	1.28	1.27	1.26	1.25	1.23	1.22	1.22
36	1.37	1.44	1.42	1.40	1.38	1.37	1.36	1.35	1.34	1.33	1.33	1.32	1.31	1.31	1.30	1.30	1.29	1.29	1.28	1.28	1.27	1.26	1.25	1.24	1.22	1.22	1.21
38	1.36	1.43	1.42	1.40	1.38	1.37	1.35	1.35	1.34	1.33	1.32	1.31	1.31	1.30	1.30	1.29	1.29	1.28	1.28	1.28	1.27	1.26	1.24	1.24	1.22	1.21	1.21
40	1.36	1.43	1.42	1.40	1.39	1.37	1.36	1.35	1.34	1.33	1.33	1.31	1.31	1.30	1.30	1.29	1.29	1.28	1.28	1.28	1.27	1.25	1.24	1.23	1.21	1.20	1.20
42	1.35	1.42	1.41	1.38	1.37	1.35	1.33	1.32	1.31	1.30	1.29	1.29	1.28	1.27	1.27	1.26	1.26	1.26	1.25	1.25	1.23	1.23	1.21	1.20	1.18	1.17	1.16
44	1.35	1.41	1.40	1.38	1.36	1.35	1.33	1.32	1.31	1.30	1.29	1.28	1.27	1.27	1.26	1.26	1.25	1.25	1.25	1.24	1.23	1.21	1.20	1.19	1.16	1.16	1.15
46	1.34	1.41	1.40	1.38	1.36	1.34	1.32	1.31	1.30	1.29	1.28	1.27	1.26	1.26	1.25	1.25	1.24	1.24	1.24	1.23	1.22	1.20	1.19	1.18	1.16	1.15	1.14
48	1.34	1.41	1.38	1.37	1.35	1.33	1.32	1.31	1.30	1.29	1.28	1.27	1.26	1.25	1.25	1.24	1.24	1.23	1.23	1.23	1.21	1.20	1.19	1.18	1.15	1.14	1.13
50	1.34	1.41	1.39	1.37	1.35	1.33	1.32	1.31	1.30	1.29	1.27	1.26	1.26	1.25	1.25	1.24	1.24	1.23	1.23	1.23	1.21	1.20	1.18	1.18	1.14	1.13	1.13
60	1.35	1.42	1.41	1.38	1.37	1.35	1.33	1.32	1.31	1.30	1.29	1.29	1.28	1.27	1.27	1.26	1.26	1.26	1.25	1.25	1.23	1.22	1.21	1.20	1.18	1.17	1.16
70	1.35	1.41	1.40	1.38	1.36	1.35	1.33	1.32	1.31	1.30	1.29	1.28	1.27	1.27	1.26	1.26	1.25	1.25	1.25	1.24	1.23	1.21	1.20	1.19	1.16	1.16	1.15
80	1.34	1.41	1.40	1.38	1.36	1.34	1.32	1.31	1.30	1.29	1.28	1.27	1.27	1.26	1.26	1.25	1.25	1.24	1.24	1.24	1.22	1.21	1.19	1.18	1.16	1.15	1.14
90	1.34	1.41	1.39	1.37	1.35	1.33	1.32	1.31	1.30	1.29	1.28	1.27	1.26	1.26	1.25	1.25	1.24	1.24	1.23	1.23	1.21	1.20	1.19	1.18	1.15	1.14	1.13
100	1.34	1.41	1.39	1.37	1.35	1.33	1.32	1.31	1.30	1.29	1.27	1.26	1.26	1.25	1.25	1.24	1.24	1.23	1.23	1.23	1.21	1.20	1.18	1.18	1.14	1.13	1.13
125	1.34	1.40	1.39	1.36	1.34	1.33	1.31	1.30	1.29	1.28	1.27	1.26	1.25	1.25	1.24	1.24	1.23	1.23	1.22	1.22	1.20	1.19	1.17	1.16	1.14	1.12	1.12
150	1.33	1.40	1.38	1.36	1.34	1.33	1.31	1.30	1.29	1.28	1.27	1.26	1.25	1.24	1.24	1.23	1.23	1.22	1.22	1.21	1.20	1.19	1.17	1.16	1.13	1.12	1.11
200	1.33	1.40	1.38	1.36	1.34	1.32	1.30	1.29	1.28	1.27	1.26	1.26	1.25	1.24	1.23	1.23	1.22	1.22	1.21	1.21	1.19	1.18	1.16	1.15	1.12	1.11	1.10
300	1.33	1.39	1.38	1.35	1.34	1.32	1.30	1.29	1.28	1.27	1.26	1.25	1.24	1.24	1.23	1.22	1.22	1.21	1.21	1.20	1.19	1.17	1.15	1.15	1.11	1.10	1.09
500	1.33	1.39	1.38	1.35	1.33	1.31	1.30	1.28	1.27	1.26	1.25	1.24	1.24	1.23	1.22	1.22	1.21	1.21	1.20	1.20	1.18	1.17	1.15	1.14	1.10	1.09	1.08
1000	1.32	1.39	1.37	1.35	1.33	1.31	1.29	1.28	1.27	1.26	1.25	1.24	1.23	1.23	1.22	1.21	1.21	1.20	1.20	1.19	1.18	1.16	1.14	1.13	1.10	1.08	1.07

DEGREES OF FREEDOM FOR DENOMINATOR

TABLE A-4 Percentiles of the *F* distribution (*continued*)

Upper 10% point of the F distribution

	DEGREES OF FREEDOM FOR NUMERATOR																										
	1	2	3	4	5	6	7	8	9	10	11	12	13	14	15	16	17	18	19	20	25	30	40	50	100	150	200
1	39.9	49.5	53.6	55.8	57.2	58.2	58.9	59.4	59.9	60.2	60.5	60.7	60.9	61.1	61.2	61.3	61.5	61.6	61.7	61.7	62.1	62.3	62.5	62.7	63.0	63.1	63.2
2	8.53	9.00	9.16	9.24	9.29	9.33	9.35	9.37	9.38	9.39	9.40	9.41	9.41	9.42	9.42	9.43	9.43	9.44	9.44	9.44	9.45	9.46	9.47	9.47	9.48	9.48	9.49
3	5.54	5.46	5.39	5.34	5.31	5.28	5.27	5.25	5.24	5.23	5.22	5.22	5.21	5.20	5.20	5.20	5.19	5.19	5.19	5.18	5.17	5.17	5.16	5.15	5.14	5.14	5.14
4	4.54	4.32	4.19	4.11	4.05	4.01	3.98	3.95	3.94	3.92	3.91	3.90	3.89	3.88	3.87	3.86	3.86	3.85	3.85	3.84	3.83	3.82	3.80	3.80	3.78	3.77	3.77
5	4.06	3.78	3.62	3.52	3.45	3.40	3.37	3.34	3.32	3.30	3.28	3.27	3.26	3.25	3.24	3.23	3.22	3.22	3.21	3.21	3.19	3.17	3.16	3.15	3.13	3.12	3.12
6	3.78	3.46	3.29	3.18	3.11	3.05	3.01	2.98	2.96	2.94	2.92	2.90	2.89	2.88	2.87	2.86	2.85	2.85	2.84	2.84	2.81	2.80	2.78	2.77	2.75	2.74	2.73
7	3.59	3.26	3.07	2.96	2.88	2.83	2.78	2.75	2.72	2.70	2.68	2.67	2.65	2.64	2.63	2.62	2.61	2.61	2.60	2.59	2.57	2.56	2.54	2.52	2.50	2.50	2.48
8	3.46	3.11	2.92	2.81	2.73	2.67	2.62	2.59	2.56	2.54	2.52	2.50	2.49	2.48	2.46	2.45	2.45	2.44	2.43	2.42	2.40	2.38	2.36	2.35	2.32	2.31	2.31
9	3.36	3.01	2.81	2.69	2.61	2.55	2.51	2.47	2.44	2.42	2.40	2.38	2.36	2.35	2.34	2.33	2.32	2.31	2.30	2.30	2.27	2.25	2.23	2.22	2.19	2.18	2.17
10	3.29	2.92	2.73	2.61	2.52	2.46	2.41	2.38	2.35	2.32	2.30	2.28	2.27	2.26	2.24	2.23	2.22	2.22	2.21	2.20	2.17	2.16	2.13	2.12	2.09	2.08	2.07
11	3.23	2.86	2.66	2.54	2.45	2.39	2.34	2.30	2.27	2.25	2.23	2.21	2.19	2.18	2.17	2.16	2.15	2.14	2.13	2.12	2.10	2.08	2.05	2.04	2.01	1.99	1.99
12	3.18	2.81	2.61	2.48	2.39	2.33	2.28	2.24	2.21	2.19	2.17	2.15	2.13	2.12	2.10	2.09	2.08	2.08	2.07	2.06	2.03	2.01	1.99	1.97	1.94	1.93	1.92
13	3.14	2.76	2.56	2.43	2.35	2.28	2.23	2.20	2.16	2.14	2.12	2.10	2.08	2.07	2.05	2.04	2.03	2.02	2.01	2.01	1.98	1.96	1.93	1.92	1.88	1.87	1.86
14	3.10	2.73	2.52	2.39	2.31	2.24	2.19	2.15	2.12	2.10	2.07	2.05	2.04	2.02	2.01	2.00	1.99	1.98	1.97	1.96	1.93	1.91	1.89	1.87	1.83	1.82	1.82
15	3.07	2.70	2.49	2.36	2.27	2.21	2.16	2.12	2.09	2.06	2.04	2.02	2.00	1.99	1.97	1.96	1.95	1.94	1.93	1.92	1.89	1.87	1.85	1.83	1.79	1.78	1.77
16	3.05	2.67	2.46	2.33	2.24	2.18	2.13	2.09	2.06	2.03	2.01	1.99	1.97	1.95	1.94	1.93	1.92	1.91	1.90	1.89	1.86	1.84	1.81	1.79	1.76	1.74	1.74
17	3.03	2.64	2.44	2.31	2.22	2.15	2.10	2.06	2.03	2.00	1.98	1.96	1.94	1.93	1.91	1.90	1.89	1.88	1.87	1.86	1.83	1.81	1.78	1.76	1.73	1.72	1.71
18	3.01	2.62	2.42	2.29	2.20	2.13	2.08	2.04	2.00	1.98	1.95	1.93	1.92	1.90	1.89	1.87	1.86	1.85	1.84	1.84	1.80	1.78	1.75	1.74	1.70	1.68	1.68
19	2.99	2.61	2.40	2.27	2.18	2.11	2.06	2.02	1.98	1.96	1.93	1.91	1.89	1.88	1.86	1.85	1.84	1.83	1.82	1.81	1.78	1.76	1.73	1.71	1.67	1.66	1.65
20	2.97	2.59	2.38	2.25	2.16	2.09	2.04	2.00	1.96	1.94	1.91	1.89	1.87	1.86	1.84	1.83	1.82	1.81	1.80	1.79	1.76	1.74	1.71	1.69	1.65	1.64	1.63
21	2.96	2.57	2.36	2.23	2.14	2.08	2.02	1.98	1.95	1.92	1.90	1.88	1.86	1.84	1.83	1.81	1.80	1.79	1.78	1.78	1.74	1.72	1.69	1.67	1.63	1.62	1.61
22	2.95	2.56	2.35	2.22	2.13	2.06	2.01	1.97	1.93	1.90	1.88	1.86	1.84	1.83	1.81	1.80	1.79	1.78	1.77	1.76	1.73	1.70	1.67	1.65	1.61	1.60	1.59
23	2.94	2.55	2.34	2.21	2.11	2.05	1.99	1.95	1.92	1.89	1.87	1.84	1.83	1.81	1.80	1.78	1.77	1.76	1.75	1.74	1.71	1.69	1.66	1.64	1.59	1.58	1.57
24	2.93	2.54	2.33	2.19	2.10	2.04	1.98	1.94	1.91	1.88	1.85	1.83	1.81	1.80	1.78	1.77	1.75	1.75	1.73	1.73	1.70	1.67	1.64	1.62	1.58	1.56	1.56
25	2.92	2.53	2.32	2.18	2.09	2.02	1.97	1.93	1.89	1.87	1.84	1.82	1.80	1.79	1.77	1.76	1.75	1.74	1.73	1.72	1.68	1.66	1.63	1.61	1.56	1.55	1.54
26	2.91	2.52	2.31	2.17	2.08	2.01	1.96	1.92	1.88	1.86	1.83	1.81	1.79	1.77	1.76	1.75	1.73	1.72	1.71	1.71	1.67	1.65	1.61	1.59	1.55	1.54	1.53
27	2.90	2.51	2.30	2.17	2.07	2.00	1.95	1.91	1.87	1.85	1.82	1.80	1.78	1.76	1.75	1.74	1.72	1.71	1.70	1.70	1.66	1.64	1.60	1.58	1.54	1.52	1.52
28	2.89	2.50	2.29	2.16	2.06	2.00	1.94	1.90	1.87	1.84	1.81	1.79	1.77	1.75	1.74	1.73	1.71	1.70	1.69	1.69	1.65	1.63	1.59	1.57	1.53	1.51	1.50
29	2.89	2.50	2.28	2.15	2.06	1.99	1.93	1.89	1.86	1.83	1.80	1.78	1.76	1.75	1.73	1.72	1.71	1.69	1.68	1.68	1.64	1.62	1.58	1.56	1.52	1.50	1.49
30	2.88	2.49	2.28	2.14	2.05	1.98	1.93	1.88	1.85	1.82	1.79	1.77	1.75	1.74	1.72	1.71	1.70	1.69	1.68	1.67	1.63	1.61	1.57	1.55	1.51	1.49	1.48
32	2.87	2.48	2.26	2.13	2.04	1.97	1.91	1.87	1.83	1.81	1.78	1.76	1.74	1.72	1.71	1.69	1.68	1.67	1.66	1.65	1.62	1.59	1.56	1.53	1.49	1.47	1.46
34	2.86	2.47	2.25	2.12	2.02	1.96	1.90	1.86	1.82	1.79	1.77	1.75	1.73	1.71	1.69	1.68	1.67	1.66	1.65	1.64	1.60	1.58	1.54	1.52	1.47	1.46	1.45
36	2.85	2.46	2.24	2.11	2.01	1.94	1.89	1.85	1.81	1.78	1.75	1.73	1.71	1.70	1.68	1.67	1.66	1.65	1.64	1.63	1.59	1.56	1.53	1.51	1.46	1.44	1.43
38	2.84	2.45	2.23	2.10	2.01	1.94	1.88	1.84	1.80	1.77	1.75	1.72	1.70	1.69	1.67	1.66	1.65	1.63	1.62	1.61	1.58	1.55	1.52	1.49	1.45	1.43	1.42
40	2.84	2.44	2.23	2.09	2.00	1.93	1.87	1.83	1.79	1.76	1.74	1.71	1.70	1.68	1.66	1.65	1.64	1.62	1.61	1.61	1.57	1.54	1.51	1.48	1.43	1.42	1.41
42	2.83	2.43	2.22	2.08	1.99	1.92	1.86	1.82	1.78	1.75	1.73	1.71	1.69	1.67	1.65	1.64	1.63	1.62	1.60	1.60	1.56	1.53	1.50	1.47	1.42	1.41	1.40
44	2.82	2.43	2.21	2.08	1.98	1.91	1.86	1.81	1.78	1.75	1.72	1.70	1.68	1.66	1.65	1.63	1.62	1.61	1.60	1.59	1.55	1.52	1.49	1.46	1.41	1.40	1.39
46	2.82	2.42	2.21	2.07	1.98	1.91	1.85	1.81	1.77	1.74	1.71	1.69	1.67	1.65	1.64	1.63	1.61	1.60	1.59	1.58	1.54	1.52	1.48	1.46	1.40	1.39	1.38
48	2.81	2.42	2.20	2.07	1.97	1.90	1.85	1.80	1.77	1.73	1.71	1.69	1.67	1.65	1.63	1.62	1.61	1.59	1.58	1.57	1.54	1.51	1.47	1.45	1.40	1.38	1.37
50	2.81	2.41	2.20	2.06	1.97	1.90	1.84	1.80	1.76	1.73	1.70	1.68	1.66	1.64	1.63	1.61	1.60	1.59	1.58	1.57	1.53	1.50	1.46	1.44	1.39	1.37	1.36
60	2.79	2.39	2.18	2.04	1.95	1.87	1.82	1.77	1.74	1.71	1.68	1.66	1.64	1.62	1.60	1.59	1.58	1.56	1.55	1.54	1.50	1.48	1.44	1.41	1.36	1.34	1.33
70	2.78	2.38	2.16	2.03	1.93	1.86	1.80	1.76	1.72	1.69	1.66	1.64	1.62	1.60	1.59	1.57	1.56	1.55	1.54	1.53	1.49	1.46	1.42	1.39	1.34	1.32	1.30
80	2.77	2.37	2.15	2.02	1.92	1.85	1.79	1.75	1.71	1.68	1.65	1.63	1.61	1.59	1.57	1.56	1.55	1.53	1.52	1.51	1.47	1.44	1.40	1.38	1.32	1.30	1.28
90	2.76	2.36	2.15	2.01	1.91	1.84	1.78	1.74	1.70	1.67	1.64	1.62	1.60	1.58	1.56	1.55	1.53	1.52	1.51	1.50	1.46	1.43	1.39	1.36	1.30	1.28	1.27
100	2.76	2.36	2.14	2.00	1.91	1.83	1.78	1.73	1.69	1.66	1.64	1.61	1.59	1.57	1.56	1.54	1.53	1.52	1.50	1.49	1.45	1.42	1.38	1.35	1.29	1.27	1.26
125	2.75	2.35	2.13	1.99	1.89	1.82	1.77	1.72	1.68	1.65	1.62	1.60	1.58	1.56	1.54	1.53	1.51	1.50	1.49	1.48	1.44	1.41	1.36	1.34	1.27	1.25	1.23
150	2.74	2.34	2.12	1.98	1.89	1.81	1.76	1.71	1.67	1.64	1.61	1.59	1.57	1.55	1.53	1.52	1.50	1.49	1.48	1.47	1.43	1.40	1.35	1.33	1.26	1.23	1.22
200	2.73	2.33	2.11	1.97	1.88	1.80	1.75	1.70	1.66	1.63	1.60	1.58	1.56	1.54	1.52	1.51	1.49	1.48	1.47	1.46	1.41	1.38	1.34	1.31	1.24	1.22	1.20
300	2.72	2.32	2.10	1.96	1.87	1.79	1.74	1.69	1.65	1.62	1.59	1.57	1.55	1.53	1.51	1.49	1.48	1.47	1.46	1.45	1.40	1.37	1.32	1.29	1.21	1.19	1.18
500	2.72	2.31	2.09	1.96	1.86	1.79	1.73	1.68	1.64	1.61	1.58	1.56	1.54	1.52	1.50	1.49	1.47	1.46	1.45	1.44	1.39	1.36	1.31	1.28	1.21	1.18	1.16
1000	2.71	2.31	2.09	1.95	1.85	1.78	1.72	1.68	1.64	1.61	1.58	1.55	1.53	1.51	1.49	1.48	1.46	1.45	1.44	1.43	1.38	1.35	1.30	1.27	1.20	1.16	1.15

DEGREES OF FREEDOM FOR DENOMINATOR

TABLE A-4 Percentiles of the *F* distribution (*continued*)

Upper 5% point of the *F* distribution

DEGREES OF FREEDOM FOR NUMERATOR

den.	1	2	3	4	5	6	7	8	9	10	11	12	13	14	15	16	17	18	19	20	25	30	40	50	100	150	200
1	161	200	216	225	230	234	237	239	241	242	243	244	245	245	246	246	247	247	248	248	249	250	251	252	253	253	254
2	18.5	19.0	19.2	19.2	19.3	19.3	19.4	19.4	19.4	19.4	19.4	19.4	19.4	19.4	19.4	19.4	19.4	19.4	19.4	19.4	19.5	19.5	19.5	19.5	19.5	19.5	19.5
3	10.1	9.55	9.28	9.12	9.01	8.94	8.89	8.85	8.81	8.79	8.76	8.74	8.73	8.71	8.70	8.69	8.68	8.67	8.67	8.66	8.63	8.62	8.59	8.58	8.55	8.54	8.54
4	7.71	6.94	6.59	6.39	6.26	6.16	6.09	6.04	6.00	5.96	5.94	5.91	5.89	5.87	5.86	5.84	5.83	5.82	5.81	5.80	5.77	5.75	5.72	5.70	5.66	5.65	5.65
5	6.61	5.79	5.41	5.19	5.05	4.95	4.88	4.82	4.77	4.74	4.70	4.68	4.66	4.64	4.62	4.60	4.59	4.58	4.57	4.56	4.52	4.50	4.46	4.44	4.41	4.39	4.39
6	5.99	5.14	4.76	4.53	4.39	4.28	4.21	4.15	4.10	4.06	4.03	4.00	3.98	3.96	3.94	3.92	3.91	3.90	3.88	3.87	3.83	3.81	3.77	3.75	3.71	3.70	3.69
7	5.59	4.74	4.35	4.12	3.97	3.87	3.79	3.73	3.68	3.64	3.60	3.57	3.55	3.53	3.51	3.49	3.48	3.47	3.46	3.44	3.40	3.38	3.34	3.32	3.27	3.26	3.25
8	5.32	4.46	4.07	3.84	3.69	3.58	3.50	3.44	3.39	3.35	3.31	3.28	3.26	3.24	3.22	3.20	3.19	3.17	3.16	3.15	3.11	3.08	3.04	3.02	2.97	2.96	2.95
9	5.12	4.26	3.86	3.63	3.48	3.37	3.29	3.23	3.18	3.14	3.10	3.07	3.05	3.03	3.01	2.99	2.97	2.96	2.95	2.94	2.89	2.86	2.83	2.80	2.76	2.74	2.73
10	4.96	4.10	3.71	3.48	3.33	3.22	3.14	3.07	3.02	2.98	2.94	2.91	2.89	2.86	2.85	2.83	2.81	2.80	2.79	2.77	2.73	2.70	2.66	2.64	2.59	2.57	2.56
11	4.84	3.98	3.59	3.36	3.20	3.09	3.01	2.95	2.90	2.85	2.82	2.79	2.76	2.74	2.72	2.70	2.69	2.67	2.66	2.65	2.60	2.57	2.53	2.51	2.46	2.44	2.43
12	4.75	3.89	3.49	3.26	3.11	3.00	2.91	2.85	2.80	2.75	2.72	2.69	2.66	2.64	2.62	2.60	2.58	2.57	2.56	2.54	2.50	2.47	2.43	2.40	2.35	2.33	2.32
13	4.67	3.81	3.41	3.18	3.03	2.92	2.83	2.77	2.71	2.67	2.63	2.60	2.58	2.55	2.53	2.51	2.50	2.48	2.47	2.46	2.41	2.38	2.34	2.31	2.26	2.24	2.23
14	4.60	3.74	3.34	3.11	2.96	2.85	2.76	2.70	2.65	2.60	2.57	2.53	2.51	2.48	2.46	2.44	2.43	2.41	2.40	2.39	2.34	2.31	2.27	2.24	2.19	2.17	2.16
15	4.54	3.68	3.29	3.06	2.90	2.79	2.71	2.64	2.59	2.54	2.51	2.48	2.45	2.42	2.40	2.38	2.37	2.35	2.34	2.33	2.28	2.25	2.20	2.18	2.12	2.10	2.10
16	4.49	3.63	3.24	3.01	2.85	2.74	2.66	2.59	2.54	2.49	2.46	2.42	2.40	2.37	2.35	2.33	2.32	2.30	2.29	2.28	2.23	2.19	2.15	2.12	2.07	2.05	2.04
17	4.45	3.59	3.20	2.96	2.81	2.70	2.61	2.55	2.49	2.45	2.41	2.38	2.35	2.33	2.31	2.29	2.27	2.26	2.24	2.23	2.18	2.15	2.10	2.08	2.02	2.00	1.99
18	4.41	3.55	3.16	2.93	2.77	2.66	2.58	2.51	2.46	2.41	2.37	2.34	2.31	2.29	2.27	2.25	2.23	2.22	2.20	2.19	2.14	2.11	2.06	2.04	1.98	1.96	1.95
19	4.38	3.52	3.13	2.90	2.74	2.63	2.54	2.48	2.42	2.38	2.34	2.31	2.28	2.26	2.23	2.21	2.20	2.18	2.17	2.16	2.11	2.07	2.03	2.00	1.94	1.92	1.91
20	4.35	3.49	3.10	2.87	2.71	2.60	2.51	2.45	2.39	2.35	2.31	2.28	2.25	2.22	2.20	2.18	2.17	2.15	2.14	2.12	2.07	2.04	1.99	1.97	1.91	1.89	1.88
21	4.32	3.47	3.07	2.84	2.68	2.57	2.49	2.42	2.37	2.32	2.28	2.25	2.22	2.20	2.18	2.16	2.14	2.12	2.11	2.10	2.05	2.01	1.96	1.94	1.88	1.86	1.84
22	4.30	3.44	3.05	2.82	2.66	2.55	2.46	2.40	2.34	2.30	2.26	2.23	2.20	2.17	2.15	2.13	2.11	2.10	2.08	2.07	2.02	1.98	1.94	1.91	1.85	1.83	1.82
23	4.28	3.42	3.03	2.80	2.64	2.53	2.44	2.37	2.32	2.27	2.24	2.20	2.18	2.15	2.13	2.11	2.10	2.08	2.06	2.05	2.00	1.96	1.91	1.88	1.82	1.80	1.79
24	4.26	3.40	3.01	2.78	2.62	2.51	2.42	2.36	2.30	2.25	2.22	2.18	2.15	2.13	2.11	2.09	2.07	2.05	2.04	2.03	1.97	1.94	1.89	1.86	1.80	1.78	1.77
25	4.24	3.39	2.99	2.76	2.60	2.49	2.40	2.34	2.28	2.24	2.20	2.16	2.14	2.11	2.09	2.07	2.05	2.04	2.02	2.01	1.96	1.92	1.87	1.84	1.78	1.76	1.75
26	4.23	3.37	2.98	2.74	2.59	2.47	2.39	2.32	2.27	2.22	2.18	2.15	2.12	2.09	2.07	2.05	2.03	2.02	2.00	1.99	1.94	1.90	1.85	1.82	1.76	1.74	1.73
27	4.21	3.35	2.96	2.73	2.57	2.46	2.37	2.31	2.25	2.20	2.17	2.13	2.10	2.08	2.06	2.04	2.02	2.00	1.99	1.97	1.92	1.88	1.84	1.81	1.74	1.72	1.71
28	4.20	3.34	2.95	2.71	2.56	2.45	2.36	2.29	2.24	2.19	2.15	2.12	2.09	2.06	2.04	2.02	2.00	1.99	1.97	1.96	1.91	1.87	1.82	1.79	1.73	1.70	1.69
29	4.18	3.33	2.93	2.70	2.55	2.43	2.35	2.28	2.22	2.18	2.14	2.10	2.08	2.05	2.03	2.01	1.99	1.97	1.96	1.94	1.89	1.85	1.81	1.77	1.71	1.69	1.67
30	4.17	3.32	2.92	2.69	2.53	2.42	2.33	2.27	2.21	2.16	2.13	2.09	2.06	2.04	2.01	1.99	1.98	1.96	1.95	1.93	1.88	1.84	1.79	1.76	1.70	1.67	1.66
32	4.15	3.29	2.90	2.67	2.51	2.40	2.31	2.24	2.19	2.14	2.10	2.07	2.04	2.01	1.99	1.97	1.95	1.94	1.92	1.91	1.85	1.82	1.77	1.74	1.67	1.64	1.63
34	4.13	3.28	2.88	2.65	2.49	2.38	2.29	2.23	2.17	2.12	2.08	2.05	2.02	1.99	1.97	1.95	1.93	1.92	1.90	1.89	1.83	1.80	1.75	1.71	1.65	1.62	1.61
36	4.11	3.26	2.87	2.63	2.48	2.36	2.28	2.21	2.15	2.11	2.07	2.03	2.00	1.98	1.95	1.93	1.92	1.90	1.88	1.87	1.81	1.78	1.73	1.69	1.62	1.60	1.59
38	4.10	3.24	2.85	2.62	2.46	2.35	2.26	2.19	2.14	2.09	2.05	2.02	1.99	1.96	1.94	1.92	1.90	1.88	1.87	1.85	1.80	1.76	1.71	1.68	1.61	1.58	1.57
40	4.08	3.23	2.84	2.61	2.45	2.34	2.25	2.18	2.12	2.08	2.04	2.00	1.97	1.95	1.92	1.90	1.89	1.87	1.85	1.84	1.78	1.74	1.69	1.66	1.59	1.56	1.55
42	4.07	3.22	2.83	2.59	2.44	2.32	2.24	2.17	2.11	2.06	2.03	1.99	1.96	1.94	1.91	1.89	1.87	1.86	1.84	1.83	1.77	1.73	1.68	1.65	1.57	1.55	1.53
44	4.06	3.21	2.82	2.58	2.43	2.31	2.23	2.16	2.10	2.05	2.01	1.98	1.95	1.92	1.90	1.88	1.86	1.84	1.83	1.81	1.76	1.72	1.67	1.63	1.56	1.53	1.52
46	4.05	3.20	2.81	2.57	2.42	2.30	2.22	2.15	2.09	2.04	2.00	1.97	1.94	1.91	1.89	1.87	1.85	1.83	1.82	1.80	1.75	1.71	1.65	1.62	1.55	1.52	1.51
48	4.04	3.19	2.80	2.57	2.41	2.29	2.21	2.14	2.08	2.03	1.99	1.96	1.93	1.90	1.88	1.86	1.84	1.82	1.81	1.79	1.74	1.70	1.64	1.61	1.54	1.51	1.49
50	4.03	3.18	2.79	2.56	2.40	2.29	2.20	2.13	2.07	2.03	1.99	1.95	1.92	1.89	1.87	1.85	1.83	1.81	1.80	1.78	1.73	1.69	1.63	1.60	1.52	1.50	1.48
60	4.00	3.15	2.76	2.53	2.37	2.25	2.17	2.10	2.04	1.99	1.95	1.92	1.89	1.86	1.84	1.82	1.80	1.78	1.76	1.75	1.69	1.65	1.59	1.56	1.48	1.45	1.44
70	3.98	3.13	2.74	2.50	2.35	2.23	2.14	2.07	2.02	1.97	1.93	1.89	1.86	1.84	1.81	1.79	1.77	1.75	1.74	1.72	1.66	1.62	1.57	1.53	1.45	1.42	1.40
80	3.96	3.11	2.72	2.49	2.33	2.21	2.13	2.06	2.00	1.95	1.91	1.88	1.84	1.82	1.79	1.77	1.75	1.73	1.72	1.70	1.64	1.60	1.54	1.51	1.43	1.39	1.38
90	3.95	3.10	2.71	2.47	2.32	2.20	2.11	2.04	1.99	1.94	1.90	1.86	1.83	1.80	1.78	1.76	1.74	1.72	1.70	1.69	1.63	1.59	1.53	1.49	1.41	1.38	1.36
100	3.94	3.09	2.70	2.46	2.31	2.19	2.10	2.03	1.97	1.93	1.89	1.85	1.82	1.79	1.77	1.75	1.73	1.71	1.69	1.68	1.62	1.57	1.52	1.48	1.39	1.36	1.34
125	3.92	3.07	2.68	2.44	2.29	2.17	2.08	2.01	1.96	1.91	1.87	1.83	1.80	1.77	1.75	1.73	1.71	1.69	1.67	1.66	1.60	1.55	1.49	1.45	1.36	1.33	1.31
150	3.90	3.06	2.66	2.43	2.27	2.16	2.07	2.00	1.94	1.89	1.85	1.82	1.79	1.76	1.73	1.71	1.69	1.67	1.66	1.64	1.58	1.54	1.48	1.44	1.34	1.31	1.29
200	3.89	3.04	2.65	2.42	2.26	2.14	2.06	1.98	1.93	1.88	1.84	1.80	1.77	1.74	1.72	1.69	1.67	1.66	1.64	1.62	1.56	1.52	1.46	1.41	1.32	1.28	1.26
300	3.87	3.03	2.63	2.40	2.24	2.13	2.04	1.97	1.91	1.86	1.82	1.78	1.75	1.72	1.70	1.68	1.66	1.64	1.62	1.61	1.54	1.50	1.43	1.39	1.30	1.26	1.23
500	3.86	3.01	2.62	2.39	2.23	2.12	2.03	1.96	1.90	1.85	1.81	1.77	1.74	1.71	1.69	1.66	1.64	1.62	1.61	1.59	1.53	1.48	1.42	1.38	1.28	1.23	1.21
1000	3.85	3.00	2.61	2.38	2.22	2.11	2.02	1.95	1.89	1.84	1.80	1.76	1.73	1.70	1.68	1.65	1.63	1.61	1.60	1.58	1.52	1.47	1.41	1.36	1.26	1.22	1.19

DEGREES OF FREEDOM FOR DENOMINATOR

651

TABLE A-4 Percentiles of the F distribution (*continued*)

Upper 2.5% point of the F distribution

DEGREES OF FREEDOM FOR NUMERATOR

df	1	2	3	4	5	6	7	8	9	10	11	12	13	14	15	16	17	18	19	20	25	30	40	50	100	150	200
1	648	800	864	900	922	937	948	957	963	969	973	977	980	983	985	987	989	990	992	993	998	1001	1006	1008	1013	1015	1016
2	38.5	39.0	39.2	39.2	39.3	39.3	39.4	39.4	39.4	39.4	39.4	39.4	39.4	39.4	39.4	39.4	39.4	39.4	39.4	39.4	39.5	39.5	39.5	39.5	39.5	39.5	39.5
3	17.4	16.0	15.4	15.1	14.9	14.7	14.6	14.5	14.5	14.4	14.4	14.3	14.3	14.3	14.3	14.2	14.2	14.2	14.2	14.2	14.1	14.1	14.0	14.0	14.0	13.9	13.9
4	12.2	10.6	9.98	9.60	9.36	9.20	9.07	8.98	8.90	8.84	8.79	8.75	8.71	8.68	8.66	8.63	8.61	8.59	8.58	8.56	8.50	8.46	8.41	8.38	8.32	8.30	8.29
5	10.0	8.43	7.76	7.39	7.15	6.98	6.85	6.76	6.68	6.62	6.57	6.52	6.49	6.46	6.43	6.40	6.38	6.36	6.34	6.33	6.27	6.23	6.18	6.14	6.08	6.06	6.05
6	8.81	7.26	6.60	6.23	5.99	5.82	5.70	5.60	5.52	5.46	5.41	5.37	5.33	5.30	5.27	5.24	5.22	5.20	5.18	5.17	5.11	5.07	5.01	4.98	4.92	4.89	4.88
7	8.07	6.54	5.89	5.52	5.29	5.12	4.99	4.90	4.82	4.76	4.71	4.67	4.63	4.60	4.57	4.54	4.52	4.50	4.48	4.47	4.40	4.36	4.31	4.28	4.21	4.19	4.18
8	7.57	6.06	5.42	5.05	4.82	4.65	4.53	4.43	4.36	4.30	4.24	4.20	4.16	4.13	4.10	4.08	4.05	4.03	4.02	4.00	3.94	3.89	3.84	3.81	3.74	3.72	3.70
9	7.21	5.71	5.08	4.72	4.48	4.32	4.20	4.10	4.03	3.96	3.91	3.87	3.83	3.80	3.77	3.74	3.72	3.70	3.68	3.67	3.60	3.56	3.51	3.47	3.40	3.38	3.37
10	6.94	5.46	4.83	4.47	4.24	4.07	3.95	3.85	3.78	3.72	3.66	3.62	3.58	3.55	3.52	3.50	3.47	3.45	3.44	3.42	3.35	3.31	3.26	3.22	3.15	3.13	3.12
11	6.72	5.26	4.63	4.28	4.04	3.88	3.76	3.66	3.59	3.53	3.47	3.43	3.39	3.36	3.33	3.30	3.28	3.26	3.24	3.23	3.16	3.12	3.06	3.03	2.96	2.93	2.92
12	6.55	5.10	4.47	4.12	3.89	3.73	3.61	3.51	3.44	3.37	3.32	3.28	3.24	3.21	3.18	3.15	3.13	3.11	3.09	3.07	3.01	2.96	2.91	2.87	2.80	2.78	2.76
13	6.41	4.97	4.35	4.00	3.77	3.60	3.48	3.39	3.31	3.25	3.20	3.15	3.12	3.08	3.05	3.03	3.00	2.98	2.96	2.95	2.88	2.84	2.78	2.74	2.67	2.65	2.63
14	6.30	4.86	4.24	3.89	3.66	3.50	3.38	3.29	3.21	3.15	3.09	3.05	3.01	2.98	2.95	2.92	2.90	2.88	2.86	2.84	2.78	2.73	2.67	2.64	2.56	2.54	2.53
15	6.20	4.77	4.15	3.80	3.58	3.41	3.29	3.20	3.12	3.06	3.01	2.96	2.92	2.89	2.86	2.84	2.81	2.79	2.77	2.76	2.69	2.64	2.59	2.55	2.47	2.45	2.44
16	6.12	4.69	4.08	3.73	3.50	3.34	3.22	3.12	3.05	2.99	2.93	2.89	2.85	2.82	2.79	2.76	2.74	2.72	2.70	2.68	2.61	2.57	2.51	2.47	2.40	2.37	2.36
17	6.04	4.62	4.01	3.66	3.44	3.28	3.16	3.06	2.98	2.92	2.87	2.82	2.79	2.75	2.72	2.70	2.67	2.65	2.63	2.62	2.55	2.50	2.44	2.41	2.33	2.30	2.29
18	5.98	4.56	3.95	3.61	3.38	3.22	3.10	3.01	2.93	2.87	2.81	2.77	2.73	2.70	2.67	2.64	2.62	2.60	2.58	2.56	2.49	2.44	2.38	2.35	2.27	2.24	2.23
19	5.92	4.51	3.90	3.56	3.33	3.17	3.05	2.96	2.88	2.82	2.76	2.72	2.68	2.65	2.62	2.59	2.57	2.55	2.53	2.51	2.44	2.39	2.33	2.30	2.22	2.19	2.18
20	5.87	4.46	3.86	3.51	3.29	3.13	3.01	2.91	2.84	2.77	2.72	2.68	2.64	2.60	2.57	2.55	2.52	2.50	2.48	2.46	2.40	2.35	2.29	2.25	2.17	2.14	2.13
21	5.83	4.42	3.82	3.48	3.25	3.09	2.97	2.87	2.80	2.73	2.68	2.64	2.60	2.56	2.53	2.51	2.48	2.46	2.44	2.42	2.36	2.31	2.25	2.21	2.13	2.10	2.09
22	5.79	4.38	3.78	3.44	3.22	3.05	2.93	2.84	2.76	2.70	2.65	2.60	2.56	2.53	2.50	2.47	2.45	2.43	2.41	2.39	2.32	2.27	2.21	2.17	2.09	2.06	2.05
23	5.75	4.35	3.75	3.41	3.18	3.02	2.90	2.81	2.73	2.67	2.62	2.57	2.53	2.50	2.47	2.44	2.42	2.39	2.37	2.36	2.29	2.24	2.18	2.14	2.06	2.03	2.01
24	5.72	4.32	3.72	3.38	3.15	2.99	2.87	2.78	2.70	2.64	2.59	2.54	2.50	2.47	2.44	2.41	2.39	2.36	2.35	2.33	2.26	2.21	2.15	2.11	2.02	2.00	1.98
25	5.69	4.29	3.69	3.35	3.13	2.97	2.85	2.75	2.68	2.61	2.56	2.51	2.48	2.44	2.41	2.38	2.36	2.34	2.32	2.30	2.23	2.18	2.12	2.08	2.00	1.97	1.95
26	5.66	4.27	3.67	3.33	3.10	2.94	2.82	2.73	2.65	2.59	2.54	2.49	2.45	2.42	2.39	2.36	2.34	2.31	2.29	2.28	2.21	2.16	2.09	2.05	1.97	1.94	1.92
27	5.63	4.24	3.65	3.31	3.08	2.92	2.80	2.71	2.63	2.57	2.51	2.47	2.43	2.39	2.36	2.34	2.31	2.29	2.27	2.25	2.18	2.13	2.07	2.03	1.94	1.91	1.90
28	5.61	4.22	3.63	3.29	3.06	2.90	2.78	2.69	2.61	2.55	2.49	2.45	2.41	2.37	2.34	2.32	2.29	2.27	2.25	2.23	2.16	2.11	2.05	2.01	1.92	1.89	1.88
29	5.59	4.20	3.61	3.27	3.04	2.88	2.76	2.67	2.59	2.53	2.48	2.43	2.39	2.36	2.32	2.30	2.27	2.25	2.23	2.21	2.14	2.09	2.03	1.99	1.90	1.87	1.86
30	5.57	4.18	3.59	3.25	3.03	2.87	2.75	2.65	2.57	2.51	2.46	2.41	2.37	2.34	2.31	2.28	2.26	2.23	2.21	2.20	2.12	2.07	2.01	1.97	1.88	1.85	1.84
32	5.53	4.15	3.56	3.22	3.00	2.84	2.71	2.62	2.54	2.48	2.43	2.38	2.34	2.31	2.28	2.25	2.22	2.20	2.18	2.16	2.09	2.04	1.98	1.93	1.85	1.82	1.80
34	5.50	4.12	3.53	3.19	2.97	2.81	2.69	2.59	2.52	2.45	2.40	2.35	2.31	2.28	2.25	2.22	2.19	2.17	2.15	2.13	2.06	2.01	1.95	1.90	1.82	1.78	1.77
36	5.47	4.09	3.50	3.17	2.94	2.78	2.66	2.57	2.49	2.43	2.37	2.33	2.29	2.25	2.22	2.20	2.17	2.15	2.13	2.11	2.04	1.99	1.92	1.88	1.79	1.76	1.74
38	5.45	4.07	3.48	3.15	2.92	2.76	2.64	2.55	2.47	2.41	2.35	2.31	2.27	2.23	2.20	2.17	2.15	2.13	2.11	2.09	2.01	1.96	1.90	1.85	1.76	1.73	1.71
40	5.42	4.05	3.46	3.13	2.90	2.74	2.62	2.53	2.45	2.39	2.33	2.29	2.25	2.21	2.18	2.15	2.13	2.11	2.09	2.07	1.99	1.94	1.88	1.83	1.74	1.71	1.69
42	5.40	4.03	3.45	3.11	2.89	2.73	2.61	2.51	2.43	2.37	2.32	2.27	2.23	2.20	2.16	2.14	2.11	2.09	2.07	2.05	1.98	1.92	1.86	1.81	1.72	1.69	1.67
44	5.39	4.02	3.43	3.09	2.87	2.71	2.59	2.50	2.42	2.36	2.30	2.26	2.22	2.18	2.15	2.12	2.10	2.07	2.05	2.03	1.96	1.91	1.84	1.80	1.70	1.67	1.65
46	5.37	4.00	3.42	3.08	2.86	2.70	2.58	2.48	2.41	2.34	2.29	2.24	2.20	2.17	2.13	2.11	2.08	2.06	2.04	2.02	1.94	1.89	1.82	1.78	1.69	1.65	1.63
48	5.35	3.99	3.40	3.07	2.84	2.69	2.56	2.47	2.39	2.33	2.27	2.23	2.19	2.15	2.12	2.09	2.07	2.05	2.02	2.01	1.93	1.88	1.81	1.77	1.67	1.64	1.62
50	5.34	3.97	3.39	3.05	2.83	2.67	2.55	2.46	2.38	2.32	2.26	2.22	2.18	2.14	2.11	2.08	2.06	2.03	2.01	1.99	1.92	1.87	1.80	1.75	1.66	1.62	1.60
60	5.29	3.93	3.34	3.01	2.79	2.63	2.51	2.41	2.33	2.27	2.22	2.17	2.13	2.09	2.06	2.03	2.01	1.98	1.96	1.94	1.87	1.82	1.74	1.70	1.60	1.56	1.54
70	5.25	3.89	3.31	2.97	2.75	2.59	2.47	2.38	2.30	2.24	2.18	2.14	2.10	2.06	2.03	2.00	1.97	1.95	1.93	1.91	1.83	1.78	1.71	1.66	1.56	1.52	1.50
80	5.22	3.86	3.28	2.95	2.73	2.57	2.45	2.35	2.28	2.21	2.16	2.11	2.07	2.03	2.00	1.97	1.95	1.92	1.90	1.88	1.81	1.75	1.68	1.63	1.53	1.49	1.47
90	5.20	3.84	3.26	2.93	2.71	2.55	2.43	2.34	2.26	2.19	2.14	2.09	2.05	2.02	1.98	1.95	1.93	1.91	1.88	1.86	1.79	1.73	1.66	1.61	1.50	1.46	1.44
100	5.18	3.83	3.25	2.92	2.70	2.54	2.42	2.32	2.24	2.18	2.12	2.08	2.04	2.00	1.97	1.94	1.91	1.89	1.87	1.85	1.77	1.71	1.64	1.59	1.48	1.44	1.42
125	5.15	3.80	3.22	2.89	2.67	2.51	2.39	2.30	2.22	2.15	2.10	2.05	2.01	1.97	1.94	1.91	1.88	1.86	1.84	1.82	1.74	1.68	1.61	1.56	1.45	1.40	1.38
150	5.13	3.78	3.20	2.87	2.65	2.49	2.37	2.28	2.20	2.13	2.08	2.03	1.99	1.95	1.92	1.89	1.87	1.84	1.82	1.80	1.72	1.67	1.59	1.54	1.42	1.38	1.35
200	5.10	3.76	3.18	2.85	2.63	2.47	2.35	2.26	2.18	2.11	2.06	2.01	1.97	1.93	1.90	1.87	1.84	1.82	1.80	1.78	1.70	1.64	1.56	1.51	1.39	1.35	1.32
300	5.07	3.73	3.16	2.83	2.61	2.45	2.33	2.23	2.16	2.09	2.03	1.99	1.95	1.91	1.88	1.85	1.82	1.80	1.77	1.75	1.67	1.62	1.54	1.48	1.36	1.31	1.28
500	5.05	3.72	3.14	2.81	2.59	2.43	2.31	2.22	2.14	2.07	2.02	1.97	1.93	1.89	1.86	1.83	1.80	1.78	1.76	1.74	1.65	1.60	1.52	1.46	1.34	1.28	1.25
1000	5.04	3.70	3.13	2.80	2.58	2.42	2.30	2.20	2.13	2.06	2.01	1.96	1.92	1.88	1.85	1.82	1.79	1.77	1.74	1.72	1.64	1.58	1.50	1.45	1.32	1.26	1.23

DEGREES OF FREEDOM FOR DENOMINATOR

652

TABLE A-4 Percentiles of the *F* distribution (*continued*)

Upper 1% point of the *F* distribution

DEGREES OF FREEDOM FOR NUMERATOR

df (denom)	1	2	3	4	5	6	7	8	9	10	11	12	13	14	15	16	17	18	19	20	25	30	40	50	100	150	200
1	4052	5000	5403	5625	5764	5859	5928	5981	6022	6056	6083	6106	6126	6143	6157	6170	6181	6192	6201	6209	6240	6261	6287	6303	6334	6345	6350
2	98.5	99.0	99.2	99.2	99.3	99.3	99.4	99.4	99.4	99.4	99.4	99.4	99.4	99.4	99.4	99.4	99.4	99.4	99.4	99.4	99.5	99.5	99.5	99.5	99.5	99.5	99.5
3	34.1	30.8	29.5	28.7	28.2	27.9	27.7	27.5	27.3	27.2	27.1	27.1	27.0	26.9	26.9	26.8	26.8	26.8	26.7	26.7	26.6	26.5	26.4	26.4	26.2	26.2	26.2
4	21.2	18.0	16.7	16.0	15.5	15.2	15.0	14.8	14.7	14.5	14.5	14.4	14.3	14.2	14.2	14.2	14.1	14.1	14.0	14.0	13.9	13.8	13.7	13.7	13.6	13.6	13.5
5	16.3	13.3	12.1	11.4	11.0	10.7	10.5	10.3	10.2	10.1	9.96	9.89	9.82	9.77	9.72	9.68	9.64	9.61	9.58	9.55	9.45	9.38	9.29	9.24	9.13	9.09	9.08
6	13.7	10.9	9.78	9.15	8.75	8.47	8.26	8.10	7.98	7.87	7.79	7.72	7.66	7.60	7.56	7.52	7.48	7.45	7.42	7.40	7.30	7.23	7.14	7.09	6.99	6.95	6.93
7	12.2	9.55	8.45	7.85	7.46	7.19	6.99	6.84	6.72	6.62	6.54	6.47	6.41	6.36	6.31	6.28	6.24	6.21	6.18	6.16	6.06	5.99	5.91	5.86	5.75	5.72	5.70
8	11.3	8.65	7.59	7.01	6.63	6.37	6.18	6.03	5.91	5.81	5.73	5.67	5.61	5.56	5.52	5.48	5.44	5.41	5.38	5.36	5.26	5.20	5.12	5.07	4.96	4.93	4.91
9	10.6	8.02	6.99	6.42	6.06	5.80	5.61	5.47	5.35	5.26	5.18	5.11	5.05	5.01	4.96	4.92	4.89	4.86	4.83	4.81	4.71	4.65	4.57	4.52	4.41	4.38	4.36
10	10.0	7.56	6.55	5.99	5.64	5.39	5.20	5.06	4.94	4.85	4.77	4.71	4.65	4.60	4.56	4.52	4.49	4.46	4.43	4.41	4.31	4.25	4.17	4.12	4.01	3.98	3.96
11	9.65	7.21	6.22	5.67	5.32	5.07	4.89	4.74	4.63	4.54	4.46	4.40	4.34	4.29	4.25	4.21	4.18	4.15	4.12	4.10	4.01	3.94	3.86	3.81	3.71	3.67	3.66
12	9.33	6.93	5.95	5.41	5.06	4.82	4.64	4.50	4.39	4.30	4.22	4.16	4.10	4.05	4.01	3.97	3.94	3.91	3.88	3.86	3.76	3.70	3.62	3.57	3.47	3.43	3.41
13	9.07	6.70	5.74	5.21	4.86	4.62	4.44	4.30	4.19	4.10	4.02	3.96	3.91	3.86	3.82	3.78	3.75	3.72	3.69	3.66	3.57	3.51	3.43	3.38	3.27	3.24	3.22
14	8.88	6.51	5.56	5.04	4.69	4.46	4.28	4.14	4.03	3.94	3.86	3.80	3.75	3.70	3.66	3.62	3.59	3.56	3.53	3.51	3.41	3.35	3.27	3.22	3.11	3.08	3.06
15	8.68	6.36	5.42	4.89	4.56	4.32	4.14	4.00	3.89	3.80	3.73	3.67	3.61	3.56	3.52	3.49	3.45	3.42	3.40	3.37	3.28	3.21	3.13	3.08	2.98	2.94	2.92
16	8.53	6.23	5.29	4.77	4.44	4.20	4.03	3.89	3.78	3.69	3.62	3.55	3.50	3.45	3.41	3.37	3.34	3.31	3.28	3.26	3.16	3.10	3.02	2.97	2.86	2.83	2.81
17	8.40	6.11	5.19	4.67	4.34	4.10	3.93	3.79	3.68	3.59	3.52	3.46	3.40	3.35	3.31	3.27	3.24	3.21	3.19	3.16	3.07	3.00	2.92	2.87	2.76	2.73	2.71
18	8.29	6.01	5.09	4.58	4.25	4.01	3.84	3.71	3.60	3.51	3.43	3.37	3.32	3.27	3.23	3.19	3.16	3.13	3.10	3.08	2.98	2.92	2.84	2.78	2.68	2.64	2.62
19	8.18	5.93	5.01	4.50	4.17	3.94	3.77	3.63	3.52	3.43	3.36	3.30	3.24	3.19	3.15	3.12	3.08	3.05	3.03	3.00	2.91	2.84	2.76	2.71	2.60	2.57	2.55
20	8.10	5.85	4.94	4.43	4.10	3.87	3.70	3.56	3.46	3.37	3.29	3.23	3.18	3.13	3.09	3.05	3.02	2.99	2.96	2.94	2.84	2.78	2.69	2.64	2.54	2.50	2.48
21	8.02	5.78	4.87	4.37	4.04	3.81	3.64	3.51	3.40	3.31	3.24	3.17	3.12	3.07	3.03	2.99	2.96	2.93	2.90	2.88	2.79	2.72	2.64	2.58	2.48	2.44	2.42
22	7.95	5.72	4.82	4.31	3.99	3.76	3.59	3.45	3.35	3.26	3.18	3.12	3.07	3.02	2.98	2.94	2.91	2.88	2.85	2.83	2.73	2.67	2.58	2.53	2.42	2.38	2.36
23	7.88	5.66	4.76	4.26	3.94	3.71	3.54	3.41	3.30	3.21	3.14	3.07	3.02	2.97	2.93	2.89	2.86	2.83	2.80	2.78	2.69	2.62	2.54	2.48	2.37	2.34	2.32
24	7.82	5.61	4.72	4.22	3.90	3.67	3.50	3.36	3.26	3.17	3.09	3.03	2.98	2.93	2.89	2.85	2.82	2.79	2.76	2.74	2.64	2.58	2.49	2.44	2.33	2.29	2.27
25	7.77	5.57	4.68	4.18	3.85	3.63	3.46	3.32	3.22	3.13	3.06	2.99	2.94	2.89	2.85	2.81	2.78	2.75	2.72	2.70	2.60	2.54	2.45	2.40	2.29	2.25	2.23
26	7.72	5.53	4.64	4.14	3.82	3.59	3.42	3.29	3.18	3.09	3.02	2.96	2.90	2.86	2.81	2.78	2.75	2.72	2.69	2.66	2.57	2.50	2.42	2.36	2.25	2.21	2.19
27	7.68	5.49	4.60	4.11	3.78	3.56	3.39	3.26	3.15	3.06	2.99	2.93	2.87	2.82	2.78	2.75	2.71	2.68	2.66	2.63	2.54	2.47	2.38	2.33	2.22	2.18	2.16
28	7.64	5.45	4.57	4.07	3.75	3.53	3.36	3.23	3.12	3.03	2.96	2.90	2.84	2.79	2.75	2.72	2.68	2.65	2.63	2.60	2.51	2.44	2.35	2.30	2.19	2.15	2.13
29	7.60	5.42	4.54	4.04	3.73	3.50	3.33	3.20	3.09	3.00	2.93	2.87	2.81	2.77	2.73	2.69	2.66	2.63	2.60	2.57	2.48	2.41	2.33	2.27	2.16	2.12	2.10
30	7.56	5.39	4.51	4.02	3.70	3.47	3.30	3.17	3.07	2.98	2.91	2.84	2.79	2.74	2.70	2.66	2.63	2.60	2.57	2.55	2.45	2.39	2.30	2.25	2.13	2.09	2.07
32	7.50	5.34	4.46	3.97	3.65	3.43	3.26	3.13	3.02	2.93	2.86	2.80	2.74	2.70	2.65	2.62	2.58	2.55	2.53	2.50	2.41	2.34	2.25	2.20	2.08	2.04	2.02
34	7.44	5.29	4.42	3.93	3.61	3.39	3.22	3.09	2.98	2.89	2.82	2.76	2.70	2.66	2.61	2.58	2.54	2.51	2.49	2.46	2.37	2.30	2.21	2.16	2.04	2.00	1.98
36	7.40	5.25	4.38	3.89	3.57	3.35	3.18	3.05	2.95	2.86	2.79	2.72	2.67	2.62	2.58	2.54	2.51	2.48	2.45	2.43	2.33	2.26	2.18	2.12	2.00	1.96	1.94
38	7.35	5.21	4.34	3.86	3.54	3.32	3.15	3.02	2.92	2.83	2.75	2.69	2.64	2.59	2.55	2.51	2.48	2.45	2.42	2.40	2.30	2.23	2.14	2.09	1.97	1.93	1.90
40	7.31	5.18	4.31	3.83	3.51	3.29	3.12	2.99	2.89	2.80	2.73	2.66	2.61	2.56	2.52	2.48	2.45	2.42	2.39	2.37	2.27	2.20	2.11	2.06	1.94	1.90	1.87
42	7.28	5.15	4.29	3.80	3.49	3.27	3.10	2.97	2.86	2.78	2.70	2.64	2.59	2.54	2.50	2.46	2.43	2.40	2.37	2.34	2.25	2.18	2.09	2.03	1.91	1.87	1.85
44	7.25	5.12	4.26	3.78	3.47	3.24	3.08	2.95	2.84	2.75	2.68	2.62	2.56	2.52	2.47	2.44	2.40	2.37	2.35	2.32	2.22	2.15	2.07	2.01	1.89	1.84	1.82
46	7.22	5.10	4.24	3.76	3.44	3.22	3.06	2.93	2.82	2.73	2.66	2.60	2.54	2.50	2.45	2.42	2.38	2.35	2.33	2.30	2.20	2.13	2.04	1.99	1.86	1.82	1.80
48	7.19	5.08	4.22	3.74	3.43	3.20	3.04	2.91	2.80	2.71	2.64	2.58	2.53	2.48	2.44	2.40	2.37	2.33	2.31	2.28	2.18	2.12	2.02	1.97	1.84	1.80	1.78
50	7.17	5.06	4.20	3.72	3.41	3.19	3.02	2.89	2.78	2.70	2.63	2.56	2.51	2.46	2.42	2.38	2.35	2.32	2.29	2.27	2.17	2.10	2.01	1.95	1.82	1.78	1.76
60	7.08	4.98	4.13	3.65	3.34	3.12	2.95	2.82	2.72	2.63	2.56	2.50	2.44	2.39	2.35	2.31	2.28	2.25	2.22	2.20	2.10	2.03	1.94	1.88	1.75	1.70	1.68
70	7.01	4.92	4.07	3.60	3.29	3.07	2.91	2.78	2.67	2.59	2.51	2.45	2.40	2.35	2.31	2.27	2.23	2.20	2.18	2.15	2.05	1.98	1.89	1.83	1.70	1.65	1.62
80	6.96	4.88	4.04	3.56	3.26	3.04	2.87	2.74	2.64	2.55	2.48	2.42	2.36	2.31	2.27	2.23	2.20	2.17	2.14	2.12	2.01	1.94	1.85	1.79	1.65	1.61	1.58
90	6.93	4.85	4.01	3.53	3.23	3.01	2.84	2.72	2.61	2.52	2.45	2.39	2.33	2.29	2.24	2.21	2.17	2.14	2.11	2.09	1.99	1.92	1.82	1.76	1.62	1.57	1.55
100	6.90	4.82	3.98	3.51	3.21	2.99	2.82	2.69	2.59	2.50	2.43	2.37	2.31	2.27	2.22	2.19	2.15	2.12	2.09	2.07	1.97	1.89	1.80	1.74	1.60	1.55	1.52
125	6.84	4.78	3.94	3.47	3.17	2.95	2.79	2.66	2.55	2.47	2.39	2.33	2.28	2.23	2.19	2.15	2.11	2.08	2.05	2.03	1.93	1.85	1.76	1.69	1.55	1.50	1.47
150	6.81	4.75	3.91	3.45	3.14	2.92	2.76	2.63	2.53	2.44	2.37	2.31	2.25	2.20	2.16	2.12	2.09	2.06	2.03	2.00	1.90	1.83	1.73	1.66	1.52	1.46	1.43
200	6.76	4.71	3.88	3.41	3.11	2.89	2.73	2.60	2.50	2.41	2.34	2.27	2.22	2.17	2.13	2.09	2.06	2.03	2.00	1.97	1.87	1.79	1.69	1.63	1.48	1.42	1.39
300	6.72	4.68	3.85	3.38	3.08	2.86	2.70	2.57	2.47	2.38	2.31	2.24	2.19	2.14	2.10	2.06	2.03	1.99	1.97	1.94	1.84	1.76	1.66	1.59	1.44	1.38	1.35
500	6.69	4.65	3.82	3.36	3.05	2.84	2.68	2.55	2.44	2.36	2.28	2.22	2.17	2.12	2.07	2.04	2.00	1.97	1.94	1.92	1.81	1.74	1.63	1.57	1.41	1.34	1.31
1000	6.66	4.63	3.80	3.34	3.04	2.82	2.66	2.53	2.43	2.34	2.27	2.20	2.15	2.10	2.06	2.02	1.98	1.95	1.92	1.90	1.79	1.72	1.61	1.54	1.38	1.32	1.28

DEGREES OF FREEDOM FOR DENOMINATOR

TABLE A-4 Percentiles of the F distribution (continued)

Upper 0.5% point of the F distribution

DEGREES OF FREEDOM FOR NUMERATOR

	1	2	3	4	5	6	7	8	9	10	11	12	13	14	15	16	17	18	19	20	25	30	40	50	100	150	200
1	••••	••••	••••	••••	••••	••••	••••	••••	••••	••••	••••	••••	••••	••••	••••	••••	••••	••••	••••	••••	••••	••••	••••	••••	••••	••••	••••
2	199	199	199	199	199	199	199	199	199	199	199	199	199	199	199	199	199	199	199	199	199	199	199	199	199	199	199
3	55.6	49.8	47.5	46.2	45.4	44.8	44.4	44.1	43.9	43.7	43.5	43.4	43.3	43.2	43.1	43.0	42.9	42.9	42.8	42.8	42.6	42.5	42.3	42.2	42.0	42.0	41.9
4	31.3	26.3	24.3	23.2	22.5	22.0	21.6	21.4	21.1	21.0	20.8	20.7	20.6	20.5	20.4	20.4	20.3	20.3	20.2	20.2	20.0	19.9	19.8	19.7	19.5	19.4	19.4
5	22.8	18.3	16.5	15.6	14.9	14.5	14.2	14.0	13.8	13.6	13.5	13.4	13.3	13.2	13.1	13.1	13.0	13.0	12.9	12.9	12.8	12.7	12.5	12.5	12.3	12.2	12.2
6	18.6	14.5	12.9	12.0	11.5	11.1	10.8	10.6	10.4	10.3	10.1	10.0	9.95	9.88	9.81	9.76	9.71	9.66	9.62	9.59	9.45	9.36	9.24	9.17	9.03	8.98	8.95
7	16.2	12.4	10.9	10.1	9.52	9.16	8.89	8.68	8.51	8.38	8.27	8.18	8.10	8.03	7.97	7.91	7.87	7.83	7.79	7.75	7.62	7.53	7.42	7.35	7.22	7.17	7.15
8	14.7	11.0	9.60	8.81	8.30	7.95	7.69	7.50	7.34	7.21	7.10	7.01	6.94	6.87	6.81	6.76	6.72	6.68	6.64	6.61	6.48	6.40	6.29	6.22	6.09	6.04	6.02
9	13.6	10.1	8.72	7.96	7.47	7.13	6.88	6.69	6.54	6.42	6.31	6.23	6.15	6.09	6.03	5.98	5.94	5.90	5.86	5.83	5.71	5.62	5.52	5.45	5.32	5.27	5.26
10	12.8	9.43	8.08	7.34	6.87	6.54	6.30	6.12	5.97	5.85	5.75	5.66	5.59	5.53	5.47	5.42	5.38	5.34	5.31	5.27	5.15	5.07	4.97	4.90	4.77	4.73	4.71
11	12.2	8.91	7.60	6.88	6.42	6.10	5.86	5.68	5.54	5.42	5.32	5.24	5.16	5.10	5.05	5.00	4.96	4.92	4.89	4.86	4.74	4.65	4.55	4.49	4.36	4.31	4.29
12	11.8	8.51	7.23	6.52	6.07	5.76	5.52	5.35	5.20	5.09	4.99	4.91	4.84	4.77	4.72	4.67	4.63	4.59	4.56	4.53	4.41	4.33	4.23	4.17	4.04	3.99	3.97
13	11.4	8.19	6.93	6.23	5.79	5.48	5.25	5.08	4.94	4.82	4.72	4.64	4.57	4.51	4.46	4.41	4.37	4.33	4.30	4.27	4.15	4.07	3.97	3.91	3.78	3.74	3.71
14	11.1	7.92	6.68	6.00	5.56	5.26	5.03	4.86	4.72	4.60	4.51	4.43	4.36	4.30	4.25	4.20	4.16	4.12	4.09	4.06	3.94	3.86	3.76	3.70	3.57	3.53	3.50
15	10.8	7.70	6.48	5.80	5.37	5.07	4.85	4.67	4.54	4.42	4.33	4.25	4.18	4.12	4.07	4.02	3.98	3.95	3.91	3.88	3.77	3.69	3.58	3.52	3.39	3.35	3.33
16	10.6	7.51	6.30	5.64	5.21	4.91	4.69	4.52	4.38	4.27	4.18	4.10	4.03	3.97	3.92	3.87	3.83	3.80	3.76	3.73	3.62	3.54	3.44	3.37	3.25	3.20	3.18
17	10.4	7.36	6.16	5.50	5.07	4.78	4.56	4.39	4.25	4.14	4.05	3.97	3.90	3.84	3.79	3.75	3.71	3.67	3.64	3.61	3.49	3.41	3.31	3.25	3.12	3.07	3.05
18	10.2	7.21	6.03	5.37	4.96	4.66	4.44	4.28	4.14	4.03	3.94	3.86	3.79	3.73	3.68	3.64	3.60	3.56	3.53	3.50	3.38	3.30	3.20	3.14	3.01	2.96	2.94
19	10.1	7.09	5.92	5.27	4.85	4.56	4.34	4.18	4.04	3.93	3.84	3.76	3.70	3.64	3.59	3.54	3.50	3.46	3.43	3.40	3.29	3.21	3.11	3.04	2.91	2.87	2.85
20	9.94	6.99	5.82	5.17	4.76	4.47	4.26	4.09	3.96	3.85	3.76	3.68	3.61	3.55	3.50	3.46	3.42	3.38	3.35	3.32	3.20	3.12	3.02	2.96	2.83	2.78	2.76
21	9.83	6.89	5.73	5.09	4.68	4.39	4.18	4.01	3.88	3.77	3.68	3.60	3.54	3.48	3.43	3.38	3.34	3.31	3.27	3.24	3.13	3.05	2.95	2.88	2.75	2.71	2.68
22	9.73	6.81	5.65	5.02	4.61	4.32	4.11	3.94	3.81	3.70	3.61	3.54	3.47	3.41	3.36	3.31	3.27	3.24	3.21	3.18	3.06	2.98	2.88	2.82	2.69	2.64	2.62
23	9.63	6.73	5.58	4.95	4.54	4.26	4.05	3.88	3.75	3.64	3.55	3.47	3.41	3.35	3.30	3.25	3.21	3.18	3.15	3.12	3.00	2.92	2.82	2.76	2.62	2.58	2.56
24	9.55	6.66	5.52	4.89	4.49	4.20	3.99	3.83	3.69	3.59	3.50	3.42	3.35	3.30	3.25	3.20	3.16	3.12	3.09	3.06	2.95	2.87	2.77	2.70	2.57	2.52	2.50
25	9.48	6.60	5.46	4.84	4.43	4.15	3.94	3.78	3.64	3.54	3.45	3.37	3.30	3.25	3.20	3.15	3.11	3.08	3.04	3.01	2.90	2.82	2.72	2.65	2.52	2.47	2.45
26	9.41	6.54	5.41	4.79	4.38	4.10	3.89	3.73	3.60	3.49	3.40	3.33	3.26	3.20	3.15	3.11	3.07	3.03	3.00	2.97	2.85	2.77	2.67	2.61	2.47	2.43	2.40
27	9.34	6.49	5.36	4.74	4.34	4.06	3.85	3.69	3.56	3.45	3.36	3.28	3.22	3.16	3.11	3.07	3.03	2.99	2.96	2.93	2.81	2.73	2.63	2.57	2.43	2.38	2.36
28	9.28	6.44	5.32	4.70	4.30	4.02	3.81	3.65	3.52	3.41	3.32	3.25	3.18	3.12	3.07	3.03	2.99	2.95	2.92	2.89	2.77	2.69	2.59	2.53	2.39	2.35	2.32
29	9.23	6.40	5.28	4.66	4.26	3.98	3.77	3.61	3.48	3.38	3.29	3.21	3.15	3.09	3.04	2.99	2.95	2.92	2.88	2.86	2.74	2.66	2.56	2.49	2.36	2.31	2.29
30	9.18	6.35	5.24	4.62	4.23	3.95	3.74	3.58	3.45	3.34	3.25	3.18	3.11	3.06	3.01	2.96	2.92	2.89	2.85	2.82	2.71	2.63	2.52	2.46	2.32	2.28	2.25
32	9.09	6.28	5.17	4.56	4.17	3.89	3.68	3.52	3.39	3.29	3.20	3.12	3.06	3.00	2.95	2.90	2.86	2.83	2.80	2.77	2.65	2.57	2.47	2.40	2.26	2.22	2.19
34	9.01	6.22	5.11	4.50	4.11	3.84	3.63	3.47	3.34	3.24	3.15	3.07	3.01	2.95	2.90	2.85	2.81	2.78	2.75	2.72	2.60	2.52	2.42	2.35	2.21	2.16	2.14
36	8.94	6.16	5.06	4.46	4.06	3.79	3.58	3.42	3.30	3.19	3.10	3.03	2.96	2.90	2.85	2.81	2.77	2.73	2.70	2.67	2.56	2.48	2.37	2.30	2.17	2.12	2.09
38	8.88	6.11	5.02	4.41	4.02	3.75	3.54	3.39	3.26	3.15	3.06	2.99	2.92	2.87	2.82	2.77	2.73	2.70	2.66	2.63	2.52	2.44	2.33	2.27	2.12	2.08	2.05
40	8.83	6.07	4.98	4.37	3.99	3.71	3.51	3.35	3.22	3.12	3.03	2.95	2.89	2.83	2.78	2.74	2.70	2.66	2.63	2.60	2.48	2.40	2.30	2.23	2.09	2.04	2.01
42	8.78	6.03	4.94	4.34	3.95	3.68	3.48	3.32	3.19	3.09	3.00	2.92	2.86	2.80	2.75	2.71	2.67	2.63	2.60	2.57	2.45	2.37	2.26	2.20	2.06	2.00	1.98
44	8.74	5.99	4.91	4.31	3.92	3.65	3.45	3.29	3.16	3.06	2.97	2.89	2.83	2.77	2.72	2.68	2.64	2.60	2.57	2.54	2.42	2.34	2.24	2.17	2.03	1.97	1.95
46	8.70	5.96	4.88	4.28	3.90	3.62	3.42	3.26	3.14	3.03	2.94	2.87	2.80	2.75	2.70	2.65	2.61	2.58	2.54	2.51	2.40	2.32	2.21	2.14	2.00	1.95	1.92
48	8.66	5.93	4.85	4.25	3.87	3.60	3.40	3.24	3.11	3.01	2.92	2.84	2.78	2.72	2.67	2.63	2.59	2.55	2.52	2.49	2.37	2.29	2.19	2.12	1.97	1.92	1.90
50	8.63	5.90	4.83	4.23	3.85	3.58	3.38	3.22	3.09	2.99	2.90	2.82	2.76	2.70	2.65	2.61	2.57	2.53	2.50	2.47	2.35	2.27	2.16	2.10	1.95	1.90	1.87
60	8.49	5.79	4.73	4.14	3.76	3.49	3.29	3.13	3.01	2.90	2.82	2.74	2.68	2.62	2.57	2.53	2.49	2.45	2.42	2.39	2.27	2.19	2.08	2.01	1.86	1.81	1.78
70	8.40	5.72	4.66	4.08	3.70	3.43	3.23	3.08	2.95	2.85	2.76	2.68	2.62	2.56	2.51	2.47	2.43	2.39	2.36	2.33	2.21	2.13	2.02	1.95	1.80	1.74	1.71
80	8.33	5.67	4.61	4.03	3.65	3.39	3.19	3.03	2.91	2.80	2.72	2.64	2.58	2.52	2.47	2.43	2.39	2.35	2.32	2.29	2.17	2.08	1.97	1.90	1.75	1.69	1.66
90	8.28	5.62	4.57	3.99	3.62	3.35	3.15	3.00	2.87	2.77	2.68	2.61	2.54	2.49	2.44	2.39	2.35	2.32	2.28	2.25	2.13	2.05	1.94	1.87	1.71	1.65	1.62
100	8.24	5.59	4.54	3.96	3.59	3.33	3.13	2.97	2.85	2.74	2.66	2.58	2.52	2.46	2.41	2.37	2.33	2.29	2.26	2.23	2.11	2.02	1.91	1.84	1.68	1.62	1.59
125	8.17	5.53	4.49	3.91	3.54	3.28	3.08	2.93	2.80	2.70	2.61	2.54	2.47	2.42	2.37	2.32	2.28	2.24	2.21	2.18	2.06	1.98	1.86	1.79	1.63	1.56	1.53
150	8.12	5.49	4.45	3.88	3.51	3.25	3.05	2.89	2.77	2.67	2.58	2.51	2.44	2.38	2.33	2.29	2.25	2.21	2.18	2.15	2.03	1.94	1.83	1.76	1.59	1.53	1.49
200	8.06	5.44	4.41	3.84	3.47	3.21	3.01	2.86	2.73	2.63	2.54	2.47	2.40	2.35	2.30	2.25	2.21	2.18	2.14	2.11	1.99	1.91	1.79	1.71	1.54	1.48	1.44
300	8.00	5.39	4.36	3.80	3.43	3.17	2.98	2.82	2.69	2.59	2.51	2.43	2.37	2.31	2.26	2.21	2.17	2.14	2.10	2.07	1.95	1.87	1.75	1.67	1.50	1.43	1.39
500	7.95	5.35	4.33	3.76	3.40	3.14	2.94	2.79	2.66	2.56	2.48	2.40	2.34	2.28	2.23	2.19	2.14	2.11	2.07	2.04	1.92	1.84	1.72	1.64	1.46	1.39	1.35
1000	7.91	5.33	4.30	3.74	3.37	3.11	2.92	2.77	2.64	2.54	2.45	2.38	2.32	2.26	2.21	2.16	2.12	2.09	2.05	2.02	1.90	1.81	1.69	1.61	1.43	1.36	1.31

DEGREES OF FREEDOM FOR DENOMINATOR

TABLE A-4 Percentiles of the *F* distribution (*continued*)

Upper 0.1% point of the F distribution

DEGREES OF FREEDOM FOR NUMERATOR (columns) / DEGREES OF FREEDOM FOR DENOMINATOR (rows)

df_2 \ df_1	1	2	3	4	5	6	7	8	9	10	11	12	13	14	15	16	17	18	19	20	25	30	40	50	100	150	200
2	999	999	999	999	999	999	999	999	999	999	999	999	999	999	999	999	999	999	999	999	999	999	999	999	999	999	999
3	167	148	141	137	135	133	132	131	130	129	129	128	128	128	127	127	127	127	127	126	126	125	125	125	124	124	124
4	74.1	61.2	56.2	53.4	51.7	50.5	49.7	49.0	48.5	48.1	47.7	47.4	47.2	46.9	46.8	46.6	46.5	46.3	46.2	46.1	45.7	45.4	45.1	44.9	44.5	44.3	44.3
5	47.2	37.1	33.2	31.1	29.8	28.8	28.2	27.6	27.2	26.9	26.6	26.4	26.2	26.1	25.9	25.8	25.7	25.6	25.5	25.4	25.1	24.9	24.6	24.4	24.1	24.0	24.0
6	35.5	27.0	23.7	21.9	20.8	20.0	19.5	19.0	18.7	18.4	18.2	18.0	17.8	17.7	17.6	17.4	17.4	17.3	17.2	17.1	16.9	16.7	16.4	16.3	16.0	15.9	15.9
7	29.2	21.7	18.8	17.2	16.2	15.5	15.0	14.6	14.3	14.1	13.9	13.7	13.6	13.4	13.3	13.2	13.1	13.1	13.0	12.9	12.7	12.5	12.3	12.2	12.0	11.9	11.8
8	25.4	18.5	15.8	14.4	13.5	12.9	12.4	12.0	11.8	11.5	11.4	11.2	11.1	10.9	10.8	10.8	10.7	10.6	10.5	10.5	10.3	10.1	9.92	9.80	9.57	9.49	9.45
9	22.9	16.4	13.9	12.6	11.7	11.1	10.7	10.4	10.1	9.89	9.72	9.57	9.44	9.33	9.24	9.15	9.08	9.01	8.95	8.90	8.69	8.55	8.37	8.26	8.04	7.96	7.93
10	21.0	14.9	12.6	11.3	10.5	9.93	9.52	9.20	8.96	8.75	8.59	8.45	8.32	8.22	8.13	8.05	7.98	7.91	7.86	7.80	7.60	7.47	7.30	7.19	6.98	6.91	6.87
11	19.7	13.8	11.6	10.3	9.58	9.05	8.66	8.35	8.12	7.92	7.76	7.63	7.51	7.41	7.32	7.24	7.17	7.11	7.06	7.01	6.81	6.68	6.52	6.42	6.21	6.14	6.10
12	18.6	13.0	10.8	9.63	8.89	8.38	8.00	7.71	7.48	7.29	7.14	7.00	6.89	6.79	6.71	6.63	6.57	6.51	6.45	6.40	6.22	6.09	5.93	5.83	5.63	5.56	5.52
13	17.8	12.3	10.2	9.07	8.35	7.86	7.49	7.21	6.98	6.80	6.65	6.52	6.41	6.31	6.23	6.16	6.09	6.03	5.98	5.93	5.75	5.63	5.47	5.37	5.17	5.10	5.07
14	17.1	11.8	9.73	8.62	7.92	7.44	7.08	6.80	6.58	6.40	6.26	6.13	6.02	5.93	5.85	5.78	5.71	5.66	5.60	5.56	5.38	5.25	5.10	5.00	4.81	4.74	4.71
15	16.6	11.3	9.34	8.25	7.57	7.09	6.74	6.47	6.26	6.08	5.94	5.81	5.71	5.62	5.54	5.46	5.40	5.35	5.29	5.25	5.07	4.95	4.80	4.70	4.51	4.44	4.41
16	16.1	11.0	9.01	7.94	7.27	6.80	6.46	6.19	5.98	5.81	5.67	5.55	5.44	5.35	5.27	5.20	5.14	5.09	5.04	4.99	4.82	4.70	4.54	4.45	4.26	4.19	4.16
17	15.7	10.7	8.73	7.68	7.02	6.56	6.22	5.96	5.75	5.58	5.44	5.32	5.22	5.13	5.05	4.99	4.92	4.87	4.82	4.78	4.60	4.48	4.33	4.24	4.05	3.98	3.95
18	15.4	10.4	8.49	7.46	6.81	6.35	6.02	5.76	5.56	5.39	5.25	5.13	5.03	4.94	4.87	4.80	4.74	4.68	4.63	4.59	4.42	4.30	4.15	4.06	3.87	3.80	3.77
19	15.1	10.2	8.28	7.27	6.62	6.18	5.85	5.59	5.39	5.22	5.08	4.97	4.87	4.78	4.70	4.64	4.58	4.52	4.47	4.43	4.26	4.14	3.99	3.90	3.71	3.65	3.61
20	14.8	9.95	8.10	7.10	6.46	6.02	5.69	5.44	5.24	5.08	4.94	4.82	4.72	4.64	4.56	4.49	4.44	4.38	4.33	4.29	4.12	4.00	3.86	3.77	3.58	3.51	3.48
21	14.6	9.77	7.94	6.95	6.32	5.88	5.56	5.31	5.11	4.95	4.81	4.70	4.60	4.51	4.44	4.37	4.31	4.26	4.21	4.17	4.00	3.88	3.74	3.64	3.46	3.39	3.36
22	14.4	9.61	7.80	6.81	6.19	5.76	5.44	5.19	4.99	4.83	4.70	4.58	4.49	4.40	4.33	4.26	4.20	4.15	4.10	4.06	3.89	3.78	3.63	3.54	3.35	3.28	3.25
23	14.2	9.47	7.67	6.70	6.08	5.65	5.33	5.09	4.89	4.73	4.60	4.48	4.39	4.30	4.23	4.16	4.10	4.05	4.00	3.96	3.79	3.68	3.53	3.44	3.25	3.19	3.16
24	14.0	9.34	7.55	6.59	5.98	5.55	5.23	4.99	4.80	4.64	4.51	4.39	4.30	4.21	4.14	4.07	4.01	3.96	3.92	3.87	3.71	3.59	3.45	3.36	3.17	3.10	3.07
25	13.9	9.22	7.45	6.49	5.89	5.46	5.15	4.91	4.71	4.56	4.42	4.31	4.22	4.13	4.06	3.99	3.94	3.88	3.84	3.79	3.63	3.52	3.37	3.28	3.09	3.03	2.99
26	13.7	9.12	7.36	6.41	5.80	5.38	5.07	4.83	4.64	4.48	4.35	4.24	4.14	4.06	3.99	3.92	3.86	3.81	3.77	3.72	3.56	3.44	3.30	3.21	3.02	2.95	2.92
27	13.6	9.02	7.27	6.33	5.73	5.31	5.00	4.76	4.57	4.41	4.28	4.17	4.08	3.99	3.92	3.86	3.80	3.75	3.70	3.66	3.49	3.38	3.23	3.14	2.96	2.89	2.86
28	13.5	8.93	7.19	6.25	5.66	5.24	4.93	4.69	4.50	4.35	4.22	4.11	4.01	3.93	3.86	3.80	3.74	3.69	3.64	3.60	3.43	3.32	3.18	3.09	2.90	2.83	2.80
29	13.4	8.85	7.12	6.19	5.59	5.18	4.87	4.64	4.45	4.29	4.16	4.05	3.96	3.88	3.80	3.74	3.68	3.63	3.59	3.54	3.38	3.27	3.12	3.03	2.84	2.78	2.74
30	13.3	8.77	7.05	6.12	5.53	5.12	4.82	4.58	4.39	4.24	4.11	4.00	3.91	3.82	3.75	3.69	3.63	3.58	3.53	3.49	3.33	3.22	3.07	2.98	2.79	2.73	2.69
32	13.1	8.64	6.94	6.01	5.43	5.02	4.72	4.48	4.30	4.14	4.02	3.91	3.81	3.73	3.66	3.60	3.54	3.49	3.44	3.40	3.24	3.13	2.98	2.89	2.70	2.64	2.60
34	13.0	8.52	6.83	5.92	5.34	4.93	4.63	4.40	4.22	4.06	3.94	3.83	3.74	3.65	3.58	3.52	3.46	3.41	3.37	3.33	3.16	3.05	2.91	2.82	2.63	2.56	2.52
36	12.8	8.42	6.74	5.84	5.26	4.86	4.56	4.33	4.14	3.99	3.87	3.76	3.67	3.59	3.51	3.45	3.40	3.34	3.30	3.26	3.10	2.98	2.84	2.75	2.56	2.49	2.46
38	12.7	8.33	6.66	5.76	5.19	4.79	4.49	4.26	4.08	3.93	3.80	3.70	3.60	3.52	3.45	3.39	3.34	3.28	3.24	3.20	3.04	2.92	2.78	2.69	2.50	2.43	2.40
40	12.6	8.25	6.59	5.70	5.13	4.73	4.44	4.21	4.02	3.87	3.75	3.64	3.55	3.47	3.40	3.34	3.28	3.23	3.19	3.14	2.98	2.87	2.73	2.64	2.44	2.38	2.34
42	12.5	8.18	6.53	5.64	5.07	4.68	4.38	4.16	3.97	3.83	3.70	3.59	3.50	3.42	3.35	3.29	3.23	3.18	3.14	3.10	2.94	2.83	2.68	2.59	2.40	2.33	2.29
44	12.4	8.12	6.48	5.59	5.02	4.63	4.34	4.11	3.93	3.78	3.66	3.55	3.46	3.38	3.31	3.25	3.19	3.14	3.10	3.06	2.89	2.78	2.64	2.55	2.35	2.28	2.25
46	12.4	8.06	6.42	5.54	4.98	4.59	4.30	4.07	3.89	3.74	3.62	3.51	3.42	3.34	3.27	3.21	3.15	3.10	3.06	3.02	2.86	2.74	2.60	2.51	2.31	2.24	2.21
48	12.3	8.00	6.38	5.50	4.94	4.55	4.26	4.04	3.85	3.70	3.58	3.48	3.38	3.31	3.24	3.17	3.12	3.07	3.02	2.98	2.82	2.71	2.56	2.47	2.28	2.21	2.17
50	12.2	7.96	6.34	5.46	4.90	4.51	4.22	4.00	3.82	3.67	3.55	3.44	3.35	3.27	3.20	3.14	3.09	3.04	2.99	2.95	2.79	2.68	2.53	2.44	2.25	2.18	2.14
60	12.0	7.77	6.17	5.31	4.76	4.37	4.09	3.86	3.69	3.54	3.42	3.32	3.23	3.15	3.08	3.02	2.96	2.91	2.87	2.83	2.67	2.55	2.41	2.32	2.12	2.05	2.01
70	11.8	7.64	6.06	5.20	4.66	4.28	3.99	3.77	3.60	3.45	3.33	3.23	3.14	3.06	2.99	2.93	2.88	2.83	2.78	2.74	2.58	2.47	2.32	2.23	2.03	1.95	1.92
80	11.7	7.54	5.97	5.12	4.58	4.20	3.92	3.70	3.53	3.39	3.27	3.16	3.07	3.00	2.93	2.87	2.81	2.76	2.72	2.68	2.52	2.41	2.26	2.16	1.96	1.89	1.85
90	11.6	7.47	5.91	5.06	4.53	4.15	3.87	3.65	3.48	3.34	3.22	3.11	3.02	2.95	2.88	2.82	2.76	2.71	2.67	2.63	2.47	2.36	2.21	2.11	1.91	1.83	1.79
100	11.5	7.41	5.86	5.02	4.48	4.11	3.83	3.61	3.44	3.30	3.18	3.07	2.99	2.91	2.84	2.78	2.73	2.68	2.63	2.59	2.43	2.32	2.17	2.08	1.87	1.79	1.75
125	11.4	7.30	5.77	4.93	4.40	4.03	3.75	3.54	3.37	3.23	3.11	3.00	2.92	2.84	2.77	2.71	2.66	2.61	2.56	2.52	2.36	2.25	2.10	2.01	1.79	1.71	1.67
150	11.3	7.24	5.71	4.88	4.35	3.98	3.71	3.49	3.32	3.18	3.06	2.96	2.87	2.80	2.73	2.67	2.61	2.56	2.52	2.48	2.32	2.21	2.06	1.96	1.74	1.66	1.62
200	11.2	7.15	5.63	4.81	4.29	3.92	3.65	3.43	3.26	3.12	3.00	2.90	2.82	2.74	2.67	2.61	2.56	2.51	2.46	2.42	2.26	2.15	2.00	1.90	1.68	1.60	1.55
300	11.0	7.07	5.56	4.75	4.22	3.86	3.59	3.38	3.21	3.07	2.95	2.85	2.76	2.69	2.62	2.56	2.50	2.46	2.41	2.37	2.21	2.10	1.94	1.85	1.62	1.53	1.48
500	11.0	7.00	5.51	4.69	4.18	3.81	3.54	3.33	3.16	3.02	2.91	2.81	2.72	2.64	2.58	2.52	2.46	2.41	2.37	2.33	2.17	2.05	1.90	1.80	1.57	1.48	1.43
1000	10.9	6.96	5.46	4.65	4.14	3.78	3.51	3.30	3.13	2.99	2.87	2.77	2.69	2.61	2.54	2.48	2.43	2.38	2.34	2.30	2.14	2.02	1.87	1.77	1.53	1.44	1.38

TABLE A-5   Values of $\frac{1}{2}\ln\frac{1+r}{1-r}=z$.

r	0.000	0.001	0.002	0.003	0.004	0.005	0.006	0.007	0.008	0.009
0.000	0.0000	0.0010	0.0020	0.0030	0.0040	0.0050	0.0060	0.0070	0.0080	0.0090
0.010	0.0100	0.0110	0.0120	0.0130	0.0140	0.0150	0.0160	0.0170	0.0180	0.0190
0.020	0.0200	0.0210	0.0220	0.0230	0.0240	0.0250	0.0260	0.0270	0.0280	0.0290
0.030	0.0300	0.0310	0.0320	0.0330	0.0340	0.0350	0.0360	0.0370	0.0380	0.0390
0.040	0.0400	0.0410	0.0420	0.0430	0.0440	0.0450	0.0460	0.0470	0.0480	0.0490
0.050	0.0501	0.0511	0.0521	0.0531	0.0541	0.0551	0.0561	0.0571	0.0581	0.0591
0.060	0.0601	0.0611	0.0621	0.0631	0.0641	0.0651	0.0661	0.0671	0.0681	0.0691
0.070	0.0701	0.0711	0.0721	0.0731	0.0741	0.0751	0.0761	0.0771	0.0782	0.0792
0.080	0.0802	0.0812	0.0822	0.0832	0.0842	0.0852	0.0862	0.0872	0.0882	0.0892
0.090	0.0902	0.0912	0.0922	0.0933	0.0943	0.0953	0.0963	0.0973	0.0983	0.0993
0.100	0.1003	0.1013	0.1024	0.1034	0.1044	0.1054	0.1064	0.1074	0.1084	0.1094
0.110	0.1105	0.1115	0.1125	0.1135	0.1145	0.1155	0.1165	0.1175	0.1185	0.1195
0.120	0.1206	0.1216	0.1226	0.1236	0.1246	0.1257	0.1267	0.1277	0.1287	0.1297
0.130	0.1308	0.1318	0.1328	0.1338	0.1348	0.1358	0.1368	0.1379	0.1389	0.1399
0.140	0.1409	0.1419	0.1430	0.1440	0.1450	0.1460	0.1470	0.1481	0.1491	0.1501
0.150	0.1511	0.1522	0.1532	0.1542	0.1552	0.1563	0.1573	0.1583	0.1593	0.1604
0.160	0.1614	0.1624	0.1634	0.1644	0.1655	0.1665	0.1676	0.1686	0.1696	0.1706
0.170	0.1717	0.1727	0.1737	0.1748	0.1758	0.1768	0.1779	0.1789	0.1799	0.1810
0.180	0.1820	0.1830	0.1841	0.1851	0.1861	0.1872	0.1882	0.1892	0.1903	0.1913
0.190	0.1923	0.1934	0.1944	0.1954	0.1965	0.1975	0.1986	0.1996	0.2007	0.2017
0.200	0.2027	0.2038	0.2048	0.2059	0.2069	0.2079	0.2090	0.2100	0.2111	0.2121
0.210	0.2132	0.2142	0.2153	0.2163	0.2174	0.2184	0.2194	0.2205	0.2215	0.2226
0.220	0.2237	0.2247	0.2258	0.2268	0.2279	0.2289	0.2300	0.2310	0.2321	0.2331
0.230	0.2342	0.2353	0.2363	0.2374	0.2384	0.2395	0.2405	0.2416	0.2427	0.2437
0.240	0.2448	0.2458	0.2469	0.2480	0.2490	0.2501	0.2511	0.2522	0.2533	0.2543
0.250	0.2554	0.2565	0.2575	0.2586	0.2597	0.2608	0.2618	0.2629	0.2640	0.2650
0.260	0.2661	0.2672	0.2682	0.2693	0.2704	0.2715	0.2726	0.2736	0.2747	0.2758
0.270	0.2769	0.2779	0.2790	0.2801	0.2812	0.2823	0.2833	0.2844	0.2855	0.2866
0.280	0.2877	0.2888	0.2898	0.2909	0.2920	0.2931	0.2942	0.2953	0.2964	0.2975
0.290	0.2986	0.2997	0.3008	0.3019	0.3029	0.3040	0.3051	0.3062	0.3073	0.3084
0.300	0.3095	0.3106	0.3117	0.3128	0.3139	0.3150	0.3161	0.3172	0.3183	0.3195
0.310	0.3206	0.3217	0.3228	0.3239	0.3250	0.3261	0.3272	0.3283	0.3294	0.3305
0.320	0.3317	0.3328	0.3339	0.3350	0.3361	0.3372	0.3384	0.3395	0.3406	0.3417
0.330	0.3428	0.3439	0.3451	0.3462	0.3473	0.3484	0.3496	0.3507	0.3518	0.3530
0.340	0.3541	0.3552	0.3564	0.3575	0.3586	0.3597	0.3609	0.3620	0.3632	0.3643
0.350	0.3654	0.3666	0.3677	0.3689	0.3700	0.3712	0.3723	0.3734	0.3746	0.3757
0.360	0.3769	0.3780	0.3792	0.3803	0.3815	0.3826	0.3838	0.3850	0.3861	0.3873
0.370	0.3884	0.3896	0.3907	0.3919	0.3931	0.3942	0.3954	0.3966	0.3977	0.3989
0.380	0.4001	0.4012	0.4024	0.4036	0.4047	0.4059	0.4071	0.4083	0.4094	0.4106
0.390	0.4118	0.4130	0.4142	0.4153	0.4165	0.4177	0.4189	0.4201	0.4213	0.4225
0.400	0.4236	0.4248	0.4260	0.4272	0.4284	0.4296	0.4308	0.4320	0.4332	0.4344
0.410	0.4356	0.4368	0.4380	0.4392	0.4404	0.4416	0.4429	0.4441	0.4453	0.4465
0.420	0.4477	0.4489	0.4501	0.4513	0.4526	0.4538	0.4550	0.4562	0.4574	0.4587
0.430	0.4599	0.4611	0.4623	0.4636	0.4648	0.4660	0.4673	0.4685	0.4697	0.4710
0.440	0.4722	0.4735	0.4747	0.4760	0.4772	0.4784	0.4797	0.4809	0.4822	0.4835
0.450	0.4847	0.4860	0.4872	0.4885	0.4897	0.4910	0.4923	0.4935	0.4948	0.4061
0.460	0.4973	0.4986	0.4999	0.5011	0.5024	0.5037	0.5049	0.5062	0.5075	0.5088
0.470	0.5101	0.5114	0.5126	0.5139	0.5152	0.5165	0.5178	0.5191	0.5204	0.5217
0.480	0.5230	0.5243	0.5256	0.5279	0.5282	0.5295	0.5308	0.5321	0.5334	0.5347
0.490	0.5361	0.5374	0.5387	0.5400	0.5413	0.5427	0.5440	0.5453	0.5466	0.5480

TABLE A-5 Values of $\frac{1}{2}\ln\frac{1+r}{1-r}$ (continued)

r	0.000	0.001	0.002	0.003	0.004	0.005	0.006	0.007	0.008	0.009
0.500	0.5493	0.5506	0.5520	0.5533	0.5547	0.5560	0.5573	0.5587	0.5600	0.5614
0.510	0.5627	0.5641	0.5654	0.5668	0.5681	0.5695	0.5709	0.5722	0.5736	0.5750
0.520	0.5763	0.5777	0.5791	0.5805	0.5818	0.5832	0.5846	0.5860	0.5874	0.5888
0.530	0.5901	0.5915	0.5929	0.5943	0.5957	0.5971	0.5985	0.5999	0.6013	0.6027
0.540	0.6042	0.6056	0.6070	0.6084	0.6098	0.6112	0.6127	0.6141	0.6155	0.6170
0.550	0.6184	0.6198	0.6213	0.6227	0.6241	0.6256	0.6270	0.6285	0.6299	0.6314
0.560	0.6328	0.6343	0.6358	0.6372	0.6387	0.6401	0.6416	0.6431	0.6446	0.6460
0.570	0.6475	0.6490	0.6505	0.6520	0.6535	0.6550	0.6565	0.6579	0.6594	0.6610
0.580	0.6625	0.6640	0.6655	0.6670	0.6685	0.6700	0.6715	0.6731	0.6746	0.6761
0.590	0.6777	0.6792	0.6807	0.6823	0.6838	0.6854	0.6869	0.6885	0.6900	0.6916
0.600	0.6931	0.6947	0.6963	0.6978	0.6994	0.7010	0.7026	0.7042	0.7057	0.7073
0.610	0.7089	0.7105	0.7121	0.7137	0.7153	0.7169	0.7185	0.7201	0.7218	0.7234
0.620	0.7250	0.7266	0.7283	0.7299	0.7315	0.7332	0.7348	0.7364	0.7381	0.7398
0.630	0.7414	0.7431	0.7447	0.7464	0.7481	0.7497	0.7514	0.7531	0.7548	0.7565
0.640	0.7582	0.7599	0.7616	0.7633	0.7650	0.7667	0.7684	0.7701	0.7718	0.7736
0.650	0.7753	0.7770	0.7788	0.7805	0.7823	0.7840	0.7858	0.7875	0.7893	0.7910
0.660	0.7928	0.7946	0.7964	0.7981	0.7999	0.8017	0.8035	0.8053	0.8071	0.8089
0.670	0.8107	0.8126	0.8144	0.8162	0.8180	0.8199	0.8217	0.8236	0.8254	0.8273
0.680	0.8291	0.8310	0.8328	0.8347	0.8366	0.8385	0.8404	0.8423	0.8442	0.8461
0.690	0.8480	0.8499	0.8518	0.8537	0.8556	0.8576	0.8595	0.8614	0.8634	0.8653
0.700	0.8673	0.8693	0.8712	0.8732	0.8752	0.8772	0.8792	0.8812	0.8832	0.8852
0.710	0.8872	0.8892	0.8912	0.8933	0.8953	0.8973	0.8994	0.9014	0.9035	0.9056
0.720	0.9076	0.9097	0.9118	0.9139	9.9160	0.9181	0.9202	0.9223	0.9245	0.9266
0.730	0.9287	0.9309	0.9330	0.9352	0.9373	0.9395	0.9417	0.9439	0.9461	0.9483
0.740	0.9505	0.9527	0.9549	0.9571	0.9594	0.9616	0.9639	0.9661	0.9684	0.9707
0.750	0.9730	0.9752	0.9775	0.9799	0.9822	0.9845	0.9868	0.9892	0.9915	0.9939
0.760	0.9962	0.9986	1.0010	1.0034	1.0058	1.0082	1.0106	1.0130	1.0154	1.0179
0.770	1.0203	1.0228	1.0253	1.0277	1.0302	1.0327	1.0352	1.0378	1.0403	1.0428
0.780	1.0454	1.0479	1.0505	1.0531	1.0557	1.0583	1.0609	1.0635	1.0661	1.0688
0.790	1.0714	1.0741	1.0768	1.0795	1.0822	1.0849	1.0876	1.0903	1.0931	1.0958
0.800	1.0986	1.1014	1.1041	1.1070	1.1098	1.1127	1.1155	1.1184	1.1212	1.1241
0.810	1.1270	1.1299	1.1329	1.1358	1.1388	1.1417	1.1447	1.1477	1.1507	1.1538
0.820	1.1568	1.1599	1.1630	1.1660	1.1692	1.1723	1.1754	1.1786	1.1817	1.1849
0.830	1.1870	1.1913	1.1946	1.1979	1.2011	1.2044	1.2077	1.2111	1.2144	1.2178
0.840	1.2212	1.2246	1.2280	1.2315	1.2349	1.2384	1.2419	1.2454	1.2490	1.2526
0.850	1.2561	1.2598	1.2634	1.2670	1.2708	1.2744	1.2782	1.2819	1.2857	1.2895
0.860	1.2934	1.2972	1.3011	1.3050	1.3089	1.3129	1.3168	1.3209	1.3249	1.3290
0.870	1.3331	1.3372	1.3414	1.3456	1.3498	1.3540	1.3583	1.3626	1.3670	1.3714
0.880	1.3758	1.3802	1.3847	1.3892	1.3938	1.3984	1.4030	1.4077	1.4124	1.4171
0.890	1.4219	1.4268	1.4316	1.4366	1.4415	1.4465	1.4516	1.4566	1.4618	1.4670
0.900	1.4722	1.4775	1.4828	1.4883	1.4937	1.4992	1.5047	1.5103	1.5160	1.5217
0.910	1.5275	1.5334	1.5393	1.5453	1.5513	1.5574	1.5636	1.5698	1.5762	1.5825
0.920	1.5890	1.5956	1.6022	1.6089	1.6157	1.6226	1.6296	1.6366	1.6438	1.6510
0.930	1.6584	1.6659	1.6734	1.6811	1.6888	1.6967	1.7047	1.7129	1.7211	1.7295
0.940	1.7380	1.7467	1.7555	1.7645	1.7736	1.7828	1.7923	1.8019	1.8117	1.8216
0.950	1.8318	1.8421	1.8527	1.8635	1.8745	1.8857	1.8972	1.9090	1.9210	1.9333
0.960	1.9459	1.9588	1.9721	1.9857	1.9996	2.0140	2.0287	2.0439	2.0595	2.0756
0.970	2.0923	2.1095	2.1273	2.1457	2.1649	2.1847	2.2054	2.2269	2.2494	2.2729
0.980	2.2976	2.3223	2.3507	2.3796	2.4101	2.4426	2.4774	2.5147	2.5550	2.5988
0.990	2.6467	2.6996	2.7587	2.8257	2.9031	2.9945	3.1063	3.2504	3.4534	3.8002

r	z
0.9999	4.95172
0.99999	6.10303

Source: Albert E. Waugh, *Statistical Tables and Problems,* McGraw-Hill Book Company, New York, 1952, Table A11, pp. 40–41, with the kind permission of the author and publisher.

Note: To obtain $\frac{1}{2}\ln(1 + r)/(1 - r)$ when r is negative, use the negative of the value corresponding to the absolute value of r; e.g., if $r = -0.242$, $\frac{1}{2}\ln(1 + 0.242)/(1 - 0.242) = -0.2469$.

TABLE A-6   Upper α point of Studentized range, $q_{k,\nu} = R/S$, $k =$ sample size for range R, $\nu =$ number of degrees of freedom for S
(entry $= q_{k,\nu,1-\alpha}$, where $P(q_{k,\nu} > q_{k,\nu,1-\alpha}) = \alpha$)

$\alpha = .05$

ν \ k	2	3	4	5	6	7	8	9	10	11	12	13	14	15	16	17	18	19	20
1	18.0	27.0	32.8	37.1	40.4	43.1	45.4	47.4	49.1	50.6	52.0	53.2	54.3	55.4	56.3	57.2	58.0	58.8	59.6
2	6.08	8.33	9.80	10.9	11.7	12.4	13.0	13.5	14.0	14.4	14.7	15.1	15.4	15.7	15.9	16.1	16.4	16.6	16.8
3	4.50	5.91	6.82	7.50	8.04	8.48	8.85	9.18	9.46	9.72	9.95	10.2	10.3	10.5	10.7	10.8	11.0	11.1	11.2
4	3.93	5.04	5.76	6.29	6.71	7.05	7.35	7.60	7.83	8.03	8.21	8.37	8.52	8.66	8.79	8.91	9.03	9.13	9.23
5	3.64	4.60	5.22	5.67	6.03	6.33	6.58	6.80	6.99	7.17	7.32	7.47	7.60	7.72	7.83	7.93	8.03	8.12	8.21
6	3.46	4.34	4.90	5.30	5.63	5.90	6.12	6.32	6.49	6.65	6.79	6.92	7.03	7.14	7.24	7.34	7.43	7.51	7.59
7	3.34	4.16	4.68	5.06	5.36	5.61	5.82	6.00	6.16	6.30	6.43	6.55	6.66	6.76	6.85	6.94	7.02	7.10	7.17
8	3.26	4.04	4.53	4.89	5.17	5.40	5.60	5.77	5.92	6.05	6.18	6.29	6.39	6.48	6.57	6.65	6.73	6.80	6.87
9	3.20	3.95	4.41	4.76	5.02	5.24	5.43	5.59	5.74	5.87	5.98	6.09	6.19	6.28	6.36	6.44	6.51	6.58	6.64
10	3.15	3.88	4.33	4.65	4.91	5.12	5.30	5.46	5.60	5.72	5.83	5.93	6.03	6.11	6.19	6.27	6.34	6.40	6.47
11	3.11	3.82	4.26	4.57	4.82	5.03	5.20	5.35	5.49	5.61	5.71	5.81	5.90	5.98	6.06	6.13	6.20	6.27	6.33
12	3.08	3.77	4.20	4.51	4.75	4.95	5.12	5.27	5.39	5.51	5.61	5.71	5.80	5.88	5.95	6.02	6.09	6.15	6.21
13	3.06	3.73	4.15	4.45	4.69	4.88	5.05	5.19	5.32	5.43	5.53	5.63	5.71	5.79	5.86	5.93	5.99	6.05	6.11
14	3.03	3.70	4.11	4.41	4.64	4.83	4.99	5.13	5.25	5.36	5.46	5.55	5.64	5.71	5.79	5.85	5.91	5.97	6.03
15	3.01	3.67	4.08	4.37	4.59	4.78	4.94	5.08	5.20	5.31	5.40	5.49	5.57	5.65	5.72	5.78	5.85	5.90	5.96
16	3.00	3.65	4.05	4.33	4.56	4.74	4.90	5.03	5.15	5.26	5.35	5.44	5.52	5.59	5.66	5.73	5.79	5.84	5.90
17	2.98	3.63	4.02	4.30	4.52	4.70	4.86	4.99	5.11	5.21	5.31	5.39	5.47	5.54	5.61	5.67	5.73	5.79	5.84
18	2.97	3.61	4.00	4.28	4.49	4.67	4.82	4.96	5.07	5.17	5.27	5.35	5.43	5.50	5.57	5.63	5.69	5.74	5.79
19	2.96	3.59	3.98	4.25	4.47	4.65	4.79	4.92	5.04	5.14	5.23	5.31	5.39	5.46	5.53	5.59	5.65	5.70	5.75
20	2.95	3.58	3.96	4.23	4.45	4.62	4.77	4.90	5.01	5.11	5.20	5.28	5.36	5.43	5.49	5.55	5.61	5.66	5.71
24	2.92	3.53	3.90	4.17	4.37	4.54	4.68	4.81	4.92	5.01	5.10	5.18	5.26	5.32	5.38	5.44	5.49	5.55	5.59
30	2.89	3.49	3.85	4.10	4.30	4.46	4.60	4.72	4.82	4.92	5.00	5.08	5.15	5.21	5.27	5.33	5.38	5.43	5.47
40	2.86	3.44	3.79	4.04	4.23	4.39	4.52	4.63	4.73	4.82	4.90	4.98	5.04	5.11	5.16	5.22	5.27	5.31	5.36
60	2.83	3.40	3.74	3.98	4.16	4.31	4.44	4.55	4.65	4.73	4.81	4.88	4.94	5.00	5.06	5.11	5.15	5.20	5.24
120	2.80	3.36	3.68	3.92	4.10	4.24	4.36	4.47	4.56	4.64	4.71	4.78	4.84	4.90	4.95	5.00	5.04	5.09	5.13
∞	2.77	3.31	3.63	3.86	4.03	4.17	4.29	4.39	4.47	4.55	4.62	4.68	4.74	4.80	4.85	4.89	4.93	4.97	5.01

Source: From pp. 176–177 of *Biometrika Tables for Statisticians*, Vol. I, by E. S. Pearson and H. O. Hartley, published by the Biometrika Trustees, Cambridge University Press, Cambridge, 1954. Reproduced with the kind permission of the authors and the publisher. Corrections of ± 1 in the last figure, supplied by James Pacheres, have been incorporated in 41 entries.

TABLE A-6 Upper α point of Studentized range (*continued*)

$\alpha = .01$

v \ k	2	3	4	5	6	7	8	9	10	11	12	13	14	15	16	17	18	19	20
1	90.0	135	164	186	202	216	227	237	246	253	260	266	272	277	282	286	290	294	298
2	14.0	19.0	22.3	24.7	26.6	28.2	29.5	30.7	31.7	32.6	33.4	34.1	34.8	35.4	36.0	36.5	37.0	37.5	37.9
3	8.26	10.6	12.2	13.3	14.2	15.0	15.6	16.2	16.7	17.1	17.5	17.9	18.2	18.5	18.8	19.1	19.3	19.5	19.8
4	6.51	8.12	9.17	9.96	10.6	11.1	11.5	11.9	12.3	12.6	12.8	13.1	13.3	13.5	13.7	13.9	14.1	14.2	14.4
5	5.70	6.97	7.80	8.42	8.91	9.32	9.67	9.97	10.2	10.5	10.7	10.9	11.1	11.2	11.4	11.6	11.7	11.8	11.9
6	5.24	6.33	7.03	7.56	7.97	8.32	8.61	8.87	9.10	9.30	9.49	9.65	9.81	9.95	10.1	10.2	10.3	10.4	10.5
7	4.95	5.92	6.54	7.01	7.37	7.68	7.94	8.17	8.37	8.55	8.71	8.86	9.00	9.12	9.24	9.35	9.46	9.55	9.65
8	4.74	5.63	6.20	6.63	6.96	7.24	7.47	7.68	7.87	8.03	8.18	8.31	8.44	8.55	8.66	8.76	8.85	8.94	9.03
9	4.60	5.43	5.96	6.35	6.66	6.91	7.13	7.32	7.49	7.65	7.78	7.91	8.03	8.13	8.23	8.32	8.41	8.49	8.57
10	4.48	5.27	5.77	6.14	6.43	6.67	6.87	7.05	7.21	7.36	7.48	7.60	7.71	7.81	7.91	7.99	8.07	8.15	8.22
11	4.39	5.14	5.62	5.97	6.25	6.48	6.67	6.84	6.99	7.13	7.25	7.36	7.46	7.56	7.65	7.73	7.81	7.88	7.95
12	4.32	5.04	5.50	5.84	6.10	6.32	6.51	6.67	6.81	6.94	7.06	7.17	7.26	7.36	7.44	7.52	7.59	7.66	7.73
13	4.26	4.96	5.40	5.73	5.98	6.19	6.37	6.53	6.67	6.79	6.90	7.01	7.10	7.19	7.27	7.34	7.42	7.48	7.55
14	4.21	4.89	5.32	5.63	5.88	6.08	6.26	6.41	6.54	6.66	6.77	6.87	6.96	7.05	7.12	7.20	7.27	7.33	7.39
15	4.17	4.83	5.25	5.56	5.80	5.99	6.16	6.31	6.44	6.55	6.66	6.76	6.84	6.93	7.00	7.07	7.14	7.20	7.26
16	4.13	4.78	5.19	5.49	5.72	5.92	6.08	6.22	6.35	6.46	6.56	6.66	6.74	6.82	6.90	6.97	7.03	7.09	7.15
17	4.10	4.74	5.14	5.43	5.66	5.85	6.01	6.15	6.27	6.38	6.48	6.57	6.66	6.73	6.80	6.87	6.94	7.00	7.05
18	4.07	4.70	5.09	5.38	5.60	5.79	5.94	6.08	6.20	6.31	6.41	6.50	6.58	6.65	6.72	6.79	6.85	6.91	6.96
19	4.05	4.67	5.05	5.33	5.55	5.73	5.89	6.02	6.14	6.25	6.34	6.43	6.51	6.58	6.65	6.72	6.78	6.84	6.89
20	4.02	4.64	5.02	5.29	5.51	5.69	5.84	5.97	6.09	6.19	6.29	6.37	6.45	6.52	6.59	6.65	6.71	6.76	6.82
24	3.96	4.54	4.91	5.17	5.37	5.54	5.69	5.81	5.92	6.02	6.11	6.19	6.26	6.33	6.39	6.45	6.51	6.56	6.61
30	3.89	4.45	4.80	5.05	5.24	5.40	5.54	5.65	5.76	5.85	5.93	6.01	6.08	6.14	6.20	6.26	6.31	6.36	6.41
40	3.82	4.37	4.70	4.93	5.11	5.27	5.39	5.50	5.60	5.69	5.77	5.84	5.90	5.96	6.02	6.07	6.12	6.17	6.21
60	3.76	4.28	4.60	4.82	4.99	5.13	5.25	5.36	5.45	5.53	5.60	5.67	5.73	5.79	5.84	5.89	5.93	5.98	6.02
120	3.70	4.20	4.50	4.71	4.87	5.01	5.12	5.21	5.30	5.38	5.44	5.51	5.56	5.61	5.66	5.71	5.75	5.79	5.83
∞	3.64	4.12	4.40	4.60	4.76	4.88	4.99	5.08	5.16	5.23	5.29	5.35	5.40	5.45	5.49	5.54	5.57	5.61	5.65

TABLE A-7 Orthogonal polynomial coefficients

k	POLYNOMIAL	X										(Σp_i^2)
		1	2	3	4	5	6	7	8	9	10	
3	Linear	−1	0	1								2
	Quadratic	1	−2	1								6
4	Linear	−3	−1	1	3							20
	Quadratic	1	−1	−1	1							4
	Cubic	−1	3	−3	1							20
5	Linear	−2	−1	0	1	2						10
	Quadratic	2	−1	−2	−1	2						14
	Cubic	−1	2	0	−2	1						10
	Quartic	1	−4	6	−4	1						70
6	Linear	−5	−3	−1	1	3	5					70
	Quadratic	5	−1	−4	−4	−1	5					84
	Cubic	−5	7	4	−4	−7	5					180
	Quartic	1	−3	2	2	−3	1					28
	Quintic	−1	5	−10	10	−5	1					252
7	Linear	−3	−2	−1	0	1	2	3				28
	Quadratic	5	0	−3	−4	−3	0	5				84
	Cubic	−1	1	1	0	−1	−1	1				6
	Quartic	3	−7	1	6	1	−7	3				154
	Quintic	−1	4	−5	0	5	−4	1				84
	Sextic	1	−6	15	−20	15	−6	1				924
8	Linear	−7	−5	−3	−1	1	3	5	7			168
	Quadratic	7	1	−3	−5	−5	−3	1	7			168
	Cubic	−7	5	7	3	−3	−7	−5	7			264
	Quartic	7	−13	−3	9	9	−3	−13	7			616
	Quintic	−7	23	−17	−15	15	17	−23	7			2184
	Sextic	1	−5	9	−5	−5	9	−5	1			264
	Septic	−1	7	−21	35	−35	21	−7	1			3,432
9	Linear	−4	−3	−2	−1	0	1	2	3	4		60
	Quadratic	28	7	−8	−17	−20	−17	−8	7	28		2,772
	Cubic	−14	7	13	9	0	−9	−13	−7	14		990
	Quartic	14	−21	−11	9	18	9	−11	−21	14		2,002
	Quintic	−4	11	−4	−9	0	9	4	−11	4		468
	Sextic	4	−17	22	1	−20	1	22	−17	4		1,980
	Septic	−1	6	−14	14	0	−14	14	−6	1		858
	Octic	1	−8	28	−56	70	−56	28	−8	1		12,870
10	Linear	−9	−7	−5	−3	−1	1	3	5	7	9	330
	Quadratic	6	2	−1	−3	−4	−4	−3	−1	2	6	132
	Cubic	−42	14	35	31	12	−12	−31	−35	−14	42	8,580
	Quartic	18	−22	−17	3	18	18	3	−17	−22	18	2,860
	Quintic	−6	14	−1	−11	−6	6	11	1	−14	6	780
	Sextic	3	−11	10	6	−8	−8	6	10	11	3	660
	Septic	−9	47	−86	92	56	−56	−42	86	−47	9	29,172
	Octic	1	−7	20	−28	14	14	−28	20	−7	1	2,860
	Novic	−1	9	−36	84	−126	126	−84	36	−9	1	48,620

TABLE A-8 Critical values for Studentized residuals and jackknife residuals, n = sample size, k = number of predictors

Column headers list JACKKNIFE k (top) over STUDENT k (bottom). The first column is JACKKNIFE k = 1 with no STUDENT k value.

$\alpha = .10$

JACKKNIFE k / STUDENT k → n ↓	1	1/2	2/3	3/4	4/5	5/6	6/7	7/8	8/9	9/10	14/15	19/20	39/40	79/80
10	3.36	3.50	3.71	4.03	4.60	5.84	9.92	63.66						
15	3.22	3.27	3.33	3.41	3.51	3.63	3.81	4.06	4.46	5.17				
20	3.20	3.22	3.23	3.29	3.33	3.37	3.43	3.50	3.58	3.69	5.60			
25	3.20	3.21	3.23	3.25	3.27	3.30	3.33	3.36	3.39	3.44	3.83	5.95		
30	3.21	3.22	3.23	3.24	3.26	3.27	3.29	3.31	3.33	3.35	3.53	3.95		
40	3.24	3.24	3.25	3.26	3.27	3.27	3.28	3.29	3.30	3.31	3.38	3.48		
60	3.30	3.30	3.30	3.31	3.31	3.31	3.32	3.32	3.32	3.33	3.35	3.38	3.66	
80	3.35	3.35	3.35	3.35	3.36	3.36	3.36	3.36	3.36	3.37	3.38	3.39	3.48	
100	3.39	3.39	3.39	3.40	3.40	3.40	3.40	3.40	3.40	3.40	3.41	3.42	3.46	3.88
200	3.54	3.54	3.54	3.54	3.54	3.54	3.54	3.54	3.54	3.54	3.54	3.55	3.55	3.58
400	3.70	3.70	3.70	3.70	3.70	3.70	3.70	3.70	3.70	3.70	3.70	3.70	3.70	3.70
800	3.86	3.86	3.86	3.86	3.86	3.86	3.86	3.86	3.86	3.86	3.86	3.86	3.86	3.86

$\alpha = .05$

JACKKNIFE k / STUDENT k → n ↓	1	1/2	2/3	3/4	4/5	5/6	6/7	7/8	8/9	9/10	14/15	19/20	39/40	79/80
10	3.83	4.03	4.32	4.77	5.60	7.45	14.09	127.32						
15	3.58	3.65	3.73	3.83	3.95	4.12	4.36	4.70	5.25	6.25				
20	3.51	3.54	3.58	3.62	3.67	3.73	3.81	3.89	4.00	4.15	6.76			
25	3.48	3.51	3.53	3.55	3.58	3.61	3.65	3.69	3.73	3.79	4.30	7.17		
30	3.48	3.49	3.51	3.52	3.54	3.56	3.58	3.60	3.63	3.66	3.88	4.42		
40	3.49	3.49	3.50	3.51	3.52	3.53	3.54	3.55	3.56	3.58	3.66	3.79		
60	3.53	3.53	3.53	3.54	3.54	3.54	3.55	3.55	3.56	3.56	3.59	3.62	3.96	
80	3.57	3.57	3.57	3.57	3.57	3.58	3.58	3.58	3.58	3.58	3.60	3.61	3.72	
100	3.60	3.60	3.60	3.60	3.61	3.61	3.61	3.61	3.61	3.61	3.62	3.63	3.68	4.19
200	3.73	3.73	3.73	3.73	3.73	3.73	3.73	3.73	3.73	3.73	3.74	3.74	3.75	3.78
400	3.87	3.87	3.87	3.87	3.87	3.87	3.87	3.88	3.88	3.88	3.88	3.88	3.88	3.88
800	4.02	4.02	4.02	4.02	4.02	4.02	4.02	4.02	4.02	4.02	4.03	4.03	4.03	4.03

$\alpha = .01$

JACKKNIFE k / STUDENT k → n ↓	1	1/2	2/3	3/4	4/5	5/6	6/7	7/8	8/9	9/10	14/15	19/20	39/40	79/80
10	5.04	5.41	5.96	6.87	8.61	12.92	31.60	636.62						
15	4.44	4.55	4.68	4.85	5.08	5.37	5.80	6.43	7.50	9.57				
20	4.23	4.29	4.35	4.42	4.50	4.60	4.72	4.86	5.05	5.29	10.31			
25	4.14	4.17	4.20	4.24	4.28	4.33	4.39	4.45	4.53	4.62	5.46	10.92		
30	4.09	4.11	4.13	4.15	4.18	4.21	4.24	4.28	4.32	4.36	4.71	5.60		
40	4.04	4.05	4.06	4.08	4.09	4.10	4.12	4.14	4.15	4.17	4.29	4.49		
60	4.03	4.03	4.04	4.04	4.05	4.05	4.06	4.06	4.07	4.07	4.12	4.17	4.67	
80	4.04	4.04	4.04	4.05	4.05	4.05	4.06	4.06	4.07	4.07	4.09	4.11	4.26	
100	4.06	4.06	4.06	4.06	4.06	4.07	4.07	4.07	4.07	4.07	4.09	4.10	4.17	4.90
200	4.15	4.15	4.15	4.15	4.15	4.15	4.15	4.15	4.15	4.15	4.15	4.16	4.17	4.21
400	4.27	4.27	4.27	4.27	4.27	4.27	4.27	4.27	4.27	4.27	4.27	4.27	4.27	4.28
800	4.40	4.40	4.40	4.40	4.40	4.40	4.40	4.40	4.40	4.40	4.40	4.40	4.40	4.40

TABLE A-9 Critical values for leverages, n = sample size, k = number of predictors

$\alpha = .10$

n \ k	1	2	3	4	5	6	7	8	9	10	15	20	40	80
10	0.626	0.759	0.847	0.911	0.956	0.984	0.997	1.000						
15	0.481	0.595	0.679	0.748	0.806	0.855	0.897	0.932	0.959	0.980				
20	0.394	0.491	0.565	0.627	0.682	0.731	0.775	0.815	0.851	0.883	0.988			
25	0.335	0.419	0.484	0.540	0.589	0.635	0.676	0.715	0.751	0.784	0.918	0.992		
30	0.293	0.366	0.424	0.474	0.519	0.560	0.599	0.635	0.669	0.701	0.837	0.937		
40	0.236	0.295	0.342	0.383	0.420	0.455	0.487	0.518	0.547	0.576	0.701	0.806		
60	0.172	0.214	0.248	0.279	0.306	0.332	0.356	0.380	0.402	0.424	0.524	0.612	0.888	
80	0.137	0.170	0.197	0.221	0.242	0.263	0.283	0.301	0.319	0.337	0.418	0.491	0.737	
100	0.114	0.141	0.164	0.183	0.201	0.219	0.235	0.250	0.266	0.280	0.348	0.410	0.625	0.941
200	0.064	0.079	0.091	0.102	0.111	0.121	0.130	0.138	0.146	0.155	0.192	0.227	0.353	0.568
400	0.036	0.043	0.050	0.055	0.060	0.065	0.070	0.075	0.079	0.083	0.104	0.122	0.190	0.311
800	0.020	0.024	0.027	0.030	0.032	0.035	0.037	0.040	0.042	0.044	0.055	0.065	0.100	0.164

$\alpha = .05$

n \ k	1	2	3	4	5	6	7	8	9	10	15	20	40	80
10	0.683	0.802	0.879	0.933	0.969	0.990	0.999	1.000						
15	0.531	0.639	0.719	0.782	0.835	0.880	0.916	0.946	0.969	0.986				
20	0.436	0.531	0.602	0.662	0.714	0.761	0.802	0.839	0.872	0.901	0.991			
25	0.372	0.454	0.518	0.573	0.621	0.665	0.705	0.742	0.776	0.807	0.931	0.994		
30	0.325	0.398	0.455	0.505	0.549	0.589	0.627	0.662	0.695	0.726	0.855	0.947		
40	0.261	0.321	0.368	0.409	0.446	0.480	0.512	0.543	0.572	0.600	0.722	0.823		
60	0.190	0.233	0.268	0.298	0.326	0.352	0.376	0.400	0.422	0.444	0.543	0.630	0.898	
80	0.151	0.185	0.212	0.236	0.258	0.279	0.299	0.318	0.336	0.353	0.435	0.508	0.751	
100	0.126	0.154	0.176	0.196	0.215	0.232	0.248	0.264	0.279	0.294	0.363	0.425	0.638	0.946
200	0.070	0.085	0.098	0.108	0.119	0.128	0.137	0.146	0.154	0.162	0.201	0.236	0.362	0.570
400	0.039	0.047	0.053	0.059	0.064	0.069	0.074	0.079	0.083	0.088	0.108	0.127	0.196	0.317
800	0.021	0.025	0.029	0.032	0.034	0.037	0.039	0.042	0.044	0.046	0.057	0.067	0.103	0.168

$\alpha = .01$

n \ k	1	2	3	4	5	6	7	8	9	10	15	20	40	80
10	0.785	0.875	0.930	0.965	0.986	0.997	1.000	1.000						
15	0.629	0.724	0.792	0.844	0.887	0.921	0.948	0.969	0.984	0.994				
20	0.524	0.612	0.677	0.731	0.777	0.817	0.852	0.883	0.910	0.933	0.996			
25	0.450	0.529	0.589	0.640	0.685	0.724	0.761	0.794	0.824	0.851	0.953	0.997		
30	0.394	0.466	0.521	0.568	0.610	0.648	0.683	0.716	0.746	0.774	0.889	0.964		
40	0.318	0.377	0.424	0.464	0.501	0.534	0.565	0.595	0.622	0.649	0.763	0.855		
60	0.231	0.275	0.310	0.341	0.369	0.395	0.420	0.443	0.465	0.487	0.584	0.668	0.917	
80	0.183	0.218	0.246	0.271	0.293	0.314	0.334	0.353	0.372	0.389	0.471	0.543	0.778	
100	0.152	0.181	0.205	0.225	0.244	0.262	0.279	0.295	0.310	0.325	0.394	0.456	0.666	0.956
200	0.085	0.100	0.113	0.124	0.135	0.145	0.154	0.163	0.172	0.180	0.219	0.255	0.383	0.598
400	0.046	0.054	0.061	0.067	0.073	0.078	0.083	0.088	0.092	0.097	0.118	0.138	0.208	0.330
800	0.025	0.029	0.033	0.036	0.039	0.041	0.044	0.046	0.049	0.051	0.062	0.073	0.110	0.175

TABLE A-10 50-percentile values of *F* distribution for Cook's d_i, n = sample size, k = number of predictors

n \ k	1	2	3	4	5	6	7	8	9	10	15	20	40	80
10	0.76	0.87	0.94	1.00	1.06	1.15	1.32	2.03						
15	0.73	0.84	0.89	0.93	0.96	0.99	1.01	1.04	1.07	1.12				
20	0.72	0.82	0.88	0.91	0.94	0.96	0.97	0.99	1.00	1.01	1.14			
25	0.71	0.81	0.87	0.90	0.92	0.94	0.96	0.97	0.98	0.99	1.03	1.15		
30	0.71	0.81	0.86	0.89	0.92	0.93	0.95	0.96	0.97	0.97	1.01	1.04		
40	0.71	0.80	0.86	0.89	0.91	0.93	0.94	0.95	0.96	0.96	0.99	1.00		
60	0.70	0.80	0.85	0.88	0.90	0.92	0.93	0.94	0.95	0.95	0.97	0.99	1.02	
80	0.70	0.80	0.85	0.88	0.90	0.91	0.93	0.94	0.94	0.95	0.97	0.98	1.00	
100	0.70	0.79	0.85	0.88	0.90	0.91	0.92	0.93	0.94	0.95	0.97	0.98	1.00	1.03
200	0.70	0.79	0.84	0.87	0.89	0.91	0.92	0.93	0.94	0.94	0.96	0.97	0.99	1.00
400	0.69	0.79	0.84	0.87	0.89	0.91	0.92	0.93	0.94	0.94	0.96	0.97	0.99	0.99
800	0.69	0.79	0.84	0.87	0.89	0.91	0.92	0.93	0.93	0.94	0.96	0.97	0.98	0.99

Appendix - Matrices and Their Relationship to Regression Analysis

B-1 Preview

Statisticians have found matrix mathematics to be a very useful vehicle for compactly presenting the concepts, methods, and formulae of regression analysis and other multivariable methods. Moreover, matrix formulation of such topics has had the important practical implication of permitting extremely efficient and accurate use of the computer for carrying out multivariable analyses on large data sets.

This appendix will summarize some of the more elementary but important notions and manipulations of matrix algebra and will use this tool to describe the general least-squares procedures of multiple regression. Admittedly, the material in this appendix is more mathematical than in the main body of the text and is not absolutely necessary knowledge for the applied user of the multivariable methods we describe. Nevertheless, the reader who is comfortable with the mathematical level used here should find the matrix formulation of regression analysis a powerful and unifying supplement that may facilitate the learning of more advanced multivariable methods.

B-2 Definitions

A *matrix* may be simply defined as a rectangular array of numbers. For example,

$$\mathbf{A} = \begin{bmatrix} 2 & 3 & 1 \\ 1 & 1 & 2 \end{bmatrix}, \quad \mathbf{B} = \begin{bmatrix} 2 & 1 \\ 3 & 2 \end{bmatrix}, \quad \mathbf{C} = \begin{bmatrix} 1 \\ 1 \\ 3 \end{bmatrix}$$

are all matrices.

The *dimensions* of a matrix are the number of rows and the number of columns that it has. For example, the matrix **A** above has two rows and three columns:

$$
\begin{array}{c}
\begin{array}{ccc}
\text{column} & \text{column} & \text{column} \\
1 & 2 & 3 \\
\downarrow & \downarrow & \downarrow
\end{array} \\
\begin{array}{l}
\text{row 1} \rightarrow \\
\\
\text{row 2} \rightarrow
\end{array}
\begin{bmatrix}
2 & 3 & 1 \\
1 & 1 & 2
\end{bmatrix}
\end{array}
$$

It is customary to say that **A** is a 2×3 matrix or to write $\mathbf{A}_{2\times3}$. The dimensions of the matrices **B** and **C** above are 2×2 and 3×1, respectively. An example of a 1×4 matrix is any matrix with one row and four columns, such as $\mathbf{D}_{1\times4} = \begin{bmatrix} -2 & 3 & 3 & 0 \end{bmatrix}$. Incidentally, matrices that contain only one row or only one column are often referred to as *vectors*; thus the matrices

$$
\mathbf{C} = \begin{bmatrix} 1 \\ 1 \\ 3 \end{bmatrix} \quad \text{and} \quad \mathbf{D} = \begin{bmatrix} -2 & 3 & 3 & 0 \end{bmatrix}
$$

are examples of a column vector and a row vector, respectively. Also, the matrix **B** is called a *square matrix* because it has the same number of rows and columns.

The numbers forming the rectangular array of a matrix are called the *elements* of the matrix. If we let a_{ij} denote the element in the ith row and jth column of the matrix **A** above,

$$
\begin{array}{ccc}
a_{11} = 2, & a_{12} = 3, & a_{13} = 1 \\
a_{21} = 1, & a_{22} = 1, & a_{23} = 2
\end{array}
$$

It is often informative to write

$$\mathbf{A}_{2\times3} = ((a_{ij}))$$

indicating that the matrix **A** (represented by a capital letter) with two rows and three columns has typical element a_{ij} (represented by the corresponding lowercase letter). Thus, if you were given that

$$\mathbf{B}_{3\times2} = ((b_{ij}))$$

where $b_{21} = 3$, $b_{31} = 2$, $b_{11} = -2$, $b_{12} = 6$, $b_{22} = 0$, $b_{32} = 1$, you should construct **B** to be

$$
\mathbf{B} = \begin{bmatrix} -2 & 6 \\ 3 & 0 \\ 2 & 1 \end{bmatrix}
$$

B-3 Matrices in Regression Analysis

Given any set of multivariable data suitable for a regression analysis, a number of key matrices can be defined that directly correspond to the basic components of the regression model being postulated. For example, consider the following $n = 12$ pairs of observations on $Y = \text{WGT}$ and $X = \text{HGT}$:

	Child											
Variable	1	2	3	4	5	6	7	8	9	10	11	12
Y (WGT)	64	71	53	67	55	58	77	57	56	51	76	68
X (HGT)	57	59	49	62	51	50	55	48	42	42	61	57

We can, in correspondence with the straight-line regression model $Y = \beta_0 + \beta_1 X + E$, define four matrices: $\mathbf{Y}$, the vector of observations on Y; $\mathbf{X}$, the matrix of independent variables; $\boldsymbol{\beta}$, the vector of parameters to be estimated; and $\mathbf{E}$, the vector of errors.

For the data given, these matrices are defined as follows:

$$
\mathbf{Y}_{12\times1} = \begin{bmatrix} 64 \\ 71 \\ 53 \\ 67 \\ 55 \\ 58 \\ 77 \\ 57 \\ 56 \\ 51 \\ 76 \\ 68 \end{bmatrix}, \quad
\mathbf{X}_{12\times2} = \begin{bmatrix} 1 & 57 \\ 1 & 59 \\ 1 & 49 \\ 1 & 62 \\ 1 & 51 \\ 1 & 50 \\ 1 & 55 \\ 1 & 48 \\ 1 & 42 \\ 1 & 42 \\ 1 & 61 \\ 1 & 57 \end{bmatrix}, \quad
\boldsymbol{\beta}_{2\times1} = \begin{bmatrix} \beta_0 \\ \beta_1 \end{bmatrix}, \quad
\mathbf{E}_{12\times1} = \begin{bmatrix} E_1 \\ E_2 \\ E_3 \\ E_4 \\ E_5 \\ E_6 \\ E_7 \\ E_8 \\ E_9 \\ E_{10} \\ E_{11} \\ E_{12} \end{bmatrix}
$$

Notice that the first column of the $\mathbf{X}$ matrix of independent variables contains only 1's. This is the general convention used for any regression model containing a constant term β_0; motivation for adopting this convention follows by imagining the β_0 term to be of the form $\beta_0 X_0$, where X_0 is a dummy variable always taking the value 1. The vectors of errors $\mathbf{E}$ contains random (and unobservable) error values, one for each pair of observations, which represent the differences between the observed Y-values and their (unknown) expected values under the given model.

B-4 Transpose of a Matrix

The transpose $\mathbf{A}'$ of a matrix $\mathbf{A}$ is defined to be that matrix whose (i, j)th element a'_{ij} is equal to the (j, i)th element of $\mathbf{A}$. For example, if

$$\mathbf{A} = \begin{bmatrix} 2 & 3 & 1 \\ 1 & 1 & 2 \end{bmatrix}$$

then

$$\mathbf{A}' = \begin{bmatrix} 2 & 1 \\ 3 & 1 \\ 1 & 2 \end{bmatrix}$$

since $a'_{11} = a_{11} = 2$, $a'_{12} = a_{21} = 1$, $a'_{21} = a_{12} = 3$, $a'_{22} = a_{22} = 1$, $a'_{31} = a_{13} = 1$, $a'_{32} = a_{23} = 2$.

Another way of looking at this is that the first column of **A** becomes the first row of **A′**, the second column of **A** becomes the second row of **A′**, and so on. Thus, if **A** is $r \times c$, then **A′** is $c \times r$.

As examples, the transposes of

$$\mathbf{A}_{3\times2} = \begin{bmatrix} 1 & 2 \\ 3 & 1 \\ 1 & 1 \end{bmatrix}, \qquad \mathbf{B}_{3\times3} = \begin{bmatrix} 1 & 0 & 2 \\ 0 & 4 & -5 \\ 2 & -5 & 3 \end{bmatrix}, \qquad \mathbf{C}_{4\times1} = \begin{bmatrix} 0 \\ 1 \\ 2 \\ -2 \end{bmatrix}$$

are

$$\mathbf{A}'_{2\times3} = \begin{bmatrix} 1 & 3 & 1 \\ 2 & 1 & 1 \end{bmatrix}, \qquad \mathbf{B}'_{3\times3} = \begin{bmatrix} 1 & 0 & 2 \\ 0 & 4 & -5 \\ 2 & -5 & 3 \end{bmatrix}, \qquad \mathbf{C}'_{1\times4} = \begin{bmatrix} 0 & 1 & 2 & -2 \end{bmatrix}$$

Also, the transpose of the matrix $\mathbf{X}_{12\times2}$ of the previous section is given by

$$\mathbf{X}'_{2\times12} = \begin{bmatrix} 1 & 1 & 1 & 1 & 1 & 1 & 1 & 1 & 1 & 1 & 1 & 1 \\ 57 & 59 & 49 & 62 & 51 & 50 & 55 & 48 & 42 & 42 & 61 & 57 \end{bmatrix}$$

Note that in the examples above, the matrix **B** is such that $\mathbf{B} = \mathbf{B}'$ (equality here means that corresponding elements are equal). A matrix satisfying this condition is said to be a *symmetric matrix*. Note that a symmetric matrix **A** must always be square, since otherwise **A** and **A′** would have different dimensions and so could not possibly be equal. A necessary and sufficient condition for the square matrix $\mathbf{A} = ((a_{ij}))$ to be symmetric is that $a_{ij} = a_{ji}$ for every $i \neq j$.

Correlation matrices such as

$$\mathbf{R}_1 = \begin{bmatrix} 1 & r_{xy} \\ r_{xy} & 1 \end{bmatrix} \qquad \text{or} \qquad \mathbf{R}_2 = \begin{bmatrix} 1 & r_{12} & r_{13} \\ r_{12} & 1 & r_{23} \\ r_{13} & r_{23} & 1 \end{bmatrix}$$

are always symmetric.

An important special case where the above condition for symmetry is satisfied is when $a_{ij} = a_{ji} = 0$ for every $i \neq j$. A square matrix having this property is said to be a *diagonal matrix*, the general form of which is given (for the 3×3 case) by

$$\begin{bmatrix} a_{11} & 0 & 0 \\ 0 & a_{22} & 0 \\ 0 & 0 & a_{33} \end{bmatrix}$$

Diagonal

The most often used diagonal matrix is the *identity matrix* **I**, which has 1's on the diagonal; for example, the 3×3 identity matrix is

$$\mathbf{I} = \begin{bmatrix} 1 & 0 & 0 \\ 0 & 1 & 0 \\ 0 & 0 & 1 \end{bmatrix}$$

We will see shortly that an identity matrix serves the same algebraic function for matrix multiplication that the number 1 serves for ordinary scalar multiplication.

B-5 Matrix Addition

The sum of two matrices, say **A** and **B**, is obtained by adding together the corresponding elements of each matrix. Clearly, such addition can be performed only when the two matrices have the same dimensions. Thus, for example, we can add the matrices

$$\mathbf{A}_{2\times3} = \begin{bmatrix} 2 & 3 & 1 \\ 1 & 1 & 2 \end{bmatrix} \quad \text{and} \quad \mathbf{B}_{2\times3} = \begin{bmatrix} 4 & 1 & 5 \\ 1 & 3 & 1 \end{bmatrix}$$

(since they have the same dimensions) to get

$$\mathbf{A}_{2\times3}\mathbf{B} = \begin{bmatrix} 2+4 & 3+1 & 1+5 \\ 1+1 & 1+3 & 2+1 \end{bmatrix} = \begin{bmatrix} 6 & 4 & 6 \\ 2 & 4 & 3 \end{bmatrix}$$

We could not, however, add the matrices **A** and **C**, where

$$\mathbf{C}_{3\times2} = \begin{bmatrix} 5 & 4 \\ 1 & 4 \\ 2 & 6 \end{bmatrix}$$

since the dimensions of these matrices are not the same.

An example of a more abstract use of matrix addition would be to sum the two 12×1 vectors

$$\mathbf{D}_{12\times1} = \begin{bmatrix} \beta_0 + 57\beta_1 \\ \beta_0 + 59\beta_1 \\ \vdots \\ \beta_0 + 57\beta_1 \end{bmatrix} \quad \text{and} \quad \mathbf{E}_{12\times1} = \begin{bmatrix} E_1 \\ E_2 \\ \vdots \\ E_{12} \end{bmatrix}$$

to obtain

$$\mathbf{D}_{12\times1}\mathbf{E} = \begin{bmatrix} \beta_0 + 57\beta_1 + E_1 \\ \beta_0 + 59\beta_1 + E_2 \\ \vdots \\ \beta_0 + 57\beta_1 + E_{12} \end{bmatrix}$$

Actually, you may recognize that each element of the matrix $\mathbf{D} + \mathbf{E}$ is obtained by substituting for X in the right side of the straight-line regression equation

$$Y = \beta_0 + \beta_1 X + E$$

each of the 12 X (HGT) values given in the data set of Section B-3.

B-6 Matrix Multiplication

Multiplication of two matrices is somewhat more complicated than addition. The first rule to remember is that the product **AB** of two matrices **A** and **B** can exist if and only if the *number of columns* of **A** *is equal to the number of rows of* **B**. Thus, if **A** is 2×3 and **B** is

3×4, the product **AB** exists since **A** has 3 columns and **B** has 3 rows. However, the product **BA** does not exist, since the number of columns of **B** (i.e., 4) is not equal to the number of rows of **A** (i.e., 2). Notationally, therefore, a matrix product can exist only if the dimensions of the matrices can be represented as follows:

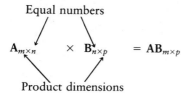

Equal numbers

$$\mathbf{A}_{m \times n} \quad \times \quad \mathbf{B}_{n \times p} \quad = \mathbf{AB}_{m \times p}$$

Product dimensions

Note also from the expression above that the dimensions $m \times p$ of the product matrix **AB** are given by the number of rows of the *pre*multiplier (i.e., **A**) and the number of columns of the *post*multiplier (i.e., **B**). For example, if **A** is 2×3 and **B** is 3×4, the dimensions of the product **AB** are 2×4.

Now, to carry out matrix multiplication, consider the two matrices

$$\mathbf{A}_{2\times3} = \begin{bmatrix} 2 & 1 & 0 \\ 0 & 3 & 1 \end{bmatrix} \quad \text{and} \quad \mathbf{B}_{3\times2} = \begin{bmatrix} 1 & -2 \\ 1 & 0 \\ 3 & 2 \end{bmatrix}$$

Since **A** is 2×3 and **B** is 3×2, the product **AB** will be a 2×2 matrix. If we let the elements of **AB** be denoted by

$$\mathbf{AB}_{2\times2} = ((c_{ij})) = \begin{bmatrix} c_{11} & c_{12} \\ c_{21} & c_{22} \end{bmatrix}$$

we can obtain the upper-left-hand corner element c_{11} by working with the first row of **A** and the first column of **B**, as follows:

$$\overset{\mathbf{A}}{\begin{bmatrix} 2 & 1 & 0 \\ 0 & 3 & 1 \end{bmatrix}} \overset{\mathbf{B}}{\begin{bmatrix} 1 & -2 \\ 1 & 0 \\ 3 & 2 \end{bmatrix}} = \overset{\mathbf{AB}}{\begin{bmatrix} (2 \times 1) + (1 \times 1) & c_{12} \\ + (0 \times 3) = 3 & \\ c_{21} & c_{22} \end{bmatrix}}$$

What we have done here is to calculate the product of each element in row 1 of **A** with the corresponding element in column 1 of **B**, and then add up these three products to obtain $c_{11} = 3$:

Column 1 of **B**

$$c_{11} = (2 \times 1) + (1 \times 1) + (0 \times 3) = 3$$

Row 1 of **A**

To find the element in the second row and first column of **AB** (i.e., c_{21}), we work with the second row of **A** and the first column of **B**, as follows:

$$
\begin{array}{ccc}
\mathbf{A} & \mathbf{B} & \mathbf{AB} \\
\begin{bmatrix} 2 & 1 & 0 \\ 0 & 3 & 1 \end{bmatrix} & \begin{bmatrix} 1 & -2 \\ 1 & 0 \\ 3 & 2 \end{bmatrix} = & \begin{bmatrix} 3 & c_{12} \\ \begin{array}{c} (0 \times 1) + (3 \times 1) \\ + (1 \times 3) = 6 \end{array} & c_{22} \end{bmatrix}
\end{array}
$$

Thus, for the element c_{21}, we find

Column 1 of **B**

$$c_{21} = (0 \times 1) + (3 \times 1) + (1 \times 3) = 6$$

Row 2 of **A**

Continuing this process, we find

$$c_{12} = (2 \times -2) + (1 \times 0) + (0 \times 2) = -4$$
$$c_{22} = (0 \times -2) + (3 \times 0) + (1 \times 2) = 2$$

Thus,

$$\mathbf{AB}_{2 \times 2} = \begin{bmatrix} 2 & 1 & 0 \\ 0 & 3 & 1 \end{bmatrix} \begin{bmatrix} 1 & -2 \\ 1 & 0 \\ 3 & 2 \end{bmatrix} = \begin{bmatrix} 3 & -4 \\ 6 & 2 \end{bmatrix}$$

In general, if $\mathbf{A}_{m \times n} = ((a_{ij}))$ and $\mathbf{B}_{n \times p} = ((b_{ij}))$, the (i, j)th element c_{ij} of the product

$$\mathbf{AB}_{m \times p} = ((c_{ij}))$$

is defined to be

$$c_{ij} = \sum_{l=1}^{n} a_{il} b_{lj} \qquad i = 1, 2, \ldots, m; \quad j = 1, 2, \ldots, p$$

Thus, as another example, if

$$\mathbf{A}_{2 \times 2} = \begin{bmatrix} -1 & 3 \\ 2 & 2 \end{bmatrix} \quad \text{and} \quad \mathbf{B}_{2 \times 1} = \begin{bmatrix} 0 \\ 1 \end{bmatrix}$$

then $m = 2$, $p = 1$, $n = 2$, and

$$c_{11} = \sum_{l=1}^{2} a_{1l} b_{l1} = (-1 \times 0) + (3 \times 1) = 3$$

$$c_{21} = \sum_{l=1}^{2} a_{2l} b_{l1} = (2 \times 0) + (2 \times 1) = 2$$

so that

$$\mathbf{AB}_{2 \times 2} = \begin{bmatrix} -1 & 3 \\ 2 & 2 \end{bmatrix} \begin{bmatrix} 0 \\ 1 \end{bmatrix} = \begin{bmatrix} 3 \\ 2 \end{bmatrix}$$

Here are a few other examples for additional practice:

1. Find **AI** and **IA**, where

$$\mathbf{A}_{2\times2} = \begin{bmatrix} -1 & 3 \\ 2 & 2 \end{bmatrix} \quad \text{and} \quad \mathbf{I}_{2\times2} = \begin{bmatrix} 1 & 0 \\ 0 & 1 \end{bmatrix}$$

2. Find **X′X**, where

$$\mathbf{X}_{3\times2} = \begin{bmatrix} 1 & 10 \\ 1 & 15 \\ 1 & 20 \end{bmatrix}$$

The answer to problem 1 is

AI = IA = A

which indicates why **I** is generally referred to as the identity matrix, since, like the scalar identity 1, the product of any square matrix (e.g., **A**) with an appropriate identity matrix (of the right dimensions) will always yield the original matrix **A**, whether **I** premultiplies or postmultiplies **A**.

The answer to problem 2 is

$$\mathbf{X'X}_{2\times2} = \begin{bmatrix} 1 & 1 & 1 \\ 10 & 15 & 20 \end{bmatrix} \begin{bmatrix} 1 & 10 \\ 1 & 15 \\ 1 & 20 \end{bmatrix} = \begin{bmatrix} 3 & 45 \\ 45 & 725 \end{bmatrix}$$

which is a symmetric matrix, as will be the case whenever any matrix is multiplied (pre or post) by its own transpose.

B-7 Inverse of a Matrix

The definition of an *inverse matrix* parallels the basic property of the *reciprocal* of an ordinary (scalar) number. That is, for any nonzero number a, its reciprocal $1/a$ satisfies the equation

$$a \times \frac{1}{a} = \frac{1}{a} \times a = 1$$

In words, when a is either pre- or postmultiplied by its reciprocal, the result is the scalar 1. Analogously, we say that a square matrix **A** has an inverse $\mathbf{A}^{-1}$ if and only if

$$\mathbf{AA}^{-1} = \mathbf{A}^{-1}\mathbf{A} = \mathbf{I}$$

That is, *the product of* **A** *by its inverse must be equal to an identity matrix.* We hasten to add at this point that *only the inverses of square matrices are being considered.* Thus, if **A** is $n \times n$, $\mathbf{A}^{-1}$ must also be $n \times n$.

We shall not describe here any of the many algorithms available for actually computing the inverse of a matrix.[1] For most practical applications, one can use standard computer packages for such computations. We wish to emphasize instead that one important attribute of inverses in statistical analyses is that their use permits the solution of matrix equations for a matrix of unknowns (e.g., the vector $\boldsymbol{\beta}$ of unknown regression parameters), in a manner similar to how division is used in ordinary algebra. Most specifically, in regression analysis, the use of an inverse is crucial because it provides a means for efficiently solving the least-squares equations, as well as providing compact formulae for additional components of the analysis (e.g., the variances of and covariances among the estimated regression coefficients).

Some examples of matrix inverses are given as follows:

$$\mathbf{A} = \begin{bmatrix} 2 & 0 & 1 \\ 1 & -1 & 2 \\ 1 & 0 & 0 \end{bmatrix} \quad \text{and} \quad \mathbf{A}^{-1} = \begin{bmatrix} 0 & 0 & 1 \\ 2 & -1 & -3 \\ 1 & 0 & -2 \end{bmatrix}$$

(The reader can check this out by multiplying $\mathbf{A}$ and $\mathbf{A}^{-1}$ together to obtain $\mathbf{I}_{3\times3}$.) The inverse of

$$\mathbf{X'X} = \begin{bmatrix} n & \sum_{i=1}^{n} X_i \\ \sum_{i=1}^{n} X_i & \sum_{i=1}^{n} X_i^2 \end{bmatrix}$$

is

$$(\mathbf{X'X})^{-1} = \begin{bmatrix} \dfrac{\sum_{i=1}^{n} X_i^2}{n \sum_{i=1}^{n} (X_i - \bar{X})^2} & \dfrac{-\bar{X}}{\sum_{i=1}^{n} (X_i - \bar{X})^2} \\[4ex] \dfrac{-\bar{X}}{\sum_{i=1}^{n} (X_i - \bar{X})^2} & \dfrac{1}{\sum_{i=1}^{n} (X_i - \bar{X})^2} \end{bmatrix}$$

B-8 Matrix Formulation of Regression Analysis

We have previously seen in Section B-3 that when fitting a straight-line model

$$Y = \beta_0 + \beta_1 X + E$$

to a set of data consisting of n pairs of observations on the variables X and Y, we can define several matrices to characterize the regression problem under consideration: $\mathbf{Y}$, the $n \times 1$

---

[1] For further details, see N. R. Draper and H. Smith, *Applied Regression Analysis* (New York: Wiley, 1966); W. Mendenhall, *Introduction to Linear Models and the Design and Analysis of Experiments* (Belmont, Calif.: Wadsworth, 1968).

vector of observations on Y; $\mathbf{X}$, the $n \times 2$ matrix of independent variables; $\boldsymbol{\beta}$, the 2×1 vector of parameters; and E, the $n \times 1$ vector of random errors.

In general, whenever we are considering a regression problem involving p independent variables using a model such as

$$Y = \beta_0 + \beta_1 X_1 + \beta_2 X_2 + \cdots + \beta_p X_p + E$$

we can analogously construct appropriate matrices based on the multivariable data set being considered. Thus, if the data on the ith individual consists of the $p + 1$ values

$$Y_i, X_{i1}, X_{i2}, \ldots, X_{ip} \qquad i = 1, 2, \ldots, n$$

the following matrices can be constructed:

$$\mathbf{Y}_{n \times 1} = \begin{bmatrix} Y_1 \\ Y_2 \\ \vdots \\ Y_n \end{bmatrix} = \begin{matrix} \text{vector of} \\ \text{observations on } Y \end{matrix}$$

$$\mathbf{X}_{n \times (p+1)} = \begin{bmatrix} 1 & X_{11} & X_{12} & \cdots & X_{1p} \\ 1 & X_{21} & X_{22} & \cdots & X_{2p} \\ \vdots & \vdots & \vdots & & \vdots \\ 1 & X_{n1} & X_{n2} & \cdots & X_{np} \end{bmatrix} = \begin{matrix} \text{matrix of} \\ \text{independent} \\ \text{variables} \end{matrix}$$

$$\boldsymbol{\beta}_{(p+1) \times 1} = \begin{bmatrix} \beta_0 \\ \beta_1 \\ \vdots \\ \beta_p \end{bmatrix} = \text{vector of parameters}$$

$$\mathbf{E}_{n \times 1} = \begin{bmatrix} E_1 \\ E_2 \\ \vdots \\ E_n \end{bmatrix} = \text{vector of random errors}$$

Using the matrices above in conjunction with the notions of matrix addition and multiplication, we can formulate the general regression model in matrix terms as follows:

$$\mathbf{Y}_{n \times 1} = \mathbf{X}_{n \times (p+1)} \boldsymbol{\beta}_{(p+1) \times 1} + \mathbf{E}_{n \times 1} \qquad (B.1)$$

This compact equation summarizes in a single statement the n equations

$$Y_i = \beta_0 + \beta_1 X_{i1} + \beta_2 X_{i2} + \cdots + \beta_p X_{ip} + E_i \qquad i = 1, 2, \ldots, n$$

Note that this equivalence follows from the following matrix calculations:

$$\mathbf{X}\boldsymbol{\beta} + \mathbf{E} = \begin{bmatrix} 1 & X_{11} & X_{12} & \cdots & X_{1p} \\ 1 & X_{21} & X_{22} & \cdots & X_{2p} \\ \vdots & \vdots & & & \vdots \\ 1 & X_{n1} & X_{n2} & \cdots & X_{np} \end{bmatrix} \begin{bmatrix} \beta_0 \\ \beta_1 \\ \vdots \\ \beta_p \end{bmatrix} + \begin{bmatrix} E_1 \\ E_2 \\ \vdots \\ E_n \end{bmatrix}$$

$$= \begin{bmatrix} \beta_0 + \beta_1 X_{11} + \beta_2 X_{12} + \cdots + \beta_p X_{1p} \\ \beta_0 + \beta_1 X_{21} + \beta_2 X_{22} + \cdots + \beta_p X_{2p} \\ \vdots \\ \beta_0 + \beta_1 X_{n1} + \beta_2 X_{n2} + \cdots + \beta_p X_{np} \end{bmatrix} + \begin{bmatrix} E_1 \\ E_2 \\ \vdots \\ E_n \end{bmatrix}$$

$$= \begin{bmatrix} \beta_0 + \beta_1 X_{11} + \beta_2 X_{12} + \cdots + \beta_p X_{1p} + E_1 \\ \beta_0 + \beta_1 X_{21} + \beta_2 X_{22} + \cdots + \beta_p X_{2p} + E_2 \\ \vdots \\ \beta_0 + \beta_1 X_{n1} + \beta_2 X_{n2} + \cdots + \beta_p X_{np} + E_n \end{bmatrix}$$

Based on the general matrix equation given by (B.1), a description of all the essential features of regression analysis can be expressed in matrix notation. In particular, the least-squares solution for the estimates of the regression coefficients in the parameter vector $\boldsymbol{\beta}$ can now be compactly written. This least-squares solution, in matrix terms, is that vector $\hat{\boldsymbol{\beta}}$ which minimizes the error sum of squares (given in matrix terms)

$$(\mathbf{Y} - \mathbf{X}\hat{\boldsymbol{\beta}})'(\mathbf{Y} - \mathbf{X}\hat{\boldsymbol{\beta}})$$

The solution to this minimization problem, which is obtained via the use of matrix calculus, yields the following easy-to-remember matrix formula:

$$\hat{\boldsymbol{\beta}} = (\mathbf{X}'\mathbf{X})^{-1}\mathbf{X}'\mathbf{Y}$$

where $\hat{\boldsymbol{\beta}}'_{1 \times (p+1)} = [\hat{\beta}_0 \quad \hat{\beta}_1 \quad \ldots \quad \hat{\beta}_p]$ denotes the vector of estimated regression coefficients.

Thus, although we have pointed out in the text the futility of *explicitly* giving the solutions to the least-squares equations for models of more complexity than a straight line, we can at least *implicitly* express these solutions in matrix notation and, because of modern computer technology, conveniently carry through with the computation of the least-squares solutions using this matrix representation.

At this point it is not our intention to carry through completely with the matrix formulation of every other aspect of a regression analysis. The reader is referred to Draper and Smith (1966; see footnote 1) for a fuller treatment of this matrix approach. However, some additional matrix results will be summarized below to give more of an indication of the utility of the matrix approach:

1. The *vector of predicted responses* is given by $\hat{\mathbf{Y}} = \mathbf{X}\hat{\boldsymbol{\beta}}$.

2. The *error sum of squares* SSE is given by SSE $= \mathbf{Y}'\mathbf{Y} - \hat{\boldsymbol{\beta}}'\mathbf{X}'\mathbf{Y}$.

3. The *regression sum of squares* is given by $\hat{\boldsymbol{\beta}}'\mathbf{X}'\mathbf{Y} - n\bar{Y}^2$.

4. The *variances of the regression coefficients*, that is, $\sigma^2_{\hat{\beta}_j}$, $j = 0, 1, 2, \ldots, p$, are given by the diagonal elements of the matrix $(\mathbf{X}'\mathbf{X})^{-1}\sigma^2$.

Finally, although we do not present the details here, any test of a (linear) statistical hypothesis concerning some subset of the regression coefficients can be formulated and carried out entirely by means of matrix operations.

Appendix - Solutions to Exercises

Chapter 3

2. nominal, ordinal, interval, ratio

3. **a.** .8413 **b.** -0.842

4. **a.** 18.475 **b.** .699

5. **a.** -1.350 **b.** .05

6. **a.** 2.51 **b.** .025

7. **a.** 0 **b.** 0 **c.** 0

8. standard normal

9. **a.** 3.0 **b.** 3 **c.** 2.8 or 3.11

10. e

11. **a.** 5.0 **b.** (187.44, 192.56)

12. $t_{0.975,27} = 2.052$

13. (24.66, 35.33)

14. 3.6858

15. **a.** significant difference **b.** significant difference

16. nonsignificant difference

17. $t_{0.995,10} = 3.169$

18. b

19. **a.** Type I error **b.** correct decision **c.** correct decision **d.** Type II error

20. a, b

21. c

22. b

23. a. $1 - \alpha$ **b.** α **c.** β **d.** $1 - \beta$

24. Accept H_0.

25. b

Chapter 5

1. c. $\hat{\beta}_0(Y|X) = -1.884,$ $\hat{\beta}_1(Y|X) = 0.235,$
$\hat{\beta}_0(Z|X) = -2.690,$ $\hat{\beta}_1(Z|X) = 0.196$

d. 95% CI for $\beta_1(Y|X)$: $0.131 < \beta_1 < 0.339$
95% CI for $\beta_1(Z|X)$: $0.191 < \beta_1 < 0.201$
For $\alpha = .05$, reject H_0: $\beta_1 = 0$ for both regression equations in favor of H_A: $\beta_1 \neq 0$, since neither CI contains the value 0.

e. For 90% confidence bands, use

$$Y \text{ on } X: \quad 0.701 + 0.235(X_0 - 11) \pm 0.884 \sqrt{0.091 + \frac{(X_0 - 11)^2}{110}}$$

$$Z \text{ on } X: \quad -0.535 + 0.196(X_0 - 11) \pm 0.052 \sqrt{0.091 + \frac{(X_0 - 11)^2}{110}}$$

for various values of X_0.

For 90% prediction bands, use

$$Y \text{ on } X: \quad 0.701 + 0.235(X_0 - 11) \pm 0.884 \sqrt{1.091 + \frac{(X_0 - 11)^2}{110}}$$

$$Z \text{ on } X: \quad -0.535 + 0.196(X_0 - 11) \pm 0.052 \sqrt{1.091 + \frac{(X_0 - 11)^2}{110}}$$

f. The straight-line regression of Z on X produces a better fit, since the relationship between Y and X is apparently not linear.

2. c. (1) $\hat{\beta}_1 = 7.024$, $\hat{\beta}_0 = 140.800$

(2) $\hat{\beta}_0$ and $\bar{Y}$(nonsmokers) have the same value. $(\hat{\beta}_0 + \hat{\beta}_1)$ and $\bar{Y}$(smokers) have the same value. The straight-line regression problem is, in this case, equivalent to the two-sample problem of comparing two population means. In fact, it can be shown by substituting $X = 0$ and $X = 1$ separately into the straight-line model that β_0 is equivalent to μ(nonsmokers) and $(\beta_0 + \beta_1)$ is equivalent to μ(smokers).

(3) Computed $T = 1.398$. Do not reject H_0 since $.10 < P < .2$.

(4) yes

d. (1) $\hat{\beta}_1 = 21.492$, $\hat{\beta}_0 = 70.576$

(3) Computed $T = 6.062$. Reject H_0 since $P < .001$.

(4) 95% CI for $\mu_{Y|\bar{X}}$: $140.99 < \mu_{Y|\bar{X}} < 148.07$

(5) For 95% prediction bands, use the following formula:

$$144.531 + 21.492(X_0 - 3.441) \pm 20.035 \sqrt{1.031 + \frac{(X_0 - 3.441)^2}{7.660}}$$

(6) Yes, since H_0 is rejected in part (3), which agrees with one's visual impression of the data.

(7) not obviously

3. **f.** straight-line-model assumption

   h. The data suggest that a concave downward parabola would provide a better fit.

4. **c.** The reliability of a predicted value $\hat{Y}$ is questionable for any X outside the range of X over which the fitted model is based.

   d. 95% CI for slope: $-0.494 < \beta_1 < -0.004$

   f. The presence of the outlier appears to make a considerable difference, since the negative straight-line relationship between DI and IQ is much more pronounced and the fit of the line to the data improves considerably when the outlier is removed.

6. **g.** Substitute various values of X_0 into the following formula:

$$519.304 - 14.041(X_0 - 9.058) \pm 28.948 \sqrt{0.077 + \frac{(X_0 - 9.058)^2}{92.424}}$$

7. **d.** Computed $T = -57.813$, which lies below $t_{17,0.005} = -2.898$. Thus, for a two-sided test, reject H_0 at $\alpha = .01$.

   f. Use for the construction of 95% confidence and prediction bands the following formulae:

   95% confidence bands: $10.175 + 0.299(X_0 - 39.737)$

$$\pm 1.711 \sqrt{0.053 + \frac{(X_0 - 39.737)^2}{4{,}473.684}}$$

   95% prediction bands: $10.175 + 0.299(X_0 - 39.737)$

$$\pm 1.711 \sqrt{1.053 + \frac{(X_0 - 39.737)^2}{4{,}473.684}}$$

9. **d.** Use the following formula to construct 99% confidence bands:

$$1.6108 - 1.785(X_0 - 0.7425) \pm 0.1762 \sqrt{0.0833 + \frac{(X_0 - 0.7425)^2}{1.1264}}$$

   e. $\hat{Y}' = 862.978 X'^{-1.785}$

   f. At maximum dosage ($X_0 = 1.15$), the 99% CI for $\mu_{Y|X_0}$ is given by

   $0.7983 < \mu_{Y|X_0=1.15} < 0.9685$

   Raising each limit to a power of 10, we obtain the following 99% CI for $\mu_{Y'|X_0'=14.125}$:

   $6.2849 < \mu_{Y'|X_0'=14.125} < 9.3004$

   At minimum dosage ($X_0 = 0.18$), the corresponding 99% CIs are

   $2.5080 < \mu_{Y|X_0=0.18} < 2.7218$

   and

   $322.107 < \mu_{Y'|X_0'=1.514} < 526.987$

g. Plot the scatter diagram of the (X', Y') pairs and draw the fitted straight line of Y' on X' on the scatter. Then compare by eye whether using Y on X gives better fit to the (X, Y) scatter than does using Y' on X' on the (X', Y') scatter. In addition, using methods described in Chapter 7, you could perform a test for lack of fit of each model and compare results.

10. c. 94% CIs for $\mu_{Y'|X}$ at $X = 5$, 4.5, and 4: $280.802 < \mu_{Y'|X=5} < 352.858$, $931.108 < \mu_{Y'|X=4.5} < 1{,}059.25$, and $2{,}795.12 < \mu_{Y'|X=4} < 3{,}512.37$

11. c. Growth rate scores for each of the three tests for each gas would provide more information.
    d. 90% CI: $2.26 < \mu_{Y|100} < 2.69$
    e. No observations were taken at $X = 200$ or more (i.e., the value 200 is outside the region of experimentation).
    f. The X-values chosen do not uniformly cover the region of experimentation (notice the big gaps between $X = 39.9$, $X = 83.8$, and $X = 131.3$).

Chapter 6

1. a. $r_{XY} = .862$, $r_{XZ} = .999$
   b. $.543 < \rho_{XY} < .960$, $.992 < \rho_{XZ} < .9995$
   c. $r_{XY}^2 = (8.166 - 2.090)/8.166 = .744$, $r_{XZ}^2 = (4.233 - 0.0071)/4.233 = .998$
   d. The regression of log dry weight on age provides a better fit, since it explains a much greater proportion of the total variation in the dependent variable than does the regression of dry weight on age.

2. d. The quantities $\hat{\beta}_1$ and r have meaning only in the context of an assumed *linear* relationship between X and Y, whereas the true model is a parabola containing no linear component.

3. a. $r = .9798$
   c. Computed $T' = 13.85$ (df = 8). Reject H_0 since $P < .001$.
   d. No observations have been taken between $X = 3$ and $X = 20$, thus making the point $(20, 20)$ an outlier in the scatter diagram. It is dangerous to extrapolate to a point far outside the range of most of the data.

4. b. 99% CI for ρ: $.4435 < \rho_{SBP,QUET} < .8923$. Since this CI does not contain 0, reject H_0: $\rho = 0$ at $\alpha = .01$.

8. a. Including outlier: $r_{DI,IQ} = -.4747$, $r_{DI,IQ}^2 = .2253$. Excluding outlier: $r_{DI,IQ} = -.8069$, $r_{DI,IQ}^2 = .6511$.
   b. Excluding outlier: computed $T = -5.291$ (df = 16). Reject H_0: $\rho = 0$ since $P < .001$.

13. b. 99% CI for ρ: $.847 < \rho_{UV} < .997$

Chapter 7

1. a.

Source	df	SS	MS	F
Regression	1	6.0748	6.0748	26.12
Residual	9	2.0936	0.2326	
Total	10	8.1684		

2. For the regression of SBP on QUET:

a.

Source	df	SS	MS	F
Regression	1	3,537.95	3,537.95	36.75
Residual	30	2,888.02	96.27	
Total	31	6,425.97		

3. For the regression of QUET on AGE:

a.

Source	df	SS	MS	F
Regression	1	4.9361	4.9361	54.36
Residual	30	2.7236	0.0908	
Total	31	7.6597		

b. Reject H_0 since $P < .001$.

12. b. Fitted straight line: $\hat{Y} = -0.4624 + 0.0358X$

c.

Source	df	SS	MS	F
Regression	1	12.8304	12.8304	632.04
Residual	18	0.3662	0.0203	

d. Computed $F = 632.04$ (1 and 18 df), $P < .001$. Conclude significant straight-line regression.

Chapter 8

1. a. (1) $\hat{Y}(\text{AGE} = 50, \text{SMK} = 1, \text{QUET} = 3.5) = 145.77$
  (2) $\hat{Y}(\text{AGE} = 50, \text{SMK} = 0, \text{QUET} = 3.5) = 135.83$
  (3) $\hat{Y}(\text{AGE} = 50, \text{SMK} = 1, \text{QUET} = 3.5) - \hat{Y}(\text{AGE} = 50, \text{SMK} = 1, \text{QUET} = 3.0) = 4.30$
b. $R^2(\text{model 1}) = .601$; $R^2(\text{model 2}) = .730$; $R^2(\text{model 3}) = .761$
c. Model 1: $H_0: \beta_1 = 0$ for the model $Y = \beta_0 + \beta_1 X_1 + E$; $F = 45.177$ (1 and 30 df); $P < .001$; reject H_0.
    Model 2: $H_0: \beta_1 = \beta_2 = 0$ for the model $Y = \beta_0 + \beta_1 X_1 + \beta_2 X_2 + E$; $F = 39.164$ (2 and 29 df), $P < .001$; reject H_0.
    Model 3: $H_0: \beta_1 = \beta_2 = \beta_3 = 0$ for the model $Y = \beta_0 + \beta_1 X_1 + \beta_2 X_2 + \beta_3 X_3 + E$; $F = 29.710$ (3 and 28 df), $P < .001$; reject H_0.
4. b. The model with X_2 and X_3, since it has the highest overall F-value.
  d. $F = 23.10$ (3 and 16 df), $P < .001$
6. a. $F = 80.87$ (3 and 21 df), $P < .001$
  b. $R^2 = .92$

7. e. X_1, because it has a higher (though negative) correlation with Y ($r = -.843$) than does X_2 ($r = .807$). Equivalently, X_1 is more highly significant from part (b) than is X_2 from part (c).

f. Best model contains both X_1 and X_2.

g. $X_1 = .043$

h. $X_1 = X_1^0 + (\hat{\beta}_2/\hat{\beta}_1)X_2^0$

Chapter 9

2. a. Testing H_0: $\rho_{YX_1} = 0$ or H_0: $\beta_1 = 0$ in the model $Y = \beta_0 + \beta_1 X_1 + E$ yields $F(X_1) = 25.758$ (1 and 20 df), $P < .001$.

Testing H_0: $\rho_{YX_2|X_1} = 0$ or H_0: $\beta_2 = 0$ in the model $Y = \beta_0 + \beta_1 X_1 + \beta_2 X_2 + E$ yields partial $F(X_2|X_1) = 6.322$ (1 and 19 df), $.01 < P < .025$.

Testing H_0: $\rho_{YX_3|X_1, X_2} = 0$ or H_0: $\beta_3 = 0$ in the model $Y = \beta_0 + \beta_1 X_1 + \beta_2 X_2 + \beta_3 X_3 + E$ yields partial $F(X_3|X_1, X_2) = 5.267$ (1 and 18 df), $.025 < P < .05$.

b. H_0: $\rho_{Y(X_2, X_3)|X_1} = 0$ or H_0: $\beta_2 = \beta_3 = 0$ in the model $Y = \beta_0 + \beta_1 X_1 + \beta_2 X_2 + \beta_3 X_3 + E$. Multiple-partial $F(X_2, X_3|X_1) = 6.505$ (2 and 18 df), $.005 < P < .01$.

c. Variables in the given ANOVA table are not entered in the necessary order. We require that X_3 enter first, and then that X_1 and X_2 enter in any order. Use the formula

$$\text{multiple-partial } F(X_1, X_2|X_3)$$
$$= \frac{[\text{regression SS}(X_1, X_2, X_3) - \text{regression SS}(X_3)]/2}{\text{MS residual } (X_1, X_2, X_3)}$$

which is distributed as an F with 2 and 18 df under H_0.

4. c. $Y = \beta_0 + \beta_1 X_1 + \beta_2 X_2 + \beta_3 X_3 + E$; $Y = \beta_0 + \beta_1 X_1 + \beta_2 X_2 + \beta_3 X_3 + \beta_4 X_1 X_3 + \beta_5 X_2 X_3 + E$; H_0: $\beta_4 = \beta_5 = 0$. Multiple-partial $F(X_1 X_3, X_2 X_3|X_1, X_2, X_3) = 0.362$ (2 and 36 df), $P > .25$.

Conclusion: The regression equation of Y on X_1 and X_2 when SPB is absent is parallel to the equation of Y on X_1 and X_2 when SPB is present (i.e., corresponding regression coefficients are the same).

d. Partial $F(X_3|X_1, X_2) = 6.855$ (1 and 38 df), $.01 < P < .025$.

Conclusion: The regression equation when SPB is absent is not coincident with the equation when SPB is present ($\alpha = .05$).

e. SPB is associated with HDL measurement, but the other two independent variables are not and so should not be included in the model.

Chapter 10

1. a. AGE

b. $r_{\text{SBP,SMK}|\text{AGE}} = .568$; $r_{\text{SBP,QUET}|\text{AGE}} = .318$. The variable SMK should be considered next, since it has the higher partial correlation. However, this variable should not be included as an important predictor of Y unless it provides significant predictive power (as measured by a test of hypothesis) over that already achieved by AGE.

c. $H_0: \rho_{SBP,SMK|AGE} = 0$; partial $F(SMK|AGE) = 13.830$ (1 and 29 df), $P < .001$

d. $r_{SBP,QUET|AGE,SMK} = .340$; partial $F(QUET|AGE, SMK) = 3.648$ (1 and 28 df), $.05 < P < .1$

e. AGE (most important), then SMK, and finally QUET. For $\alpha = .05$, only AGE and SMK would be considered important. For $\alpha = .01$, only AGE would be considered important enough to be included in the model.

f. $r^2_{SBP(QUET,SMK)|AGE} = .4009$; multiple-partial $F(SMK, QUET|AGE) = 9.371$ (2 and 28 df), $P < .001$

2. a. For $H_0: \rho_{YX_3|X_1} = 0$, $T = 1.152$ (85 df); $P > .20$; do not reject H_0.
   For $H_0: \rho_{YX_3|X_1,X_2} = 0$, $T = 1.117$ (84 df); $P > .20$; do not reject H_0.

b.

Source	df	SS	MS
X_1	1	0.8087	0.8087
$X_2|X_1$	1	0.0344	0.0344
$X_3|X_1, X_2$	1	0.2958	0.2958
Residual	84	19.9111	0.2370

c. $H_0: \rho_{Y(\text{interaction terms})|\text{main effects}} = 0$; multiple-partial $F(\text{interaction terms}|\text{main effects}) = 0.7488$ (25 and 54 df), $P > .25$

d. overall $F(X_1) = 1.600$ (1 and 86 df), $.10 < P < .25$
   overall $F(\text{main effects}) = 1.4435$ (8 and 79 df), $.10 < P < .25$
   overall $F(\text{main effects} + \text{interactions}) = 0.8894$ (33 and 54 df), $P < .25$

e. No evidence of a relationship has been found.

3. a. Compute and interpret $r_{XY|Z}$, and test $H_0: \rho_{XY|Z} = 0$.

b. (1) $r_{YX_1|X_2} = .028$; $r_{YX_1|X_3} = .184$; $r_{YX_1|X_2,X_3} = -.1265$.

   (2) All these partial correlations are small, suggesting a spurious correlation of .35 between unemployment level and respiratory morbidity rate. The tests of hypotheses concerning the significance of these partial correlations are all nonsignificant:
   Testing $H_0: \rho_{XY_1|X_2} = 0$ yields $T = 0.131$ (22 df), $P > .8$.
   Testing $H_0: \rho_{YX_1|X_3} = 0$ yields $T = 0.878$ (22 df), $P > .2$.
   Testing $H_0: \rho_{YX_1|X_2,X_3} = 0$ yields $T = -0.584$ (21 df), $P > .4$.

8. a. $r_{YX_2|X_1} = .764$; $r_{YX_3|X_1} = -.727$

b. X_2 should be entered next, provided that it significantly adds to the prediction of Y.

c. Computed $T = 3.349$ (8 df), $.01 < P < .02$ (two-sided)

d. $r^2_{Y(X_2,X_3)|X_1} = .7134$; multiple-partial $F(X_2, X_3|X_1) = 8.713$ (2 and 7 df), $.01 < P < .025$

Chapter 11

1. a. $WGT = \beta_0 + \beta_1 HGT + \beta_2 AGE + \beta_3 (AGE)^2 + E$

b. $\hat{\beta}_1[\text{model with HGT, AGE and } (AGE)^2] = 0.724$; $\hat{\beta}_1(\text{model with HGT only}) = 1.073$. These two regression coefficients are meaningfully different. Therefore, there is confounding due to AGE and/or $(AGE)^2$ using the regression coefficient as the criterion.

c. Yes. The regression coefficient $\hat{\beta}_1$ changes from 0.724 for the initial (full) model to 0.722 when $(AGE)^2$ is dropped from the model. This is a very slight change.

Consequently, $(AGE)^2$ does not need to be controlled in order to control confounding.

d. No. $(AGE)^2$ may be dropped because it does not meaningfully increase precision once AGE and HGT are already in the model. The partial $F[(AGE)^2|HGT,AGE] = 0.01$, which is clearly nonsignificant. Also, the 95% confidence interval for $\hat{\beta}_1$ is given by

$$\hat{\beta}_1 \pm 2.262 S_{\hat{\beta}_1} = (.097, 1.351)$$

for the full model, whereas the corresponding confidence interval is $(0.132, 1.312)$ for the model that contains HGT and AGE but not $(AGE)^2$. The latter confidence interval is narrower than the former; thus dropping $(AGE)^2$ from the model increases the precision.

e. $Y = \beta_0 + \beta_1 HGT + \beta_2 AGE + E$. The variable AGE cannot be dropped because of confounding. Also, the partial F statistic $F(AGE|HGT)$ equals 4.79, which is of borderline significance $(.05 < P < .1)$; thus AGE also helps precision.

f. $Y = \beta_0 + \beta_1 HGT + \beta_2 AGE + \beta_3 (AGE)^2 + \beta_4 (HGT \times AGE) + \beta_5 [HGT \times (AGE)^2] + E$

g. Test $H_0: \beta_4 = \beta_5 = 0$ in the modified (interaction) model. Use a multiple-partial F statistic

$$F[HGT \times AGE, HGT \times (AGE)^2|HGT, AGE, (AGE)^2]$$

which has 2 and $(n - 6)$ df.

2. a. There is confounding because $\hat{\beta}_1$ changes when X_2 is dropped from the model.
    b. There is no confounding because $r_{YX_1|X_2} = r_{YX_1}$.
    c. Different conclusions about confounding are possible depending on which measure of association is used.

3. a. There is no confounding because $\hat{\beta}_1$ does not change when X_2 is dropped from the model.
    b. There is confounding because $r_{YX_1|X_2} \neq r_{YX_1}$.
    c. Same answer as in Problem 2(c).

4. f. No, since we do not know how $\hat{\beta}_2$ changes when X_1 is removed from the model.

5. c. $H_0: \rho_{Y(X_1X_3, X_2X_3)|X_1, X_2, X_3} = 0$; multiple-partial $F(X_1X_3, X_2X_3|X_1, X_2, X_3) = 0.362$ (2 and 36 df), $P > .25$.
      Conclusion: The regression equation of Y on X_1 and X_2 when SPB is absent is parallel to the equation of Y on X_1 and X_2 when SPB is present (i.e., corresponding regression coefficients are the same).
    d. $H_0: \rho_{YX_3|X_1X_2} = 0$; partial $F(X_3|X_1, X_2) = 6.855$ (1 and 38 df), $.01 < P < .025$.
      Conclusion: The regression equation when SPB is absent is not coincident with the equation when SPB is present $(\alpha = .05)$.

Chapter 12

1. a. $r^2 = .7442$, $F = 26.18$, $P = .0006$, significant
    d. For studentized residuals, Shapiro–Wilks' test of normality $P = .295$, not significant. For jackknife residuals, $P = .039$, but see comment in 1(f), noting

$n = 11$. Largest studentized residual is 2.35, which is less than 3.50 (critical value from Appendix Table A-8 for $\alpha = .05$, $N = 15$). Largest jackknife residual is 3.57, which is less than 3.90. Note: Since the largest jackknife is not significantly unusual, can use next largest sample size (15) without interpolating.

e. No particular observation raises questions. However, plot of residuals versus predictor is obviously curvilinear, indicating need for different model.

f. Residuals are t distributed, as well as correlated. As sample size increases, t goes to Z, so problem lessens quickly.

2. a. $r^2 = .8000$, $F = 36.00$, $P = .0002$, significant

d. Studentized residuals give Shapiro–Wilks' $P = .843$, and jackknife residuals give $P = .963$; both are not significant. Largest studentized residual is -1.73 and largest jackknife residual is -2.00; each is far from required value in Appendix Table A-8.

e. No particular observation raises questions. However, plot of residuals against predictor gives pattern of two straight lines, one below AGE = 12 and one above. May alternatively be interpreted as random scatter, assuming these are hypothetical data.

f. Logarithmic transformation of weight seems superior, but residual plot is still troublesome.

3. a. $r^2 = .1853$, $F = 4.093$, $P = .0582$

d. Studentized Shapiro–Wilks' $P < .01$, jackknife Shapiro–Wilks' $P < .01$. For studentized residuals, critical value is 3.51 ($\alpha = .05$), compared with 2.206 for largest value. Similarly, for jackknife residuals, critical value is 3.54 compared with observed value of -1.914.

e. No particular observation raises questions. However, residual distribution is skewed. Large values either show pattern, or randomness, possible with small sample and weak or nonexistent relationship in plot of residuals versus predictor.

4. a. $r^2 = .2253$, $F = 4.65$, $P = .0465$

d. Studentized Shapiro–Wilks' $P = .091$; jackknife Shapiro–Wilks' $P < .01$. For studentized residuals, Appendix Table A-8 gives .05 critical values of 3.58 and 3.51 for $n = 15$ and $n = 20$. Since the largest studentized residual is 3.14, the test for an outlier being present is not significant. For jackknife, corresponding values are 3.65, 3.54, and 4.90, which is significant, indicating the need for further attention.

e. Observation 17 (DI = 39.60, IQ = 134) seems detached from the other data (see plot of residuals versus predictor). The data should be checked again for accuracy. Results could be computed without the observation and might change, thereby casting uncertainty on any interpretation.

15. a. $r^2 = .9497$, $F = 1095.17$, $P < .0001$

d. Kolmogorov test of normality gives $P < .01$ for both studentized and jackknife residuals. Critical value is 3.53 ($\alpha = .05$) or 4.03 ($\alpha = .01$) for both types of residuals (Appendix Table A-8), versus 3.65 and 4.13 observed maximum values.

e. One observation is highlighted. The plot of residuals versus predictor shows marked heterogeneity and also curvilinearity, indicating specification-of-model problems.

16. a. $r^2 = .9813$, $F = 3046.52$, $P < .0001$

  d. Kolmogorov test gives $P = .105$ (studentized) and .093 (jackknife). Largest studentized residual is 2.81, and largest jackknife is 3.00; both are far from the critical value.

  e. No observation seems troublesome. Plot of residuals versus predictor have achieved homogeneity, but some small curvilinearity (quadratic) is present. May not matter since $r^2 = .9813$, so only 2% of variance is left.

  f. Logarithms are far superior. Obvious only by residual analysis since $r^2 = .9497$ for original data.

17. a. $r^2 = .9165$, $F = 636.825$, $P < .001$

  d. Kolmogorov test of normality gives $P < .01$ for both studentized and jackknife residuals. Largest jackknife residual value is 4.05, which is greater than the .05 critical value of 3.53.

  e. Same answer as 15(e).

18. a. $r^2 = .9816$, $F = 3086.84$, $P < .0001$

  d. Kolmogorov test of normality gives $P > .15$ for both studentized and jackknife residuals. Largest values are 2.41 and 2.51, which are even less bothersome than in Problem 17.

  e. No observation is troublesome. Plot of residual versus predictor shows homogeneity and no obvious pattern. A periodic sine wave pattern does not seem plausible for these nonordered data.

22. a. $R^2 = .8463$, $F_{0.99,10,35} = 3.215 < 19.27$, significant

  b. .8867, .8875, .9550, .9745, .9290, .9809, .4952, .9606, .9585, .4962

  c. All are at least mildly suspicious, except MAXH and WIND. AVST is the worst.

Chapter 13

1. c.

Source		df	SS	MS	F
Regression		1	12.7050	12.7050	43.69
Residual	lack of fit	4	4.4196	1.1049	57.03
	pure error	12	0.2325	0.0194	
Total		17	17.3571		

d.

Source		df	SS	MS	F
Regression	degree 1 (X)	1	12.7050	12.7050	43.69
	degree 2 (X²|X)	1	3.9051	3.9051	78.46
Residual	lack of fit	3	0.5145	0.1715	8.85
	pure error	12	0.2325	0.0194	
Total		17	17.3571		

e. $r_{XY}^2 = 0.732$; R^2(quadratic regression of Y on X) $= .957$

f. Test for significance of straight-line regression of Y on X: $F = 12.705/$ $0.2908 = 43.69$ (1 and 16 df), $P < .001$; reject H_0. Test for adequacy of straight-line model: $F = 1.1049/0.0194 = 57.03$ (4 and 12 df), $P < .001$; reject H_0; conclude straight-line model is not adequate.

g. Test for significance of quadratic regression: $F = 8.3051/0.0498 = 166.77$ (2 and 15 df), $P < .001$; reject H_0. Test for addition of X^2 term: partial $F(X^2|X)$ $= 3.9051/0.0498 = 78.42$ (1 and 15 df), $P < .001$; reject H_0. Test for adequacy of quadratic model: $F = 0.1715/0.0194 = 8.85$ (3 and 12 df), $.001 < P$ $< .005$; reject H_0; that is, conclude quadratic model is not adequate.

h. Test for significance of straight-line regression of ln Y on X: $F = 4550.35$, $P <$ $.001$; reject H_0. Test for straight-line model adequacy of ln Y on X: $F = 0.6819$ (4 and 12 df), $P > .25$; conclude that model is adequate.

i. R^2(straight-line regression of ln Y on X) $= .996$; R^2(quadratic regression of Y on X) $= .957$. The straight-line fit of ln Y on X provides a better fit to the data than the quadratic model of Y on X.

j. (1) Homoscedasticity assumption appears to be much more reasonable when using ln Y on X than when using Y on X.

(2) Straight-line regression of ln Y on X is preferred: It gives a higher R^2, is an adequate model, satisfies assumption of homoscedasticity, and provides a better graphical fit.

k. Independence assumption would be violated.

7. b. Straight-line regression: $F = 8.218$ (1 and 10 df), $.01 < P < .025$. Addition of X^2: $F = 2.250$ (1 and 9 df), $.10 < P < .25$. Addition of X^3: $F = 12.451$ (1 and 8 df); $.005 < P < .01$.

c. The cubic model is best for several reasons: It appears to fit the scatter better, the addition of X^3 to the model is highly significant, and the change in R^2 is considerable when going from degree 1 ($R^2 = .451$) to degree 2 ($R^2 = .561$) to degree 3 ($R^2 = .828$).

Chapter 14

1. a. smokers: $\widehat{SBP} = 79.26 + 20.12QUET$
   nonsmokers: $\widehat{SBP} = 49.31 + 26.30QUET$

b. t test for equal slopes (one-sided): $T = 0.892$, $.1 < P < .2$; do not reject H_0.

c. t test for equal intercepts (two-sided): $T = -1.248$, $.2 < P < .4$; do not reject H_0.

d. Since the tests for slope and intercept did not lead to rejection, we conclude that the two lines are coincident (recognizing, nevertheless, that a more efficient test is possible using multiple regression analysis).

2. e. Z test for equal correlations (two-sided): $Z = 0.594$, $P = .56$; do not reject H_0. The test for equality of slopes is not equivalent to the test for equality of correlations; that is, equal slopes do not imply equal correlations.

6. a. males: $\hat{Y} = 13.767 + 14.966X$
   females: $\hat{Y} = 15.656 + 12.735X$

b. t test for equal slopes (two-sided): $T = 0.82$ (28 df), $.4 < P < .6$; do not reject H_0: "lines are parallel."

c. 99% CI for $\beta_{1M} - \beta_{1F}$: $-5.264 < \beta_{1M} - \beta_{1F} < 9.726$

d. t test for equal intercepts (two-sided): $T = 0.69$ (28 df), $.4 < P < .6$; do not reject H_0: "equal intercepts." Since both the test for equal intercepts and the test for equal slopes are not rejected, conclude that there is no evidence that the lines are not coincident.

7. d. 95% CI for $\mu_{Y^A|15} - \mu_{Y^B|15}$: $4.201 < \mu_{Y^A|15} - \mu_{Y^B|15} < 4.799$

8. a. SBP $= \beta_0 + \beta_1 \text{QUET} + \beta_2 \text{SMK} + \beta_3(\text{QUET} \times \text{SMK}) + E$, where

$$\text{SMK} = \begin{cases} 1 & \text{if smoker} \\ 0 & \text{if nonsmoker} \end{cases}$$

smokers: SBP $= (\beta_0 + \beta_2) + (\beta_1 + \beta_3)\text{QUET} + E$
nonsmokers: SBP $= \beta_0 + \beta_1\text{QUET} + E$

b. smokers: $\widehat{\text{SBP}} = 79.253 + 20.118\text{QUET}$
nonsmokers: $\widehat{\text{SBP}} = 49.312 + 26.303\text{QUET}$

c. H_0: $\beta_3 = 0$ (parallelism): partial $F(\text{QUET} \times \text{SMK}|\text{QUET, SMK}) = 0.796$ (1 and 28 df), $P > .25$

H_0: $\beta_2 = \beta_3 = 0$ (coincidence): multiple-partial $F(\text{SMK, QUET} \times \text{SMK}|\text{QUET}) = 4.03$ (2 and 28 df), $.025 < P < .05$

d. Test for parallelism gives the same result as previously obtained. Test for coincidence gives a different result for $\alpha = .05$.

10. a. males: $\widehat{\text{SBPSL}} = (\hat{\beta}_0 + \hat{\beta}_2) + (\hat{\beta}_1 + \hat{\beta}_4)\text{SBP1} + (\hat{\beta}_3 + \hat{\beta}_6)\text{RW}$
$\qquad\qquad + (\hat{\beta}_5 + \hat{\beta}_7)(\text{SBP1} \times \text{RW})$
females: $\widehat{\text{SBPSL}} = (\hat{\beta}_0 - \hat{\beta}_2) + (\hat{\beta}_1 - \hat{\beta}_4)\text{SBP1} + (\hat{\beta}_3 - \hat{\beta}_6)\text{RW}$
$\qquad\qquad + (\hat{\beta}_5 - \hat{\beta}_7)(\text{SBP1} \times \text{RW})$

b. Variables are entered in the given ANOVA table in an inappropriate order for making the required tests. To carry out either the test for parallelism or the test for coincidence, the correct order of entry should be X_1, X_3, and X_5 followed by X_2, X_4, X_6, and X_7.

In testing for parallelism (H_0: $\beta_4 = \beta_6 = \beta_7 = 0$), the test statistic to be used is the multiple-partial $F(X_4, X_6, X_7|X_1, X_2, X_3, X_5)$, which has the F distribution with 3 and 96 df under H_0.

In testing for coincidence (H_0: $\beta_2 = \beta_4 = \beta_6 = \beta_7 = 0$), the test statistic to be used is the multiple-partial $F(X_2, X_4, X_6, X_7|X_1, X_3, X_5)$, which has the F distribution with 4 and 96 df under H_0.

13. a. $R = 1$ and TD $= 1$: $\widehat{\text{SBPSL}} = (\hat{\beta}_0 + \hat{\beta}_2 + \hat{\beta}_4 + \hat{\beta}_9) + \hat{\beta}_1\text{SBP1} + (\hat{\beta}_6 + \hat{\beta}_{11})\text{RW}$
$R = 0$ and TD $= 1$: $\widehat{\text{SBPSL}} = (\hat{\beta}_0 + \hat{\beta}_3 + \hat{\beta}_4 + \hat{\beta}_7) + \hat{\beta}_1\text{SBP1} + (\hat{\beta}_6 + \hat{\beta}_{13})\text{RW}$
$R = -1$ and TD $= 1$: $\widehat{\text{SBPSL}} = (\hat{\beta}_0 - \hat{\beta}_2 - \hat{\beta}_3 + \hat{\beta}_4 - \hat{\beta}_7 - \hat{\beta}_9) + \hat{\beta}_1\text{SBP1}$
$\qquad\qquad + (\hat{\beta}_6 - \hat{\beta}_{11} - \hat{\beta}_{13})\text{RW}$
$R = 1$ and TD $= 0$: $\widehat{\text{SBPSL}} = (\hat{\beta}_0 + \hat{\beta}_2 + \hat{\beta}_5 + \hat{\beta}_{10}) + \hat{\beta}_1\text{SBP1} + (\hat{\beta}_6 + \hat{\beta}_{12})\text{RW}$
$R = 0$ and TD $= 0$: $\widehat{\text{SBPSL}} = (\hat{\beta}_0 + \hat{\beta}_3 + \hat{\beta}_5 + \hat{\beta}_8) + \hat{\beta}_1\text{SBP1} + (\hat{\beta}_6 + \hat{\beta}_{14})\text{RW}$
$R = -1$ and TD $= 0$: $\widehat{\text{SBPSL}} = (\hat{\beta}_0 - \hat{\beta}_2 - \hat{\beta}_3 + \hat{\beta}_5 - \hat{\beta}_8 - \hat{\beta}_{10}) + \hat{\beta}_1\text{SBP1}$
$\qquad\qquad + (\hat{\beta}_6 - \hat{\beta}_{12} - \hat{\beta}_{14})\text{RW}$

$$R = 1 \text{ and } TD = -1: \widehat{\text{SBPSL}} = (\hat{\beta}_0 + \hat{\beta}_2 - \hat{\beta}_4 - \hat{\beta}_5 - \hat{\beta}_9 - \hat{\beta}_{10}) + \hat{\beta}_1 \text{SBP1}$$
$$+ (\hat{\beta}_6 - \hat{\beta}_{11} - \hat{\beta}_{12})\text{RW}$$
$$R = 0 \text{ and } TD = -1: \widehat{\text{SBPSL}} = (\hat{\beta}_0 + \hat{\beta}_3 - \hat{\beta}_4 - \hat{\beta}_5 - \hat{\beta}_7 - \hat{\beta}_8) + \hat{\beta}_1 \text{SBP1}$$
$$+ (\hat{\beta}_6 - \hat{\beta}_{13} - \hat{\beta}_{14})\text{RW}$$
$$R = -1 \text{ and } TD = -1: \widehat{\text{SBPSL}} = (\hat{\beta}_0 - \hat{\beta}_2 - \hat{\beta}_3 - \hat{\beta}_4 - \hat{\beta}_5 + \hat{\beta}_7 + \hat{\beta}_8 + \hat{\beta}_9 + \hat{\beta}_{10})$$
$$+ \hat{\beta}_1 \text{SBP1} + (\hat{\beta}_6 + \hat{\beta}_{11} + \hat{\beta}_{12} + \hat{\beta}_{13} + \hat{\beta}_{14})\text{RW}$$

  b. $H_0: \beta_{11} = \beta_{12} = \beta_{13} = \beta_{14} = 0$; multiple-partial $F = 1.409$ (4 and 89 df), $.1 < P < .25$

  c. multiple-partial $F = 1.102$ (8 and 89 df), $P > .25$

  d. $H_0: \beta_2 = \beta_3 = \beta_4 = \beta_5 = \beta_7 = \beta_8 = \beta_9 = \beta_{10} = \beta_{11} = \beta_{12} = \beta_{13} = \beta_{14} = 0$. Use

$$\text{multiple-partial } F = \frac{[\text{regression SS(full model)} - \text{regression SS}(X_1, X_6)]/12}{\text{MS residual (full model)}}$$

which has 12 and 89 df under H_0.

14. a. location 1: $|\widehat{d_{ij}}| = -0.42 + 0.16(X_{ij1} + X_{ij2})$

   location 2: $|\widehat{d_{ij}}| = 2.01 - 0.04(X_{ij1} + X_{ij2})$

  b. partial $F = 41.64$ (1 and 16 df), $P < .001$

  c. multiple-partial $F = 25.21$ (2 and 16 df), $P < .001$

15. a. $Y = \beta_0 + \beta_1 X + \beta_2 X^2 + \beta_3 Z + \beta_4 ZX + \beta_5 ZX^2 + E$, where

$$Z = \begin{cases} 0 & \text{if medium A} \\ 1 & \text{if medium B} \end{cases}$$

  b. medium A: $\hat{Y}_A = 0.4946 + 0.00537X + 0.00022X^2$
     medium B: $\hat{Y}_B = 0.3509 + 0.00660X + 0.00021X^2$

  c. parallelism: multiple-partial $F(ZX, ZX^2 | X, X^2, Z) = 0.006$ (2 and 34 df), $P > .25$

  d. coincidence: multiple-partial $F(Z, ZX, ZX^2 | X, X^2) = 4.08$ (3 and 34 df), $.01 < P < .025$

  e. No. A test to determine whether a quadratic term is needed in the model for each medium could be made by assessing whether the addition of X^2 and ZX^2 to a model containing X, Z, and ZX is significant. This would require results for a regression run based on a reduced model containing only X, Z, and ZX, but the results for such a run have not been presented.

Chapter 15

1. a. $\text{SBP} = \beta_0 + \beta_1(\text{QUET}) + \beta_2(\text{SMK}) + E$

  b. smokers: $\overline{\text{SBP}}(\text{adj}) = 148.547$; $\overline{\text{SBP}}(\text{unadj}) = 147.823$
     nonsmokers: $\overline{\text{SBP}}(\text{adj}) = 139.977$; $\overline{\text{SBP}}(\text{unadj}) = 140.800$

  c. $H_0: \beta_2 = 0$; partial $F(\text{SMK}|\text{QUET}) = 7.326$ (1 and 29 df), $.01 < P < .025$

  d. $2.094 < \beta_2 < 15.048$

4. a. $Y = \beta_0 + \beta_1 X + \beta_2 Z_1 + \beta_3 Z_2 + E$ where $\begin{cases} Z_1 = 1 & \text{if group 1 and 0 otherwise} \\ Z_2 = 1 & \text{if group 2 and 0 otherwise} \end{cases}$

  b. $Y = \beta_0 + \beta_1 X + \beta_2 Z_1 + \beta_3 Z_2 + \beta_4 XZ_1 + \beta_5 XZ_2 + E$
     $H_0: \beta_4 = \beta_5 = 0$; multiple-partial $F(XZ_1, XZ_2 | X, Z_1, Z_2) = 4.1312$ (2 and 24 df), $.025 < P < .05$; do not reject H_0 at $\alpha = .01$.

   c. group 1: $\bar{Y}(\text{adj}) = 4.40$; $\bar{Y}(\text{unadj}) = 4.10$
     group 2: $\bar{Y}(\text{adj}) = 4.80$; $\bar{Y}(\text{unadj}) = 4.86$
     group 3: $\bar{Y}(\text{adj}) = 4.74$; $\bar{Y}(\text{unadj}) = 5.00$
     Testing for differences among the adjusted mean scores: $H_0: \beta_2 = \beta_3 = 0$; multiple-partial $F(Z_1, Z_2|X) = 0.637$ (2 and 26 df), $P > .25$.

5. a. $\text{VIAD} = \beta_0 + \beta_1 \text{IQM} + \beta_2 \text{IQF} + \beta_3 Z + E$
   b. male experimenter: $\overline{\text{VIAD}}(\text{adj}) = -3.295$; $\overline{\text{VIAD}}(\text{unadj}) = -3.00$
     female experimenter: $\overline{\text{VIAD}}(\text{adj}) = 1.901$; $\overline{\text{VIAD}}(\text{unadj}) = 1.60$
   c. $H_0: \beta_3 = 0$; partial $F(Z|\text{IQM}, \text{IQF}) = 2.751$ (1 and 16 df), $.10 < P < .25$
   d. $-1.446 < \beta_3 < 11.838$

7. a. $\bar{Y}_{\text{EXP}}(\text{adj}) = 13.75$; $\bar{Y}_{\text{CON}}(\text{adj}) = 12.68$
   b. Partial $F(Z|X) = 0.3356$ (1 and 34 df), $P > .25$
   c. Two-sample t test comparing mean difference scores for each group. Computed $T = 1.324$ (35 df), $.2 < P < .4$.
   d. The two tests are not equivalent: Regression model for covariance analysis is $Y = \beta_0 + \beta_1 X + \beta_2 Z + E$, whereas the regression model for the t test is $Y - X = \beta_0^* + \beta_1^* Z + E$.
   e. Lack-of-fit $F = 6.357$ (30 and 4 df), $.025 < P < .05$. Suggests (at $\alpha = .05$) that the covariance model is not completely appropriate.

Chapter 16

1. a. $\widehat{\text{WGT}} = 37.600 + 0.053(\text{AGE} \times \text{HGT})$, $R^2 = .754$
   b. same answer as in part (a)
   c. $\widehat{\text{WGT}} = 6.553 + 0.722\text{HGT} + 2.05\text{AGE}$, $R^2 = .780$

2. a. $\widehat{\text{SBP}} = 48.050 + 1.709\text{AGE} + 10.294\text{SMK}$
   b. same answer as in part (a)
   c. same answer as in part (a)

Variables in Model	QUET	AGE	SMK	QUET, AGE	QUET, SMK	AGE, SMK	QUET, AGE, SMK
R^2	.551	.601	.061	.641	.641	.730	.761

   d. For $\alpha = .05$, add AQ only to get

$$\widehat{\text{SBP}} = 79.096 + 0.482\text{AGE} + 9.819\text{SMK} + 0.186\text{AQ}$$

     for which $R^2 = .773$.

4. a. $\hat{Y} = -34.073 + 1.224X_2 + 4.399X_3$, $R^2 = .802$
   b. same answer as in part (a)
   c. Best one-variable solution involves X_3, with $R^2 = .748$.
     Best two-variable solution involves X_2 and X_3, with $R^2 = .802$.
     The three-variable solution yields $R^2 = .818$.

Chapter 17

1. a. $\bar{Y}_1 = 7.5$, $\bar{Y}_2 = 5.0$, $\bar{Y}_3 = 4.333$, $\bar{Y}_4 = 5.167$, $\bar{Y}_5 = 6.167$
     $S_1 = 1.6432$, $S_2 = 1.2649$, $S_3 = 1.0328$, $S_4 = 1.4720$, $S_5 = 2.0412$

b.

Source	df	SS	MS	F
Treatments	4	36.467	9.1167	3.90
Error	25	58.500	2.3400	
Total	29	94.967		

c. $F = 9.1167/2.3400 = 3.90$ (4 and 25 df), $.01 < P < .025$. Conclude that the treatments have significantly different effects at $\alpha = .025$ but not at $\alpha = .01$.

d. Estimates of effects $(\bar{Y}_i - \bar{Y})$: 1.867 for treatment 1, -0.633 for 2, -1.300 for 3, -0.467 for 4, 0.533 for 5. Also, $\sum_{i=1}^{5} (\bar{Y}_i - \bar{Y}) = 0$.

e. $Y = \beta_0 + \alpha_1 X_1 + \alpha_2 X_2 + \alpha_3 X_3 + \alpha_4 X_4 + E$. For

$$X_i = \begin{cases} 1 & \text{for treatment } i \\ 0 & \text{otherwise} \end{cases} \quad i = 1, 2, 3, 4$$

the regression coefficients may be expressed as follows: $\beta_0 = \mu_5$, $\alpha_1 = \mu_1 - \mu_5$, $\alpha_2 = \mu_2 - \mu_5$, $\alpha_3 = \mu_3 - \mu_5$, and $\alpha_4 = \mu_4 - \mu_5$. For

$$X_i = \begin{cases} -1 & \text{for treatment } 5 \\ 1 & \text{for treatment } i \\ 0 & \text{otherwise} \end{cases} \quad i = 1, 2, 3, 4$$

the regression coefficients may be expressed as follows: $\beta_0 = \mu$, $\alpha_1 = \mu_1 - \mu$, $\alpha_2 = \mu_2 - \mu$, $\alpha_3 = \mu_3 - \mu$, and $\alpha_4 = \mu_4 - \mu$. Also, $-(\alpha_1 + \alpha_2 + \alpha_3 + \alpha_4) = \mu_5 - \mu$.

f. Ranking the sample means yields $\bar{Y}_1 > \bar{Y}_5 > \bar{Y}_4 > \bar{Y}_2 > \bar{Y}_3$, so the order of comparisons is 1 versus 3 (significant), 1 versus 2 (nonsignificant), and so on, with the remaining comparisons nonsignificant. The confidence intervals for the first two of these comparisons using an overall confidence level of 95% are:

Scheffé: $0.24 < \mu_1 - \mu_3 < 6.10$; $-0.43 < \mu_1 - \mu_2 < 5.43$

Tukey: $0.57 < \mu_1 - \mu_3 < 5.76$; $-0.10 < \mu_1 - \mu_2 < 5.10$

LSD: $0.34 < \mu_1 - \mu_3 < 6.00$; $-0.33 < \mu_1 - \mu_2 < 5.33$

Conclude with an overall $\alpha = .05$ that treatments 1 and 3 are significantly different, but that all other pairwise comparisons are nonsignificant. Schematically:

Of the three methods, Scheffé's method gives the widest interval, the LSD method the next to widest, and Tukey's method the narrowest interval.

2. a.

Source	df	SS	MS	F
Between cities	2	26	13.0	4.89
Within cities	6	16	2.67	
Total	8	42		

    f. $-6.59 < \mu_I - \mu_{II} < -1.41$

3. a.

Source	df	SS	MS	F
Labs	2	18.7	9.35	1.75
Error	9	48.0	5.33	
Total	11	66.7		

    c. $Y_{ij} = \mu + T_i + E_{ij}$; $i = $ I, II, III; $j = 1, 2, 3, 4$: T_i and E_{ij} are independent random variables satisfying $T_i \sim N(0, \sigma_T^2)$ and $E_{ij} \sim N(0, \sigma^2)$.

    d. A measure of repeatability is σ^2, which is estimated by MSE = 5.33; a measure of reproducibility is $\sigma^2 + 4\sigma_T^2$, which is estimated by MS(labs) = 9.35.

Source	df	SS	MS	F
Between groups	2	6,796.87	3,398.435	32.94
Within groups	197	20,323.00	103.162	
Total	199	27,119.87		

6. c. Use a t statistic of the form

$$T = \frac{\bar{Y}_{\text{COMBINED}} - \bar{Y}_{\text{NO-DIFF}}}{\sqrt{\text{MSE}(\frac{1}{125} + \frac{1}{75})}} \sim t_{197} \quad \text{under } H_0$$

7. b. The validity of the independence assumption might be questioned, since the same high schools were considered at different points in time (although different students were involved at each time point).

    c.

Source	df	SS	MS	F
Years	2	551.33	275.67	2.28
Error	12	1,453.60	121.13	
Total	14	2,004.93		

8. c. 95% Scheffé confidence interval: $(\bar{Y}_i - \bar{Y}_j) \pm 3.38$ for $i, j = $ A, B, C. Conclude that A is significantly better than C, that B is significantly better than C, but that A and B are not significantly different.

    d. The tests performed in (b) and (c) address only the issue of whether the persons differ from each other in ESP ability, not whether one or more actually has significant ESP ability. Consideration of the latter issue would require a test of H_0: $\mu = 13$ versus H_A: $\mu > 13$ for each person.

10. a.

Source	df	SS	MS	F
Locations	3	30.00	10.00	13.33 (.001 < P < .005)
Error	8	6.00	0.75	
Total	11	36.00		

Conclusion: Significant differences (at $\alpha = .05$) among the four locations.
b. $F = 39.2$ (1 and 8 df), $P < .001$
c. $F = 44.55$ (1 and 10 df), $P < .001$
d. $-3\mu_1 - \mu_2 + \mu_3 + 3\mu_4 = -3[\beta_0 + \beta_1(0)] - [\beta_0 + \beta_1(10)] + [\beta_0 + \beta_1(20)]$
$$+ 3[\beta_0 + \beta_1(30)]$$
$$= 100\beta_1$$
e. Different denominator mean squares were used; if there is no lack of fit, both are estimates of σ^2. Modification would involve using the mean square for pure error, which is equivalent to the ANOVA error mean square, in the denominator.
f. $F = 0.40$ (2 and 8 df), $P > .25$
g. same value as obtained in part (f)

Chapter 18

1. a. Rats are blocks; chemicals are treatments.

b.

Source	df	SS	MS	F
Chemicals	2	25.0	12.500	7.00
Rats	7	18.5	2.643	1.48
Error	14	25.0	1.786	
Total	23	68.5		

c. Test for difference in chemical effects: $F = 12.5/1.786 = 7.00$ (2 and 14 df), $.005 < P < .01$; reject H_0.
d. 98% CI for $\mu_I - \mu_{II}$: $-3.0 < \mu_I - \mu_{II} < 0.5$
e. $R^2 = .635$
f. Fixed-effects ANOVA model: $Y_{ij} = \mu + \tau_i + \beta_j + E_{ij}$; $i = $ I, II, III; $j = 1, 2, \ldots,$ 8 ($\tau_i = i$th chemical effect, $\beta_j = j$th rat effect)
Regression model: $Y = \mu + \alpha_1 X_1 + \alpha_2 X_2 + \sum_{j=1}^{7} \beta_j Z_j + E$, where

$$X_1 = \begin{cases} 1 & \text{if chemical I} \\ 0 & \text{if chemical II} \\ -1 & \text{if chemical III} \end{cases} \quad X_2 = \begin{cases} 0 & \text{if chemical I} \\ 1 & \text{if chemical II} \\ -1 & \text{if chemical III} \end{cases}$$

$$Z_j = \begin{cases} -1 & \text{if rat 8} \\ 1 & \text{if rat } j \ (j = 1, 2, \ldots, 7) \\ 0 & \text{otherwise} \end{cases}$$

2. a. Let n denote the number of individuals in each group, let

$$G = \begin{cases} 1 & \text{if group 1} \\ -1 & \text{if group 2} \end{cases}$$

and let

$$B_j = \begin{cases} -1 & \text{if block } n \\ 1 & \text{if block } j \quad j = 1, 2, \ldots, n - 1 \\ 0 & \text{otherwise} \end{cases}$$

Then the following six regression models correspond to approaches (1) through (6):

(1) $SP2 - SP1 = \beta_0 + \sum_{j=1}^{n-1} \beta_j B_j + \tau G + E$

(2) $SP2 - SP1 = \beta_0 + \sum_{j=1}^{n-1} \beta_j B_j + \gamma SP1 + \tau G + E$

(3) $SP2 = \beta_0 + \sum_{j=1}^{n-1} \beta_j B_j + \gamma SP1 + \tau G + E$

(4) $SP2 - SP1 = \beta_0 + \alpha_1 AGE + \alpha_2 SEX + \tau G + E$, where

$$SEX = \begin{cases} 1 & \text{if male} \\ -1 & \text{if female} \end{cases}$$

(5) $SP2 - SP1 = \beta_0 + \alpha_1 AGE + \alpha_2 SEX + \gamma SP1 + \tau G + E$

(6) $SP2 = \beta_0 + \alpha_1 AGE + \alpha_2 SEX + \gamma SP1 + \tau G + E$

b. For each of the models above, the null hypothesis of interest is $H_0: \tau = 0$.

(1)

Source	df	MS
Blocks	$n - 1$	MSB
Group|blocks	1	MSG
Error	$n - 1$	MSE
Total	$2n - 1$	

critical region: $F = MSG/MSE \geq F_{1,n-1,1-\alpha}$

(2)

Source	df	MS
Blocks	$n - 1$	MSB
SP1|blocks	1	MS1
Group|blocks, SP1	1	MSG
Error	$n - 2$	MSE
Total	$2n - 1$	

critical region: $F = MSG/MSE \geq F_{1,n-2,1-\alpha}$

(3)

Source	df	MS
Blocks	$n - 1$	MSB
SP1|blocks	1	MS1
Group|blocks, SP1	1	MSG
Error	$n - 2$	MSE
Total	$2n - 1$	

critical region: $F = MSG/MSE \geq F_{1,n-2,1-\alpha}$

(4)

Source	df	MS
AGE	1	MSA
SEX|AGE	1	MSS
Group|SEX, AGE	1	MSG
Error	$2n - 4$	MSE
Total	$2n - 1$	

critical region: $F = \text{MSG}/\text{MSE} \geq F_{1,2n-4,1-\alpha}$

(5)

Source	df	MS
AGE	1	MSA
SEX|AGE	1	MSS
SP1|AGE, SEX	1	MS1
Group|SEX, AGE, SP1	1	MSG
Error	$2n - 5$	MSE
Total	$2n - 1$	

critical region: $F = \text{MSG}/\text{MSE} \geq F_{1,2n-5,1-\alpha}$

(6)

Source	df	MS
AGE	1	MSA
SEX|AGE	1	MSS
SP1|AGE, SEX	1	MS1
Group|SEX, AGE, SP1	1	MSG
Error	$2n - 5$	MSE

critical region: $F = \text{MSG}/\text{MSE} \geq F_{1,2n-5,1-\alpha}$

   c. Using either model (2) or model (3) would be preferable to using model (1), unless it can be assumed that SP2 − SP1 is not significantly correlated with SP1. However, to be on the safe side, controlling for SP1 is suggested; in any case, if SP1 is not an important covariate, this will be detected by the stepwise-regression runs for approaches (2) and (3).

3. b. The model chosen depends primarily on how one prefers to treat the copper-ion-concentration variable. If treated nominally as in part (a), only differences among the five concentration levels treated as (qualitative) groups can be detected. If treated continuously as in part (b), a trend (i.e., a dose–response type of relationship) between copper ion concentration and BOD level can possibly be quantified; in part (b), this dose–response relationship is modeled as a straight line for each sample. These remarks suggest that the model in part (b) is probably preferable, since copper ion concentration is actually a continuous variable.

   c. There appears to be a decreasing trend in the means, suggesting that increasing the copper ion concentration magnifies the retardation effect on the bacterial action.

d. $F = 2{,}299.17/114.22 = 20.13$ (4 and 8 df), $P < .001$; reject H_0.

f. $T = 2.090$ (7 df), $.05 < P < .1$ (two-sided test); do not reject H_0 at $\alpha = .05$.

6. d. No, since any or all of the companies may be discriminatory even though they do not differ in the extent of that discrimination. One would need to test whether any given company has a rate discrepancy that is significantly different from some standard rate (which corresponds to no discrimination).

e. Both the assumptions of normality and variance homogeneity are in question. Alternative analysis procedures would entail using nonparametric ANOVA techniques (e.g., see Siegel, 1956) or making a suitable data transformation (see Chapter 12).

Chapter 19

1. b. Both factors should be considered fixed.

d.

| | Epinephrine | | |
Levorphanol	$-$	$+$	Marginals
$-$	2.25	5.28	3.77
$+$	1.75	2.57	2.16
Marginals	2.00	3.93	2.96

The presence of levorphanol appears to reduce stress, whereas the presence of epinephrine appears to increase stress. In the presence of epinephrine, levorphanol reduces stress by an average of 2.71 units, whereas in the absence of epinephrine, levorphanol reduces stress by only 0.50 unit (suggesting an interaction effect).

e.

Source	df	SS	MS	F (fixed effects)
Levorphanol	1	12.832	12.832	12.59
Epinephrine	1	18.586	18.586	18.24
Interaction	1	6.161	6.161	6.05
Error	16	16.298	1.019	

f. Main effect of levorphanol: $F = 12.832/1.019 = 12.59$ (1 and 16 df), $.001 < P < .005$; the presence of levorphanol significantly reduces stress.

Main effect of epinephrine: $F = 18.586/1.019 = 18.24$ (1 and 16 df), $P < .001$; the presence of epinephrine significantly increases stress.

Interaction: $F = 6.161/1.019 = 6.05$ (1 and 16 df), $.025 < P < .05$; the interaction effect as described in part (d) is significant at $\alpha = .05$ but not at $\alpha = .025$.

2. a. The classification depends on the purpose of the investigation, although in the most reasonable classification, "specialty" is fixed and "city" is random.

b.

	City 1	City 2	City 3	Marginals
Ped	83.58	83.98	83.83	83.79
Ob-Gyn	79.03	77.20	77.35	77.86
Db-Hyp	62.95	62.75	58.50	61.40
Marginals	75.18	74.64	73.22	74.35

There appears to be a highly significant main effect of "specialty," with pediatric FNPs being most competent, Ob-Gyn FNPs next most competent, and Db-Hyp FNPs least competent. There is no main effect of "city," and apparently there is no interaction effect.

c.

Source	F (both fixed)	F (both random)	F (SPEC fixed, CITY random)	F (SPEC random, CITY fixed)
Specialty	$15.02_{(2,27)}$	$187.04_{(2,4)}$	$187.04_{(2,4)}$	$15.02_{(2,27)}$
City	$0.11_{(2,27)}$	$1.42_{(2,4)}$	$0.11_{(2,27)}$	$1.42_{(2,4)}$
Interaction	$0.08_{(4,27)}$	$0.08_{(4,27)}$	$0.08_{(4,27)}$	$0.08_{(4,27)}$

d. The tests for each factor classification scheme all give the same results, which support the interpretation given in part (b).

e. Use $(\bar{Y}_{\text{Ped}} - \bar{Y}_{\text{Ob-Gyn}}) \pm \sqrt{(3-1)F_{2,27,0.95}\text{MSE}(\frac{1}{12} + \frac{1}{12})}$, which gives $-.4554 < \mu_{\text{Ped}} - \mu_{\text{Ob-Gyn}} < 11.41$.

f. $Y = \beta_0 + \beta_1 S_1 + \beta_2 S_2 + \beta_3 C_1 + \beta_4 C_2 + \beta_5 S_1 C_1 + \beta_6 S_1 C_2 + \beta_7 S_2 C_1 + \beta_8 S_2 C_2 + E$, where

$$S_1 = \begin{cases} -1 & \text{if Db-Hyp} \\ 1 & \text{if Ped} \\ 0 & \text{otherwise} \end{cases} \qquad S_2 = \begin{cases} -1 & \text{if Db-Hyp} \\ 1 & \text{if Ob-Gyn} \\ 0 & \text{otherwise} \end{cases}$$

$$C_j = \begin{cases} -1 & \text{if city 3} \\ 1 & \text{if city } j \qquad j = 1, 2 \\ 0 & \text{otherwise} \end{cases}$$

$\hat{\beta}_0 = 74.35$, $\hat{\beta}_1 = 9.44$, $\hat{\beta}_2 = 3.51$, $\hat{\beta}_3 = 0.833$, $\hat{\beta}_4 = 0.292$

4. a.

College Type	Male	Female	Marginals
≥75% male coed	39.3	29.7	34.49
<75% male coed	26.7	26.9	26.80
Not coed	34.0	31.7	32.85
Marginals	33.33	29.43	31.38

The table of means suggests that males are generally more sexist than females and that coed colleges with a large proportion of females have less sexism than the other types of colleges considered. Furthermore, and perhaps most important, it appears that males are much more sexist than females in coed schools with relatively few females, that there is little difference in attitudes in coed schools with relatively many females, and that males are somewhat more sexist than females in colleges that are not coed (i.e., there is an interaction effect).

5. a. Yes. There is an apparent same-direction interaction, which is reflected in the fact that the difference in mean waiting times between suburban and rural court locations is larger for state 1 than for state 2.

6. c.

Source	F (both fixed)	F (both random)	F (INST fixed, METH random)	F (INST random, METH fixed)
Instructor	$0.0004_{(1,96)}$	$0.0004_{(1,1)}$	$0.0004_{(1,1)}$	$0.0004_{(1,96)}$
Method	$5.996_{(1,96)}$	$5.742_{(1,1)}$	$5.996_{(1,96)}$	$5.742_{(1,1)}$
Interaction	$1.044_{(1,96)}$	$1.044_{(1,96)}$	$1.044_{(1,96)}$	$1.044_{(1,96)}$

d. student knowledge of subject matter at beginning of the experiment, IQ, mathematics aptitude, attitudes toward subject matter, and so on

e. analysis-of-covariance model: $Y = \beta_0 + \beta_1 C + \beta_2 I + \beta_3 M + \beta_4 IM + E$, where

$$I = \begin{cases} 1 & \text{if instructor A} \\ -1 & \text{if instructor B} \end{cases} \qquad M = \begin{cases} 1 & \text{if self-instruction method} \\ -1 & \text{if lecture method} \end{cases}$$

9. a. "Species" fixed, "locations" random (unless, for "locations," only the four chosen are of interest).

b.

Source	F (mixed)	F (both fixed)
Species	$9.140_{(2,6)}$	$7.972_{(2,48)}$
Locations	$0.944_{(3,48)}$	$0.944_{(3,48)}$
Interaction	$0.872_{(6,48)}$	$0.872_{(6,48)}$

mixed: species $(.01 < P < .025)$, locations $(P > .25)$, interaction $(P > .25)$
fixed: species $(.001 < P < .005)$, locations $(P > .25)$, interaction $(P > .25)$

Chapter 20

1. b.

Source	df	SS	MS
Modern rank	2	687.72	343.86
Traditional rank	2	33.123	16.561
Interaction	4	4,369.7	1,092.4
Error	21	1,592.5	75.833

(*Note*: Harmonic mean $= 3.1034$.)
$F(\text{modern}) = 343.86/75.833 = 4.53$ (2 and 21 df), $.01 < P < .025$; significant at .025
$F(\text{traditional}) = 16.561/75.833 = .218$ (2 and 21 df), $P > .25$; not significant
$F(\text{interaction}) = 1,092.4/75.833 = 14.41$ (4 and 21 df), $P < .001$; highly significant

Keep in mind that finding such a significant interaction effect somewhat negates the importance of any main-effect tests.

c. $Y = \beta_0 + \alpha_1 X_1 + \alpha_2 X_2 + \beta_1 Z_1 + \beta_2 Z_2 + \gamma_{11} X_1 Z_1 + \gamma_{12} X_1 Z_2 + \gamma_{21} X_2 Z_1 + \gamma_{22} X_2 Z_2 + E$, where

$$X_i = \begin{cases} -1 & \text{if LO modern rank} \\ 1 & \text{if modern rank } i \qquad i = 1, 2 \\ 0 & \text{if otherwise} \end{cases}$$

$$Z_j = \begin{cases} -1 & \text{if LO traditional rank} \\ 1 & \text{if traditional rank } j \qquad j = 1, 2 \\ 0 & \text{if otherwise} \end{cases}$$

($i = 1$ for HI modern rank and $i = 2$ for MED modern rank; $j = 1$ for HI traditional rank and $j = 2$ for MED traditional rank)

d. *"Modern" main effect*:

(1) $F(X_1, X_2) = \dfrac{977.70/2}{235.5172}$

$= 2.076$ (2 and 27 df), $.10 < P < .25$; not significant

(2) $F(X_1, X_2 | Z_1, Z_2) = \dfrac{871.92/2}{246.5377}$

$= 1.768$ (2 and 25 df), $.10 < P < .25$; not significant

"Traditional" main effect:

(1) $F(Z_1, Z_2) = \dfrac{301.30/2}{260.5690}$

$= 0.578$ (2 and 27 df), $P > .25$; not significant

(2) $F(Z_1, Z_2 | X_1, X_2) = \dfrac{195.52/2}{246.5377}$

$= 0.397$ (2 and 25 df), $P > .25$; not significant

Interaction:

$$F(X_1 Z_1, X_1 Z_2, X_2 Z_1, X_2 Z_2 | X_1, X_2, Z_1, Z_2) = \dfrac{4{,}570.85/4}{75.8333}$$

$= 15.069$, $P > .001$; highly significant

Regression analysis agrees well with unweighted-means analysis.

e. $Y = \beta_0 + \beta_1 X_1 + \beta_2 X_2 + \beta_3 X_1 X_2 + E$, where, for example,

$$X_1 = \begin{cases} 0 & \text{if LO modern rank} \\ 1 & \text{if MED modern rank} \\ 2 & \text{if HI modern rank} \end{cases} \qquad X_2 = \begin{cases} 0 & \text{if LO traditional rank} \\ 1 & \text{if MED traditional rank} \\ 2 & \text{if HI traditional rank} \end{cases}$$

The difficulty arises with regard to assigning numerical values to the categories of each factor. The coding scheme for X_1 and X_2 given here assumes that the categories are equally spaced, which may not be the case.

2. **a.** $Y = \beta_0 + \beta_1 X_1 + \beta_2 X_2 + \beta_3 X_1 X_2 + E$
   b. $Y = \beta_0 + \alpha_1 A + \beta_1 C + \gamma_{11} AC + E$, where

$$A = \begin{cases} -1 & \text{if high attitude} \\ 1 & \text{if low attitude} \end{cases} \qquad C = \begin{cases} -1 & \text{if high communication} \\ 1 & \text{if low communication} \end{cases}$$

This corresponds to a two-way ANOVA model with fixed effects.

   c. A choice between the models given in parts (a) and (b) should depend on the confidence one has in the sensitivity of the measurements. If both the communication and attitude scores range from 0 to 100, for example, with meaningful interpretations for scores differing by 2 or 3 points, the model in part (a) would probably be best. On the other hand, if only high and low scores can be discriminated for the attitude and communication factors, the model in part (b) is preferable.

7. **b.** Two-way tables of sample means:

	Season		
Residence	Summer	Winter	Marginals
Rural	2.72	4.81	3.76
Urban	3.39	4.07	3.73
Marginals	3.06	4.44	

	Region			
Residence	East	Central	West	Marginals
Rural	3.78	3.80	3.71	3.76
Urban	3.52	3.83	3.85	3.73
Marginals	3.65	3.82	3.78	

	Region			
Season	East	Central	West	Marginals
Summer	3.07	2.95	3.15	3.06
Winter	4.23	4.68	4.42	4.44
Marginals	3.65	3.82	3.78	

These tables indicate that there is a seasonal main effect (i.e., more TV viewing in the winter), but that there are no main effects of the other two factors. There is also an apparent interaction between "season" and "residence" such that seasonal differences are much greater for rural residents than for urban residents. Furthermore, there appears to be no "residence" × "region" interaction (since appropriate mean differences are quite small). There is a slight hint of a "season" × "region" interaction, since the summer–winter difference is somewhat larger in the central part of the state than in the other two regions. Finally, there appears to be some evidence of a three-way interaction, since there is no apparent two-way interaction between "region" and "season" for rural residents but there is somewhat of a "region" × "season" interaction for urban residents.

c.

Source	df (num., den.)	F	P
RESID	1, 468	0.13	$P > .25$
REGION	2, 468	1.25	$P > .25$
SEASON	1, 468	224.73	$P < .001$
RESID × REGION	2, 468	1.63	$.10 < P > .25$
RESID × SEASON	1, 468	59.20	$P < .001$
REGION × SEASON	2, 468	3.54	$.025 < P < .05$
RESID × REGION × SEASON	2, 468	3.30	$.025 < P < .05$

The main effect of SEASON and the RESID × SEASON interaction effect are both very highly significant. The REGION × SEASON interaction and the triple interaction are significant at the .05 level.

d. $Y = \beta_0 + \beta_1 R + \beta_2 S + \beta_3 L_1 + \beta_4 L_2 + \beta_5 RS + \beta_6 RL_1 + \beta_7 RL_2 + \beta_8 SL_1 + \beta_9 SL_2 + \beta_{10} RSL_1 + \beta_{11} RSL_2 + E$, where

$$R = \begin{cases} 1 & \text{if rural} \\ -1 & \text{if urban} \end{cases} \qquad S = \begin{cases} 1 & \text{if winter} \\ -1 & \text{if summer} \end{cases}$$

$$L_1 = \begin{cases} 1 & \text{if east} \\ 0 & \text{if central} \\ -1 & \text{if west} \end{cases} \qquad L_2 = \begin{cases} 0 & \text{if east} \\ 1 & \text{if central} \\ -1 & \text{if west} \end{cases}$$

8. a. Factors are sex (fixed) and treatment (fixed).

b.

Sex	Control	Estrogen	Total
Male	$n_{11} = 12$	$n_{12} = 12$	$n_{1.} = 24$
Female	$n_{21} = 8$	$n_{22} = 8$	$n_{2.} = 16$
Total	$n_{.1} = 20$	$n_{.2} = 20$	$n_{..} = 40$

$n_{1.}n_{.1}/n_{..} = 12 = n_{11}$; $n_{1.}n_{.2}/n_{..} = 12 = n_{12}$; $n_{2.}n_{.1}/n_{..} = 8 = n_{21}$; $n_{2.}n_{.2}/n_{..} = 8 = n_{22}$

c.

Source	df	SS	MS	F
Treatment	1	536.5562	536.5562	36.429 $(P < .001)$
Sex	1	19.7227	19.7227	1.339 $(P > .25)$
Interaction	1	1.3500	1.3500	0.092 $(P > .25)$
Error	36	530.2412	14.7289	

d.

Source	df	SS	MS	F
Treatment	1	504.60	504.60	34.259 $(P < .001)$
Sex	1	19.723	19.723	1.339 $(P > .25)$
Interaction	1	1.350	1.350	0.092 $(P > .25)$
Error	36	530.2412	14.7289	

(*Note*: SS terms have each been multiplied by the harmonic mean of the n_{ij}'s, 9.60.)

e. $Y = \beta_0 + \beta_1 S + \beta_2 T + \beta_3 ST + E$, where

$$S = \begin{cases} -1 & \text{if male} \\ 1 & \text{if female} \end{cases} \qquad T = \begin{cases} -1 & \text{if control} \\ 1 & \text{if estrogen} \end{cases}$$

f. The regression results demonstrate that in the proportional-cell-frequency case, it makes no difference to the analysis which main-effect variable is entered first. Thus the methodology for equal-cell-number two-way ANOVA is appropriate.

Chapter 21

1. $n = 30$ (6 age–sex groups $\times$ 5 years)

2. Here, $i = 5$ and $k = 1962 - 1960 = 2$, so that

$$E(Y_{52}) = \ell_{52}\lambda_{52} = \ell_{52}e^{\alpha_5 + 2\beta}$$

3. Log rate changes linearly with time. In particular, for the ith group,

$$\ln \lambda_{ik} = \alpha_i + \beta k$$

so that α_i is the intercept and β is the slope of the straight line relating the response $\ln \lambda_{ik}$ to the time variable $k = [\text{year}] - 1960$.

4. Model (1) assumes no interaction between age–sex group and time in the sense that the change in log rate over time (as measured by β) does not depend on i. Since $\ln \lambda_{ik} = \alpha_i + \beta k$, a graph of $\ln \lambda_{ik}$ versus k for each i would plot as a series of parallel straight lines, that is, lines all with the same slope (β) but possibly different intercepts (the α_i's). A lack of parallelism would reflect interaction between age–sex groups and time because the change in log rate over time would be different for different age–sex groups.

5. $\ln \text{IDR}_{ik} = \ln \lambda_{ik} - \ln \lambda_{10} = (\alpha_i + \beta k) - (\alpha_1 + \beta \cdot 0) = (\alpha_i - \alpha_1) + \beta k$, so that

$$\text{IDR}_{ik} = e^{\alpha_i - \alpha_1}e^{\beta k}$$

Note that this is a function of both age–sex group (i) and time (k).

6. An appropriate model is

$$E(Y_{ik}) = \ell_{ik}\lambda_{ik}$$

where

$$\ln \lambda_{ik} = \sum_{i=1}^{6} \alpha_i A_i + \beta k + \sum_{i=1}^{5} \gamma_i(A_i k)$$

For age–sex group i, then,

$$\ln \lambda_{ik} = \alpha_i + \beta k + \gamma_i k$$
$$= \alpha_i + (\beta + \gamma_i)k$$
$$= \alpha_i + \delta_i k$$

where $\delta_i = \beta + \gamma_i$. Hence, the slope for group i, namely δ_i, is now a function of i.

Only when all six δ_i's (or, equivalently, all six γ_i's) are equal will the six straight lines be parallel.

7. **a.** Yes, since $D(\hat{\boldsymbol{\beta}})_{(1)} - D(\hat{\boldsymbol{\beta}})_{(3)} = 300 - 175 = 125$, which is highly significant when compared to appropriate upper-tail χ^2-values with $29 - 24 = 5$ df.

   b. Yes, since $D(\hat{\boldsymbol{\beta}})_{(3)} - D(\hat{\boldsymbol{\beta}})_{(4)} = 175 - 60 = 115$, which is highly significant when compared to appropriate upper-tail χ^2-values with $24 - 23 = 1$ df.

   c. No, since $D(\hat{\boldsymbol{\beta}})_{(4)} - D(\hat{\boldsymbol{\beta}})_{(5)} = 60 - 59 = 1$, which is clearly not significant when compared to appropriate upper-tail χ^2-values with $23 - 22 = 1$ df.

   d. Yes, since $D(\hat{\boldsymbol{\beta}})_{(4)} - D(\hat{\boldsymbol{\beta}})_{(7)} = 60 - 20 = 40$, which is highly significant when compared to appropriate upper-tail χ^2-values with $23 - 18 = 5$ df.

   e. H_0 is rejected, since $D(\hat{\boldsymbol{\beta}})_{(4)} - D(\hat{\boldsymbol{\beta}})_{(6)} = 60 - 25 = 35$, which is highly significant when compared to appropriate upper-tail χ^2-values with $23 - 22 = 1$ df.

   f. Only models (6) and (7) are candidates to be the final model. Model (6) has a deviance of 25 based on 22 df, indicating a good fit to the data. Model (7) has a deviance of 20 based on 18 df, also indicating a good fit. All other candidate models have significant lack of fit. Note that $D(\hat{\boldsymbol{\beta}})_{(6)} - D(\hat{\boldsymbol{\beta}})_{(7)} = 25 - 20 = 5$, which is clearly not significant when compared to appropriate upper-tail χ^2-values based on $22 - 18 = 4$ df. Hence, model (6) certainly fits the data as well as model (7), and it has the important advantage that it characterizes very specifically the type of interaction present in the data (namely, that the group 1 slope is different from the slope common to the other five groups). Hence, model (6) is our choice as the final model.

   g. For model (6),

$$\text{pseudo-}R^2 = \frac{300 - 25}{300} = \frac{275}{300} = .917$$

which is indicative of a good model.

   h. (1) $\hat{\beta} \pm 1.96[\widehat{s.e.}(\hat{\beta})] = 0.50 \pm 1.96(0.20) = 0.50 \pm 0.392$, or $(0.108, 0.892)$

      (2) The point estimate of δ_1 is $\hat{\delta}_1 = \hat{\beta} + \hat{\gamma}_1 = 0.50 - 3.00 = -2.50$. The variance of the estimator $\hat{\delta}_1$ is

$$\text{Var}(\hat{\delta}_1) = \text{Var}(\hat{\beta}) + \text{Var}(\hat{\gamma}_1) + 2\text{Cov}(\hat{\beta}, \hat{\gamma}_1)$$

so that the estimated variance of $\hat{\delta}_1$ is

$$\widehat{\text{Var}}(\hat{\delta}_1) = (0.20)^2 + (0.50)^2 + 2(-0.10) = .09$$

Finally, an approximate 95% CI for δ_1 is:

$$\hat{\delta}_1 \pm 1.96\sqrt{\widehat{\text{Var}}(\hat{\delta}_1)} = -2.50 \pm 1.96\sqrt{.09} = -2.50 \pm 0.588,$$
$$\text{or} \quad (-3.088, -1.912)$$

These two confidence intervals suggest that log rate increases linearly ($\beta > 0$) for groups 2 through 6, but decreases linearly ($\delta_1 < 0$) for group 1.

Chapter 22

1. **b.** two-response, no factor $\rightarrow$ independence
one-response, one-factor $\rightarrow$ homogeneity

c.

	Response Category ($r = 6$)						
	1	2	3	4	5	6	
	East and Approve	East and Disapprove	East and No Opinion	South and Approve	South and Disapprove	South and No Opinion	Total
One subpopulation ($s = 1$)	$n_{11} = 500$ Π_{11}	$n_{12} = 350$ Π_{12}	$n_{13} = 150$ Π_{13}	$n_{14} = 400$ Π_{14}	$n_{15} = 300$ Π_{15}	$n_{16} = 300$ Π_{16}	$n_{1.} = 2000$

e. H_0: $f_1(\Pi) = \ln \Pi_{11} + \ln \Pi_{16} - \ln \Pi_{14} - \ln \Pi_{13} = 0$
$\quad\quad f_2(\Pi) = \ln \Pi_{12} + \ln \Pi_{16} - \ln \Pi_{15} - \ln \Pi_{13} = 0$
so that $f_1(\Pi) = \sum_{\gamma=1}^{6} k_{1\gamma} \ln a_\gamma(\Pi)$, where $a_\gamma(\Pi) = \Pi_{1\gamma}$, $\gamma = 1, 2, \ldots, 6$, $k_{11} = 1$, $k_{12} = 0$, $k_{13} = -1$, $k_{14} = -1$, $k_{15} = 0$, and $k_{16} = 1$; and $f_2(\Pi) = \sum_{\gamma=1}^{6} k_{2\gamma} \ln a_\gamma(\Pi)$, where $a_\gamma(\Pi) = \Pi_{1\gamma}$, $\gamma = 1, 2, \ldots, 6$, $k_{21} = 0$, $k_{22} = 1$, $k_{23} = -1$, $k_{24} = 0$, $k_{25} = -1$, and $k_{26} = 1$.

h. H_0: $a_1(\Pi) = \Pi_{11} - \Pi_{21} = 0$ and $a_2(\Pi) = (\Pi_{12} - \Pi_{22}) = 0$, so that

$$a_1(\Pi) = \sum_{i=1}^{2} \sum_{j=1}^{3} a_{1,ij} \Pi_{ij} \quad \text{where} \quad a_{1,11} = 1, \quad a_{1,21} = -1, \quad \text{all other } a_{1,ij} = 0$$

$$a_2(\Pi) = \sum_{i=1}^{2} \sum_{j=1}^{3} a_{2,ij} \Pi_{ij} \quad \text{where} \quad a_{2,12} = 1, \quad a_{2,22} = -1, \quad \text{all other } a_{2,ij} = 0$$

i. $\chi^2 = 64.96$ with 2 df, $P < .0005$; therefore, reject H_0 of no association (i.e., there is strong evidence of either dependence or nonhomogeneity, depending on the sampling scheme used).

j. Usual χ^2 of 64.96 is close to GSK χ^2-values of 62.82 and 67.14.

4. a.

	Subpopulation ($s = 12$)		Category of Response ($r = 2$)	
Subpopulation	Year of Graduation	MD Type	Respondent	Nonrespondent
1	Before 1940	Infect. dis. ped.	Π_{11} ($n_{11} = 5$)	Π_{12} ($n_{12} = 5$)
2	Before 1940	General pract.	Π_{21} ($n_{21} = 38$)	Π_{22} ($n_{22} = 33$)
3	Before 1940	Family phys.	Π_{31} ($n_{31} = 56$)	Π_{32} ($n_{32} = 86$)
4	1940–1950	Infect. dis. ped.	Π_{41} ($n_{41} = 9$)	Π_{42} ($n_{42} = 10$)
5	1940–1950	General pract.	Π_{51} ($n_{51} = 85$)	Π_{52} ($n_{52} = 58$)
6	1940–1950	Family phys.	Π_{61} ($n_{61} = 42$)	Π_{62} ($n_{62} = 51$)
7	1951–1960	Infect. dis. ped.	Π_{71} ($n_{71} = 23$)	Π_{72} ($n_{72} = 14$)
8	1951–1960	General pract.	Π_{81} ($n_{81} = 102$)	Π_{82} ($n_{82} = 60$)
9	1951–1960	Family phys.	Π_{91} ($n_{91} = 81$)	Π_{92} ($n_{92} = 61$)
10	After 1960	Infect. dis. ped.	$\Pi_{10,1}$ ($n_{10,1} = 8$)	$\Pi_{10,2}$ ($n_{10,2} = 2$)
11	After 1960	General pract.	$\Pi_{11,1}$ ($n_{11,1} = 49$)	$\Pi_{11,2}$ ($n_{11,2} = 21$)
12	After 1960	Family phys.	$\Pi_{12,1}$ ($n_{12,1} = 27$)	$\Pi_{12,2}$ ($n_{12,2} = 35$)

b. $a_\gamma(\Pi) = \Pi_{\gamma 1}$, $\gamma = 1, 2, \ldots, 12$

c. $a_1(p) = \frac{5}{10} = 0.5000$, $a_2(p) = \frac{36}{71} = 0.5352$, $a_3(p) = \frac{56}{142} = 0.3944$, $a_4(p) = \frac{9}{19} = 0.4652$, $a_5(p) = \frac{85}{143} = 0.5944$, $a_6(p) = \frac{42}{93} = 0.4516$, $a_7(p) = \frac{23}{37} = 0.6241$, $a_8(p) = \frac{102}{162} = 0.6296$, $a_9(p) = \frac{81}{142} = 0.5704$, $a_{10}(p) = \frac{8}{10} = 0.8000$, $a_{11}(p) = \frac{49}{70} = 0.7000$, $a_{12}(p) = \frac{27}{62} = 0.4355$

d. $\Pi_{\xi 1} = \mu + x_1\alpha_1 + x_2\alpha_2 + x_3\beta_1 + x_4\beta_2 + x_5\beta_3 + x_1x_3\gamma_{13} + x_1x_4\gamma_{14} + x_1x_5\gamma_{15} + x_2x_3\gamma_{23} + x_2x_4\gamma_{24} + x_2x_5\gamma_{25}$, where

$$x_1 = \begin{cases} 1 & \text{if infect. dis. ped.} \\ 0 & \text{otherwise} \end{cases} \qquad x_2 = \begin{cases} 1 & \text{if gen. ped.} \\ 0 & \text{otherwise} \end{cases}$$

$$x_3 = \begin{cases} 1 & \text{if grad. before 1940} \\ 0 & \text{otherwise} \end{cases} \qquad x_4 = \begin{cases} 1 & \text{if grad. 1940–1950} \\ 0 & \text{otherwise} \end{cases}$$

$$x_5 = \begin{cases} 1 & \text{if grad. 1951–1960} \\ 0 & \text{otherwise} \end{cases}$$

e.

Source	df	χ^2	P
MD type	2	12.630	$.001 < P < .005$
Year of graduation	3	9.669	$.01 < P < .05$
Interaction	6	6.732	$.30 < P < .40$

Conclude that "MD type" is strongly significant, that "year of graduation" is mildly significant, and that there is no interaction effect.

f. MD type: $d = 2$; H_0: $\alpha_1 = 0$, $\alpha_2 = 0$.
Year of graduation: $d = 3$; H_0: $\beta_1 = 0$, $\beta_2 = 0$, $\beta_3 = 0$.
Interaction: $d = 6$; H_0: $\gamma_{13} = 0$, $\gamma_{14} = 0$, $\gamma_{15} = 0$, $\gamma_{23} = 0$, $\gamma_{24} = 0$, $\gamma_{25} = 0$.

g. $\Pi_{\xi 1} = \mu + x_1\alpha_1 + x_2\alpha_2 + x_3\beta_1 + x_4\beta_2 + x_5\beta_3$

5. b. $a_\gamma(\Pi) = \Pi_{\gamma 1}$, $\gamma = 1, 2, \ldots, 12$

e.

Source	df	χ^2	P
W	1	7.356	$.005 < P < .01$
E	2	0.156	$P > .9$
S	1	2.705	$P \approx .1$
$W \times E$	2	2.832	$P \approx .25$
$W \times S$	1	8.402	$.001 < P < .005$
$E \times S$	2	2.285	$.3 < P < .4$
$W \times E \times S$	2	1.611	$.4 < P < .5$

Only the main effect of W and the $W \times S$ interaction effect are strongly significant; also, the main effect of S is significant at the $\alpha = .1$ level. Probably the most important finding is that the difference in the likelihood of getting byssinosis between dusty and not dusty workplaces is far greater for smokers than for nonsmokers, which is reflected in the highly significant $W \times S$ interaction effect.

f. $\Pi_{\xi 1} = \mu + x_1\alpha + x_2\beta_1 + x_3\beta_2 + x_4\gamma + x_1x_2\delta_{12} + x_1x_3\delta_{13} + x_1x_4\delta_{14}$

g.

Source	df	χ^2	P
W	1	20.619	$P < .0005$
E	2	1.017	$P \approx .60$
S	1	0.625	$.40 < P < .5$
$W \times E$	2	12.758	$.001 < P < .005$
$W \times S$	1	13.324	$P < .0005$
Goodness of fit	4	4.414	$.3 < P < .4$

W and $W \times S$ are strongly significant as before, but for this model $W \times E$ is also significant. Also, the fit of the model is adequate since the goodness-of-fit test is not significant.

h.

Predicted Response	Subpopulation											
	1	2	3	4	5	6	7	8	9	10	11	12
p_{i1}	.0556	.1288	.1500	.2388	.1111	.2715	.0118	.0103	.0095	.0121	.0102	.0196
$\hat{p}_{i1}$	.0456	.1402	.1454	.2400	.1447	.2393	.0093	.0118	.0090	.0115	.0128	.0153

i. Based on the results in part (g), one might try the model

$$\Pi_{\xi 1} = \mu + x_1\alpha + x_1 x_2 \delta_{12} + x_1 x_3 \delta_{13} + x_1 x_4 \delta_{14}$$

Nevertheless, based on the results in part (h), the model defined in part (f) provides an adequate fit.

Chapter 23

1. a. Use the two-sample t test:

$$T = \frac{\bar{X}_1 - \bar{X}_2}{S_p\sqrt{1/n_1 + 1/n_2}} \sim t_{n_1+n_2-2} \quad \text{under } H_0$$

SBP: $T = 2.201$ with 23 df, $.02 < P < .05$
CHOL: $T = 1.322$ with 23 df, $.1 < P < .2$
AGE: $T = 1.617$ with 23 df, $.1 < P < .2$

b. $Y = 0.4$ for noncases and $Y = -0.6$ for cases.

c. $c^* = 0.205849$

d. $\ell = -0.0683\text{SBP} + 0.0118\text{CHOL} - 0.0677\text{AGE}$
$b_0 = -\frac{1}{2}(\bar{\ell}_1 + \bar{\ell}_2) = 10.9545$
$\ell^* = 10.9545 - 0.0683\text{SBP} + 0.0118\text{CHOL} - 0.0677\text{AGE}$
 (1) If $\ell^* \geq -0.4055$, assign to NCHD group; if $\ell^* < -0.4055$, assign to CHD group.
 (2) If $\ell^* \geq 0$, assign to NCHD group; if $\ell^* < 0$, assign to CHD group.

e. Noncase 1: $\ell^* = 1.3661$; assign to NCHD group.
 Case 10: $\ell^* = -1.5792$; assign to CHD group.

f. There is not a clear-cut choice; the first table has more incorrect classifications of actual CHD cases than the second table ($\frac{2}{15} = 0.1333$, $\frac{5}{10} = 0.5000$), whereas the second table has more incorrect classifications of noncases than the first table ($\frac{4}{15} = 0.2667$, $\frac{3}{10} = 0.3000$). The investigator would have to decide which type of error is more serious.

g. Apparent error rates: $\frac{4}{15} = 0.2667$, $\frac{3}{10} = 0.3000$. Probabilities of misclassification: $\hat{P}_1 = \phi(-0.5061) = \hat{P}_2 = .3050$.

h. $F = 1.871$ with 3 and 21 df, $.1 < P < .2$; conclude that there are no significant differences between the two groups based on a model involving SBP, CHOL, and AGE.

i. Only SBP makes a significant contribution to discrimination ($P < .05$); therefore, recommend a model involving only this variable.

j. $c^* = 0.21547$, $b_1 = -0.06804$, $b_0 = -\frac{1}{2}(\bar{\ell}_1 + \bar{\ell}_2) = 10.1288$

   Rule: If $\ell^* = (10.1288 - 0.06804\text{SBP})$ is greater than 0, assign to NCHD group; if $\ell^* < 0$, assign to CHD group. To compare this model with the one containing all three variables, examine the misclassification rates.

k. $$P(\ell^*) = \frac{1}{1 + \exp(-10.1288 + 0.06804\text{SBP})}$$

   where $1 - P(\ell^*)$ denotes the probability (or risk) of becoming a CHD case. risk for noncase 1 $= .2802$; risk for case 10 $= .7369$

2. a. $\ell = -1.7881$ (max. breadth) $+ 6.7966$ (basialv. length)
      $+0.6289$ (nasal height) $+ 4.6848$ (basibr. height)

   b. $b_0 = -1,074.5625$
      Rule: Assign to period I if $\ell^* = -1,074.5625 + \ell > 0$; assign to period II otherwise.

   c. $F = 11.730$ with 4 and 5 df, $.005 < P < .01$

   d. Period I, no. 1: $\ell^* = 10.522$; assign to period I.
      Period II, no. 1: $\ell^* = -18.659$; assign to period II.

   e. Perfect discrimination is achieved.

4. Both variables: $c^* = 0.1273$; $\ell = -0.0035\text{MW} + 0.2442\text{MAGE}$
   MW only: $c^* = 0.1294$; $\ell = -0.0223\text{MW}$

5. $\ell^* = 4.1468 + 0.72106\text{DEF} - 1.70867\text{ENV} - 0.66671\text{RACE} - 0.1806\text{ECON}$

6. $\ell^* = 6.17532 - 3.22708\text{TD} + 1.60927\text{WR} + 1.16119\text{HS} - 0.51035\text{AD}$

8. b. X_1 (significant), X_3 (significant), X_2 (nonsignificant)

   c. 1.0735

   d. For $\ell = 1.25614X_1 + 0.79186X_2 + 2.63219X_3$, assign to species 1 if $\ell \geq 1.0735$ and assign to species 2 if $\ell \leq 1.0735$. It is not necessary to compute c^* unless the a priori probabilities and/or costs of misclassification are such that $\ln(p_2c_{21}/p_1c_{12})$ is nonzero.

   e. Cutoff point: 0.6678. Using $\ell = 1.25118X_1 + 3.36758X_3$, assign to species 1 if $\ell \geq 0.6678$ and assign to species 2 if $\ell \leq 0.6678$.

   f. $F = 50.82$ (2 and 21 df), $P < .001$

   g. The rule involving only X_1 and X_3 is more appropriate, since it is simpler and discriminates as well as the rule involving all three variables.

Chapter 24

1. a. T   b. T   c. F   d. F   e. F   f. F   g. F   h. F
   i. F   j. T   k. T   l. F   m. F   n. T   o. F   p. T
   q. T   r. F   s. F   t. T   u. T   v. T   w. F   x. T
   y. F

2. a. small sample size
   b. Rotated factors have a much simpler structure than do the principal components.
   c. possibly items 12 and 13
   d. factor 1: 1, 2, 3, 4, 6, 10, 11
   factor 2: 1, 5, 7, 8, 9

3. a. The three dimensions were theoretically conceived to overlap somewhat.
   b. Factor 1 = "value," factor 2 = "confidence," factor 3 = "interest." Suggested clusters are as follows:

4. c.

	PML		IML
Item	Factor Weight	Item	Factor Weight
1	0.271	6	0.405
2	0.519	7	0.229
3	0.447	8	0.423
4	0.409	9	0.580
5	0.540	10	0.517

   d. person no. 1: PIML score = $6.311 - 0.797 = 5.514$
   person no. 2: PIML score = $1.841 - 1.957 = -0.116$
   person no. 3: PIML score = $-1.034 - 6.046 = -7.080$

5. factor 1: hostility, suspiciousness, uncooperativeness
   factor 2: anxiety, guilt feelings, depressive mood
   factor 3: conceptual disorganization, hallucinatory behavior, unusual thought content

Index